Second Edition

Multimedia Image and Video Processing

IMAGE PROCESSING SERIES

Series Editor: Phillip A. Laplante, Pennsylvania State University

Published Titles

Second Edition

Multimedia Image and Video Processing

Edited by
Ling Guan
Yifeng He
Sun-Yuan Kung

 CRC Press
Taylor & Francis Group
Boca Raton London New York

CRC Press is an imprint of the
Taylor & Francis Group, an **informa** business

CRC Press
Taylor & Francis Group
6000 Broken Sound Parkway NW, Suite 300
Boca Raton, FL 33487-2742

First issued in paperback 2017

© 2012 by Taylor & Francis Group, LLC
CRC Press is an imprint of Taylor & Francis Group, an Informa business

No claim to original U.S. Government works
Version Date: 20120103

ISBN 13: 978-1-138-07253-4 (pbk)
ISBN 13: 978-1-4398-3086-4 (hbk)

Visit the Taylor & Francis Web site at
http://www.taylorandfrancis.com

and the CRC Press Web site at
http://www.crcpress.com

Contents

Part I Fundamentals of Multimedia

Part II Methodology, Techniques, and Applications: Coding of Video and Multimedia Content

Part III Methodology, Techniques, and Applications: Multimedia Search, Retrieval, and Management

Part IV   Methodology, Techniques, and Applications: Multimedia Security

Part V   Methodology, Techniques, and Applications: Multimedia Communications and Networking

Part VI   Methodology, Techniques, and Applications: Architecture Design and Implementation for Multimedia Image and Video Processing

List of Figures

Preface

We have witnessed significant advances in multimedia research and applications due to the rapid increase in digital media, computing power, communication speed, and storage capacity. Multimedia has become an indispensable aspect in contemporary daily life, and we can feel its presence in many applications ranging from online multimedia search, Internet Protocol Television (IPTV), and mobile multimedia, to social media. The proliferation of diverse multimedia applications has been the motivating force for the research and development of numerous paradigm-shifting technologies in multimedia processing.

This book documents the most recent advances in multimedia research and applications. It is a comprehensive book, which covers a wide range of topics including multimedia information mining, multimodal information fusion and interaction, multimedia security, multimedia systems, hardware for multimedia, multimedia coding, multimedia search, and multimedia communications. Each chapter of the book is contributed by prominent experts in the field. Therefore, it offers a very insightful treatment on the topic.

This book includes an Introduction and 28 chapters. The Introduction provides a comprehensive overview on recent advances in multimedia research and applications. The 28 chapters are classified into 7 parts. Part I focuses on Fundamentals of Multimedia, and Parts II through VII focus on Methodology, Techniques, and Applications.

Part I includes Chapters 1 through 6. Chapter 1 provides an overview of multimedia standards including video coding, still image coding, audio coding, multimedia interface, and multimedia framework. Chapter 2 provides the fundamental methods for histogram processing, image enhancement, and feature extraction and classification. Chapter 3 gives an overview on the design of an efficient application-specific multimedia architecture. Chapter 4 presents the architecture for a typical multimedia information mining system. Chapter 5 reviews the recent methods in multimodal information fusion and outlines the strength and weakness of different fusion levels. Chapter 6 presents bidirectional, human-to-computer and computer-to-human, affective interaction techniques.

Part II focuses on coding of video and multimedia content. It includes Chapters 7 through 10. Chapter 7 is a part overview, which provides a review on various multimedia coding standards including JPEG, MPEG-1, MPEG-2, MPEG-4, H.261, H.263, and H.264. Chapter 8 surveys the recent work on applying distributed source coding principles to video compression. Chapter 9 reviews a number of important 3D representation formats and the associated compression techniques. Chapter 10 gives a detailed description to Audio Video Coding Standard (AVS) developed by the China Audio Video Coding Standard Working Group.

Part III focuses on multimedia search, retrieval, and management. It includes Chapters 11 through 14. Chapter 11 is a part overview which provides the research trends in the area of multimedia search and management. Chapter 12 reviews the recent work on video modeling and retrieval including semantic concept detection, semantic video retrieval, and interactive video retrieval. Chapter 13 presents a variety of existing techniques for image retrieval, including visual feature extraction, relevance feedback, automatic image annotation, and large-scale visual indexing. Chapter 14 describes three basic components: content structuring and organization, data cleaning, and summarization, to enable management of large digital media archival.

Part IV focuses on multimedia security. It includes Chapters 15 through 18. Chapter 15 is a part overview which reviews the techniques for information hiding for digital media, multimedia forensics, and multimedia biometrics. Chapter 16 provides a broad view of biometric systems and the techniques for measuring the system performance. Chapter 17 presents the techniques in watermarking and fingerprinting for multimedia protection. Chapter 18 reviews content-based fingerprinting approaches that are applied to images and videos.

Part V focuses on multimedia communications and networking. It includes Chapters 19 through 22. Chapter 19 is a part overview, which discusses several emerging technical challenges as well as research opportunities in next-generation networked mobile video communication systems. Chapter 20 presents a two-tier proxy-based peer-to-peer (P2P) live streaming network, which consists of a low-delay high-bandwidth proxy backbone and a peer-level network. Chapter 21 presents the recent studies on exploring the scalability of scalable video coding (SVC) and the quality of service (QoS) provided by the IEEE 802.11e to improve performance for video streaming over wireless local area networks (WLANs). Chapter 22 provides a review of recent advances on optimal resource allocation for video communications over P2P streaming systems, wireless ad hoc networks, and wireless visual sensor networks.

Part VI focuses on architecture design and implementation for multimedia image and video processing. It includes Chapters 23 through 25. Chapter 23 presents the methodology for concurrent optimization of both algorithms and architectures. Chapter 24 introduces dataflow-based methods for efficient parallel implementations of image processing applications. Chapter 25 presents the design issues and methodologies of application-specific instruction set processor (ASIP) for video processing.

Part VII focuses on multimedia systems and applications. It includes Chapters 26 through 28. Chapter 26 presents the design and implementation of a mixed-reality environment for learning. Chapter 27 reviews the recent methods for converting conventional monocular video sequences to stereoscopic or multiview counterparts for display using 3D visualization technology. Chapter 28 presents a Second Life (SL) HugMe prototype system that bridges the gap between virtual and real-world events by incorporating interpersonal haptic communication system.

The target audience of the book includes researchers, educators, students, and engineers. The book can be served as a reference book in the undergraduate or graduate courses on multimedia processing or multimedia systems. It can also be used as references in research of multimedia processing and design of multimedia systems.

Ling Guan
Yifeng He
Sun-Yuan Kung

Acknowledgments

First, we would like to thank all the chapter contributors, without whom this book would not exist. We would also like to thank the chapter reviewers for their constructive comments. We are grateful to Nora Konopka, Jessica Vakili, and Jennifer Stair of Taylor & Francis, LLC, and S.M. Syed of Techset Composition, for their assistance in the publication of the book. Finally, we would like to give special thanks to our families for their patience and support while we worked on the book.

Introduction: Recent Advances in Multimedia Research and Applications

Guo-Jun Qi, Liangliang Cao, Shen-Fu Tsai, Min-Hsuan Tsai, and Thomas S. Huang

CONTENTS

0.1 Overview

In the past 10 years, we have witnessed the significant advances in multimedia research and applications. Amount of new technologies have been invented for various fundamental multimedia research problems. They are helping the computing machines better perceive, organize, and retrieve the multimedia content. With the rapid development of multimedia hardware and software, nowadays we can easily make, access and share considerable multimedia contents, which could not be imagined only 10 years before. All of these result in many urgent technical problems for effectively utilizing the exploding multimedia information, especially for efficient multimedia organization and retrieval in different levels from personal photo albums to web-scale search and retrieval systems. We look into some of these edge cutting techniques arising in the past few years, and in brevity summarize how they are applied to the emerging multimedia research and application problems.

0.2 Advances in Content-Based Multimedia Annotation

Content-based multimedia annotation has been attracting great effort, and significant progress has been made to achieve high effectiveness and efficiency in the past decade. A large number of machine learning and pattern recognition algorithms have been introduced and adopted for improving annotation accuracy, among which are support vector machines (SVMs), ensemble methods (e.g., AdaBoost), and semisupervised classifiers. To apply these classic classification algorithms for annotation task, multimodality algorithms have been invented to fuse different kinds of feature cues, ranging from color, texture, and shape features to the popular scale-invariant feature transform (SIFT) descriptors. They make use of the complementary information across different feature descriptors to enhance annotation accuracy. On the other hand, recent results show that the annotation tasks do not exist independently across different multimedia concepts, but the annotation tasks of these concepts are strongly correlated with each other. This idea yields many new annotation algorithms which explore intrinsic concept correlations. In this section, we first review some basic annotation algorithms which have been successfully applied for annotation tasks, followed by some classic algorithms that fuse the different feature cues and explore the concept correlations.

0.2.1 Typical Multimedia Annotation Algorithms

Kernel methods and ensemble classifiers have gained great success since they are proposed in the late 1990s. As the typical discriminative models, they become prevailing in real annotation systems [62]. Generally speaking, when there are enough training samples, discriminative models result in more accurate classification results than generative models [66]. On the contrary, generative models, including naive Bayes, Bayesian Network, Gaussian Mixture Model (GMM), Hidden Markov Model (HMM), and many graphical models, are also widely applied to multimedia annotation and the results showed they can complement with the discriminative models to improve the annotation accuracy. For example, Refs. [71] and [105] use two-dimensional dependency-tree model hidden HMMs and GMMs respectively to represent each image adapting from universal background models in the first step. Then discriminative SVMs are built upon the kernel machines comparing the similarity of these image representations.

In case of small number of training samples, semisupervised algorithms are more effective for annotation. They explore the distribution of testing samples so that more robust annotation results can be achieved, avoiding from overfitting into training set of small size. Zhou et al. [104] proposes to combine partial label to estimate the score of each sample. The similar idea is also developed by Zhu et al. [107], where the scores on labeled samples are fixed as that in training sert and the resulted method corresponds to the harmonic solution of un-normalized graph Laplacian.

0.2.2 Multimodality Annotation Algorithms

Efficiently fusing a set of multimodal features is one of the key problems in multimedia annotation. The weights of different features often vary for each annotation task, or even change from one multimedia object to another. This propels us to develop sophisticated modality fusion algorithms. A common approach to fusing multiple features is to use different kernels for different features and then combine them by a weighted summation, which is so called multiple kernel learning (MKL) [5,46,75]. In MKL, the weight for the different feature does not depend on the multiple objects and remains the same across all the samples. Consequently, such a linear weighting approach does not describe possible nonlinear relationships among different types of features.

Recently, Gnen and Alpaydin [29] proposed a localized weighting approach to MKL by introducing a weighting function for the samples that is assumed to either a linear or quadratic function of the input sample. Cao et al. [11] proposed an alternative Heterogeneous Feature Machine (HFM) that builds a kernel logistic regression (LR) model based on similarities that combine different features and distance metrics.

0.2.3 Concept-Correlative Annotation Algorithms

The goal of multimedia annotation is to assign a set of labels to multimedia documents based on their semantic content. In many cases, multiple concepts can be assigned to one multimedia document simultaneously. For example, in many online video/image sharing web sites (e.g., Flickr, Picasa, and YouTube), most of the multimedia documents have more than one tags manually labeled by users. It results in a multilabel multimedia annotation problem that is more complex and challenging compared to multiclass annotation problem. This is because the annotations of multiple concepts are not independent but strongly correlated with each other. Evidences have shown exploring the label correlations plays the key role to improve the annotation results. Naphade et al. [63] proposes a probabilistic Bayesian Multinet approach that explicitly models the relationship between the multiple concepts through a factor graph upon the underlying multimedia ontology semantics. Wu et al. [97] uses an ontology-based multilabel learning algorithm for multimedia concept detection. Each concept is first independently modeled by a classifier, and then a predefined ontology hierarchy is leveraged to improve the detection accuracy of each individual classifier. Smith and Naphade [85] presents a two-step Discriminative Model Fusion approach to mine the unknown or indirect relationship to specific concepts by constructing model vectors based on detection scores of individual classifiers. SVM is then trained to refine the detection results of the individual classifiers. Alternative fusion strategy can also be used, for example, Hauptmann et al. [34] proposed to use LR to fuse the individual detections. Users were involved in their approach to annotate a few concepts for extra video clips, and these manual annotations are then utilized to help infer and improve detections of other concepts.

Although it is intuitively correct that contextual relationship can help improve detection accuracy of individual detectors, experimental results have shown that such improvement is not always stable, and the overall performance can even be worse than individual detectors alone. It is due to the fact that these algorithms are built on top of the independent binary detectors with a second step to fuse them. However, the output of the individual independent detectors can be unreliable and therefore their detection errors can propagate to the fusion step. To address the difficulties faced in the first and second paradigms, a new paradigm of an integrated multilabel annotation is proposed. It simultaneously models both the individual concepts and their interactions in a single formulation. Qi et al. [70] proposes a structure SVM-based algorithm that encodes the label correlations as well as feature representations in a unifying kernel space. The concepts for each multimedia document can be efficiently inferred by a graphical model with effective approximation algorithms, such as Loopy Belief Propagation and Gibbs Sampling. Alternatively Wang et al. [90] proposes a semisupervised method which models the label correlation by a concept graph on the whole dataset, and then GraphCut is leveraged to annotate the unlabeled documents.

Multilabel algorithms are also extended to explore more realistic scenarios for the annotation task. Recently, Zhou et al. [106] proposes a new multi-instance multilabel paradigm to combine bag-of-words feature representations with multilabel models so that the instance and label redundancies can be fully investigated in the same framework. In another research orientation, Qi et al. [71] proposes a multilabel active learning approach that utilizes the label correlations to save the human labor of collecting training samples. It only requests manual annotation of a small portion of key concepts and the other concepts could be readily inferred by their correlations to the annotated ones. The result shows that the algorithm can significantly reduce labeling effort by a user interface which organizes the human annotation by concept.

0.3 Advances in Constructing Multimedia Ontology

One of the essential goals in the multimedia community is to bridge the semantic gaps between the low-level feature and high-level image semantics. Many efforts have been made to introduce multimedia semantics to close the gap. Due to its explicit representation, ontology, a formal specification of the domain knowledge that consists of concepts and relationships between concepts, has become one of the most important ingredients in multimedia analysis and has been deeply prompted recently.

In a typical ontology, concepts are represented by terms, while in a multimedia ontology concepts might be represented by terms or multimedia entities (images, graphics, video, audio, segments, etc.) [67]. In this section, we mainly focus on the recent advance of the ontologies regarding multimedia terms (i.e., typical ontology).

There are mainly two active research areas related to multimedia ontology: ontology construction and ontological inference.

0.3.1 Construction of Multimedia Ontologies

It is never an easy task to construct ontology for the purpose of ontological inference in multimedia. Indeed, there are quite a few challenges, for example, limited resources to extract concept relations, vague concepts in multimedia domains.

There have been much effort made to construct multimedia ontologies manually for some small and specific domains or for the proof-of-concept purpose. For example, Chai et al. [16] utilizes manually constructed ontologies to organize the domain knowledge and provide explicit and conceptual annotation for personal photos. Similarly in Schreiber et al. [82] used a domain-specific ontology for the animal domain which was constructed manually to provide knowledge describing the photo subject matter vocabulary.

As manual construction of ontology is too expensive, if not infeasible, recent efforts on building the multimedia ontologies have been shifted to leverage existing lexical resource such as WordNet or Cyc as a reference or a starting point. Benitez and Chang [7] proposed an automatic concept extraction technique that first extracts semantic concepts from annotated image collections with a disambiguation process, which is followed by a semantic relationship discovery based on WordNet. Deng et al. [23] constructed the ImageNet dataset with the aid of the semantic structure of WordNet, mainly focusing on the taxonomy (i.e., the 'is-a' or 'subclass-of' relationships).

On the other hand, Jaimes and Smith [41] argued that both manual and automatic construction of ontologies are not feasible for large-scale ontology and domains that require domain-specific decisions. Therefore they proposed to build data-driven multimedia ontologies semiautomatically with text mining techniques and content-based image retrieval tools.

0.3.2 Ontological Inference

The explicit representation of domain knowledge in multimedia ontologies enables inference for new concept based on concepts and their relationships. It is due to this capability that more and more research work incorporate ontologies into all kinds of multimedia tasks, such as image annotation, retrieval and video event detection, in the past few years.

In general multimedia ontologies may exist in the form of relational graph, where the vertices in the graph represent the concepts and the edges between vertex represent the relationships among corresponding vertices. In [57], Marszalek and Schmid used binary classifiers associated with the edge of the graph, that is, the relations, to compute conditional likelihood for each concept. Then a most probable path-based decision is made based on the marginal likelihood, which attains the maximum of all the path likelihoods from the root concept to the target concept. More recently, the generative probabilistic model, for example, Bayesian Network, has been adopted to explore the conditional dependence among concepts and perform the probabilistic inference for high-level concept extraction [26,53,64].

On the other hand, in order to reduce the complexity of the inference problem, it has been proposed to consider only part of the relationships such as 'is-a' and/or 'part-of' so that the ontologies can be formulated in tree- or forest hierarchies. For example, in the earlier work of Chua et al. [19], the ontology was constructed in a tree hierarchy that being used to represent the domain knowledge in their mapping algorithm. Similarly Park et al. [68] used an animal ontology as the animal taxonomy in their semantic inference process for annotating images. Wu Coworkers [87] proposed an ontology-based multiclassification algorithm that learns semantically meaningful influence paths based on predefined forest structured ontology hierarchies for video concept detection.

0.4 Advances in Sparse Representation and Modeling for Multimedia

In the past decade, the amount and dimensionality of multimedia data have grown larger and higher, respectively, due to the advances of data storage and Internet technology. The ability to handle such enormous data becomes critical in many multimedia applications . Sparse representation arises as a way to explore and exploit the underlying low dimensional structure of high dimensional data, which can yield much compact representation of multimedia data for effective retrieval and indexing. In particular, suppose for some application the observed data $\mathbf{y}$ is presumably a linear combination of a small set of columns of dictionary matrix $\mathbf{A}$, then sparse representation seeks the sparse coefficients $\mathbf{x}$ such that $\mathbf{y} = \mathbf{A}\mathbf{x}$.

0.4.1 Computation

Given the observed data $\mathbf{y}$ and dictionary $\mathbf{A}$, it has been proved that finding the sparsest $\mathbf{x}$ is itself a NP-complete problem and difficult to approximate [4], implying any deterministic algorithm would most likely require more than exponential time to get such $\mathbf{x}$, unless $P = NP$. Although discouraging as it seems, researchers proceeded to find out that instead of minimizing ℓ^0 norm of $\mathbf{x}$, that is the number of nonzero elements of $\mathbf{x}$, minimizing ℓ^1 norm of $\mathbf{x}$, that is the sum of absolute values of $\mathbf{x}$ would often yield the sparsest one [27]. Finding $\mathbf{x}$ with minimum ℓ^1 norm and constraint $\mathbf{y} = \mathbf{A}\mathbf{x}$ can be cast as a linear programming problem for which various polynomial time algorithms have been well-studied. Because of this advance on computation, pursuit of $\mathbf{x}$ which is truly the sparsest is no longer impractical, and researchers can try it out and get rid of the less sparse $\mathbf{x}$ yielded by previous approximation algorithms.

Technically, Donoho [27] defines the equivalent breakdown point (EBP) of a matrix $\mathbf{A}$ as the maximum number k such that if $\mathbf{y} = \mathbf{A}\mathbf{x}_0$ for some $\mathbf{x}_0$ with less than k nonzero entries, then the minimal ℓ^1 norm solution $\hat{\mathbf{x}}$ is equal to that sparse generator $\mathbf{x}_0$. Roughly speaking, EBP characterizes condition for a given dictionary $\mathbf{A}$ where the sparsest representation can be computed efficiently by linear programming algorithms. An upper bound on EBP(A) is derived from the theory of centrally neighborly polytopes: EBP(A) $\leq \lfloor (m + 1)/3 \rfloor$, where m is the height of A.

0.4.2 Application

Sparse representation has been applied in some tasks such as face recognition and foreground detection. The key to applying sparsity is to find a suitable model for the application where the assumption of sparsity holds.

0.4.2.1 Face Recognition

Wright et al. [96], proposed a novel way of creating and exploiting sparsity to do face recognition. Assuming the test face is a linear combination of the training faces of the same object, then it can be regarded as a sparse linear combination of all training faces in the database. With controlled pixel alignment, image orientation, and frontal face images, the proposed method outperforms the state-of-the-art ones.

Moreover, this model was extended to handle occlusion, and the resulting algorithm is shown to be able to cope with occlusion induced by sunglasses and scarf. Specifically, occlusion is modeled by an error component $\mathbf{e}$, that is, $\mathbf{y} = \mathbf{A}\mathbf{x} + \mathbf{e} = [\mathbf{A}\,\mathbf{I}][\mathbf{x}\,\mathbf{e}]^T = \mathbf{B}\mathbf{w}$, where $\mathbf{B} = [\mathbf{A}\,\mathbf{I}]$, $\mathbf{w} = [\mathbf{x}\,\mathbf{e}]^T$, and further assuming that $\mathbf{e}$ is sparse as well. This model can thus deal with occlusion of arbitrary magnitude as long as the number of occluded pixels does not exceed certain threshold. Given that EBP is bounded by EBP(A) $\leq \lfloor (m + 1)/3 \rfloor$, the model is able to handle up to occlusion of about 33% of the entire image. However, in their experiments it was noticed that the algorithm works well way beyond 33% occlusion. Wright and Ma [95] then further investigated the cause of the phenomenon and thereafter developed a "bouquet" signal model for this particular application to explain, where "bouquet" refers to face images as they all lie in a very narrow range in the high dimensional space, just like bouquet.

0.4.2.2 Video Foreground Detection

Similar idea has also been applied in video foreground detection [25], where the dictionary consists of background video frames, and the algorithm identifies the occluded pixels in the input frame as foreground pixels. This approach avoids statistical modeling foreground and background pixels and parameter estimation while utilizing the underlying simple foreground/background structure.

0.4.3 Robust Principal Component Analysis

In this section, we make a detour to a somewhat related topic: robust principal component analysis (Robust PCA). It is a more general treatment of huge amount of high dimensional data. Given a large data matrix M, Robust PCA seeks decomposition $M = L_0 + S_0$ where L_0 has low rank and S_0 is sparse. Like $\ell^1 - \ell^0$ equivalence in sparse representation, minimizing the sum of singular values of L plus the sum of absolute values of S under the constraint $M = L + S$ often yields a low rank L and sparse S [9]. This again encourages people to pursue the decomposition and apply it to various tasks.

Robust PCA can be applied to face recognition and foreground detection as well [9]. Moreover, in Latent Semantic Indexing, it can be used to decompose the document versus term matrix M into sparse term S_0 and low rank term L_0, where L_0 captures common words used in all documents and S_0 captures the few key words that best distinguish each document from others [9]. Similarly, collaborative filtering (CF) is a process of jointly obtaining multidimensional attributes of objects of interest, for example, millions of user rankings of a large set of films, and joint annotations for an image database over a huge annotation vocabulary.

0.5 Advances in Social Media

The center concept governing the community media is that users play the central role in retrieving, indexing, and mining media content. The basic idea is quite different from the traditional content-centric multimedia system. The web sites providing community media are not solely operated by the the owners but by millions of amateur users who provide, share, edit, and index these media content. In this section, we will review recent research advancement in community media system from two aspects: (1) retrieval and indexing system for community media based on user-contributed tags and (2) community media recommendation by mining user ratings on media content.

0.5.1 Retrieval and Search for Social Media

Recent advances in internet speed and easy-to-use user interfaces provided by some web companies, such as Flickr, Corbis, and Facebook, have significantly prompted image sharing, exchange and propagation among user. Meanwhile, the infrastructures of image-sharing social networks make it easier for users to attach tags to images than before. These huge amount of user tags enable better understanding of the associated images and provide many research opportunities to boost image search and retrieval performance. On the other hand, the user tags somehow reflect the users' intentions and subjectivities and therefore can be leveraged to build a user-driven image search system.

To develop a reliable retrieval system for community media based on these user contributed tags, two basic problems must be resolved. First of all, the user tags are often quite noisy or even semantically meaningless [18]. More specifically, the user tags are known to be ambiguous and limited in terms of completeness, and overly personalized [31,58]. This is not surprising because of the uncontrolled nature of social tagging and the diversity of knowledge and cultural background of the users [47]. To guarantee a satisfactory retrieval performance, tag denoising methods are required to refine these tags before they can be used for retrieval and indexing. Some examples of tag denoising methods are as follows: Tang et al. [86] proposes to construct an intermediate concept space from user tags which can be used as medium to infer and detect more generic concepts of interest in future. Weinberger et al. [94] proposes a probabilistic framework to resolve ambiguous tags which are likely to occur but appear in different context with the help of human effort. In the meantime, there exist many tag suggestion methods [2,59,84,98] which help users annotate community media with most informative tags, and avoid meaningless or low-quality tags. For all these methods, tag suggestion systems are involved to actively guide users to provide high-quality tags based on tags co-occurrence relations.

Secondly, the tags associated with an image are generally in a random order without any importance or relevance information, which limits the effectiveness of these tags in search and other applications. To overcome this problem, Liu et al. [50] proposes a tag ranking scheme that aims to automatically rank the tags associated with a given image according to their relevance to the image content. This tag ranking system estimates the initial relevance scores for the tags based on probability density estimations, followed by a random walk over a tag similarity graph to refine the relevance scores. Another method was proposed in [47] that learns tag relevance by accumulating votes from visually similar neighbors. Treated as tag frequency, these learned tag relevance is seamlessly embedded into tag-based social image retrieval paradigms.

Many efforts have been made on developing the multimedia retrieval systems by mining the user tags. Semantic distance metric can be set up from these web images and their associated user tags, which can be directly applied to retrieve web images by examples at semantic level rather than at visual level [73]. Meanwhile Wang et al. [91] proposed a novel attempt at model-free image annotation, which is a data-driven approach to annotating images by the returned search results based on user tags and surrounding text. Since no training data set is required, their approach enables annotating with unlimited vocabulary and is highly scalable and robust to outliers.

0.5.2 Multimedia Recommendation

Developing recommendation systems for community media has attracted many attentions with the popularity of Web 2.0 applications, such as Flickr, YouTube, and Facebook. Users give their own comments and rates on multimedia items, such as images, amateur videos, and movies. However, only a small portion of multimedia items have been rated and thus the available user ratings are quite sparse. Therefore, an automatic recommendation system is desired to be able to predict users' ratings on multimedia items so that they can easily find the interesting images, videos and movies from shared multimedia contents.

Recommendation systems measure the user interest in given items or products to provide personalized recommendations based on user's taste [6,37]. It becomes more and more important to enhance user's experience and loyalty by providing them with the most appropriate products in e-commerce web sites (such as Amazon, eBay, Netflix, TiVo, and Yahoo!), which bring many interests in designing a user-satisfied recommendation system.

Currently, existing recommendation systems can be categorized into two different types. The content-based approach [1] creates a profile for each user or product which depicts its nature. User profiles can be described by their historical rating records on movies, personal information (such as their ages, genders, occupations) and their movie types of interest. Meanwhile, movie profiles can be represented by other features, such as their titles, release date, and movie genres (e.g., action, adventure, animation, comedy). The obtained profiles allow programs to quantify the association between users and products.

The other popular recommendation systems only rely on the past user ratings on the products with no need to create explicit profiles. This method is known as CF [30] which analyzes relationships between users and interdependencies among products. In other words, it aims at predicting an user's ratings based on users' ratings on the same set of multimedia items. The only information used in CF is the historical behavior of users, such as their previous transactions or the way they rate products. The CF method can also be cast as two primary approaches—the neighborhood approach and the latent factor models. Neighborhood methods compute the relationships between users [8,36,76] or items [24,48,81] or combination thereof [89] to predict the preference of a user to a product. On the other hand, latent factor models transform both users and items into the same latent factor space and measure their interactions in this space directly. The most representative methods of latent factor models are singular value decomposition (SVD) [69]. Evaluations on the recommendation systems suggest SVD methods have gained the state-of-the-art performance among many other methods [69].

Some public data sets are available for comparison purpose among different recommendation systems. Among them, the most exciting and popular one is the Netflix data set for movie recommendation at http://www.netflixprize.com. The Netflix data set contains more than 100 million ratings on near 18 thousand movie titles from over 480 thousand randomly-chosen, anonymous customers. These users' ratings were collected between October 1998 and December 2005, and they are able to represent the users' trend and preference during this period. The ratings are given on a scale from one to five stars. The date of each rating as well as the title and year of release for each movie are provided. No other data, such as customer or movie information, were employed to compute Cinematch's accuracy values used in this contest. In addition to the training data set, a qualifying test set is provided with over 2.8 million customer/movie pairs and the rating dates. These pairs were selected from the most recent ratings from a subset of the same customers in the training data set, over a subset of the same movies. Netflix offers a Grand Prize with $1,000,000$ and Progress Prizes with $50,000$. On September 2009, Netflix announced the latest Grand Prize winner as team BellKor's Pragmatic Chaos [45], whose result achieves the root mean squared error (RMSE) 0.8567 on the test subset—a 10.06% improvement over the Cinematch's score on the test subset, which uses straightforward statistical linear models with a lot of data conditioning. The next Grand Prize winner will have to improve RMSE by at least 10% compared to BellKor's Pragmatic Chaos algorithm.

0.6 Advances in Distributed Multimedia Mining

The explosion of multimedia data has made it impossible for single PCs or small computer clusters to store, index, or understand real multimedia information networks. In the

United Kingdom, there are about 4.2 million surveillance cameras, which means, there is one surveillance camera for every 14 residents. On the other hand, both videos and photos have become prevalent on popular websites such as YouTube and Facebook. Facebook has collected the largest photo bank in the history (15 billion photos in total, with an increasing rate of 220 million new photos per week). It has become a serious challenge to manage or process such an overwhelming amount of multimedia files. Fortunately, we are entering the era of "cloud computing," which provides the potential of processing huge multimedia and building large-scale intelligent interface to help us understand and manage the media content.

Cloud computing, conceptually speaking, describes the new computing interface whereby details are abstracted from the users who no longer have need of, expertise in, or control over the technology infrastructure "in the cloud" that supports them. Cloud computing system includes a huge data storage center and compute cycles nearby. It constitutes front ends for users to submit their jobs using the service provided. It incorporates multiple geographically distributed sites where the sites might be constructed with different structure and services. From the developer's viewpoint, cloud computing also reduces developing efforts. For example, working on MapReduce (a programming paradigm in cloud computing) is much easier than using classical parallel message passing interface. In the past few years, many successful cloud computing systems have been constructed, including Amazon EC2, Google App, IBM SmarterPlanet, and Microsoft Windows Azure.

Cloud computing is still in the growing stage, and most of the existing work on this topic are related to documentation data. Designing a new paradigm especially for multimedia community is a brave but untested idea. On the other hand, computing high dimension features has been a classical problem for decades.

MapReduce is a new parallel programming paradigm developed in Google [22] for large scale data processing. In recent years, some researchers applied MapReduce paradigm to many machine learning and multimedia processing problems. MapReduce framework has also been applied to many graph mining problems. Haque and Chokkapu [33] reformulate the PageRank algorithm with MapReduce framework. This work is further extended by Kang et al. [43] who implemented graph mining library, PEGASUS, on Hadoop platform. They first introduce a GIM-V framework for matrix-vector multiplication and show how it can be applied to graph mining algorithms such as diameter estimation, the PageRank estimation, random walk with restart calculation, and finding connected-components. In another work, Husain et al. [40] propose a framework to store and retrieve a large number of resource description framework (RDF) data for semantic web systems. RDF is a standard model for data interchanging on the Web, and it is essential to process RDF efficiently for semantic web system. They suggest a devised schema for RDF data for Hadoop and show how to answer queries with the new framework.

Many research works also use Hadoop for more general applications. Chu et al. [17] showed that many popular machine learning algorithms, such as weighted linear regression, naive bayes, and PCA rely on statistics which can be computed by a Map and Reduce procedure directly (single-pass learning). Some other algorithms, such as K-means, EM, Neural Network, LR and SVM, can be computed by iterative methods, of which each iteration is composed by a Map procedure and a Reduce procedure (iterative learning). This idea is further examined by Qillick et al., who compare in details the cost of Hadoop overhead and the benefit of distributing the task to a large cluster. For single pass algorithms, we need to put in more efforts to reduce the Hadoop overhead cost. For iterative methods, the map stage of each iteration depends upon the parameters generated by the reduce phase of the previous iteration, where the benefit of distributive computing is more attractive.

In recent years, there have been more and more efforts in designing MapReduce based machine learning algorithms. Newman et al. [65] employs MapReduce to inferring latent Dirichlet allocation. Ye et al. [101] implement decision trees on Hadoop. Gonzalez et al. [32] discuss how to implement parallel Belief Propagation using Splash approach. Cardona et al. [14] implement probabilistic neural network with MapReduce framework and show the performance with simulation. Many algorithms have been implemented in Apache Mahout library [56], including K-Nearest Neighbor (KNN), naive Bayes, expectation maximization (EM), Frequent Pattern (FP)-growth, Kmeans, mean-shift, Latent Dirichlet Allocation, and etc. Yan et al. [99] recognized the overhead of MapReduce paradigm for multimedia classification problem. They proposed a new algorithm called robust subspace bagging, which builds ensemble classifiers in different dataset partitions. The basic idea is first to train multiple classification models using bootstrapped samples, and add such model into the ensemble classifier when it improves the recognition accuracy. These new algorithms alleviate the cost of exploring the whole dataset, however, at a cost of much slower testing speed since the ensemble model has to evaluate multiple base classifiers when evaluating each testing samples.

Despite all these work, MapReduce is still not efficient for multimedia data analysis. Since MapReduce is originally designed for processing texts, for new tasks involving high-dimensional numerical data, the overhead of MapReduce becomes huge in the following aspects:

- In MapReduce, each individual task is designed to handle 16M to 64M samples, which is often not efficient for high dimensional data.
- When the dimension becomes high, loading cost will increase. In many large scale multimedia applications, the bottleneck lies in not only computing but also loading process.
- When the size of the data is huge, it is expensive to transfer data over the network. Moreover, sorting/merging the results in reduce step will become slow.

We are expecting more progress to be obtained in these new directions.

0.7 Advances in Large-Scale Multimedia Annotation and Retrieval

Annotating and retrieving large-scale multimedia corpus are critical to real-world multimedia information system. Efficiently working on a corpus of millions or even billions of multimedia documents is one of the most important criteria to evaluate the system. On the server end, the system should be able to access, organize, and index the whole corpus dynamically with constantly new data. On the front user end, the system shall respond to user queries on time without too long delay. All these impose requirements on effective data structure for multimedia documents and the associated annotation and retrieval algorithms. In past few years, locality-sensitive hashing [3] has been widely accepted as such data structure to represent and index the multimedia documents in the corpus. It uses a family of hashing functions to map each document to an integer number so that the similar images/videos could have the same hashing value with high probability. Meanwhile, those dissimilar documents could probably have a distinct hashing value. Since hashing function can be efficiently computed on each multimedia document, the desired documents with similar content can be retrieved efficiently given the user queries.

Considering the dynamic nature of multimedia database, new images and videos are constantly accumulated into the system and must be processed in a real time manner. Online and/or adaptive algorithms are developed to handle the new data and update the corresponding annotation and retrieval models. An example of adaptive learner for video annotation problem is given in [100]. In this work, the squared Euclidean distance between the parameters of two successive linear models is minimized while the new coming training data can be best classified. Qi et al. [72] proposes an online algorithm which deals with correlative multimedia annotation problem. Kullback-Leibler divergence is minimized between two consecutive probabilistic models, while the first two moments of statistics on new training examples and models comply with each other with minimal differences. Besides the online learner which involves a sequence of single training examples, Davis et al. [21] proposed another form of online learner which updated a metric model with a pair of new examples. In their method, the Bregman divergence over the positive definite convex cone is used to measure the progress between two successive metric models parameterized by Mahalonobis matrix. Meanwhile, the squared loss between the prediction Mahalonobis distance and the target distance is used to measure the correctiveness of new model on each trial.

0.8   Advances in Geo-Tagged Social Media

Geographical information is longitude-latitude pair to represent the locations where the images are taken. In recent years, the use of geographical information has become more and more popular. With the advance in low-cost Global Positioning System (GPS) chips, cell phones, and cameras become equipped with GPS receivers and thus are able to record the locations while taking pictures. Many online communities, such as Flickr and Google Earth, allow users to specify the location of their shared images either manually through placement on a map or automatically using image metadata embedded in the image files. At the end of 2009, there have been nearly 100 million geo-tagged images in Flickr with millions of images added each month. Given the geographical information associated with images, it becomes possible to infer the image semantics with geographical information, or to estimate the geographical information from visual content. By exploring the rich media such as user tags, satellite images, and wikipedia knowledge, we can leverage the visual and geographical information for many novel applications.

To give a clear overview, we roughly group the research of geo-tagged social media into two groups: to estimate the geographical information from general images and to utilize geographical information for better understanding of the image semantics. Although there are also a lot of work of using geo-tagged information to build different multimedia systems [10,74,92], in this survey we focus on the research problems instead of applications.

One interesting question is whether we can look at an image to estimate its geographical locations even when they are not provided. As evidenced by the success of Google Earth, there is great need for such geographic information among the mass. Hays and Efros [35] are among the first to consider the problem of estimate the location of a single image using only its visual content. Their results show the approach is able to locate about a quarter of the images in the test images to within a small country ($\sim$750 km) of their true location.

Motivated by Hays and Efras [35], Gallagher et al. [28] incorporate textual tags to estimate the geographical locations of images. Cao et al. [12] combine tags with visual information for annotation using a novel model named logistic canonical correlation regression which explores the canonical correlations between geographical locations, visual content, and community tags. Crandall et al. [20] argue that it is difficult to estimate the exact location at which a photo was taken and they only estimate the approximate location of a novel photo. Their results show that at the landmark scale, both the text annotations and visual content perform better than chance while at the metropolitan scale, only the text annotations perform better than chance. In a recent research work [103] supported by Google, Zhen et al. built a web-scale landmark recognition engine named "Tour the world" using 20 million GPS-tagged photos of landmarks together with online tour guide web pages. The experiments demonstrate that the engine can deliver satisfactory recognition performance with high efficiency. However, it is still an open question whether it is possible to recognize nonlandmark location reliably.

Geographical information can also help to obtain a better understanding of the image semantics. Naaman et al. [61] proposed a system to suggest candidate identity labels based on the metadata of a photo including its time stamp and location. The system explores the information of events and locations, and the co-occurrence statistics of people. However, image analysis techniques are not used in the system. After Naaman's work, more research works aim to understand semantics from both visual information and geographical collections. Joshi and Luo [42] propose to explore the geographical information systems (e.g., Google Map) database using a given geographical location. They use descriptions of small local neighborhoods to form bags of geo-tags as their representation and demonstrate that the context of geographical location is a strong cue for the visual event/activity recognition. Yu and Luo [102] propose another way to leverage nonvisual contexts such as location and time stamp, which can be obtained through picture metadata automatically. Both visual and nonvisual context information are fused using a probabilistic graphical model to improve the accuracy of object region recognition. Luo et al. [55] explore satellite images corresponding to picture location data and investigate their novel uses to recognize the picture-taking environment, as if through a third eye above the object. This satellite information is combined with classical vision-based event detection methods. The fusion of the complementary views (photo and satellite) achieves significant performance improvement over the ground view baseline.

0.9 Advances in Multimedia Applications

In this section, we look into a set of multimedia applications that make high impact on our everyday lives, including personal photo album, automatic video editing, multimedia advertisement, and photo tourism. A lot of multimedia processing and analysis techniques mentioned above have been applied into these applications and gain the potential of commercial goals.

Proliferation of high quality digital cameras at consumer prices resulted in an explosion of personal digital photo collection in the past ten years. It becomes often for a home user to take thousands of digital photos each year. These images are largely un-labeled, and often have minimal organization done by the user. Managing, accessing, and making use of this personal photo collection has become a challenging task. In personal photo albums,

Both identity of people and number of people in photos are important cues to users in how they remember, describe, and search personal photos [60]. There is a large body of research on face recognition. Subjects in personal photos within a collection are usually related by family ties or friendship. Discovering and understanding the relationship among identified people in personal collections has significant application impact. Sharing of personal photos (e.g., on social networking sites such as Facebook) makes it possible to discover relations beyond an individual's collection. On the other hand, personal collections are generally captured using a small number of cameras with time stamp (and sometimes GPS information). Typically, users store these photos in folders labeled with time and event information [44]. Event and location information are also among the top cues people remember and use when searching for photos [60]. Due to the availability and importance of time and location information for personal collections, both cues are used in photo clustering and annotation work for personal photos.

To ease the access to the rich media information, automatic multimedia editing and summarization is an effective tool to help users understand and quickly locate the desired images and/or video clips in a large scale multimedia corpus. For example, Hua et al. [39] proposes an optimization-based automatic video editing system to summarize home videos into informative short summarizations. Salient video clips are detected and extracted, and the abstracted video sequences are aligned with the tempo of incident music so that users can better enjoy the edited video content. The resulting video editing system has been successfully deployed in a industry software—Movie Maker. Meanwhile, recent popularity of video sharing web sites make the organization and summarization of the web videos more interesting and challenging. Hong et al. [38] constructs an event-driven video summarization system and users can better perceive large scale web video corpus organized by events.

With the giant progress of multimedia search technology, more and more industrial systems begin to take multimedia information into account when developing search engines. Although none of the major search engines (e.g., Google, Microsoft Bing Search) utilized visual feature before 2008, nowadays there have been quite a few attempts to develop large scale visual search systems. TwitPic was launched to allow users to post pictures to follow Twitter post. Tiltomo was developed to search the Flickr dataset based on tags and similar themes. Tineye and GazoPa allow users to provide their own pictures and find similar peers in the internet. Such similar image searching functions have also been supported by Google Image and Bing Image. Moreover, Google has built a beta version of "Swirl" search, which organizes the image search results into groups by hierarchically clustering the visual features. In addition, more and more companies have target the searching problem in mobile platform, and quite a few systems have been developed including Google Goggles, kooaba, snaptell, etc. Another group of companies focus on vertical visual search, which considers a specific segment of visual search, for example, Paperboy considers on searching news articles or books, while Plink focuses on searching works of art.

References

1. G. Adomavicius and A. Tuzhilin. Toward the next generation of recommender systems: A survey of the state-of-the-art and possible extensions. *IEEE Transactions on Knowledge and Data Engineering*, 17(6): 734–749, 2005.

2. M. Ames and M. Naaman. Why we tag: Motivations for annotation in mobile and online media. In *Proceedings of the SIGCHI Conference on Human Factors in Computing Systems*, 2007.

3. A. Andoni and P. Indyk. Near-optimal hashing algorithms for approximate nearest neighbor in high dimensions. *Communications of the ACM*, 51(1):117–122, 2008.

4. E. Amaldi and V. Kann. On the approximability of minimizing nonzero variables or unsatisfied relations in linear systems. *Theoretical Computer Science*, 209, 237–260, 1998.

5. F. Bach, G. R. G. Lanckriet, and M. I. Jordan. Multiple kernel learning, conic duality, and the smo algorithm. In *International Conference on Machine Learning*, 2004.

6. R. M. Bell, Y. Koren, and C. Volinsky. Modeling relationships at multiple scales to improve accuracy of large recommender systems. In *ACM SIGKDD Internal Conference on Knowledge Discovery and Data Mining*, August 2007.

7. A. B. Benitez and S.-F. Chang. Semantic knowledge construction from annotated image collections. In *Proceedings of IEEE International Conference on Multimedia*, pp. 26–29, 2002.

8. J. S. Breese, D. Heckerman, and C. Kadie. Empirical analysis of predictive algorithms for collaborative filtering. In *Proceedings of UAI*, 1998.

9. E. J. Cands, X. Li, Y. Ma, and J. Wright. Robust principal component analysis? Technical Report No. 2009-13, Department of Statistics, Stanford University, December 2009.

10. L. Cao, J. Luo, A. Gallagher, X. Jin, J. Han, and T. Huang. A worldwide tourism recommendation system based on geotagged web photos. In *International Conference on Acoustics, Speech, and Signal Processing (ICASSP)*, 2010.

11. L. Cao, J. Luo, F. Liang, and T. S. Huang. Heterogeneous feature machines for visual recognition. In *International Conference on Computer Vision*, 2009.

12. L. Cao, J. Yu, J. Luo, and T. Huang. Enhancing semantic and geographic annotation of web images via logistic canonical correlation regression. In *Proceedings of the ACM International Conference on Multimedia*, pp. 125–134, 2009.

13. L. Cao, J. Yu, J. Luo, and T. S. Huang. Enhancing semantic and geographic annotation of web images via logistic canonical correlation regression. In *Proceedings of the 17th ACM International Conference on Multimedia*, pp. 125–134. ACM Press, 2009.

14. K. Cardona, J. Screvan, M. Georgiopoulos, and G. Anagnostopoulos. A grid based system for data mining using mapreduce. 2007.

15. Y. Chai, X.-Y. Zhu, and J. Jia. Ontoalbum: An ontology based digital photo management system. In A. C. Campilho and M. S. Kamel, ed. *ICIAR*, volume 5112 of Lecture Notes in Computer Science, pp. 263–270. Springer, 2008.

16. Y.-M. Chai, X.-Y. Zhu, and J.-P. Jia. Ontoalbum: An ontology based digital photo management system. In *ICIAR '08: Proceedings of the 5th international conference on Image Analysis and Recognition*, pp. 263–270, 2008.

17. C.-T. Chu, S. K. Kim, Y.-A. Lin, Y.Y. Yu, G. Bradski, and A. Y. Ng. Map-reduce for machine learning on multicore. *Advances in Neural Information Processing Systems*, 2006.

18. T.-S. Chua, J. Tang, R. Hong, H. Li, Z. Luo, and Y. Zheng. Nus-wide: A real-world web image database from national university of singapore. In *Proceedings of ACM International Conference on Image and Video Retrieval*, 2009.

19. T.-.S Chua, K. C. Teo, B. C. Ooi, and K. L. Tan. Using domain knowledge in querying image databases. In *International Conference on Multimedia Modeling*, pp. 12–15, 1996.

20. D. Crandall, L. Backstrom, D. Huttenlocher, and J. Kleinberg. Mapping the world's photos. In *International Conference on World Wide Web*, pp. 761–770, 2009.

21. J. V. Davis, B. Kulis, P. Jain, S. Sra, and I. S. Dhillon. Information-theoretic metric learning. In *Proceedings of International Conference on Machine Learning*, 2007.

22. J. Dean and S. Ghemawat. Mapreduce: Simplified data processing on large clusters. *OSDI*, 2004.

23. J. Deng, W. Dong, R. Socher, L.-J. Li, K. Li, and L. Fei-Fei. Imagenet: A large-scale hierarchical image database. *Computer Vision and Pattern Recognition*, 2009.

24. M. Deshpande and G. Karypis. Item-based top-n recommendation algorithms. *ACM Transactions on Information Systems*, 22(1):143–177, 2004.

25. M. Dikmen and T. S. Huang. Robust estimation of foreground in surveillance videos by sparse error estimation. *International Conference on Pattern Recognition*, 2008.

26. Z. Ding and Y. Peng. A probabilistic extension to ontology language owl. In *Proceedings of the 37th Hawaii International Conference on System Sciences (HICSS-37)*, Big Island, 2004.

27. D. Donoho. For most large underdetermined systems of equations, the minimal ℓ^1-norm near-solution approximates the sparsest near-solution. *Communications on Pure and Applied Mathematics*, 59(7): 907–934, 2006.

28. A. Gallagher, D. Joshi, J. Yu, and J. Luo. Geo-location inference from image content and user tags, pp. 55–62, Miami, FL, June 2009.

29. M. Gnen and E Alpaydin. Localized multiple kernel learning. In *International Conference on Machine Learning*, 2008.

30. D. Goldberg, D. Nichols, B. M. Oki, and D. Terry. Using collaborative filtering to weave an information tapestry. *Communications of the ACM*, pp. 61–70, 1992.

31. S. A. Golder and B. A. Huberman. Usage patterns of collzaborative tagging systems. *Journal of Information Science*, 32(2):198–208, 2006.

32. J. Gonzalez, Y. Low, and C. Guestrin. Residual splash for optimally parallelizing belief propagation. AISTATS, 2009.

33. A.-ul Haque and V. Chokkapu. The web laboratory: Pagerank calculation using map reduce. 2008.

34. A. Hauptmann, M.-Y. Chen, and M. Christel. Confounded expectations: Informedia at TRECVID 2004. In *TREC Video Retrieval Evaluation Online Proceedings*, 2004.

35. J. Hays and A. A. Efros. Im2gps: Estimating geographic information from a single image. In *IEEE Conference on Computer Vision and Pattern Recognition*, 2008.

36. J. L. Herlocker, J. A. Konstan, A. Borchers, and J. Riedl. An algorithmic framework for performing collaborative filtering. In *Proceedings of the Annual International ACM SIGIR Conference on Research and Development in Information Retrieval*, 1999.

37. J. L. Herlocker, J. A. Konstan, L. G. Terveen, and J. T. Riedl. Evaluating collaborative filtering recommender systems. *ACM Transactions on Information Systems*, 22(1): 5–53, 2004.

38. R. Hong, J. Tang, H.-K. Tan, S. Yan, C.-W. Ngo, and T.-S. Chua. Event driven summarization for Web videos, in *Proceedings of the first SIGMM workshop on Social media*, pp. 43–48, Beijing, China, 2009.

39. X.-S. Hua, L. Lu, and H.-J. Zhang. Optimization-based automated home video editing system *IEEE Transaction on Circuits and Syst. for Video Technology*. 14(5): 572–583, 2004.

40. M. F. Husain, P. Doshi, L. Khan, and B. M. Thuraisingham. Storage and retrieval of large rdf graph using hadoop and mapreduce. In *CloudCom*, Lecture Notes in Computer Science. Springer, 2009.

41. A. Jaimes and J. R. Smith. Semi-automatic, data-driven construction of multimedia ontologies. In *ICME*, pp. 781–784, 2003.

42. D. Joshi and J. Luo. Inferring generic places based on visual content and bag of geotags. In *ACM Conference on Content-Based Image and Video Retrieval*, 2008.

43. U. Kang, C. E. Tsourakakis, and C. Faloutsos. Pegasus: A peta-scale graph mining system—implementation and observations. *IEEE International Conference on Data Mining*, 2009.

44. D. S. Kirk, A. J. Sellen, C. Rother, and K. R. Wood. Understanding photowork. In *Proceedings of CHI 2006, ACM Press*, pp. 761–770. ACM Press, 2006.

45. Y. Koren. The bellkor solution to the netflix grand prize. Technical report, Netflix, Inc, August 2009.

46. G. Lanckriet, N. Cristianini, P. Bartlett, L. El Ghaoui, and M. Jordan. Learning the kernel matrix with semidefinite programming. *The Journal of Machine Learning Research*, 5:27–72, 2004.

47. X. Li, C. G. Snoek, and M. Worring. Learning tag relevance by neighbor voting for social image retrieval. In *MIR '08: Proceedings of the 1st ACM International Conference on Multimedia Information Retrieval*, 2008.

48. G. Linden, B. Smith, and J. York. Amazon.com recommendations: Item-to-item collaborative filtering. *IEEE Internet Computing*, 2003.

49. N. Littlestone. *Mistake Bounds and Logarithmic Linear-Threshold Learning Algorithms*. PhD Thesis, University of California Santa Cruz, 1989.

50. D. Liu, X.-S. Hua, L. Yang, M. Wang, and H.-J. Zhang. Tag ranking. In *Proceedings of International Conference on World Wide Web*, 2009.
51. Y. Liu, B. Gao, T.-Y. Liu, Y. Zhang, Z. Ma, S. He, and H. Li. Browserank: Letting web users vote for page importance. In *International ACM SIGIR Conference on Research and Development in Information Retrieval*, pp. 451–458, 2008.
52. Y. Liu, D. Xu, I. Tsang, and J. Luo. Using large-scale web data to facilitate textual query based retrieval of consumer photos. In *Proceedings of the 17th ACM International Conference on Multimedia*, pp. 55–64. ACM, 2009.
53. Y. Liu, J. Zhang, Z. Li, and D. Tjondronegoro. High-level concept annotation using ontology and probabilistic inference. In *ICIMCS*, pp. 97–101, 2009.
54. A. C. Loui and A. E. Savakis. Automatic image event segmentation and quality screening for albuming applications. In *IEEE International Conference on Multimedia and Expo*, pp. 1125–1128.
55. J. Luo, J. Yu, D. Joshi, and W. Hao. Event recognition: Viewing the world with a third eye. In *ACM International Conference on Multimedia*, pp. 1071–1080, 2008.
56. Mahout library. http://lucene.apache.org/mahout.
57. M. Marszalek and C. Schmid. Semantic hierarchies for visual object recognition. *IEEE Conference on Computer Vision and Pattern Recognition, 2007. CVPR '07*, pp. 1–7, June 2007.
58. K. Matusiak. Towards user-centered indexing in digital image collections. *OCLC Systems and Services*, 22(4):283–298, 2006.
59. G. Mishne. Autotag: A collaborative approach to automated tag assignment for weblog posts. In *Proceedings of the 15th International Conference on World Wide Web*, 2006.
60. M. Naaman, S. Harada, and Q. Wang. Context data in geo-referenced digital photo collections. In *Proceedings of the 12th ACM International Conference on Multimedia*, pp. 196–203. ACM Press, 2004.
61. M. Naaman, Y. Song, A. Paepcke, and H. Garcia-Molina. Automatic organization for digital photographs with geographic coordinates. In *International Conference on Digital Libraries*, 7, pp. 53–62, 2004.
62. M. Naphade, L. Kennedy, J. Kender, S. Chang, J. Smith, P. Over, and A. Hauptmann. A light scale concept ontology for multimedia understanding for TRECVID 2005. *IBM Research Report RC23612 (W0505-104)*, 2005.
63. M. Naphade, I. Kozintsev, and T. Huang. Factor graph framework for semantic video indexing. *IEEE Transaction on CSVT*, 12(1): 40–52, 2002.
64. M. R. Naphade, T. Kristjansson, B. Frey, and T. S. Huang. Probabilistic multimedia objects (multijects): A novel approach to video indexing and retrieval in multimedia systems. In *ICIP*, pp. 536–540, 1998.
65. D. Newman, A. Asuncion, P. Smyth, and M. Welling. Distributed inference for latent dirichlet allocation. *Advances in Neural Information Processing Systems*, 2007.
66. A. Ng and M. Jordan. On discriminative vs. generative classifiers: A comparison of logistic regression and naive bayes. In *Advances in Neural Information Processing Systems*, 2002.
67. M. Pagani. *Encyclopedia of Multimedia Technology and Networking*. Information Science Reference—Imprint of: IGI Publishing, Hershey, PA, 2008.
68. K.-W. Park, J.-W. Jeong, and D.-H. Lee. Olybia: ontology-based automatic image annotation system using semantic inference rules. In *DASFAA'07: Proceedings of the 12th International Conference on Database Systems for Advanced Applications*, pp. 485–496, 2007.
69. A. Paterek. Improving regularized singular value decomposition for collaborative filtering. In *Proceedings of KDD Cup and Workshop*, August 2007.
70. G.-J. Qi, X.-S. Hua, Y. Rui, J. Tang, T. Mei, and H.-J. Zhang. Correlative multi-label video annotation. In *Proceedings of the 14th Annual ACM International Conference on Multimedia*, Augusberg, Germany, 2007. ACM.
71. G.-J. Qi, X.-S. Hua, Y. Rui, J. Tang, Z.-J. Zha, and H.-J. Zhang. A joint appearance-spatial distance for kernel-based image categorization, In *IEEE Conference on Computer Vision and Pattern Recognition*, Anchorage, Alaska, June 24–26, 2008.

72. G.-J. Qi, X.-S. Hua, Y. Rui, J. Tang, and H.-J. Zhang. Two-dimensional multi-label active learning with an efficient online adaptation model for image classification. *IEEE Transactions on Pattern Analysis and Machine Intelligence*, 31(10):1880–1897, 2009.

73. G.-J. Qi, X.-S. Hua, and H.-J. Zhang. Learning semantic distance from community-tagged media collection. In *Proceedings of ACM International Conference on Multimedia*, 2009.

74. T. Quack, B. Leibe, and L. Van Gool. World-scale mining of objects and events from community photo collections. *ACM Conference on Image and Video Retrieval*, pp. 47–56, 2008.

75. A. Rakotomamonjy, F. Bach, S. Canu, and Y. Grandvalet. SimpleMKL. *Journal of Machine Learning Research*, 9:2491–2521, 2008.

76. P. Resnick, N. Iacovou, M. Suchak, P. Bergstrom, and J. Riedl. Grouplens: An open architecture for collaborative filtering of netnews. In *Proceedings of ACM CSCW*, 1994.

77. S. Roweis and L. Saul. Nonlinear dimensionality reduction by locally linear embedding. *Science*, 290(5500): 2323–2326, 2000.

78. X. Rui, M. Li, Z. Li, W.-Y. Ma, and N. Yu. Bipartite graph reinforcement model for web image annotation. In *ACM International Conference on Multimedia*, pp. 585–594, 2007.

79. Y. Rui, T. Huang, and S.-F. Chang. Image retrieval: Current techniques, promising directions, and open issues. *Journal of Visual Communication and Image Representation*, 10: 39–62, 1999.

80. D. Rumelhart, G. Hinton, and R. Williams. *Parallel Data Processing*, volume 1, chapter Learning Internal Representation by Error Propagation, pp. 318–362. The M.I.T. Press, Cambridge, MA, 1986.

81. B. Sarwar, G. Karypis, J. Konstan, and J. Riedl. Item-based collaborative filtering recommendation algorithms. In *Proceedings of WWW*, 2001.

82. A. T. (Guus) Schreiber, B. Dubbeldam, J. Wielemaker, and B. Wielinga. Ontology-based photo annotation. *IEEE Intelligent Systems*, 16(3):66–74, 2001.

83. S. Shalev-Shwartz, Y. Singer, and N. Srebro. Pegasos: Primal estimated sub-gradient solver for svm. In *Proceedings of Internatioanl Conference on Machine Learning*, 2007.

84. B. Sigurbjorsnsson and B. van Zwol. Flickr tag recommendation based on collective knowledge. In *Proceedings of the 17th International Conference on World Wide Web*, 2008.

85. J. R. Smith and M. Naphade. Multimedia semantic indexing using model vectors. In *Proceedings of IEEE International Conferences on Multimedia and Expo*, 2003.

86. J. Tang, S. Yan, R. Hong, G.-J. Qi, and T.-S. Chua. Inferring semantic concepts from community-contributed images and noisy tags. In *Proceedings of ACM International Conference on Multimedia*, 2009.

87. B. L. Tseng, Y. Wu and J. R. Smith. Ontology-based multi-classification learning for video concept detection. In *ICME*, pp. 1003–1006, 2004.

88. H. Wang, X. Jiang, L.-T. Chia, and A.-H. Tan. Wikipedia2onto — adding wikipedia semantics to web image retrieval. In *WebSci'09: Society on-Line*, 2009.

89. J. Wang, A. P. D. Vries, and M. J. T. Reinders. Unifying user-based and item-based collaborative filtering approaches by similarity fusion. In *Proceedings of SIGIR*, 2006.

90. J. Wang, Y. Zhao, X. Wu, and X.-S. Hua. A transductive multi-label learning approach for video concept detection. *Pattern Recognition*, 44: 2274–2286, 2011.

91. X.-J. Wang, L. Zhang, X. Li, and W.-Y. Ma. Annotating images by mining image search results. *IEEE Transactions on Pattern Analysis and Machine Intelligence*, 30(11):1919–1932, 2008.

92. Y. Wang, J. Yang, W. Lai, R. Cai, L. Zhang, and W. Ma. Exploring traversal strategy for efficient web forum crawling. In *Proceedings of SIGIR*, 2008.

93. M. K. Warmuth and D. Kuzmin. Randomized online pca algorithms with regret bounds that are logarithmic in the dimension. *Journal of Machine Learning Research*, 9:2287–2320, 2008.

94. Q. Weinberger, M. Slaney, and R. V. Zwol. Resolving tag ambiguity. In *Proceedings of International ACM Conference on Multimedia*, 2008.

95. J. Wright and Y. Ma. Dense error correction via L1-minimization. *IEEE Transactions on Information Theory*, 2010.

96. J. Wright, A. Yang, A. Ganesh, S. Sastry, and Y. Ma. Robust face recognition via sparse representation. In *International Conference on Computer Vision*, 2009.

97. Y. Wu, B. Tseng, and J. Smith. Ontology-based multi-classification learning for video concept detection. In *Proceedings of IEEE International Conference on Multimedia and Expo*. IEEE, 2004.

98. Z. Xu, Y. Fu, J. Mao, and D. Su. Towards the semantic web: Collaborative tag suggestions. In *Collaborative Web Tagging Workshop at WWW 2006*, 2006.

99. R. Yan, M.-O. Fleury, M. Merler, A. Natsev, and J. R. Smith. Large-scale multimedia semantic concept modeling using robust subspace bagging and mapreduce. In *ACM Workshop on Large-Scale Multimedia Retrieval and Mining*, pp. 35–42, 2009.

100. J. Yang, R. Yan, and A. Hauptmann. Cross-domain video concept detection using adaptive svms. In *ACM Conference on Multimedia*, 2007.

101. J. Ye, J.-H. Chow, J. Chen, and Z. Zheng. Stochastic gradient boosted distributed decision trees. In *CIKM '09*, 2009.

102. J. Yu and J. Luo. Leveraging probabilistic season and location context models for scene understanding. In *International Conference on Content-Based Image and Video Retrieval*, pp. 169–178, 2008.

103. Y. Zheng, M. Zhao, Y. Song, H. Adam, U. Buddemeier, A. Bissacco, F. Brucher, T. Chua, and H. Neven. Tour the World: Building a web-scale landmark recognition engine. *IEEE Conference on Computer Vision and Pattern Recognition*, 2009.

104. D. Zhou, O. Bousquet, T. Lal, J. Weston, and B. Scholkopf. Learning with local and global consistency. *Advances in Neural Information Processing Systems (NIPS)*, 16:321–328, 2004.

105. X. Zhou, N. Cui, Z. Li, F. Liang, and T. S. Huang. Hierarchical Gaussianizatin for image classification, In *IEEE Conference on Computer Vision*, 2009.

106. Z.-H. Zhou and M.-L. Zhang. Multi-Instance Multi-Label Learning with Application to Scene Classification. In *NIPS*, 2006.

107. X. Zhu, J. Laerty, and Z. Ghahramani. Semi-supervised learning: From Gaussian fields to Gaussian processes. *ICML*, 2003.

Editors

Dr. Ling Guan is a Tier I Canada research chair in multimedia and computer technology and a professor of electrical and computer engineering at Ryerson University, Canada. He received his PhD from University of British Columbia in 1989. Dr. Guan has been conducting research in image and multimedia signal processing and communications, and published extensively in the field. He serves/served on the editorial boards of numerous international journals, including *IEEE Transactions on Multimedia, IEEE Transactions on Circuit and Systems for Video Technology, IEEE Transactions on Neural Networks, IEEE Transactions on Evolutionary Computations, IEEE Signal Processing Magazine*, and *IEEE Computational Intelligence Magazine*. Dr. Guan also guest-edited special issues for *Proceedings of the IEEE* (September 1999) and a dozen other international journals. He served as the general chair of the 2006 *IEEE International Conference on Multimedia and Expo*, Toronto, Canada, and the general co-chair of the 2008 *ACM International Conference on Image and Video Retrieval*, Niagara Falls, Canada. In 2000, Dr. Guan played the leading role in the inauguration of *IEEE Pacific-Rim Conference on Multimedia* and served as the founding general chair. Dr. Guan is a Fellow of the IEEE, a Fellow of the Engineering Institute of Canada, and a Fellow of the Canadian Academy of Engineering. He is an IEEE Circuits and System Society Distinguished Lecturer (2010–2011) and a recipient of the 2005 IEEE Transactions on Circuits and Systems for Video Technology Best Paper Award.

Dr. Yifeng He has been an assistant professor in the Department of Electrical and Computer Engineering, Ryerson University, Canada, since July 2009. He received his PhD from Ryerson University, Canada, in 2008. Dr. He was a research intern at Microsoft Research Asia in 2007, and a postdoctoral fellow at Ryerson University from 2008 to 2009. His research interests include Peer-to-Peer (P2P) video streaming, wireless video streaming, and resource allocation for multimedia communications. Dr. He has published over 40 research papers in the field of multimedia communications. He is the associate editor of the *International Journal of Digital Multimedia Broadcasting* and the *Journal of Communications and Information Sciences*. Dr. He is the recipient of 2008 Canada Governor General's Gold Medal, and 2007 Best Paper Award of Pacific-rim Conference on Multimedia (PCM).

Dr. Sun-Yuan Kung is a professor at the Department of Electrical Engineering in Princeton University. His research areas include VLSI (very large-scale integration) array processors, system modeling and identification, machine learning, wireless communication, sensor array processing, multimedia signal processing, and genomic signal processing and data mining. He was a founding member of several Technical Committees (TC) of the IEEE Signal Processing Society, and was appointed as the first associate editor in VLSI area (1984) and later the first associate editor in neural network (1991) for the *IEEE Transactions on Signal Processing*. He has been a Fellow of IEEE since 1988. He served as a member of the Board of Governors of the IEEE Signal Processing Society (1989–1991). Since 1990, he has been the editor-in-chief of the *Journal of VLSI Signal Processing Systems*. He was a recipient of IEEE Signal Processing Society's Technical Achievement Award for the contributions on "parallel processing and neural network algorithms for signal processing" (1992); a distinguished lecturer of IEEE Signal Processing Society (1994); a recipient of IEEE Signal Processing Society's

Best Paper Award for his publication on principal component neural networks (1996); and a recipient of the IEEE Third Millennium Medal (2000). He has authored and coauthored more than 400 technical publications and numerous textbooks, including *VLSI and Modern Signal Processing*, Prentice-Hall (1985), *VLSI Array Processors'*, Prentice-Hall (1988); *Digital Neural Networks'*, Prentice-Hall (1993); *Principal Component Neural Networks'*, John Wiley (1996); and *Biometric Authentication: A Machine Learning Approach*, Prentice-Hall (2004).

Contributors

S. K. Alamgir Hossain
School of Electrical Engineering and
    Computer Science
University of Ottawa
Ottawa, Canada

Dimitrios Androutsos
Department of Electrical and Computer
    Engineering
University of Ryerson
Toronto, Canada

Oscar Au
Department of Electronic and Computer
    Engineering
Hong Kong University of Science
    and Technology
Hong Kong, China

Shuvra S. Bhattacharyya
Department of Electrical and Computer
    Engineering
University of Maryland
College Park, Maryland

David Birchfield
School of Computing and Informatics
Arizona State University
Tempe, Arizona

Winslow Burleson
School of Computing and Informatics
Arizona State University
Tempe, Arizona

Jianfei Cai
School of Computer Engineering
Nanyang Technological University
Singapore

Hong Cao
Institute for Infocomm Research
    Technology
Singapore

S.-H. Gary Chan
Department of Computer Science and
    Engineering
The Hong Kong University of Science
    and Technology
Hong Kong, China

Chang Wen Chen
Department of Computer Science
    and Engineering
State University of New York
    at Buffalo
Buffalo, New York

Liang-Gee Chen
Department of Electrical Engineering
National Taiwan University
Taipei, Taiwan

Tung-Chien Chen
Department of Electrical Engineering
National Taiwan University
Taipei, Taiwan

Tat-Seng Chua
Institute for Infocomm Research
    Technology
Singapore

Tzu-Der Chuang
Department of Electrical Engineering
National Taiwan University
Taipei, Taiwan

Marius D. Cordea
University of Ottawa
Ottawa, Canada

Lionel Daniel
Department of Multimedia
    Communications EURECOM
Sophia Antipolis, France

Antitza Dantcheva
Department of Multimedia
   Communications
EURECOM
Sophia Antipolis, France

Igor Dolgov
School of Computing and Informatics
Arizona State University
Tempe, Arizona

Jean-Luc Dugelay
Department of Multimedia
   Communications
EURECOM
Sophia Antipolis, France

Nesli Erdogmus
Department of Multimedia
   Communications
EURECOM
Sophia Antipolis, France

Mani Malek Esmaeili
Department of Electrical and Computer
   Engineering
University of British Columbia
Vancouver, Canada

Mehrdad Fatourechi
Department of Electrical and Computer
   Engineering
University of British Columbia
Vancouver, Canada

Chuan Heng Foh
School of Computer Engineering
Nanyang Technological University
Singapore

Wen Gao
Institute of Digital Media
Peking University
Beijing, China

Ling Guan
Department of Electrical and Computer
   Engineering
Ryerson University
Toronto, Canada

Sarah Hatton
School of Computing and
   Informatics
Arizona State University
Tempe, Arizona

Yifeng He
Department of Electrical and Computer
   Engineering
Ryerson University
Toronto, Canada

Xian-Sheng Hua
Microsoft Corporation

April Khademi
Department of Electrical and Computer
   Engineering
University of Toronto
Toronto, Canada

Sung Dae Kim
School of Electrical and Computer
   Engineering
Ajou University
Suwon, Korea

Neslihan Kose
Department of Multimedia
   Communications EURECOM
Sophia Antipolis, France

Alex C. Kot
School of Electrical and Electronic
   Engineering
Nanyang Technological University
Singapore

Sridhar Krishnan
Department of Electrical and Computer
   Engineering
Ryerson University
Toronto, Canada

Sun Yuan Kung
Department of Electrical Engineering
Princeton University
Princeton, New Jersey

Gwo Giun (Chris) Lee
Department of Electrical Engineering
National Cheng Kung University
Tainan, Taiwan

Xiaoli Li
Department of Electrical and Computer
    Engineering
Ryerson University
Toronto, Canada

He Yuan Lin
Department of Electrical Engineering
National Cheng Kung
    University
Tainan, Taiwan

Xudong Lv
Department of Electrical and Computer
    Engineering
University of British Columbia
Vancouver, Canada

Ngok-Wah Ma
Department of Electrical and Computer
    Engineering
Ryerson University
Toronto, Canada

Siwei Ma
Institute of Digital Media
Peking University
Beijing, China

Wei-Ying Ma
Microsoft Research Asia

A. S. M. Mahfujur Rahman
School of Electrical Engineering and
    Computer Science
University of Ottawa
Ottawa, Canada

Brandon Mechtley
School of Computing and Informatics
Arizona State University
Tempe, Arizona

M. Colleen Megowan-Romanowicz
School of Computing and Informatics
Arizona State University
Tempe, Arizona

Rui Min
Department of Multimedia
    Communications EURECOM
Sophia Antipolis, France

Alan R. Moody
Department of Medical Imaging
Sunnybrook Health Sciences Center
Toronto, Canada

Chong-Wah Ngo
Deptartment of Computer Science
City University of Hong Kong
Hong Kong, China

Zefeng Ni
School of Computer Engineering
Nanyang Technological University
Singapore

Yaqing Niu
School of Information Engineering
Communication University of China
Beijing, China

Emil M. Petriu
School of Electrical Engineering and
    Computer Science
University of Ottawa
Ottawa, Canada

Raymond Phan
Department of Electrical and Computer
    Engineering
Ryerson University
Toronto, Canada

William Plishker
Department of Electrical and Computer
    Engineering
University of Maryland
College Park, Maryland

Gang Qian
School of Computing and Informatics
Arizona State University
Tempe, Arizona

Dongni Ren
Department of Computer Science
    and Engineering
The Hong Kong University of Science
    and Technology
Hong Kong, China

Richard Rzeszutek
Department of Electrical and Computer
    Engineering
Ryerson University
Toronto, Canada

A. El Saddik
School of Electrical Engineering and
    Computer Science
University of Ottawa
Ottawa, Canada

Chung-Ching Shen
Department of Electrical and Computer
    Engineering
University of Maryland
College Park, Maryland

Huifang Sun
Mitsubishi Electric Research Laboratories
Cambridge, Massachusetts

Myung Hoon Sunwoo
School of Electrical and Computer
    Engineering
Ajou University
Suwon, Korea

Song Tan
Department of Computer Science
City University of Hong Kong
Hong Kong, China

Harvey Thornburg
School of Computing and Informatics
Arizona State University
Tempe, Arizona

Carmelo Velardo
Department of Multimedia
    Communications EURECOM
Sophia Antipolis, France

Anastasios N. Venetsanopoulos
Department of Electrical and Computer
    Engineering
Ryerson University

and

Department of Electrical and Computer
    Engineering
University of Toronto
Toronto, Canada

Anthony Vetro
Mitsubishi Electric Research
    Laboratories
Cambridge, Massachusetts

Yongjin Wang
Department of Electrical and Computer
    Engineering
Ryerson University
Toronto, Canada

Z. Jane Wang
Department of Electrical and Computer
    Engineering
University of British Columbia,
Vancouver, Canada

Rabab K. Ward
Department of Electrical and Computer
    Engineering
University of British Columbia
Vancouver, Canada

Bin Wei
AT&T Labs Research
Florham Park, New Jersey

Thomas E. Whalen
Communications Research
    Centre
Ottawa, Canada

Zixiang Xiong
Department of Electrical and Computer
  Engineering
Texas A&M University
College Station, Texas

Huijuan Yang
Institute for Infocomm Research
Agency for Science, Technology
  and Research
Singapore

Linjun Yang
Microsoft Research Asia
Beijing, China

Jin Yuan
School of Computing National University
  of Singapore
Singapore

Bing Zeng
Department of Electrical and Electronic
  Engineering
Hong Kong University of Science
  and Technology
Hong Kong, China

Zheng-Jun Zha
School of Computing
National University of Singapore
Singapore

Hong-Jiang Zhang
Microsoft Corporation
Beijing, China

Lei Zhang
Microsoft Research Asia
Beijing, China

Li Zhang
Institute of Digital Media
Peking University
Beijing, China

Qin Zhang
School of Information Engineering
Communication University of China
Beijing, China

Ruofei Zhang
Yahoo! Labs
Sunnyvale, California

Yu Zhang
School of Computer Engineering
Nanyang Technological University
Singapore

Zhongfei (Mark) Zhang
Department of Computer Science
State University of New York
  at Binghamton
Binghamton, New York

and

Department of Information Science and
  Electronics Engineering
Zhejiang University
Zhejiang, China

Debin Zhao
Harbin Institute of Technology
Harbin, China

Xuran Zhao
Department of Multimedia
  Communications
EURECOM
Sophia Antipolis, France

Yisu Zhao
School of Electrical Engineering and
  Computer Science
University of Ottawa
Ottawa, Canada

Yan-Tao Zheng
Institute for Infocomm Research
Singapore

Part I

Fundamentals of Multimedia

1

Emerging Multimedia Standards

Huifang Sun

CONTENTS

1.1 Introduction

Significant progress in digital audio/video processing and communication technology has enabled the dream of many applications such as High-Definition Television (HDTV) broadcasting, digital versatile disk (DVD) storage, high-quality real-time audio/video streaming over various networks, and recent 3DTV. Among these technologies, digital audio/video coding or compression technique has assumed an important role to bridge the gap between huge amount of video data required for the transmission/storage and limited bandwidth of communication channels or limited size of storages. For most applications, a huge number of receivers are used. Therefore, the multimedia standards including audio/video/image coding standards are needed and they are very important for the multimedia industry.

Since the early 1990s, several multimedia standards including digital audio/video/image coding standards have been successfully developed. For developing these standards, especially for video coding standards, two international organizations have taken important roles and led the actions. These two organizations are the Moving Pictures Expert Group (MPEG) which belongs to International Organization for Standardization (ISO) and International Electrotechnical Commission (IEC), and the Video Coding Expert Group (VCEG) of the International Telecommunication Union (ITU). The video coding standards developed by ISO/IEC include MPEG-1 [1], MPEG-2 [2], and MPEG-4 [3], the still image coding, JPEG [4] and JPEG-2000 [5], as well as multimedia interface standard MPEG-7 [6] and Multimedia Framework standard MPEG-21 [7]. The standards developed by ITU include H.261 [8], H.263 [9], and H.264/AVC [10]. The H.262 is the same as MPEG-2, which is a joint standard of MPEG and ITU. The H.264/AVC video coding standard is developed by the joint video team (JVT) of MPEG and VCEG which is also referred to as the MPEG-4 Part 10 [10]. These standards have found many successful applications. The applications can be mainly classified into two aspects: digital television (DTV) and digital telephony. With the advances of computer technologies, personal computers (PCs) have become the most flexible and most inexpensive platform for digital video communication and players. Currently, a joint collaborative team on video coding (JCT-VC) of MPEG and VCEG is making efforts to develop a next-generation video coding which is referred to as high-efficiency video coding (HEVC).

From the history of multimedia standards development, it can be summarized that a successful multimedia standard, which is widely accepted by the industry and extensively used in many applications, should have several important features. The first important feature is that the standard should adopt the most advanced techniques developed at that time. These techniques are able to solve the most difficult problems to meet the immediate needs from the industry. For example, during the development of Advanced Television Systems Committee (ATSC) HDTV standard, the industry was required to provide full-quality HDTV service in a single 6 MHz channel. The function of compression layer has to compress the raw data from about 1 Gbps to the data rate of ~19 Mbps to satisfy the 6 MHz spectrum bandwidth requirement. This goal has been achieved by using the main profile and high level of MPEG-2 video coding standard and advanced digital modulation technique. The MPEG-2 video standard uses hybrid coding techniques, which mainly combine the discrete cosine transform (DCT) with motion-compensated predictive coding and other techniques to provide 50:1 compression ratio at the broadcast quality. Therefore, the MPEG-2 becomes the basis of the standard for DTV. Also the techniques adopted by the standard should consider both facts, that is, good coding performance and less computational complexity. Specially, the industry has to consider the implementation issues before accepting a standard. Sometimes, the techniques adopted by standards may be a trade-off between performance and computational complexity. Some techniques have significant technical innovation but may not be adopted by the previous standard due to the implementation difficulty at that time. But it may be adopted by the later standard or extension of standard with advances of computer and VLSI technology. Many examples about this can be found in the standard of H.264/AVC [10,11].

The second important feature is that the audio/video coding standards should be independent of transmission and storage media. The compressed video and audio bitstreams can be used for different applications. The bitstreams can be transmitted through broadcasting, computer network, wireless network, and other media. They can also be stored in different kinds of storage. In order to achieve this goal, a system layer standard has been developed. System standards specify the packetization, multiplexing, and packet header syntax for delivering or storing the audio- and video-compressed bitstreams in different media.

Another feature the standards should have is compatibility. Compatibility is important for manufacturers. If a new standard can decode the bitstreams of prior standards, it is easy to let new products be introduced to the market. For example, since the MPEG-2 decoders are able to decode the MPEG-1 bitstreams, the DVD receivers, which use the MPEG-2 standard, can handle the video CDs (VCDs), which contain MPEG-1-compliant bitstreams. Compatibility is a very important feature for the DTV and DVD industry, which cannot upgrade the hardware easily. But compatibility is not essential for some applications, for instance, in the computer industry. The most PCs have big memory and high-speed CPU, which allow PCs to use software decoders. The software decoders would easily be upgraded, or multiple decoder software can be installed in the PCs. For example, the MPEG-4 and H.264/AVC are not backward compatible with MPEG-2 and H.263, respectively. Also, it should be noted that with rapidly advanced VLSI technology, it is not difficult to provide the decoders which can decode multistandards. For example, the Blu-ray standard uses several video coding standards including MPEG-2, H.264/AVC, and VC-1. The transcoder is also a solution.

Before we describe in detail the multimedia standards, a very important concept should be noted, that is, the standard should only define the syntax and semantics of the compressed bitstream. This means that a standard defines only the decoder process. How to encode and generate the bitstream is not the normative part of a standard. In other words, the standard does not define the encoder. However, during the development of the standard, both the encoder and the decoder are needed to verify the syntax and semantics of the bitstreams. Therefore, most standards provide the reference software encoder as an informative part of the standard which is usually not optimized. In this sense, the encoding algorithms are open for competition. Different encoding algorithms may generate the bitstreams with different coding performance and different computational complexity. Only one thing is common for all kinds of encoders, the generated bitstreams must be standard compliant. Therefore, the encoder manufacturers can compete on the items of coding performance and complexity. On the contrary, decoder manufacturers need to produce standard-compliant decoders, but can compete not only on the price but also on additional features such as postprocessing and error concealment capability.

The another concept that should also be mentioned is that for multimedia applications the source information includes not only video and image but also audio as well as data. The different sources of information are all converted into the digital format referred to as elementary bitstreams at the source coding layer. Then these elementary bitstreams are mixed together to a new bitstream at the system layer with additional information such as information for time synchronization. This new format of information, binary bitstream, is a revolutionary change in the multimedia industry, since the digitized information format, that is, the bitstream, can be decoded by not only the traditional consumer electronic products such as television and any kind of video recorder but also the digital computers through different transmission media.

Subsequently, we describe the technical details of major multimedia coding and multimedia interface as well as multimedia framework standards.

1.2 Digital Video Coding

1.2.1 MPEG-1 and MPEG-2 Video

The MPEG-1 was completed in 1991 [1]. The target application of MPEG-1 is digital storage media, CD-ROM, at bit rates up to 1.5 Mbps. The MPEG-2 [2] is also referred to as H.262, as a joint standard of MPEG and ITU. The MPEG-2 was completed in 1994. It is an extension of MPEG-1 and allows for greater input format flexibility and higher data rates for both HDTV and Standard Definition Television (SDTV). The US ATSC DTV standard and European DTV standard DVB both use the MPEG-2 as their source-coding standard but use different transmission systems. At the system layer, the DTV uses Transport Stream, which is designed for the lossy transmission environment. The MPEG-2 is also used for DVD. The data format of DVD is MPEG-2 Program Stream which is designed for the clear transmission environment. Since DVD can provide SDTV quality, which is much better than the traditional analog VCR or digital VCD.

As we know, the MPEG-2 video coding has the feature of being backward compatible with MPEG-1. This means that the MPEG-2 decoders can decode the MPEG-1 bitstreams. It turns out to be that most of the decoders in the market are MPEG-2-compliant decoders. For simplicity, we introduce the technical details of MPEG-1 and then describe the enhanced features of MPEG-2, which MPEG-1 does not have. Since most video coding standards have a similar structure, we shall provide the technical details with regard to MPEG-1 video and will only describe the difference or enhancements with MPEG-1 video for other coding standards.

The MPEG-1 video algorithm is mainly based on DCT coding and interframe motion compensation. A typical MPEG-1 video encoder structure is shown in Figure 1.1.

The MPEG-1 video allows only progressive pictures, but offers great flexibility in size, up to 4095×4095 pixels. The coder itself is optimized to the extensively used video Source Input Format (SIF) picture format, which has luminance resolution of 360×240 pixels/frame at 30 frames/second or 360×288 pixels/frame at 25 frames/second. The SIF is a simple derivative of the ITU-R 601 video format for DTV applications. According to ITU-R 601, a color video source has three components, a luminance component (Y) and two chrominance components (Cb and Cr) which are in the 4:2:0 subsampling format. In order to remove both spatial and temporal redundancy, the video sequence is first divided into groups of pictures (GOPs). Each GOP may include three types of pictures or frames: intracoded (I) picture or frame, predictive-coded (P) picture or frame, and bidirectionally predictive-coded (B) picture or frame. I-pictures are coded by intraframe techniques only, with no need for previous information. In other words, I-pictures are self-sufficient. They are used as anchors for forward and/or

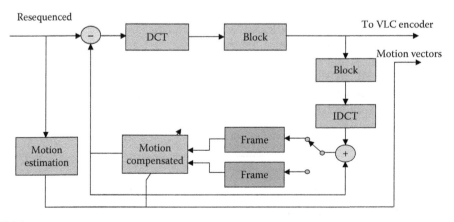

FIGURE 1.1
Typical MPEG.1 encoder structure.

backward prediction, and also they can serve as points of random access. P-pictures are coded using one-directional motion-compensated prediction from a previous anchor frame, which could be either I-picture or P-picture. The distance between two nearest I-frames is denoted by N, which is the size of GOPs. The distance between two nearest anchor frames is denoted by M. Parameter N and M both are user selectable, which are selected by the user during encoding. A larger number of N and M will increase the coding performance but cause error propagation or drift. Usually, N is chosen from 12 to 15 and M from 1 to 3 in MPEG-1 and MPEG-2 video. If M is selected to be 1, this means that no B-picture will be used, which can be used for low-delay applications. Finally, the B-pictures can be coded using predictions from either past or future anchor frames (I or P), or both. Regardless of the type of frame, each frame may be divided into slices; each slice consists of several macroblocks (MBs). There is no rule to decide the slice size. A slice could contain all MBs in a row of a frame or all MBs of a frame. Smaller slice size is favorable for the purpose of error resilience, but will decrease coding performance due to higher overhead. It should be noted that if B-pictures are used as in Figure 1.2a, the transmission or encoding order and display order will be different as shown in Figure 1.2b and c, respectively. Since the encoding order is different from the display order, the input sequence has to be reordered for encoding. A typical example of a GOP structure is shown as in Figure 1.2.

As it is shown in Figure 1.1, the DCT coding is used to remove the intraframe redundancy and the motion compensation is used to remove the interframe redundancy. In order to remove the spatial redundancy in the input video, the video frame is first decomposed to MBs. An MB contains a 16×16 luminance (Y) component and spatially corresponding two 8×8 croma (Cb and Cr) components for 4:2:0 video format. The luminance component of an MB is further divided into four 8×8 blocks. Therefore, an MB contains four luminance blocks and two chrominance blocks (for 4:2:0 sampling format). The DCT operation is then performed on each block as shown in Figure 1.1. The original spatial domain data are then transferred to the frequency domain for each block. In the original spatial domain all pixels are equally important, but the transformed coefficients are no longer equally important since the lower-order coefficients contain more energy than higher-order coefficients from statistics for most natural videos. Therefore, the video data can be coded more efficiently in the frequency domain than the spatial pixel domain.

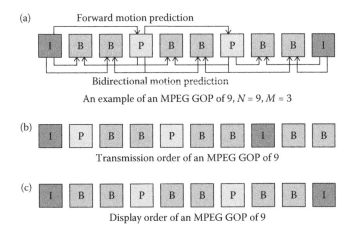

(a) Forward motion prediction

Bidirectional motion prediction

An example of an MPEG GOP of 9, $N = 9$, $M = 3$

(b) Transmission order of an MPEG GOP of 9

(c) Display order of an MPEG GOP of 9

FIGURE 1.2
(a) An example of an MPEG GOP of 9, $N = 9$, $M = 3$. (b) Transmission order of an MPEG GOP of 9 and (c) display order of an MPEG GOP of 9.

The motion-compensated prediction is used to remove the interframe redundancy. The fundamental model used assumes that a translational motion can approximate the motion of a block. If all elements in a video scene are approximately spatially displaced, the motion between frames can be described by a limited number of motion vectors. It is obvious that it would be too expensive to transmit motion vector for each pixel in the block. Actually, limited motion vectors for each block would be sufficient since the spatial correlation between adjacent pixels is usually very high. The MPEG-1/2 video uses the MB structure for motion compensation, that is, for each 16×16 MB only one or sometimes two motion vectors are transmitted. The motion vectors for any block are found within a search window that can be up to 512 pixels in each direction. Also, the matching can be done at half-pixel accuracy, where the half-pixel values are computed by averaging the full-pixel values. All MBs in the I-frame are coded in intramode without motion compensation. The MBs in P- and B-frames can be coded in several modes. The mode decision is usually made by the optimized rate-distortion process with the price of increasing the encoding complexity. Within each slice, the values of motion vectors and DC values of each MB are coded using differential pulse-code modulation (DPCM). The structure of MPEG coder implies that if an error occurs within I-frames, it will be propagated through all frames in the GOPs. Similarly, an error in a P-frame will affect the related P- and B-frames, while B-frame errors will be isolated.

The transform coefficients are then quantized. During the process of quantization a weighted quantization matrix is used. The function of quantization matrix is to quantize high frequencies with coarser quantization steps that will suppress high frequencies with no subjective degradation, thus taking advantage of human visual perception characteristics. The bits saved for coding high frequencies are used for lower frequencies to obtain better subjective coded images. Two quantizer weighting matrices are used in MPEG-2 video, an intraquantizer weighting matrix and a nonintraquantizer weighting matrix; the later is more flat since the energy of coefficients in interframe coding is more uniformly distributed than in intraframe coding.

The transformed coefficients are processed in zigzag order since most of the energy is usually concentrated in the lower-order coefficients. The zigzag ordering of elements in an 8×8 matrix as shown in Figure 1.3 allows for a more efficient run-length coder.

Finally, the zigzagged run-Zoro pairs of intra-/interquantized coefficients are coded by the VLC (variable length coding) codes. The VLC tables are obtained by statistically

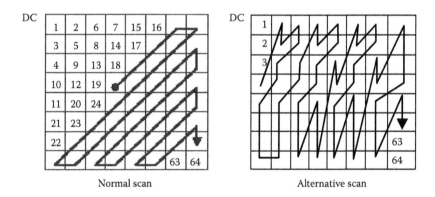

Normal scan   Alternative scan

FIGURE 1.3
Two zigzag scan methods for MPEG-2 video coding. (From MPEG-2 Test model 5, *ISO/IEC/JTC1/SC29/WG11*, April, 1993. With permission.)

optimizing a large number of training video sequences and are included in the MPEG-2 specification. The same idea is applied to code the DC values, motion vectors, and other information. Therefore, the MPEG-1/2 video standard contains a number of VLC tables. After coding, all the information is converted into binary bits. The MPEG video bitstream consists of several well-defined layers with headers and data fields. These layers include sequence, GOP, picture, slice, MB, and block. In the sequence layer, the header contains the information of picture size, frame rate, bit rate, buffering requirement, and programmable coding parameters. The header of the GOP layer provides information about random access and time code. In the picture layer, the picture header contains the timing information, bufferfullness, temporal reference, and coding type (I, P, or B). In the slice layer, the header provides the intraframe addressing information and coding reinitialization for error resilience. The header of the MB contains information about the basic coding structure, coding mode, motion vectors, and quantization. The lowest layer comprises blocks that contain DCT coefficients. The syntax elements contained in the headers and the amount of bits defined for each element can be found in the standard.

1.2.2 MPEG-2 Enhancements

The basic coding structure of MPEG-2 video is the same as that of MPEG-1 video, that is, the intraframe and interframe DCT coding with I-, P-, and B-pictures is used; while in the P- and B-pictures the motion-compensated prediction is used. The most important enhancements of MPEG-2 over MPEG-1 video coding include:

- Field/frame prediction modes for supporting the interlaced video input
- Field/frame DCT coding syntax
- Downloadable quantization matrix and alternative scan order
- Scalability extension

The above enhancement items are all coding performance improvements related to the support of interlaced material. There are also several noncompression enhancements, which include:

- Syntax to facilitate 3:2 pull-down in the decoder
- Pan and scan codes with 1/16-pixel resolution
- Display flags indicating chromaticity, subcarrier amplitude, and phase (for NTSC/PAL/SECAM source material)

Technical details of the enhancement items can be found in reference [2].

1.2.3 H.261 and H.263

As mentioned previously, the ITU VCEG developed the video coding standards H.261 [6] and H.263 [7]. The H.261 was published by the ITU (International Telecom Union) in 1990, and it is mainly used for Integrated Services Digital Network (ISDN) video conferencing.

The H.261 has many common features with the MPEG-1 video. However, since they target different applications, there exist many differences between the two standards such as data rates, picture quality, end-to-end delay, and others.

The major similarity between H.261 and MPEG-1 can be described as follows.

First, the key coding algorithms of H.261 and MPEG-1 are very similar. Both H.261 and MPEG-1 use the DCT-based block coding to remove intraframe redundancy and use the motion-compensated predictive coding to remove interframe redundancy.

Second, both standards are used to code the similar video format. H.261 is mainly used to code Common Interchange Format (CIF) which has luminance resolution of 352×288 pixels/frame at 30 frames/second and the chrominance has half the luminance resolution in both vertical and horizontal dimensions, or Quarter Common Interchange Format (QCIF). Format QCIF spatial resolution for teleconference application. MPEG-1 uses CIF, SIF, or MPEG-2 is used for DTV and DVD.

The main differences between the H.261 and MPEG-1 include the following H.261 uses only I- and P-MBs but no B-MBs while MPEG-1 uses three MB types, I-, P-, B-MBs (I-MB is an intraframe-coded MB, P-MB is a predictive-coded MB, and B-MB is a bidirectionally coded MB), also three picture types, I-, P-, and B-pictures as defined in the MPEG-1 standard.

There is a constraint of H.261 that for every 132 interframe-coded MBs, which correspond to four group of blocks (GOBs) or to one-third of CIF pictures, it requires at least one intraframe-coded MB. In order to obtain better coding performance at low-bit-rate applications, most encoding schemes of H.261 prefer not to use intraframe coding on all the MBs of a picture but only few MBs in every picture with a rotational scheme. MPEG-1 uses the GOPs' structure, where the size of GOPs (the distance between two I-pictures) is not specified.

The end-to-end delay is not a critical issue for MPEG-1, but is critical for H.261. The video encoder and video decoder delays of H.261 need to be known to allow audio compensation delays to be fixed when H.261 is used in interactive applications. This will allow lip synchronization to be maintained.

The accuracy of motion compensation in MPEG-1 is up to a half-pixel, but only a full pixel is used in H.261. However, H.261 uses a loop-filter to smooth the previous frame. This filter attempts to minimize the prediction error.

In H.261, a fixed picture aspect ratio of 4:3 is used. In MPEG-1, several picture aspect ratios can be used and the picture aspect ratio is defined in the picture header.

Finally, in H.261 the encoded picture rate is restricted to allow up to three skipped frames. This would allow the control mechanism in the encoder to have some flexibility for controlling the encoded picture quality and at the same time for satisfying the buffer regulation. Although MPEG-1 has no restriction on skipped frames, the encoder usually does not perform frame skipping. Rather, the syntax for B-frames is exploited, as B-frames require much fewer bits than P-pictures.

The H.263 is based on the H.261 framework and was completed in 1996. The H.263 encoder structure is similar to that of the H.261 encoder. The main components of an H.263 encoder include DCT-based block transform, motion-compensated prediction, block quantization, and VLC. However, there is no loop filter in an H.263 encoder. For H.263 encoder, each picture is partitioned into GOBs. A GOB contains multiple number of 16 lines, $k*16$ lines, depending on the picture format ($k = 1$ for sub-QCIF, QCIF; $k = 2$ for 4CIF; $k = 4$ for 16CIF). Each GOB is divided into MBs that are the same as in H.261, each MB consists of four 8×8 luminance blocks and two 8×8 chrominance blocks. The H.263 video includes more computationally intensive and efficient algorithms to increase the coding performance. It is used for video conferencing with analog telephone lines, desktop computer, and mobile terminals connected to the Internet. After the baseline of H.263 was approved, more than 15 additional optional modes have been added. These optional modes provide the tools for improving coding performance for very low-bit-rate applications and addressing the needs of mobile video and other noisy transmission environments.

These new features for improving coding efficiency for very low-bit-rate applications include picture-extrapolating motion vectors (or unrestricted motion vector mode), motion compensation with half-pixel accuracy, advanced prediction (which includes variable-block size motion compensation and overlapped block motion compensation), syntax-based arithmetic coding, and PB-frame mode.

1.2.4   MPEG-4 Part 2 Video

MPEG-4 was completed in 2000 [3,13]. It is the first object-based video coding standard and is designed to address the highly interactive multimedia applications. The MPEG-4 can provide tools for efficient coding, object-based interactivity, object-based scalability, and error resilience. The MPEG-4 provides tools not only for coding natural videos, but also for synthetic video and audio, still image, and graphics.

The major feature of MPEG-4 is to provide the technology for object-based compression, which is capable of separately encoding and decoding video objects (VOs). To clearly explain the idea of object-based coding, we should review the set of VO-related definitions. An image scene may contain several objects. The time instant of each VO is referred to as the VO plane. The concept of a VO provides a number of functionalities of MPEG-4 which are either impossible or very difficult in MPEG-1 or MPEG-2 video coding. Each VO is described by the information of texture, shape, and motion vectors. The video sequence can be encoded in a way that will allow the separate decoding and reconstruction of the objects and allow the editing and manipulation of the original scene by simple operation on the compressed bitstream domain. The feature of object-based coding is also able to support functionality such as warping of synthetic or natural text, textures, images, and video overlays on reconstructed VOs.

Since MPEG-4 aims at providing coding tools for multimedia environment, these tools not only allow one to efficiently compress natural VOs, but also compress synthetic objects, which are a subset of the larger class of computer graphics. The tools of MPEG-4 video include:

- Motion estimation and compensation
- Texture coding
- Shape coding
- Sprite coding
- Interlaced video coding
- Wavelet-based texture coding
- Generalized temporal and spatial as well as hybrid scalability
- Error resilience

The technical details of these tools can be found in [3,7].

1.2.5   MPEG-4 Part 10 Advanced Video Coding/H.264

H.264 is also referred to as MPEG-4 Part 10 Advanced Video Coding [10,11] which was developed by the JVT of ISO and ITU. The goal of H.264/AVC is to provide very high coding performance, which is much higher than MPEG-2 and MPEG-4 at a large range of bit rates. To achieve this goal, the H.264/AVC uses many new tools, which are different or modified from MPEG-2 and MPEG-4 Part 2, for improving coding performance. These tools include multiple reference frames that can be up to 15 frames, multiple motion compensation modes with block size from 16×16 to 4×4, up to quarter-pixel accuracy motion compensations, small block-size transformation, in-loop deblocking filter, adaptive arithmetic entropy coding, and many other tools.

Based on the conventional block-based motion-compensated hybrid video coding concepts with many new tools, the H.264/AVC provides approximately 50% bit-rate savings from equivalent perceptual quality relative to the performance of prior standards. This has been shown from extensive simulation results. As we have mentioned that the superior coding performance of H.264/AVC is obtained because many new features are incorporated. With high coding efficiency, the H.264/AVC can provide technical solutions for many applications. However, it will not be that easy to replace the current existing standard such as MPEG-2 with H.264/AVC in some applications such as in the area of the DTVs. However, it may be used for new application areas, such as nest-generation broadcasting, video storage on optical and magnetic devices, Blu-ray disk, mobile video transmission, and video streaming over different media. On the contrary, to achieve the high coding efficiency, H.264/AVC has to use a lot of new tools or modified tools from existing standards that substantially increases the complexity of the codec; the complexity would be about four times higher for the decoder and nine times higher for the encoder compared with MPEG-2 video coding standard. However, with fast advances of the semiconductor technique, the silicon solution can alleviate the problem of high complexity.

Compared to other existing video coding standards, the basic coding structure of H.264/AVC is similar, which is the structure with the motion-compensated transform coding. The block diagram of H.264/AVC video encoder is shown in Figure 1.4.

Except for many common tools, the H.264/AVC includes many highlighted features that are enabled to greatly improve the coding efficiency and increase the capability of error robustness and the flexibility for operation over a variety of network environments. Features for improving coding efficiency can be classified into two parts: the first is to improve the accuracy of prediction for the picture to be encoded and the second includes the method of transform and entropy coding. Several tools have been adopted in H.264/AVC to improve inter- and intraprediction, which are briefly summarized as follows.

Variable block size for motion compensation with small block sizes is used, in which seven selections of block sizes are used for motion compensation in H.264/AVC, among which the smallest block size for luma motion compensation can be as small as 4×4.

The quarter-pixel accurate motion compensation is adopted in H.264/AVC. The quarter-pixel accurate motion compensation has been used in the advanced profile of MPEG-4 Part 2, but H.264/AVC further reduces the complexity of the interpolation process.

Multiple reference pictures for motion compensation and weighted prediction are used to predict the P-pictures and B-pictures. The number of reference pictures can be up to 15 for level 3.0 or lower and four reference pictures for levels higher than 3.0. When the multiple reference pictures are used for motion compensation prediction, the contribution of prediction from different references should be weighted and offset by amounts specified by the encoder. This can greatly improve coding efficiency for those scenes that contain fades.

Directional spatial prediction for intracoding is adopted for further improving coding efficiency. In this technique, the intracoded regions are predicted with the references of the previously coded areas, which can be selected from different spatial directions. In such a

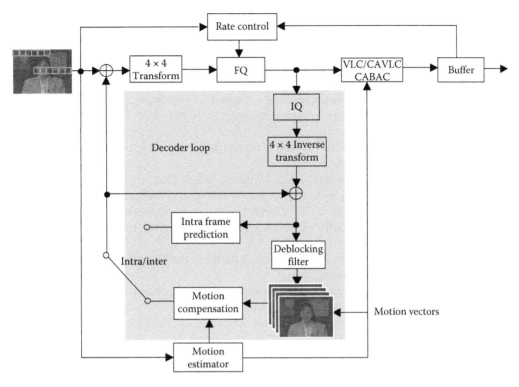

FIGURE 1.4
Block diagram of an H.264 encoder.

way the edges of the previously decoded areas of the current picture can be extrapolated into the current intracoded regions.

Skip mode in P-picture and direct mode for B-picture are used to alleviate the problem for using too many bits for coding motion vectors in the interframe coding. H.264/AVC uses the skip mode for P-pictures and direct mode for B-pictures. In these modes, the reconstructed signal is obtained directly from the reference frame with the motion vectors derived from previously encoded information by exploiting either spatial (for skip mode) or temporal (for direct mode) correlation of the motion vectors between adjacent MBs or pictures. In such a way, bits saving for coding motion vectors can be achieved.

The use of loop deblocking filters is another feature that is used to reduce the block artifacts and improve both objective and subjective video quality. The difference from MPEG-1/2 is that in H.264/AVC the deblocking filter is brought within the motion compensation loop, so that it can be used for improving the interframe prediction and therefore improving the coding efficiency.

H.264/AVC uses a small transform block size of 4×4 instead of 8×8 as in most video coding standards. The merit of using the small transform block size is able to encode the picture in a more local adaptive fashion, which would reduce the coding artifacts such as ringing noise. However, the problem of using small transform block size may cause coding performance degradation due to the correlations of a large area, which may not be exploited for certain pictures. H.264/AVC uses two ways to alleviate this problem, one is by using a hierarchical transform to extend the effective block size of nonactive chroma information to an 8×8 block, and another is by allowing the encoder to select a special coding type of intracoding, which enables the extension of the length of the luma transform for nonactive area to a 16 × 16 block size. Also, for a high profile of H.264/AVC, the 8 × 8 transform is used to address the above problems. As mentioned previously, the basic function of integer transform used in H.264/AVC do not have an equal norm. To solve this problem, quantization table size has been increased.

Two very powerful entropy coding methods, content-adaptive variable length coding and content-adaptive binary arithmetic coding are used in H.264/AVC for further improving coding performance.

In H.264/AVC, several tools have been adopted for increasing the capability of error robustness.

Flexible slice size allows the encoder to adaptively select the slice size for increasing the capability of error robustness. Flexible macroblock ordering (FMO) allows partitioning the MBs into slices in a flexible order. Since each slice is an independently decodable unit, the FMO can significantly enhance error robustness by managing the spatial relationship between the MBs in the slice.

There are also several features that are used to increase the flexibility for operation over a variety of network environments.

The parameter set structure is used to provide a more flexible way to protect the key header information and increase the error robustness.

The Network Abstraction Layer (NAL) unit syntax structure allows for carrying video content in a manner appropriate for each specific network in a customized way.

Arbitrary slice ordering is used to improve end-to-end delay in real-time application, particularly for the applications on the Internet protocol networks.

Switching P and switching I slices are new slice types. They are specially encoded slices that allow efficient switching between video bitstreams and efficient random access for video decoders. This feature can be used for efficiently switching a decoder to decode different bitstreams with different bit rates, recovery from errors, and trick modes.

An overview of H.264/AVC video coding standard can be found in [11] and the detailed specification can be found in [10]. The technical details of the above tools are described in the following sections.

1.2.6 New Video Coding Standard, HEVC

Recently, The MPEG and ITU started a joint effort again for developing a new video coding standard. At the *86th MPEG Meeting* in Busan 2008, MPEG determined the need for a next generation of video compression technology. The next-generation video coding standard would be intended mainly for high-quality applications, by providing performance in terms of coding efficiency at higher resolutions, with applicability for entertainment-quality services such as high-definition mobile, home cinema, and ultrahigh-definition TV [14]. A call for evidence [15] was issued that allowed proponents to report about the existence of such technologies. The response to this CfE was evaluated at the *89th MPEG Meeting* in July 2009. Although MPEG and VCEG could independently create separate next-generation video coding standards, two new standards of similar functionalities might not be welcomed by industry. Based on the previous success in jointly creating H.264/AVC, future collaboration was established at this meeting and a JCT-VC was created.

The target of new coding standard should be capable of providing a bit-rate reduction of at the same subjective quality compared to H.264/AVC high profile as used in these applications. More specifically, HEVC should be capable of operating with a complexity ranging from 50% to three times of H264/AVC High Profile. When the HEVC is operated at a complexity of 50% compared to H.264/AVC High Profile, it should provide a 25% bit-rate savings compared to H.264/MPEG-4 AVC High Profile at equivalent subjective quality [16].

The call for proposals (CfP) has been issued in the Kyoto *91st MPEG Meeting*, January 2010 [17]. The HEVC work officially started. There are in total 27 responses to CfP and results of subjective tests for these responses were reported in the July 2010 *93rd MPEG Meeting* [18]. The tools proposed in the proposals with top performance have been suggested to the test model under consideration (TMuC) [19]. In the TMuC, several new coding tools can be summarized as follows:

1. *Intraprediction*: In [20], H.264 intraprediction is enhanced with additional bidirectional intraprediction (BIP) modes, where BIP combines prediction blocks from two prediction modes using a weighting matrix.

2. *Interprediction*: To further improve interprediction efficiency, finer fractional motion prediction and better motion vector prediction were proposed. Increasing the resolution of the displacement vector from 1/4-pixel to 1/8-pixel to obtain higher efficiency of the motion compensated prediction is suggested in [21]. In [22], a competing framework for better motion vector coding and Skip mode is proposed, where both spatial and temporal redundancies in motion vector fields are captured. Moreover, [23] suggests extending the MB size up to 64×64 so that new partition sizes $64 \times 64, 64 \times 32, 32 \times 64, 32 \times 32, 32 \times 16$, and 16×32 can be used. Instead of using the fixed interpolation filter from H.264/AVC, adaptive interpolation filters (AIFs) are proposed, such as 2D AIF [24], separable AIF [25], directional AIF [26], Enhanced AIF [27], and enhanced directional AIF [28].

3. *Quantization*: To achieve better quantization, optimized quantization decision at the MB level and at different coefficient positions are proposed. Rate Distortion Optimized Quantization (RDOQ), which performs optimal quantization on an MB, was added to the JM reference software. RDOQ does not require a change of H.264/AVC decoder syntax. More recently, [29] gives an improved, more efficient RDOQ implementation. In [30], Adaptive Quantization Matrix Selection, a method deciding the best quantization matrix index, where different coefficient positions can have different quantization steps, is proposed to optimize the quantization matrix at an MB level.

4. *Transform*: For motion partitions bigger than 16×16, a 16×16 transform is suggested in addition to 4×4 and 8×8 transforms [23]. Moreover, transform coding is not always a must. In [31], it is proposed that for each block of the prediction error, either standardized transform coding or spatial domain coding can be adaptively chosen.

5. *In-Loop Filter*: Besides the deblocking filter, an additional adaptive loop filter (ALF) is added to improve coding efficiency by applying filters to the deblocking-filtered picture. Two different ALF techniques are adopted so far: Quadtree-based adaptive loop filter [32] and Block-based adaptive loop filter [33].

6. *Internal Bit Depth Increase*: By using 12 bits of internal bit depth for 8-bit sources, so that the internal bit depth is greater than the external bit depth of the video codec, the coding efficiency can be further improved [34].

Besides the techniques listed above, there are some noticeable contributions such as decoder side motion estimation for B-frame motion vector decision, which improves coding efficiency by saving bits on B-frame motion vector coding. Also, some new techniques are under investigation and will be presented in the future meeting.

In the October 2010 *94th MPEG Meeting*, the HEVC Test Model HM1 [35] was created based on the TMuC. The coding tools in HM1 are divided into two categories, high efficiency (HE) and low complexity (LC).

The coding tools included in both HE and LC are as follows:

- Coding units 8×8 up to 64×64 in tree structure
- Prediction units
- Transform block size of 4×4 to 32×32 samples
- Angular intraprediction up to 34 directions
- Advanced motion vector prediction
- Deblocking

The tools only in HE include:

- Transform unit tree (3 level maximum)
- 12-Tap DCT-based interpolation filter
- CABAC entropy coding
- Internal bit-depth increase (4 bits)
- ALF

The tools only in LC include:

- Transform unit tree (2 level maximum)
- 6-Tap directional interpolation filter
- Low-complexity entropy coding (LCEC) phase 2
- Transform precision extension (4 bits)

Several tools have been removed from TMuC which include:

- Asymmetric motion partition
- Geometric partition
- Adaptive intrasmoothing
- Combined intraprediction
- Planar prediction for intra-, edge-based prediction, interleaved motion vector prediction, adaptive motion vector resolution, motion vector prediction scaling
- PU (prediction unit)-based merging
- Partition-based illumination compensation
- MDDT (Mode Dependent Direction Al Transform)

The tools removed from TMuC and new proposed tools will be evaluated in the future meetings and the TMuC will become HEVC Test Model (HM). The final decision for the adoption of tools by the HM will depend on the coding performance and complexity.

In summary, so far the coding efficiency has been improved up to 40%. The new results will be reported in the near future. The current timetable for the HEVC is as follows.

2012 February: Committee Draft (CD)
2012 July: Draft of International Standard (DIS)
2013 January: Final Draft of International Standard (FDIS)

1.3 Still Image Coding

1.3.1 Joint Photographic Experts Group

The joint committee of ISO/IEC JTC1 and ITU-T (formerly CCITT), Joint Photographic Experts Group (JPEG), was created in the middle of the 1980s. The JPEG has developed the joint international standards JPEG and JPEG-2000 for the compression of still images. Officially, JPEG [4] is the ISO/IEC international standard 10918-1; digital compression and coding of continuous-tone still images, or the ITU-T recommendation T.81. JPEG became an international standard in 1992. The JPEG standard includes two working modes: lossy coding and lossless coding. The lossy coding is a DCT-based coding scheme which is the baseline of JPEG and is sufficient for many applications. However, to meet the needs of applications that cannot tolerate loss such as coding of medical images; a lossless coding scheme is needed. The lossless coding mode of JPEG is based on predictive coding scheme.

The coding structure of the baseline algorithm of JPEG is similar to the intraframe coding of MPEG-1; both are DCT-based block coding. However, the JPEG baseline algorithm includes two coding modes: the sequential DCT-based coding mode and the progressive DCT-based coding mode. In the sequential DCT-based coding mode, an image is first partitioned into blocks of 8×8 pixels. The blocks are then converted into the transform domain with DCT. The transformed coefficients are quantized and entropy-coded to the bitstream. The difference between sequential coding and progressive coding modes is that in the sequential DCT-based coding the encoding process is performed according to the scanning order from left to right and top to bottom. In the progressive DCT-based coding mode, the DCT coefficients are first stored in a buffer before the encoding is performed. The DCT coefficients in the buffer are then encoded by a multiple scanning process. In each scan, the quantized DCT coefficients are partially encoded by either spectral selection or successive approximation. In the method of spectral selection, the quantized DCT coefficients are first divided into multiple spectral bands according to a zigzag order. In each scan, a specified band is then encoded. In the method of successive approximation, a specified number of most significant bits of the quantized coefficients are first encoded and the least significant bits are then encoded in subsequent scans. Therefore, the difference between sequential coding and progressive coding can be summarized as follows. In sequential coding an image is encoded part by part according to the scanning order, while in the progressive coding an image is encoded by a multiple scanning process and in each scan the full image is encoded to a certain quality level.

The lossless coding scheme is used to satisfy some special applications such as medical applications. The lossless coding mode of JPEG is based on a predictive coding scheme. In this scheme, three nearest-neighboring pixels are used to predict the pixel to be coded. The prediction difference is entropy coded using either Huffman coding or arithmetic coding. It should be noted that the prediction difference in this scheme is not quantized, the coding is lossless.

Except for the two baseline modes and lossless mode, the JPEG also supports the hierarchical coding mode. In this mode an image is first spatially down-sampled to a multilayered pyramid which results in a set of frames. To code this set of frames the predictive scheme is applied to obtain the differences between coded frame and reference frame. The reference frame is the reconstructed previous frame in the decoder. The difference can be coded with DCT-based coding, lossless coding, or DCT-based coding with the final lossless process. Of course, the down-sampling and up-sampling processing has to use the antialiasing filters. The hierarchical coding mode provides a progressive presentation similar to the progressive DCT-based coding mode, but it can provide multiresolution applications such as in the scalable video coding, which the progressive DCT-based mode cannot.

It should be noted that the extended version of JPEG, the Motion-JPEG, can also be used for video coding. In Motion-JPEG, each frame in the video is encoded and stored as a still image with the JPEG format which may be easier for editing video for some applications.

1.3.2 JPEG-2000

JPEG-2000 is also the standard for still image coding, which is also jointly developed by ISO/IEC and ITU-T in 2001 as the international standard ISO/IEC 15444-1 or ITU-T Recommendation T.800 [36,37]. The reason to develop JPEG-2000 is to meet some advanced requirements of many of today's applications such as Internet, wireless devices, digital cameras, image scanning, and client/server imaging which the JPEG is not so easy to satisfy. In order to satisfy the advanced applications, the JPEG-2000 is not just optimized for coding efficiency, also includes the function of scalability and interoperability. The main difference between JPEG-2000 and conventional JPEG is that the JPEG-2000 uses the wavelet transform as a core technology while the conventional JPEG uses traditional DCT-based coding technology. The encoding process of JPEG-2000 is shown in Figure 1.5.

The input image is first decomposed into rectangular nonoverlapped blocks that are called "tiles" in JPEG-2000. The tiles are the basic units of the original or reconstructed image and they are compressed independently. The use of tiling can reduce the memory requirements since the tiles are reconstructed independently and can be used for decoding specific parts of image instead of the whole image. The selection of tile size is the trade-off between memory size and coding performance. It is obvious that the smaller tile uses less memory

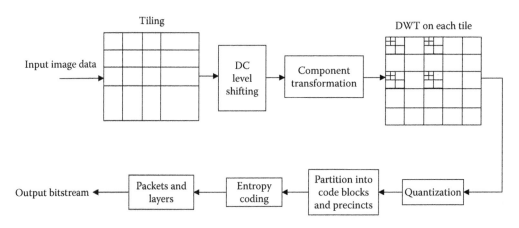

FIGURE 1.5
Encoding processing of JPEG-2000.

size but causes more image quality degradation, especially, in the low-bit-rate cases. The tiling process is followed by the DC level shifting. The DC level shifting actually converts an unsigned representation to a two's complement representation, or vice versa. Component transformations can improve compression and allow for visually relevant quantization. The discrete wavelet transform (DWT) is then applied to each tile of the image. The JPEG-2000 supports both irreversible and reversible DWT. After the wavelet transformation, a quantizer is applied to all transformed coefficients. After quantization, each subband is divided into nonoverlapped rectangular blocks. Each code block is coded by the bit plane method with entropy coding. Entropy coding uses an arithmetic coding system that compresses binary symbols relative to an adaptive probability model associated with each of 18 different coding contexts. Additionally, a header is added at the beginning of the bitstream for describing the original image and the various decomposition and coding styles that are used to locate, extract, decode, and reconstruct the image with the desired resolution, fidelity, regions of interest or other characteristics.

Finally, several advantages of JPEG-2000 over JPEG are summarized as follows:

- Better image quality at the same bit rate or comparable image quality at 25–35% bits saving.
- Good image quality can be achieved at very high compression ratios, such as over 80:1.
- Have low-complexity option mode for devices with limited resources.
- With JPEG-2000, the image that best matches the target device can be extracted from a single compressed bitstream on a server.

1.4 Digital Audio Coding

1.4.1 Introduction

It is the same as for digital video coding systems an audio codec consists of an audio encoder and an audio decoder. The encoder is to receive the audio signal and to compress the input audio data to a binary output that is referred to as a bitstream, while the decoder receives the compressed bitstream and reconstructs a perceptually identical copy of the input audio signal. During the past two decades, several audio coding standards have been developed which include MPEG-1 audio, MPEG-2 audio, MPEG-4 audio, and AC-3, as well as others. The principle of audio coding standards is well presented in the references [1–3, 38]. In the following sections we give a brief introduction of these standards.

1.4.2 MPEG-1 Audio

The MPEG-1 audio is the first high-quality audio compression standard used for the applications of storage media. Later the MPEG-1 audio was tested by ITU-R and was recommended for broadcasting applications. The block diagram of MPEG-1 audio encoder and decoder are shown in the Figure 1.6a and b, respectively.

In the encoder, the input audio signal is fed to the time-to-frequency (TF) mapping block and the psychoacoustic model block. The outputs of these blocks are sent to the allocation

(a)

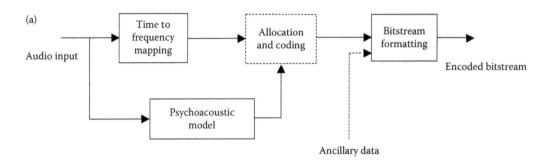

(b)

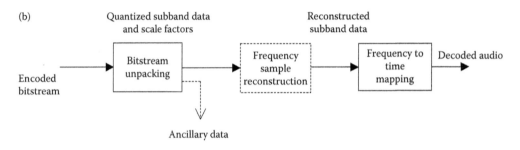

FIGURE 1.6
(a) MPEG-1 audio encoder. (b) MPEG-1 audio decoder. (Adapted from ISO/IEC JTC1 IS 11172 (MPEG-1), Coding of moving picture and coding of continuous audio for digital storage media up to 1.5 Mbps, 1992.)

and coding block. The bitstream is formatted with the quantized data with side information and ancillary data. The decoder performs in the reverse way to reconstruct the audio signal.

The MPEG-1 audio standard consists of three layers with different complexity and subjective performance. Layer I is the simplest with lower performance and Layer III is more complicated with the highest performance among the three layers. The TF mapping in Layers I and II uses a 32-PQMF (Pseudo-QMF (Quadrature Mirror Filter)) and a uniform midtread quantizer while in Layer III the output of the PQMF is sent to a Modified Discrete Cosine Transform (MDCT) stage and the filter bank is adaptive instead of static as in Layers I and II. The psychoacoustic model used in Layer III is also quite different with the one used in Layers I and II. Therefore, the performance of Layer III is much better than Layers I and II. The MPEG-1 audio coding Layer III is referred to as MP3 which is very famous and widely used in consumer electronics.

The MPEG-1 audio has four operation modes: mono, stereo, dual channel, and joint stereo. In the joint stereo mode, higher compression ratio can be achieved by using either the correlation between the left and right channels or the irrelevancy of the phase difference between two channels through some intelligent exploitation.

1.4.3 MPEG-2 Audio

The advanced features of MPEG-2 audio compared with MPEG-1 audio include the follows. The first is that the MPEG-2 audio supports multichannel applications, up to 5.1 channels including five full-bandwidth channels of the 3/2 stereo, plus an optional low-frequency enhancement channel. This feature can be used not only for audio-only applications, but also for the applications of HDTV and DVD. The second is that the MPEG-2 audio includes two coding standards: the Backward Compatible (BC) standard which is the backward compatibility with MPEG-1 audio, and the Advanced Audio Coding (AAC) standard which does not have the feature of backward compatibility.

The feature of backward compatibility of MPEG-2 BC with MPEG-1 audio allows the MPEG-1 audio decoder enabling to decode the bitstream of MPEG-2 BC audio. The differences between two standards are located in sampling frequency field, a bit-rate index field, and a psychoacoustic model used in bit allocation tables. Since the MPEG-2 BC audio has the feature of lower sampling rates, it is possible to compress two-channel audio signals to bit rates less than 64 kb/s with good quality.

MPEG-2 BC can provide good audio quality at data rates of 640–896 kb/s for five full-bandwidth channels, while MPEG-2 AAC targets at less than half of that data rate, 320 kb/s or lower, with the same audio quality. The MPEG-2 AAC does not have the constraints of backward compatibility with NPEG-1 audio as MPEG-2 BC does. The MPEG-2 AAC contains many coding tools including gain control, filter bank, prediction, quantization and coding, noiseless coding, bitstream multiplexing, temporal noise shaping (TNS), Mid/Side (M/S) stereo coding, and intensity stereo coding. The function of gain control is to divide the input signal into four equally spaced frequency bands, which are then flexibly encoded to fit into a variety of sampling rates, and also to alleviate the pre-echo effect. In the filter bank, the signals are converted from the time domain to the frequency domain. The prediction tool exploits the correlation between adjacent frames and increases the coding performance. The quantization and coding tools are similar to the ones in MPEG-1 Layer III. The TNS tool is used to control the temporal shape of the quantization noise. Intensity stereo coding tool tries to reduce perceptually irrelevant information by combining multiple channels in high-frequency regions into a single channel. The M/S stereo coding tool is used to code the sum and difference of left and right channels instead of coding them separately, in such a way that the correlation between two channels can be efficiently exploited to increase the coding efficiency.

MPEG-2 AAC standard is defined to three profiles: main profile, the low-complexity profiles and the sampling rate scalable (SRS) profile for applications with different requirements of coding performance and complexity. To achieve these requirements each profile contains different sets of coding tools. The main profile provides the highest coding efficiency by using a full set of the tools with the exception of the gain control tool. In the low-complexity profile, the prediction and preprocessing tools are not used and the TNS is limited which can reduce the requirements on the memory and computing power and favor for some applications. The SRS profile requires the gain control, but does not require the prediction tool; it offers a scalable complexity by allowing partial decoding of a reduced audio bandwidth.

1.4.4 MPEG-4 Audio

The MPEG-4 standard is an object-based coding method which targets at functionalities such as high coding efficiency, object-based scalability, error resiliency coding, and many others. The MPEG-4 audio targets at broader applications than MPEG-1 and MPEG-2 audio do. The applications of MPEG-4 audio include telephony, mobile communication, digital broadcasting, Internet, and interactive multimedia. In order to address these applications, MPEG-4 audio consists of two types of audio coding: synthetic and natural. The synthetic coding contains the tools for creating symbolically defined music and speech which include Musical Instrument Digital Interface and Text-to-Speech systems. The synthetic coding also includes the tools for the 3D localization of sound which allow the creation of artificial sound environments using artificial and natural sources. The natural audio coding is optimized at bit rates ranging from 2 kb/s up to 64 kb/s.

For satisfying different applications, the MPEG-4 audio consists of three types of codec. The first codec is the parametric codec which is used for speech coding at lower bit rates between 2 and 4 kb/s, but even lower bit rate such as 1.2 kb/s in average can be achieved if variable rate coding mode is enabled. The second codec is the code-excited linear predictive codec which is used at the bit rates in the range of 4–24 kbps for general audio coding such as speaking voice over background music. The third codec is the TF codec which includes MPEG-2 AAC and vector-quantizer-based tools for the higher-bit-rate applications.

1.4.5   Dolby AC-3

Dolby AC-3 [39] is adopted as the audio standard for the ATSC HDTV system in the United States in early 1990s. It has also been adopted for DVD films and DVB (Digital Video Broadcasting) standards. Dolby AC-3 can be used from 1 channel up to 5.1 channels including left, right, center, left-surrounding, right-surrounding, and low-frequency enhancement channels.

The AC-3 algorithm is similar to the MPEG-1 and MPEG-2 audio that are based on the perceptual coding principles. In the AC-3 encoder the audio signal is first grouped into blocks of 512 PCM time samples. The blocks of signal are first transformed to frequency domain by the MDCT. The transformed coefficients are then divided into nonuniform subbands the resolution of which is selected by dynamically adapting the filter bank block size depending on the nature of the input signal. The transformed coefficients of each subband are converted into a floating-point representation where each coefficient is described by an exponent and a mantissa. Each exponent is then encoded by a strategy based on the required time and frequency resolution. The encoded exponents are sent to the psychoacoustic model which calculates the perceptual resolution according to the encoded exponents and proper perceptual parameters. Finally, the mantissas are quantized according to the perceptual resolution and the bit allocation output.

From above coding process of AC-3 it can be seen that the AC-3 audio standard has several important features. One feature is that the AC-3 is very flexible for the applications with different bit rates and it can process multiple channels as a single ensemble. Therefore, AC-3 can encode multiple audio channels into a bitstream at the bit rates from 32 to 640 kbps. Another feature is that due to the close relationship among exponent coding, psychoacoustic models, and the bit allocation, the encoded exponents can provide an estimate of the spectral envelope which can be used in the psychoacoustic model to determine the mantissa quantization. Therefore, the AC-3 decoder can obtain the quantizer information from the decoded exponents and limited perceptual parameters without the need of side information. Of course, this approach cannot provide a detailed psychoacoustic analysis which would affect the coding efficiency. Therefore, there exists a trade-off between the number of bits for transmitting side information and the constraint psychoacoustic precision for the coding efficiency of AC-3.

1.5   MPEG-7: Multimedia Interface Standard

MPEG-7 is not a coding standard and it is the standard of Multimedia Content Description Interface, which is an ISO/IEC standard developed by MPEG [6]. The objectives of MPEG-7 include:

- Describe main issues about the content (low-level characteristics, structure, models, collections, etc.).
- Index a big range of applications.
- Audiovisual information that MPEG-7 deals with are audio, voice, video, images, graphs, and three-dimensional (3D) models.
- Inform about how objects are combined in a scene.
- Independence between description and the information itself.

To achieve the above goals, several MPEG-7 tools have been developed. These tools include: descriptors (Ds), description schemes (DSs), description definition language (DDL), and systems tools. The D is used to represent a syntactically and semantically defined feature of multimedia objects. It should be noted that a multimedia object can be described by several Ds from different aspects. The DSs are used to specify the structure and semantics of the relations between Ds and DSs. DDL is based on the Extensible Markup Language used to define the structural relations between Ds. It allows the creation and modification of DSs and also the creation of new Ds. The functions of system tools deal with the issues of binarization, synchronization, transport, storage of Ds, and protection of intellectual property. The relations between these tools are described in Figure 1.7 [6].

The MPEG-7 defines a huge number of Ds, such as visual Ds, audio Ds, and others. The visual Ds include:

- *Color Descriptors*: Color Space, Color Quantization, Dominant Color, Scalable Color, Color Layout, Color Structure, Group of Frame Color, and Group of Picture Color.
- *Texture Descriptors*: Homogeneous Texture, Texture Browsing, Edge Histogram.
- *Motion Descriptors*: Camera Motion, Motion Trajectory, Parametric Motion, Motion Activity.

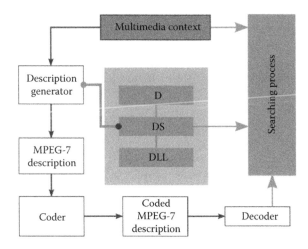

FIGURE 1.7
Relations between tools of MPEG.7. (Adapted from ISO/IEC JTC1 15938 (MPEG-7), Information Technologulti-media Content Description Interface—Multimedia Description Schemes, 2001.)

- *Shape Descriptors*: Region Shape, Contour Shape, Shape 3D.
- *Localization Descriptors*: Region Locator, Spatio-Temporal Locator.
- Face Recognition Descriptor.

The audio Ds include:

- *Basic Descriptors*: Audio Waveform and Audio Power
- *Basic Spectral Descriptors*: AudioSpectrumEnvelope, AudioSpectrumCentroid Descriptor, AudioSpectrumSpread and AudioSpectrumFlatness
- *Signal Parameters Descriptors*: AudioFundamentalFrequency, AudioHarmonicity
- *Timbral Temporal Descriptors*: LogAttackTime, TemporalCentroid
- *Tembral Spectral Descriptors*: SpectralCentroid, HarmonicSpectralCentroid, HarmonicSpectralDeviation, HarmonicSpectralSpread, HarmonicSpectralVariation
- *Spectral Basis Descriptors*: AudioSpectrumBasis, AudioSpectrumProjection

MPEG-7 audio has also included five specialized high-level tools that are used to exchange some generality for descriptive richness. The Musical Instrument Timbre description tools are used to describe perceptual features of instrument sounds. The Melody description tools are used for monophonic melodic information to facilitate efficient, robust, and expressive melodic similarity matching. The General Sound Recognition and Indexing description tools are for indexing and categorization of general sounds with applications to sound effect. Finally, the Spoken Content Description tools can be used for a detailed description of words spoken within an audio stream. The details of other parts of MPEG-7 can be found in [6].

1.6 MPEG-21: Multimedia Framework Standard

The MPEG-21 standard is referred to as Multimedia Framework and it is developed by MPEG as ISO/IEC 21000 [7]. The MPEG-21 is the standard which aims at developing the technology for supporting users to exchange, access, consume, trade, and otherwise manipulate the fundamental unit in the multimedia framework. The fundamental unit of multimedia unit is defined as Digital Items in the MPEG-21 standard. The Digital Item is the fundamental unit of multimedia for distribution and transaction. In practice, any multimedia content used in MPEG-21 is referred to as a Digital Item. The multimedia content used in MPEG-21 can be videos, audio tracks, images, metadata such as MPEG-7 Ds, structures (describing the relationship between resources), and any combination of the above items. The MPEG-21 has also defined another important concept, which of users interacting with the digital items. These two concepts give the overall picture of MPEG-21. With these two concepts, MPEG-21 defines a normative open framework for multimedia delivery and consumption which can be used by all the parties in the chain of delivery and consumption. To achieve this purpose, the MPEG-21 developed many parts in the standard. In total there are 20 parts so far:

Part 1—Vision, Technologies and Strategy
Part 2—Digital Item Declaration (DID)

Part 3—Digital Item Identification (DII)
Part 4—Intellectual Property Management and Protection (IPMP) Components
Part 5—Rights Expression Language (REL)
Part 6—Rights Data Dictionary (RDD)
Part 7—Digital Item Adaptation (DIA)
Part 8—Reference Software
Part 9—File Format
Part 10—Digital Item Processing (DIP)
Part 11—Evaluation Tools for Persistent Association
Part 12—Test Bed for MPEG-21 Resource Delivery
Part 13—Scalable Video Coding (moved out of MPEG-21)
Part 14—Conformance
Part 15—Event Reporting (ER)
Part 16—Binary Format
Part 17—Fragment Identification of MPEG Resources
Part 18—Digital Item Streaming
Part 19—Multimedia Value Chain Ontology
Part 20—Contract Expression Language

The MPEG-21 aims at many applications. We take universal multimedia access (UMA) application as an example which is strongly related to the MPEG-21 part 7. The concept of the UMA has two aspects. From the user's side, UMA allows users access to a rich set of multimedia content through various connections such as Internet, Optical Ethernet, Wireless, Cable, Satellite, terrestrial broadcasting, and others, with different terminal devices. From the content or service provider's side, UMA promises to deliver timely multimedia contents with various formats for a wide range of receivers that have different capabilities and are connected through various access networks. The major purpose for UMA is to fix the mismatch between the content formats, the conditions of transmission networks,

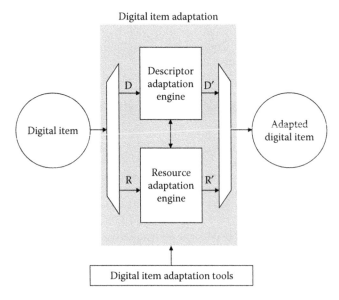

FIGURE 1.8
Illustration of MPEG.21 DIA.

and the capability of receiving terminals. A mechanism for adaptation has to be created for this purpose and can be considered into the following two ways. One way is to adapt the content to fit the playback environment and the other is to adapt playback environment to accommodate the existing input contents. The MPEG-21, especially Part 7 of MPEG-21, DIA, aims at fixing these gaps between elements. The conceptual architecture of DIA is illustrated in Figure 1.8.

From this architecture it can be found that the adapted Digital Item that meets the content playback environment is obtained from the original input Digital Item through either D adaptation engine or resource adaptation engine. DIA aims at providing the standardized descriptions and tools that can be used by these adaptation engines [40]. The details of other parts of MPEG-21 can be found in [7].

1.7 Summary

In this chapter, the multimedia standards including video coding, still image coding, audio coding, multimedia interface, and multimedia framework have been introduced. Due to limited space, we have given a brief introduction. For the details of the standards, readers can refer the listed references or specifications of the official standard documents. Some standards such as VC-1 video coding and other audio coding standards are not included in this chapter since they have similar codec structure and readers can easily understand from their specifications. Finally, the author extends his gratitude to many MPEG and VCEG colleagues who provided the references and figures.

References

1. ISO/IEC JTC1 IS 11172 (MPEG-1), Coding of moving picture and coding of continuous audio for digital storage media up to 1.5 Mbps, 1992.
2. ISO/IEC JTC1 IS 13818 (MPEG-2), Generic coding of moving pictures and associated audio, 1994.
3. ISO/IEC JTC1 IS 14386 (MPEG-4), Generic coding of moving pictures and associated audio, 2000.
4. Digital compression and coding of continuous-tone still images—Requirements and Guidelines, ISO//IEC International Standard 10918-1, CCITT T.81, September, 1992.
5. JPEG-2000 Verification Model 4.0, SC29WG01 N1282, April 22, 1999.
6. ISO/IEC JTC1 15938 (MPEG-7), Information Technology—Multimedia Content Description Interface—Multimedia Description Schemes, 2001.
7. Requirements Group, *MPEG-21 Overview v5*, J. Bormans, ed., *ISO/IEC JTC1/SC29/WG11 N4801*, Fairfax, USA, May 2002.
8. ITU-T Recommendation H.261, Video Codec for Audiovisual Services at px64 Kb/s, March 1993.
9. ITU-T Recommendation H.263, Video Coding for Low Bit Rate Communication, Draft H.263, May 2, 1996.
10. ISO/IEC 14496-10 AVC or ITU-T Rec. H.264, September 2003.
11. T. Wiegand, G. J. Sullivan, G. Bjontegaard and A. Luthra, Overview of the H.264/AVC video coding standard, *IEEE Transactions on Circuits and Systems for Video Technology*, 13(7), 560–576, 2003.

12. MPEG-2 Test model 5, *ISO/IEC/JTC1/SC29/WG11*, April, 1993.
13. ISO/IEC WG11 MPEG Video Group, MPEG-4 Video Verification Model version 16.0, *ISO/IEC JTC1/SC29/WG11 MPEG00/N3312*, Noordwijkerhout, March, 2000.
14. ISO/IEC JTC 1/SC 29/WG 11 N10117, Busan (Korean), 2008.
15. ISO/IEC JTC 1/SC 29/WG 11 N10553, London (UK), July 2009.
16. G.J. Sullivan, Informal report of VCEG actions at SG 16 meeting Jan/Feb 2009, ITU-T Q.6/SG16 VCEG, VCEG-AK04, Yokohama, Japan, 2009.
17. ISO/IEC JTC 1/SC 29/WG 11 N11113, Kyoto (Japan), January 2010.
18. S. Klomp and J. Ostermann, Response to call for evidence in HVC: Decoder-side motion estimation for improved prediction, *ISO/IEC JTC 1/SC 29/WG 11 M16570*, London, UK, 2009.
19. JCT-VC A205, Test model under consideration (TMuC), April 2010.
20. Y. Ye and M. Karczewicz, Improved intra coding, *ITU-T Q.6/SG16 VCEG, VCEG-AG11*, Shenzhen, China, 2007.
21. J. Ostermann and M. Narroschke, Motion compensated prediction with 1/8-pel displacement vector resolution, *ITU-T Q.6/SG16 VCEG, VCEG-AD09*, Hangzhou, China, 2006.
22. G. Laroche, J. Jung, and B. Pesquet-Popescu, RD Optimized coding for motion vector predictor selection, *IEEE Transactions on Circuits and Systems for Video Technology*, 18, 1681–1691, 2008.
23. JCT-VC C402, High Efficiency Video Coding Test Model 1 (HM1), October, 2010.
24. Y. Vatis and J. Ostermann, Prediction of P- and B-frames using a two-dimensional non-separable adaptive Wiener interpolation filter for H.264/AVC, *ITU-T Q.6/SG16 VCEG, VCEG-AD08*, Hangzhou, China, 2006.
25. S. Wittmann and T. Wedi, Simulation results with separable adaptive interpolation filter, *ITU-T Q.6/SG16 VCEG, AG10*, Shenzhen, China, 2007.
26. D. Rusanovskyy, K. Ugur, and J. Lainema, Adaptive interpolation with directional filters, *ITU-T Q.6/SG16 VCEG, AG21*, Shenzhen, China, 2007.
27. Y. Ye and M. Karczewicz, Enhanced adaptive interpolation filter, *ITU-T SG16/Q.6, doc. T05-SG16-C-0464*, Geneva, Switzerland, 2008.
28. T. Arild Fuldseth, D. Rusanovskyy, K. Ugur, and J. Lainema, Low complexity directional interpolation filter, *ITU-T Q.6/SG16 VCEG, VCEG-AI12*, Berlin, Germany, 2008.
29. M. Karczewicz, Y. Ye, and I. Chong, Rate distortion optimized quantization, *ITU-T Q.6/SG16 VCEG, VCEG-AH21*, Antalya, Turkey, 2008.
30. A. Tanizawa and T. Chujoh, Adaptive quantization matrix selection, ITU-T Q.6/SG16 VCEG, D.266, 2006.
31. M. Narroschke and H.G. Musmann, Adaptive prediction error coding in spatial and frequency domain for H.264/AVC, *ITU-T Q.6/SG16 VCEG, VCEG-AB06*, Bangkok, Thailand, 2006.
32. T. Chujoh, N. Wada, and G. Yasuda, Quadtree-based adaptive loop filter, *ITU-T Q.6/SG16 VCEG, C.181*, 2009.
33. G. Yasuda, N. Wada, T. Watanabe, and T. Yamakage, Block-based adaptive loop filter, *ITU-T Q.6/SG16 VCEG, VCEG-AI18*, Berlin, Germany, 2008.
34. T. Chujoh and R. Noda, Internal bit depth increase for coding efficiency, *ITU-T Q.6/SG16 VCEG, VCEG-AE13*, Marrakech, MA, 2007.
35. P. Chen, Y. Ye, and M. Karczewicz, Video coding using extended block sizes, *ITU-T SG16/Q6, doc. C-123*, 2009.
36. ISO/IEC 15444-1 or ITU-T Rec. T.800, March 2001.
37. A. Skodras, C. Christopoulos, and T. Ebrabimi, The JPEG2000 still image compression standard, *IEEE Signal Processing Magazine*, 18(5), 36–58, 2001.
38. M. Bosi and R. E. Goldberg, *Introduction to Digital Audio Coding and Standards*, Kluwer Academic Publisher, Boston/Dordrecht/London, 2003.
39. ATSC A/52/10, Digital Audio Compression Standard (AC-3), December 1995.
40. ISO/IEC 21000-7:2004 (E), Information Technology—Multimedia Framework—Part 7: Digital Item Adaptation, 2004.

2

Fundamental Methods in Image Processing

April Khademi, Anastasios N. Venetsanopoulos, Alan R. Moody, and Sridhar Krishnan

CONTENTS

Multimedia is a term that collectively describes the variety of data sources or media content available today, including audio, still images, video, animation, text, and so on. As most of this content is in digital format, much research is dedicated to the investigation of automated algorithms to manipulate and analyze these multimedia signals. For example, segmentation algorithms can be used to detect tumors in medical images and the volume of each tumor can be automatically computed [1], or a subject's face can be detected in each frame of a video sequence for biometric identification [2]. Other applications include classification-based techniques, such as computer-aided diagnosis (CAD) (automatic diagnosis support) [3], and content-based image retrieval (automated tagging/retrieving images based on a query) [4]. Automated techniques are becoming the preferred method of image analysis, since they are faster, more reliable, and produce repeatable and objective results, in comparison with traditional human-based analysis.

Although processing techniques have advanced for all types of content, the development of image processing algorithms is of fundamental value. They form the basis of several computer vision operations, as many methodologies are first developed for images, and then extended to the realm of video processing (images as a function of time) or volumetric time series data (image volumes collected for several time instances). Since there are so many applications for image processing by itself, and because of their importance in the development of multidimensional data processors, this chapter focuses on some of the fundamental image processing methods for grayscale images.

Any chapter on fundamental methods in image processing would be incomplete without a discussion on histogram-based methods. Consequently, Section 2.1 is dedicated to the image's histogram, including the definition of the histogram, histogram statistics, and modification of the histogram to achieve segmentation and contrast enhancement (CE). The last topic in this section touches upon a fuzzy edge measure which is based on the histogram of the *gradient* values (instead of *graylevel* values). As is shown in later sections, this fuzzy edge strength metric has great utility in image enhancement applications.

Section 2.2 is based on the types of degradations present in images and ways to reduce these effects via denoising or enhancement. An initial model for image artifacts is given, and the effect of the additive noise field is considered. Two nonparametric statistical tests are presented to analyze the stationarity and spatial correlatedness of this additive noise field for exploratory purposes. The remainder of this section focuses on CE and edge-preserving smoothing where the performance of these algorithms are objectively quantified by two image quality assessment (IQA) techniques.

Section 2.3 focuses on a classification system that discriminates between images based on texture. The feature extraction problem is formulated first, highlighting the importance of designing a robust feature set (scale, translation, and rotation invariance). To quantify textural characteristics, wavelet analysis is employed to get a rich description of the space-localized events. A robust multiscale feature extraction scheme is then presented, which mimics human texture perception for image discrimination. This section concludes with information on the classifier used, as well as results on three medical image databases.

2.1 Histogram Processing

A large family of image processing algorithms are based on modification of the graylevel histogram. Because of its wide applicability, and basis for many algorithms, several fundamental image processing methods based on the histogram are covered here. This introductory section begins with a general definition of the histogram (first-order probability density function (PDF)) and continue with the methods used to robustly estimate the PDF. This section concludes with the definition of histogram statistics, histogram thresholding for image segmentation, CE via histogram equalization (HE) and a robust edge strength measure that is obtained with the histogram of the *gradient* values.

2.1.1 Histogram Definition

Consider an input grayscale image $g(x, y)$, where $(x, y) \in \mathbf{Z}^2$ are the spatial coordinates of each pixel value. The histogram of $g(x, y)$ measures the number of occurrences of each graylevel $g \in [0, \ldots, L-1]$ in the image, as in

$$h(g) = n_g, \tag{2.1}$$

where n_g is the number of pixels with graylevel g and L is the number of intensity levels in the image. Normalization of this histogram results in the first-order discrete PDF

$$p_G(g) = \frac{n_g}{MN}, \tag{2.2}$$

where MN is the total number of pixels in the image (M rows, N columns) such that $\sum_{g=0}^{L-1} p(g) = 1$. The horizontal axis of the histogram is the response variable (graylevel g), and the vertical axis is the frequency count, as shown for the brain magnetic resonance imaging (MRI) in Figure 2.1. The result is one of the most basic descriptions of the graylevel distribution and despite its simplicity, it has been used successfully in many applications, such as segmentation [5], classification [6], and quantification and summarization [7].

In addition to aiding automated solutions, the histogram is also a very useful visualization tool as it highlights the underlying properties of the image data. For example, visual inspection of the histogram in Figure 2.1b yields a number of qualitative observations. First, there are two image classes which are the largest in the image (in terms of area), as indicated by the two largest peaks in the histogram. These large peaks correspond to the classes with the most number of pixels in the image (hence area). The largest peak corresponds to the background (pure noise, no signal) and the second smaller peak corresponds to the brain tissue class. Therefore, the histogram highlights the approximate graylevels of our classes. Moreover, the relative spread of these classes indicates that there must be a noise source present, since there is ambiguity in defining boundaries between tissue classes (as opposed to each class being comprised of a single value in the noise-free case).

Although the histogram visualizes the empirical density of the sample, a common challenge is the subjective choices of the bin width and number of bins [8]. Consider Figure 2.1a

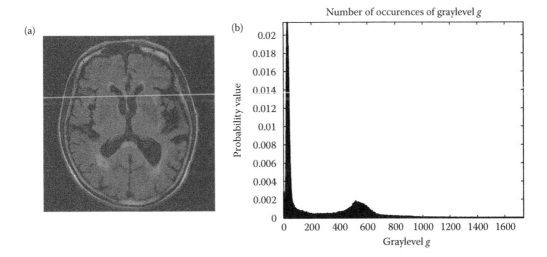

FIGURE 2.1
Histogram example with L number of bins. (a) FLAIR MRI (brain). (b) PDF $p_G(g)$ of (a).

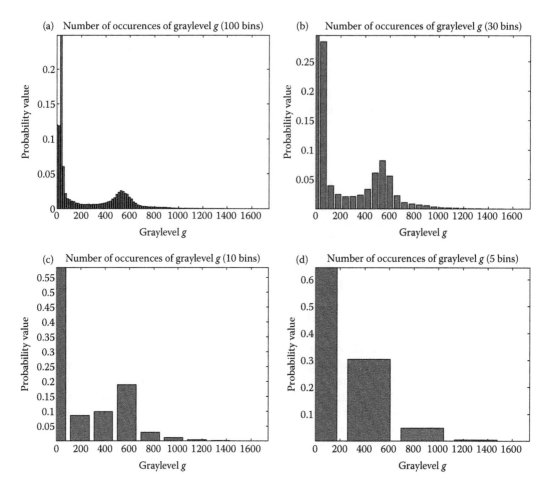

FIGURE 2.2
Example histograms with varying number of bins (bin widths). (a) 100 bins, (b) 30 bins, (c) 10 bins, (d) 5 bins.

again, and the histograms in Figure 2.2 which are the result of varying the total number of bins. As can be seen, by decreasing the number of bins, the corresponding bin widths are also increased, resulting in a series of images that highlight subtle nuances (Figure 2.2a) and gross features (Figure 2.2d). It then becomes difficult to decide which representation is yielding a more appropriate representation of the intrinsic characteristics of the data. To achieve a smoother, objective and more flexible representation of the graylevel histogram (PDF), which does not depend on bin width choices and number of bins, a Kernel Density Approximation (KDA) [9] may be used.

2.1.2 Robust Density Approximation

From the empirical histogram, KDA methods use a kernal to uniformly spread out the weight of a single observation throughout the given interval for PDF estimation [8]. The result is the summation of a series of kernels for all sample data, which smoothly approximates the shape of a histogram. For graylevels $G_1, G_2, \ldots, G_n$ the kernel density estimate is

$$\hat{p}_G(g) = \frac{1}{nh_n} \sum_{i=1}^{n} K\left(\frac{g - g_i}{h_n}\right),$$
(2.3)

where $G_i = g_i, i = 1, \ldots, n$, $K(\cdot)$ is the kernel function, and h_n is the scale parameter of $K(\cdot)$. The result is a robust representation of the histogram which is nonparametric by nature—it does not require any parameter selection including the number of bins, or the bin start position. Moreover, the representation is smoother than the empirical one, which opens up the possibility of processing the histogram itself—that is, taking the derivative for peak finding algorithms.

These advantages are demonstrated with the following example. Consider a random sample of data comprised of 60 samples, where 30 come from a centered normal distribution ($N(0,1)$) and the other 30 from a noncentered normal distribution of $N(5,1)$. Using 10 bins, the histogram of these data was calculated and is shown in Figure 2.3a. As can be seen, the empirical histogram is discrete in appearance and the true characteristics of the data are not represented correctly (i.e., it is not apparent that the underlying random variables are normal). However, as shown in Figure 2.3b, a KDA of this same dataset with a Gaussian kernel of $h_n = 1$ localizes the Gaussian random variables in a smooth manner. Therefore, the kernel density estimator offers a more robust, smooth and accurate representation of the graylevel distribution when compared with that of the traditional histogram.

Computing the kernel density estimate of some random data requires the selection of a kernel or windowing function. Some commonly used windows are box, triangle, normal, and Epanechnikov, as shown in Figure 2.4. In general, smoother kernels reconstruct a smoother density function—see the kernel density estimate for each of the other windows (box, triangle, and Epanechnikov) in Figure 2.5. The box (least smooth window) provides the coarsest approximation.

In the kernel density estimate in Figure 2.3b, since the variance of the data was known, the scale of the kernel h_n was chosen accordingly. However, this information is mostly unavailable and an incorrect kernel width will lead to erroneous results. A small value for h_n, while highlighting subtle nuances in the main part of density, also results in noisy tail ends [8] and an overall noisy appearance, making the distribution spiky and hard to

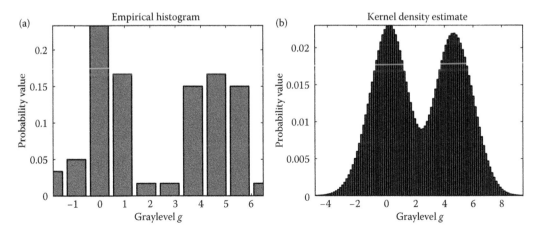

FIGURE 2.3
Empirical histogram and KDA estimate of two random variables, $N(0,1)$ and $N(5,1)$. (a) Histogram. (b) KDA.

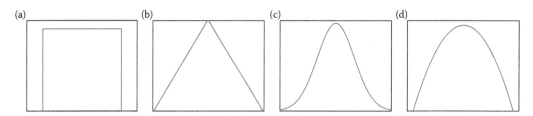

FIGURE 2.4
Types of kernels for KDA. (a) Box, (b) triangle, (c) Gaussian, and (d) Epanechnikov.

interpret. Conversely, a large bandwidth h_n will smooth the tail ends, but in general will also over-smooth the density masking the underlying structure of the data.

As the kernel width is a critical parameter in determining the shape of the final PDF, an objective method should be used to find the optimal scale of the window, h_n^*. One such method determines the optimal parameter through the optimization of the mean integrated

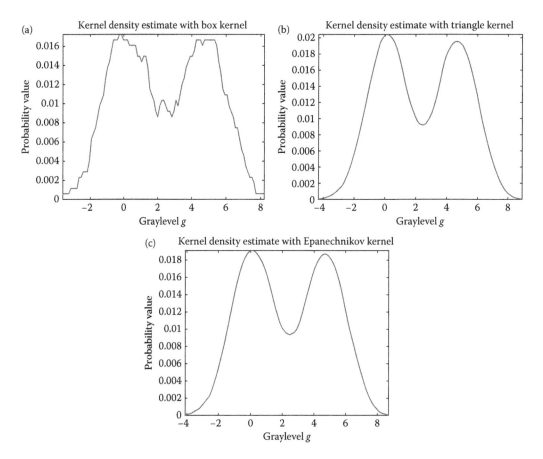

FIGURE 2.5
KDA of random sample $(N(0,1) + N(5,1))$ for box, triangle, and Epanechnikov kernels. (a) Box, (b) triangle, and (c) Epanechnikov.

squared error (MISE)

$$MISE = \mathbf{E}\left(\int (p(g) - \hat{p}(g))^2 \, dg\right), \tag{2.4}$$

$$= \int \text{Bias}^2(\hat{p}(g)) \, dg + \int \text{Var}^2(\hat{p}(g)) \, dg. \tag{2.5}$$

Substituting known expressions for the bias and variance, and solving for the optimal scale parameter h_n^* results in the following relation [8]:

$$h_n^* = \left(\frac{R(K)}{\sigma_K^4 R(p'(g))}\right)^{1/5} n^{-1/5}, \tag{2.6}$$

where $R(K) = \int K(u)^2 \, du$. Consequently, KDA methods have the advantage of automatically determining the bin width, bin start, and scale parameter in an objective way, while constructing a smooth and representative PDF, which is easier to interpret and manipulate.

2.1.3 Histogram Statistics

In the previous subsections, methods to compute the graylevel PDF were shown. Summary statistics may be computed from these distributions to gain insight into the underlying characteristics of the data. As these measures are computed from the graylevel histogram, such descriptions shed light into the the global properties of the image.

One of the most common metrics computed is the mean intensity value, which is computed from the expectation of graylevel values

$$\mu_G = \mathbf{E}[G], \tag{2.7}$$

$$\mu_g = \sum_{i=0}^{L-1} i \times p_G(i).$$

The mean quantifies the most likely pixel value to occur (on average), where $p_G(g)$ may be computed from either the traditional histogram definition or from a kernel density estimator. To measure the relative spread of the distribution, the variance can be computed by the expectation of the squared deviation of graylevels from the mean μ_G, as in

$$\sigma_G^2 = \mathbf{E}[(G - \mu_G)^2], \tag{2.8}$$

$$\sigma_G^2 = \sum_{g=0}^{N-1} (g - \mu_G)^2 \times p_G(g).$$

Consider the example of brain shown in Figure 2.6a and its corresponding histogram in Figure 2.6b, where the mean and variance are marked by vertical lines. As can be seen, the mean value is located near the largest peak in the histogram, which corresponds to the brain region. This concurs with our observation that the brain tissue intensities are the ones that are most frequently occurring. The variance gives an indication as to the relative spread of the graylevel distribution, in comparison with the mean value.

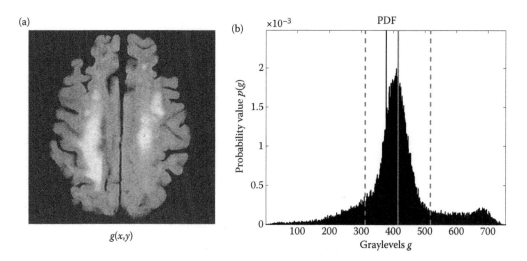

FIGURE 2.6
Example image and its corresponding histogram with mean and variance indicated. (a) $g(x,y)$. (b) PDF $p_G(g)$ of (a).

Many other statistical metrics, such as skewness and kurtosis, are used to describe the shape of the intensity PDF. Skewness is defined as the

$$s_G = \mathbf{E}\left[\left(\frac{X-\mu}{\sigma}\right)^3\right] = \frac{\mu_3}{\sigma^3}, \tag{2.9}$$

and the kurtosis is

$$\kappa_G = \frac{\mu_4}{\sigma^4} - 3, \tag{2.10}$$

where μ_3 and μ_4 are the third and fourth moment about the mean, respectively, and σ is the standard deviation. In general, skew indicates which tail of the distribution is longer, and where majority of the distribution is concentrated. On the other hand, kurtosis measures the "peakedness" of the probability distribution. These metrics dictate where most of the graylevels are concentrated, and how distributed this peak is they.

Order statistics have also been used over the years and give a statistical summary of the distribution based on percentiles (i.e., median). Other measures such as correlation and covariance may be used to examine the similarity of two images' intensity distributions. All these measures describe the global phenomena of the image and have been used successfully in image processing applications, such as the brain extraction algorithm, which collects data statistics from the histogram to find a starting point for the segmentation [5].

2.1.4 Histogram Equalization for Contrast Enhancement

The distribution of the graylevel PDF also dictates the contrast in the image. *Contrast* refers to the difference in visual properties that makes an object distinguishable from other objects and/or background. A low-contrast image has a washed-out appearance and it is difficult to distinguish the objects from one another (they have similar intensity and the histogram is peaked and concentrated around middle graylevel values [7]). Conversely, a high-contrast image has a histogram that is distributed over the entire range of graylevels, which offers a better representation of image objects and detail since the difference in the objects' intensities is larger. As a result, the "ideal" high-contrast image can be said to have a graylevel PDF that is uniformly distributed [7].

Given a low-contrast image, and the knowledge that the PDF of a high-contrast image is uniform, HE techniques [7] were developed to convert a low-contrast image into one of high contrast by reshaping its histogram to resemble a uniform PDF. In the continuous case, this is achieved by equalizing, or flattening/spreading the frequency values of the PDF via the integration operator

$$q_G(g) = \int_{t=0}^{g} p_G(t)\, dt, \tag{2.11}$$

where $q_G(g)$ is the equalized histogram and t is a dummy variable used for integration. Frequencies that lie close together will dramatically be stretched out by integration. For discrete values, summations are used instead of integrals:

$$q_G(g) = \sum_{i=0}^{g} p_G(i). \tag{2.12}$$

Unlike its continuous counterpart, it cannot be proven that this discrete transformation will always yield a uniformly distributed graylevel PDF in the output image, but it does have the tendency of spreading the histogram so that the graylevels span a fuller range [7]. Through this operation, the intensities can be better distributed throughout the histogram permitting areas of lower local contrast to gain a higher contrast.

To examine the benefit of HE for CE purposes, consider the mammogram lesions in Figure 2.7a and the corresponding HE versions in Figure 2.7b. The original images are

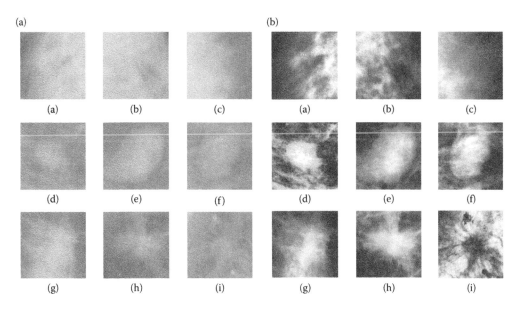

FIGURE 2.7
HE techniques applied to mammogram lesions. (a) Original. (b) Histogram equalized.

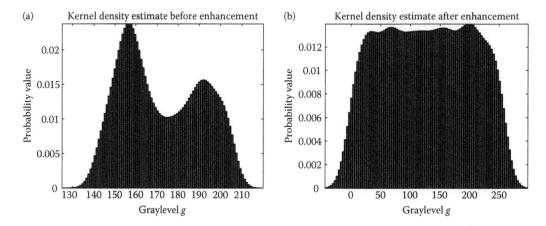

FIGURE 2.8
The KDA of lesion "(e)" in Figure 2.7, before and after enhancement. Note that after equalization, the histogram resembles a uniform PDF. (a) Before equalization. (b) After equalization.

of low contrast and the boundary of lesions are extremely hard to detect. However, the enhanced lesions are much more discernable and there is a greater discrimination between image objects (pathology and normal breast anatomy). The graylevel PDF before and after equalization for one of the lesions is shown in Figure 2.8, demonstrating how the original bimodal histogram has been transformed, now resembling a uniform distribution.

Although in some applications such HE techniques work, in many images they do not. If there are large variations in the image (on a gross scale), and many graylevels are represented in the original histogram, the maximum enhancement result cannot be achieved. Although modifications have been proposed (local HE, adaptive HE), histogram-based techniques have limited utility for all types of images. In Section 2.1.5, a more advanced CE technique is explored (non-histogram based).

2.1.5 Histogram Thresholding for Image Segmentation

In many images, the intensity classes or objects comprise unique peaks or modes in the histogram (PDF). In these cases, from an inspection of the histogram, a global threshold T may be selected and applied to the image to generate a binary mask $B(x, y)$ of the segmentation

$$B(x, y) = \begin{cases} 1, & \text{if } g(x, y) > T, \\ 0, & \text{otherwise,} \end{cases} \tag{2.13}$$

which by multiplying with the original image can be used to select the pixels above intensity T, as in

$$g_T(x, y) = B(x, y) * g(x, y), \tag{2.14}$$

where $g_T(x, y)$ is the segmented image. The threshold T may be found visually (which is subjective and time consuming), or it may be found automatically.

Recall the example in Figure 2.1, where the mean and variance were localized on the histogram. If the task is segmentation of the brain tissue class, the mean and variance can be used to determine a threshold automatically. As the brain tissue class generates the largest peak in the histogram, a threshold of $T = \mu_G \pm \beta * \sigma_G$ should approximately segment the brain tissue class. Using $\beta = 1$ to examine the intensity values within a single standard deviation from the mean, the thresholding rule becomes

$$B(x,y) = \begin{cases} 1, & \text{if } g(x,y) > \mu_G - \sigma_G, \\ 1, & \text{if } g(x,y) < \mu_G + \sigma_G, \\ 0, & \text{otherwise.} \end{cases} \tag{2.15}$$

Figure 2.9 contains the result of using this mask to find the thresholded image, as well as the image of the remaining pixels generated by $1 - B(x,y)$. As can be seen, this threshold separates the brain matter from the lesions. However, there is some residual brain pixels in the periphery that have been missed.

Simple and intuitive, global thresholding methods provide a fast approach to image segmentation. However, the choice of the threshold is very sensitive and small deviations can drastically change the result. For example, if β was chosen to be 2 in the previous example, the segmentation results would be completely different. Perhaps this would correct the misclassified pixels on the edge of the brain but would also likely mis-segment the boundaries of the lesions.

In other images, especially in those with low signal-to-noise ratios, the histogram becomes distorted and image classes are no longer easily differentiable. These effects become worse as the noise and/or degradations are increased; a pixel's intensity no longer represents the expected intensity of the object. The result is a nontrivial thresholding task, because it is difficult to determine where the threshold should be placed. To combat this, an automatic histogram thresholding algorithm known as Otsu's thresholding [10] can be used.

Otsu's thresholding is a neat segmentation algorithm which directly operates on the PDF $p_G(g)$ and requires no *a priori* information, except for the number of classes to segment. In a two-class problem, Otsu's algorithm assumes that the histogram has a bimodal distribution, and finds the threshold that minimizes the weighted within-class variance between these two modes (which is the same as maximizing the weighted between-class variance). The weighted within-class variance, which is a function of the location of threshold t, is

$$\sigma_\omega^2(t) = q_1(t)\sigma_1^2(t) + q_2(t)\sigma_2^2(t), \tag{2.16}$$

(a)   (b)   (c)

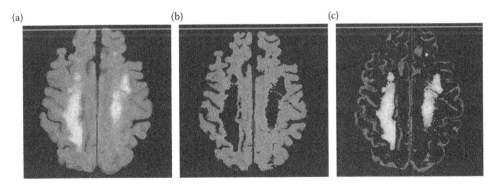

FIGURE 2.9
Image segmentation based on global histogram thresholding. (a) Original. (b) $B(x,y) * g(x,y)$. (c) $(1 - B(x,y)) * g(x,y)$.

where $q_1(t)$ and $q_2(t)$ are the class probabilities which may be estimated as

$$q_1(t) = \sum_{g=1}^{t} p_G(g),$$ (2.17)

$$q_2(t) = \sum_{t+1}^{L-1} p_G(g).$$ (2.18)

The two classes, separated by threshold t have class means that are given by

$$\mu_1(t) = \sum_{g=1}^{t} \frac{g \times p_G(g)}{q_1(t)},$$ (2.19)

$$\mu_2(t) = \sum_{t+1}^{L-1} \frac{g \times p_G(g)}{q_2(t)}.$$ (2.20)

Solving for the individual class variances yields

$$\sigma_1^2(t) = \sum_{g=1}^{t} [g - \mu_1(t)]^2 \frac{p(g)}{q_1(t)},$$ (2.21)

$$\sigma_2^2(t) = \sum_{t+1}^{L-1} [g - \mu_2(t)]^2 \frac{p(g)}{q_2(t)}.$$ (2.22)

All these variables are used to find the optimal threshold t^* that minimizes Equation 2.16. The threshold that minimizes this equation is the one that maximally separates the two classes, while ensuring that the within-class scatter is minimal. In an essence, we are finding the most optimum separation of the graylevels, in terms of class separability. The threshold is easily found with a recursive algorithm, which may exploit the relationships between the within-class and between-class variance to generate a recursion relation that permits a much faster calculation [10]. Moreover, this technique is easily extended to a multithresholding case [10] for multiclass problems.

Applying a three-class multilevel Otsu on the original image of Figure 2.9a, generates the segmentation result shown in Figure 2.10. As can be seen, the three classes are robustly

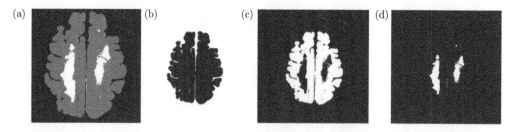

FIGURE 2.10
The result of a three-class Otsu segmentation on the image of Figure 2.6a. The left image is the segmentation result of all three classes (each class is assigned a unique intensity value). The images on the left are binary segmentations for each tissue class $B(x, y)$. (a) Otsu segmentation. (b) Background class. (c) Brain class. (d) Lesion class.

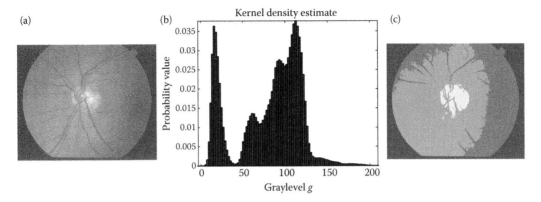

FIGURE 2.11
Otsu's segmentation on retinal image showing several misclassified pixels. (a) Original. (b) PDF $p_G(g)$ of (a). (c) Otsu segmentation.

detected (background, brain, and lesion), without requiring any choice for the optimal threshold, as it is found automatically.

Of course depending on the application, Otsu's method will not always render such an optimal segmentation result. For example, when applied to the retinal image of Figure 2.11 (four classes), several pixels are misclassified. This is because the histogram of the image strongly disobeys the assumption of modalness, which is the basis of the Otsu algorithm. Moreover, there is a heavy shading artifact present, which strongly skews the intensity distribution of the tissue classes.

2.1.6 Edge Measures from Gradient Histogram

So far, the section on the histogram focused on the distribution of the *graylevel* values. However, in this concluding section, the histogram of another variable is discussed. In particular, the gradient magnitude is examined and it will be shown that the PDF of this quantity may be used to obtain a robust, fuzzy edge strength measure. The gradient localizes the edge information in the image and is computed via a discrete version of the two-dimensional (2D) continuous gradient [7]. There have been some issues in the past presented for the discrete gradient, namely the nullspace issue [11] as well as its inability to classify all significant edges as the same class.

To combat these downfalls, the edge content of the image can be found with a fuzzy technique based on the cumulative distribution function (CDF) of the gradient [11,12]. This is a nonlinear, fuzzy mapping that results in a robust quantification of the "certainty of edge presence." Consider an image with intensities $y(x_1, x_2)$, $(x_1, x_2) \in \mathbf{Z}^2$, which may simply be denoted as y. To compute the fuzzy edge strength measure, first, the traditional magnitude of the gradient, g, is estimated

$$g = \|\nabla y\| = \sqrt{\left|\frac{\partial y}{\partial x_1}\right|^2 + \left|\frac{\partial y}{\partial x_2}\right|^2}, \tag{2.23}$$

where a discrete operator, such as Sobel may be used. Then from this, the PDF of the gradient $p_G(g)$ is computed, and based on this PDF, the CDF of the gradient magnitude ρ_k

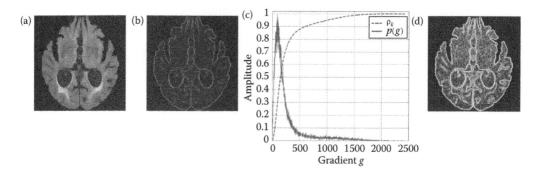

FIGURE 2.12
Example FLAIR with WML, gradient image, and fuzzy edge mapping functions. (a) $y(x_1, x_2)$. (b) $g(x_1, x_2) = \|\nabla y\|$. (c) ρ_k and $p_G(g)$. (d) $\rho_k(x_1, x_2)$.

is found [11] using

$$\rho_k = \mathrm{Prob}(g \leq g_k),\tag{2.24}$$

$$\rho_k = \sum_{g=0}^{g_k} p_G(g),$$

where $\rho_k \in [0,1]$. It is a "fuzzyfication" of the edge information because it determines the membership, or likeliness that a pixel belongs to the edge class. For example, $\rho_k = 1$ indicates with the most certainty that the current pixel belongs to an edge.

Consider Figure 2.12, which contains a brain-extracted FLAIR MRI with white matter lesions (WMLs) $y(x_1, x_2)$, the PDF of the gradient $p_G(g)$ and the fuzzy edge measure ρ_k, as well as these quantities mapped back to the spatial domain. Quantifying the edge information in images may be regarded as a classification problem where a pixel is either: (1) on an edge (object boundary) or (2) in a flat region (interior of image object). Thus, a useful edge feature should robustly classify both the lesion's and brain's boundaries as having a significant edge presence. As can be seen in the gradient image $g(x_1, x_2)$, the magnitude of the WML's edges are much lower than the amplitude of the brain-background edge, thus classifying only the brain-background edge as significant.

As shown in Figure 2.12c, majority of the edge information is comprised of a small gradient value (created by low-contrast edges, such as noise and/or white matter tracts). Since majority of the edges are small and the edge map is cumulative, the final mapping ρ_k assigns large and similar values to significant edges, despite them occurring over a wide range of g and with few occurrences. Thus, such a nonlinear mapping function has the effect of grouping significant edges, while separating them from the irrelevant edges (such as noise). Sections 2.2.4 and 2.2.5 show the utility of such a measure for the enhancement and denoising of images.

2.2 Image Degradations and Enhancement

Images with inherent artifacts may cause slight difficultly for human interpretation, but they can cause significant challenges for computing devices. A computer can easily tell the difference between two pixels' intensity values, even if they differ from one another only by a few graylevel values. As a result, images with degradations can generate erroneous results in intensity-based segmentation schemes [6], misclassified pixels in automated tissue classification algorithms [13], inaccurate three-dimensional (3D) reconstructions, and so on. To combat the effects of such artifacts and to make way for robust image analysis, much research has been conducted to examine the noise fields and types of degradations that are possible in images. Mathematical modeling, simulation, and many other methods have been used to characterize the image noise fields. Understanding in a quantitative sense how specific artifacts affect the intensity distribution has given way to automated artifact reduction schemes for the preprocessing of images.

As a result of the importance of noise and noise reduction schemes in image processing, Section 2.2 is devoted to these topics. Initially, a noise degradation model is presented, followed by two nonparametric statistical tests that may be used to explore the spatial correlation and nonstationarity of the additive noise field. To quantify the performance of image denoisers, IQA techniques are presented next. The remaining sections deal with ways to suppress this noise, via CE and an edge-preserving smoothing filter.

2.2.1 Noise Degradation Model

All natural images are degraded by artifacts that cause challenges in image processing algorithms. Considering an undistorted, "clean" image $f(x,y)$, many types of imaging artifacts can be described by the following relation:

$$g(x,y) = f(x,y) \times \beta(x,y) + n(x,y), \tag{2.25}$$

where $g(x,y)$ is the distorted image, $\beta(x,y)$ is a multiplicative noise source, and $n(x,y)$ is an additive source of noise. The multiplicative noise can be some shading artifact [7] or it can be a smoothly varying bias field as is common in MRI [14]. The additive noise fields are the most common types of noise sources found in images and are usually due to acquisition noise, variations in the detector sensitivity, environmental variations, the discrete nature of radiation, transmission or quantization errors, and so on.

These noise sources create challenges in automated analysis techniques because they alter the underlying true image classes. Resultantly, many research works are dedicated to modeling these artifacts and developing methods to reduce or remove the image degradations to estimate the noise-free $f(x,y)$. For example, many works use Gaussian mixture models to mathematically characterize image classes in the presence of Gaussian noise for image classification [15] and segmentation [16]. Other noise distributions create larger challenges, but have been handled by several authors; see [17] for descriptions on how Rician and Poisson distributions are modeled for image processing. Multiplicative bias fields have been modeled by smoothly varying polynomials [18] or low-frequency fields, and deconvolution methods [14], wavelet-based methods, and other iterative procedures [19] have been used to remove this field.

As additive noise is the most common type of noise source that is combated in images of all kinds, this section focuses on the additive noise field $n(x,y)$ versus the multiplicative one. To examine how the noise field $n(x,y)$ creates challenges in image processing consider the following simulated $T1$- and $T2$-weighted brain magnetic resonance (MR) images and their corresponding histograms in Figure 2.13. These images are generated (simulated) without noise and are used in many research works to validate algorithms [13,20] since the ground truth is known. As shown by the histograms, the intensity distribution profile is a multimodal distribution with sharp peaks corresponding to the intensities of the tissue

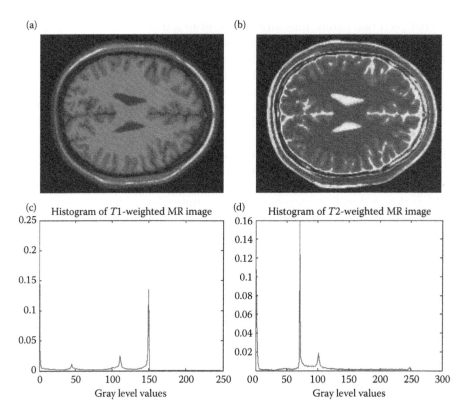

FIGURE 2.13

T1- and T2-weighted MR images (1 mm slice thickness) of the brain and corresponding histograms. Images are from BrainWeb database; see http://www.bic.mni.mcgill.ca/brainweb/. (a) T1-weighted MRI. (b) T2-weighted MRI. (c) Histogram of Figure 2.13a. (d) Histogram of Figure 2.13b. (D.L. Collins. et al., Design and construction of a realistic digital brain phantom, *IEEE Transactions on Medical Imaging*, 17, 463–468, 1998. © (1998) IEEE.)

classes. Since there is no additive noise in these images, there is minimal spread in the intensity distribution of tissue classes.*

Figure 2.14 contains the same images as shown in Figure 2.13, except with 9% noise added.[†] The histograms of these images are also included in the figure; note how each class blends with adjacent classes, which is a stark contrast to the isolated peaks in the histograms of the original images (Figure 2.13c and d). This variance or spread in intensity values of tissue classes is caused by acquisition noise.

As shown by the histograms of Figure 2.14c and d, the intensity profile is significantly altered by the addition of acquisition noise. The tissue classes are no longer clearly delineated in the histogram, and moreover, the overall uniformity of the tissues is significantly altered by this noise. This causes significant challenges in automated algorithms, such as texture classifiers, since the noise adds "erroneous" texture to each of the classes, or threshold-based segmentation schemes since locating the threshold can be difficult.

---

* Tissue classes are not delta functions (single intensity values) due to another artifact, known as partial volume averaging [16], which causes smearing between tissue classes.

[†] BrainWeb represents noise as a percentage, which is the ratio of the standard deviation of the Gaussian noise added to each channel to the signal amplitude for a reference tissue.

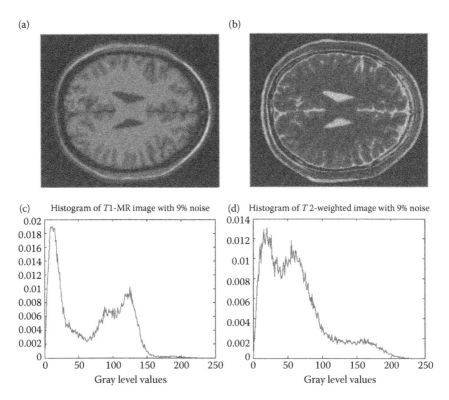

FIGURE 2.14

$T1$- and $T2$-weighted MR images (1 mm slice thickness) with 9% noise and corresponding histograms. Images are from BrainWeb database; see http://www.bic.mni.mcgill.ca/brainweb/. (a) $T1$-weighted MRI with 9% noise. (b) $T2$-weighted MRI with 9% noise. (c) Histogram of Figure 2.14a. (d) Histogram of Figure 2.14b. (D.L. Collins. et al., Design and construction of a realistic digital brain phantom, *IEEE Transactions on Medical Imaging*, 17, 463–468, 1998. © (1998) IEEE.)

2.2.2 Nonparametric Tests for Correlation and Stationarity

Many times, quantifying the characteristics of the noise field $n(x, y)$ requires good knowledge of the image acquisition process. Understanding the acquisition process in image processing is becoming ever more important, since many noise reduction tasks achieve optimal results when they exploit the distributional and statistical properties of the image, which are directly correlated to how the image was generated. However, when the image acquisition/generation process is not known, it is difficult to make any *a priori* assumptions about the data. This is the case with some commercial MR scanners, since the MR image generation process is proprietary [22]. Unknown properties create challenges in algorithm design since the underlying characteristics of the data are unknown and so few data assumptions can be made.

Consequently, for images generated from unknown processes or proprietary algorithms, or for general exploratory analysis, tests which can quantify the noise characteristics would be of value. Of particular interest is the spatial correlation and 2D nonstationarity of the noise field $n(x, y)$, as this topic has received little attention in the research community and two new tests have been proposed in the literature [22]. Statistical tests to determine whether nonstationarity or spatial correlation exists in the image is useful since as appropriate models can then be chosen to incorporate or exploit these effects. For example, Samsonov and

Johnson [23] propose a nonlinear anisotropic diffusion filter which adapts to the image's nonstationarity to maximize the reduction in noise.

Since the distribution of the noise may not be known ahead of time, stationarity or any other "nice" property like normality cannot be assumed. Consequently, any statistical method used for 2D exploratory analysis must not incorporate any of these assumptions in the test. To achieve this, nonparametric tests are utilized, since they do not rely on any parameters, or assumptions regarding the test statistic's and data's distribution. They are flexible approaches which are very general, thus allowing minimal information to be known about the data *a priori*. Section 2.2.2.1 details these methods.

2.2.2.1 2D Spatial Correlation Test

To examine whether correlation exists between pixels in an image, at first glance, one would consider computing the 2D-autocorrelation function. However, conventional auto-correlation estimates do not apply to nonstationary data [24]; thus, for generalized spatial dependency analysis, a more robust spatial correlation test is required.

To quantify the presence or absence of spatial correlation, several other statistical tests were considered, such as the Spearman's rank coefficient [8]. Although this test is non-parametric, it considers correlation between neighboring *regions*, and does not quantify the amount of correlation between neighboring *pixels*.

For the analysis of the spatial correlatedness of any random 2D data, the following elaborates on a technique known as Mantel's test for clustering. This test was developed in 1967 to determine clustering trends of disease in terms of geographic location (space) and time [25,26]. It is nonparametric by nature, since no assumptions are made about the underlying trends in the data.

The extension of Mantel's test to images is called the 2D spatial correlation test (2DSCT), and it uses pixel locations instead of geographical location (space), and graylevels instead of time to quantify the correlation or dependence between pixel values. Consider the location or indices of the pixel data to be $s = [x, y]'$ where x and y are the vertical and horizontal coordinates on the sampling lattice. Additionally, let the value of the pixel (or intensity) at these indices be $Z(s)$. If no correlation exists, the fact that two pixels s_1 and s_2 occur close to one another has no bearing on the pixel values $Z(s_1)$ and $Z(s_2)$ at these points. However, if correlation exists, proximity in space is coupled with similarity of attribute (intensity) values [26].

To quantify this effect, the image pixels are visualized in a 3D space $(x, y, Z(x, y))$ and clusterability, in terms of space and graylevel proximity, can be determined. For example, consider Figure 2.15, which contains a Gaussian-distributed image (noncorrelated noise), a colored (correlated) noise image, and the 3D representation of both. As can be seen in the 3D images, for correlated noise, points that are close in the (x, y)-plane have similar values of $Z(x, y)$, generating a cluster in this space-graylevel location. Conversely, for uncorrelated noise (Figure 2.15b), there is no coupling between spatial location and graylevel proximity (and therefore no clustering). This clustering in 3D is related to the correlation between pixel values [26].

Assume that the image consists of N pixels $(Z(s_1), Z(s_1), \ldots, Z(s_N))$. If correlation or spatial dependance exists between the pixel values, spatial coordinates that are close to one another will be coupled with intensity values that are also similar to each other. To statistically model this clustering process, distance matrices $\mathbf{W}_{ij}$ and $\mathbf{U}_{ij}$ can be used to describe the spatial proximity of s_i to s_j and the closeness of graylevel $Z(s_i)$ to $Z(s_j)$, respectively, as in

$$\mathbf{W}_{ij} = \|s_i - s_j\|, \tag{2.26}$$

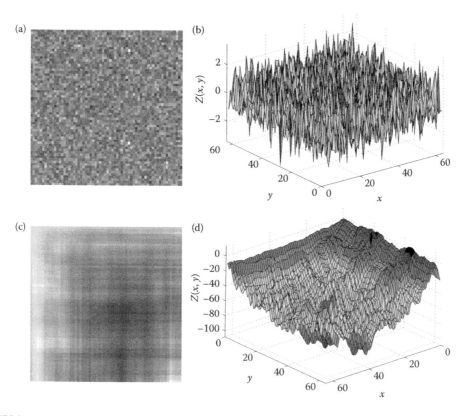

FIGURE 2.15
(Un)correlated noise sources and their 3D surface representation. (a) 2D Guassian IID noise. (b) Surface representation of Figure 2.15a. (c) 2D Colored noise. (d) Surface representation of Figure 2.15c.

and

$$\mathbf{U}_{ij} = |Z(s_i) - Z(s_j)|, \tag{2.27}$$

for all i, j, where i and j correspond to integer positions on a sampling lattice.

Using these matrices, Mantel defined a test statistic M_2, which quantifies spatial correlation between pixels:

$$M_2 = \sum_{i=1}^{N} \sum_{j=1}^{N} \mathbf{W}_{ij} \mathbf{U}_{ij}. \tag{2.28}$$

Using M_2's distribution, it is possible to perform a hypothesis test, where the hypothesis of no correlation may be rejected if the found M_2^{obs} is unusual enough (sufficiently extreme).

As the distribution of M_2 is not known ahead of time, Monte Carlo simulations can be used to generate the empirical distribution for M_2, by reordering the data several times and computing the test statistic for each of the rearrangements. For a significance level of 1%, the number of rearrangements must exceed 999 [26]. With such a high level of samples generated, it is then possible to use the empirically generated distribution to find the parameters for a z-test.

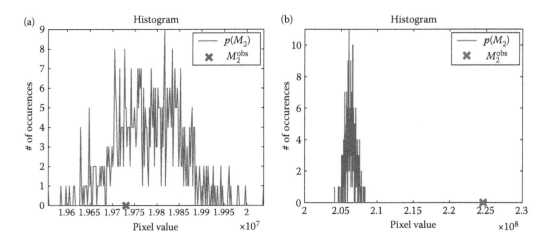

FIGURE 2.16
Empirically found M_2 distribution and the observed M_2^{obs} for uncorrelated and correlated 2D data of Figure 2.15.
(a) $p(M_2)$ and M_2^{obs} for Figure 2.15a. (b) $p(M_2)$ and M_2^{obs} for Figure 2.15c.

For the 2D noise sources shown in Figure 2.15, the M_2 distribution determined by Monte Carlo methods is shown in Figure 2.16. The corresponding M_2 computed for the observed (original data), M_2^{obs}, is shown as an X along the M_2 axis. As can be easily seen, the M_2^{obs} is sufficiently extreme from the center of the PDF of M_2 ($p(M_2)$) for the colored noise scenario. Consequently, the null hypothesis of no correlation must be rejected in this case. Conversely, the M_2^{obs} for the IID data (uncorrelated) falls near the center of the distribution, indicating with high likelihood that the null hypothesis of no correlation between pixels can be accepted.

To examine the performance of such 2DSCT, several colored images were simulated [22] and used for experimental purposes. The simulated images were 32×32 in size, and were generated from normal $N(\mu, \sigma)$ and uniformly $U(a, b)$ distributed variables. There are two parameters that must be tuned for each of these distributions which were determined by randomly sampling the following sets of possible values for each of the parameters:

$$\mu \in \{0, 1, 2, \ldots, 100\},$$

$$\sigma \in \{1, 1.1, 1, 2, \ldots, 10\},$$

$$a \in \{0, 1, 2, \ldots, 100\},$$

$$b \in \{1, 2, \ldots, 100\}.$$

Note that a is always less than b and the mean of the uniform and normal random variables were set to zero. The randomly selected parameters are shown in Table 2.1. These variables were used to generate colored images according to [22], and are shown in Figure 2.17.

The colored noise sources shown in Figure 2.17 were used to test the 2DSCT and the results are shown in Table 2.2, where h is the hypothesis test result (if $h = 0$ the hypothesis of no spatial correlation is accepted, whereas it is rejected when $h = 1$) and p is the associated p-value found at a confidence level of 0.05. This table contains the results from testing both the original IID (uncorrelated) random variables, alongside the colored images. As can be seen, correlation was correctly detected in every case for the colored images and the p-value

TABLE 2.1

Parameters Used for Generating the Nonstationary and Correlated Noise Sources

	Correlated Noise					Nonstationary Noise			
	μ	σ	a	b		μ	σ	a	b
Img1	0	7.4	43	−a	Img11	79	9.9	55	94
Img2	0	2.4	86	−a	Img12	34	8.8	16	73
Img3	0	1.0	26	−a	Img13	22	3.4	70	72
Img4	0	1.4	83	−a	Img14	84	3.6	75	88
Img5	0	4.8	36	−a	Img15	14	8.5	54	86
Img6	0	4.9	79	−a	Img16	10	5.9	4	37
Img7	0	6.3	30	−a	Img17	32	4.2	66	73
Img8	0	5.1	7	−a	Img18	88	7.1	3	6
Img9	0	4.2	67	−a	Img19	51	4.1	25	29
Img10	0	8.6	37	−a	Img20	53	2.8	6	15

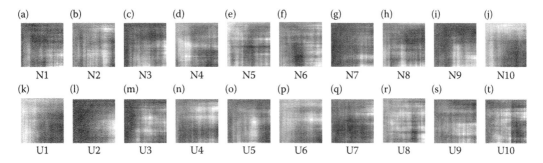

FIGURE 2.17

Correlated 2D variables generated from normally (N) and uniformly (U) distributed random variables. Parameters used to simulate the random distributions are shown in Table 2.1.

TABLE 2.2

Results for 2DSCT

	Gaussian				Uniform			
	Uncorrelated		Correlated		Uncorrelated		Correlated	
	h	p	h	p	h	p	h	p
Img1	0	0.52	1	0	0	0.23	1	0
Img2	0	0.99	1	2.1 e − 33	0	0.43	1	0
Img3	1	0.02	1	0	0	0.96	1	0
Img4	0	0.84	1	0	0	0.30	1	0
Img5	0	0.11	1	2.2 e − 30	0	0.20	1	8.8 e − 38
Img6	0	0.07	1	2.0 e − 23	0	0.86	1	0
Img7	0	0.85	1	0	0	0.17	1	0
Img8	0	0.55	1	0	0	0.56	1	6.1 e − 14
Img9	0	0.14	1	0	1	0.04	1	0
Img10	0	0.31	1	0	0	0.23	1	0.0001

is very low (almost zero) for all cases as well. However, two of the uncorrelated cases were incorrectly classified as correlated.

On visual inspection, the misclassified, uncorrelated noise seems to be relatively unremarkable. It is difficult to judge whether there was a spatial pattern randomly generated in the supposed uncorrelated images that is not visible to the eye, or whether the test is sensitive to particular parameters of the underlying distribution of the images. Recall that statistical tests allow us to state a conclusion *with great confidence*; not with 100% certainty, indicating that misclassifications are possible. More results of the test are shown in [22], and future works are going to increase the sample size, number of distribution parameters as well as the image size itself.

2.2.2.2 2D Reverse Arrangement Test

To date, exploration of nonstationarity has been mainly investigated for one-dimensional (1D) signals. For example, statistical tests such as the runs test, reverse arrangements (RA), and modified reverse arrangement [27] tests have been used to detect signal nonstationarity or drift phenomena [22]. These tests, which have been successfully applied to electroencephalography signals [27] as well as sensor drift as a function of time [28], are used to understand/verify assumptions about the data, or to develop accurate data models. However, since such techniques are strictly developed for 1D signals and do not directly apply to images, the following sections will discuss a 2D nonparametric test for image nonstationarity based on the work in [22].

In the 1D reverse arrangement test (1DRAT), the total number of *reverse arrangements* (RA) present in the data are used to determine whether a nonstationary trend exists. For example, consider a 1D signal $Z(t)$, with N time points, $t = (t_1, t_2, \ldots, t_N)$. An RA occurs when the following equality holds true: $Z(t_j) > Z(t_i)$ with $i < j$ for a single instance of i and j. Thus, if a future data point $Z(t_j)$ has a value greater than the current data sample $Z(t_i)$, a reversal or RA is detected. For example, consider Figure 2.18, which contains 1D

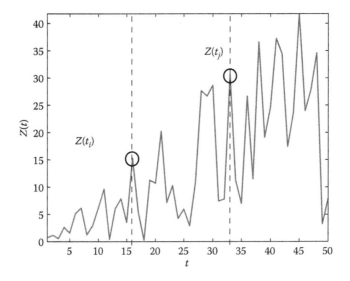

FIGURE 2.18
1D nonstationary data.

sequence that exhibits nonstationarity. Two data points, at two time instants t_i and t_j, are circled. In this case, an RA exists at point t_i, when considering point t_j, since $Z(t_i) < Z(t_j)$.

The total number of RAs at data point i, for all $j > i$, is used to determine whether nonstationarity exists or not, that is, a 1D sequence with a linearly increasing trend will have a large (more than average) number of RAs per data point. Therefore, such techniques have the ability to detect the presence of a monotonic trend in the data (which is related to nonstationarity [27]).

For the 2DRAT, similar principles are used, except that the number of RAs are found using a 2D causal grid. The 2D causal grid which is used is illustrated in Figure 2.19, which shows the image at various stages of the algorithm. The black box is the current pixel being processed and the gray area contains the permissable pixels values used to detect an RA.

More formally, given an intensity image $Z(x, y)$, an RA is detected when a future pixel $Z(x_j, y_n)$ is greater than the current pixel being processed $Z(x_i, y_m)$ with $i < j, m < n$, for some i, j, m, n. These pixels, according to a causal grid definition, are the future pixels of $Z(x_i, y_m)$. The total number of RAs at pixel (x_i, y_m), denoted $r(x_i, y_m)$, may be found by counting the number of times $Z(x_j, y_n)$ is greater than the current pixel $Z(x_i, y_m)$ for all $j > i, n > m$:

$$r(x_i, y_m) = \# \, Z(x_j, y_n) > Z(x_i, y_m), \quad \text{where}, i < j, m < n, \tag{2.29}$$

The sum of all the RAs $r(x_i, y_m)$ found by the above equation is the test statistic R^* for the image (this is the same as the 1D RA test) and is found by

$$R^* = \sum_i \sum_m r(x_i, y_m). \tag{2.30}$$

Since the distribution of the RAs is not known, the Monte Carlo methods described in Section 2.2.2.1 are used to generate the distribution of R ($p(R)$), and the parameters for a Z-test are computed again. If R^* is sufficiently extreme, the null hypothesis of stationarity is rejected. Randomization would ruin any trend or drift that was originally in the graylevel values. Therefore, an extreme value of R^* indicates nonstationarity.

To demonstrate this, consider the images Figure 2.20. Figure 2.20a contains stationary noise, generated from IID uniform random variables and Figure 2.20c nonstationary noise, with an obvious trend in the data (which is related to the way the statistics of each random variable changes). The histogram plots adjacent to these images contain $p(R)$, the distribution of R, and the total number of RAs R^* for the original image. As can be seen by these

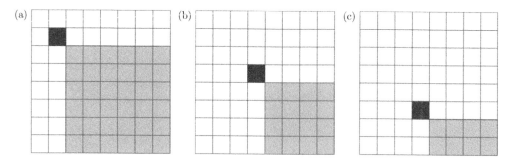

FIGURE 2.19
Grid for 2D-extension of RA test. (a), (b), and (c) show several examples of different spatial locations where the number of RAs are computed.

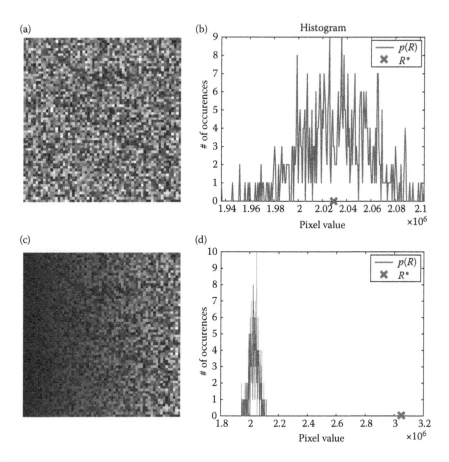

FIGURE 2.20
Empirically found distribution of R and the observed R^* for 2D stationary and nonstationary data. (a) IID stationary noise. (b) $p(R)$ and R^* of (a). (c) Nonstationary noise. (d) $p(R)$ and R^* of (c).

plots, the nonstationary noise scenario results in an extreme value of R^*, thereby causing us to reject the null hypothesis of stationarity. Conversely, the stationary image data generate a value for R^* which is near the middle of $p(R)$, thus allowing us to accept the null hypothesis of stationarity with high probability. Results on nonstationary, 2D data are shown in [22].

To test the 2DRAT, nonstationary images were generated using the methods in [22] with the distributions of Table 2.1. These images are shown in Figure 2.21. Note that the mean of these distributions are nonzero. Alongside the results for the stationary IID random variables, the results for the simulated nonstationary images are shown in Table 2.3. As can be seen, all images were correctly classified—the null hypothesis of stationarity was rejected in 100% of the nonstationary cases whereas the null hypothesis was accepted for all the stationary cases. Therefore, the 2DRAT quantifies image nonstationarity with good accuracy. Additional test results are shown in [22].

2.2.3 Image Quality Metrics

In the previous sections, a noise model for images as well as ways to analyze the additive noise field for correlation and nonstationarity were presented. In the following sections,

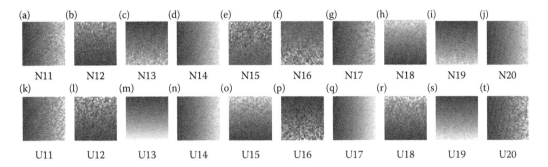

FIGURE 2.21
Nonstationary 2D variables generated from normally (N) and uniformly (U) distributed. Parameters (μ, σ) and (a, b) used to simulate the underlying distributions are shown in Table 2.1.

TABLE 2.3

Results of 2DRAT

	Gaussian				Uniform			
	Stationary		Nonstationary		Stationary		Nonstationary	
	h	p	h	p	h	p	h	p
Img11	0	0.20	1	9.7 e − 181	0	0.36	1	2.1 e − 180
Img12	0	0.91	1	1.7 e − 146	0	0.17	1	1.2 e − 106
Img13	0	0.87	1	1.2 e − 185	0	0.75	1	4.0 e − 228
Img14	0	0.41	1	7.1 e − 221	0	0.27	1	5.6 e − 222
Img15	0	0.17	1	6.8 e − 66	0	0.86	1	1.0 e − 191
Img16	0	0.92	1	1.8 e − 63	0	0.28	1	3.4 e − 078
Img17	0	0.41	1	2.7 e − 180	0	0.20	1	3.5 e − 225
Img18	0	0.42	1	2.7 e − 209	0	0.53	1	7.2 e − 161
Img19	0	0.30	1	1.2 e − 204	0	0.92	1	1.0 e − 233
Img20	0	0.73	1	1.0 e − 217	0	0.21	1	1.5 e − 217

methods to reduce this additive noise source in images will be explored and applied on real images. In order to quantify the performance of these algorithms, metrics that *objectively* assess the quality of the noise suppressed image of fundamental importance are that is, how much improvement is achieved with the image enhancement scheme and are details left in tact? To achieve this, a series of mathematical measurements that quantify the quality of the image (in an objective manner) may be utilized. Such methods are covered in this section and then are applied in Sections 2.2.4 and 2.2.5 to quantify the performance of the proposed noise suppression and CE schemes.

In general, image quality metrics may be broadly classified into two groups: (1) reference-based IQA and (2) nonreference-based IQA techniques, which differ based on the number of images required to do the computation. Since both denoising and enhancement applications have access to the original image, as well as the processed one, reference-based IQA techniques are preferred.

Reference-based IQA methods are a family of *full reference* metrics; that is, they are dependent on a secondary image to perform the assessment. For example, the mean square error (MSE) is a full reference-based IQA technique that compares the original image $g(x, y)$

with the approximation $\hat{g}(x, y)$, by computing the mean-squared difference between the two images. The MSE has been used in several works with some success [29]. Although simple to apply, it has been shown that the MSE is not a good indicator of algorithm performance as it does not correlate well to visual judgement of image quality [29]. Consequently, edge details in the output image may not be pronounced, artifacts may be more noticeable than measured, and so on. As visual appearance of the processed image is a good indicator of algorithm performance, reference-based methods have been working to model human image quality perception to judge the image quality of the processed image. The following methods provide robust alternatives to the MSE which are in tune with human perception.

2.2.3.1 IQA for Image Denoising

A reference-based image quality metric can quantify the noise reduction achieved by quantifying an image characteristic before and after noise reduction. The edge quality is one example, since common low-pass filtering is infamous for suppressing the noise while removing the useful details such as edges as well. Therefore, comparing the integrity of the edges before denoising and after would quantify the performance of the algorithm in terms of retaining edges.

To quantify the performance of the filter in Section 2.2.4 objectively, the technique in [12] and [29] is utilized. It is based on a scatter plot of the gradient magnitude of the original image (x-axis) versus the gradient magnitude of the filtered image (y-axis) (see Figure 2.22 for an example scatterplot of a filtered image).

Sharpened and smoothed pixels are grouped into two sets: A—pixels that have been smoothed (below the line $y = x$) and B—pixels that have been sharpened (above the line $y = x$). With fitting techniques, a line $y = mx + y_o$ is computed to describe the trends of the data in sets A and B. The slope of the lower line (m_A) dictates the smoothing induced by the filter and the slope of the upper line (m_B) is the sharpening or enhancement of the filter [29].

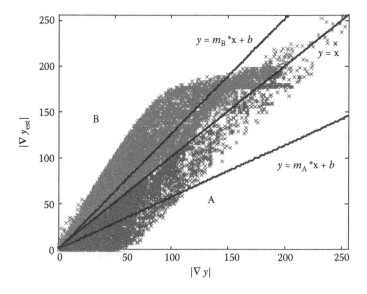

FIGURE 2.22
Scatterplot of gradient magnitude images of original image (x-axis) and reconstructed version (y-axis).

To account for the number of pixels in each set, each slope is weighted by the relative number of points to generate the following two filter descriptors [29]:

$$\text{Smoothness} = (m'_A - 1) * \frac{|A|}{|A| + |B|},$$

$$\text{Sharpness} = (m_B - 1) * \frac{|B|}{|A| + |B|},$$

where both measures are $\in [0, \infty)$ and $m'_A = 1/m_A$. These metrics are the attenuation factor of noise in flat regions (smoothness) and the edge amplification factor (sharpness).

2.2.3.2 *Image Quality Assessment for Contrast Enhancement*

Later in this section, a CE scheme will be presented which requires validation of its performance. A full reference approach is an attractive choice since two images are available to compute the metric from the original image as well as the enhanced image output. Therefore, the contrast of the images may be compared before and after enhancement.

In line with human contrast perception, some authors have proposed local contrast measures to quantify image contrast. One example of such a measure is presented in [30] and [12], which computes a local contrast value $c(x, y)$, by taking the difference of the mean values in two rectangular windows centered on a pixel (x, y) as in

$$c(x, y) = \frac{|\mu_O - \mu_N|}{|\mu_O + \mu_N|}, \qquad (2.31)$$

where μ_O is the mean value of a center region (a 3×3 region containing the 8-connect neighbors of pixel (x, y)) and μ_N is the mean graylevel value of the pixels in a 7×7 neighborhood region surrounding the 3×3 window. As contrast is often defined as the difference in mean luminance between an object and its surrounding, such a definition of local contrast is in line with human contrast perception. If the Object (O) has a mean intensity that is equal to that of the neighboring region (N), then there is no contrast between these objects and it is likely that both O and N are from the same region, with $c(x, y) \sim 0$. If the mean of the O region is significantly larger than the mean of the N region, the local contrast approaches a maximum, $c(x, y) \sim 1$.

This contrast feature may be computed on both the enhanced and original image and compared to quantify the contrast improvement. One such method, known as the contrast improvement ratio (CIR) [12,30], computes the squared error between the contrast before and after enhancement, and is found to be as follows:

$$CIR = \frac{\sum_{(x,y) \in \mathbf{R}} |c(x, y) - \tilde{c}(x, y)|^2}{\sum_{(x,y) \in \mathbf{R}} c(x, y)}, \qquad (2.32)$$

where c and $\tilde{c}$ represent the local contrast before and after CE, respectively. The CIR looks at the average difference in contrast between the original and new output.

2.2.4 Image Denoising

In this section, we deal with ways to decrease the effect of the additive noise field on an image $y(x_1, x_2)$. As stated in the preamble, such additive degradation can be caused by many effects and is present in nearly all images, causing significant challenges in automated techniques. Therefore, methods to reduce this artifact is of paramount importance in ensuring that image processors return the most accurate and robust results.

There exists many image processing methods to reduce additive noise, such as spatial filtering [7] or wavelet-based denoising [31], and so on. Many of these algorithms have downfalls in the sense that either there is an oversmoothing of important details [32], or assumed *a priori* models are inaccurate causing difficulties in noise modeling, threshold selection, and so on.

To this end, the current section discusses an edge-preserving smoothing filter, which retains the important image details, while smoothing in the flat regions. Such filters have been gaining more attention lately, as they combat the downfalls of traditional filters. The most popular edge-preserving smoothing algorithm to date is the bilateral filter [33], where filtering is completed via convolution with a kernel that is the product of two Gaussians:

$$BF[y] = \frac{1}{W} \sum_{q \in S} G_{\sigma_s}(||p - q||) G_{\sigma_r}(||y_p - y_q||) y_q, \qquad (2.33)$$

where y is the noisy image, W is a normalizing factor, p, q are spatial locations, G_{σ_s} and G_{σ_r} are the spatial and range Gaussians, with standard deviations of σ_s and σ_r, respectively. The range Gaussian has the effect of probing for "edgey" regions and turns filtering off when an edge is detected. It gives way to Gaussian smoothing in flat regions (inside image objects) and little-to-no filtering on the object boundaries (edge preservation). The result is an image with much of the noise removed, while the boundaries are retained. To avoid fine tuning the parameters, default parameters may be used: $\sigma_r = 10\% y_{\max}$ and $\sigma_s = 1/16 * \min(M, N)$, where $y_{\max}$ is the maximum intensity of the image, and M, N are the dimensions of the rows and columns. Two example images, with the corresponding bilateral filtered results are shown in Figure 2.23, demonstrating both smoothing and edge retainment.

Although this technique is widely used, a major issue is the setting of optimal parameters. Default parameters could not adequately retain the edges in all types of images. In this section we discuss a new edge-preserving filter which does not require any parameter tuning and will show that it outperforms the bilateral filter for FLAIR MRIs with WMLs. This work is based on [34].

To perform edge-preserving smoothing on the FLAIR MRIs, an initial edge estimate is required. The fuzzy edge estimate ρ_k, described in Section 2.1.6, will be utilized. Recall that it is based on the empirical CDF of the gradient magnitude and $\rho_k \in [0, 1]$. It is a "fuzzyfication" of the edge information because it determines the membership, or likeliness that a pixel belongs to the edge class.

(a)   (b)   (c)   (d)

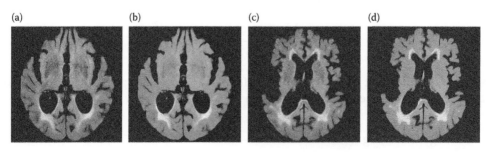

FIGURE 2.23
Bilaterally filtered examples. (a) Original. (b) Bilaterally filtered. (c) Original. (d) Bilaterally filtered.

As discussed in Section 2.1.6, the final mapping ρ_k assigns large and similar values to high gradient values (significant edges), despite them occurring over a wide range of g and with few occurrences. This allows automated algorithms to deduce that both WMLs and brain boundaries have significant edge presence. Thus, such a nonlinear mapping function has the effect of separating relevant from irrelevant edges.

Although the edge map has good localization of significant edges, it is also noisy. To improve this estimate, intensity information is coupled with the fuzzy edge measure through the conditional PDF of fuzzy edge strength ρ_k, for a particular intensity y as in

$$p_{P|Y}(P = \rho_k|Y = y) = \frac{\# \text{ pixels with } P = \rho_k|Y = y}{\# \text{ pixels with } Y = y}, \qquad (2.34)$$

which may be denoted as $p_{P|Y}(\rho_k|y)$. This PDF quantifies the distribution of the edge information for a specific graylevel value y. The interior of the brain and WML classes are defined by low- and high-intensity values, respectively. Since these regions are flat (no large edges), there is clustering in the PDF in low-edge values over these intensity ranges (classes). Across anatomical boundaries (high-contrast edges), clustering of high-edge values arise in the PDF for class-specific intensity values. Therefore, to gain a more descriptive and useful representation of the edge information in the image, an averaging operation may be used.

The conditional expectation operator is used to find this new and improved edge map, which is a global representation of the edge information in the image. Conditional expectation offers the best prediction of the fuzzy edge measure ρ_k given that the intensity is y according to the MSE criteria

$$\mu_{\rho_k|y}(y) = \mathrm{E}\{P = \rho_k|Y = y\}, \qquad (2.35)$$

$$= \sum_{\forall \rho_k} \rho_k * p_{P|Y}(\rho_k|y), \qquad (2.36)$$

which is an intensity-specific (global) approximation of the edge content.

Using the new edge map, the image is reconstructed with different levels of quality based on preset amounts of image detail (edges/noise) included in the reconstruction. Since this profile is approximating the global edge content of the image, it is used as an estimate of the image's gradient *as a function of the graylevel y*:

$$\mu_{\rho_k|y}(y) \simeq y'(y), \qquad (2.37)$$

which is integrated to find the estimated intensity Y_{rec}:

$$Y_{\text{rec}}(y) = \sum_{n=0}^{y} y'(n), \qquad (2.38)$$

$$Y_{\text{rec}}(y) = \sum_{n=0}^{y} \mu_{\rho_k|y}(n). \qquad (2.39)$$

For inclusion of significant edge magnitudes only, the feature map can be refined via thresholding with τ

$$\mu_{\rho_k|y}^{\tau}(y) \to \mu_{\rho_k|y}(y) \geq \tau, \qquad (2.40)$$

and the enhanced image is reconstructed with the modified edge map

$$Y_{\text{rec}}^{\tau} = \sum_{n=0}^{y} \mu_{\rho_k|y}^{\tau}(n). \tag{2.41}$$

Varying the threshold τ includes more or less details in the final image reconstruction, generating images with varying qualities. The results of reconstructing the brain shown in Figure 2.23a are shown in Figure 2.24. Images reconstructed from low thresholds still contain noise due to minimal noise removal. With higher thresholds, less amounts of irrelevant details are included in the reconstruction creating a smoother image (image objects appear cartoon-like).

　　Larger thresholds cause the difference between the mean intensities (class separability) of WMLs and brain tissues to be reduced (large and similar edge values are used to reconstruct both the brain and WML). This causes the reconstruction to be smooth (minimal intraclass scatter) at the expense of reduced contrast between WML and brain classes (minimal inter-class separation). Additionally, an even higher threshold utilizes only the largest edges in the image for reconstruction which causes the WMLs to be missed completely (see Figure 2.24d). Consequently, an appropriate threshold must be selected that achieves maximal smoothing while maintaining class separation between the WMLs and brain tissues.

　　Quantitative metrics will be used to select the optimal threshold. Recall that smoothing and contrast are opposing forces (over-smoothing decreases separability between the WML and brain intensities, or as contrast increases, the smoothness of our image decreases). We may use these two features (randomness S and contrast C) to find the optimal threshold τ^*

$$\tau^* = \min_{\tau} \frac{S(Y_{\text{rec}}^{\tau}(x_1, x_2))}{C(Y_{\text{rec}}^{\tau}(x_1, x_2))}, \tag{2.42}$$

where τ^* is the threshold that *maximizes contrast* between the WML and normal brain and *minimizes the randomness* (or noise) in the reconstruction.

　　The randomness metric S is 1 for maximum randomness, and 0 for complete smoothness. It is computed based on the homogeneity H of the image

$$S = 1 - H, \tag{2.43}$$

where the traditional co-occurrence matrix-based definition of homogeneity is used.

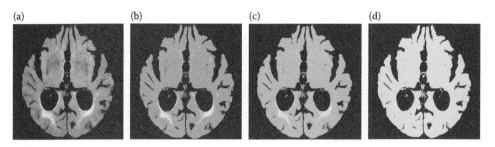

FIGURE 2.24
Image reconstruction of example shown in Figure 2.23a. (a) $Y_{\text{rec}}^{0.35}$, (b) $Y_{\text{rec}}^{0.50}$, (c) $Y_{\text{est}}^{0.58}$, and (d) $Y_{\text{rec}}^{0.70}$.

The contrast measure C is based on the difference between the most frequently occurring intensity in the brain tissue class ($y_{peak} = \arg\max(Hist(Y))$) and the maximum intensity ($y_{max} = \max(Y)$) (WML class) as in

$$C = \frac{y_{max} - y_{peak}}{y_{max}}, \tag{2.44}$$

which is normalized to set $C \in [0,1]$. A maximum contrast situation receives a $C = 1$ value, while a minimum value of $C = 0$ occurs when the WML and brain classes are no longer differentiable.

In total, 14 volumes (patients) were used to test the image denoiser, resulting in a database of 97 images (slices). The only preprocessing applied is automatic brain extraction (skull stripping) [12].

Each image was reconstructed based on the optimal threshold found by Equation 2.42. The results for Figure 2.23a and another example are shown in Figure 2.25. The objective function exhibits a local minima for medium-valued τ ($\tau \sim 0.5$), indicating that this threshold noise is removed, while maintaining the contrast between the WML and brain in an optimal manner. For low τ values, the objective function is approximately constant and similar to the initial value ($\tau = 0$). This indicates that there is minimal removal of noise. In the other extreme, as the threshold increases beyond the local minimum, the objective function increases rapidly. This occurs when the threshold is too high and removes the WML from the reconstruction. In this scenario, contrast approaches zero causing a dramatic increase in the objective function, thereby ensuring that this threshold which removes the WML is not selected. Thus, the objective function delivers a robust method for suppressing noise while maintaining separability between the two tissue classes (WML and brain).

Figure 2.25 also the histogram before and after noise reduction. The histogram is much more peaked in the denoised case and it appears that the brain is almost represented by a single graylevel value indicating maximal denoising. Since the output image is reconstructed based on a global edge map, such a technique is performing edge-preserving smoothing as strong edge strengths are retained by the reconstruction, while small edges in flat regions are suppressed.

To quantify the performance of this filter objectively, the technique described in Section 2.2.3 is utilized. Both the Smoothness and Sharpness metrics were computed and compared against those found with the bilateral filter. To determine the improvement in performance, the smoothness and sharpness measures of the bilaterally filtered image were subtracted from those found with the reconstruction method, followed by a normalization by the respective reconstructed filter results. These normalized differences, for all 97 images are shown in Figures 2.26a and b. As shown by these results, only few values are negative, indicating that both the smoothing and sharpening indices from the proposed work are higher in most cases. On average, the reconstructed images provided attenuation of noise in flat regions with a smoothness of 44.39% higher than the bilateral filter, along with a simultaneously edge enhancement increase of 34.14% over the bilateral filter on average.

The few cases where the sharpening or smoothing index of the bilateral filter was better corresponded to images with ringing artifacts (caused by patient movement). The proposed algorithm enhances the ringing artifacts (since they are of high-intensity value) which creates a reduction in the smoothing index and superficial changes of the sharpness index in low gradient regions. Future works will utilize a knowledge-based system to remove such artifacts, since the location of them is known (periphery of brain).

The proposed edge-preserving smoothing method supplies a near ideal tissue model with each tissue class being represented by few graylevel values. Future works will investigate its

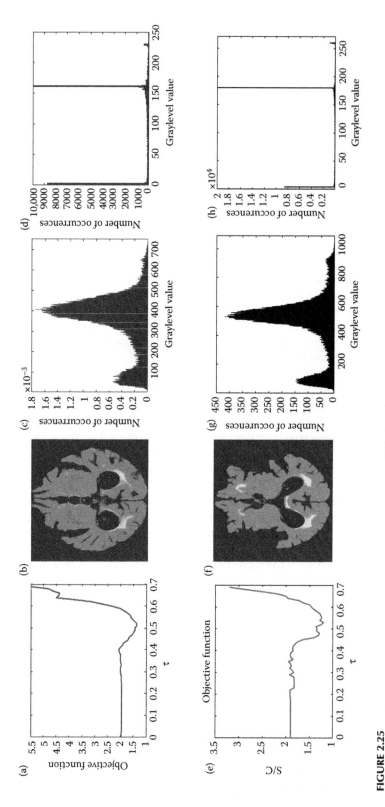

FIGURE 2.25

Reconstruction example ($\tau^* = 0.51$ and $\tau^* = 0.53$, respectively). (a) $\frac{S(Y_{rec}^\tau(x_1,x_2))}{C(Y_{rec}^\tau(x_1,x_2))}$. (b) $Y_{rec}^{0.51}$. (c) Hist(Y). (d) Hist $\left(Y_{rec}^{0.51}\right)$. (e) $\frac{S(Y_{rec}^\tau(x_1,x_2))}{C(Y_{rec}^\tau(x_1,x_2))}$. (f) $Y_{rec}^{0.53}$. (g) Hist(Y). (h) Hist $\left(Y_{rec}^{0.53}\right)$.

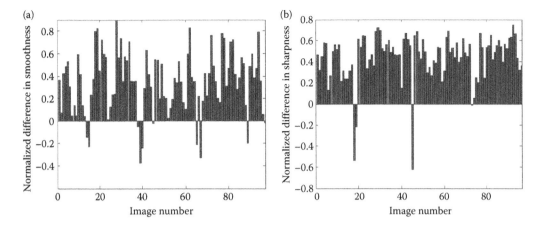

FIGURE 2.26
Normalized differences in smoothness and sharpness, between the proposed method and the bilateral filter. (a)
Smoothness. (b) Sharpness.

efficacy as a preprocessor for WML segmentation and validation studies will be conducted
to determine the optimal configuration. Other issues that will be addressed are whether the
features C and S were the best parameters to determine the objective function. Additionally,
the computational complexity of the algorithm will have to be improved in future versions as finding the objective function right now requires the reconstruction of 100 images
($\tau = \{0, 0.01, 0.02, \ldots, 1\}$) and the calculation of C and S for each of the images.

2.2.5 Contrast Enhancement

Although many imaging modalities produce images of superior quality, noise and/or
low-contrast objects can cause challenges in automated recognition tasks. As discussed
in Section 2.1.4, to increase the contrast of an image, HE techniques have been proposed
and used for many years. However, such a technique has limited application and often
does not work in all images. To combat this, many other techniques have been proposed
in the literature, see [30] for a comprehensive overview. Yang and co-workers attempted to
overcome over-enhancement issues common with adaptive HE schemes, by proposing a
local bihistogram equalization technique [35]. Results for cardiac MRI look promising and
preform better than traditional HE-based algorithms. Vidaurrazaga et al. [36] propose a
wavelet-based enhancement approach for medical images, which simultaneously shrinks
noisy wavelet coefficients, while enhancing image details by linearly combining wavelet
coefficients from all scales (with more importance placed on lower scales).

Each CE technique aims to exploit some image property to create a larger separation
between classes (objects), which ultimately increases the contrast of the image. Although
many are successful in the given task, for best enhancement performance, such algorithms
must be specifically designed for the modality [30]. To this end, this work discusses an
enhancement scheme first presented in [12], which is tuned to the characteristics of FLAIR-
weighted MRI of patients with WMLs.

Generalized CE algorithm design can be coarsely broken into two stages (apart from
preprocessing): (1) discriminatory feature design and (2) CE transfer function generation.
The first part of the design involves finding image features that discriminate between

anatomical objects. Then using these features, the contrast between the image objects are maximized by an enhancement function.

The intensity of pixels are itself a discriminatory feature, since it can differentiate between the brain and WML (visually). However, a single feature is not enough on its own and must be combined with another image characteristic to maximize differences between the image classes. To enhance the intensity features, a measure of the edge strength is incorporated.

The fuzzy edge measure described in Section 2.1.6 is used as an edge strength measure. As discussed in the previous section, like tissues are represented by similar intensities. Since edges which describe the boundary between two image objects are (intensity) dependent, edge magnitudes are also class dependent and thus carry discriminant information.

The intensity and edge features are combined into a single descriptor which estimates the global fuzzy edge content. This global edge profile, which removes the variance caused by noise in the original edge estimate, uniquely describes both the intensity and fuzzy edge information for enhanced discrimination between image objects. The global edge estimate, $\mu_\rho(y)$, is computed by averaging all the fuzzy edge values ρ_k at a particular intensity y

$$\mu_\rho(y) = \frac{1}{N} \sum_{\forall \rho_k | y} \rho_k, \tag{2.45}$$

where N is the number of ρ_k values at a given intensity y. The result is an average edge profile that estimates the edge content as a function of intensity value y. This edge measure describes the overwhelming trend in edge values, and in effect is denoising the original edge estimate ρ_k. Consider the image in Figure 2.23a, and its corresponding cluster plot of fuzzy edge strength ρ_k versus intensity y in Figure 2.27a, as well as the average edge profile $\mu_\rho(y)$ in Figure 2.27b. As can be seen by the global edge measure, the new estimate retains the overall trend of the data and is represented by a "noisefree" 1D curve. The global edge feature mapped back to the spatial domain is also shown in Figure 2.27c. Flat regions are replaced by low and similar values (little-to-no edge information) and the edgey regions, such as the brain-background border and the WML boundaries, are highlighted with a large value (high certainty of edge presence).

Another example is shown in Figure 2.28. As can be seen by $\mu_\rho(x_1, x_2)$, the global edge feature is robustly estimating the edge content in the image. Again, boundaries between objects are represented by a large magnitude (both edge classes are detected with high

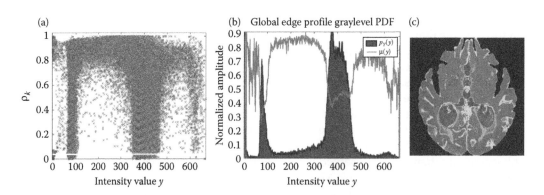

FIGURE 2.27

Fuzzy edge strength ρ_k versus intensity y for the image in Figure 2.23a. (a) ρ_k vs. y, (b) $\mu_\rho(y)$, and (c) $\mu_\rho(x_1, x_2)$.

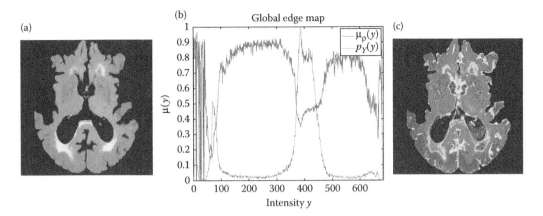

FIGURE 2.28
Original image $y(x_1, x_2)$, global edge profile $\mu_\rho(y)$ and global edge values mapped back to spatial domain $\mu_\rho(x_1, x_2)$. (a) $y(x_1, x_2)$, (b) $\mu_\rho(y)$, and (c) $\mu_\rho(x_1, x_2)$.

values), whereas flat regions are represented by a low and constant value. Thus, the new edge map is providing discriminatory information about the image classes.

As shown by the edge profiles, the brain-background edge $\mu_\rho(y)$, for low y values, is large and similar in value to the $\mu_\rho(y)$ of the WML boundaries (high y values). The fact that the brain and WML occur at different intensities (with high-edge value) will be exploited to maximize the contrast between these two classes.

To increase the separation between these two classes, the $\mu_\rho(y)$ function can be amplified for high-intensity values (WML boundaries), and attenuated over low-intensity values (the brain-background edge). To achieve this, the global edge profile $\mu_\rho(y)$ is weighted by a weighting function $w(y)$, to result in a modified transform $c(y)$

$$c(y) = w(y) * \mu_\rho(y). \tag{2.46}$$

In this work, we consider a linear, fuzzy weighting function, $w(y) \in [0, 1]$, where the attenuation of μ_ρ in low intensities (brain boundary) and amplification of μ_ρ in high intensities (WML boundaries) can be achieved by the following mapping:

$$w(y) = \frac{y}{y_{max}}. \tag{2.47}$$

Equation 2.47 quantifies the belongingness of a pixel to the bright pixel class (i.e., WMLs). When $w(y) = 1$, the pixel belongs to the bright pixel class with 100% certainty, whereas when $w(y) = 0$, there is no chance that this pixel belongs to the bright pixel class. All other values ($0 < w(y) < 1$), the further the intensity is from y_{max}, the less likely it is. This weighting factor $w(y)$ increases discrimination between edgy objects, in a class-dependant manner, since the edge strength value $\mu_\rho(y)$ is weighted by the normalized intensity value that it occurs at.

Multiplying $w(y)$ by the edge profile $\mu_\rho(y)$ results in a modified transfer function $c(y)$

$$c(y) = \frac{y}{y_{max}} * \mu_\rho(y) \tag{2.48}$$

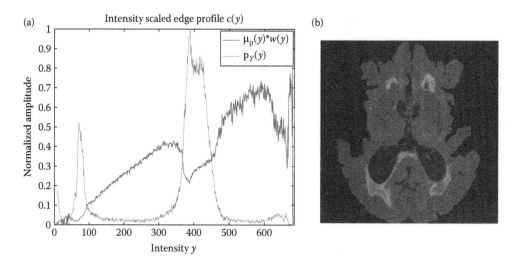

FIGURE 2.29
Modified transfer function $c(y)$ with original graylevel PDF $p_Y(y)$, and the resultant image, $c(x_1, x_2)$. (a) $c(y)$ and $p_Y(y)$ and (b) $c(x_1, x_2)$ of (b).

which amplifies large edges caused by WMLs and attenuates large edges caused by the brain-background edge. The new function provides an enhanced description of the tissue classes and an example is shown in Figure 2.29. The brain-background edge is suppressed and the WML borders are enhanced. These two features (intensity and weighted-edge strength) uniquely describe the boundaries of image objects.

To complete the enhancement, the interior of objects must be "filled in" since they were detected with low-edge values (and hence are of lower values than their counterpart boundaries). To enhance the image objects, the scaled average CDF is summed to a particular graylevel y to render the CE function y_{CE} for an input intensity y:

$$y_{CE}(y) = \sum_{i=0}^{y} \frac{i}{y_{max}} * \mu_\rho(i). \tag{2.49}$$

This transformation is a fuzzification describing the membership of a pixel to the WML class. The largest values of $y_{CE}(y)$ are assigned to the center of the WML (brightest position) indicating with the highest certainty that the WML is present. Lower values are assigned to the lesion boundaries, and even lower values are assigned to the brain region. As the brain-edge was suppressed, the WML and brain regions are further separated in intensities in the reconstructed image.

The final CE mapping function $y_{CE}(y)$ is shown in Figure 2.30, along with the contrast-enhanced output. The beginning portion of the enhancement function is piecewise linear with an overall decrease in intensity values (and range) in the output image (brain boundary and background is suppressed). In the middle of the transfer function, there is a nonlinear "dip" which occurs in the brain region, indicating that the brain intensities are also being suppressed in a nonlinear manner (in comparison to the WML). The transfer function after the nonlinear dip is steep—thus leading to an enhancement of the WML.

To test the enhancement scheme, five patients with WML were chosen, resulting in 18 images (slices) in total. The results for several images are shown in Figure 2.31. As can be seen by these results, the enhanced images offer much more discrimination between

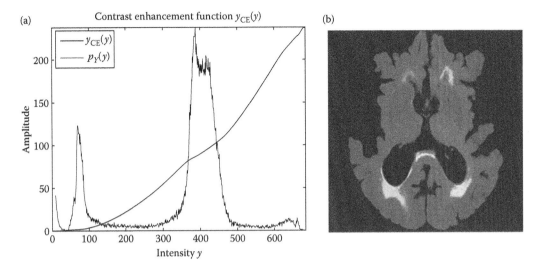

FIGURE 2.30
CE transfer function and contrast-enhanced image. (a) $y_{CE}(y)$ and $p_Y(y)$. (b) $y_{CE}(x_1, x_2)$ of (b).

WML and the normal brain tissue. To objectively assess the performance of the CE scheme, the CIR described in Section 2.2.3.2 is used. As this work focuses on CE of WMLs, it is not practical to use the entire image to compute the CIR (it would quantify global enhancement, versus local enhancement). Therefore, to measure the local enhancement of the lesions, several small regions from each image were chosen to compute the CIR (typically, they were approximately 25×25 in size). Each region contained normal brain tissue and WML. These values are shown in Table 2.4, and the average value over 18 images is 41.1%. As shown by the results (both in terms of CIR and visual appearance), the CE scheme performs very well.

To demonstrate the robustness of such a CE scheme as a preprocessor for segmentation, a threshold-based segmenter was applied to the images, and are also shown in Figure 2.31. The threshold was automatically set to the sum of the mean and standard deviation of the CE image.

Other modifications of the CE scheme are possible. For example, perhaps a nonlinear scaling function could be applied (instead of linear scaling) to the average CDF to further separate the WML and normal brain matter. Additionally, other features could be extracted instead of the mean CDF value per intensity y. These are the subjects of future works.

2.3 Feature Extraction and Classification

Computing devices are becoming an integral part of our daily lives and in many times, these algorithms are designed to mimic human behavior. In fact, this is the major motivation of many computer vision systems; to understand and analyze image content in the same fashion as humans do. Since texture has been shown to be an important feature for discrimination in images [37], understanding how humans perceive texture provides important clues into how a computer vision system should be designed. This section focuses on the

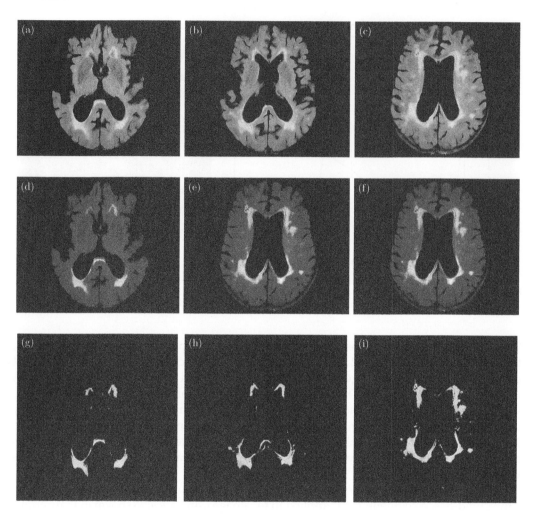

FIGURE 2.31
Original, contrast-enhanced images and WML segmentation. (a–c) Original. (d–f) Enhanced. (g–i) Segmentation.

design and implementation of a feature extraction and classification application, which is based on human texture perception. This work is described in more detail in [3,37–39].

When a surface is viewed, the human visual system can discriminate between textured regions quite easily. To describe how the human visual system can easily differentiate between textures, Julesz [40] defined textons, which are elementary units of texture.

TABLE 2.4

CIR results for 18 images

Im	CIR (%)	Im	CIR (%)	Im	CIR (%)	Im	CIR (%)
1	10.1	6	39.2	11	36.1	16	69.1
2	26.0	7	61.0	12	37.3	17	11.7
3	21.0	8	31.2	13	44.1	18	15.0
4	23.2	9	49.3	14	33.6	—	—
5	20.4	10	23.1	15	189	—	—

Textured regions can be decomposed using these textons, which include elongated blobs, lines, terminators, and more. It was found that the frequency content, scale, orientation, and periodicity of these textons can provide important clues on how to differentiate between two or more textured areas [40]. Therefore, to create a system which mimics human understanding of texture, it is necessary that the analysis system can detect these properties of the elementary units of texture (texture markers). To accommodate this, a system that detects textured events based on their scale, frequency, and orientation will be designed.

2.3.1 Feature Extraction and Invariance

To describe the textured events of images, a feature extraction scheme will be used. The extracted features are fed into a classifier, which arrives at a decision regarding the class of the image. Let $\mathcal{X} \subset \mathbf{R}^n$ represent the signal space which contains all images with the dimensions of $n = N \times N$. Since the images contained within $\mathcal{X}$ can be expected to have a very high dimensionality, using all these samples to arrive at a classification result would be prohibitive [41]. Furthermore, the original image space $\mathcal{X}$ is also redundant, which means that all the image samples are not necessary for classification. Therefore, to gain a more useful representation, a *feature extraction operator f* may map the subspace $\mathcal{X}$ into a feature space $\mathcal{F}$

$$f : \mathcal{X} \rightarrow \mathcal{F}, \tag{2.50}$$

where $\mathcal{F} \subset \mathbf{R}^k$, $k \leq n$ and a particular sample in the feature space may be written as a feature vector: $F = \{F_1, F_2, F_2, \ldots, F_k\}$. If $k < n$, the feature space mapping would also result in a dimensionality reduction.

Although it is important to choose features which provide the maximum discrimination between textures, it is also important that these features are robust. A feature is robust if it provides consistent results across the entire application domain [42]. To ensure robustness, the numerical descriptors should be rotation, scale, and translation invariant. In other words, if the image is rotated, scaled, or translated, the extracted features should be insensitive to these changes, or it should be a rotated, scaled, or translated version of the original features, but not modified [43]. This would be useful for classifying unknown image samples since these test images will not have structures with the same orientation and size as the images in the training set [44]. By ensuring invariant features, it is possible to account for the natural variations and structures within the images.

If a feature is extracted from a transform domain, it is also important to investigate the invariance properties of the transform since any invariance in this domain also translates to an invariance in the features.

2.3.2 Multiresolution Analysis

Julesz formalized the theory of human texture perception in terms of several features which are necessary for texture discrimination. If an image analysis system is to understand image content (in terms of texture) in a similar fashion as humans do, then it is necessary to utilize feature extractors which quantify these textural characteristics.

To analyze the scale and frequency properties of textural elements, nonstationary image analysis is an attractive choice since texture is comprised of space-localized frequency events. In particular, multiresolutional analysis techniques will be used to capture these nonstationary events, since the wavelet basis has excellent joint space–frequency resolution [43]. The diffusion of textural features or events will occur across subbands, which allows features to be captured not only *within* subbands, but also *across* subbands.

For an example of the localization properties of wavelets in an image, as well as the textural differences between images, consider Figure 2.32. The normal image's decomposition exhibits an overly homogeneous appearance of the wavelet coefficients in the HH, HL, and LH bands (which reflects the uniform nature of the original image). The decomposition of the abnormal image shows that each of the higher frequency subbands localizes the pathology, which appears as heterogeneous textured blobs (high-valued wavelet coefficients) in the center of the subband. This illustrates how the discrete wavelet transform (DWT) can localize the textural differences in images and also how multiscale texture may be used to discriminate between images.

Another benefit of wavelet analysis is that the basis functions are scale invariant. Scale-invariant basis functions will give rise to a localized description of the texture elements, regardless of their size or scale, that is, coarse texture can be made up of large textons, while fine texture is comprised of smaller elementary units.

Although the DWT is scale invariant, it is well known that the DWT is shift-variant [43], that is, the coefficients of a circularly shifted image are not translated versions of the original image's coefficients. Shift-variance causes significant challenges in feature extraction and therefore, for shift-invariant features, a shift-invariant discrete wavelet transform (SIDWT) is utilized.

To achieve translation invariance, the 2D version of Belkyns's SIDWT is employed [3,45]. This algorithm computes the DWT for all circular translates of the image, in a computationally efficient manner. Coifman and Wickerhauser's best basis algorithm is employed [41] to ensure that the same set of coefficients are chosen, regardless of the input shift. The result is a shift-invariant representation.

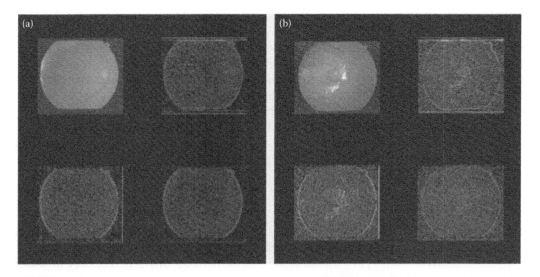

FIGURE 2.32
One level of DWT decomposition of retinal images. (a) Normal image decomposition; (b) decomposition of the retinal images with diabetic retinopathy. CE was performed in the higher frequency bands (HH, LH, HL) for visualization purposes.

2.3.3 Multiscale Texture Analysis

Now that a transformation has been employed which can robustly localize the scale-frequency properties of the textured elements in the images, it is important to design an analysis scheme which can quantify such textured events. To do this, this work proposes the use of a multiscale texture analysis scheme. Extracting features from the wavelet domain will result in a *localized* texture description, since the DWT has excellent space-localization properties.

To extract texture-based features, normalized graylevel co-occurrence matrices (GCMs) are used. Let each entry of the normalized GCM be represented as a probability distribution $p(l_1, l_2, d, \theta)$, where l_1 and l_2 are two graylevel values at a distance d and angle θ.

In the wavelet domain, GCMs are computed from each scale j at several angles θ. They are computed at multiple angles since, as Julesz discussed, orientation is also an important factor for texture discrimination. Each subband isolates different frequency components—the HL band isolates horizontal edge components, the LH subband isolates horizontal edges, the HH band captures the diagonal high-frequency components, and the LL band contains the low-pass-filtered version of the original. Consequently, to capture these oriented texture components, the GCM is computed at 0° in the HL band, 90° in the LH subband, 45° and 135° in the HH band, and 0°, 45°, 90°, and 135° in the LL band to account for any directional elements which may still be present in the low-frequency subband. Moreover, $d = 1$ for fine texture analysis.

From these GCMs, homogeneity h and entropy e are computed for each decomposition level using Equations 2.51 and 2.52. Homogeneity (h) describes how uniform the texture is and entropy (e) is a measure of nonuniformity or the complexity of the texture.

$$h(\theta) = \sum_{l_1=0}^{L-1} \sum_{l_2=0}^{L-1} p^2(l_1, l_2, d, \theta) \tag{2.51}$$

$$e(\theta) = -\sum_{l_1=0}^{L-1} \sum_{l_2=0}^{L-1} p(l_1, l_2, d, \theta) \log_2(p(l_1, l_2, d, \theta)) \tag{2.52}$$

These features describe the relative uniformity of textured elements in the wavelet domain.

For each decomposition level j, more than one directional feature is generated for the HH and LL subbands. The features in these subbands are averaged so that features are not biased to a particular orientation of texture and the representation will offer some rotational invariance. The features generated in these subbands (HH and LL) are shown below (note that the quantity in parentheses is the angle at which the GCM was computed):

$$\tilde{h}_{HH}^j = \frac{1}{2}\left(h_{HH}^j(45°) + h_{HH}^j(135°)\right),$$

$$\tilde{e}_{HH}^j = \frac{1}{2}\left(e_{HH}^j(45°) + e_{HH}^j(135°)\right),$$

$$\tilde{h}_{LL}^j = \frac{1}{4}\left(h_{LL}^j(0°) + h_{LL}^j(45°) + h_{LL}^j(90°) + h_{LL}^j(135°)\right),$$

$$\tilde{e}_{LL}^j = \frac{1}{4}\left(e_{LL}^j(0°) + e_{LL}^j(45°) + e_{LL}^j(90°) + e_{LL}^j(135°)\right).$$

As a result, for each decomposition level j, two feature sets are generated:

$$F_h^j = [h_{HL}^j(0°), h_{LH}^j(90°), \tilde{h}_{HH}^j, \tilde{h}_{LL}^j],$$ (2.53)

$$F_e^j = [e_{HL}^j(0°), e_{LH}^j(90°), \tilde{e}_{HH}^j, \tilde{e}_{LL}^j],$$ (2.54)

where $\tilde{h}_{HH}^j$, $\tilde{h}_{LL}^j$, $\tilde{e}_{HH}^j$, and $\tilde{e}_{LL}^j$ are the averaged texture descriptions from the HH and LL band previously described and $h_{HL}^j(0°)$, $e_{HL}^j(0°)$, $h_{LH}^j(90°)$, and $e_{LH}^j(90°)$ are homogeneity and entropy texture measures extracted from the HL and LH bands. Since directional GCMs are used to compute the features in each subband, the final feature representation is not biased for a particular orientation of texture and may provide a semirotational invariant representation.

2.3.4 Classifier

After the multiscale texture features have been extracted, a pattern recognition technique is necessary to correctly classify the features. A large number of test samples are required to evaluate a classifier with low error (misclassification) rates since a small database will cause the parameters of the classifiers to be estimated with low accuracy. This requires the image database to be large, which is not always be the case—especially for medical image databases. If the extracted features are strong (i.e., the features are mapped into nonoverlapping clusters in the feature space) the use of a simple (linear) classification scheme will be sufficient in discriminating between classes.

To satisfy the above criteria, linear discriminant analysis (LDA) will be the classification scheme used in conjunction with the *Leave One Out Method* (LOOM). In LOOM, one sample is removed from the whole set and the discriminant functions are derived from the remaining $N - 1$ data samples and the left out sample is classified. This procedure is completed for all N samples. LOOM will allow the classifier parameters to be estimated with least bias [46].

2.3.5 Results on Medical Image Databases

Consider Figure 2.33, which contains several healthy and diseased medical images. On visual inspection, it is obvious that texture discriminates between the normal and abnormal images. Pathological regions from the retinal and small bowel images contain inhomogeneous, oriented texture regions, whereas the normal images are relatively uniform (homogeneous). Considering the mammographic lesions, it is easy to see that the malignant lesion contains oriented, nonuniform spicules, whereas the benign mass is relatively smooth. Consequently, this work investigates human texture perception for image understanding and aims to model human texture analysis for CAD.

The classification performance of the proposed system is evaluated for three types of imagery:

1. *Small bowel images:* 41 normal and 34 abnormal (submucosal masses, lymphomas, jejunal carcinomas, multifocal carcinomas, polypoid masses, Kaposi's sarcomas, etc.)
2. *Retinal images:* 38 normal, 48 abnormal (exudates, large drusens, fine drusens, choroidal neovascularization, central vein and artery occlusion, arteriosclerotic retinopathy, histoplasmosis, hemicentral retinal vein occlusion, and more)
3. *Mammograms:* 35 benign and 19 malignant lesions

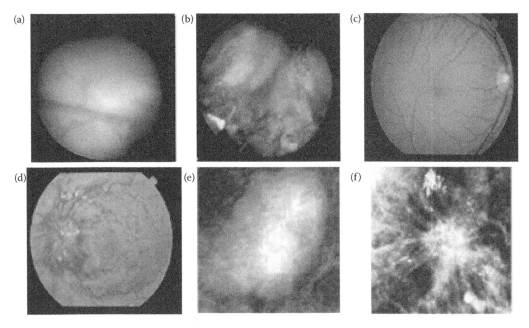

FIGURE 2.33
(**See color insert.**) Medical images exhibiting texture. (a) Normal small bowel, (b) small bowel lymphoma, (c) normal retinal image, (d) central retinal vein occlusion, (e) benign lesion, and (f) malignant lesion. CE was performed on (e) and (f) for visualization purposes.

The image specifications are shown in Table 2.5 and example images were shown earlier in Figure 2.33. Only the luminance plane was utilized for the color images (retinal and small bowel), in order to examine the performance of grayscale-based features. Furthermore, in the mammogram images, only a 128×128 region of interest is analyzed which contains the candidate lesion (to strictly analyze the textural properties of the lesions). Features were extracted from the higher levels of decomposition (the last three levels were not included as further decomposition levels contain subbands of 8×8 or smaller, resulting in skewed probability distribution (GCM) estimates). Therefore, the extracted features are F_e^j and F_h^j for $j = \{1, 2, \ldots, J\}$, where J is the number of decomposition levels minus three.

In order to find the optimal subfeature set, an exhaustive search was performed (i.e., all possible feature combinations were tested using the proposed classification scheme). For the small bowel images, the optimal classification performance was achieved by combining homogeneity features from the first and third decomposition levels with entropy from the

TABLE 2.5

Medical Image Specifications

Small Bowel	Retinal	Mammogram
Color (24 bpp)	Color (24 bpp)	Grayscale (8 bpp)
Lossy (.jpeg)	Lossy (.jpeg)	Raw (.pgm)
256×256	700×605	1024×1024

first decomposition level (see [39] for more details):

$$F_h^1 = [h_{HL}^1(0°), h_{LH}^1(90°), \widetilde{h}_{HH}^1, \widetilde{h}_{LL}^1], \tag{2.55}$$

$$F_h^3 = [h_{HL}^3(0°), h_{LH}^3(90°), \widetilde{h}_{HH}^3, \widetilde{h}_{LL}^3], \tag{2.56}$$

$$F_e^1 = [e_{HL}^1(0°), e_{LH}^1(90°), \widetilde{e}_{HH}^1, \widetilde{e}_{LL}^1]. \tag{2.57}$$

The optimal feature set for the retinal images were found to be homogeneity features from the fourth decomposition level with entropy from the first, second, and fourth decomposition levels (see [3] for more details):

$$F_h^4 = [h_{HL}^4(0°), h_{LH}^4(90°), \widetilde{h}_{HH}^4, \widetilde{h}_{LL}^4], \tag{2.58}$$

$$F_e^1 = [e_{HL}^1(0°), e_{LH}^1(90°), \widetilde{e}_{HH}^1, \widetilde{e}_{LL}^1], \tag{2.59}$$

$$F_e^2 = [e_{HL}^2(0°), e_{LH}^2(90°), \widetilde{e}_{HH}^2, \widetilde{e}_{LL}^2], \tag{2.60}$$

$$F_e^4 = [e_{HL}^4(0°), e_{LH}^4(90°), \widetilde{e}_{HH}^4, \widetilde{e}_{LL}^4]. \tag{2.61}$$

Finally, the optimal feature set for the mammographic lesions were found by combining homogeneity features from the second decomposition level with entropy from the fourth decomposition level:

$$F_h^2 = [h_{HL}^2(0°), h_{LH}^2(90°), \widetilde{h}_{HH}^2, \widetilde{h}_{LL}^2], \tag{2.62}$$

$$F_e^4 = [e_{HL}^4(0°), e_{LH}^4(90°), \widetilde{e}_{HH}^4, \widetilde{e}_{LL}^4]. \tag{2.63}$$

Using the above features in conjunction with LOOM and LDA, the classification results for the small bowel, retinal, and mammogram images are shown as a confusion matrix in Tables 2.6, 2.7, and 2.8, respectively.

Using the generalized framework for image classification based on texture, a total of 75 small bowel images were correctly classified at an average rate of 85%, 86 retinal images had an average classification accuracy of 82.2%, and the mammogram lesions (54) were classified correctly 69% on average. The classification results are quite high, considering that the system was not tuned for a specific modality and was strictly reliant on textural

TABLE 2.6

Results for Small Bowel Image Classification

	Normal	Abnormal
Normal	35 (85%)	6 (15%)
Abnormal	5 (15%)	29 (85%)

TABLE 2.7

Results for Retinal Image Classification

	Normal	Abnormal
Normal	30 (79%)	8 (21%)
Abnormal	7 (14.6%)	41 (85.4%)

TABLE 2.8

Results for Mammogram ROI Classification

	Benign	Malignant
Benign	28 (80%)	7 (20%)
Malignant	8 (42%)	11 (58%)

differences between cases. The system performed well, even though (1) pathologies came in various orientations, (2) pathologies arose in a variety of locations in the image, (3) the masses and lesions were of various sizes and shapes, and (4) there was no restriction on the type of pathology for the retinal and small bowel images. This is a direct consequence of the robust feature design (translation, scale, and semirotational invariance).

Although the classification results are high, any misclassification can be accounted to cases where there is a lack of statistical differentiation between the texture uniformity of the pathologies. Additionally, normal tissue can sometimes assume the properties of abnormal regions; for example, consider a normal small bowel image which has more than the average amount of folds. This may be characterized as nonuniform texture and consequently would be misclassified. In a normal retinal image, if the patient has more than the average number of vessels in their eye, this may be detected as oriented or heterogeneous texture and could be misclassified. Moreover, when considering the mammogram lesions, the normal breast parenchyma is overlapping with the lesions and also assumes some textural properties itself. In order to improve the performance of the mammogram lesions, a segmentation step could be applied prior to feature extraction.

Another important consideration arises from the database sizes. As was stated in Section 2.3.4, the number of images used for classification can determine the accuracy of the estimated classifier parameters. Since only a modest number of images were used, misclassification could result due to the lack of proper estimation of the classifiers parameters (although the scheme tried to combat this with LOOM). This could be the case for the mammogram lesions especially, since the number of benign lesions outnumbered the malignant lesions by almost double—this could have caused difficulties in classification parameter accuracy. Additionally, finding the right trade-off between the number of features and database size is an ongoing research topic and is yet to be perfectly defined [46].

References

1. J. Suri, S. Singh, and L. Reden, Computer vision and pattern recognition techniques for 2-D and 3-D MR cerebral cortical segmentation (part I): A state-of-the-art review, *Pattern Analysis and Applications*, 5(1), 46–76, 2002.
2. B. Froba and C. Kublbeck, Robust face detection at video frame rate based on edge orientation features, in *IEEE Conference on Face and Gesture Recognition*, Washington, DC, May 2002, pp. 342–347.
3. A. Khademi and S. Krishnan, Shift-invariant discrete wavelet transform analysis for retinal image classification, *Medical and Biological Engineering and Computing (MBEC) Journal*, 5(12), 2011–2022, 2007.
4. Q. Iqbal and J.K. Aggarwal, Combining structure, color and texture for image retrieval: A performance evaluation, in *International Conference on Pattern Recognition*, Quebec City, Canada, August 2002, Vol. 2, pp. 438–443.

5. S.M. Smith, Fast robust automated brain extraction, *Human Brain Mapping*, 17(3), pp. 143–155, 2002.
6. C.R. Jack, P.C. O'Brien, D.W. Rettman, M.M. Shiung, Y.C. Xu, R. Muthupillai, A. Manduca, R. Avula, and B.J. Erickson, Flair histogram segmentation for measurement of leukoaraiosis volume, *Journal of Magnetic Resonance Imaging*, 14(6), pp. 668–676, 2001.
7. R. Gonzalez and R.E. Woods, *Digital Image Processing (2nd Edition)*, Prentice-Hall, Inc., NJ, 2001.
8. P. Kvam and B. Vidakovic, *Nonparametric Statistics with Applications to Science and Engineering*, John Wiley & Sons, Inc., Hoboken, NJ, 2007.
9. A. Khademi, A. Venetsanopoulos, and A.R. Moody., Edge-based partial volume averaging estimation in FLAIR MRI with white matter lesions, in *IEEE EMBC* (Engineering in Medicine and Biology Conference), Buenos Aires, Argentina, pp. 1–4, 2010.
10. N. Otsu, A tlreshold selection method from gray-level histograms, *IEEE Transactions on SMC*, 9, 1408–1419, 1979.
11. P. Meer and B. Georgescu, Edge detection with embedded confidence, *IEEE Transactions on PAMI*, 23(12), 1351–1365, 2001.
12. A. Khademi, A. Venetsanopoulos, and A.R. Moody, Automatic contrast enhancement of white matter lesions in FLAIR MRI, in *IEEE ISBI* (International Symposium on Biomedical Imaging), Boston, Massachusetts, pp. 322–325, 2009.
13. M.B. Cuadra, B. Platel, E. Solanas, T. Butz, and J.-Ph. Thiran, Validation of tissue modelization and classification techniques in $T1$-weighted mr brain images, *MICCAI* (Medical Image Computing and Computer-Assisted Intervention), 2488, 290–297, 2002.
14. J. Sled, A. Zijdenbos, and A. Evans, A nonparametric method for automatic correction of intensity nonuniformity in mri data, *IEEE Transactions on Medical Imaging*, 17(1), 87–96, 1998.
15. L.P. Clarke, R.P. Velthuizen, S. Phuphanich, J.D. Schellenberg, J.A. Arrington, and M. Silbiger, MRI: Stability of three supervised semgentation algorithms, *Magnetic Resonance Imaging*, 11(1), 95–106, 1993.
16. M.B. Cuadra, L. Cammoun, T. Butz, O. Cuisenaire, and J.-P. Thiran, Comparison and validation of tissue modelization and statistical classification methods in $T1$-weighted MR brain images, *IEEE Transactions on Medical Imaging*, 24(12), 1548–1565, 2005.
17. P. Gravel, G. Beaudoin, and J.A. De Guise, A method for modeling noise in medical images, *IEEE Transactions on Medical Imaging*, 23, 1221–1232, 2004.
18. M. Styner, C. Brechbuhler, G. Szckely, and G. Gerig, Parametric estimate of intensity inhomogeneities applied to MRI, *IEEE Transactions on Medical Imaging*, 19, 153–165, 2000.
19. H. Cheng and F. Huang, Magnetic resonance imaging image intensity correction with extrapolation and adaptive smoothing, *Magnetic Resonance in Medicine*, 55, 959–966, 2006.
20. A. Noe and J.C. Gee, Partial volume segmentation of cerebral MRI scans with mixture model clustering, *Information Processing in Medical Imaging—Lecture Notes in Computer Science*, 2082, 423–430, 2001.
21. D.L. Collins, A.P. Zijdenbos, V. Kollokian, J.G. Sled, N.J. Kabani, C.J. Holmes, and A.C. Evans, Design and construction of a realistic digital brain phantom, *IEEE Transactions on Medical Imaging*, 17, 463–468, 1998.
22. A. Khademi, D. Hosseinzadeh, A. Venetsanopoulos, and A. Moody, Nonparametric statistical tests for exploration of correlation and nonstationarity in images, in *DSP Conference*, Santorini, Greece, pp. 1–6, 2009.
23. A. Samsonov and C. Johnson, Noise-adaptive nonlinear diffusion filtering of MR images with spatially varying noise levels, *Magnetic Resonance in Medicine*, 52, 798–806, 2004.
24. W. Wierwille and J. Knight, Off-line correlation analysis of nonstationary signals, *IEEE Transactions on Computers*, 17(6), 525–536, 1968.
25. N. Mantel, The detection of disease clustering and a generalized regression approach, *Cancer Research*, 27(2), 209–220, 1967.
26. O. Schabenberger and C. Gotway, *Statistical Methods for Spatial Data Analysis*, Chapman & Hall CRC, Boca Raton, FL, 2005.
27. T. Becka, T. Housha, J. Weirb, J. Cramerc, V. Vardaxisb, G. Johnsona, J. Coburnd, M. Maleka, and M. Mielkea, An examination of the runs test, reverse arrangements test, and modified reverse

arrangements test for assessing surface emg signal stationarity, *Journal of Neuroscience Methods*, 156(1–2), 242–248, 2006.

28. P. Cappa, S. Silvestri, and S. Sciuto, On the robust utilization of non-parametric tests for evaluation of combined cyclical and monotonic drift, *Measurement Science and Technology*, 12, 1439–1444, 2006.

29. J. Dijk, D. de Ridder, P.W. Verbeek, J. Walraven, I.T. Young, and L.J. Van Vliet, A quantitative measure for the perception of sharpening and smoothing in images, in *5th Advanced School for Computing and Imaging*, Delft, the Netherlands, pp. 291–298, 1999.

30. C. Wyatt, Y.-P. Wang, M. T. Freedman, M. Loew, and Y. Wang, *Biomedical Information Technology*, Chapter 7, pp. 165–169, Elsevier, USA, 2007.

31. R.D. Nowak, Wavelet-based Rician noise removal for magnetic resonance imaging, *IEEE Transactions on Image Processing*, 8, 1408–1419, 1999.

32. J.C. Wood and K.M. Johnson, Wavelet packet denoising of magnetic resonance images: Importance of Rician noise at low SNR, *Magnetic Resonance in Medicine*, 41, 631–635, 1999.

33. S. Paris, P. Kornprobst, J. Tumblin, and F. Durand, A gentle introduction to bilateral filtering and its applications, in *IEEE CVPR*, Cairo, Egypt, 2008, 1–130.

34. A. Khademi, A. Venetsanopoulos, and A.R. Moody, Image reconstruction for noise suppression in FLAIR MRI with white matter lesions, *IEEE Signal Processing Letters*, 17(12), 989–992, 2010.

35. Y.-C. Fan H.-Y. Yang, Y.-C. Lee, and H.-W. Taso, A novel algorithm of local contrast enhancement for medical image, in *IEEE Nuclear Science Symposium*, Honolulu, Hawaii, 2007, 3951–3954.

36. M. Vidaurrazaga, L.A. Diago, and A. Cruz, Contrast enhancement with wavelet transform in radiololgical images, in *IEEE EMBS*, Chicago, Illinois, 2000, 1760–1763.

37. A. Khademi, Multiresolutional analysis for classification and compression of medical images, MS thesis, Ryerson University, Canada, 2006.

38. A. Khademi and S. Krishnan, Medical image texture analysis: A case study with small bowel, retinal and mammogram images, in *IEEE CCECE* (Canadian Conference on Electrical and Computer Engineering), Niagara Falls, Canada, May 2008, 1–6.

39. A. Khademi and S. Krishnan, Multiresolution analysis and classification of small bowel medical images, in *IEEE EMBC*, Lyon, France, August 2007, 1–4.

40. B. Julesz, Textons, the elements of texture perception, and their interactions, *Nature*, 290(5802), 91–97, 1981.

41. R.R. Coifman and N. Saito, Local discriminant bases and their applications, *Journal of Mathematical Imaging and Vision*, 5, 337–358, 1995.

42. S.E. Umbaugh, Y.S. Wei, and M. Zuke, Feature extraction in image analysis a program for facilitating data reduction in medical image classification, *IEEE Engineering in Medicine and Biology Magazine*, 16, 62–73, 1997.

43. S. Mallat, *Wavelet Tour of Signal Processing*, Academic Press, USA, 1998.

44. M.M. Leung and A.M. Peterson, Scale and rotation invariant texture classification, in *Conference Record of The Twenty-Sixth Asilomar Conference on Signals, Systems and Computers*, Pacific Grove, California, October 1992, Vol. 1, 461–465.

45. J. Liang and T.W. Parks, Image coding using translation invariant wavelet transforms with symmetric extensions, *IEEE Transactions on Image Processing*, 7, 762–769, 1998.

46. K. Fukunaga and R.R. Hayes, Effects of sample size in classifier design, *IEEE Transactions on Pattern Analysis and Machine Intelligence*, 11(8), 873–885, 1989.

3

Application-Specific Multimedia Architecture

Tung-Chien Chen, Tzu-Der Chuang, and Liang-Gee Chen

CONTENTS

3.1   Introduction

Multimedia applications, such as radio, audio, camera phone, digital camera, camcorder, and mobile broadcasting TV, are increasingly popular as the technologies of image sensor, communication, VLSI manufacture, and video coding standards have made great progress. Among various multimedia applications, video and image applications are always attractive, but the required transmission bandwidth or storage size is also much larger than others. Hence many image and video coding standards such as JPEG, JEPG-2000, MPEG series, and H.26x-series are established to compress image and video data in order to save the required transmission bandwidth or the storage size.

However, the toughest challenge of image and video compression is real-time processing. For HDTV applications, several tera-operations/instructions per second (TOPS) of computing power and several tera-bytes per second (TB/s) of memory access are demanded. The required resources of real-time coding are far beyond the capabilities of today's general purpose processors. For mobile applications with much smaller image sizes, power consumption is the most critical issue. If a processor-based software implementation is adopted, the operating frequency of the processor should be as high as hundreds of megaHertz, which will violate the strict power constraints. Although the complexity of mobile applications is much lower than that of HDTV applications, real-time compression is still a heavy burden for processors. Hence, application-specific multimedia architecture plays a role in multimedia applications. Hardware-oriented algorithms and efficient VLSI architectures for application-specific multimedia architecture are urgently needed.

In this chapter, we take the latest and the most powerful video coding standard, the H.264/AVC [1], as design example to illustrate the design issues and methodologies of application-specific multimedia architecture. In a video coding system, temporal prediction is always the processing bottleneck. More than half, sometimes even up to 90% of computing power and memory access are dominated by temporal prediction. Therefore, in this chapter, we will focus on how to design an efficient VLSI architecture for temporal prediction of H.264/AVC to meet the requirement of huge computation complexity and memory bandwidth, which are the most critical issues for the hardware design of a video coding system. This chapter is organized as follows. Section 3.2 is a brief introduction of design issue of application-specific multimedia architecture. Section 3.3 shows the system-level design of a H.264/AVC video encoder. Sections 3.4 through 3.6 discuss the temporal prediction architecture and data reuse method of H.264/AVC. Section 3.4 is a comprehensive discussion about design issues and architecture exploration of the most important function in video coding, the integer motion estimation. In Section 3.5, data bandwidth problem and related design methodologies are discussed. In Section 3.6, the architecture of new coding tools proposed by H.264/AVC, the fractional motion estimation, is introduced. Finally, Section 3.7 summarizes this chapter.

3.2 Design Issues of Application-Specific Multimedia Architecture

3.2.1 Multimedia Architecture

A general architecture of multimedia applications system is shown in Figure 3.1, which is composed of many modules, such as network/communication, audio processor, central controller, co-processor, and video processor. Each module is responsible for one task of a multimedia system. Among different tasks, video/image processing is the core technique, and the toughest challenge of video/image coding systems is real-time processing due to its huge computation complexity and ultra large memory access.

Since video/image processing is the key component, we focus on the hardware design of video/image processing in this chapter. A general architecture of video/image processing is shown in Figure 3.2a, which consists four modules, processing elements, on-chip memory, system bus, and external memory. If we want to implement a video/image processing algorithm into this architecture, there are four design challenges that we need to take care.

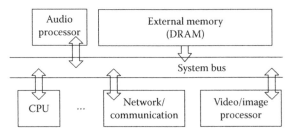

FIGURE 3.1
A general architecture of multimedia applications system.

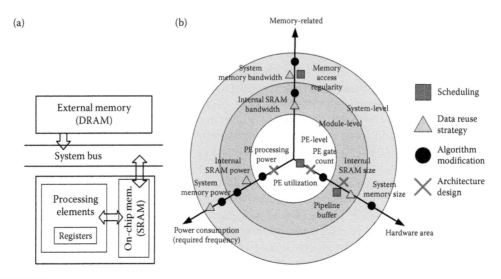

FIGURE 3.2
(a) The general architecture and (b) hardware design issues of the video/image processing engine.

The first one is the architecture design for processing elements. We require a good architecture to satisfy the required computation complexity with a smaller area cost. The second one is data reuse strategy which optimizes memory access behavior to solve the large memory bandwidth problem. The on-chip memory arrangement is the third one. On-chip memory is used to be data buffer and it has two requirements. One is the memory access requirement of processing elements, and the other is the memory size requirement of data reuse strategy. A good arrangement should satisfy these two constraints at the same time. The scheduling of external memory access is the fourth design challenges. In general, irregular memory access in external memory reduces the efficiency of memory access and system bandwidth, so the scheduling to guarantee regular memory access is important.

These four design challenges are encountered from the module-level viewpoint. From the system-level viewpoint, the critical issues are processing speed, power consumption, hardware area, system memory bandwidth (external memory bandwidth) and memory storage (on-chip memory size and external memory size). The relationship between the four design challenges and four system issues is shown in Figure 3.2b. These issues can be further discussed in three design-level, system-level, module-level, and PE-level. The architecture design is related to hardware area and power consumption, the scheduling is the tradeoff between control complexity and system memory bandwidth, and the data reuse strategy is the tradeoff between system memory bandwidth, hardware area, power, and external memory size. In addition, the adopted algorithm always decides the basic requirement of memory requirement, hardware area, and power.

3.2.2 Design Issues

3.2.2.1 Processing Speed

For image and video applications, the speed requirement is usually the most urgent issue. Actually, for on-line video applications, the first priority of an implementation is to meet the real-time specification.

Since there is a large amount of data to be processed within a tight time constraint, general processors, which execute the computations sequentially, cannot afford such high computational load. Only raising the working frequency is not a good approach to solve the problem. Working fast but not efficiently is of course not an optimal way. Also, high-power consumption is a problem at higher frequency especially for battery-powered portable appliances. Therefore, for image and video processing designs, hardware architecture designers usually do not pursue very high operating frequencies. On the contrary, designers look for more efficient architectures to be operated at lower clock rate. SIMD (single instruction, multiple data), VLIW (very-long instruction word), array processor, and dedicated parallel architecture are explored for higher computational efficiency.

3.2.2.2 Area

IC designers always look for compact architectures since smaller die area means lower cost. A more compact design will have better competitiveness. In a dedicated accelerator design, parallel architecture is usually adopted to achieve the required specification while lowering the required working frequency. Area is used to trade with frequency.

Parallel architecture results in higher area cost. Therefore, hardware utilization is an important index that designers should take care of and check. Simply duplicating multiple processing elements and memories may not be an optimal way if some of these resources are with low hardware utilization rate. The optimization goal should be a just enough parallelism with as higher as possible hardware utilization. Algorithm-level optimization, hardware sharing and folding are example techniques for chip area optimization.

3.2.2.3 Power

The power issue becomes one of the most important issue in multimedia applications. As with the process development, a chip can provide more and more transistors, but the allowed power consumption in a chip will not increase. The emerging portable multimedia devices ask for more restricted power consumption. In addition, the heat due to high-power consumption will also cause the reliability problem.

To cope with the power problem, a design should be carefully examined not only in architecture level but also in algorithm and circuit levels. Actually, the higher-level optimization usually provides more gain in power saving. The development of fast algorithms with lower complexity, and algorithms with lower data bandwidth are examples of this. Designers have to understand the algorithm characteristics by detailed analysis.

At architecture level, designers have to look into the detailed operations of each module or even each gate, and its power consumption behavior. High hardware utilization architecture will be more power-efficient since the power will not be wasted on idle gates. Memory is also a big source of power consumption. Therefore, memory hierarchy and arrangement will play an important role on a low-power design. At last, the algorithm and architecture characteristics can be combined with circuit level techniques such as clock gating, dynamic voltage/frequency scaling, multiple voltage domain, and so on.

3.2.2.4 Memory Bandwidth and Storage

Memory bandwidth and on-chip memory capacity are limiting factors for many multimedia applications. Today, in many designs, on-chip memory has already occupied more than 50% of total chip area. Good memory management and area-, power-, and yield-efficient

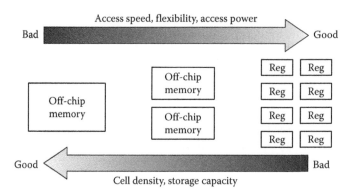

FIGURE 3.3
Memory hierarchy: trade-offs and characteristics.

memory implementations, become important for a successful multimedia architecture solution.

The memory management is to provide an efficient memory hierarchy that consists of off-chip memory, on-chip memory, and registers [2], as shown in Figure 3.3. Different memory types have different features. Off-chip memory, usually implemented by DRAM, offers a large amount of storage size but consumes the most power. The off-chip memory and I/O access may dominate the power budget. The on-chip memory, usually implemented by SRAM, provides faster access and less power consumption than the off-chip memory, but the memory cell size is much larger. Registers is faster than on-chip memory and can be accessed with more flexibility. However, the size of a register is the largest. In general, registers are more suitable for the implementation of smaller-size buffers.

Memory management can be organized from two different levels: algorithm level and architecture-level. The algorithm-level memory hierarchy optimization is to modify the algorithm of video/image processing system improve architecture performance, such as power, bandwidth, and area. On the other hand, the architecture-level memory organization can optimize the memory hierarchy from modifying hardware architecture, such as applying different reuse schemes to reduce memory storage size or adopting a cache to reduce memory bandwidth.

3.3   System Architecture of H.264/AVC Codec

3.3.1   Instruction Profiling

Before introducing the hardware architecture, we first do instruction profiling of an video/image processing system. Instruction profiling is based on a reduced instruction set computing (RISC) platform. It can be viewed as a hardware architecture with only one processing element (PE), or an "extremely folded" architecture. This data is valuable for software implementation, hardware implementation, and software/hardware partitioning. For software implementation, it can be used to find the critical module to be optimized. For hardware implementation, it can be used to find the parallelism requirement for each module in order to achieve the given specification. As for software/hardware partitioning, it can be used to roughly find some critical modules that need to be implemented in

TABLE 3.1

Instruction Analysis of an H.264/AVC Baseline Profile Encoder

Functions	Arithmetic		Controlling		Data Transfer		
	MIPS	%	MIPS	%	MIPS	MB/s	%
Integer-pel motion estimation	95,491.9	78.31	21,915.1	55.37	116,830.8	365,380.7	77.53
Fractional-pel motion estimation	21,396.6	17.55	14,093.2	35.61	30,084.9	85,045.7	18.04
Fractional-pel interpolation	558.0	0.46	586.6	1.48	729.7	1067.6	0.23
Lagrangian mode decision	674.6	0.55	431.4	1.09	880.7	2642.6	0.56
Intra prediction	538.0	0.44	288.2	0.73	585.8	2141.8	0.45
Variable length coding	35.4	0.03	36.8	0.09	44.2	154.9	0.03
Transform and quantization	3223.9	2.64	2178.6	5.50	4269.0	14,753.4	3.13
Deblocking	29.5	0.02	47.4	0.12	44.2	112.6	0.02
Total	121,948.1	100.00	39,577.3	100.00	153,469.3	471,299.3	100.00

Note: The parameters are CIF, 30 frames/s, 5 reference frames, ± 16-pel search range, $QP = 20$, and low complexity mode decision. MIPS stands for million instructions per second.

hardware and some modules whose complexity is small enough to be handled by software executed on a given processor.

In this chapter, we take H.264/AVC as an example. The instruction profiling results prepared by Chen [3] are shown in Table 3.1. The simulation model is based on standard-compatible software, and the platform is a SunBlade 2000 workstation with a 1.015 GHz Ultra Sparc II CPU and 8 Gbytes RAM. Arithmetic, controlling, and data transfer instructions are separated in this table. It can be observed that motion estimation, including integer-pel motion estimation, fractional-pel motion estimation, and fractional-pel interpolation in the table, takes up more than 95% of the computation in the whole encoder, which is a common characteristic in all video encoders. Among motion estimation, integer-pel motion estimation, which is implemented with a full search algorithm, plays the most important role. The total required computing complexity for a H.264 encoder is more than 300 giga instructions per second (GIPS), which is hardly achieved by existing processors. Therefore, an application-specific hardware architecture is necessary for H.264. Furthermore, the amount of data transfer is more than 460 Gbytes/s, which cannot be achieved by existing memory systems. Most of the data transfer comes from motion estimation. Therefore, the memory architecture of motion estimation must be carefully designed. Note that the simulation model used for this profiling is not optimized software. Hence, some data above may be larger than those of commercial products. However, this can still be valuable information for hardware architecture design.

3.3.2 Conventional Two-Stage Macroblock Pipelining

For MPEG-1/2/4 series encoders, since the required number of cycles for ME for one macroblock is larger than the summation of those of other modules, two-stage pipelining is usually used. Figure 3.4 shows the conventional two-stage macroblock (MB) pipelining architecture. The processor is responsible for MB-level hardware control and other high-level tasks. Accelerators of motion estimation (ME) and block engine (BE) are connected to the bus to speed up the encoding operations. When one MB is processed at the ME stage, its previous MB is handled by the BE stage. Two MBs are simultaneously processed. The prediction tasks and texture coding tasks are thus clearly separated. The mode decision

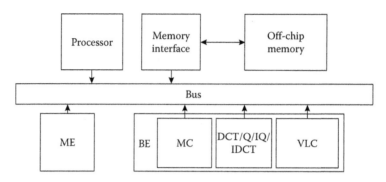

FIGURE 3.4
Conventional two-stage macroblock pipelining architecture.

result is fetched by the processor from ME to BE via bus. To reduce the traffic on system bus, the motion compensation (MC), differential pulse code modulation (DPCM) loop, and variable length coding (VLC) modules are linked with shared memories. The ME module requires the most computation and bandwidth.

3.3.3 Design Challenges of Two-Stage Macroblock Pipelining Architecture for H.264/AVC Encoder

The conventional two-stage MB pipelining architecture is widely adopted in prior video encoding hardware designs, such as MPEG-1/2/4 series encoders. However, three main problems will be encountered if the two-stage MB pipelining is directly applied to H.264/AVC encoding. The first problem is the low utilization or high frequency. The prediction in H.264/AVC includes integer motion estimation (IME), fractional motion estimation (FME), and intra prediction (IP). IME demands the most computation and memory bandwidth. FME is over 100 times more complex than that of prior standards due to the multiple reference frames (MRF), variable block sizes (VBS), quarter-pixel accuracy, more precise distortion evaluation, and mode decision subsequent to FME. In addition, FME cannot be parallelized with IME for the same MB. IP has 17 modes in H.264/AVC baseline profile and is also very time-consuming. Putting IME, FME, and IP in the same MB pipeline stage leads to very low utilization. Even if the resource is shared for the three prediction tasks, the operating frequency will be high due to a large number of required cycles for an MB in the prediction stage. The second problem is the required high bandwidth. Because of MRF and VBS, the required bus bandwidth increases a lot for motion compensation (MC) and transmitting mode decision results. The third problem is about the feasibility. IP utilizes neighboring reconstructed pixels to generate predictors. Therefore, IP cannot be separated with the DPCM loop (DCT/Q/IQ/IDCT). Conventional pipelining violates the data dependency between IP and DPCM. Therefore, new MB pipelining scheme and efficient module designs are required for H.264/AVC encoding systems.

3.3.4 Four-Stage Macroblock Pipelining Architecture

To overcome the design challenges mentioned in previous sections, Huang [4] proposed a four-stage MB pipelining architecture for H.264/AVC encoder. The system architecture is shown in Figure 3.5. The encoder contains a main controller and five engines for integer

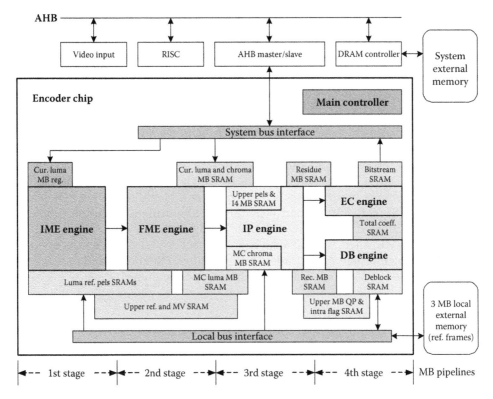

FIGURE 3.5
Block diagram of the four-stage MB pipelining H.264/AVC encoding system.

motion estimation (IME), fractional motion estimation (FME), intra prediction (IP), entropy coding (EC), and deblocking (DB). Reference frames are buffered at the external memory via local bus. The maximum bandwidth requirements are 40 and 240 MB/s for the system bus and local bus, respectively, for the HDTV720p specification.

The system is characterized by the four-stage MB pipelining. As mentioned before, in H.264/AVC encoding, IME has the most computation and memory requirements. FME is 100 times more complex than that of prior standards, and it cannot be parallelized with IME for the same MB. IP is also very time-consuming. Moreover, it is difficult to execute IME, FME, and IP operations on the same circuits. Even if hardware resource sharing of the three prediction tasks is achieved, the required operating frequency will be very high. Hence, the prediction engine is partitioned into three stages, and EC/DB is placed at the 4th stage to generate bitstream and reference frames. MB data propagate through IME, FME, IP, and EC/DB, and four MBs are simultaneously processed. The throughput is thus approximately doubled. IP must also integrate forward/inverse transform/quantization because reconstructed neighboring pixels are necessary to generate predictors. In this architecture, processing cycles of the four stages are balanced to achieve high utilization, and local data transfer between stages is used to reduce bus traffic. This four-stage MB pipelining architecture establishes a milestone for H.264/AVC encoder design. Several advanced H.264/AVC encoder designs [5,6] are based on this four-stage MB pipelining architecture with modifications.

3.4 Integer Motion Estimation Architecture

Motion-compensated transform coding has been adopted by all of the existing international video coding standards, such as the MPEG-1/2/4 series and the H.26x series. Motion estimation (ME) removes temporal redundancy within frames and thus provides coding systems with high compression ratio. Block matching approach is mostly selected as the ME module in video codecs and is also adopted in all existing video coding standards because of its simplicity and good performance. In MPEG-1/2/4 series, the fix block size motion estimation (FBSME) is adopted, while the variable block size motion estimation (VBSME) is proposed in H.264/AVC. VBSME is one kind of block-based ME algorithms and it can provide more accurate prediction, compared to traditional FBSME. However, VBSME increases the difficulty of hardware architecture design. In this section, we first introduce the algorithm of ME, and we will have a comprehensive discussion on full search architecture of FBSME. Then, we will analyze the impact of VBSME on hardware architectures. By classifying inter-level or intra-level architectures, data flows, and the seven loops of ME, the impact of VBSME on hardware architectures is analyzed.

3.4.1 Motion Estimation

3.4.1.1 Introduction of Motion Estimation

As mentioned before, block matching algorithm (BMA) is usually adopted in motion estimation due to its simplicity and good performance. Among various BMAs, full search is usually preferred because it can provide the best quality. The computation flow of full search is as follows. The current frame is divided into many small macroblocks (MBs), and each MB in the current frame (current MB) is matched in the searching range of the reference frame by calculating the distortion. Figure 3.6 shows the spatial relationship between the current MB and the searching region. The most commonly used distortion model is the sum of absolute differences (SAD), which can be written as

$$SAD(k,p) = \sum_{j=1}^{B_V} \sum_{i=1}^{B_H} Distortion(i,j,k,p), \tag{3.1}$$

$$Distortion(i,j,k,p) = |cur(i,j) - ref(i+k,p+j)|, \tag{3.2}$$

$$-P_H \leq k < P_H, \quad -P_V \leq p < P_V, \tag{3.3}$$

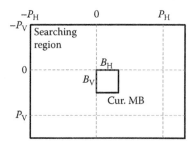

FIGURE 3.6
The spatial relationship between the current macroblock and the searching range.

```
for Number-of-Frame (Frame-level loop)

    for Number-of-MBv (MB-level loop in the vertical direction)
    for Number-of-MBH (MB-level loop in the horizontal direction)

        for Number-of-SRv (SR-level loop in the vertical direction)
        for Number-of-SRH (SR-level loop in the horizontal direction)

            for Number-of-CBv (CurBlock-level loop in the vertical direction)
            for Number-of-CBH (CurBlock-level loop in the horizontal direction)
                ......
                ...
            end of Number-of-CBH
            end of Number-of-CBv

        end of Number-of-SRH
        end of Number-of-SRv

    end of Number-of-MBH
    end of Number-of-MBv

end of Number-of-Frame
```

FIGURE 3.7
The procedure of ME in a video coding system for a sequence.

where B_H (B_V) is the horizontal (vertical) size of the current MB, the searching range is $[-P_H, P_H)$ and $[-P_V, P_V)$ in the horizontal and vertical directions, (k, p) is one position in the searching range (searching candidate), its corresponding block is called the reference block, $cur(i, j)$ is the pixel value in the current MB (current pixel), $ref(i + k, j + p)$ is the pixel value in the reference block (reference pixel), $Distortion(i, j, k, p)$ is the difference between the current pixel $cur(i, j)$ and the reference pixel $ref(i + k, j + p)$, and $SAD(k, p)$ is the summation of all distortions in the current MB for the searching candidate (k, p). The row SAD is the summation of B_H distortions in a row, and the column SAD is the summation of B_V distortions in a column. After examining all searching candidates, the searching candidate with the smallest SAD is selected as the best reference block, and the associated (k, p) is the motion vector of this current MB.

For a video sequence, the procedure of full search in a video coding system can be decomposed into seven loops, as shown in Figure 3.7. The first loop (*Frame-level loop*) is the number of frames in a video sequence, and the second and third loops (*MB-level loop*) represent the number of current MBs in one frame. The fourth and fifth loops (*SR-level loop*) exhibit the number of searching candidates in the searching region, and the last two loops (*CurBlock-level loop*) describe the number of pixels in one current MB for the computation of SAD.

3.4.1.2 *Variable Block Size Motion Estimation*

In block-based motion estimation algorithms, VBSME is a newly developed coding technique, and it can provide more accurate predictions compared to traditional FBSME. With FBSME, if a MB consists of two objects with different motion directions, the coding performance of this MB is worse. On the other hand, for the same condition, as such MB can be divided into smaller blocks in order to fit the different motion directions with VBSME. Hence, the coding performance is improved. VBSME has been adopted in the latest video coding standards, including H.263 [7], MPEG-4 [8], and H.264/AVC [1]. As shown in Figure 3.8, in H.264/AVC, a MB with variable block size can be divided into seven kinds

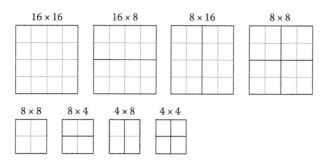

FIGURE 3.8
Block partition of H.264/AVC variable block size.

of blocks including 4×4, 4×8, 8×4, 8×8, 8×16, 16×8, and 16×16. Although VBSME can achieve a higher compression efficiency, its prices are the huge computation complexity and the increase of the difficulty in ME hardware implementation.

Traditional ME hardware architectures are designed for FBSME, and they can provide different advantages in some specific system issues, such as memory bitwidth, area, latency, throughput, and so on. Since many hardware architectures for FBSME are proposed and VBSME becomes important, the main purpose of this section is not only to make a study of FBSME architectures but also to analyze the impact of VBSME on hardware architectures. Beyond, we provide a comprehensive comparison for system designers to select the optimal tradeoff.

3.4.2 Exploration of Full Search Architectures

In this section, eight representative works of ME hardware architectures for FBSME are introduced. They are the works of Yang et al. [9], Yeo and Hu [10], Lai and Chen [11], Komarek and Pirsch [12], Vos and Stegherr [13], and Hsieh and Lin [14]. These six architectures are significant works, and many hardware architectures are proposed based on them. For example, Reference [15] is the extension of [9]. Reference [16] is proposed based on [13]. Reference [17] combines [14] with multilevel successive elimination algorithm [18,19]. Reference [20] is the extension of [12]. Furthermore, two more architectures, propagate partial SAD proposed by Huang et al. [21] and SAD tree proposed by Chen et al. [22], are also introduced and added into comparisons.

Traditional ME hardware architectures can be roughly classified into two categories. One is an inter-level architecture, where each processing element (PE) is responsible for one SAD of a specific searching candidate, as shown in (3.1), and the other is an intra-level architecture, where each PE is responsible for the distortion of a specific current pixel in the current MB for all searching candidates, as shown in (3.2). That is, inter-level architectures perform the computations of *SR-level loop* in parallel but execute the computations of *CurBlock-level loop* sequentially. Contrarily, intra-level architectures perform the operations of *SR-level loop* sequentially but execute the computations of *CurBlock-level loop* in parallel. Besides pure inter-/intra-level architectures, there are other kinds of architectures such as AS2 in [12] and tree-based architecture in [23], which are hybrids of inter-level and intra-level architectures. For the sake of simplicity, we only discuss the pure inter-/intra-level architectures. The others can be easily extended based on our analysis. In the following discussion, we use the classification of inter-level or intra-level architectures, the data flows, and the seven loops of ME to describe the characteristics of each hardware architecture. The block size is

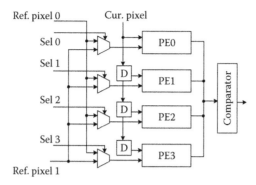

FIGURE 3.9
The hardware architecture of 1DInterYSW, where $N = 4$, $P_h = 2$, and $P_v = 2$.

assumed to be $N \times N$, and the searching range is $[-P_h, P_h)$ and $[-P_v, P_v)$ in the horizontal and vertical directions.

3.4.2.1 Work of Yang, Sun, and Wu

Yang et al. implemented the first VLSI motion estimator [9], as shown in Figure 3.9, which is a 1-D inter-level hardware architecture (1DInterYSW). 1DInterYSW computes the searching candidates in the fifth loop (*Horizontal SR-level loop*) in parallel and performs others sequentially, so the number of PEs is equal to the number of searching candidates in the horizontal direction, $2P_h$. In 1DInterYSW, reference pixels are broadcasted into all PEs. By selection signals, the corresponding reference pixel is selected and input into each PE. Current pixels are propagated with propagation registers, and the partial SADs are stored in registers of PEs. In each cycle, each PE computes the distortion and accumulates the SAD of a searching candidate. In this architecture, the most important concept is data broadcasting. With broadcasting technique, the memory bitwidth which is defined as the number of bits for the required reference data in one cycle is reduced significantly, although some global routings are required.

3.4.2.2 Work of Yeo and Hu

Figure 3.10 shows a 2-D inter-level hardware architecture which is proposed by Yeo and Hu (2DInterYH) [10]. 2DInterYH is similar to 1DInterYSW but computes all the searching candidates in the *SR-level loop* in parallel. Therefore, 2DInterYH consists of $2P_h \times 2P_v$ PEs. Reference pixels are broadcasted into PEs, and current pixels are propagated with propagation registers. The partial SADs are stored and accumulated in PEs. Because of broadcasting reference pixels in both directions for data reuse, the number of PEs has to match the MB size. Hence the searching range should be partitioned into $(2P_h/N) \times (2P_v/N)$ regions, and each region is computed by a set of $N \times N$ PEs. In 2DInterYH, the reference pixels are broadcasted in two directions at the same time, which can increase the data reuse.

3.4.2.3 Work of Lai and Chen

Lai and Chen also propose another 1-D PE array that implemented a 2-D inter-level architecture with two data interlacing reference arrays (2DInterLC) [11], as shown in Figure 3.11.

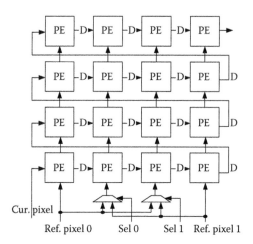

FIGURE 3.10
The hardware architecture of 2DInterYH, where $N = 4$, $P_h = 2$, and $P_v = 2$.

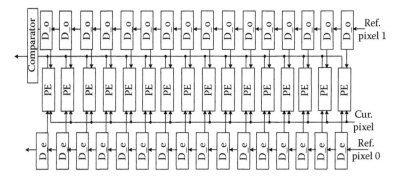

FIGURE 3.11
The hardware architecture of 2DInterLC, where $N = 4$, $P_h = 2$, and $P_v = 2$.

2DInterLC also computes all the searching candidates in the *SR-level loop* in parallel, and it is also similar to 2DInterYH except two aspects. Reference pixels are propagated with propagation registers, and current pixels are broadcasted into PEs. The partial SADs are still stored in registers of PEs. In addition, in 2DInterLC, reference pixels are loaded into propagation registers before computing SADs. The latency of loading reference pixels can be reduced by partitioning the searching range in 2DInterLC. For example, the searching range can be partitioned into $(2P_h/N) \times (2P_v/N)$ parts for a shorter latency.

3.4.2.4 Work of Vos and Stegherr

A 2-D intra-level architecture is proposed by Vos and Stegherr (2DIntraVS) [13], as shown in Figure 3.12, which performs the computation of the *CurBlock-level loop* in parallel. In 2DIntraVS, the number of PEs is equal to the block size, and each PE is corresponding to a current pixel. Current pixels are stored in PEs, and reference pixels are propagated with propagation registers. The important concept of 2DIntraVS is the scanning order in searching candidates, snake scan. In order to realize it, a lot of propagation registers are used to store reference pixels, and the data in propagation registers can be shifted in

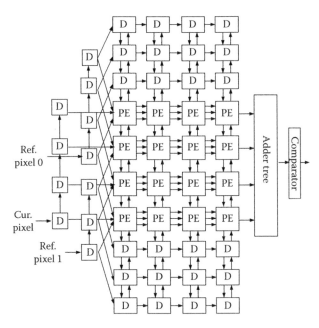

FIGURE 3.12
The hardware architecture of 2DIntraVS, where $N = 4$, $P_h = 2$, and $P_v = 2$.

upward, downward, and right directions. These propagation registers and the long latency for loading reference pixels are the tradeoffs for the reduction of memory usages. The computation flow is as follows. First, the distortion is computed in each PE, and N partial row SADs are propagated and accumulated in the horizontal direction. Second, an adder tree is used to accumulate the N row SADs to be SAD. The accumulations of row SADs and SAD are done in one cycle, so no partial SAD is required to be stored.

3.4.2.5 Work of Komarek and Pirsch

Komarek and Pirsch contribute a detailed systolic mapping procedure by the dependence graph (DG) [12]. By different DGs, including different scheduling and projections, different systolic hardware architectures can be derived. AB2 (2DIntraKP) is a 2-D intra-level architecture, as shown in Figure 3.13. Because of 2-D intra-level architecture, 2DIntraKP also performs the computation of the *CurBlock-level loop* in parallel. Current pixels are stored in corresponding PEs. Reference pixels are propagated PE by PE in the horizontal direction. The N partial column SADs are propagated and accumulated in the vertical direction, first. After the vertical propagation, these N column SADs are propagated in the horizontal direction. In each PE, the distortion of a current pixel in the current MB is computed and added with the partial column SAD which is propagated in PEs from top to bottom in the vertical direction. In the horizontal propagation, these N column SADs are accumulated one by one by N adders and $2N$ registers.

3.4.2.6 Work of Hsieh and Lin

Hsieh and Lin propose another 2-D intra-level hardware architecture with searching range buffer (2DIntraHL) [14], as shown in Figure 3.14. 2DIntraHL consists of N PE arrays in the

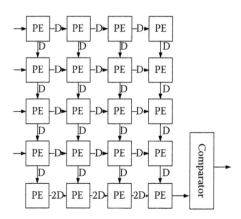

FIGURE 3.13
The hardware architecture of 2DIntraKP, where $N = 4$, $P_h = 2$, and $P_v = 2$.

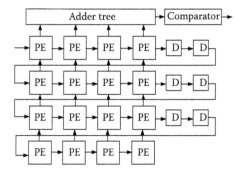

FIGURE 3.14
The hardware architecture of 2DIntraHL, where $N = 4$, $P_h = 2$, and $P_v = 2$.

vertical direction, and each PE array is composed of N PEs in a row. In 2DIntraHL, reference pixels are propagated with propagation registers one by one, which can provide the advantages of serial data input and increasing the data reuse. Current pixels are still stored in PEs. The N partial column SADs are propagated in the vertical direction from bottom to up. In each computing cycle, each PE array generates N distortions of a searching candidate and accumulates these distortions with N partial column SADs in the vertical propagation. After the accumulation in the vertical direction, N column SADs are accumulated in the top adder tree in one cycle. The longer latency for loading reference pixels and large propagation registers are the penalties for the reduction of memory bandwidth and memory bitwidth.

3.4.2.7 Work of Huang et al., Propagate Partial SAD

Huang et al. proposes a *Propagate Partial SAD* architecture [21], which is also a 2-D intra-level architecture. Figure 3.15a and b show the concept and hardware architecture of *Propagate Partial SAD*, respectively. The architecture is composed of N PE arrays with 1-D adder tree in the vertical direction. Current pixels are stored in each PE, and two sets of N continuous reference pixels in a row are broadcasted to N PE arrays at the same time. In each PE array with 1-D adder tree, N distortions are computed and summed by 1-D adder tree to generate

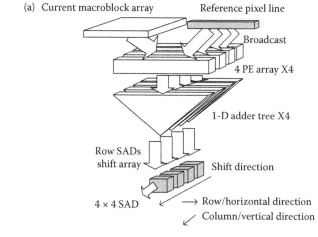

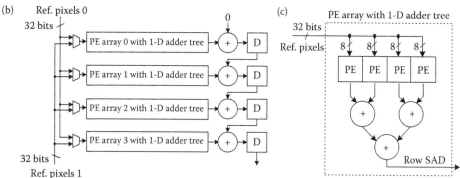

FIGURE 3.15
(a) The concept, (b) the hardware architecture, and (c) the detailed architecture of PE array with 1-D adder tree, of *Propagate Partial SAD*, where $N = 4$.

one row SAD, as shown in Figure 3.15c. The row SADs are accumulated and propagated with propagation registers in the vertical direction, as shown on the right-hand part of Figure 3.15b.

The reference data of searching candidates in the even and odd columns are input through *Ref. Pixels* 0 and *Ref. Pixels* 1, respectively. After initial cycles, the SAD of the first searching candidate in the zeroth column is generated, and the SADs of the other searching candidates are sequentially generated in the following cycles. When computing the last $N - 1$ searching candidates in each column, the reference data of searching candidates in the next column are started to be input through another reference input. Then the hardware utilization is 100% except the initial latency. In *Propagate Partial SAD*, by broadcasting reference pixel rows and propagating partial row SADs in the vertical direction, it provides the advantages of fewer reference pixel registers and a shorter critical path.

3.4.2.8 Work of Chen et al., SAD Tree

Figure 3.16a shows the concept of SAD tree [22], which is proposed by Chen et al. *SAD Tree* is a 2-D intra-level architecture and consists of a 2-D PE array and one 2-D adder tree with propagation registers, as shown in Figure 3.16b and c. Current pixels are stored in

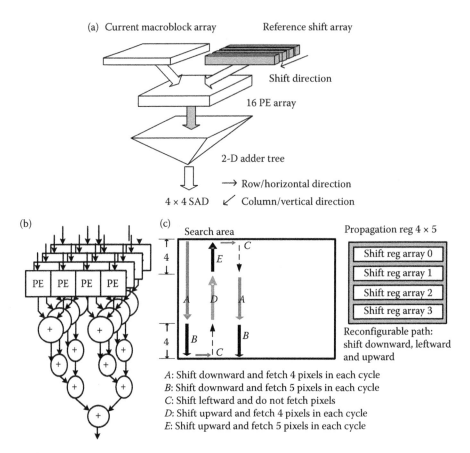

FIGURE 3.16
(a) The concept, (b) the hardware architecture, and (c) the scan order and memory access, of *SAD Tree*, where
$N = 4$.

each PE, and reference pixels are stored in propagation registers for data reuse. In each
cycle, $N \times N$ current and reference pixels are input to PEs. Simultaneously, N continuous
reference pixels in a row are input into propagation registers to update reference pixels.
In propagation registers, reference pixels are propagated in the vertical direction row by
row. In *SAD Tree* architecture, all distortions of a searching candidate are generated in the
same cycle, and by an adder tree, $N \times N$ distortions are accumulated to derive the SAD in
one cycle.

In order to provide a high utilization and data reuse, snake scan is adopted and reconfigurable data path propagation registers are developed in *SAD Tree*, as shown in Figure 3.16c,
which consists of five basic steps from A to E. The first step, A step, fetches N pixels in a
row and the shift direction of propagation registers is downward. When calculating the last
N candidates in a column, extra one reference pixel is required to be input, that is, B step.
When finishing the computation of one column, the reference pixels in the propagation
registers are shifted left in step C. Because the reference data have already been stored in
the propagation registers, the SAD can be directly calculated. The next two steps, step D
and E, are the same as to step A and B except that the shift direction is upward. After finishing the computation of one column in the searching range, we execute step C and then
go back to step A. This procedure iterates, until all searching candidates in the searching

range have been calculated. By snake scan and reconfigurable propagation registers, *SAD Tree* not only can maximize the data reuse between two successive searching candidates but also can approach 100% hardware utilization.

3.4.3 The Impact of Supporting VBSME

There are many methods to support VBSME in hardware architectures. For example, we can increase the number of PEs or the operating frequency to do ME for different block sizes. One of them is to reuse the SADs of the smallest blocks, which are the blocks partitioned with the smallest block size, to derive the SADs of larger blocks. By this method, the overhead of VBSME is only a slight increase of gate count, and the other factors, like frequency, hardware utilization, memory usage, and so on, are the same as those of FBSME. When this method is adopted, the circuit for the SAD calculation is the only difference in hardware designs between FBSME and VBSME. Hence the impact of VBSME on hardware architectures depends on the data flow of partial SADs. In inter-level architectures, the partial SADs are stored in registers of PEs, which is called storing in registers of PEs. In intra-level architectures, there are two kinds of data flows of partial SADs, called propagating with propagation registers and no partial SADs. In the following, the register overhead of VBSME with three different data flows is analyzed. We assume that the size of a MB is $N \times N$, and it can be divided into $n \times n$ smallest blocks of size $(N/n) \times (N/n)$.

3.4.3.1 Data Flow I: Storing in Registers of PEs

In inter-level architectures, each PE is responsible for computing the distortion and accumulating the SAD of a searching candidate, as shown in Figure 3.17a. The partial SADs are stored in registers of PEs. When supporting VBSME, the number of partial SADs is increased from one to n^2. In order to store these partial SADs, more data buffers are required in each PE, as shown in Figure 3.17b. In addition, there are extra two n^2-to-1 and 1-to-n^2 multiplexers in each PE for the selection of partial SADs. All PEs of inter-level architectures, including 1DInterYSW, 2DInterYH, and 2DInterLC, should be replaced with that in Figure 3.17b to support VBSME. The number of bits for the data buffer in each PE is increased from $\log_2 N^2 + 8$ to $n^2 \times (\log_2(N/n)^2 + 8)$, where N^2 or $(N/n)^2$ is the number of pixels in

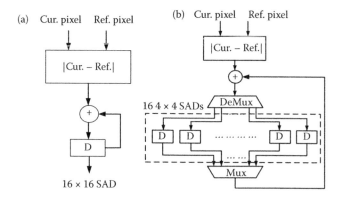

FIGURE 3.17
The hardware architecture of inter-level PE with data flow I for (a) FBSME, where $N = 16$; (b) VBSME, where $N = 16$ and $n = 4$.

one block, and 8 is the wordlength of one pixel. For instance, if a MB is 16×16 and can be divided into 16 4×4 blocks, the size of data buffer is increased from 16 bits to 16×12 bits in one PE.

3.4.3.2 Data Flow II: Propagating with Propagation Registers

In intra-level architectures, partial SADs can be accumulated and propagated with propagation registers. Each PE computes the distortion of one corresponding current pixel in the current MB. By propagation adders and registers, the partial SAD is accumulated with these distortions. The hardware architecture of *Propagate Partial SAD* is a typical example, as shown in Figure 3.15b, where the partial SADs are propagated in the vertical direction. When supporting VBSME, more propagation registers are required to store partial SADs of the smallest blocks. In each propagating direction, the number of propagation registers is n times that in the original architecture, for the n smallest blocks in the other direction. For example, in Figure 3.18, because there are four smallest blocks in the horizontal direction, four partial SADs of the smallest blocks have to be propagated in the vertical direction at the same time in order to reuse them. Therefore, the propagation registers are duplicated four copies, and the number of propagation registers increases from 16 to 64.

Furthermore, some extra delay registers are required in order to synchronize the timing of the SADs of the smallest blocks, as shown in Figure 3.18. In each propagating direction, the number of delay registers is equal to $n \times (n(n-1)/2) \times (N/n)$. That is, in Figure 3.18, there are 4 delay register arrays. In each delay register array, the top smallest block requires $(4-1) \times 16/4$ delay registers, the second smallest block requires $(4-2) \times 16/4$ delay registers, the third smallest block requires $(4-3) \times 16/4$ delay registers, and the bottom smallest block does not require delay registers. Totally, there are $4 \times (4(4-1)/2) \times (16/4)$ delay registers. In addition to *Propagate Partial SAD*, 2DIntraHL also propagates the partial SADs in the vertical direction, and the partial SADs in 2DIntraKP are propagated in two directions. In these three intra-level architectures, extra propagation and delay registers are required in their propagating directions, when VBSME is supported.

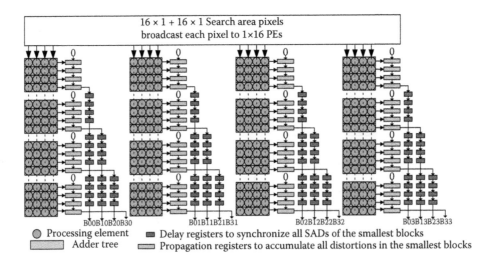

FIGURE 3.18
The hardware architecture of *Propagate Partial SAD* with Data Flow II for VBSME, where $N = 16$ and $n = 4$.

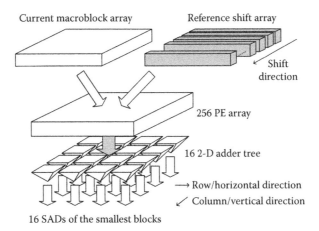

Current macroblock array   Reference shift array

Shift direction

256 PE array

16 2-D adder tree

→ Row/horizontal direction
↙ Column/vertical direction

16 SADs of the smallest blocks

FIGURE 3.19
The hardware architecture of *SAD Tree* with Data Flow III for VBSME, where $N = 16$ and $n = 4$.

3.4.3.3 Data Flow III: No Partial SADs

In intra-level architectures, it is possible that no partial SADs are required to be stored, such as *SAD Tree* and 2DIntraVS. In *SAD Tree*, each PE computes the distortion of one current pixel for a searching candidate, and the total SAD is accumulated by an adder tree in one cycle, as shown in Figure 3.16a. Because there is no partial SAD in this architecture, there are no registers overhead to store partial SADs when supporting VBSME. The adder tree is the one to be reorganized to support VBSME, as shown in Figure 3.19. That is, 2-D adder tree is partitioned into several parts to get the SADs of the smallest blocks first and further reuse these SADs to derive those of large blocks. Although there is no additional register overhead, some extra adders may be required for the reorganization to support VBSME, and this still induces the increase of the required area.

3.4.4 Experimental Results and Discussion

In this section, we discuss the performances of these eight hardware architectures. First of all, we summarize the characteristics of eight ME hardware architectures in Tables 3.2 and 3.3. In Table 3.2, the number of PEs, required cycles, and latency are formulated to show the degree of parallelism and utilization, and the data flow of partial SADs are listed to categorize the impact of supporting VBSME in each hardware. In Table 3.3, current buffer, reference buffer, and memory bitwidth are used to evaluate the tradeoff between data buffer and memory usage. Note that because we reuse the SADs of the smallest blocks to derive the SADs of larger blocks, the impact of VBSME on hardware architectures is only the increase of chip area. The other factors are the same for FBSME and VBSME.

Besides the theoretical analysis in Tables 3.2 and 3.3, an real case is also given to provide a practical comparison in Table 3.4. The specifications of ME are as follows. The MB size is 16×16, and the searching range is $P_h = 64$ and $P_v = 32$. The frame size is D1 size, 720×480. In Table 3.4, Verilog-HDL and SYNOPSYS Design Compiler with ARTISAN UMC $0.18um$ cell library are used to implement each hardware architecture. Because the timing of the critical path in some architectures is too long, which means the maximum operating frequency is limited without modifying the architecture, the frame rate is set as only 10 frames per second (fps). The discussion of these experimental results is given below.

TABLE 3.2

Parallelism, Cycle, Latency, and Data Flow of Architectures

Name	No. of PEs	Operating Cycles (Cycles/Macroblock)	Latency (Cycles)	Data Flow
1DInterYSW [9]	$2P_h$	$N^2 \times 2P_v + 2P_h$	N^2	Data Flow I
2DInterYH [10]	$2P_h \times 2P_v$	$2N^2$	N^2	Data Flow I
2DInterLC [11]	$2P_h \times 2P_v$	$2N^2$	$2N^2$	Data Flow I
2DIntraVS [13]	N^2	$2P_h \times 2P_h + N \times 2P_v$	$N \times 2P_v$	Data Flow III
2DIntraKP [12]	N^2	$2P_v \times (N + 2P_h) + N$	$3N$	Data Flow II
2DIntraHL [14]	N^2	$(2P_v + N - 1) \times (2P_h + N - 1)$	$2N + (N - 1)$ $\times (2P_h + N - 2)$	Data Flow II
Propagate Partial SAD [21]	N^2	$2P_h \times 2P_v + N - 1$	N	Data Flow II
SAD Tree [22]	N^2	$2P_h \times 2P_v + N - 1$	N	Data Flow III

Note: The analysis is based on a video coding system with macroblock pipelining architecture.

TABLE 3.3

Data Buffer and Memory Bitwidth of Architectures

Name	Current Buffer (Pixels)	Reference Buffer (Pixels)	Memory Bitwidth (Bits/Cycle)
1DInterYSW [9]	$2P_h - 1$	—	$(2P_h/N + 2) \times 8$
2DInterYH [10]	$N^2 - 1$	—	$(2(2P_v/N) \times$ $(2P_h/N) + 1) \times 8$
2DInterLC [11]	—	$2P_h \times 2P_v$	$(2(2P_v/N) \times$ $(2P_h/N) + 1) \times 8$
2DIntraVS [13]	N^2	$N \times (4P_v + N - 2)$	$\{1, (2)^*\} \times 8$
2DIntraKP [12]	N^2	$N \times (N - 1)$	$N \times 8$
2DIntraHL [14]	N^2	$N^2 + (N - 1) \times$ $(2P_h - 2)$	1×8
Propagate Partial SAD [21]	N^2	—	$\{N, (2N)^*\} \times 8$
SAD Tree [22]	N^2	$N \times (N + 1)$	$\{N, (N + 1)^*\} \times 8$

Note: $(.)^*$ is the worst case.

3.4.4.1 Required Frequency and Area

The required frequency is dominated by the degree of parallelism in a hardware architecture. The smaller the degree of parallelism is, the higher the required frequency is. In Tables 3.2 and 3.4, because the degree of parallelism in 1DInterYSW is the smallest, the required frequency is the highest. On the contrary, because 2DInterYH and 2DInterLC have the largest degrees of parallelism among eight hardware architectures, their required frequencies are the smallest.

There are two columns of chip area in Table 3.4. One is for FBSME and the other is for VBSME. The area consists of PE array, Current Buffer and Reference Buffer. Therefore, for FBSME, the area of 1DInterYSW is the smallest. The area of 2DInterLC is larger than that of 2DInterYH because of the huge reference buffer, as shown in Table 3.3. Similarly, the area of 2DIntraVS is also larger than that of *SAD Tree*, because large propagation registers exist in

TABLE 3.4

Comparison of Architectures for FBSME and VBSME

Name	Area (FBS) (kgates)	Area (VBS) (kgates)	Freq. (MHz)	Bitwidth (Bits)	Bandwidth (kbits)	Latency (Cycles)	Util. (%)
1DInterYSW [9]	61.9	359.6	222.7	80	1290	256	99.2
2DInterYH [10]	2907.0	20,422.0	6.9	520	260	256	50.0
2DInterLC [11]	4055.0	21,647.0	6.9	520	260	512	50.0
2DIntraVS [13]	301.3	318.7	127.7	16	90	1024	88.9
2DIntraKP [12]	108.8	159.1	123.7	128	1146	48	89.3
2DIntraHL [14]	231.9	254.6	152.5	8	90	2162	72.5
Propagate Partial SAD [21]	66.6	81.5	110.8	256	1259	16	99.8
SAD Tree [22]	88.4	88.6	110.8	136	1,044	16	99.8

Note: The specification is D1 size, 10 fps, and the searching range is [−64, 64) in a video coding system with macroblock pipelining architecture.

2DIntraVS. The impact of supporting VBSME is apparently observed in the other column of area, the area for VBSME, in Table 3.4. Among these eight hardware architectures, all inter-level architectures with Data Flow I increase gate count dramatically. The chip area is five times of that in FBSME at least. In intra-level architectures with Data Flow II, the increase of gate count is much smaller, and the increasing ratio is from 9.8% to 46.3%. If the data flow of partial SADs is Data Flow III in intra-level architectures, the area overheads are 0.2% and 5.8%, for *SAD Tree* and 2DIntraVS, respectively.

Due to the characteristics of inter-level architectures, the chip area overhead of inter-level architectures for VBSME is large. In three inter-level architectures, the overhead of 2DInterLC is the smallest, because of a lot of propagation registers in 2DInterLC compared to other architectures. The alike condition occurs when we compare the performances of 2DIntraHL and *Propagate Partial SAD*. Therefore, the chip area overhead of supporting VBSME in 2DIntraHL is smaller than that in *Propagate Partial SAD*. In three intra-level architectures with Data Flow II, 2DIntraKP has the largest chip area overhead, because its partial SADs are propagated in two directions. The others only propagate their partial SADs in one direction. In the intra-level architectures with Data Flow III, 2DIntraVS and *SAD Tree*, there is no partial SAD to be stored, and only some extra adders are required for VBSME. Hence their increase ratios are very small. The reason why the chip area overhead of 2DIntraVS is larger than that of *SAD Tree* is a longer critical path in 2DIntraVS. Finally, among the eight hardware architectures, the chip area of *Propagate Partial SAD* is the smallest, and *SAD Tree* has the smallest overhead, when VBSME is supported.

3.4.4.2 Latency

The latency is defined as *The number of start-up cycles that a hardware takes to generate the first SAD*. The latency is more important for a video coding system than for a single ME module, because the latency affects the effect of parallel computation. In a video coding system, we usually use a large degree of parallelism to achieve realtime computation. However, if a module has a long latency and it cannot be shortened by parallel architectures, the effect of parallel computation is reduced. That is, a shorter latency is better for video coding systems. There are two factors to affect the latency. One is the type of a hardware architecture. In inter-level architectures, the latency is at least $N \times N$, as shown in Tables 3.2

and 3.4. Conversely, there is no constraint in intra-level architectures. The other factor to affect the latency is the memory bitwidth and reference buffer, as shown in Table 3.3. If there is a large reference buffer but fewer memory bitwidth, for example, 2DIntraVS or 2DIntraHL, the architecture takes more initial cycles to load reference data into the reference buffer. Compared to these hardware architectures, the other intra-level architectures, such as *Propagate Partial SAD* and *SAD Tree*, have shorter latencies.

3.4.4.3   Utilization

To evaluate the hardware efficiency, we defined the utilization as *Computing Cycles/Operating Cycles for a MB*. Then, the utilization is dominated by the operating cycles. The operating cycles include three parts, latency, computing cycles, and bubble cycles. Computation cycles are the number of cycles when we can get one SAD at least. That is, if the utilization is 100%, we can get one SAD at least in each cycle. Fewer operating cycles let the penalty of the latency be apparent. The more bubble cycles are, the lower the utilization is. 2DInterYH and 2DInterLC are two examples which have low utilizations because of their fewer operating cycles, as shown in Tables 3.2 and 3.4. In *Propagate Partial SAD* and *SAD Tree*, there are shorter latencies and no bubble cycles, so their utilizations can achieve 99.8%.

3.4.4.4   Memory Usage

Memory usage consists of two parts, memory bitwidth and memory bandwidth. Memory bitwidth is defined as *The Number of Bits which a Hardware Has to Access From Memory in Each Cycle*, and memory bandwidth is re-defined as *The Number of Bits which a Hardware Has to Access From Memory for a MB*. Memory bandwidth affects the loading of system bus without on-chip memory or the power of on-chip memory, and memory bitwidth is the key to the data arrangement of on-chip memories. Memory bitwidth and bandwidth depend on the data reuse scheme and operating cycles. From Table 3.3, because 2DIntraHL and 2DIntraVS have larger reference buffers to reuse reference pixels, the required memory bitwidths and bandwidths are fewer. In 2DInterYH and 2DInterLC, because of their high degrees of parallelism as shown in Table 3.2, the large memory bitwidths are required. But the memory bandwidths are much fewer because of fewer operating cycles. The data reuse schemes in 2DIntraKP, *Propagate Partial SAD*, and *SAD Tree* are similar, and the differences in these three architectures are resulted from the different data reuse schemes when changing columns.

3.4.5   Architecture Design of Integer Motion Estimation in H.264/AVC

Based on the above analysis, Chen et al. present the first Integer Motion Estimation (IME) hardware for H.264/AVC [22]. Figure 3.20 shows the overall IME architecture, which mainly comprises eight parallel PE-Array of 2-D intra-level SAD Trees to fit the huge computation complexity of HDTV720p 30 frame/s realtime encoding. The Current MB (CMB) is stored in *Cur. MB Reg*. The reference pixels are read from external memory and stored in *Luma Ref. Pels SRAMs*. Each PE array and its corresponding 2-D SAD tree compute the 41 SADs of VBS for one searching candidate at each cycle. Therefore, eight horizontally adjacent candidates are processed in parallel. All SAD results of VBS are input to the *Comparator Tree Array*. Each comparator tree finds the smallest SAD among the eight search points and updates the best MV for a certain block-size.

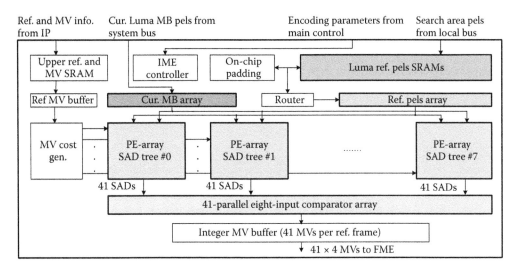

FIGURE 3.20
Block diagram of the IME engine. It mainly consists of eight PE-Array SAD Trees. Eight horizontally adjacent candidates are processed in parallel. (C.-Y. Chen et al. Analysis and architecture design of variable block-size motion estimation for H.264/AVC. *IEEE Transactions on Circuits and Systems I: Regular Papers*, 53(3):578–593, 2006. © (2006). IEEE.)

Figure 3.21 shows the M-parallel PE-array SAD Tree architecture. A horizontal row of reference pixels, which are read from SRAMs, is stored and shifted downward in *Ref. Pels Reg. Array*. When one candidate is processed, 256 reference pixels are required. When eight horizontally adjacent candidates are processed in parallel, not (256×8) but $(256 + 16 \times 7)$ reference pixels are required. In addition, when the ME process is changed to the next eight candidates, most data can be reused in Ref. Pels Array. This parallel architecture achieves inter-candidate data reuse in both horizontal and vertical directions and reduce the on-chip SRAM bandwidth.

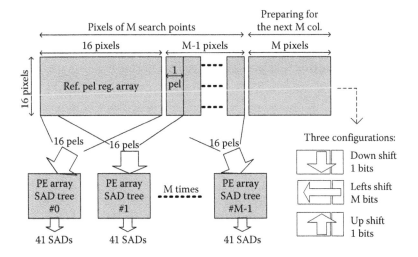

FIGURE 3.21
M-parallel PE-array SAD Tree architecture. The inter-candidate data reuse can be achieved in both horizontal and vertical directions with *Ref. Pels Reg. Array*, and the on-chip SRAM bandwidth is reduced.

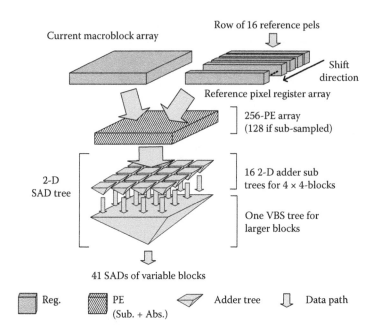

Current macroblock array

Row of 16 reference pels

Shift direction

Reference pixel register array

256-PE array
(128 if sub-sampled)

2-D
SAD tree

16 2-D adder sub
trees for 4 × 4-blocks

One VBS tree for
larger blocks

41 SADs of variable blocks

Reg.   PE   Adder tree   Data path
            (Sub. + Abs.)

FIGURE 3.22
PE-array SAD Tree architecture. The cost of 16 4 × 4 blocks are separately summed up by 16 2-D Adder sub-trees
and then reduced by one VBS Tree for larger blocks.

Figure 3.22 shows the architecture of *PE-Array of 2-D intra-level SAD Tree*. The cost of 4 × 4 blocks are separately summed up by 16 2-D *Adder Sub-trees*, and then reused by one *VBS Tree* for larger blocks. This is so-called intra-candidate data reuse. All 41 SADs for one candidate are simultaneously generated and compared with the 41 best costs. No intermediate data are buffered. Therefore, this architecture can support VBS without any partial SAD registers.

3.5   Multiple Reference Frame Motion Estimation

Multiple reference frames motion estimation (MRF-ME) is an important tool for H.264/ AVC [24]. For MRF-ME, the high computation complexity and large system bandwidth requirement are two main challenges for hardware implementation. In the previous section, many parallel architectures are introduced to provide significant computation for integer motion estimation (IME). After the parallel processing, the problem of high computation complexity can be solved. For a VLSI system where the system bandwidth is usually limited, the next challenge is to reduce the large requirement of the system bandwidth. In tradition, the data of current macroblock (MB) and search window (SW) are loaded from system memory and then buffered in on-chip SRAMs or registers. By means of local data reuse (DR), the system bandwidth requirement can be greatly reduced. Four DR strategies [25] have been proposed with different tradeoffs between system bandwidth and local memory size. However, these schemes cannot efficiently support the MRF-ME in H.264/AVC. The required system bandwidth and local memory size are linearly increased with the number of reference frames. In this session, a new frame-level DR scheme will be introduced [26].

With the frame-level rescheduling, the data of one loaded SW can be reused by multiple current MBs in different original frames for the MRF-ME, and the system bandwidth is greatly reduced. The required local memory size is almost the same with the previous design supporting only one reference frame.

3.5.1 Conventional Data Reuse Scheme

In ME, in order to find the best matched candidate of one current MB, SWs for multiple reference frames have to be searched. The bandwidth between system memory and ME core is very heavy if all required pixels are loaded from system memory. It consumes too much power and is not achievable in the current VLSI system since the system bandwidth is limited. The common solution is to design the local buffers to store reusable data. By means of local memory access, the system bandwidth can be greatly reduced.

Figure 3.23 shows the operation loops of MRF-ME for H.264/AVC. In the second loop, the SWs of the neighboring current MBs for one reference frame are considerably overlapped. So are the reference pixels of neighboring candidates for one current MB in the fifth loop. Four DR schemes have been proposed with different tradeoffs between local memory size and system bandwidth and are indexed from level-A to level-D [25,27]. Level-A requires the smallest local memory size and the highest system bandwidth, while level-D has the largest local memory size and the lowest system bandwidth.

Figure 3.24a describes the level-C DR scheme, which is generally adopted nowadays and will be used as the benchmark to explain our framework. The level-C scheme reuses

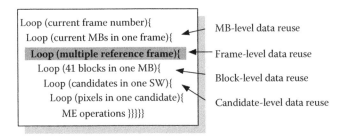

FIGURE 3.23
The operation loops of MRF-ME for H.264/AVC.

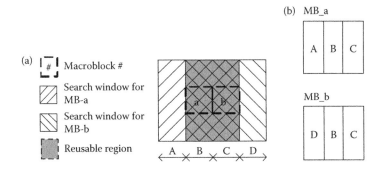

FIGURE 3.24
The level-C data reuse scheme. (a) There are overlapped region of SWs for horizontally adjacent MBs; (b) the physical location to store SW data in local memory.

the horizontally overlapped region between two SWs of the neighboring current MBs. As Figure 3.24b shows, the SW of $MB - b$ includes the area of B,C, and D. When the ME of $MB - a$ is finished and the $MB - b$ is going to be processed, only the reference pixels in D are loaded to replace A in the local memory.

3.5.2   System Bandwidth Issue in MRF-ME

In H.264/AVC, to support MRF-ME with level-C DR scheme, multiple SW memories are implemented as shown in Figure 3.25. There are four SW memories, and each SW memory will be independently loaded and updated as Figure 3.24. This can be referred as multiple reference frames single current MB (MRSC) scheme.

Table 3.5 shows the requirement of system bandwidth and local memory size for MRSC scheme. The hardware cost is almost proportional to the maximum reference frame number. The large system bandwidth requirement increases the power consumption and becomes the bottleneck of the whole system. In addition, the increased memory size increases the silicon area and cost. Recently, the block-level data reuse scheme can be utilized for the fourth loop in Figure 3.23 [21,28]. The distortion costs of the smallest 4×4 blocks can be computed first. The costs of larger blocks can be on-line calculated by summing up the corresponding 4×4 costs. However, this reuse scheme can only reduce the computation complexity but the system bandwidth. A new DR scheme is demanded for the MRF-ME loop.

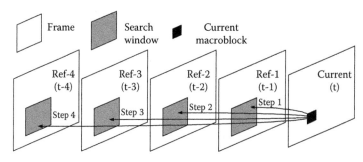

FIGURE 3.25
The MRSC scheme for MRF-ME requires multiple SWs memories. The reference pixels of multiple reference frames are loaded independently according to the level-C data reuse scheme.

TABLE 3.5

System Bandwidth and SW Memory Size for MRF-ME with Level-C MRSC Scheme

System Bandwidth (MByte/s)		Local Memory Size (kByte)	
MPEG-1/2/4	H.264	MPEG-1/2/4	H.264
(Ref. = 1)	(Ref. = 4)	(Ref. = 1)	(Ref. = 4)
60.13	209.43	6.656	25.86

Note: SDTV (720 × 480, 30 fps), search range = [−32, +31].

3.5.3 Frame-Level Data Reuse Scheme for MRF-ME

3.5.3.1 Frame-Level Data Reuse

To further reduce the system bandwidth and the local memory requirement, a single Reference frame Multiple Current MBs (SRMC) scheme is proposed. The SRMC scheme exploits the frame-level DR in the multiple reference frame loop as shown in Figure 3.23. This new frame-level DR scheme is orthogonal to the traditional candidate-level and MB-level DR schemes. That is, the SRMC scheme can be jointly used with any of the four conventional DR schemes.

Figure 3.26 shows the concept of frame-level DR in SRMC scheme. The reconstructed frame at time slot t-4, is the first reference frame of the original frame at time slot t-3. It is also the second, third, and fourth reference frame of original frames at time slot t-2, t-1, and t, respectively. Therefore, when the SW in the first previous reconstruction frame of one current MB is loaded to local memory, it can also be reused by the current MBs at the same location of the following original frames.

In the MRSC scheme, one current MB is loaded only one time, and one reference SW needs to be loaded several times. In SRMC scheme, one current MB is loaded several times while one reference SW only needs to be loaded once. Since the SW is much larger than one MB, both the bandwidth and memory size can be largely reduced. When the frame size and the Search Range (SR) are increased, the benefit of the SRMC scheme will become more obvious. For HDTV video specifications, after the SRMC scheme is applied for MRF-ME, the required hardware resource is almost the same with the hardware resource supporting only one reference frame.

3.5.3.2 Frame-Level Rescheduling

In order to achieve the frame-level DR, the ME procedures of one current MB in different reference frames are processed at different time slots. Therefore, the frame-level rescheduling is designed to rearrange the first three loops in Figure 3.23. Figure 3.27 shows the original schedule of the MRSC scheme along with the rearranged schedule of the SRMC scheme for MRF-ME. For the simplification, it is assumed that there are six MBs in each frame and five P-frames to be coded. The maximum reference frame number is four. Three indices are used to explain one ME cube—a ME procedure of one current MB for one reference. The first, second, and third indices stand for the absolute time information of the processed current MB, the absolute time information of the corresponding SW, and the current MB index in

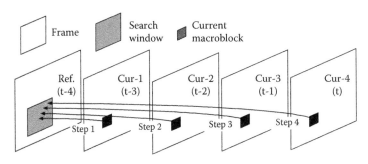

FIGURE 3.26
The SRMC scheme can exploit the frame-level DR for MRF-ME. Only single SW memory is required.

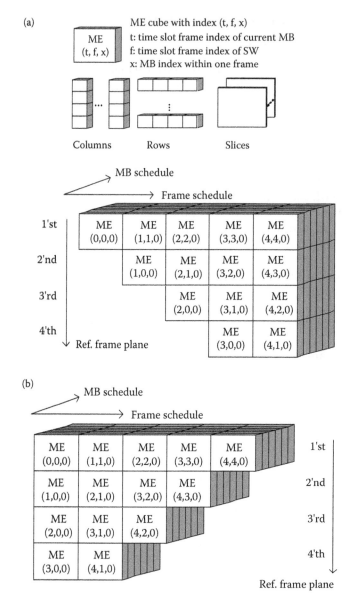

FIGURE 3.27
Schedule of MB tasks for MRF-ME; (a) the original (MRSC) version; (b) the proposed (SRMC) version.

one original frame, respectively. The ME cubes are processed sequentially along *Ref frame plane*, *MB schedule*, and then *Frame schedule* axes.

Figure 3.27a represents the original schedule of MRSC scheme. The ME cubes is performed reference frame by reference frame of one current MB, MB by MB in one frame, and then frame by frame, just like the reference software [29]. The MB cubes in one vertical column stand for the ME tasks of one current MB in different reference frames. According to the indices, the first, second, third and fourth ME cubes represent the *step*-1, *step*-2, *step*-3, and *step*-4 searching processes in Figure 3.25. Note that the ME cubes in one vertical column have the same t index and different f index.

Figure 3.27b represent the rearranged schedule for SRMC scheme. The second, third, and fourth horizontal rows of each slice are shifted leftward for one, two, and three frame slots, respectively. Now, the first, second, third, and fourth ME cubes in one vertical column represent the *step*-1, *step*-2, *step*-3, and *step*-4 searching processes in Figure 3.26. The ME cubes in one vertical column have the same index of *f* and different index of *t*. Therefore, one SW data can be reused by multiple current MBs for different reference frames, and the MRF-ME is still successively achieved.

3.5.3.3   Mode Decision in the SRMC Scheme

The issues of the mode decision are needed to be reconsidered after the frame-level rescheduling in the proposed SRMC scheme. In reference software, the Lagrangian cost function takes MV costs into consideration. The MV of each block is generally predicted by the medium values of MVs from the left, top, and top right neighboring blocks. Not until the modes of neighboring blocks or MBs are decided can the Motion Vector Predictor (MVP) of the current block or MB become available. The rate term of the Lagrangian cost function can be computed only after that MVs of the neighboring blocks or MBs are decided from the solution space of variable blocks and multiple reference frames.

To support the block-level data reuse scheme in hardware, the distortion values of all candidates for all variable blocks must be stored for the following Lagrangian mode decision flow. This flow can hardly be implemented in hardware due to its large data size. The modified MVP [3] is adopted for all 41 blocks in one MB as shown in Figure 3.28. The exact MVPs of variable blocks, which are ideally the medium of MVs of the left, top, and top-right blocks, are changed to the medium of MVs of the left, top, and top-right MBs. For example, the exact MVP of the C22 4 × 4-block is the medium of the MVs of C12, C13, and C21. The MVPs of all 41 blocks are changed to the medium of MV0, MV1, and MV2 in order to facilitate the parallel processing for the block-level data reuse.

Furthermore, in the SRMC scheme, the ME tasks in different reference frames are processed at different time slots. The MV information in different reference frames cannot be jointly considered because the information is needed to be stored in the frame basis and the large storage space is required. A two-stage mode decision method can be applied to deal with this problem. The mode decision flow is divided into partial mode decision (PMD) and final mode decision (FMD) as shown in Table 3.6. The PMD is responsible for

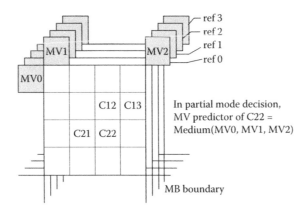

FIGURE 3.28
Estimated MVPs in PMD for Lagrangian mode decision.

TABLE 3.6

Two-Stage Mode Decision

	Partial Mode Decision (PMD)	**Final Mode Decision (FMD)**
Process	On-line/Dedicated hardware	Off-line/System RISC
Task	Decide the best matches of 41 blocks in each reference frame	Decide the best combination from all block modes in all reference frames

on-line deciding the best matches of 41 variable blocks for each reference frame. Since the exact MVPs of the current MB cannot be calculated on-line with the distortion costs, these MVPs are estimated according to the information within one reference frame as shown in Figure 3.28. The MVs and the distortion costs of these suboptimal results will be written to the external memory. After the PMD results of all reference frames are generated for a certain current MB, the FMD uses system RISC to decide the best combination of variable blocks in different reference frames. At this time, the exact MV costs can be sequentially generated.

3.5.4 Architecture Design

Figure 3.29 shows the architecture of H.264/AVC ME engine using the SRMC scheme. Different from the MRSC scheme, only one SW memory is required to support MRF-ME. The ME core computes the distortion values of the candidates, and the PMD engine on-line decides the MVs of variable blocks according to the estimated MVPs. Both the full-search or fast-search ME algorithms can be implemented as the ME core. The PMD results are buffered at system memory, and then the RISC performs FMD.

Figure 3.30 shows the schedule. Referred to Figure 3.26, the SW at the frame marked as t-4 is loaded to SW buffer first. Then, the ME task of the current MB in the frame marked as t-3 will be performed. After that, the FMD of this current MB is then done by RISC after the PMD results are written out. At the same time, the current MBs at the same location of the following frames marked as t-2, t-1, and t are processed one after another. Although multiple current MBs are loaded, only one current MB buffer is required. Note that FMD can also be implemented as dedicated hardware with the same schedule. Another system bandwidth to load back the PMD results is the penalty in this case.

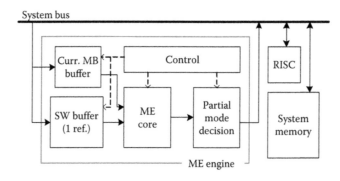

FIGURE 3.29
Proposed architecture with SRMC scheme.

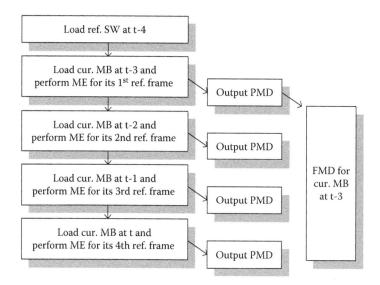

FIGURE 3.30
The schedule of SRMC scheme in the proposed framework.

3.5.5 Performance Evaluation

3.5.5.1 Encoding Performance

Figure 3.31 shows the rate-distortion efficiency of the reference software and the proposed framework. Four sequences with different characteristics are used for the experiment. Foreman has lots of deformations and media motions. Mobile has complex textures and regular motions. Akiyo is a still scene, while Stefan has large motions. The encoding parameters are baseline profile, IPPP... structure, four reference frames, ±16-pel search range, and low complexity mode decision. In the proposed framework, two algorithm modifications are involved compared with reference software. First, to facilitate the parallel processing for VBS-ME, the modified MVP is adopted for the block-level data reuse. Second, to reduce the system bandwidth for MRF-ME, the SRMC scheme are proposed with two-stage mode decision for the frame-level data reuse.

According to the simulation results, the proposed framework has similar compression performances compared to the reference software except for Foreman. For Foreman sequence in lower bitrates, upto 0.3 dB quality drop is induced compared with the reference software. This is mainly caused by the difference between estimated MVP and exact MVP. However, compared to the previous hardwired encoder where the modified MVP is applied [3], only 0.12 dB quality drop is induced by the proposed two-stage mode decision. In addition, for applications with higher bitrates, the distortion part dominates the Lagrangian cost function, and the compression performance of the proposed framework is approach to the idea case.

3.5.6 System Bandwidth and Memory Size

In this section, we will show the efficiency of the proposed SRMC DR scheme in terms of system bandwidth and memory size requirement. The level-C DR scheme is used as original MRSC DR scheme for the comparison. The framework with two-stage mode decision flow is used. The PMD is done by the dedicated hardware while the FMD is handled by the

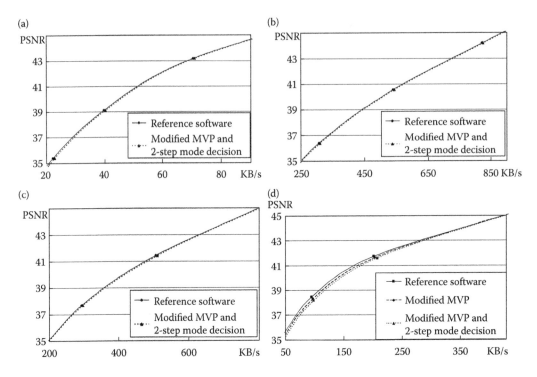

FIGURE 3.31
The rate-distortion efficiency of the reference software and the proposed framework. Four sequences with different characteristics are used for the experiment. Foreman has lots of deformation with media motions. Mobile has complex textures and regular motion. Akiyo has the still scene, while Stefan has large motions. The encoding parameters are baseline profile, IPPP... structure, CIF, 30 frames/s, 4 reference frames, ±16-pel search range, and low complexity mode decision. (a) Akiyo (CIF, 30 fps); (b) Mobile (CIF, 30 fps); (c) Stefan (CIF, 30 fps); (d) Foreman (CIF, 30 fps).

RISC. The required bus bandwidth and memory size requirement of the MRSC schedule are calculated as follows:

$$BW_{MRSC} = BW_{one\_ref} \times num\_ref + BW_{one\_cur\_MB}$$

$$mem_{MRSC} = mem_{one\_SW} \times num\_ref + mem_{one\_cur\_MB}$$

The required bus bandwidth and memory size of the rearranged SRMC schedule are calculated as follows:

$$BW_{SRMC} = BW_{one\_ref} + BW_{one\_cur\_MB} \times num\_ref + BW_{PMD}$$

$$mem_{SRMC} = mem_{one\_SW} + mem_{one\_cur\_MB}$$

The BW_{PMD} includes the MVs and the matching costs of variable blocks. The bus bandwidth requirement of BW_{PMD} is relatively small. Table 3.7 summarizes the performance evaluation for three cases. The proposed SRMC scheme can save about 75% of on-chip memory size and 35.4–62.6% of system bandwidth compared to the conventional MRSC scheme. The BW_{one_ref} increases with larger search range. Therefore, the proposed SRMC scheme has better performance for the videos with larger frame sizes that inherently require larger search ranges.

TABLE 3.7

Performance of the Proposed SRMC Scheme

	System Bandwidth (MByte/s)			Local Memory Size (kByte)		
	MRSC Scheme	SRMC Scheme	Saved (%)	MRSC Scheme	SRMC Scheme	Saved (%)
Case A	46.97	25.63	−35.4	11.78	2.56	−78.3
Case B	209.43	97.21	−53.6	25.86	6.65	−74.3
Case C	973.82	364.28	−62.6	83.20	20.99	−74.8

Note: Case A: CIF (352 × 288, 30 fps), search range = [−16, +15], 5 reference frames
Case B: SDTV (720 × 480, 30 fps), search range = [−32, +31], 4 reference frames
Case C: HDTV (1280 × 720, 30 fps), search range = [−64, +63], 4 reference frames

3.6 Fractional Motion Estimation

Fractional motion estimation (FME) with the rate-distortion constrained mode decision can improve the rate-distortion efficiency by 2–6 dB in peak signal-to-noise ratio (PSNR). However, it comes with considerable computation complexity. Acceleration by dedicated hardware is a must for real-time applications. The main difficulty for FME hardware implementation is parallel processing under the constraint of the sequential flow and data dependency. In this section, a systematic method is presented to efficiently project the FME algorithm into VLSI hardware. Based on the reference software, the FME algorithm is formulated as the nested-loop structure. Then the feasibility of parallelization and the data locality is explored by use of the loop analysis. Afterwards the data reuse (DR) techniques along with the corresponding hardware are designed to achieve realtime processing and to reduce memory access power. Finally the implementation results are also included.

3.6.1 FME Algorithm

3.6.1.1 Functionality Overview

H.264/AVC can save 25–45% and 50–70% bitrates compared with MPEG-4 [8] and MPEG-2 [30], respectively. The improvement in compression performance mainly comes from the new prediction tools and the significantly larger computation complexity is the penalty. In H.264/AVC, the inter prediction, or motion estimation (ME), can be divided into two parts—Integer ME (IME) and Fractional ME (FME). The IME searches for the initial prediction in coarse resolution, while the FME refines this result to the best match in fine resolution. The FME in H.264/AVC contributes 2–6 dB of rate-distortion efficiency in PSNR but consumes 45% of the run-time in H.264/AVC inter prediction [31].

In H.264/AVC, FME supports quarter-pixel accuracy with variable block sizes (VBS) and multiple reference frames (MRF). For the MRF-ME shown in Figure 3.32, more than one prior reconstructed frames can be used as reference frames. This tool is effective for the uncovered backgrounds, repetitive motions, and highly textured areas [32]. For the VBS-ME, the block size in H.264/AVC ranges from 16 × 16 to 4 × 4 luminance samples. As shown in Figure 3.33, the luminance component of each Macroblock (MB) can be selected from four kinds of partitions: 16 × 16, 16 × 8, 8 × 16, and 8 × 8. For the partition 8 × 8, each 8 × 8 block can be further split into four kinds of sub-partitions: 8 × 8, 8 × 4, 4×8, and 4 × 4. This tree-structured partition leads to a large number of possible combinations

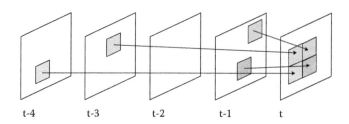

FIGURE 3.32
Multiple reference frame motion estimation.

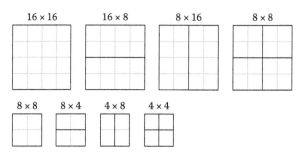

FIGURE 3.33
Variable block size motion estimation.

within each MB. In general, large blocks are appropriate for homogeneous areas, and small partitions are beneficial for textured area and objects of variant motions. The accuracy of motion compensation is in quarter-pixel resolution. It can significantly improve compression performance especially for pictures with complex texture. A six-tap FIR filter is applied for half-pixel generation as shown in Figure 3.34a, and another bilinear filter for quarter pixel generation as shown in Figure 3.34b.

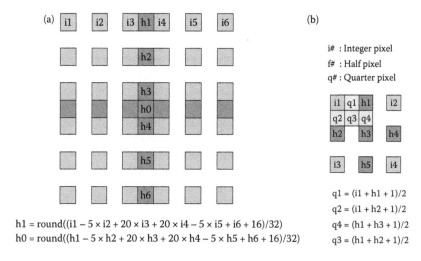

$$h1 = round((i1 - 5 \times i2 + 20 \times i3 + 20 \times i4 - 5 \times i5 + i6 + 16)/32)$$
$$h0 = round((h1 - 5 \times h2 + 20 \times h3 + 20 \times h4 - 5 \times h5 + h6 + 16)/32)$$

$$q1 = (i1 + h1 + 1)/2$$
$$q2 = (i1 + h2 + 1)/2$$
$$q4 = (h1 + h3 + 1)/2$$
$$q3 = (h1 + h2 + 1)/2$$

FIGURE 3.34
Interpolation scheme for luminance component: (a) 6-tap FIR filter for half pixel interpolation. (b) Bilinear filter for quarter pixel interpolation.

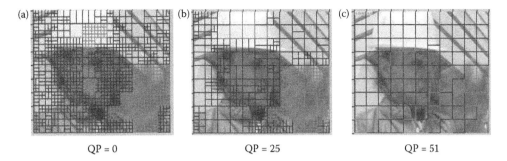

QP = 0   QP = 25   QP = 51

FIGURE 3.35
Best partition for a picture with different quantization parameters (black block: inter block, gray block: intra block).

The mode decision algorithm is left as an open issue in H.264/AVC standard. In the reference software [29], the Lagrangian cost function is adopted. Given the quantization parameter QP and the Lagrange parameter λ_{MODE} (a QP dependent variable), the Lagrangian mode decision for a macroblock MB_k proceeds by minimizing

$$J_{MODE}(MB_k, I_k | QP, \lambda_{MODE})$$
$$= Distortion(MB_k, I_k | QP) + \lambda_{MODE} \cdot Rate(MB_k, I_k | QP)$$

where the MB mode I_k denotes all possible coding modes and MVs. The best MB mode is selected by considering both the distortion and rate parts. Due to the large computation complexity and sequential issues in the high complexity mode of H.264/AVC, it is less suitable for real-time applications. In this design, we focus on low complexity mode decision. The distortion is evaluated by the sum of absolute transformed differences (SATD) between the predictors and the original pixels. The rate is estimated by the number of bits to code the header information and the MVs. Figure 3.35 shows the best partition for a picture with different quantization parameters. With larger QP, the mode decision tends to choose the larger block or the modes with less overhead in the MB header. In contrast, when the QP is small, it tends to choose the smaller block for more accurate prediction.

3.6.1.2 FME Procedure in Reference Software

Figures 3.36 and 3.37 show the refinement flow and procedure of FME in the H.264/AVC reference software [29], respectively. To find the sub-pixel MV refinement of each block, a

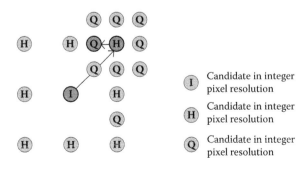

Candidate in integer
pixel resolution

Candidate in integer
pixel resolution

Candidate in integer
pixel resolution

FIGURE 3.36
FME refinement flow for each block and sub-block.

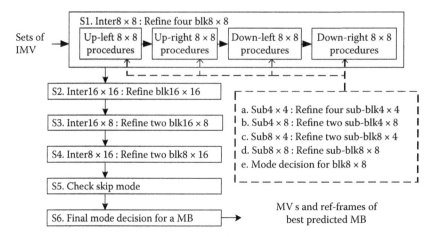

FIGURE 3.37
FME procedure of Lagrangian inter mode decision in H.264/AVC reference software.

two-step refinement is adopted for every block. In Figure 3.36, the half-pixel MV refinements are performed around the best integer search positions, I, from integer ME results. The search range of half-pixel MV refinements are $\pm1/2$ pixel along both horizontal and vertical directions. The quarter-pixel ME, as well, is then performed around the best half search position with $\pm1/4$ pixel search range. Each refinement has nine candidates, including the refinement center and its eight neighborhood, for the best match.

In Figure 3.37, the best MB mode is selected from five candidates: Inter8 × 8, Inter16 × 16, Inter16 × 8, Inter8 × 16, and skip mode, denoted as $S1$–$S5$. In $S1$ procedure, each 8 × 8 block should find its best sub-MB mode from four choices: Sub4 × 4, Sub4×8, Sub8 × 4, and Sub8 × 8, denoted as *a–d*. Thus, nine sub-blocks are processed for each 8 × 8 block, and a total of 41 blocks and sub-blocks are involved per reference frame. The inter mode decision is done after all costs are computed in quarter-pel precision in all reference frames. Note that the sub-blocks in each 8 × 8 block should be within the same reference frame.

Based on the reference software, the matching cost of each candidate is calculated as the flow shown in Figure 3.38. The reference pixels are interpolated to produce the fractional pixels for each searching candidate. Afterwards, the residues are generated by subtracting the corresponding fractional pixels from the current pixels. Then, the absolute values of the 4 × 4-based Hadamard transformed residues are accumulated as the distortion cost. It is called the sum of absolute transformed differences (SATD). The final matching cost is calculated by adding the SATD with the MV cost. Taking MV cost into consideration improves the compression performance for VBSME, but brings many data dependencies among blocks. Because of the MV predictor of the block is defined by its neighboring blocks in H.264/AVC standard, the cost can be correctly derived only after the prediction modes of the neighboring blocks are determined.

3.6.2   FME Loop Decomposition and Data Reuse

3.6.2.1   FME Operation Loop

Based on the FME algorithm in the reference software, we can divide the entire procedure of Figures 3.36 through 3.38 into six iteration loops as shown in Figure 3.39. The first two loops are the current frames and MB in one current frame to be coded. The next two loops are the

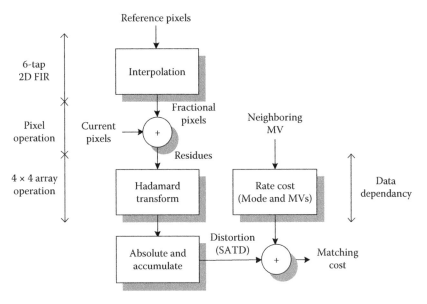

FIGURE 3.38
The matching cost flowchart of each candidate.

```
Loop 0 (current frame index)
   Loop 1 (MB index in current frame)
      Loop 2 (block index)
         Loop 3 (reference frame index)
            Loop 4 (candidate index)
               Loop 5 (pixel index in a candidate)
               {
                   Interpolation operations;
                   SATD operations;
               }
```

FIGURE 3.39
Nested loops of fractional motion estimation.

prediction modes of the variable blocks and multiple reference frames. The fourth loop is the refinement process of half-pixel refinement followed by the quarter-pixel refinement. Each refinement flow has 3×3 candidates to be searched. The last loop is the pixel iteration within one candidate, and the iteration number ranges from 16 × 16 to 4 × 4 depending on the block size. The main tasks inside the most inner loops are the interpolation and SATD calculation.

3.6.2.2 Parallel Hardware and Data Reuse

For the realtime constraint, some of the loops must be unrolled and efficiently mapped into the parallel hardware. During the parallelization of the algorithm, the temporal and spatial data locality should be explored in order to achieve DR. If the two processes has data locality and are performed in parallel, the datum can be only accessed once and then shared between the processing units (PU). Hence, the amount of data access is reduced as well as

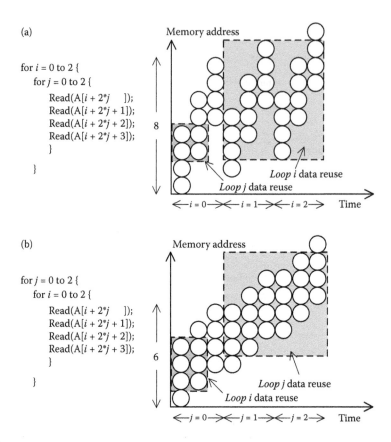

FIGURE 3.40
Data reuse exploration with loop analysis. (a) Original nested loops; (b) *Loop i* and *Loop j* are interchanged.

the power consumption. A simplified example is shown in Figure 3.40a. There are two loops in the algorithm. Inside the *Loop j*, four consecutive data are read. Among them, two data are overlapped in the two consecutive processes of *Loop j*. Therefore, *Loop j* data locality is found, and it is a candidate for DR. In addition, there are also several overlapped data in two consecutive processes of *Loop i*, and thus *Loop i* data locality is another candidate for DR.

The techniques of rescheduling and hardware-oriented algorithm modification can be applied to improve the DR in parallel hardware. In some algorithms, the original processing flow cannot achieve efficient DR. In this case, loop interchange may be an effective way to improve DR efficiency. Here comes an example. With the original schedule in Figure 3.40a, two overlapped data in the consecutive processes of *Loop j* can be on-line reused if parallel processing is applied. However, if *Loop i* and *Loop j* are interchanged. Three overlapped data in the consecutive processes of *Loop i* can be reused as shown in Figure 3.40b. As a result, the DR efficiency in the lower loop becomes better. The SRMC scheme in Section 3.5.3 is another example of improving the DR efficiency by the frame-level re-scheduling.

In some algorithms, there are data dependencies inside a loop. The tasks must be processed sequentially, and parallel processing is not applicable. Sometimes there are data dependencies between two loops in some algorithms. The processing order is fixed and cannot be re-scheduled. In those cases, algorithm modification is required for improvement of DR efficiency, and it may lead to some penalties. In a video coding system, the penalty may be the coding performance degradation. Therefore, there will be a trade-off between coding performance and power in this case. An example can also be found in Section 3.5.3. To achieve the frame-level re-scheduling in the SRMC scheme, the two-stage mode decision is used and results in a 0.3 dB quality drop in PSNR.

3.6.2.3 FME Loop Decomposition

According to the nested loops of FME algorithm in Figure 3.39, the loop analysis about the data reuse is as follows.

Loop 2: For the loop of block index, 41 initial MVs of variable blocks may point to different positions in the search window (SW) after the IME. Therefore, there is few data locality between blocks. If the parallelization is applied in this loop, the reference pixels of VBS must be read in parallel and the required memory bitwidth of SW becomes too large. Therefore, the parallel processing cannot be efficiently applied here for DR.

Loop 3: For the reference frame loop, the costs of a certain block on different reference frames can be processed independently. This loop can be unrolled for parallel processing by duplicating the basic FME PUs and the corresponding SRAMs multiple times for multiple reference frames. The required on-chip SW memory size and the amount of off-chip memory access is proportional to the number of reference frames in this case. Inter-frame DR can be achieved by use of the frame-level rescheduling— the interchange between the *Loop 0* and *Loop 3*. Multiple current MBs of different current frames can reuse the data inside the SW of one reference frame. In this way, only one SW buffer is required, and the external memory BW is greatly reduced. This scheme is called SRMC scheme. In this section, we will emphasis less on this scheme, and refer to Section 3.5.3 for details.

Loop 5: Here, *Loop 5* is discussed before the *Loop 4*. For the on-line interpolation, the 6-tap interpolation filter increases the memory access requirement of FME engine. For example, a 6×6 window of reference data are required to generate an (H, H) interpolated pixel. Note that "H" indicate the position of the half pixel in coordinate. DR can be explored from the neighboring interpolated pixels at the same search candidate. For two horizontally adjacent (half, half) interpolated pixels, 6×5 pixels in the overlapped region of the two interpolation windows could be reused as shown in Figure 3.41a. For an (H, H) search candidate with the 4×4 block-size, a 9×9 window of reference pixels are enough to generate all the interpolated pixels as shown in Figure 3.41b. In this case, the required memory access is reduced from 576 ($4 \times 4 \times 6 \times 6$) to 81 ($9 \times 9$). This technique reuses the reference data of the neighboring interpolated pixels within the same candidate, and is called inter-pixel or intra-candidate DR of FME.

Loop 4: DR can also be explored between the neighboring search candidates for FME. Take the half-pel refinement of a 4×4 block for example. There are nine candidates for the half-pixel refinement, comprising one (I, I), two (H, I), two (I, H), and four (H, H) candidates. Note that "I" indicates the position of the integer pixel in coordinate. With intra-candidate DR, the (I, I), (H, I), (I, H), and (H, H) candidates require $4 \times 4, 4 \times 9, 9 \times 4$, and 9×9 interpolation windows of reference data, respectively. However, many parts of those interpolation windows are overlapped as shown in Figure 3.42. If the nine search candidates are processed in parallel, the

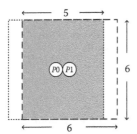
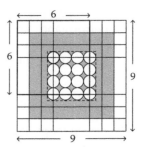

P0 : Interpolated pixel 0   ☐ : Interpolation window of P0   ○ : Interpolated pixel   ☐ : 4 × 4 interpolated block
P1 : Interpolated pixel 1   ☐ : Interpolation window of P1   ☐ : Interpolation window for a pixel

FIGURE 3.41
Intra-candidate data reuse for fractional motion estimation. (a) Reference pixels in the overlapped (gray) interpolation windows for two horizontally adjacent interpolated pixels P0 and P1 can be reused; (b) Overlapped (gray) interpolation windows data reuse for a 4 × 4 interpolated block. Totally, 9 × 9 reference pixels are enough with the technique of intra-candidate data reuse.

reference data read from memory can be shared with each other. Finally, a 10×10 window of reference data is enough to compute the matching costs of all the nine candidates for half-pixel refinement. In this case, the required memory access is reduced from 484 ($4 \times 4 + 2 \times 4 \times 9 + 2 \times 9 \times 4 + 4 \times 9 \times 9$) to 100 ($10 \times 10$). The technique reuses the reference data for neighboring search candidates and is called inter-candidate DR.

The FME algorithm in the reference software is a two-stage refinement process. The quarter-pixel refinement is based on the best matching of the result of the half-pixel refinement. Therefore, the data dependencies is involved in this loop, and the tasks must be processed sequentially. Different searching patterns or refinement criteria may be applied to break this data dependency and facilitate the parallel processing. It may affect the coding performance and the careful evaluation is required.

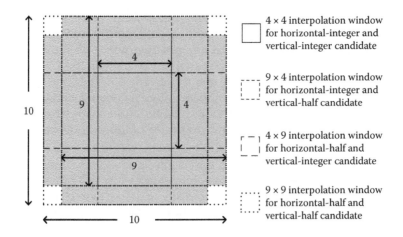

FIGURE 3.42
Inter-candidate data reuse for half-pel refinement of fractional motion estimation. The overlapped (gray) region of interpolation windows can be reused to reduce memory access.

3.6.3 FME Architecture Design

3.6.3.1 Basic Architecture for Half-Than-Quarter Refinement FME Algorithm

Based on the previous analysis, the basic hardware architecture of FME engine is shown in Figure 3.43 [33]. The 4 × 4 block size is the smallest element of VBS, and the SATD computation is also based on 4 × 4 blocks. In addition, all other larger block-sizes can be decomposed into several 4 × 4-elements with the same MV. Therefore, a 4 × 4-block processing unit (PU) is designed and reused for larger blocks by means of the folding. There are nine 4 × 4-block PUs to process nine candidates around the refinement center for the half-pixel and quarter-pixel refinement. Each 4 × 4-block PU is responsible for the residue generation and Hadamard transformation. The interpolation engine generates the half or quarter reference pixels. The interpolated pixels are shared by all 4 × 4-block PUs to achieve intra-candidate and inter-candidate DR. The inputs of the interpolation engine are the parallel integer pixels from the SW SRAMs. In the following, we will use the bottom-up order to introduce these hardware modules.

The architecture of each 4 × 4 PU is shown in Figure 3.44. Four subtractors generate four residues in each cycle and transmit them to the 2-D Hadamard transform unit. The 2-D Hadamard transform unit [34] contains two 1-D transform units and one transposed register array. The first 1-D transform unit filters the residues row by row in each 4 × 4 block, while the second 1-D transform unit processes column by column. The data path of the transposed

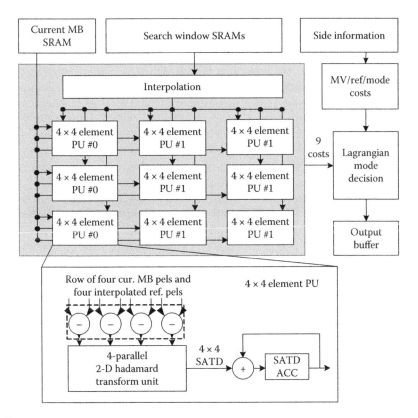

FIGURE 3.43
Hardware architecture for fractional motion estimation engine.

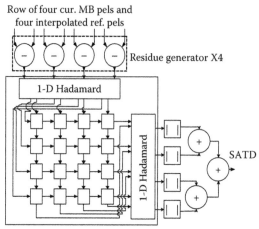

Row of four cur. MB pels and
four interpolated ref. pels

Residue generator X4

1-D Hadamard

1-D Hadamard

SATD

2-D Hadamard transform unit

FIGURE 3.44
Block diagram of 4 × 4-block PU.

registers can be configured as rightward shift or downward shift. The two configurations interchange with each other every four cycles. First, the rows of 1-D transformed residues of the first 4 × 4 block are written into transpose registers horizontally. After four cycles, the columns of 1-D transformed residues are read vertically for the second 1-D Hadamard transform. Meanwhile, the rows of 1-D transformed residues of the second 4 × 4 block are written into transposed registers vertically. In this way, the Hadamard transform unit is fully pipelined with residue generators. The latency of the 2-D transform is four cycles, and there is no bubble cycle for processing the adjacent 4 × 4 blocks for larger blocks.

Figure 3.45a shows the parallel architecture of 2-D interpolation engine. The operations of 2-D FIR filter are decomposed into two 1-D FIR filters with the interpolation shifting buffer. A row of 10 horizontally adjacent integer pixels are input to interpolate five horizontal half pixels simultaneously. These five half pixels and six neighboring integer pixels are latched and shifted downward in the "V-IP Unit" as shown in Figure 3.45b. After the latency of six cycles, the 11 vertical filters generate 11 vertical half pixels by filtering the corresponding columns of six pixels within the "V-IP Units." The dotted rectangle in the bottom of Figure 3.45b represents all predictors needed by residue generation in half-pixel refinement for each cycle. As for quarter-pixel refinement, another bilinear filtering engine with input from the dotted rectangle is enabled for quarter-pixels generation.

In this design, a folding technique is applied for larger blocks to iteratively utilize the interpolation circuits and 4 × 4-block PUs. An efficient scheduling is designed along with the advanced memory arrangement in order to reuse the data. The basic processing scheduling is shown in Figure 3.46a. The 4 × 4 blocks are strung up in the vertical direction. The reference pixels are read row by row from SW SRAMs and input to the interpolation engine. Therefore, vertically inter-block DR can be obtained as depicted in Figure 3.47a. For the next column of 4 × 4 blocks, reference pixels should be re-reloaded, and no DR can be achieved horizontally. With the basic flow of FME, DR of 4 × 8 and 8 × 16 block-sizes is not efficient. Another advanced scheduling can be applied as shown in Figure 3.46b. For 4 × 8 and 8 × 16 block sizes FME, the 4 × 4 sub blocks can be strung up in the horizontal direction. In this case, reference pixels should be read column by column from SW SRAMs,

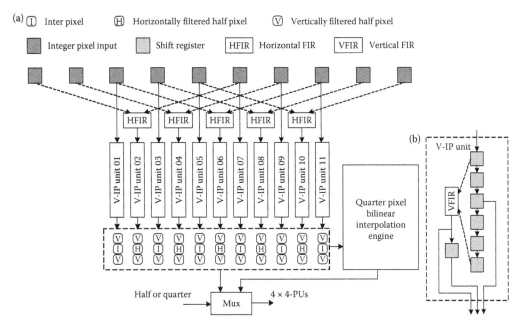

FIGURE 3.45
Block diagram of interpolation engine.

and the inter-block DR can be achieved horizontally as shown in Figure 3.47b. The memory access can be further reduced.

In order to support the inter block DR for both the vertical and horizontal directions, the memory access capability of consecutively horizontal or vertical pixels is required. We will start from the traditional memory arrangement for SW SRAMs. The pixel location in the physical SW is shown in Figure 3.48a. The conventional data arrangement is shown in Figure 3.48b. Horizontally adjacent pixels are arranged in different banks of SW SRAMs. The first column of reference pixels, $A1$–$A8$, are placed in the bank $M1$. The second column

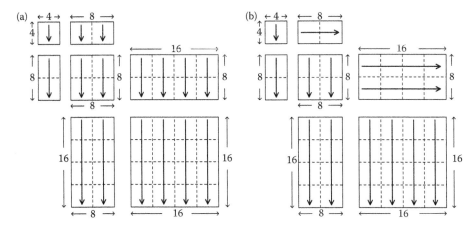

FIGURE 3.46
Hardware processing flow of variable-block size fractional motion estimation. (a) Basic flow; (b) advanced flow.

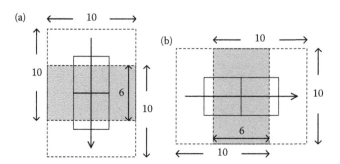

FIGURE 3.47
Inter-4 × 4-block interpolation window data reuse. (a) Vertical data reuse, (b) horizontal data reuse.

of pixels, B1–B8, are placed in the bank M2, and so on. If there are eight banks of SRAM, the ninth column of pixels are placed in the first bank $M1$. In this way, a row of reference pixels, like A5–H5, can be read in parallel. However, a column of reference pixels, like C1–C8, cannot be accessed at the same time because they are located in the same bank $M3$. This is called 1-D random access.

The new ladder-shaped SW data arrangement technique is depicted in Figure 3.48c. After the traditional arrangement, the second and third rows are rotated rightward by one and two pixels. The other rows are also rotated in the same manner. In this way, the reference pixels of A5–H5 and C1–C8 are both put in different banks of SRAMs and can be accessed in one cycle. Therefore, a row or a column of pixels can both be accessed in parallel, and 2-D random access is achieved. As a result, the proposed data arrangement can enhance random access capability of SW SRAMs, and improve DR efficiency.

Here, a factor of AAP_{FME} is defined to stand for average accessed pixels for one FME half-pixel candidate, to indicate the DR efficiency of a FME engine. AAP_{FME} can be computed as

$$AAP_{FME} = \frac{\text{Total numbers of accessed pixels}}{9 \ (\#. \ \text{of candidate}) \times 7 \ (\#. \ \text{of block size})} \quad (3.4)$$

The amount of memory access for different block-sizes and AAP_{FME} for different DR techniques are listed in Table 3.8. The presented architecture with the advanced processing flow

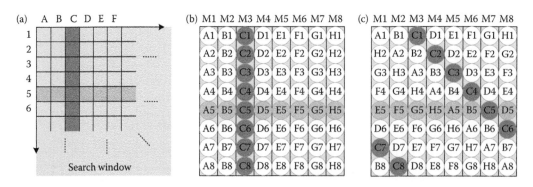

FIGURE 3.48
Search Window SRAMs data arrange. (a) Physical location of reference pixels in the search window; (b) traditional data arrangement with 1-D random access; (c) proposed ladder-shaped data arrangement with 2-D random access.

TABLE 3.8

Memory Access Requirement for Half-Pel Refinement of Fractional
Motion Estimation

Block Size	(1)	(2)	(3)	(4)	(5)
16×16	24,832	2482	484	880	880
16×8	24,832	2852	616	1120	880
8×16	24,832	2852	704	880	880
8×8	24,832	3272	784	1120	1120
8×4	24,832	4112	1120	1600	1120
4×8	24,832	4112	1120	1120	1120
4×4	24,832	5152	1600	1600	1600
Total	173,824	24,834	6428	8320	7600
AAP_{FME}	2759.11	394.19	102.03	132.06	120.63
Ratio	100%	14.29%	3.70%	4.79%	4.37%

Note: (1) No data reuse.
  (2) Only intra-candidate data reuse.
  (3) Full intra-candidate and inter-candidate data reuse.
  (4) Proposed architecture with the basic processing flow.
  (5) Proposed architecture with the advanced processing flow.

can achieve efficient DR and reduce the memory BW to 4.37% which is very close to the lower bound of 3.70% with fully inter- and intra-candidate DR.

3.6.3.2 Advanced Architecture for One-Pass Refinement FME Algorithm

In the reference software, half-pixel candidates around the best integer-pixel candidate are first refined. Then, the quarter-pixel candidates around the best half-pixel candidate are further refined to find the best MV. As a result, 17 candidates are searched to find the best matching candidates. In H.264/AVC, a 6-tap interpolation filter is adopted to generate the half-pixel reference data from integer-pixel reference data. The quarter-pixel reference data are generated from the standard defined neighboring half-pixel reference data with bilinear filter. SATD (Sum of Absolute Transformed Difference) is used as the matching cost of FME and defined as follows. The difference values of the current block and the interpolated reference block are computed first and then processed with Hadamard transform (HT). The resulting data are called transformed residues. At last, the absolute values of the transformed residues are accumulated to generate the SATD cost.

The required interpolation windows for half-pixel and quarter-pixel candidates are overlapped. If the conventional two-step algorithm is adopted, the overlapped reference data are loaded twice and wasteful. Therefore, we propose a hardware-oriented one-pass algorithm. The main concept is that the half-pixel and quarter-pixel candidates are processed simultaneously to share the memory access data and thus reduce data access power. There are 49 fractional-pixel candidates in all for FME, comprising of 1 integer-pixel, 8 half-pixel, and 40 quarter-pixel candidates as shown in Figure 3.49b. If all the candidates are searched, the computation is 2.88 (49/17) times of the two-step algorithm. However, according to the simulation, 87% of the best matching candidates are located at the central 25 candidates as shown in Figure 3.49b. Therefore, the marginal 24 candidates are skipped to save the computation.

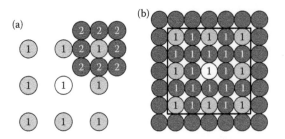

FIGURE 3.49
Illustration of fractional motion estimation algorithm. The white circles are the best integer-pixel candidates. The light-gray circles are the half-pixel candidates. The dark-gray circles are the quarter-pixel candidates. The circles labeled "**1**" and "**2**" are the candidates refined in the first and second passes, respectively. (a) Conventional two-step algorithm; (b) Proposed one-pass algorithm. The 25 candidates inside the dark square are processed in parallel.

In order to further reduce the data processing power of the quarter-pixel candidates, the linearity of HT is utilized. Equation (3.5) shows the linearity of HT. $HT(\cdot)$ means the HT function. a and b are scalars, and **A** and **B** are two 4×4 blocks of reference data. In Equation (3.6), **Q** is a 4×4 block of a quarter-pixel candidate and it is bilinearly interpolated from two 4×4 blocks (**A** and **B**) of half-pixel candidates. $Round(\cdot)$ means the rounding function. In Equation 3.7, **U** is the 4×4 current block. Due to the linearity, the transformed residues of **Q** can be approximated by the bilinear interpolation of the transformed residues of **A** and **B** (only the rounding effect is nonlinear). With the approximation, data processing power for HT of all quarter-pixel candidates is saved. The simulation results of the proposed one-pass FME algorithm are shown in Figure 3.50. Compared to the two-step algorithm in the reference software [29], the performance degradation is only 0.06 dB quality drop in average for CIF (352×288) 30 fps video encoding. In addition, the required memory access of the proposed one-pass algorithm is the same with the only-half refinement algorithm (i.e., MVs are only refined to the half-pixel precision), but the rate-distortion performance is much better.

$$HT(a\,\mathbf{A} + b\,\mathbf{B}) = a\,HT(\mathbf{A}) + b\,HT(\mathbf{B}) \tag{3.5}$$

$$\mathbf{Q} = \text{Round}\left(\frac{\mathbf{A} + \mathbf{B}}{2}\right) \tag{3.6}$$

$$HT(\mathbf{Q} - \mathbf{U}) = HT\left(\text{Round}\left(\frac{\mathbf{A} + \mathbf{B}}{2}\right) - \mathbf{U}\right)$$

$$\approx HT\left(\text{Round}\left(\frac{\mathbf{A} - \mathbf{U} + \mathbf{B} - \mathbf{U}}{2}\right)\right) \tag{3.7}$$

$$\approx \text{Round}\left(\frac{HT(\mathbf{A} - \mathbf{U}) + HT(\mathbf{B} - \mathbf{U})}{2}\right)$$

The corresponding FME architecture is shown in Figure 3.51 [35]. This architecture is based on the previous subsection with additional PUs in the light-gray box. At first, the reference pixels are read from search window memory and input to the "6-tap interpolation filter" to generate the half-pixel reference data. The reference and the current MB data are input to "4×4 block PE" to compute the transformed residues. The absolute values of the transformed residues are then accumulated with the MV costs to generate the matching

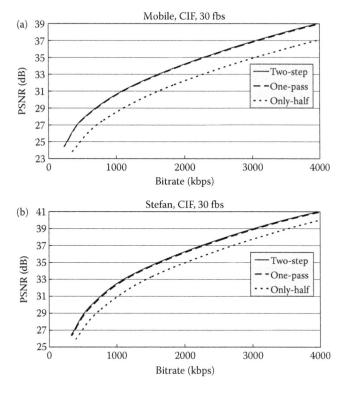

FIGURE 3.50
Rate-distortion performance of the proposed one-pass FME algorithm. The solid, dashed, and dotted lines show the performance of the two-step algorithm in the reference software, the proposed one-pass algorithm, and the algorithm with only half-pixel refinement.

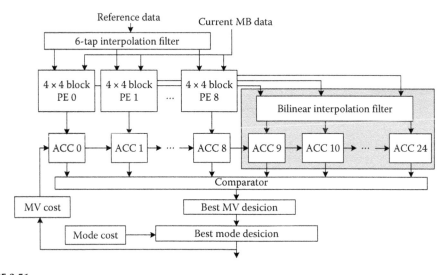

FIGURE 3.51
Architecture of fractional motion estimation. The processing engines on the left side are used to generate the matching costs of integer-pixel and half-pixel candidates. The transformed residues are reused to generate the matching costs of quarter-pixel candidates with the processing engines inside the light-gray box on the right side. Then, the 25 matching costs are compared to find the best MV.

costs of 1 integer-pixel and 8 half-pixel candidates. On the other hand, the transformed residues of half-pixel candidates are reused and input to "Bilinear Interpolation Engine" to generate the approximate transformed residues and the matching costs of 16 quarter-pixel candidates with Equation 3.7. The final 25 matching costs are compared to find the best fractional MV. After comparing matching costs of different inter-modes, the best inter-mode can be decided. With the proposed architecture and one-pass algorithm, a large amount of data access power can be saved. In addition, the throughput of the FME engine is doubled compared to the previous architecture with the conventional two-step algorithm.

3.7   Summary

Multimedia applications are more and more popular as the technologies have made great progress. Among various multimedia applications, video and image applications are always attractive, but the real-time processing of these applications becomes a tough challenge due to its huge computation complexity and memory bandwidth. To support these attractive multimedia applications, application-specific multimedia architecture development is one of the most important issues. In this chapter, the design issues of multimedia architecture are described. By taking the H.264/AVC video encoder as example, system-level design considerations are introduced. For the key component of H.264/AVC system, the temporal prediction architectures, different implementation strategies are surveyed and analyzed. From these architecture designs, how to support huge computation complexity with efficient parallel architecture, and how to reduce system bandwidth with novel data reuse method are illustrated. This chapter is intended to provide an overview, from theory to practice, on how to design an efficient application-specific multimedia architecture. These analysis method and design concept can be applied to different application-specific multimedia architecture.

References

1. *Advanced Video Coding for Generic Audiovisual Services*. ITU-T Rec. H.264 and ISO/IEC 14496-10 (MPEG-4 AVC), ITU-T and ISO/IEC JTC 1, May 2003.
2. F. Catthoor, S. Wuytack, E. De Greef, F. Balasa, L. Nachtergaele, and A. Vandecappele. *Custom Memory Management Methodology: Exploration of Memory Organization for Embedded Multimedia System Design*. Kluwer Academic Publishers, Massachusetts, USA, January 1998.
3. T.-C. Chen, S.-Y. Chien, Y.-W. Huang, C.-H. Tsai, C.-Y. Chen, T.-W. Chen, and L.-G. Chen. Analysis and architecture design of an HDTV720p 30 frames/s H.264/AVC encoder. *IEEE Transactions on Circuits and Systems for Video Technology*, 16(6):673–688, 2006.
4. Y.-W. Huang, T.-C. Chen, C.-H. Tsai, C.-Y. Chen, T.-W. Chen, C.-S. Chen, C.-F. Shen et al., A 1.3tops H.264/AVC single-chip encoder for HDTV applications. In *Proceedings of IEEE International Solid-State Circuits Conference*, San Francisco, California, pp. 128–129, 588, 2005.
5. Z. Liu, Y. Song, M. Shao, S. Li, L. Li, S. Ishiwata, M. Nakagawa, S. Goto, and T. Ikenaga. A 1.41w H.264/AVC real-time encoder soc for hdtv1080p. In *Proceedings of IEEE Symposium on VLSI Circuit*, Honolulu, Oahu, Hawaii, pp. 12–13, June 2007.

6. Y.-K. Lin, D.-W. Li, C.-C. Lin, T.-Y. Kuo, S.-J. Wu, W.-C. Tai, W.-C. Chang, and T.-S. Chang. A 242 mw 10 mm^2 1080p H.264/AVC high-profile encoder chip. In *Proceedings of IEEE International Solid-State Circuits Conference*, San Francisco, California, pp. 314–315, 615, 2008.

7. *Video Coding for Low Bit Rate Communication.* ITU-T Rec. H.263 version 3, ITU-T, November 2000.

8. *Coding of Audio-Visual Objects–Part 2: Visual.* ISO/IEC 14496-2 (MPEG-4 Visual), ISO/IEC JTC 1, April 1999.

9. K. M. Yang, M. T. Sun, and L. Wu. A family of VLSI designs for the motion compensation block-matching algorithm. *Transactions on Circuits and System*, 36(2):1317–1325, 1989.

10. H. Yeo and Y. H. Hu. A novel modular systolic array architecture for full-search block matching motion estimation. *IEEE Transactions on Circuits and Systems for Video Technology*, 5(5):407–416, 1995.

11. Y. K. Lai and L. G. Chen. A data-interlacing architecture with two-dimensional data-reuse for full-search block-matching algorithm. *IEEE Transactions on Circuits and Systems for Video Technology*, 8(2):124–127, 1998.

12. T. Komarek and P. Pirsch. Array architectures for block matching algorithms. *Transactions on Circuits and System*, 36(2):1301–1308, 1989.

13. L. D. Vos and M. Stegherr. Parameterizable VLSI architectures for the full-search block-matching algorithm. *Transactions on Circuits and System*, 36(2):1309–1316, 1989.

14. C. H. Hsieh and T. P. Lin. VLSI architecture for block-matching motion estimation algorithm. *IEEE Transactions on Circuits and Systems for Video Technology*, 2(2):169–175, 1992.

15. J. F. Shen, T. C. Wang, and L. G. Chen. A novel low-power full search block-matching motion estimation design for H.263+. *IEEE Transactions on Circuits and Systems for Video Technology*, 11(7):890–897, 2001.

16. N. Roma and L. Sousa. Efficient and configurable full-search block-matching processors. *IEEE Transactions on Circuits and Systems for Video Technology*, 12(12):1160–1167, 2002.

17. V. L. Do and K. Y. Yun. A low-power VLSI architecture for full-search block-matching motion estimation. *IEEE Transactions on Circuits and Systems for Video Technology*, 8(4):393–398, 1998.

18. X. Q. Gao, C. J. Duanmu, and C. R. Zou. A multilevel successive elimination algorithm for block matching motion estimation. *IEEE Transactions on Image Processing*, 9(3):501–504, 2000.

19. M. Brünig and W. Niehsen. Fast full-search block matching. *IEEE Transactions on Circuits and Systems for Video Technology*, 11(2):241–247, 2001.

20. S. F. Chang, J. H. Hwang, and C. W. Jen. Scalable array architecture design for full search block matching. *IEEE Transactions on Circuits and Systems for Video Technology*, 5(4):332–343, 1995.

21. Y.-W. Huang, T.-C. Wang, B.-Y. Hsieh, and L.-G. Chen. Hardware architecture design for variable block size motion estimation in MPEG-4 AVC/JVT/ITU-T H.264. In *Proceedings of IEEE International Symposium on Circuits and System*, Rio de Janeiro, Brazil, pp. 796–799, 2003.

22. C.-Y. Chen, S.-Y. Chien, Y.-W. Huang, T.-C. Chen, T.-C. Wang, and L.-G. Chen. Analysis and architecture design of variable block-size motion estimation for H.264/AVC. *IEEE Transactions on Circuits and Systems I: Regular Papers*, 53(3):578–593, 2006.

23. H. M. Jong, L. G. Chen, and T. D. Chiueh. Parallel architectures for 3-step hierarchical search block-matching algorithm. *IEEE Transactions on Circuits and Systems for Video Technology*, 4(4):407–416, 1994.

24. T. Wiegand, X. Zhang, and B. Girod. Long-term memory motion-compensated prediction. *IEEE Transactions on Circuits and Systems for Video Technology*, 9:70–84, 1999.

25. J.-C. Tuan, T.-S. Chang, and C.-W. Jen. On the data reuse and memory bandwidth analysis for full-search block-matching VLSI architecture. *IEEE Transactions on Circuits and Systems for Video Technology*, 12:61–72, 2002.

26. T.-C. Chen, C.-Y. Tsai, Y.-W. Huang, and L.-G. Chen. Single reference frame multiple current macroblocks scheme for multiple reference frame motion estimation in H.264/AVC. *IEEE Transactions on Circuits and Systems for Video Technology*, 17(2):242–247, 2007.

27. M.-Y. Hsu. Scalable module-based architecture for MPEG-4 BMA motion estimation. Master's thesis, National Taiwan University, June 2000.

28. S. Y. Yap and J. V. McCanny. A VLSI architecture for variable block size video motion estimation. *IEEE Transactions on Circuit and System II*, 51:384–389, 2004.

29. *Joint Video Team Reference Software JM8.5*. http://bs.hhi.de/suehring/tml/download/, September 2004.

30. *Generic Coding of Moving Pictures and Associated Audio Information–Part 2: Video*. ITU-T Rec. H.262 and ISO/IEC 13818-2 (MPEG-2 Video), ITU-T and ISO/IEC JTC 1, May 1996.

31. T.-C. Chen, H.-C. Fang, C.-J. Lian, C.-H. Tsai, Y.-W. Huang, T.-W. Chen, C.-Y. Chen, Y.-H. Chen, C.-Y. Tsai, and L.-G. Chen. Algorithm analysis and architecture design for HDTV applications. *IEEE Circuits and Devices Magazine*, 22(3):22–31, 2006.

32. Y. Su and M.-T. Sun. Fast multiple reference frame motion estimation for H.264. In *Proceedings of IEEE International Conference on Multimedia and Expo (ICME'04)*, 2004.

33. Y.-H. Chen, T.-C. Chen, S.-Y. Chien, Y.-W. Huang, and L.-G. Chen. VLSI architecture design of fractional motion estimation for H.264/AVC. *Journal of Signal Processing Systems*, 53(3):335–347, 2008.

34. T.-C. Wang, Y.-W. Huang H.-C. Fang, and L.-G. Chen. Parallel 4 × 4 2D transform and inverse transform architecture for MPEG-4 AVC/H.264. In *Proceedings of IEEE International Symposium on Circuits and Systems (ISCAS'03)*, Bangkok, Thailand, pp. 800–803, 2003.

35. Y.-H. Chen, T.-C. Chen, C.-Y. Tsai, S.-F. Tsai, and L.-G. Chen. Algorithm and architecture design of power-oriented H.264/AVC baseline profile encoder for portable devices. *IEEE Transactions on Circuits and Systems for Video Technology*, 19(8):1118–1128, 2009.

4

Multimedia Information Mining

Zhongfei (Mark) Zhang and Ruofei Zhang

CONTENTS

4.1 Introduction

Multimedia information mining is a recently emerged, very interdisciplinary and multidisciplinary area involving three independent research fields: *multimedia, information retrieval,* and *data mining.* However, multimedia information mining is *not* a research area that simply combines the research of the three fields together. Instead, multimedia information mining research focuses on the theme of merging the research of the three fields together to exploit the synergy among the three fields to promote understanding of and to enhance knowledge discovery and information retrieval in multimedia. Multimedia information mining is highly related to another recently emerging area called *multimedia data mining* [1] that synergistically relies on the state-of-the-art research in multimedia and data mining but at the same time fundamentally differs from either multimedia or data mining or a simple combination of the two fields. Specifically, multimedia information mining encompasses all the research in multimedia data mining in addition to the added focus on information retrieval in the context of multimedia.

Information retrieval is a well-established research field and the early work in this field dates back to the time around the middle of the last century. The information retrieval research concerns science and/or methodologies in searching documents or information within the documents [2]. Historically, the study of information retrieval mainly focuses on the text document retrieval, and thus, in the literature, information retrieval is also often referred to as text retrieval or document retrieval. With the emergence of the Internet, the recent research of information retrieval has expanded from the traditional pure text retrieval to also incorporating multimedia (e.g., imagery or video) information retrieval with the typical context of and applications in the Web information retrieval.

On the contrary, multimedia and data mining are two very interdisciplinary and multidisciplinary fields. Both fields started in the early 1990s with only a very short history. Therefore, both fields are relatively young in comparison with the field of information retrieval. On the other hand, with substantial application demands, both fields have undergone independent and simultaneous rapid developments in recent years.

Multimedia is a very diverse, interdisciplinary, and multidisciplinary research field.* The word *multimedia* refers to a combination of multiple media types together. Due to the advanced development of the computer and digital technologies in the early 1990s, multimedia began to emerge as a research field [3,4]. As a research field, multimedia refers to the study and development of an effective and efficient multimedia system targeting a specific application. In this regard, the research in multimedia covers a very wide spectrum of subjects, ranging from multimedia indexing and retrieval, multimedia databases, multimedia networks, multimedia presentation, multimedia quality of services, multimedia usage and user study, to multimedia standards, to name just a few.

While the field of multimedia is so diverse with many different subjects, those that are related to multimedia information mining mainly include multimedia indexing and retrieval, multimedia databases, and multimedia presentation [5–7]. Today, it is well known that multimedia information is ubiquitous and is often required, if not necessarily essential, in many applications. This phenomenon has made multimedia repositories widespread and extremely large. There are tools for managing and searching within these collections, but the need for tools to extract hidden useful knowledge embedded within multimedia collections is becoming pressing and central to many decision-making applications. For example, it is highly desirable for developing the tools needed today for discovering relationships between objects or segments within images, classifying images based on their content, extracting patterns in sound, categorizing speech and music, and recognizing and tracking objects in video streams.

At the same time, researchers in multimedia information systems, in the search for techniques for improving the indexing and retrieval of multimedia information, are in search of new methods for discovering indexing information. A variety of techniques, from machine learning, statistics, databases, knowledge acquisition, data visualization, image analysis,

---

* Here we are only concerned with a research field; multimedia may also be referred to industries and even social or societal activities.

high-performance computing, and knowledge-based systems, have been used mainly as research handcraft activities. The development of multimedia databases and their query interfaces recalls again the idea of incorporating multimedia information mining methods for dynamic indexing.

On the other hand, data mining is also a very diverse, interdisciplinary, and multi-disciplinary research field. The terminology *data mining* refers to knowledge discovery. Originally, this field began with knowledge discovery in databases. However, data mining research today has been advanced far beyond databases [8,9]. This is due to the following reasons. First, today's knowledge discovery research requires more than ever the advanced tools and theory beyond the traditional database research, noticeably mathematics, statistics, machine learning, and pattern recognition. Second, with the fast explosion of the data storage scale and the presence of multimedia data almost everywhere, it is not enough for today's knowledge discovery research to just focus on the structured data in the traditional databases; instead, it is common to see that the traditional databases have evolved into data warehouses, and the traditional structured data have evolved into more nonstructured data such as imagery data, time-series data, spatial data, video data, audio data, and more general multimedia data. Adding into this complexity is the fact that in many applications these nonstructured data do not even exist in a more traditional "database" anymore; they are just simply a collection of the data, even though many times people still call them databases (e.g., image database, video database).

Examples are the data collected in fields such as art, design, hypermedia and digital media production, case-based reasoning and computational modeling of creativity, including evolutionary computation, and medical multimedia data. These exotic fields use a variety of data sources and structures, interrelated by the nature of the phenomenon that these structures describe. As a result there is an increasing interest in new techniques and tools that can detect and discover patterns that lead to new knowledge in the problem domain where the data have been collected. There is also an increasing interest in the analysis of multimedia data generated by different distributed applications, such as collaborative virtual environments, virtual communities, and multiagent systems. The data collected from such environments include a record of the actions in them, a variety of documents that are part of the business process, asynchronous threaded discussions, transcripts from synchronous communications, and other data records. These heterogeneous multimedia data records require sophisticated preprocessing, synchronization, and other transformation procedures before even moving to the analysis stage.

Consequently, with the independent and advanced developments of the research fields in information retrieval, multimedia, and data mining, with today's explosion of the data scale and the existence of the pluralism of the data media types, it is natural to evolve into this new area called *multimedia information mining*. While it is presumably true that multimedia information mining involves the research among the three fields, the research in multimedia information mining refers to the synergistic application of knowledge discovery and information retrieval theory and techniques in a multimedia database or collection. As a result, "inherited" from these parent fields, multimedia information mining by nature is also an interdisciplinary and multidisciplinary area; the research in multimedia information mining heavily relies on the research from many other fields, noticeably from mathematics, statistics, machine learning, computer vision, and pattern recognition. Figure 4.1 illustrates the relationships among these interconnected fields.

While we have clearly given the working definition of multimedia information mining as an emerging, active research area, due to historic reasons, it is helpful to clarify several misconceptions and to point out several pitfalls at the beginning.

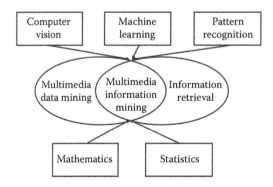

FIGURE 4.1
Relationships among the interconnected fields to multimedia information mining.

- *Information* versus *Data*: Jain [6] gives a clear conceptual description of the difference between information and data. Essentially, information is what we are actually concerned about and data are simply the symbolic representation of what we are actually concerned about. Consequently, when we say information mining, we must proceed with data mining in order to discover knowledge from the information.

- *Multimedia Information Mining* versus *Multimedia Data Mining*: Based on the working definition of multimedia information mining given at the beginning of this chapter as well as the difference between information and data, multimedia information mining is more broadly defined than multimedia data mining and essentially incorporates all the research aspects of multimedia data mining in addition to the added focus on multimedia information retrieval. However, in the literature, the two terminologies are often used interchangeably.

- *Multimedia Indexing and Retrieval* versus *Multimedia Information Mining*: It is well known that in the classic data mining research, the pure text retrieval as the classic information retrieval focus is *not* considered as part of data mining, as there is no knowledge discovery involved [9]. However, in multimedia information mining, when it comes to the scenarios of multimedia indexing and retrieval, it is considered as part of multimedia information mining. The reason is that a typical multimedia indexing and/or retrieval system reported in the recent literature often contains a certain level of knowledge discovery such as feature selection, dimensionality reduction, concept discovery, as well as mapping discovery between different modalities (e.g., imagery annotation, where a mapping from an image to textual words is discovered and word-to-image retrieval, where a mapping from a textual word to images is discovered). On the other hand, if a multimedia indexing or retrieval system uses a "pure" indexing system such as the text-based indexing technology employed in many commercial imagery/video/audio retrieval systems on the Web, this system is not considered as a multimedia information mining system.

- *Database* versus *Data Collection*: In a classic database system, there is always a database management system to govern all the data in the database. This is true for the classic, structured data in the traditional databases. However, when the data become nonstructured data, in particular, multimedia data, often we do not have such a management system to "govern" all the data in the collection. Typically, we simply have a whole collection of multimedia data, and we expect to develop an indexing/retrieval system or other data mining system on top of this data collection. For historic reasons, in many literature references, we still use the terminology of "database" to refer to such a multimedia data collection, even though this is different from the traditional, structured database in concept.

- *Multimedia Data* versus *Single Modality Data*: Although "multimedia" refers to the multiple modalities and/or multiple media types of data, conventionally in the area of multimedia, multimedia indexing and retrieval also includes the indexing and retrieval of a single, nontext modality of data, such as image indexing and retrieval, video indexing and retrieval, and audio indexing and retrieval. Consequently, in multimedia information mining, we follow this convention to include the study of any knowledge discovery dedicated to any single modality of data as part of the multimedia data mining or multimedia information mining research. Therefore, studies in image data mining, video data mining, and audio data mining alone are considered as part of the multimedia data mining or multimedia information mining area.

Multimedia information mining, although still in its early booming stage as an area that is expected to have a further development, has already found enormous application potential in a wide spectrum covering almost all the sectors of society, ranging from people's daily lives to economic development to government services. This is due to the fact that in today's society almost all the real-world applications often have data with multiple modalities, from multiple sources, and in multiple formats. For example, in homeland security applications, we may need to mine data from an air traveler's credit history, traveling patterns, photo pictures, and video data from surveillance cameras in the airport. In the manufacturing domains, business processes can be improved if, for example, part drawings, part descriptions, and part flow can be mined in an integrated way instead of separately. In medicine, a disease might be predicted more accurately if the magnetic resonance imaging (MRI) imagery is mined together with other information about the patient's condition. Similarly, in bioinformatics, data are available in multiple formats.

The rest of the chapter is organized as follows. In the next section, we give the architecture for a typical multimedia information mining system or methodology in the literature. Then in order to showcase a specific multimedia information mining system and how it works, we present an example of a specific method on concept discovery through image annotation in a multimodal database in the following section. Finally, Section 4.4 concludes this chapter.

4.2 A Typical Architecture of a Multimedia Information Mining System

A typical multimedia information mining system, or framework, or method always consists of the following three key components. Given the raw multimedia data, the very first step for mining the multimedia information is to convert a specific raw data collection (or a database) into a representation in an abstract space which is called the feature space. This process is called feature extraction. Consequently, we need a feature representation method to convert the raw multimedia data to the features in the feature space, before any mining activities are able to be conducted. This component is very important as the success of a multimedia information mining system to a large degree depends on how good the feature representation method is. The typical feature representation methods or techniques are taken from the classic computer vision research, pattern recognition research, as well as multimedia information indexing and retrieval research in multimedia field.

Since knowledge discovery is an intelligent activity, like other types of intelligent activities, multimedia information mining requires the support of a certain level of knowledge. Therefore, the second key component is the knowledge representation, that is, how to effectively represent the required knowledge to support the expected knowledge discovery activities in a multimedia database. The typical knowledge representation methods used in the multimedia information mining literature are directly taken from the general knowledge representation research in artificial intelligence field with possible special consideration in multimedia information mining problems such as spatial constraints-based reasoning.

Finally, we come to the last key component—the actual mining or learning theory and/or technique to be used for the knowledge discovery in a multimedia database. In the current literature of multimedia information mining, there are mainly two paradigms of the learning or mining theory/techniques that can be used separately or jointly in a specific multimedia information mining application. They are *statistical learning theory* and *soft computing theory*, respectively. The former is based on the recent literature on machine learning and in particular statistical machine learning, whereas the latter is based on the recent literature on soft computing such as fuzzy logic theory. This component typically is the core of the multimedia information mining system.

In addition to the three key components, in many multimedia information mining systems, there are user interfaces to facilitate the communications between the users and the mining systems. Like the general data mining systems, for a typical multimedia information mining system, the quality of the final mining results can be judged only by the users. Hence, it is necessary in many cases to have a user interface to allow a communication between the users and the mining systems and the evaluations of the final mining quality; if the quality is not acceptable, the users may need to use the interface to tune different parameter values of a specific component used in the system, or even to change different components, in order to achieve better mining results, which may go into an iterative process until the users are happy with the mining results.

Figure 4.2 illustrates this typical architecture of a multimedia information mining system.

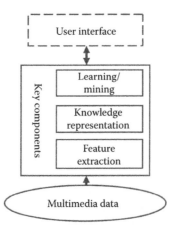

FIGURE 4.2
The typical architecture of a multimedia information mining system.

4.3 An Example: Concept Discovery in Multimodal Data

In this section, as an example to showcase the research as well as the technologies developed in multimedia information mining, we address the multimedia database modeling problem in general and, in particular, focus on developing a semantic concept discovery methodology in a multimodal database through image annotation to address effective semantics-intensive mining and retrieval problems.

Specifically, in this section, we describe a probabilistic semantic model in which the visual features and the textual words are connected via a hidden layer which constitutes the semantic concepts to be discovered to explicitly exploit the synergy between the two modalities; the association of visual features and the textual words is determined in a Bayesian framework such that the confidence of the association can be provided; and extensive evaluations on a large-scale, visually and semantically diverse image collection crawled from the Web are reported to evaluate the prototype system based on the model. In the developed probabilistic model, a hidden concept layer which connects the visual features and the word layer is discovered by fitting a generative model to the training images and annotation words. An expectation—maximization (EM)-based iterative learning procedure is developed to determine the conditional probabilities of the visual features and the textual words, given a hidden concept class. Based on the discovered hidden concept layer and the corresponding conditional probabilities, the image annotation and the text-to-image retrieval are performed using the Bayesian framework. The evaluations of the prototype system on 17,000 images and 7736 automatically extracted annotation words from the crawled Web pages for multimodal image data mining and retrieval have indicated that the model and the framework are superior to a state-of-the-art peer system in the literature.

4.3.1 Background

Efficient access to a multimedia database requires the ability to search and organize multimedia information. In traditional image retrieval, users have to provide examples of images that they are looking for. Similar images are found based on the match of image features. Even though there have been many studies on this traditional image retrieval paradigm, empirical studies have shown that using image features solely to find similar images is usually insufficient due to the notorious *semantic gap* between low-level features and high-level semantic concepts [10]. As a step further to reduce this gap, region-based features (describing object–level features), instead of raw features of the whole image, to represent the visual content of an image are proposed [11–13].

On the other hand, it is well observed that often imagery does not exist in isolation; instead, typically there is rich collateral information coexisting with image data in many applications. Examples include the Web, many domain-archived image databases (in which there are annotations to images), and even consumer photo collections. In order to further reduce the semantic gap, recently multimodal approaches to image data mining and retrieval have been proposed in the literature [14] to explicitly exploit the redundancy coexisting in the collateral information to the images. In addition to the improved mining and retrieval accuracy, a benefit for the multimodal approaches is the added querying modalities. Users can query an image database by imagery, by a collateral information modality (e.g., text), or by any combination.

In this section, we describe a probabilistic semantic model and the corresponding learning procedure to address the problem of automatic image annotation and show its application to multimodal image data mining and retrieval. Specifically, we use the developed probabilistic semantic model to explicitly exploit the synergy between the different modalities of the imagery and the collateral information. In this work, we focus only on a specific collateral modality—text. The model may be generalized to incorporate other collateral modalities. Consequently, the synergy here is explicitly represented as a hidden layer between the imagery and the text modalities. This hidden layer constitutes the concepts to be discovered through a probabilistic framework such that the confidence of the association can be provided. An EM-based iterative learning procedure is developed to determine the conditional probabilities of the visual features and the words, given a hidden concept class. Based on the discovered hidden concept layer and the corresponding conditional probabilities, the image-to-text and text-to-image retrievals are performed in a Bayesian framework.

In the existing image data mining and retrieval literature, COREL data have been extensively used to evaluate the performance [15–18]. It has been argued [19] that the COREL data are much easier to annotate and retrieve due to their small number of concepts and small variations of the visual content. In addition, the relatively small number (1000–5000) of the training images and test images typically used in the early literature further makes the problem easier and the evaluation less convictive. In order to truly capture the difficulties in real scenarios such as Web image data mining and retrieval and to demonstrate the robustness and the promise of the developed model and the framework in these challenging applications, we have evaluated the prototype system on a collection of 17,000 images with the automatically extracted textual annotations from various crawled Web pages. We have shown that the developed model and framework work well on this scale of a very noisy image dataset and substantially outperform the state-of-the-art peer system MBRM [17].

The specific contributions of this work include:

1. We develop a probabilistic semantic model in which the visual features and textual words are connected via a hidden layer to constitute the concepts to be discovered to explicitly exploit the synergy between the two modalities. An EM-based learning procedure is developed to fit the model to the two modalities.

2. The association of visual features and textual words is determined in a Bayesian framework such that the confidence of the association can be provided.

3. Extensive evaluations on a large-scale collection of visually and semantically diverse images crawled from the Web are performed to evaluate the prototype system based on the model and the framework. The experimental results demonstrate the superiority and the promise of the approach.

4.3.2 Probabilistic Semantic Model

To achieve automatic image annotation as well as multimodal image data mining and retrieval, a probabilistic semantic model is developed for the training imagery and the associated textual word annotation dataset. The probabilistic semantic model is developed by the EM technique to determine the hidden layer connecting image features and textual words, which constitutes the semantic concepts to be discovered to explicitly exploit the synergy between the imagery and text.

4.3.2.1 Probabilistically Annotated Image Model

First, a word about notation: $f_i, i \in [1, N]$ denotes the visual feature vector of images in the training database, where N is the size of the image database. $w^j, j \in [1, M]$ denotes the distinct textual words in the training annotation word set, where M is the size of annotation vocabulary in the training database.

In the probabilistic model, we assume the visual features of images in the database, $f_i = [f_i^1, f_i^2, \ldots, f_i^L], i \in [1, N]$, are known i.i.d. samples from an unknown distribution. The dimension of the visual feature is L. We also assume that the specific visual feature annotation word pairs $(f_i, w^j), i \in [1, N], j \in [1, M]$ are known i.i.d. samples from an unknown distribution. Furthermore, we assume that these samples are associated with an unobserved *semantic concept* variable $z \in Z = \{z_1, \ldots, z_K\}$. Each observation of one visual feature $f \in F = \{f_i, f_2, \ldots, f_N\}$ belongs to one or more concept classes z_k, and each observation of one word $w \in V = \{w^1, w^2, \ldots, w^M\}$ in one image f_i belongs to one concept class. To simplify the model, we have two more assumptions. First, the observation pairs (f_i, w^j) are generated independently. Second, the pairs of random variables (f_i, w^j) are conditionally independent, given the respective hidden concept z_k,

$$P(f_i, w^j | z_k) = p_{\mathcal{F}}(f_i | z_k) P_{\mathcal{V}}(w^j | z_k). \tag{4.1}$$

The visual feature and word distributions are treated as a randomized data generation process, described as follows:

- Choose a concept with probability $P_{\mathcal{Z}}(z_k)$.
- Select a visual feature $f_i \in F$ with probability $P_{\mathcal{F}}(f_i | z_k)$.
- Select a textual word $w^j \in V$ with probability $P_{\mathcal{V}}(w^j | z_k)$.

As a result, one obtains an observed pair (f_i, w^j), while the concept variable z_k is discarded. The graphic representation of this model is depicted in Figure 4.3.

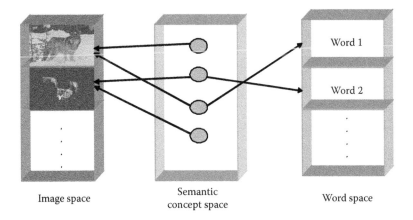

Image space   Semantic concept space   Word space

FIGURE 4.3
Graphic representation of the model developed for the randomized data generation for exploiting the synergy between imagery and text.

Translating this process into a joint probability model results in the expression

$$P(f_i, w^j) = P(w^j)P(f_i|w^j)$$

$$= P(w^j) \sum_{k=1}^{K} P_{\mathcal{F}}(f_i|z_k)P(z_k|w^j). \tag{4.2}$$

Inverting the conditional probability $P(z_k|w^j)$ in Equation 4.2 with the application of Bayes' rule results in

$$P(f_i, w^j) = \sum_{k=1}^{K} P_{\mathcal{Z}}(z_k)P_{\mathcal{F}}(f_i|z_k)P_{\mathcal{V}}(w^j|z_k). \tag{4.3}$$

The mixture of Gaussian [20] is assumed for the feature-concept conditional probability $P_{\mathcal{F}}(\bullet|Z)$. In other words, the visual features are generated from K Gaussian distributions, each one corresponding to a z_k. For a specific semantic concept variable z_k, the conditional pdf of visual feature f_i is

$$p_{\mathcal{F}}(f_i|z_k) = \frac{1}{(2\pi)^{L/2}|\sum_k|^{1/2}} e^{-\frac{1}{2}(f_i - \mu_k)^T \sum_k^{-1}(f_i - \mu_k)}, \tag{4.4}$$

where $\sum_k$ and μ_k are the covariance matrix and mean of the visual features belonging to z_k, respectively. The word–concept conditional probabilities $P_{\mathcal{V}}(\bullet|Z)$, that is, $P_{\mathcal{V}}(w^j|z_k)$ for $k \in [1, K]$, are estimated through fitting the probabilistic model to the training set.

Following the likelihood principle, one determines $P_{\mathcal{F}}(f_i|z_k)$ by the maximization of the log-likelihood function

$$\log \prod_{i=1}^{N} p_{\mathcal{F}}(f_i|Z)^{u_i} = \sum_{i=1}^{N} u_i \log \left(\sum_{k=1}^{K} P_{\mathcal{Z}}(z_k)p_{\mathcal{F}}(f_i|z_k) \right), \tag{4.5}$$

where u_i is the number of the annotation words for image f_i. Similarly, $P_{\mathcal{Z}}(z_k)$ and $P_{\mathcal{V}}(w^j|z_k)$ can be determined by the maximization of the log-likelihood function

$$\mathcal{L} = \log P(F, V) = \sum_{i=1}^{N} \sum_{j=1}^{M} n(w_i^j) \log P(f_i, w^j), \tag{4.6}$$

where $n(w_i^j)$ denotes the weight of annotation word w^j, that is, the occurrence frequency, for image f_i.

4.3.2.2 EM-Based Procedure for Model Fitting

From Equations 4.2, 4.5, and 4.6, we derive that the model is a statistical mixture model [21], which can be resolved by applying the EM technique [22]. The EM alternates in two steps: (i) an expectation (E) step where the posterior probabilities are computed for the hidden variable z_k, based on the current estimates of the parameters and (ii) a maximization (M) step, where parameters are updated to maximize the expectation of the complete-data likelihood $\log P(F, V, Z)$, given the posterior probabilities computed in the previous E-step. Thus, the probabilities can be iteratively determined by fitting the model to the training image database and the associated annotations.

Applying Bayes' rule to Equation 4.3, we determine the posterior probability for z_k under f_i and (f_i, w^j):

$$p(z_k|f_i) = \frac{P_{\mathcal{Z}}(z_k)p_{\mathcal{F}}(f_i|z_k)}{\sum_{t=1}^{K} P_{\mathcal{Z}}(z_t)p_{\mathcal{F}}(f_i|z_t)}, \tag{4.7}$$

$$P(z_k|f_i, w^j) = \frac{P_{\mathcal{Z}}(z_k)P_{\mathcal{Z}}(f_i|z_k)P_{\mathcal{V}}(w^j|z_k)}{\sum_{t=1}^{K} P_{\mathcal{Z}}(z_t)P_{\mathcal{F}}(f_i|z_t)P_{\mathcal{V}}(w^j|z_t)}. \tag{4.8}$$

The expectation of the complete-data likelihood $\log P(F, V, Z)$ for the estimated $P(Z|F, V)$ derived from Equation 4.8 is

$$\sum_{(i,j)=1}^{K} \sum_{i=1}^{N} \sum_{j=1}^{M} n(w_i^j) \log [P_{\mathcal{Z}}(z_{i,j})p_{\mathcal{F}}(f_i|z_{i,j})P_{\mathcal{V}}(w^j|z_{i,j})]P(Z|F, V), \tag{4.9}$$

where

$$P(Z|F, V) = \prod_{s=1}^{N} \prod_{t=1}^{M} P(z_{s,t}|f_s, w^t).$$

In Equation 4.9, the notation $z_{i,j}$ is the concept variable that associates with the feature–word pair (f_i, w^j). In other words, (f_i, w^j) belongs to concept z_t, where $t = (i, j)$.

Similarly, the expectation of the likelihood $\log P(F, Z)$ for the estimated $P(Z|F)$ derived from Equation 4.7 is

$$\sum_{k=1}^{K} \sum_{i=1}^{N} \log(P_{\mathcal{Z}}(z_k)p_{\mathcal{F}}(f_i|z_k))p(z_k|f_i). \tag{4.10}$$

Maximizing Equations 4.9 and 4.10 with Lagrange multipliers to $P_{\mathcal{Z}}(z_l)$, $p_{\mathcal{F}}(f_u|z_l)$, and $P_{\mathcal{V}}(w^v|z_l)$, respectively, under the following normalization constraints

$$\sum_{k=1}^{K} P_{\mathcal{Z}}(z_k) = 1, \sum_{k=1}^{K} P(z_k|f_i, w^j) = 1 \tag{4.11}$$

for any f_i, w^j, and z_l, the parameters are determined as

$$\mu_k = \frac{\sum_{i=1}^{N} u_i f_i p(z_k|f_i)}{\sum_{s=1}^{N} u_s p(z_k|f_s)}, \tag{4.12}$$

$$\sum_k = \frac{\sum_{i=1}^{N} u_i p(z_k|f_i)(f_i - \mu_k)(f_i - \mu_k)^T}{\sum_{s=1}^{N} u_s p(z_k|f_s)}, \tag{4.13}$$

$$P_{\mathcal{Z}}(z_k) = \frac{\sum_{j=1}^{M} \sum_{i=1}^{N} u(w_i^j)P(z_k|f_i, w^j)}{\sum_{j=1}^{M} \sum_{i=1}^{N} n(w_i^j)}, \tag{4.14}$$

$$P_{\mathcal{V}}(w^j|z_k) = \frac{\sum_{i=1}^{N} n(w_i^j)P(z_k|f_i, w^j)}{\sum_{u=1}^{M} \sum_{v=1}^{N} n(w_v^u)P(z_k|f_v, w^u)}. \tag{4.15}$$

Alternating Equations 4.7 and 4.8 with Equations 4.12 through 4.15 defines a convergent procedure to a local maximum of the expectation in Equations 4.9 and 4.10.

4.3.2.3 Estimating the Number of Concepts

The number of concepts, K, must be determined in advance for the EM model fitting. Ideally, we intend to select the value of K that best agrees to the number of the semantic classes in the training set. One readily available notion of the fitting goodness is the log-likelihood. Given this indicator, we can apply the minimum description length principle [23] to select among values of K. This can be done as follows [23]: choose K to maximize

$$\log(P(F, V)) - \frac{m_K}{2} \log(MN), \tag{4.16}$$

where the first term is expressed in Equation 4.6 and m_K is the number of free parameters needed for a model with K mixture components. In our probabilistic model, we have

$$m_K = (K - 1) + K(M - 1) + K(N - 1) + L^2 = K(M + N - 1) + L^2 - 1.$$

As a consequence of this principle, when models with different values of K fit the data equally well, the simpler model is selected. In the experimental database reported in Section 4.3.4, K is determined through maximizing Equation 4.16.

4.3.3 Model-Based Image Annotation and Multimodal Image Mining and Retrieval

After the EM-based iterative procedure converges, the model fitting to the training set is obtained. The image annotation and multimodal image mining and retrieval are conducted in a Bayesian framework with the determined $P_{\mathcal{Z}}(z_k)$, $p_{\mathcal{F}}(f_i|z_k)$, and $P_{\mathcal{V}}(w^j|z_k)$.

4.3.3.1 Image Annotation and Image-to-Text Querying

The objective of image annotation is to return words which best reflect the semantics of the visual content of images. In this developed approach, we use a joint distribution to model the probability of an event that a word w^j belonging to semantic concept z_k is an annotation word of image f_i. Observing Equation 4.1, the joint probability is

$$P(w^j, z_k, f_i) = P_{\mathcal{Z}}(Z_k) p_{\mathcal{F}}(f_i|z_k) P_{\mathcal{V}}(w^j|z_k). \tag{4.17}$$

Through applying Bayes' law and the integration over $P_{\mathcal{Z}}(z_k)$, we obtain the following expression:

$$
\begin{aligned}
P(w^j|f_i) &= \int P_{\mathcal{V}}(w^j|z) p(z|f_i)\, \mathrm{d}z \\
&= \int P_{\mathcal{V}}(w^j|z) \frac{p_{\mathcal{F}}(f_i|z) P(z)}{p(f_i)}\, \mathrm{d}z \\
&= E_z \left\{ \frac{P_{\mathcal{V}}(w^j|z) p_{\mathcal{F}}(f_i|z)}{p(f_i)} \right\},
\end{aligned}
\tag{4.18}
$$

where

$$p(f_i) = \int p_{\mathcal{F}}(f_i|z) P_{\mathcal{Z}}(z)\, \mathrm{d}z = E_z\{p_{\mathcal{F}}(f_i|z)\}. \tag{4.19}$$

In the above equations, $E_z\{\bullet\}$ denotes the expectation over $P(z_k)$, the probability of semantic concept variables. Equation 4.18 provides a principled way to determine the probability of word w^j for annotating image f_i. With the combination of Equations 4.18 and 4.19, the automatic image annotation can be solved fully in the Bayesian framework.

In practice, we derive an approximation of the expectation in Equation 4.18 by utilizing the Monte Carlo sampling [24] technique. Applying Monte Carlo integration to Equation 4.18 derives

$$P(w^j|f_i) \approx \frac{\sum_{k=1}^{K} P_{\mathcal{V}}(w^j|z_k)p_{\mathcal{F}}(f_i|z_k)}{\sum_{h=1}^{K} p_{\mathcal{F}}(f_i|z_h)}$$

$$= \sum_{k=1}^{K} P_{\mathcal{V}}(w^j|z_k)x_k, \tag{4.20}$$

where $x_k = p_{\mathcal{F}}(f_i|z_k)/\sum_{h=1}^{K} p_{\mathcal{F}}(f_i|z_h)$. The words with the top highest $P(w^j|f_i)$ are returned to annotate the image. Given this image annotation scheme, the image-to-text querying may be performed by retrieving documents for the returned words based on the traditional text retrieval techniques.

4.3.3.2 Text-to-Image Querying

The traditional text-based image retrieval systems, for example, Google image search, solely use textual information to index images. It is well known that this approach fails to achieve a satisfactory image retrieval, which actually has motivated the content-based image indexing research. Based on the model obtained in Section 4.3.2 to explicitly exploit the synergy between imagery and text, we develop an alternative and much more effective approach using the Bayesian framework to image data mining and retrieval given a text query.

Similar to the derivation in Section 4.3.3.1, we retrieve images for word queries by determining the conditional probability $P(f_i|w^j)$:

$$P(f_i|w^j) = \int P_{\mathcal{F}}(f_i|z)P(z|w^j)dz$$

$$= \int P_{\mathcal{V}}(w^j|z)\frac{p_{\mathcal{F}}(f_i|z)P(z)}{P(w^j)}dz$$

$$= E_z\left\{\frac{P_{\mathcal{V}}(w^j|z)p_{\mathcal{F}}(f_i|z)}{P(w^j)}\right\}. \tag{4.21}$$

The expectation can be estimated as follows:

$$P(f_i|w^j) \approx \frac{\sum_{k=1}^{K} P_{\mathcal{V}}(w^j|z_k)p_{\mathcal{F}}(f_i|z_k)}{\sum_{h=1}^{K} P_{\mathcal{V}}(w^j|z_h)}$$

$$= \sum_{k=1}^{K} p_{\mathcal{F}}(f_i|z_k)y_k, \tag{4.22}$$

where $y_k = P_{\mathcal{V}}(w^j|z_k)/\sum_{h} P_{\mathcal{V}}(w^j|z_h)$. The images in the database with the top highest $P(f_i|w^j)$ are returned as the querying result for each query word.

4.3.4   Experiments

We have implemented the approach in a prototype system. The architecture of the prototype system is illustrated in Figure 4.4. The system supports both image-to-text (i.e., image annotation) and text-to-image queryings.

4.3.4.1   Dataset and Feature Sets

It has been noted that the datasets used in the existing automatic image annotation systems [15–18] fail to capture the difficulties inherent in many real-world image databases. Two issues are taken into the consideration in the design of the experiments reported in this section. First, the commonly used COREL database is much easier for image annotation and retrieval due to its limited semantics conveyed and small variations of the visual content. Second, the typical small scales of the datasets reported in the existing literature are far away from being realistic in all the real-world applications. To address these issues, we decide not to use the COREL database in the evaluation of the prototype system; instead, we evaluate the system on a collection of a large-scale real-world dataset automatically crawled from the Web. The web pages crawled are from the Yahoo! photos Web site; then the images and the surrounding text describing the images' content are extracted from the blocks containing the images by using the VIPS algorithm [25]. The surrounding text is processed using the standard text processing techniques to obtain the annotation words. Apart from the images and the annotation words, the weight of each annotation word for the images is computed by using a scheme incorporating TF, IDF, and the tag information in VIPS, and is

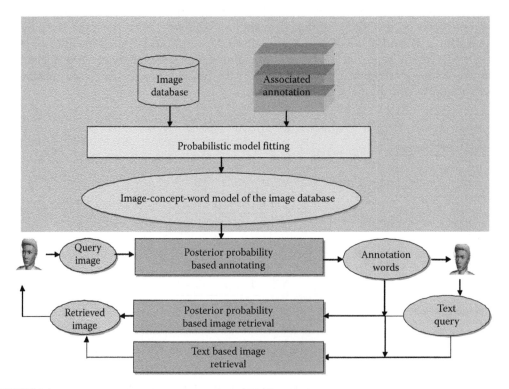

FIGURE 4.4
(**See color insert**.) The architecture of the prototype system.

People (6), mountain (6), sky (5)

FIGURE 4.5
An example of image and annotation word pairs in the generated database. The number following each word is the corresponding weight of the word.

normalized to range (0,10]. The image–annotation word pairs are stemmed and manually cleaned before using as the training database for the model fitting and testing. The data collection consists of 17,000 images and 7736 stemmed annotation words. Among them, 12,000 images are used as the training set and the remaining 5000 images are used for the testing purpose. Compared with images in COREL, the images in this set are more diverse both on semantics and on visual appearance, which reflects the true nature of image search in many real-world applications. Figure 4.5 shows an image example with the associated annotation words in the generated database.

The focus of this section is not on image feature selection and the developed approach is independent of any visual features. For implementation simplicity and easy comparison purposes, similar features used in [17] are used in the prototype system. Specifically, a visual feature is a 36-dimensional vector, consisting of 24 color features (autocorrelogram computed over eight quantized colors and three Manhattan Distances) and 12 texture features (Gabor energy computed over three scales and four orientations).

4.3.4.2 Evaluation Metrics

To evaluate the effectiveness and the promise of the prototype system for multimodal image data mining and retrieval, the following performance measures are defined:

- Hit-Rate3 (HR3): the average rate of at least one word in the ground truth of a test image is returned in the top three returned words for the test set.
- Complete-Length (CL): the average minimum length of the returned words which contain all the ground truth words for a test image for the test set.
- Single-Word-Query-Precision (SWQP(n)): the average rate of the relevant images (here "relevant" means that the ground truth annotation of this image contains the query word) in the top n returned images for a single-word query for the test set.

HR3 and CL measure the accuracy of image annotation (or the image-to-text querying); the higher the HR3, and/or the lower the CL, the better the annotation accuracy. SWQP(n) measures the precision of the text-to-image querying; the higher the SWQP(n), the better the text-to-image querying precision.

Furthermore, we also measure the image annotation performance by using the annotation recall and precision defined in [17]. $recall = B/C$ and $precision = B/A$, where A is the number of the images automatically annotated with a given word in the top-10-returned-word list; B is the number of the images correctly annotated with that word in the top-10-returned-word list; and C is the number of the images having that word in the ground truth annotation. An ideal image annotation system would have a high average annotation recall and annotation precision simultaneously.

4.3.4.3 Results of Automatic Image Annotation

The interface of the prototype system for automatic image annotation is shown in Figure 4.6. In this system, words and their confidence scores (conditional probabilities) are returned to annotate images upon users' querying.

Applying the method of estimating the number of the hidden concepts described in Section 4.3.2.3 to the training set, the number of the concepts is determined to be 262. Compared with the number of the images in the training set, 12,000, and the number of the stemmed and cleaned annotation words, 7736, the number of the semantic concept variables is far less. In terms of the computational complexity, the model fitting is computation intensive; it takes 45 h to fit the model to the training set on a Pentium IV 2.3 GHz computer with 1 GB memory. Fortunately, this process is performed offline and only once. For online image annotation and single-word image querying, the response time is acceptable (<1 s).

To show the effectiveness and the promise of the probabilistic model in image annotation, we have compared the accuracy of the developed method with that of MBRM [17].

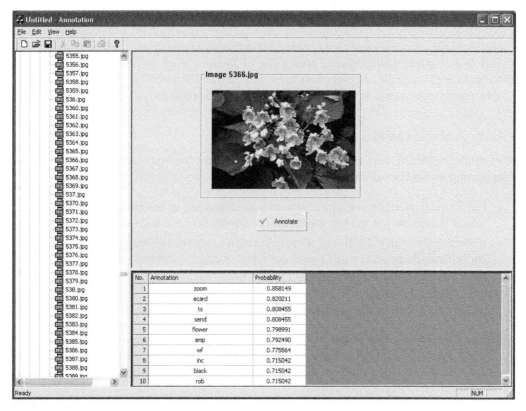

FIGURE 4.6
(**See color insert.**) The interface of the automatic image annotation prototype.

TABLE 4.1

Comparisons between the Examples of the Automatic Annotations Generated by the Developed Prototype System and MBRM

Systems	MBRM	Our Prototype
	Animal, water, wolf, house, tiger	Wolf, winter, wild, animal, stone
	Male-face, hair, people, bear, sky	Male-face, people, hair, man, monologue
	Bird, grass, leopard, sail, cuckoo	Bird, cuckoo, yellow, sand, sky
	Flower, red, tree, meadow, outdoor	Flower, red, azalea, leaf, landscape
	Desert, beach, mummy, building, church	Pyramid, Egypt, desert, mummy, beach

Source: R. Zhang et al., A probabilistic semantic model for image annotation and multi-modal image retrieval. In *Proceedings of IEEE International Conference on Computer Vision*, 2005. © (2005) IEEE Computer Society Press; With kind permission from Springer-Verlag Press: *ACM Multimedia Systems Journal*, A probabilistic semantic model for image annotation and multi-modal image retrieval. 12(1):27–33, 2006, R. Zhang et al.

In MBRM, the word probabilities are estimated using a multiple Bernoulli model, and no association layer between visual features and words is used. We compare the developed approach with MBRM because MBRM reflects the performance of the state-of-the-art automatic image annotation research. In addition, since the same image visual features are used in MBRM, a fair comparison of the performance is expected. Table 4.1 shows examples of the automatic annotation obtained by the developed prototype system and MBRM on the test image set. Here the top five words (according to probability) are taken as the automatic annotation of an image. The performance comparison demonstrated in the table clearly indicates that the developed system performs better than MBRM.

The systematic evaluation results are shown for the test set in Table 4.2. The results are reported for all (7736) words in the database. The developed approach clearly outperforms MBRM. As shown, the average recall improves by 48% and the average precision improves by 69%. The multiple Bernoulli generation of the words in MBRM is artificial and the association of the words and features is noisy. On the contrary, in the developed model no explicit word distribution is assumed, and the synergy between the visual features and the words exploited by the hidden concept variables reduces the noise substantially. We believe that these reasons account for the better performance of the developed approach. We note that certain returned words with top rank from the developed system for a given image query are found semantically relevant by subjective examinations, although they

TABLE 4.2

Performance Comparison on the Task of Automatic Image Annotation on the Test Set

Models	MBRM	Developed Model
HR3	0.56	0.83
CL	1265	574
# Words with *recall* > 0	3295	6078
Results on all 7736 words		
Average per-word recall	0.19	0.28
Average per-word precision	0.16	0.27

Source: R. Zhang et al., A probabilistic semantic model for image annotation and multi-modal image retrieval. In *Proceedings of IEEE International Conference on Computer Vision*, 2005. © (2005) IEEE Computer Society Press; With kind permission from Springer-Verlag Press: *ACM Multimedia Systems Journal*, A probabilistic semantic model for image annotation and multi-modal image retrieval, 12(1):27–33, 2006, R. Zhang et al.

are not contained in the ground truth annotation of the image. We did not count these words in the computation of the performance in Table 4.2. Consequently, the HR3, recall, and precision in the table are actually underestimated while the CL is overestimated for the developed system.

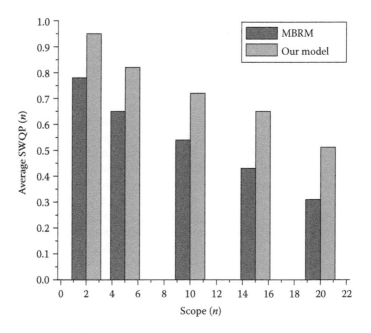

FIGURE 4.7
Average SWQP(n) comparisons between MBRM and the developed approach. (R. Zhang et al., A probabilistic semantic model for image annotation and multi-modal image retrieval. In *Proceedings of IEEE International Conference on Computer Vision*, 2005. © (2005) IEEE Computer Society Press; With kind permission from Springer-Verlag Press: *ACM Multimedia Systems Journal*, A probabilistic semantic model for image annotation and multi-modal image retrieval. 12(1):27–33, 2006, R. Zhang et al.)

4.3.4.4 Results of Single-Word Text-to-Image Querying

The single-word text-to-image querying results on a set of 500 randomly selected query words are shown in Figure 4.7. The average SWQP (2,5,10,15,20) values of the developed system and those of MBRM are reported. A returned image is considered as relevant to the single-word query if this word is contained in the ground truth annotation of the image. It is shown that the performance of the developed probabilistic model has higher overall SWQP than that of MBRM. It is also noticeable that when the scope of the returned images increases, the SWQP(n) in the developed system attenuates more gracefully than that in MBRM, which is another advantage of the developed model.

4.3.4.5 Results of Image-to-Image Querying

From Equations 4.18 and 4.21, it is clear that if we have an image query q_f, based on Equation 4.18, we immediately generate the top m annotation words based on the probability $P(w^j|q_f)$. For each of the m annotation words, based on Equation 4.21, we immediately generate an image list based on the probability $P(f_i|w^j)$. Finally, we merge the m ranked lists as the final retrieval result based on the posterior probability $P(f_i|w^j)P(w^j|q_f)$. Clearly, for a general query consisting of words and images, each component of the query may be individually processed and the final retrieval may be obtained by merging all the retrieved lists together based on the posterior probability. For the reference purpose, we call this general indexing and retrieval method UPMIR, standing for *Unified Posterior-based Multimedia Information Retrieval*. For the image-to-text annotation and single-word text-to-image retrieval, we have reported the evaluations of UPMIR against MBRM. Since the originally reported MBRM method did not include the image-to-image scenario, we report the image-to-image evaluations for UPMIR against UFM [12], a well-known image-to-image retrieval

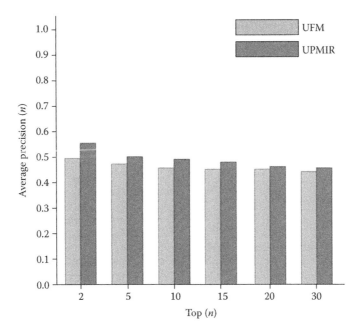

FIGURE 4.8
Precision comparison between UPMIR and UFM.

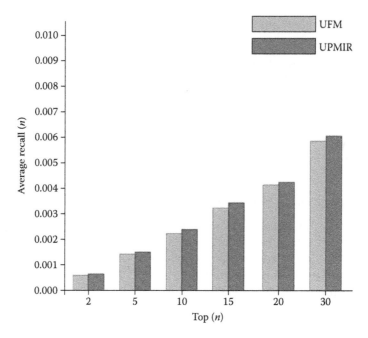

FIGURE 4.9
Recall comparison between UPMIR and UFM.

system. Figures 4.8 and 4.9 document the averaged precision and recall as the performance comparison with UFM using the image-to-image querying mode for the 600 query images on the same evaluation dataset. It is clear that with the pure image querying mode, UPMIR performs at least the same as UFM and in most cases better than UFM (e.g., with the top two images retrieved, UPMIR has a 10% better retrieval precision than UFM). To further demonstrate that UPMIR is also more efficient than UFM in image retrieval, we note that UPMIR and UFM are both implemented and evaluated in the same environment of a Pentium IV 2.26 GHz CPU with 1 GB memory. Given the scale of 17,000 images and 7736 word vocabulary, the average response time for each query for UPMIR is 0.936 s, while that for UFM is 9.14 s. Clearly, UPMIR beats UFM substantially. This is due to the fact that UPMIR has much lower complexity than UFM, as UFM is a region-based approach and for each comparison between two images UFM requires a combinatorial complexity, while for UPMIR it is only a constant complexity.

4.3.4.6 *Results of Performance Comparisons with Pure Text Indexing Methods*

Since the UPMIR performance is biased toward the text component querying, we intend to experimentally justify and demonstrate that UPMIR still offers a better image retrieval than a pure text indexing scheme. For this purpose, we manually evaluate UPMIR using the pure text querying mode (i.e., single-word text-to-image querying mode) against Google and Yahoo!. We randomly select 20 words out of the 7736 word vocabulary and use each of them as a pure text query to pose to UPMIR, Google image search, and Yahoo! image search, respectively. We manually examine the precisions. Figure 4.10 clearly demonstrates that UPMIR outperforms Google and Yahoo! for different numbers of the top images retrieved.

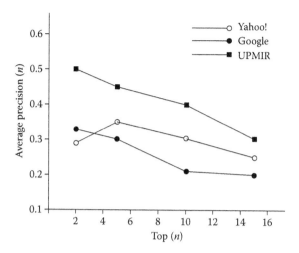

FIGURE 4.10
Average precision comparison among UPMIR, Google Image Search, and Yahoo! Image Search.

Since we do not have access to Google or Yahoo! image databases, though the comparing databases are different in size and content, this is the best we can do to compare their performances. The purpose is to show that a multimodal image mining and retrieval system such as UPMIR still has clear advantages for image retrieval over a pure text-based indexing system.

4.4 Summary

In this chapter we have introduced the new, emerging area called multimedia information mining. We have given a working definition of what this area is about; we have corrected a few misconceptions that typically exist in the related research communities; and we have given a typical architecture for a multimedia information mining sytem or methodology. Finally, in order to showcase what a typical multimedia information mining system does and how it works, we have given an example of a specific method for semantic concept discovery in an image annotation and multimodal retrieval application.

Multimedia information mining, despite being a new and emerging area, has undergone an independent and rapid development over the last few years. For example, a systematic introduction to the related multimedia data mining area may be found in [1] as well as the references of this chapter.

Acknowledgments

This work is supported in part by the US National Science Foundation through grants IIS-0535162, IIS-0812114, and CCF-1017828, and the National Basic Research Program of China (973 Program) through grants 2012CB316406. Any opinions, findings, and conclusions or recommendations expressed in this material are those of the authors and do not necessarily reflect the views of the funding agencies. Zhen Guo of Yahoo! Labs; Mingjing Li, Wei-Ying Ma, and Hong-Jiang Zhang of Microsoft Research contributed to part of this work to a certain degree.

References

1. Z. Zhang and R. Zhang. *Multimedia Data Mining—A Systematic Introduction to Concepts and Theory.* Taylor & Francis, Boca Raton, FL, 2008.
2. G. Salton. *Automatic Text Processing: The Transformation, Analysis, and Retrieval of Information by Computer.* Addison-Wesley, 1989.
3. B. Furht, ed. *Multimedia Systems and Techniques.* Kluwer Academic Publishers, 1996.
4. R. Steinmetz and K. Nahrstedt. *Multimedia Fundamentals—Media Coding and Content Processing.* Prentice-Hall PTR, 2002.
5. C. Faloutsos, R. Barber, M. Flickner, J. Hafner, W. Niblack, D. Petkovic, and W. Equitz. Efficient and effective querying by image content. *Journal of Intelligent Information Systems,* 3(3/4):231–262, 1994.
6. R. Jain. Infoscopes: Multimedia information systems. In B. Furht, ed., *Multimedia Systems and Techniques.* Kluwer Academic Publishers, 1996.
7. V. S. Subrahmanian. *Principles of Multimedia Database Systems.* Morgan Kaufmann, 1998.
8. C. Faloutsos. *Searching Multimedia Databases by Content.* Kluwer Academic Publishers, 1996.
9. J. Han and M. Kamber. *Data Mining—Concepts and Techniques.* 2nd edition. Morgan Kaufmann, 2006.
10. A. W. M. Smeulders, M. Worring, S. Santini, A. Gupta, and R. Jain. Content-based image retrieval at the end of the early years. *IEEE Transactions on Pattern Analysis and Machine Intelligence,* 22: 1349–1380, 2000.
11. C. Carson, M. Thomas, S. Belongie, J. M. Hellerstein, and J. Malik. Blobworld: A system for region-based image indexing and retrieval. In *The 3rd International Conference on Visual Information System Proceedings,* pp. 509–516, Amsterdam, Netherlands, June 1999.
12. Y. Chen and J. Z. Wang. A region-based fuzzy feature matching approach to content-based image retrieval. *IEEE Transactions on PAMI,* 24(9):1252–1267, 2002.
13. J. Z. Wang, J. Li, and G. Wiederhold. Simplicity: Semantics-sensitive integrated matching for picture libraries. *IEEE Transactions on PAMI,* 23(9), 2001.
14. Z. Zhang, R. Zhang, and J. Ohya. Exploiting the cognitive synergy between different media modalities in multimodal information retrieval. In *the IEEE International Conference on Multimedia and Expo (ICME'04),* Taipei, Taiwan, July 2004.
15. K. Barnard, P. Duygulu, N. de Freitas, D. Blei, and M. I. Jordan. Matching words and pictures. *Journal of Machine Learning Research,* 3:1107–1135, 2003.
16. P. Duygulu, K. Barnard, J. F. G. d. Freitas, and D. A. Forsyth. Object recognition as machine translation: Learning a lexicon for a fixed image vocabulary. In *the 7th European Conference on Computer Vision,* Vol. IV, pp. 97–112, Copenhagon, Denmark, 2002.
17. S. L. Feng, R. Manmatha, and V. Lavrenko. Multiple bernoulli relevance models for image and video annotation. In *the International Conference on Computer Vision and Pattern Recognition,* Washington, DC, June, 2004.
18. J. Li and J. Z. Wang. Automatic linguistic indexing of pictures by a statistical modeling approach. *IEEE Transactions on PAMI,* 25(9), 2003.
19. T. Westerveld and A. P. de Vries. Experimental evaluation of a generative probabilistic image retrieval model on "easy" data. In *The SIGIR Multimedia Information Retrieval Workshop 2003,* August 2003.

20. W. R. Dillon and M. Goldstein. *Multivariate Analysis, Mehtods and Applications.* John Wiley and Sons, New York, 1984.
21. G. Mclachlan and K. E. Basford. *Mixture Models.* Marcel Dekker, Inc., Basel, New York, 1988.
22. A. Dempster, N. Laird, and D. Rubin. Maximum likelihood from incomplete data via the em algorithm. *Journal of the Royal Statistical Society, Series B*, 39(1):1–C38, 1977.
23. J. Rissanen. *Stochastic Complexity in Statistical Inquiry.* World Scientific, 1989.
24. G. Fishman. *Monte Carlo Concepts, Algorithms and Applications.* Springer-Verlag, 1996.
25. D. Cai, S. Yu, J.-R. Wen, and W.-Y. Ma. VIPS: A vision-based page segmentation algorithm. *Microsoft Technical Report* (MSR-TR-2003-79), 2003.
26. R. Zhang, Z. Zhang, M. Li, W.-Y. Ma, and H.-J. Zhang. A probabilistic semantic model for image annotation and multi-modal image retrieval. In *Proceedings of IEEE International Conference on Computer Vision*, 2005.
27. R. Zhang, Z. Zhang, M. Li, W.-Y. Ma, and H.-J. Zhang. A probabilistic semantic model for image annotation and multi-modal image retrieval. *ACM Multimedia Systems Journal*, 12(1):27–33, 2006.

5

Information Fusion for Multimodal Analysis and Recognition

Yongjin Wang, Ling Guan, and Anastasios N. Venetsanopoulos

CONTENTS

5.1 Introduction

The advances in sensing technology and the proliferation of multimedia content have motivated the design and development of computationally efficient and economically feasible multimodal systems for a broad spectrum of applications. Multimedia contain a combination of information content from different media sources in various content forms, such as audio, video, image, and text, each of which can be deemed as a modality in a multimodal multimedia representation. The integration of multimodal data contains more information about the semantics presented in the media, and may provide a more complete and effective pattern representation. The objective of information fusion is to combine different modalities for accomplishing more accurate pattern analysis and recognition.

In the past two decades, significant improvement has been accomplished in various facets of multimedia. In particular, the development of intelligent multimedia systems have drawn increasingly extensive interest in both research and industrial sectors, in a plethora of applications such as security and surveillance, video conferencing, video streaming, education and training, health care, database management, and human–computer interaction. Human-centered computing (HCC), which embraces applications that are associated with direct interaction between human and computing technology, is an emerging field that aims at bridging the gaps between the various disciplines that are associated with the design of computing systems that support people's activities. The ultimate goal in the design and development of an intelligent HCC system is to create a more secure, friendly, convenient, and efficient living and working environment for people.

A great deal of effort has been put into the design of intelligent HCC systems in the past, under a wide variety of application contexts. However, most of the existing works focus on one modality only. Such unimodal systems usually afford low-level performance due to the drastic variation and noisy nature of the data. The main characteristics of human communication are the multiplicity and multimodality of communication channels [18]. Examples of human communication channels include auditory channels that carry speech and vocal intonation, and visual channels that carry facial expressions, gestures, and gait. In many real-life communication scenarios, different channels are activated at the same time, and thus the communication is very flexible and robust. The information presented in different channels, when being utilized for the description of the same perception, may provide complementary characteristics, and it is highly probable that a more complete, precise, and discriminatory representation can be derived by incorporating multiple media sources [19].

The design of a multimodal system is critically dependent on the characteristics of the data as well as the requirement of the application. Many multimodal systems have been introduced after Bolt's pioneering work in voice and gesture [10]. A set of key points that need to be considered include [33]: information sources, feature extraction method, fusion level, fusion strategy, system architecture, and whether any background knowledge needs to be embedded. The selection of information sources is determined by the requirements of the application. In general, the selected modalities should be capable of providing discriminatory patterns for classification, can be collected quantitatively, and from a practical point of view, the system should be of low cost and computationally efficient. Some applications may have specific qualifications, such as the universality, stability, and user-acceptability requirements for biometrics [27]. For example, a multimodal emotion recognition system may utilize speech and video for security and consumer service-related applications, and can be possibly integrated with some biological signals, such as electrocardiogram (ECG) and electroencephalography (EEG) in a medical environment. A multimodal biometric system can utilize biometric traits such as fingerprint and iris for authentication, but face and gait may be more appropriate for surveillance applications.

The effective integration of multimodal information is a challenging problem, with the major difficulties primary to the identification of the complementary relationship between different modalities, and the effective fusion of information from multiple channels. In general, the fusion strategy should be capable of taking full advantage of information collected from multiple sources and bearing a better description of the intended perception. An ill-designed multimodal system will possibly produce degraded performance and lowered feasibility. This chapter introduces the fundamentals of information fusion, reviews the state of the art of fusion strategies, and outlines the issues and challenges that are associated with the design of a multimodal system. Furthermore, we introduce kernel-based fusion methods, and examine their feasibility in an audiovisual-based human emotion recognition problem.

The remainder of this chapter is organized as follows. Section 5.2 provides a review of multimodal information fusion and related works according to their levels of fusion. Section 5.3 introduces kernel-based fusion strategies. The proposed bimodal emotion recognition system is outlined in Section 5.4, followed by experimental evaluation presented in Section 5.5. A summary is provided in Section 5.6.

5.2   Information Fusion Levels and Strategies

A pattern recognition system usually consists of four basic modules: sensor, feature extraction, classification, and decision. Accordingly, it is possible to combine the information collected from different modalities at any of the modules. That is, the fusion of the multimodal information can be performed at the sensor level, feature level, score level, or decision level, as depicted in Figure 5.1. The selection of an information fusion level is dependent on the specific application context and requirement. In some cases, a hybrid fusion strategy which involves the combination of different fusion levels may also be used to take advantage of the benefits of individual levels. In the following, we discuss the fusion strategies at each of the levels.

5.2.1   Sensor-Level Fusion

Sensor-level fusion involves the combination of raw data/samples before performing feature extraction. It is the earliest stage of information fusion. A simple example is to average the intensity values of multiple sensed images of the same object, and the subsequent feature extraction and classification are based on the resulting one single composite image. Fusion at the sensor level has been mainly investigated in object detection and tracking, remote sensing image fusion, Global Positioning System (GPS) and Inertial Navigation System (INS) navigation, and multisensor biometric recognition [23].

In [8], many multisensor-based fusion methods have been introduced for target tracking applications. The extended Kalman filter (EKF), which utilizes the multiple raw measurements from multiple sensors for the estimation of the states of the target, is one of the most widely employed techniques. Sasiadek [47] presented a review of different approaches for fusing GPS and INS sensors. Generally, the fusion methods can be categorized as probabilistic model, least-square-based technique (e.g., variants of EKF, optimal theory),

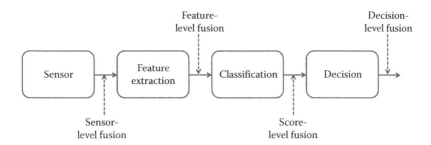

FIGURE 5.1
Multimodal information fusion levels.

and intelligent fusion (e.g., fuzzy logic, neural network, genetic algorithm). Lee et al. [35] combined multiple snapshots of a face that has been taken by cameras positioned at different locations to generate three-dimension (3D) facial expressions. Bowyer et al. [11] have combined multiple two-dimension (2D), 3D, and infrared face images for biometric recognition, and demonstrated improved recognition performance.

The raw data that are acquired directly from the sensor contain the richest information content. It will provide the most accurate analysis and recognition results if the information can be appropriately utilized. Information fusion at the sensor level also provides storage advantage since only one or few signal samples are generated. However, the raw measurements usually contain various noise due to imperfect acquisition conditions. Therefore, enhancement or quality improvement procedure is required. One major limitation of sensor-level fusion is that it requires that the samples from different sensors should be compatible [29], that is, in the same or integratable format, for example, multiple image samples of a face or a fingerprint. But for applications that involve the combination of signals with disparate characteristics, such as a face and a fingerprint, or a speech signal and an image, it is difficult to perform fusion at the sensor level.

5.2.2 Feature-Level Fusion

Feature-level fusion combines the extracted features from each modality through a fusion strategy. The resulting features, which are usually represented as a vector, are taken as the input to a classifier for recognition purpose. Examples of classification algorithms include nonparametric techniques such as k-nearest neighbors, Parzen window; stochastic methods such as Bayesian classifier, maximum likelihood, and hidden Markov model (HMM); soft computing techniques such as neural networks and support vector machines (SVM); and other nonmetric-based approaches such as decision trees and rule-based methods [15].

The extraction of features that can truly capture the universal characteristics of the intended perception, as well as offering distinctive representation against other perceptions, is of fundamental importance in a pattern recognition problem. In general, a preprocessing step is first applied to the original signal to reduce noise or detect the region of interest. For example, speech noise reduction to remove the background noise, and face detection to localize the facial region. Feature extraction techniques are then employed on the preprocessed signal to extract a set of attributes for representing the characteristics of the original signal. For instance, the Mel-frequency cepstral coefficient (MFCC) and linear predictive coding features for speech, color and shape features for images, and motion vectors for video sequence. Note that depending on the specific problem, even for the same type of signal, different feature extractors are needed. For example, prosodic features are usually believed to be the major indicators of human emotion in speech [53], while phonetic features are usually used for speech recognition [6]. In addition, when a large number of features are extracted, a feature selection/dimensionality reduction procedure might be necessary to reduce the dimensionality of the feature space, whilst selecting those most discriminatory ones. Representative techniques of dimensionality reduction include the unsupervised principal component analysis (PCA), the supervised linear discriminant analysis, and their variants. Example methods of feature selection are exhaustive search, branch and bound, forward selection and backward elimination, and genetic algorithm.

Many feature-level fusion-based works have been proposed on different pattern analysis problems, using various signal modalities and feature extractors. One simple and widely adopted approach of feature-level fusion is to concatenate the attributes from multiple modalities into one vector. Representative works along this line include biometric recognition [13,34,45], audiovisual emotion recognition [54]. For video-based analysis, a popular approach is the graphical model-based methods to capture the temporal structure and ensemble characteristics. Graphical models (e.g., HMM and its variants) represent the observations or states of different modalities as the nodes, and use the edges to represent their probabilistic dependencies [4]. Some representative works include audiovisual-based video shot classification [56], speech recognition [28,41], speaker localization [22,42], biometric recognition [7], object tracking [5], and human emotion recognition [59]. More recently, a few works have also been proposed to perform correlation analysis using canonical correlation analysis (CCA) [12,46,52] and its kernelized version KCCA [60]. Such methods identify a set of optimal transformations, either linearly or nonlinearly, to maximize the correlation between different modalities. They are capable of finding the features that contribute to the intended perception in all the modalities, and have demonstrated improved performance for their specific analysis problems.

Feature-level fusion provides several advantages. First, the features contain rich information about the perception of interest. Second, an appropriate designed feature extractor can reduce the nonrelevant or redundant information presented in the original signal. One major drawback of fusion at this level is due to the problem of "curse of dimensionality," which is usually computationally expensive and requires a large set of training data [32]. Furthermore, the fusion problem may be difficult due to the disparate characteristics of features extracted from different modalities, such as the minute points for fingerprints and the PCA features for face images, which have different numerical range and are usually evaluated using different metrics.

5.2.3 Score-Level Fusion

Score-level fusion combines the classification scores generated from multiple classifiers using multiple modalities. Each modality functions as an expert system with its own procedure of feature extraction and classification. The scores obtained from each modality provides a measure of similarity between an input sample and those belonging to a certain class. The integrated score values from multiple modalities may offer a better classification performance by considering multiple aspects of a perception. A score normalization process is usually required to scale the scores generated by different modalities in the same range, such that no single modality will overpower the others, and the significance of each individual modality is leveraged in the final decision. Typical score normalization techniques include min–max and z-score. A detailed study on score normalization method can be found in [26].

Generally, score-level fusion can be grouped into two categories: rule-based methods and pattern classification approaches. In the rule-based scheme, the multiple normalized classification scores are combined by using some predefined rules, such as maximum, minimum, sum [24], weighted sum [43], product [3], logistic regression [38], and majority voting [31]. Alternatively, the score-level fusion can be conducted in a pattern classification sense in which the scores are taken as features into a classification algorithm. Examples of pattern classification-based score-level fusion include fusing audio, visual, and text for semantic indexing [1], audiovisual-based emotion recognition [39], and multimodal biometrics [2].

The design of a score-level fusion-based multimodal system is flexible and scalable, since each individual modality/source can be processed independently to find the best feature representation and classifier of that modality with respect to the specific analysis problem. Particularly, it provides a feasible solution for many applications in which the recognition task needs to be carried out even when only one modality presents. For example, for audiovisual-based emotion analysis, when the audio channel is silent, the recognition scores from the visual channel can still be used for recognition. In addition, fusion at the score level is simple and easy to combine and understand. However, the scores represent the property of an input sample with respect to the class information, but do not provide sufficient information about the original signal characteristic, that is, it contains limited information. When different modalities are independent of each other, score-level fusion may provide a more effective solution. But for dependent sources, it will be hard to model the dependency between different modalities using the scores.

5.2.4 Decision-Level Fusion

Decision-level fusion generates the final results based on the decision from multiple modalities. Each modality has standalone feature extraction, classification, and decision modules, and makes independent local decision about the class that an input sample belongs to. It is the latest level of fusion, and is usually performed by rule-based methods, such as AND/OR [25], and majority voting [61]. For example, in a biometric verification system, the AND rule will output an acceptance if and only if all the modalities make the same decision. On the other hand, the OR rule will decide a match if any of the modalities produces a match. The majority voting produces a decision based on the majority of individual matches. It is usually used for applications with more than two modalities or classifiers. Similar to the score-level fusion, information fusion at the decision level is flexible for system design, with the integrity of individual expert systems being well preserved. However, the decisions are usually in binary zeros or ones. The fusion at this level is very rigid due to the limited information left.

5.2.5 Remarks

The selection of a fusion level is dependent on the characteristics of the data and the requirements of the application problem. Generally, sensor-level fusion might be more appropriate for applications that involve a large number of sensors capturing the same concept or perception. For example, multiple images collected from multiple cameras positioned at different locations to capture the same object. These images can be fused to generate a single 3D image. Feature-level fusion is one of the most effective solutions, particularly when the multiple sources are not independent. It preserves the important information, and can be applied to identify the complementary relationship between different modalities. On the other hand, score-level fusion might be a more appropriate choice for applications with independent sources, such as the fusion of fingerprint and face images for biometric recognition. Decision-level fusion, although contains the least information, can be used in applications that employ independently developed single-modality systems. For example, in many commercial biometrics-related product, only the decisions are available. In such cases, a decision-level fusion can be devised to incorporate products from different venders without worrying about their compatibility.

5.3 Kernel-Based Information Fusion

Despite the extensive studies and progress made in the past, information fusion is still a difficult problem far from being solved. We believe that one of the major issues in information fusion is to identify the complementary and associative relationship between different modalities. In this section, we introduce kernel-based fusion methods for addressing these problems.

5.3.1   Kernel Machine Technique

Kernel method is considered as one of the most important tools in the design of nonlinear feature extraction and classification techniques, and has demonstrated its success in various pattern analysis problems. The premise behind the kernel method is to find a nonlinear mapping from the original space ($\mathbb{R}^J$) to a higher-dimensional kernel feature space $\mathbb{F}^F$ by using a nonlinear function $\phi(\cdot)$, that is,

$$\phi : \mathbf{z} \in \mathbb{R}^J \to \phi(\mathbf{z}) \in \mathbb{F}^F \quad J < F \leq \infty. \tag{5.1}$$

In the kernel feature space $\mathbb{F}^F$, the pattern distribution is expected to be simplified so that better classification performance can be achieved by applying traditional linear methodologies [40]. In general, the dimensionality of the kernel space is much larger than that of the original input space, sometimes even infinite. Therefore, an explicit determination of the nonlinear mapping ϕ is either difficult or intractable. Fortunately, with the so-called "kernel trick," the nonlinear mapping can be performed implicitly in the original space $\mathbb{R}^J$ by replacing dot products of the feature representations in $\mathbb{F}^F$ with a kernel function defined in $\mathbb{R}^J$. Let $\mathbf{z}_i \in \mathbb{R}^J$ and $\mathbf{z}_j \in \mathbb{R}^J$ represent two vectors in the original feature space, the dot product of their feature representations $\phi(\mathbf{z}_i) \in \mathbb{F}^F$, $\phi(\mathbf{z}_j) \in \mathbb{F}^F$ can be computed by a kernel function $\kappa(\cdot)$ defined in $\mathbb{R}^J$, that is,

$$\phi(\mathbf{z}_i) \cdot \phi(\mathbf{z}_j) = \kappa(\mathbf{z}_i, \mathbf{z}_j) \tag{5.2}$$

The function selected as the kernel function should satisfy the Mercer's condition, that is, positive semidefinite [40]. Some commonly used kernel functions include the linear kernel $\kappa(\mathbf{z}_i, \mathbf{z}_j) = <\mathbf{z}_i, \mathbf{z}_j>$, the Gaussian kernel $\kappa(\mathbf{z}_i, \mathbf{z}_j) = \exp(\frac{-||\mathbf{z}_i - \mathbf{z}_j||^2}{2\sigma^2})$, and the polynomial kernel $\kappa(\mathbf{z}_i, \mathbf{z}_j) = (<\mathbf{z}_i, \mathbf{z}_j> +1)^d$.

Kernel principal component analysis (KPCA) [48] is one of the most typical kernel techniques. KPCA is an implementation of the traditional PCA algorithm in the kernel feature space. Let C denote the number of classes and C_i the number of samples of the ith class; then the covariance matrix $\tilde{\mathbf{S}}_t$ defined in $\mathbb{F}^F$ could be expressed as follows:

$$\tilde{\mathbf{S}}_t = \frac{1}{N} \sum_{i=1}^{C} \sum_{j=1}^{C_i} \left(\phi(\mathbf{z}_{ij}) - \bar{\phi}\right) \left(\phi(\mathbf{z}_{ij}) - \bar{\phi}\right)^T \tag{5.3}$$

where $\bar{\phi} = \frac{1}{N} \sum_{i=1}^{C} \sum_{j=1}^{C_i} \phi(\mathbf{z}_{ij})$ is the mean of training samples in $\mathbb{F}^F$. The KPCA subspace is spanned by the first d significant eigenvectors of $\tilde{\mathbf{S}}_t$, denoted as $\tilde{\mathbf{W}}_{\text{KPCA}}$, corresponding to d largest eigenvalues, that is, $\tilde{\mathbf{W}}_{\text{KPCA}} = [\tilde{\mathbf{w}}_1, \ldots, \tilde{\mathbf{w}}_d]$ and $\tilde{\lambda}_1 > \tilde{\lambda}_2, \ldots, \tilde{\lambda}_d$, where $\tilde{\mathbf{w}}_k$ is the eigenvector and $\tilde{\lambda}_k$ is the corresponding eigenvalue. The nonlinear mapping to a high-dimensional subspace can be solved implicitly using the kernel trick. Given n samples

$\{\mathbf{z}_1, \mathbf{z}_2, \ldots, \mathbf{z}_n\}$, the kernel matrix K can be computed as

$$
K = \begin{bmatrix}
\kappa(\mathbf{z}_1, \mathbf{z}_1) & \kappa(\mathbf{z}_1, \mathbf{z}_2) & \cdots & \kappa(\mathbf{z}_1, \mathbf{z}_n) \\
\kappa(\mathbf{z}_2, \mathbf{z}_1) & \kappa(\mathbf{z}_2, \mathbf{z}_2) & \cdots & \kappa(\mathbf{z}_2, \mathbf{z}_n) \\
\vdots & \vdots & \ddots & \vdots \\
\kappa(\mathbf{z}_n, \mathbf{z}_1) & \kappa(\mathbf{z}_n, \mathbf{z}_2) & \cdots & \kappa(\mathbf{z}_n, \mathbf{z}_n)
\end{bmatrix}
\tag{5.4}
$$

We can center the kernel matrix by

$$
\tilde{K} = K - I_n K - K I_n - I_n K I_n,
\tag{5.5}
$$

where I_n is an $n \times n$ matrix with the elements set to $1/n$. Then the solution to the above problem can be formulated as the following eigenvalue problem,

$$
\tilde{K} \tilde{\mathbf{u}}_i = \tilde{\lambda}_i \tilde{\mathbf{u}}_i.
\tag{5.6}
$$

For a testing sample $\mathbf{z}'$, its projection on the ith eigenvector is computed as

$$
\tilde{z}'_i = \tilde{\mathbf{u}}_i^T \begin{bmatrix}
\kappa(\mathbf{z}', \mathbf{z}_1) \\
\kappa(\mathbf{z}', \mathbf{z}_2) \\
\vdots \\
\kappa(\mathbf{z}', \mathbf{z}_n)
\end{bmatrix}
\tag{5.7}
$$

5.3.2 Kernel Matrix Fusion

In the kernel-based method, the kernel matrix is an embedding of the original features to the kernel feature space, with each entry representing a certain notion of similarity between two specific patterns in a higher-dimensional space. It is possible to combine different kernel matrices using algebraic operations, such as addition and multiplication. Such operations still preserve the positive semidefiniteness of the kernel matrix. In many multimodal analysis problems, different channels capture different aspects of the same semantic. The kernel matrix derived from each channel therefore provides a partial description of the specific information of one sample to the others. The kernel fusion technique allows for integrating the respective kernel matrices to identify their joint characteristic pertaining to the associated semantic, which may provide a more discriminatory representation. It incorporates features of disparate characteristics into a common format of kernel matrices, and therefore provides a viable solution of fusing heterogenous data. In addition, kernel-based fusion offers a flexible solution since different kernel functions can be used for different modalities.

In this chapter, we examine and compare two basic algebraic operations, weighted sum and multiplication. Let $\mathbf{x}_1, \mathbf{x}_2, \ldots, \mathbf{x}_n$ and $\mathbf{y}_1, \mathbf{y}_2, \ldots, \mathbf{y}_n$ denote the information extracted two different modalities, respectively, then the kernel matrices K^x and K^y can be computed as $K^x_{ij} = \kappa^x(\mathbf{x}_i, \mathbf{x}_j)$ and $K^y_{ij} = \kappa^y(\mathbf{y}_i, \mathbf{y}_j)$. The fused kernel matrix K^f can be computed as

$$
K^f_{ij} = a K^x_{ij} + b K^y_{ij}, \quad a + b = 1,
\tag{5.8}
$$

$$
K^f_{ij} = K^x_{ij} \times K^y_{ij},
\tag{5.9}
$$

Substituting the fused kernel K^f into Equations 5.5 and 5.6, we can compute the KPCA eigenvectors $(\tilde{\mathbf{u}}_1, \tilde{\mathbf{u}}_2, \ldots, \tilde{\mathbf{u}}_d)$. For a pair of testing sample $(\mathbf{x}', \mathbf{y}')$, the projection to the ith eigenvector in the joint space will be computed as follows:

$$\tilde{z}'_i = \tilde{\mathbf{u}}_i^T \begin{bmatrix} a\kappa^x(\mathbf{x}', \mathbf{x}_1) + b\kappa^y(\mathbf{y}', \mathbf{y}_1) \\ a\kappa^x(\mathbf{x}', \mathbf{x}_2) + b\kappa^y(\mathbf{y}', \mathbf{y}_2) \\ \vdots \\ a\kappa^x(\mathbf{x}', \mathbf{x}_n) + b\kappa^y(\mathbf{y}', \mathbf{y}_n) \end{bmatrix}, \tag{5.10}$$

$$\tilde{z}'_i = \tilde{\mathbf{u}}_i^T \begin{bmatrix} \kappa^x(\mathbf{x}', \mathbf{x}_1) \times \kappa^y(\mathbf{y}', \mathbf{y}_1) \\ \kappa^x(\mathbf{x}', \mathbf{x}_2) \times \kappa^y(\mathbf{y}', \mathbf{y}_2) \\ \vdots \\ \kappa^x(\mathbf{x}', \mathbf{x}_n) \times \kappa^y(\mathbf{y}', \mathbf{y}_n) \end{bmatrix}. \tag{5.11}$$

5.3.3 Kernel Canonical Correlation Analysis

CCA [21] is a statistical tool to find projection directions such that the correlation between two multidimensional variables can be maximized. Let $X \in \mathbb{R}^{n \times p}$ and $Y \in \mathbb{R}^{n \times q}$ denote two mean centralized matrices with each row being a feature vector, n represents the total number of samples, and p and q represent the dimensionality of the two feature streams, respectively. The corresponding rows of X and Y, denoted as $\{(\mathbf{x}_1, \mathbf{y}_1), (\mathbf{x}_2, \mathbf{y}_2), \ldots, (\mathbf{x}_n, \mathbf{y}_n)\}$, constitute pairs of representations with respect to certain semantic, from two different modalities. The linear CCA method tries to find projection matrices $\alpha = (\alpha_1, \alpha_2, \ldots, \alpha_d)$ and $\beta = (\beta_1, \beta_2, \ldots, \beta_d)$, $d \leq \min(p, q)$, through optimization of the following criterion:

$$\max_{\alpha, \beta} \frac{\alpha^T C_{xy} \beta}{\sqrt{\alpha^T C_{xx} \alpha} \sqrt{\beta^T C_{yy} \beta}}, \tag{5.12}$$

where $C_{xx} = X^T X$, $C_{yy} = Y^T Y$, and $C_{xy} = X^T Y$ denote the within- and between-set covariance matrices, respectively. This can be solved by applying a series of optimization problem successively to determine the basis vector pairs $\{(\alpha_1, \beta_1), (\alpha_2, \beta_2), \ldots, (\alpha_d, \beta_d)\}$:

$$\max_{\alpha_i, \beta_i} \alpha_i^T C_{xy} \beta_i, \tag{5.13}$$

subject to: $\alpha_i^T C_{xx} \alpha_i = 1, \beta_i^T C_{yy} \beta_i = 1$.

By using the Lagrangian method, the maximization of the above equation can be solved by the following eigenvalue problem:

$$C_{xx}^{-1} C_{xy} C_{yy}^{-1} C_{yx} \alpha_i = \lambda^2 \alpha_i$$

$$C_{yy}^{-1} C_{yx} C_{xx}^{-1} C_{xy} \beta_i = \lambda^2 \beta_i \tag{5.14}$$

The CCA approach can be kernelized to KCCA using the kernel trick, in which the optimization criterion becomes

$$\max_{\alpha, \beta} \frac{\alpha^T K^x K^y \beta}{\sqrt{\alpha^T (K^x)^2 \alpha} \sqrt{\beta^T (K^y)^2 \beta}}, \tag{5.15}$$

where K^x and K^y denote the kernel matrices. Similar to the CCA method, this can be solved as a generalized eigenvalue problem. However, when K^x or K^y are not invertible, the optimization of the above equation will lead to degenerate solutions. To solve this problem, a regularized KCCA can be performed by maximizing the following criterion [9]:

$$\max_{\alpha, \beta} \frac{\alpha^T K^x K^y \beta}{\sqrt{\alpha^T((1-\tau)(K^x)^2 + \tau K^x)\alpha}\sqrt{\beta^T((1-\tau)(K^y)^2 + \tau K^y)\beta}}, \tag{5.16}$$

where $0 \leq \tau \leq 1$ is a regularization parameter.

For a pair of feature vectors $\mathbf{x}'$ and $\mathbf{y}'$ with kernel functions $\kappa^x(\cdot)$ and $\kappa^y(\cdot)$ respectively, their projections onto the ith eigenvectors of α and β can be computed as

$$\tilde{x}'_i = \alpha_i^T \begin{bmatrix} \kappa^x(\mathbf{x}', \mathbf{x}_1) \\ \kappa^x(\mathbf{x}', \mathbf{x}_2) \\ \vdots \\ \kappa^x(\mathbf{x}', \mathbf{x}_n) \end{bmatrix}, \quad \tilde{y}'_i = \beta_i^T \begin{bmatrix} \kappa^y(\mathbf{y}', \mathbf{y}_1) \\ \kappa^y(\mathbf{y}', \mathbf{y}_2) \\ \vdots \\ \kappa^y(\mathbf{y}', \mathbf{y}_n) \end{bmatrix} \tag{5.17}$$

5.4 Emotion Recognition System

Emotion plays an important role in our daily social interactions and activities. Machine recognition of human emotion has been an increasingly important problem for accomplishing the goal of building a more natural and friendly communication interface between humans and computers. It can find applications in a wide spectrum of context such as intelligent human–computer interaction, security and surveillance, consumer relations, entertainment, and education.

The affective intention of an individual can be inferred from different sources such as voice, facial expressions, body language, semantic meaning of the speech, ECG, and EEG. Among various modalities, vocal and facial expressions are two of the most natural, passive, and noninvasive types of traits, and they are primaries to the objectives of an emotion recognition system in the field of human–computer interaction. Moreover, they can be easily captured by low-cost sensing devices, which makes them economically more feasible for potential deployment in a wide range of applications.

Many tentative solutions have been proposed in the literature for machine recognition of human emotion. The majority of the early works focus on either speech alone or facial expression only. However, due to the drastic variation and the noise presented in the signals, such unimodal systems usually afford low-recognition performance. The inefficiency of unimodel approaches can be partially addressed by incorporating multimodal information. De Silva et al. [49] first reported human perception of emotion, and demonstrated that some emotions (e.g., sadness, fear) are audio dominant, some (e.g., happiness, surprise, anger) are visual dominant, while others (e.g., disgust) are mixed dominant. Their results actually echo the intrinsic properties of human emotion, and imply the possibility of improving recognition performance by the integration of audiovisual information. De Silva and Ng [50] proposed a rule-based fusion at the decision level to combine a prosodic feature-based audio system and a local feature-based video system. Go et al. [17] presented a rule-based method for fusion of audio and video recognition results, with both using wavelet-based techniques for feature extraction. A semantic data fusion model at the decision level was introduced in

[14], where prosodic features are used for audio representation and an active appearance model for facial expression analysis. Kanluan et al. [30] used prosodic and spectral features for audio representation, 2D discrete Cosine transform on predefined image blocks for visual feature extraction, and a weighted linear combination for fusion at the decision level. Score-level fusion was investigated in [39] where a Gaussian mixture model is used for audio and video feature extraction, and a Bayesian classifier and an SVM are employed for classifying the scores. Han et al. [20] also presented an SVM-based score-level fusion scheme with prosodic and local features for audio and visual channels, respectively. Several HMM-based methods have been introduced in [51,57,59], with different approaches for feature extraction. Wang and Guan [55] proposed a hierarchical multi-classifier scheme for the analysis of combinations of different classes, and the fusion of multimodal data was performed at the feature level. A survey of the state of the art in emotion recognition are discussed in [58].

Existing works have demonstrated that the performance of emotion recognition systems can be improved by integrating audio and visual information. However, it is far from a solved problem due to the limits in the accuracy, efficiency, and generality of the proposed systems. In addition, most of these works treat audio and visual channels as independent sources, and fail to identify their joint characteristics and dependency. In this section, we present the proposed audiovisual-based multimodal emotion recognition system using kernel-based fusion. The proposed system analyzes the audio and visual information presented in a short time window **w**, and the features from successive windows are then modeled by an HMM for capturing the change of audio and visual information with respect to time. In the following, we detail the feature extraction, fusion, and classification methods.

5.4.1 Feature Extraction

5.4.1.1 Audio Feature Representation

For the purpose of obtaining a better representation of the patterns embedded in the original signal, it is necessary to perform noise reduction first, such that the negative impact due to the "hiss" of the recording machine and background noise can be minimized. In this work, we perform noise reduction using a wavelet coefficient thresholding methods on every windowed audio signal of size **w** [53]. Traditional low-pass filtering methods, which are linear time invariant, can blur the sharp features in a signal. For wavelets, which are nonlinear functions, the amplitude, instead of the location, of the Fourier spectra, differs from that of the noise. If a signal has energy concentrated in a small number of wavelet coefficients, their values will be large in comparison with the noise that has its energy spread over a large number of coefficients. These localizing properties of the wavelet transform allow the thresholding of the wavelet coefficients to remove the noise without blurring the features in the original signal.

Subsequently, we perform spectral analysis on the noise-reduced speech signals to extract audio features. The spectral analysis method is only reliable when the signal is stationary, that is, the statistical characteristics of a signal are invariant with respect to time. For speech, this holds only within the short time intervals of articulatory stability, during which a short-time analysis can be performed by windowing a signal into a succession of windowed sequences, called frames. These speech frames can then be processed individually. In this work, we use a Hamming window of size 512 points, with 50% overlap between adjacent windows.

The pitch, energy, and MFCC features are then extracted from each frame. The pitch is estimated based on the Fourier analysis of the logarithmic amplitude spectrum of the signal [53]. By taking the logarithm of the frequency spectrum, the low-frequency components of the spectrum are exaggerated. If the log amplitude spectrum contains many regularly spaced harmonics, then the Fourier analysis of the spectrum will show a peak corresponding to the spacing between the harmonics: that is the fundamental frequency. The energy features are extracted in the time domain and represented in decibel (dB). MFCC is a powerful analytical tool to mimic the behavior of human ears by applying cepstral analysis, and has demonstrated great success in the field of speech recognition. In this work, the implementation of MFCC feature extraction follow the same procedure as described in [6]. The pitch, energy, and the first 13 MFCCs are then extracted from each frame, and the features of successive frames within **w** are concatenated as the audio features.

5.4.1.2 Visual Feature Representation

For visual feature representation, we perform feature extraction on the middle frame of the corresponding audio time window **w**. We first detect the face region from the image frame using an HSV (hue, saturation, and value) color model-based method [55]. The HSV color model corresponds closely to the human intuition on color. The RGB (red, green, and blue) components of an image are first converted to the HSV color space, followed by a planar envelop approximation method [16] to approximate the human skin color. Subsequently, morphological operations are applied to remove the spurious nonskin regions, such as small isolated blobs and narrow belts, whose color falls into the skin color space.

The resulting facial image is normalized to a size of 64 × 64. A Gabor filter bank of five scales and eight orientations is then applied for feature extraction. Using Gabor wavelet features to represent facial expression has been explored and shown to be very efficient in the literature [36]. It allows a description of spatial frequency structure in the image while preserving information about spatial relations. In this work, the Gabor filter bank is implemented using the algorithm proposed in [37]. Due to the large dimensionality of the Gabor coefficients, we downsample each subband to a size of 32 × 32, and then perform dimensionality reduction on all the down-sampled Gabor coefficients using the PCA method.

5.4.2 Fusion Strategies

The kernel matrix fusion approach finds a joint subspace for different modalities, and outputs one feature vector. Figure 5.2 depicts a block diagram of kernel matrix fusion-based

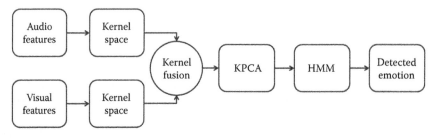

FIGURE 5.2
Block diagram of kernel matrix fusion-based system.

system. During the training process, for each of the windows, a separate kernel matrix is computed from the extracted audio and visual features, respectively. Algebraic operations are then performed on the two matrices (as in Equations 5.8 and 5.9), and the eigenvectors of the joint subspace can be computed as in Equation 5.6. For a new input, the projection is performed according to Equations 5.10 and 5.11, and the resulting features are considered as the final features of the current window.

To capture the statistical dependence across successive windows and identify the inherent temporal structure of the features, an HMM [44] is employed to characterize the distribution of an video sequence. The projected features of each window are considered as the observation of an HMM, of which the probability density functions (pdfs) given a state is modeled using Gaussian mixtures. The output of the HMM contains the likelihood of the sample with respect to different classes, and the decision can be made based on the maximum likelihood value.

The KCCA approach produces two projections, one for each modality. Therefore, the fusion can be performed at either feature or score levels, as shown in Figures 5.3 and 5.4 respectively. In feature-level fusion, the KCCA-projected features for audio and visual channels are concatenated into one stream and the HMM outputs the likelihood of the sample with respect to different classes. In score-level fusion, an HMM is constructed for the KCCA-projected audio and visual channels, respectively, and the outputs of each modality, that is, the likelihood values, are considered as the classification scores. Based on the obtained scores, we adopt a rule-based fusion approach, in which the score values of different modalities are linearly combined through the equal-weight sum fusion. A sample is classified into a category according to

$$\hat{l} = \arg\max_{i}(0.5s_i^a + 0.5s_i^v), \quad i = 1, \ldots, C, \tag{5.18}$$

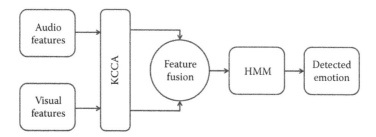

FIGURE 5.3
Block diagram of KCCA-based fusion at the feature level.

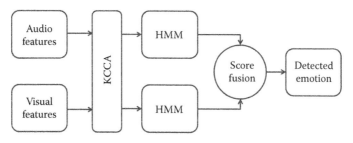

FIGURE 5.4
Block diagram of KCCA-based fusion at the score level.

where $\hat{l}$ presents the predicted label, C is the number of classes, and s_i^a and s_i^v denote the ith scores from audio and visual channels, respectively.

5.5   Experimental Results

To evaluate the effectiveness of the proposed solution, we conduct experiments on the RML (Ryerson Multimedia research Laboratory) emotion database [55], which consists of video samples from eight human subjects, speaking six languages, and expressing six basic emotions (anger, disgust, fear, happiness, sadness, and surprise). The samples were recorded at a sampling rate of 22,050 Hz, and a frame rate of 30 fps. In our experiments, we selected 400 video clips, each of which is then truncated to 2-s-long such that both audio and visual information present. The video sample is then segmented into 10 uniform windows, and the dimensionality of audio and visual features are set to 240 and 200, respectively. The evaluation is based on cross-validation, where each time 75% of samples are randomly selected for training and the rest for testing. This process is repeated 10 times, and the average of the results is reported.

In our experiments, we normalize each audio and visual attribute to zero mean and unit variance. The number of hidden states of the HMM is set to 3 and a Gaussian mixture with three components is chosen to model the pdf of an observation given a state. We have conducted extensive experiments to select the kernel functions and corresponding parameters, and the reported results are based on a Gaussian kernel with $\sigma = 10$. The results for audio, visual, and audiovisual information are based on the original features. For the weighted sum-based kernel matrix fusion, the parameter a is set to 0.1–0.9 with a step size of 0.1. For KCCA, the projected dimensionality is set to the same value for both audio and visual channels, and the regularization parameter is set to $\tau = 0.2$–1.0 for comparison.

Figure 5.5 depicts the recognition performance of kernel matrix fusion, in comparison with audio-only, visual-only, and concatenation-based audiovisual feature-level fusion. By using the weighted sum rule at $a = 0.4$, it produces the best overall recognition accuracy of 82.22%, which significant outperforms other methods. When the original features are utilized, the concatenation-based feature-level fusion actually exhibits degraded performance compared to visual features only. This demonstrates that the disparate characteristics of different modalities may produce negative impact to the recognition performance. On the other hand, the introduced method is capable of integrating heterogeneous data through the fusion of the kernel matrix, and providing improved performance. In addition, it also provides efficient dimensionality reduction, compared to the original dimensionality of $240 + 200 = 440$ for audio and visual features.

Figures 5.6 and 5.7 show the experimental results of KCCA-based fusion at the feature and score levels, respectively. For the original features, the score-level fusion combines the classification scores, which represent the likelihood of being classified as a certain semantic, and provides improved performance than single modalities only. The linear CCA method produces lower-recognition performance. This demonstrates that the linear approaches fail to identify the true relationship between the two subsets of features. The kernel-based nonlinear methods, KCCA, which effectively capture the intrinsic nonlinearity between two modalities, significantly improve the recognition accuracy, using both fusion strategies.

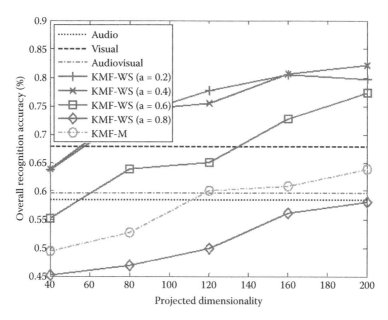

FIGURE 5.5
Experimental results of kernel matrix fusion (KMF)-based method (weighted sum (WS), multiplication (M)).

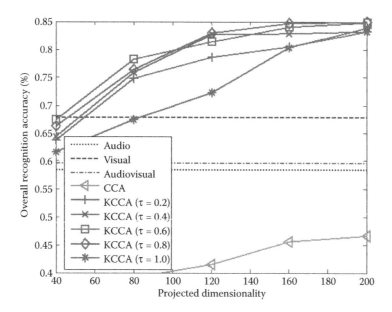

FIGURE 5.6
Experimental results of KCCA-based fusion at the feature level.

Overall, the KCCA fusion at the feature level obtains the best recognition accuracy of about 85%. Comparing with kernel matrix fusion method which finds a subspace for describing the joint characteristics of different modalities, the KCCA approach seeks for two distinct sets of projections, one for each modality, such that the correlation between these two sets of features can be maximized. It can better identify the associated information across

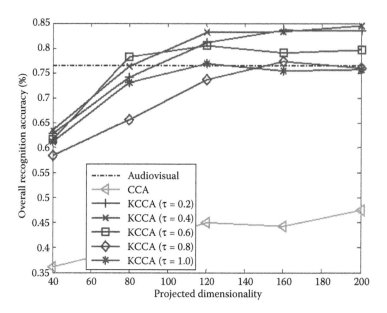

FIGURE 5.7
Experimental results of KCCA-based fusion at the score level.

multiple modalities with respect to the presented semantic, and therefore offers superior performance. On the other hand, the kernel matrix fusion approach only integrate the information by combining their respective kernel matrix, but without identifying the relationship between different modalities.

5.6 Conclusions

This chapter has presented a review of the state of the art in multimodal information fusion, and outlined the strength and weakness of different fusion levels. We have introduced kernel-based fusion by examining kernel matrix fusion and KCCA-based methods. Kernel-based fusion methods provide efficient solutions for modeling the complex and nonlinearity properties of the features, combining multimodal data of disparate characteristics, and identifying the relationship between different modalities. The effectiveness of the introduced solutions is demonstrated through extensive experimentation in an audiovisual-based emotion recognition problem.

Information fusion is an intrinsically complex problem. Despite the great advances achieved in the past, a more comprehensive investigation is needed to address some fundamentally important issues, such as how to evaluate the amount of additional information provided by integrating multiple modality, how to utilize the beneficial information while removing the redundant or negative information, and, how to measure the dependency between different modalities with respect to a classification problem. Most of the existing works focus on their specific application scenarios, which lack generality. We hope that this chapter will provide the readers with some insights into the design and development of multimodal systems, and motivate the research community to further explore this important and promising area in detail.

References

1. W. H. Adams, G. Iyengar, M. R. Naphade, C. Neti, H. J. Nock, and J. R. Smith. Semantic indexing of multimedia content using visual, audio and text cues. *EURASIP Journal on Applied Signal Processing*, 2:170–185, 2003.
2. J. F. Aguilar, J. O. Garcia, D. G. Romero, and J. G. Rodriguez. A comparative evaluation of fusion strategies for multimodal biometric verification. In *Proceedings of International Conference on Video-Based Biometric Person Authentication*, Springer Lecture Notes in Computer Science, Vol. 2688, pp. 830–837, 2003.
3. F. M. Alkoot and J. Kittler. Experimental evaluation of expert fusion strategies. *Pattern Recognition Letters*, 20:1361–1369, 1999.
4. P. K. Atrey, M. A. Hossain, A. E. Saddik, and M. S. Kankanhalli. *Multimodal Fusion for Multimedia Analysis: A Surevey*, Multimedia Systems, Vol. 16, pp. 345–379. Springer, 2010.
5. M. J. Beal, N. Jojic, and H. Attias. A graphical model for audio-visual object tracking. *IEEE Transactions on Pattern Analysis and Machine Intelligence*, 25(7): 828–836, 2003.
6. C. Becchetti and L. P. Ricotti. *Speech Recognition: Theory and C++ Implementation*. Wiley, Hoboken, New Jersey, 1999.
7. S. Bengio. Multimodal authentication using asynchronous hmms. In *Proceedings of International Conference on Audio and Video Based Biometric Person Authentication*, pp. 770–777, Guildford, UK, 2003.
8. S. Blackman and R. Popoli. *Design and Analysis of Modern Tracking Systems*. Artech House, Norwood, Massachusetts, 1999.
9. M. B. Blaschko and C. H. Lampert. Correlational spectral clustering. In *Proceedings of the IEEE Conference on Computer Vision and Pattern Recognition*, Anchorage, Alaska, 2008.
10. R. A. Bolt. Put that there: Voice and gesture at the graphic interface. *Computer Graphics*, 14(3): 262–270, 1980.
11. K. W. Bowyer, K. I. Chang, P. J. Flynn, and X. Chen. Face recognition using 2-d, 3-d, and infrared: Is multimodal better than multisample? *Proceedings of the IEEE*, 94(11):2000–2012, 2006.
12. H. Bredin and G. Chollet. Audio-visual speech synchrony measure for talking-face identity verification. In *Proceedings of IEEE International Conference on Speech, Acoustics, and Signal Processing*, Honolulu, Hawaii, 2007.
13. G. Chetty and M. Wagner. Audio-visual multimodal fusion for biometric person authentication and liveness verification. In *Proceedings of the 2005 NICTA-HCSNet Multimodal User Interaction Workshop*, Sydney, Australia, 2005.
14. D. Datcu and L. J. M. Rothkrantz. Semantic audio-visual data fusion for automatic emotion recognition. In *Proceedings of Euromedia*, Porto, Portugal, 2008.
15. R. O. Duda, P. E. Hart, and D. G. Stork. *Pattern Classification*. Wiley-Interscience, 2000.
16. C. Garcia and G. Tziritas. Face detection using quantized skin color regions merging and wavelet packet analysis. *IEEE Transactions on Multimedia*, 1:264–277, 1999.
17. H. Go, K. Kwak, D. Lee, and M. Chun. Emotion recognition from the facial image and speech signal. In *Proceedings of SICE Annual Conference*, Vol. 3, pp. 2890–2895, 2003.
18. L. Guan, Y. Wang, and Y. Tie. Toward natural and efficient human computer interaction. In *Proceedings of IEEE International Conference on Multimedia and Expo*, New York, USA, pp. 1560–1561, 2009.
19. L. Guan, Y. Wang, R. Zhang, Y. Tie, A. Bulzachi, and M. T. Ibrahim. Multimodal information fusion for selected multimedia applications. *International Journal of Multimedia Intelligence and Security*, 1(1):5–32, 2010.

20. M. Han, J. H. Hus, and K. T. Song. A new information fusion method for bimodal robotic emotion recognition. *Journal of Computers*, 3(7):39–47, 2008.
21. D. R. Hardoon, S. Szedmak, O. Szedmak, and J. Shawe-taylor. Canonical correlation analysis: An overview with application to learning methods. In *Technical Report CSD-TR-03-02*. Department of Computer Science, Royal Holloway, University of London, 2003.
22. J. Hershey, H. Attias, N. Jojic, and T. Krisjianson. Audio visual graphical models for speech processing. In *Proceedings of IEEE International Conference on Speech, Acoustics, and Signal Processing*, Montreal, Quebec, Canada, 2004.
23. D. Huang, H. Leung, and W. Li. Fusion of dependent and independent biometric information sources. *Contract Report*, DRDC Ottawa, (2), March 2005.
24. M. Indovina, U. Uludag, R. Snelick, A. Mink, and A. Jain. Multimodal biometric authentication methods: A cots approach. In *Proceedings of 2003 Workshop on Multimodal User Authentication*, Santa Barbara, California, pp. 99–106, 2003.
25. S. A. Israel, W. T. Todd Scruggs, W. J. Worek, and J. M. Irvine. Fusing face and ECG for personal identification. In *Proceedings of the 32nd Applied Imagery Pattern Recognition Workshop*, pp. 226–231, 2003.
26. A. K. Jain, K. Nandakumar, and A. Ross. Score normalization in multimodal biometric systems. *Pattern Recognition*, 38(12):2270–2285, 2005.
27. A. K. Jain, A. Ross, and S. Prabhakar. An introduction to biometric recognition. *IEEE Transactions on Circuits and Systems for Video Technology*, 14:4–20, 2004.
28. N. Joshi and L. Guan. Feature fusion applied to missing data ASR with the combination of recognizers. *The Journal of Signal Processing Systems*, 58:359–370, 2010.
29. T. Joshi, S. Dey, and D. Samanta. Multimodal biometrics: State of the art in fusion techniques. *International Journal of Biometrics*, 1(4):393–417, 2009.
30. I. Kanluan, M. Grimm, and Kroschel K. Audio-visual emotion recognition using an emotion space concept. In *Proceedings of 16th European Signal Processing Conference*, Lausanne, Switzerland, 2008.
31. J. Kittler, J. Matas, K. Jonsson, and M. U. Ramos Sanchez. Combining evidence in personal identity verification systems. *Pattern Recognition Letters*, 18:845–852, 1997.
32. J. Kludas, E. Bruno, and S. M. Maillet. Information fusion in multimedia information retrieval. In *Proceedings of 5th International Workshop on Adaptive Multimedia Retrieval: Retrieval, User, and Semantics*, pp. 147–159, Paris, France, 2007.
33. M. M. Kokar, J. A. Tomasik, and J. Weyman. Formalizing classes of information fusion systems. *Information Fusion*, 5(4):189–202, 2004.
34. A. Kumar and D. Zhang. Personal recognition using hand shape and texture. *IEEE Transactions on Image Processing*, 15(8):2454–2461, 2006.
35. H. C. Lee, E. S. Kim, G. T. Hur, and H. Y. Choi. Generation of 3d facial expressions using 2d facial image. Computer and Information Science, *ACIS International Conference on Computer and Information Science*, pp. 228–232, Jeju Island, South Korea, 2005.
36. M. Lyons, S. Akamatsu, M. Kamachi, and J. Gyoba. Coding facial expressions with gabor wavelets. In *Proceedings of the Third IEEE International Conference on Automatic Face and Gesture Recognition*, pp. 200–205, Nara, Japan, 1998.
37. B. S. Manjunath and W. Y. Ma. Texture features for browsing and retrieval of image data. *IEEE Transactions on Pattern Analysis and Machine Intelligence*, 18:837–842, 1996.
38. G. L. Marcialis and F. Roli. Fingerprint verification by fusion of optical and capacitive sensors. *Pattern Recognition Letters*, 25:1315–1322, 2004.
39. A. Metallinou, S. Lee, and S. Narayanan. Audio-visual emotion recognition using gaussian mixture models for face and voice. In *Proceedings of 10th IEEE International Symposium on Multimedia*, pp. 250–257, Berkeley, California, 2008.
40. K. R. Muller, S. Mika, G. Ratsch, K. Tsuda, and B. Scholkopf. An introduction to kernel-based learning algorithms. *IEEE Transactions on Neural Networks*, 12(2):181–201, 2001.
41. A. V. Nefian, L. Liang, X. Pi, X. Liu, and K. Murphye. Dynamic bayesian networks for audio-visual speech recognition. *EURASIP Journal on Applied Signal Processing*, pp. 1–15, 2002.

42. H. J. Nock, G. Iyengar, and C. Neti. Assessing face and speech consistency for nomologue detection in video. In *Proceedings of ACM International Conference on Image and Video Retrieval*, London, UK, 2002.
43. M. G. K. Ong, T. Connie, A. B. T. Jin, and D. N. C. Ling. A single-sensor hand geometry and palmprint verification system. In *Proceedings of the 2003 ACM SIGMM Workshop on Biometrics Methods and Applications*, pp. 100–106, Berkeley, California, 2003.
44. L. R. Rabiner. A tutorial on hidden markov models and selected applications in speech recognition. In *Proceedings of the IEEE*, 77: 257–286, 1989.
45. A. Ross and R. Govindarajan. Feature level fusion using hand and face biometrics. In *Proceedings of SPIE Conference on Biometric Technology for Human Identification II*, pp. 196–204, Orlando, USA, 2005.
46. M. E. Sargin, Y. Yemez, E. Erzin, and A. M. Tekalp. Audiovisual Synchronization and Fusion Using Canonical Correlation Analysis. *IEEE Transactions on Multimedia*, 9(7):1396–1403, 2007.
47. J. Z. Sasiadek. Sensor fusion. *Annual Reviews in Control*, 26:203–228, 2002.
48. B. Scholkopf, A. Smola, and K. R. Muller. Nonlinear component analysis as a kernel eigenvalue problem. *Neural Computation*, 10(5):1299–1319, 1998.
49. L. C. De Silva, T. Miyasato, and R. Nakatsu. Facial emotion recognition using multimodal information. In *Proceedings of IEEE International Conference on Information, Communications and Signal Processing*, pp. 397–401, Singapore, 1997.
50. L. C. De Silva and P. C. Ng. Bimodal emotion recognition. In *Proceedings of 4th IEEE International Conference on Automatic Face and Gesture Recognition*, pp. 332–335, 2000.
51. M. Song, C. Chen, and You M. Audio-visual based emotion recognition using tripled hidden markov model. In *Proceedings of IEEE International Conference on Acoustics, Speech, and Signal Processing*, Vol. 5, pp. 877–880, Montreal, Quebec, Canada, 2004.
52. Q. Sun, S. Zeng, Y. Liu, P. Heng, and D. Xia. A new method of feature fusion and its application in image recognition. *Pattern Recognition*, 36(12):2437–2448, 2005.
53. Y. Wang and L. Guan. An investigation of speech-based human emotion recognition. In *Proceedings of IEEE International Workshop on Multimedia Signal Processing*, pp. 15–18, Siena, Italy, 2004.
54. Y. Wang and L. Guan. Recognizing human emotion from audiovisual information. In *Proceedings of IEEE International Conference on Acoustics, Speech, and Signal Processing*, pp. 1125–1128, Philadelphia, Pennsylvania, 2005.
55. Y. Wang and L. Guan. Recognizing human emotional state from audiovisual signals. *IEEE Transactions on Multimedia*, 10(5):936–946, 2008.
56. Y. Wang, Z. Liu, and J. C. Huang. Multimedia content analysis: Using both audio and visual clues. *IEEE Signal Processing Magazine*, pp. 12–36, 2000.
57. Z. Zeng, Y. Hu, G. I. Roisman, Z. Wen, Y. Fu, and T. S. Huang. Audio-visual spontaneous emotion recognition. In *Artificial Intelligence for Human Computing*, Springer Lecture Notes in Computer Science, T. S. Huang et al. (Eds), Vol. 4451, pp. 72–90, 2007.
58. Z. Zeng, M. Pantic, G. I. Roisman, and T. S. Huang. A survey of affect recognition methods: audio, visual, and spontaneous expressions. *IEEE Transactions on Pattern Analysis and Machine Intelligence*, 31(1):39–58, 2009.
59. Z. Zeng, J. Tu, B. Pianfetti, and T. S. Huang. Audio-visual affective expression recognition through multi-stream fused HMM. *IEEE Transactions on Multimedia*, 10(4):570–577, 2008.
60. J. Zhao, Y. Fan, and W. Fan. Fusion of global and local features using KCCA for automatic target recognition. In *Proceedings of International Conference on Image and Graphics*, Xi'an, China, 2009.
61. Y. A. Zuev and S. K. Ivanon. The voting as a way to increase the decision reliability. *Journal of the Franklin Institute*, 336(2):361–378, 1999.

6

Multimedia-Based Affective Human–Computer Interaction

Yisu Zhao, Marius D. Cordea, Emil M. Petriu, and Thomas E. Whalen

CONTENTS

As computers are becoming an essential part of humans' daily life, they are expected to offer more natural interaction modalities between humans and computers, better tuned to the affective computing paradigm [1–4]. This chapter discusses bidirectional, human-to-computer and computer-to-human, affective interaction techniques developed by authors, with special attention paid to the human emotion communication based on facial expression, head movement, and eye gaze attributes.

6.1 Introduction

While computers are becoming smarter and provide human interfaces that accommodate a wider range of human capabilities [5,6], there are still many challenges to be overcome in order to achieve a truly natural and transparent human–computer interaction (HCI).

Human beings can recognize a wide range of visual, auditory, olfactory, gustatory, and haptic stimuli, and classify these stimuli according to abstract patterns [7]. While communicating with each other, human beings are adept at empathic communications, understanding vocal and facial emotions, and understanding body language. Human beings are also dexterous. They can manipulate a wide variety of objects, regardless of variations in their shape or texture.

Figure 6.1 illustrates the main HCI devices for the three main human sensing modalities: visual, auditory, and haptic, which already are commercially available. The olfactory and gustatory HCI devices, which are still in an early development phase, are not shown.

Human beings are very high-bandwidth creatures. The human visual system is capable of perceiving more than a hundred megabits of information per second. The largest human sense organ, the skin, is capable of perceiving nearly that much as well. The human auditory system does not have nearly as much capacity as the visual system, but does add another significant source of input.

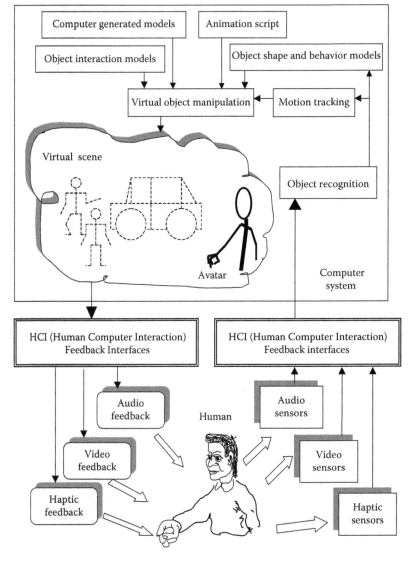

FIGURE 6.1
HCI devices for three main human sensing modalities: audio, video, and haptic.

People communicate through body language which includes gestures, facial expressions, and eye movements, all of these being very important nonverbal human communication modalities.

Human beings are the only animals with a small colored iris embedded in a white sclera, making it possible for other people to detect very small changes in the direction of a person's gaze. Moreover, there is evidence that people make important judgments about other people based on microexpressions, fleeting movements of the facial muscles that are barely noticeable.

A specialized form of gesture is drawing and writing in which a pen, or keyboard, is moved by the gesture to form persistent images or linguistic tokens. Existing computer interfaces, even graphical user interfaces on high-resolution monitors, only make use of a small fraction of the bandwidth that is available to human beings. Once again, we rely on the intelligence of the human being to learn to use nonstandard interfaces and understand the ambiguous information that they are presented with.

As computing intelligence expands at an ever-increasing rate, it is expected that the current HCI interfaces will evolve to more user-friendly human–computer symbiosis interfaces [5]. A thorough review of the emotion detection method can be found in [8,9].

6.2 Emotional HCI-Based on Combined Facial Expression, Head Movement, and Eye Gaze Analysis

Human beings are emotional. They can vary the characteristics of response, depending on the global state of each individual. Depending on emotional state, the human may trade speed of response for complexity of the analysis of different options; trade accuracy for force of response; or even trade inaction for action.

Human emotions are usually recognized using a combination of facial expressions, head movements, and eye gaze.

Facial expression refers to the physical muscle movement change in human face. There are six universal facial expressions: happiness, anger, sadness, surprise, disgust, and fear. However, facial expressions cannot provide enough information to infer the real affective states of a person. For example, a person may be very happy or very sad within, however, expressionless on the face. Under such circumstances, the facial expression of that person is natural, while the emotion of that person is very happy or very sad. Another example is that a smiling face usually means happiness. However, if a person is smiling to you while shaking his/her head, this may not mean that he/she is in a mood of happiness. In this case, head movement plays a leading role in emotion recognition than facial expression. Although facial expression shows happiness, the actual emotion may more likely be disagreement or denial.

6.2.1 Facial-Expression Recognition

An early scientific study on facial expression was *The Expression of the Emotions in Man and Animals* published by Charles Darwin in the mid-nineteenth century [10].

Almost 100 years later, Ekman and Friesen, two sign communication psychologists, developed the anatomically oriented *Facial Action Coding System* (FACS) based on numerous experiments with facial muscles. Ekman and Friesen reduced the 50,000 distinguishable expression space to a comprehensive system, which could distinguish all possible facial expressions by using a combination of a limited set of only 46 Action Units (AUs).. The AU refers to a basic visually distinguishable facial movement, which cannot be decomposed into smaller units [11–13]. An AU can be controlled by one or more muscles.

Facial expression analysis, recognition, and interpretation represent challenging tasks for image processing, pattern recognition, and computer vision. There are a number of surveys [14,15] discussing in detail the automated facial expression analysis in images and video.

A facial expression analyzer has three functional modules [14–16]:

1. Face acquisition
2. Feature extraction
3. Expression classification

Face acquisition automatically locates the face area in the input image or sequence of images. In the case of static images the process is referred to as *detecting* a face. In the case of sequence of images the process is referred to as *tracking* a face. Once the face is found, the facial area is further processed to extract facial features.

Facial features can be extracted from facial images using three types of methods [14]:

1. Analytic or feature based [17,18]
2. Holistic or model based [19–21]
3. Hybrid or analytic-to-holistic [22–26]

Feature-based methods localize the facial features of an analytic face model in the input image or track them in the image sequence. Model-based methods fit a holistic face model to the face in the input image or track it in the image sequence. Deformable templates in general and active models (Active Shape Model (ASM), Active Appearance Model (AAM)) [27] in particular represent good examples of the holistic approach to extract facial features.

Facial expressions can be classified either in the AU space, or in terms of the prototypical (basic) emotional expressions: happiness, surprise, anger, sadness, fear, and disgust. There are several issues to keep in mind when discussing the expression classification [15]: (1) as there is no unique description based on Aus, all the FACS rely on Ekman's linguistic description of the six basic facial expressions and (2) each expression depends on the subject's physiognomy and the timing and intensity of facial deformation. These factors make automated expression recognition a difficult task. The most used classification techniques are [15]:

1. Model based [19–21,23,24]
2. Neural networks [22,26]
3. Rule based [17]

The methods used for facial expression recognition can also be classified from the way images are processed in time:

1. Static: expressions are recognized in a single frame [17,19–23,26]
2. Dynamic: expressions are recognized from image sequences [18,24,25,28]

6.2.1.1 Databases

One item used in all emotion recognition research, for static as well as for sequences of images, is the *testbed database*.

One such database, JAFFE [29], contains images of 10 Japanese female frontal face images with different facial expressions (Figure 6.2).

Another well-known database is the Facial Expression and Emotion Database of the FG-NET (Face and Gesture Recognition Network) consortium [30]. It contains image sequences to assist research on human facial expressions (Figure 6.2). The main characteristics of the database are:

1. Allowance of spontaneous and natural expressions
2. No restrictions in head movement
3. Good lighting conditions and constant background

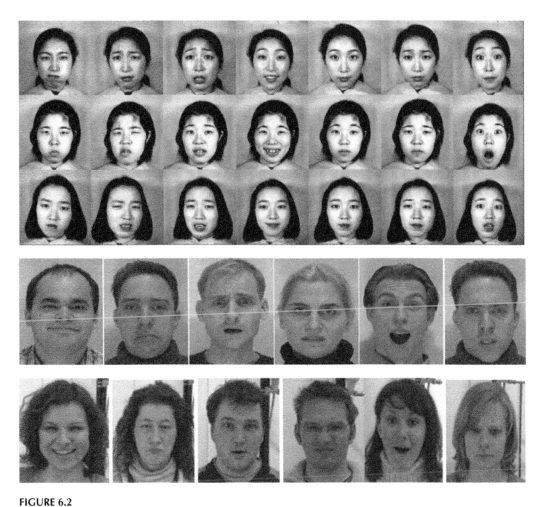

FIGURE 6.2
Examples of emotional facial expressions from JAFFE (first three rows), MMI (fourth row), and FG-NET (last row) databases.

The database contains images collected from 19 individuals, each performing all six basic expressions and the neutral sequence three times. Each sequence starts with a neutral expression, it develops without following any pattern, namely it can increase or decrease in intensity at any time, then ending again with a neutral face.

Another database available for research is the MMI (Man Machine Interaction) Facial Expression Database [31]. It provides a large testbed for research on automated facial expression analysis, and was already used in Pantic's [17] method validation. The database is composed of more than 1500 static images and image sequences of faces display-ing the prototypic expressions, as well as single and multiple AU-based expressions (Figure 6.2). The database includes 19 different persons, ranging in age from 19 to 62, with European, Asian, or South American ethnic background. The main characteristics of the database are:

1. 79 staged emotional and AU expressions in frontal and profile views
2. Allowance of minimal head movement
3. Good lighting conditions and cluttered background

6.2.1.2 Static Facial-Expression Recovery

Cordea et al. [32] introduced the three-dimensional (3D) Anthropometric Muscle-Based Active Appearance Model (AMB AAM) for modeling the shape and appearance of human faces using a constrained 3D AAM. This model allows modeling the shape and appearance of human faces using a generic 3D wireframe model of the face. The shape of the model is described by two sets of controls: (1) the muscle actuators to model facial expressions and (2) the anthropometrical controls to model different facial types. The 3D face deformations are defined on the anthropometric-expression space, which is derived from Waters' muscle-based face model [33], FACS [11], and anthropometric research [34].

6.2.1.3 Dynamic Facial-Expression Recovery

As the head movement and facial expressions are combined in images, a real-time tracking system has first to separate and recover the head's rigid movements and the nonrigid facial expressions. In order to do this, Cordea et al. [35,36], used an Extended Kalman Filter (EKF) to recover the head pose (global motion) and a 3D AMB AAM to recover the facial expressions (local motion) [32].

When using AMB AAM to recover the facial expressions, each image is analyzed and transformed into a 3D-wireframe head model, controlled by seven pairs of muscles: *zygo-matic major, anguli depressor, frontalis inner, frontalis outer, frontalis major, lateral corrugator, nasi labii*, and the *jaw drop*, shown in red in Figure 6.3.

Anthropometric deformations are chosen in conformity with anthropometrical statistics, to provide a way to deform the 3D generic mesh into a personal one. The deformation parameters are used to make the jaw wider, the nose-bridge narrower, and so on.

Figures 6.4 and 6.5 illustrate the performance of the facial expression recognition for two cases, person dependent and person independent (i.e., person not present in training set), respectively.

While, as expected, the person-dependent cases have a better recognition rate, 86.1%, the system has a respectable 76.2% recognition rate in person-independent cases, which could be quite acceptable for some practical applications.

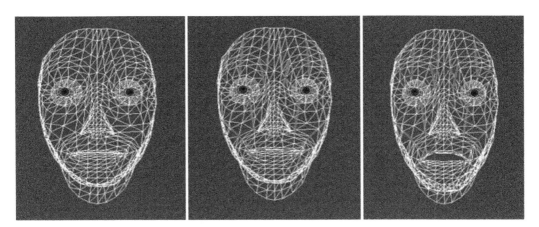

FIGURE 6.3
Muscle-controlled 3D wireframe head model.

Au	Signification	No.	Correct	False	Missed	Confused	Recognition rate
0	Neutral	20	19	1	0	0	95%
1	Inner brow raiser	24	21	0	3	0	87.5%
2	Outer brow raiser	52	41	1	9	1	78.8%
4	Brow lowerer	42	41	0	0	1	97.6%
12	Lip corner puller	51	48	3	0	0	94.1%
15	Lip corner depressor	18	17	0	1	0	94.4%
26	Jaw drop	74	55	0	19	0	74.3%
Total		281	242	5	32	2	86.1%
False alarm: 1.7%, Missed: 11.3%							

FIGURE 6.4
Person-dependent recognition of facial expressions for faces from the MMI database. (From M. Pantic et al., Web-based database for facial expression analysis, *Proceedings of IEEE Conference on Multimedia and Expo*, Amsterdam, the Netherlands, pp. 317–321, 2005. Available at http://www.mmifacedb.com. © (2005) IEEE. With permission.)

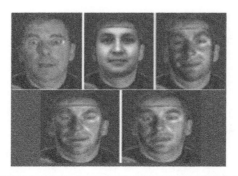

Au	Signification	No.	Correct	False	Missed	Confused	Recognition rate
0	Neutral	20	17	3	0	0	85.0%
1	Inner brow raiser	10	8	0	2	0	80.0%
2	Outer brow raiser	27	20	1	5	1	74.0%
4	Brow lowerer	24	21	1	1	1	87.5%
12	Lip corner puller	17	13	0	4	0	76.4%
15	Lip corner depressor	13	10	1	2	0	76.9%
26	Jaw drop	24	14	0	10	0	58.3%
Total		135	103	6	24	2	76.2%
False alarm: 4.4%, Missed: 17.7%							

FIGURE 6.5
Person-independent recognition of facial expressions for faces from the MMI database. (From M. Pantic et al., Web-based database for facial expression analysis, *Proceedings of IEEE Conference on Multimedia and Expo*, Amsterdam, the Netherlands, pp. 317–321, 2005. Available at http://www.mmifacedb.com. © (2005) IEEE. With permission.)

The architecture of the dynamic facial-expression recovery system [35,36], illustrated in Figure 6.6, consists of two main modules: (1) *tracking module*, delivering the 2D measurements $p_i(u_i, v_i)$ of the tracked features, where $i = 1, \ldots, m$, and m is the number of measurement points and (2) *estimator module* for the estimation of 3D motion and expressions, delivering a state vector:

$$s = (t_x, t_y, t_z, \alpha, \beta, \lambda, f, a_j)$$

where $(t_x, t_y, t_z, \alpha, \beta, \lambda)$ are the 3D camera/object relative translation and rotation, f is the camera focal length, and a_j are the 3D model's muscles that generate facial expressions, where $j = 1, \ldots, n$, and n is the number of muscles.

The set of 2D features to be tracked is obtained by projecting their corresponding 3D model points $P_i(X_i, Y_i, Z_i)$, where $i = 1, \ldots, m$, and m is the number of tracked features.

Tracking allows estimating the motion while locating the face and coping with different degrees of rigid motion and nonrigid deformations. Since the global and local motion search spaces are separated, the features to be tracked must originate from rigid regions of the face.

The main purpose of the EKF is to filter and predict the rigid and nonrigid facial motion. It consists of two stages: time updates (or prediction) and measurement updates (or correction). EKF provides iteratively an optimal estimate of the current state using the current state measurements, and producing a future state estimate based on the underlying state model.

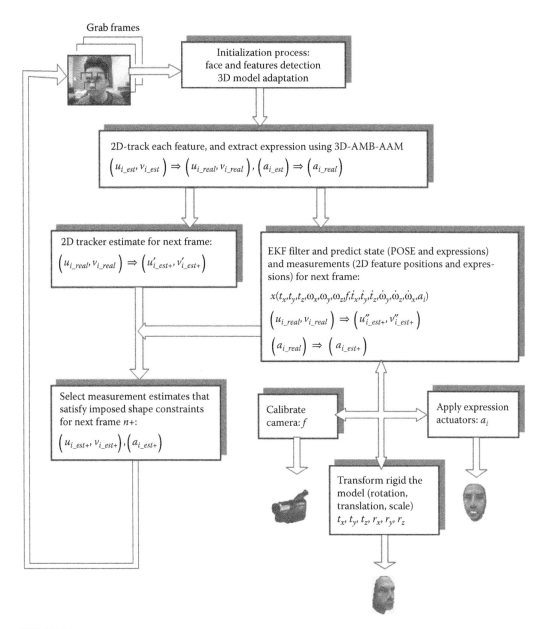

FIGURE 6.6
Visual tracking and recognition of facial expression. (From M. D. Cordea et al., Three-dimensional head tracking and facial expression recovery using an anthropometric muscle-based active appearance model, *IEEE Transactions on Instrumentation and Measurement*, 57(8), 1578–1588, 2008. © (2008) IEEE. With permission.)

The EKF state and measurement equations are

$$s(k+1) = As(k) + \xi(k) \text{ and } m(k) = Hs(k) + \eta(k),$$

where s is the state vector s (*translation, rotation, velocity, focal_length, actuators*) that contains the relative 3D camera–object translation, rotation and their velocities, camera focal length, and the actuators/muscles, m is the measurement vector, A is the state transition matrix, H is the Jacobian that relates state to measurement, and $\xi(k)$ and $\eta(k)$ are error terms modeled as Gaussian white noise.

6.2.2 Head Movement Recognition

The goal of this section is to determine whether head movement direction is nodding, shaking, or stationary. The general procedure is shown in Figure 6.7.

The nostril detection and tracking method is motivated by [37,38]. We extend the method in [37] by first applying the Harris corner detection algorithm [39] to automatically detect nostril point candidates instead of manually labeling them. After nostril detection, the x- and y-coordinates of nostril are used to analyze the direction of head movement. Suppose the coordinate of nostril is (x_{n-1}, y_{n-1}) and (x_n, y_n) in two consecutive frames, respectively, then $|y_n - y_{n-1}| \gg |x_n - x_{n-1}|$ indicate head nods and $|y_n - y_{n-1}| \ll |x_n - x_{n-1}|$ indicate head shakes. If both $|y_n - y_{n-1}|$ and $|x_n - x_{n-1}|$ are lower than a certain threshold, then stationary status is occurred. GentleBoost-based pattern analyzer [40] is used in the classification stage.

In our head movement detection, the linguistic variables are *High-Frequency-Shaking*, *Low-Frequency Nodding*, and so on. We define the quantized input variables using the following formulas:

$$Head\text{-}movement = 0.5 + (\text{sgn}) \times v_{\text{normal}},$$

$$\text{sgn} = \begin{cases} 1, & noding \\ -1, & shaking \end{cases},$$

$$v_{\text{normal}} = \frac{v_{hm}}{2 \times MaxSpeed}, \quad hm = nod \text{ or } shake,$$

where *Head-movement* is the quantified value for linguistic variable inputs, v_{hm} is the speed of a certain head nod or head shake movement, and *MaxSpeed* is the maximal speed of the

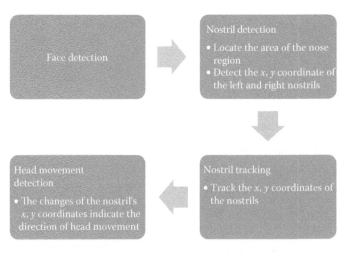

FIGURE 6.7
General steps of proposed head movement detection.

head movement. After repeated testing, we find the maximum speed that human beings can achieve for both head nods and shakes is 3 cycle/s. Therefore, 3 would be a reasonable value for *MaxSpeed*.

6.2.3 Eye Gaze Recognition

We propose an eye gaze recognition method in this section. The goal of this function is to determine whether the human subject's gaze direction is direct or averted. This is done using a series of frontal face images, detected in previous stages, following the process shown in Figure 6.8.

A straightforward eye location method is based on the assumption that for most of the cases, the face skin is brighter than facial features such as the eyes, lips, and nostrils, an observation that is used for the construction of a binary face mask where these eyes and nostrils are clearly identified. The coordinates of the nostrils and eyes are recovered for each frame using the first known frontal view as reference.

Since there are only two eye gaze states to identify, direct gaze and averted gaze, the problem is solved in a straightforward way by comparing the face image to a known direct gaze image as the template in order to evaluate the possible gaze state.

The eye gaze evaluation algorithm uses the four parameters r_R, r_L, α, and β, where

$$r_R = \frac{|AC|}{|CD|}, \quad r_L = \frac{|BD|}{|CD|}$$

are defined in the geometrical model of the eyes and nostrils, as shown in Figure 6.9. These parameters are immediately recovered once the positions of the eye pupils and nostrils are identified in the frontal view of the human face from the sequence of video frames.

These parameters are then used to define a global parameter $D = WT \cdot P$, where WT is a weight vector $WT = [w_1, w_2, w_3, w_4]$, and $P = [\alpha, \beta, r_R, r_L]^T$.

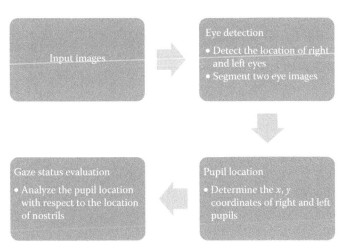

FIGURE 6.8
General steps of proposed eye gaze detection.

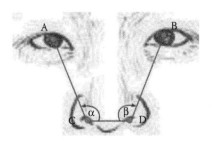

FIGURE 6.9
Geometrical eye and nostril model.

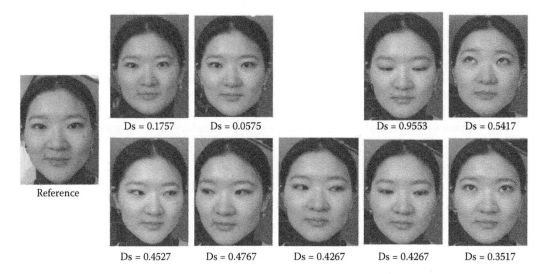

Reference

Ds = 0.1757   Ds = 0.0575   Ds = 0.9553   Ds = 0.5417

Ds = 0.4527   Ds = 0.4767   Ds = 0.4267   Ds = 0.4267   Ds = 0.3517

FIGURE 6.10
Example of gaze detection based on the $|D - D_0|$ global parameter difference.

A fuzzified *Eye Gaze* evaluation parameter is then calculated as follows:

$$Eye\ Gaze = \begin{cases} 1, & \overline{Ds}_{normal} = 0 \\ 1 - \overline{Ds}_{normal}, & otherwise \end{cases} \quad \text{where } \overline{Ds}_{normal} \text{ is the normalized average } Ds,$$

where $Ds = D - D_0$ is the difference between the current value and the reference value (D_0) calculated for the reference frame in which the subject assumes the direct gaze.

Illustrating this method, Figure 6.10 shows the normalized Ds values for different eye gaze directions. The two direct gaze examples on the top left images have lower Ds value than the other averted gaze examples.

6.2.4 Emotion Evaluation

Human emotions greatly depend on the intensity of facial expressions, the frequency and orientation of head movements, and the direction of eye gazes.

An efficient procedure for the evaluation of expressions could be implemented in a human-like fashion using a fuzzy inference procedure fashion [41] based on a combination of facial expressions, frequency of head movements, and direction of eye gaze.

Figure 6.11 illustrates the taxonomy of the human head typical attributes that are used as input fuzzy sets for the fuzzy inference engine that evaluates the facial emotions. These 17 basic head attributes, including 10 facial expressions, 5 head motions, and 2 eye gazes, can be used as vocabulary of a fuzzy-type *human-head language* for nonverbal communication.

The linguistic attributes and the set of rules used for classifying the state of emotion are defined by human experience common sense and collective wisdom.

The emotion evaluation system is based on a standard Mamdani fuzzy inference engine [41] shown in Figure 6.12 having five fuzzy inputs (happiness, anger, sadness, head movement, and eye gaze) and three fuzzy outputs.

The system, based on a standard Mamdani fuzzy inference engine [41], is extendable allowing the addition of more fuzzy variables, changing their fuzzy membership functions, and modifying the fuzzy rules.

Figure 6.13 shows the fuzzy membership functions for the three facial expression input variables: *happiness*, *anger*, and *sadness*, and for the head movement and eye gaze input variables.

The emotional states were divided into three fuzzy outputs *emotion set-A*, *emotion set-B*, and *emotion set-C* defined by fuzzy membership functions as shown in Figure 6.14.

The first emotional output, *set-A*, is characterized by four fuzzy membership functions: *thinking*, *decline*, *admire*, and *agree*. The second emotional output, *set-B*, is characterized by three fuzzy membership functions: *disagree*, *annoyed*, and *resentment*. The third emotional output, *set-C*, is characterized by three fuzzy membership functions: *disappointed*, *guilty*, and *distressed*.

FIGURE 6.11
Taxonomy of the *human-head language* attributes.

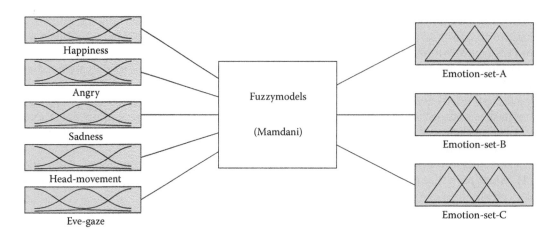

FIGURE 6.12
Fuzzy inferences system for multimodal emotion evaluation.

The *fuzzy inference rule base* consists of the following 11 rules based on human expert-opinions:

Rule 1: IF (*Happiness is Very-Happy*) AND (*Head-Movement is High-Frequency-Nodding*) AND (*Eye-Gaze is Direct-Gaze*) THEN (*Emotion-Set-A is Agree*)

Rule 2: IF (*Happiness is Happy*) AND (*Head-Movement is Low-Frequency-Nodding*) AND (*Eye-Gaze is Direct-Gaze*) THEN (*Emotion-Set-A is Admire*)

Rule 3: IF (*Happiness is Little-Happy*) AND (*Head-Movement is Low-Frequency-Shaking*) AND (*Eye-Gaze is Direct-Gaze*) THEN (*Emotion-Set-A is Decline*)

Rule 4: IF (*Happiness is Maybe-Happy*) AND (*Head-Movement is Low-Frequency-Nodding*) AND (*Eye-Gaze is Avert-Gaze*) THEN (*Emotion-Set-A is Thinking*)

Rule 5: IF (*Happiness is Maybe-Happy*) AND (*Head-Movement is Stationary*) AND (*Eye-Gaze is Avert-Gaze*) THEN (*Emotion-Set-A is Thinking*)

Rule 6: IF (*Angry is Very-Angry*) AND (*Head-Movement is Low-Frequency-Shaking*) AND (*Eye-Gaze is Direct-Gaze*) THEN (*Emotion-Set-B is Resentment*)

Rule 7: IF (*Angry is Maybe-Angry*) AND (*Head-Movement is High-Frequency-Shaking*) AND (*Eye-Gaze is Direct-Gaze*) THEN (*Emotion-Set-B is Disagree*)

Rule 8: IF (*Sadness is Very-Sad*) AND (*Head-Movement is Low-Frequency-Shaking*) AND (*Eye-Gaze is Avert-Gaze*) THEN (*Emotion-Set-C is Distressed*)

Rule 9: IF (*Sadness is Sad*) AND (*Head-Movement is High-Frequency-Nodding*) AND (*Eye-Gaze is Avert-Gaze*) THEN (*Emotion-Set-C is Guilty*)

Rule 10: IF (*Sadness is Maybe-Sad*) AND (*Head-Movement is Low-Frequency-Shaking*) AND (*Eye-Gaze is Direct-Gaze*) THEN (*Emotion-Set-C is Disappointed*)

Rule 11: IF (*Angry is Angry*) AND (*Head-Movement is High-Frequency-Shaking*) AND (*Eye-Gaze is Avert-Gaze*) THEN (*Emotion-Set-B is Annoyed*)

The fuzzy-inference engine is easily extendable allowing the addition of more fuzzy variables, changing their fuzzy membership functions, and modifying the fuzzy rules in order to accept other human body language attributes, such as hand gestures [42], body postures, and so on for a more complex and refined emotion evaluation [43,44].

The resulting multimodal fuzzy emotion evaluation system based on facial expression, head movement, and eye gaze attributes was validated using 88 video clips under different background conditions.

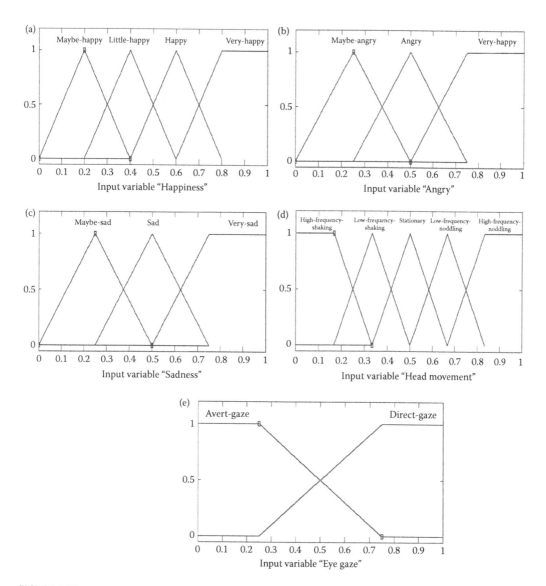

FIGURE 6.13
Fuzzy membership functions for the five input variables. (a) Happiness, (b) anger, (c) sadness, (d) head-movement, and (e) eye-gaze.

Figure 6.15 shows as an example a sequence of images, captured from a video clip at 250 ms intervals, of a female subject showing the *admire* emotional output state.

6.3 Emotional Computer–Human Interaction

Computers can communicate to humans nonverbal human-style emotional messages using animated human avatars able to display a wide range of facial expressions.

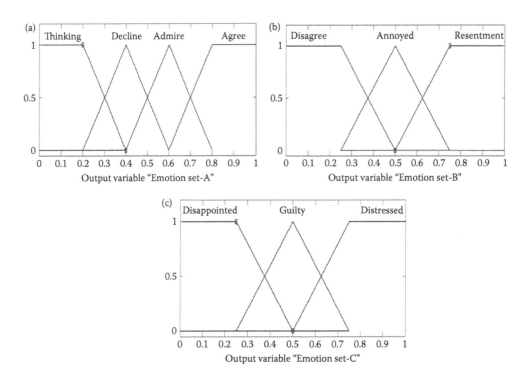

FIGURE 6.14
Fuzzy membership functions for the three output variables. (a) Emotion set-A, (b) emotion set-B, and (c) emotion set-C.

FIGURE 6.15
Image sequence of female subject showing the *admire* emotion state.

Human face serves to identify a person, reflect an individual's emotional state, allowing other people to relate to that person. Therefore, the control mechanism used to stimulate facial expressions plays an important role in the facial animation framework.

Facial Animation Using Shape Interpolation: This facial animation technique originates from the early work of Parke [45]. The process interpolates among a set of 3D key facial postures obtained by different computer vision methods (stereophotogrammetry, structured light, 3D-digitizers, etc.). This set of facial postures has the property of topological equivalence of the wireframe facial models. Once a complete set is derived, then in between interpolation of the (x, y, z) vertex coordinates can be computed. Parke's algorithm generates the frames between two chosen key-pose expressions. This approach requires complete specification of the model geometry for each facial expression, the result being a very labor-intensive technique.

DeGraf and Wahrman [46] used puppetry together with shape interpolation to animate expressive characters. Unfortunately, this technique is limited to the matrix of facial postures, which prompted the development of parameterization schemes.

Facial Animation Using Key-Node Parameterization: In his PhD dissertation [47], Parke pioneered the idea of the animation of facial expressions based on a parametric control algorithm. This parameterization reduces the face model to a small set of control parameters that can be "hard-wired" into a particular facial geometry. These parameters are only loosely based on the dynamics of the face. For example, the expression on the eyebrow from surprise involves the manipulation of five or six vertices of the facial geometry. Parke proposed two major categories of facial parameters [47]: *conformation parameters*, which control the structure of a face, and *expression parameters*, which control its emotional content. Interpolating between key-node positions updates the node positions after a change of parameter values.

6.3.1 Physically Based Animation

Physically based animation attempts to represent the shape and dynamic changes of a face by modeling the underlying properties of facial tissue and muscle actions. Most of these models are based on mass-spring meshes and spring lattices, with muscle actions approximated by a variety of force functions [48,49].

Platt and Balder [48] proposed a *muscle-based animation* technique using a multidomain model of the human face consisting of skin nodes, muscle nodes, and bone nodes. In order to animate this face model, they used a physically based elastic behavior connecting 3D facial nodes by springs. Applying forces to this elastic skin mesh via the underlying muscles generates a facial expression. For example, when a single muscle fiber contracts, a force is applied to a muscle point in the direction of its bone point. The force is then reflected along all arcs adjacent to the point. These reflected forces are then applied to their corresponding adjacent points. In this fashion, a force is propagated out from the initiating point across the face. Based on earlier successful applications of physics-based modeling of deformable objects in computer graphics [50,51], Terzopoulos and Waters extended this method to the animation of human faces [49,52]. They proposed a deformable face using facial muscles and forward dynamic skin models based on the biological structure of human skin [49]. Physically based models are computationally expensive and difficult to control with force-based functions.

Waters [33] introduced a simplified version of the physically based animation using only facial muscles for the activation of the neighboring nodes in a facial mesh. The model mimics at a simple level the action of three primary muscle groups of the face (Figure 6.16). It uses a simple spring-mass model for muscles, which have vector properties and are independent of the underlying bone structure.

Waters' method offers distinct advantages: (1) independence of particular facial geometry, (2) yields a reduced set of facial parameters (muscle activation values) convenient for low-bit-rate communication, and (3) low computation time.

Each muscle functions independently. The muscles control parameters are (1) location on bone and skin, (2) zones of influence, and (3) type of contraction profile (linear, nonlinear).

Muscles are mapped to AUs in FACS. Consequently, orchestrating a proper sequence of AU muscle contractions can create desired facial expressions. Muscle control parameters

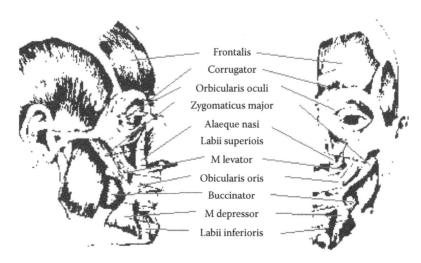

FIGURE 6.16
Facial muscles. (From Waters, K, *Computer Graphics*, 21(4), 1987. With permission.)

are then scripted into a keyframe animation system. At each frame in the sequence, muscles are contracted that in turn deform the facial geometry. The resulting geometry is rendered in accordance with viewpoint, light source, and skin reflectance information.

6.3.2   3D Head and Facial Animation

Figure 6.17 presents the architecture of the 3D head and facial animation system developed by Cordea [53,54].

Facial expressions are described using the FACS. It supports 46 AUs, among which 37 are muscle-controlled and 11 do not involve facial muscles. This allows for a muscle-level programming of specific facial expressions, as shown in Figure 6.18.

Two types of "controllers" control the whole facial deformation at the low level: (1) *muscle* controllers and (2) *jaw*, *eyelids*, and *eyes* controllers.

Eyes are constructed as spheres with iris and pupil. The eyes rotate simultaneously in their coordinate system. Eyelids are wireframe models, which can blink during the animation.

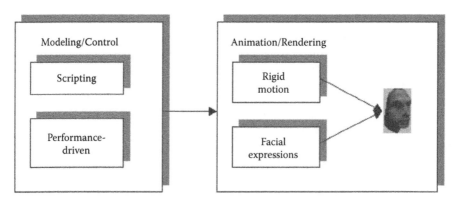

FIGURE 6.17
The architecture of the 3D head and facial animation system.

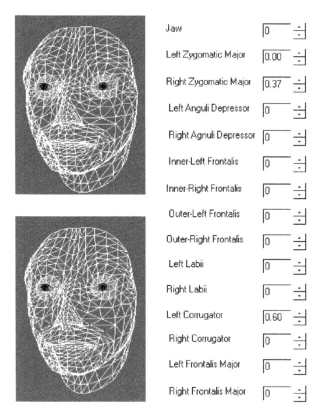

Jaw	0
Left Zygomatic Major	0.00
Right Zygomatic Major	0.37
Left Anguli Depressor	0
Right Agnuli Depressor	0
Inner-Left Frontalis	0
Inner-Right Frontalis	0
Outer-Left Frontalis	0
Outer-Right Frontalis	0
Left Labii	0
Right Labii	0
Left Corrugator	0.60
Right Corrugator	0
Left Frontalis Major	0
Right Frontalis Major	0

FIGURE 6.18
The muscle control of the wireframe model of the face.

Using combinations of muscle actuators, it is possible to create a wide range of emotional expressions, as, for instance, the six primary facial expressions that communicate anger, disgust, fear, happiness, sadness, and surprise, as shown in Figure 6.19.

Speech-Driven Lip Animation: Bondy et al. [55] present a speech-driven lip animation method based on the strong correlation that exists between the shape of the speaker's lips contour and the speaker's voice. This allows for the speech to directly drive the shape of the lips contour in the model-based animation of the speaker's face.

This animation technique uses a parametric speaker-dependent model of the speaker's lips and a Linear Predictive Coding (LPC) analysis and LPC cepstral coefficients for the recognition of the speech. The 7-parameter model adequately describes the shape of the lips used to animate the lips of the speaker's face model.

6.4   Conclusion

This chapter presents bidirectional, human-to-computer and computer-to-human, affective interaction techniques developed by authors. The human-to-computer interaction modality, described in Section 6.2, is based on a combination of head movement, eye gaze, and facial expression recognition techniques. The computer-to-human interaction modality described

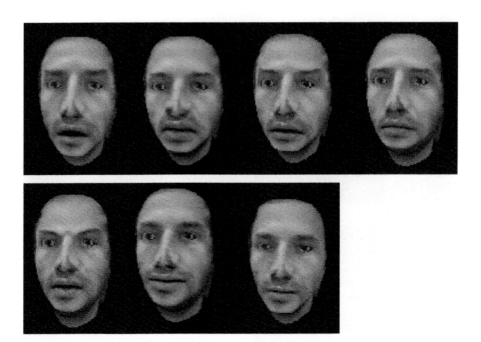

FIGURE 6.19
Fundamental facial expressions generated by the 3D muscle-controlled facial animation system: surprise, disgust, fear, sadness, anger, happiness, and neutral position. (From M. D. Cordea, *Real Time 3D Head Pose Recovery for Model Based Video Coding*, MASc thesis, University of Ottawa, ON, Canada, 1999. With permission.)

in Section 6.3 allows the computer to communicate to humans nonverbal human-style emotional messages using animated human avatars able to display a wide range of facial expressions.

The authors gratefully acknowledge the contributions and assistance of past and current colleagues from the University of Ottawa: Michel Bondy, David Chen, Miriam Goubran, Muhammad Abid, Feng Shi, and Xin Wang.

References

1. R. Picard, *Affective Computing*, Cambridge, MA: MIT Press, 1997.
2. R. A. Calvo, S. D'Mello, Affect detection: An interdisciplinary review of models, methods, and their applications, *IEEE Transactions on Affect Computing*, 1, 18–37, 2010.
3. T. Balomenos, A. Raouzaiou, S. Ioannou, A. Drosopoulos, K. Karpouzis, S. Kollias, Emotion analysis in man-machine interaction system, *LNCS*, 3361, 318–328, 2005.
4. R. W. Picard, E. Vyzas, J. Healey, Toward machine emotional intelligence: Analysis of affective physiological state, *IEEE Transactions on Pattern Analysis and Machine Intelligence*, 23(10), 1175–1191, 2001.
5. R. W. Picard, Human-computer coupling, *Proceedings of IEEE*, 86(8), 1803–1807, 1998.
6. D. P. Mahoney, Better than real, *Computer Graphics World*, 22, 32–40, February 1999.
7. E. M. Petriu, T. E. Whalen, Computer-controlled human operators, *IEEE Instrumentation and Measurement Magazine*, 5(1), 35–38, 2002.
8. R. A. Calvo, S. D'Mello, Affect detection: An interdisciplinary review of models, methods, and their applications, *IEEE Transactions on Affect Computing*, 1, 18–37, 2010.

9. Z. Zeng, M. Pantic, G. I. Roisman, T. S. Huang, A survey of affect recognition methods: Audio, visual, and spontaneous expressions, *IEEE Transactions on Pattern Analysis and Machine Intelligence*, 31(1), 39–58, 2009.

10. C. R. Darwin, *The Expression of Emotions in Man and Animals*, New York: Appleton, 1896.

11. P. Ekman, W. Friesen, *Facial Action Coding System: A Technique for the Measurement of the Facial Movement*, Palo Alto: Consulting Psychologists Press, 1977.

12. P. Ekman, W. V. Friesen, *EMFACS-7: Emotional Facial Action Coding System*, Unpublished manuscript, University of California, San Francisco, 1983.

13. P. Ekman, W. Friesen, *Manual for the Facial Action Coding System*, Palo Alto: Consulting Psychologists Press, 1986.

14. B. Fasel, J. Luettin, Automatic facial expression analysis: A survey, *Pattern Recognition*, 36(1), 259–275, 2003.

15. M. Pantic, L. J. M. Rothkrantz, Automatic analysis of facial expressions: The state of the art, *IEEE Transactions on Pattern Analysis and Machine Intelligence*, 22(12), 1424–1445, 2000.

16. Y. Tian, T. Kanade, J. Cohn, *Handbook of Face Recognition*, Heidelberg: Springer-Verlag, 2003.

17. M. Pantic, L. J. M. Rothkrantz, Expert system for automatic analysis of facial expression, *Image and Vision Computing Journal*, 18(11), 881–905, 2000.

18. J. F. Cohn, A. J. Zlochower, J. J. Lien, T. Kanade, Feature-point tracking by optical flow discriminates subtle differences in facial expression, *Proceedings of International Conference on Automatic Face and Gesture Recognition*, Nara, Japan, pp. 396–401, 1998.

19. G. J. Edwards, T. F. Cootes, C. J. Taylor, Face recognition using active appearance models, *Proceedings of the European Conference on Computer Vision*, Freiburg, Germany, 2, 581–695, 1998.

20. H. Hong, H. Neven, C. von der Malsburg, Online facial expression recognition based on personalized galleries, *Proceedings of International Conference on Automatic Face and Gesture Recognition*, Nara, Japan, pp. 354–359, 1998.

21. C. L. Huang, Y. M. Huang, Facial expression recognition using model-based feature extraction and action parameters classification, *Journal of Visual Communication and Image Representation*, 8(3), 278–290, 1997.

22. M. Yoneyama, Y. Iwano, A. Ohtake, K. Shirai, Facial expressions recognition using discrete hopfield neural networks, *Proceedings of International Conference on Information Processing*, Santa Barbara, CA, 3, 117–120, 1997.

23. M. J. Lyons, J. Budynek, S. Akamatsu, Automatic classification of single facial images, *IEEE Transactions on Pattern Analysis and Machine Intelligence*, 21(12), 1357–1362, 1999.

24. I. Essa, A. Pentland, Coding, analysis interpretation, recognition of facial expressions, *IEEE Transactions on Pattern Analysis and Machine Intelligence*, 19(7), 757–763, 1997.

25. M. Wang, Y. Iwai, M. Yachida, Expression recognition from time-sequential facial images by use of expression change model, *Proceedings of International Conference on Automatic Face and Gesture Recognition*, Nara, Japan, pp. 324–329, 1998.

26. Z. Zhang, M. Lyons, M. Schuster, S. Akamatsu, Comparison between geometry-based and gabor wavelets-based facial expression recognition using multi-layer perceptron, *Proceedings of International Conference on Automatic Face and Gesture Recognition*, Nara, Japan, pp. 454–459, 1998.

27. T. F. Cootes, C. J. Taylor, D. Cooper, J. Graham, Active shape models—Their training and application, *Computer Vision and Image Understanding*, 1(61), 38–59, 1995.

28. M. J. Black, Y. Yacoob, Recognizing facial expressions in image sequences using local parameterized models of image motion, *International Journal of Computer Vision*, 25(1), 23–48, 1997.

29. M. J. Lyons, S. Akamatsu, M. Kamachi, J. Gyoba, Coding facial expressions with Gabor wavelets, *Proceedings of IEEE International Conference on Automatic Face and Gesture Recognition*, Nara, Japan, pp. 200–205, 1998.

30. FGnet-IST-2000-26434, Face and Gesture Recognition Working Group. Available at http://www.mmk.ei.tum.de/~waf/fgnet/feedtum.html

31. M. Pantic, M. F. Valstar, R. Rademaker, L. Maat, Web-based database for facial expression analysis, *Proceedings of IEEE Conference on Multimedia and Expo*, Amsterdam, the Netherlands, pp. 317–321, 2005. Available at http://www.mmifacedb.com

32. M. D. Cordea, E. M. Petriu, A 3-D anthropometric-muscle-based active appearance model, IEEE Transactions on Instrumentation and Measurement, 55(1), 91–98, 2006.
33. Waters, K, A muscle model for animating 3D facial expressions, *Computer Graphics*, 21(4), 1987.
34. L. Farkas, *Anthropometry of the Head and Face*, Raven Press, New York, 1994.
35. M. D. Cordea, E. M. Petriu, D. C. Petriu, Three-dimensional head tracking and facial expression recovery using an anthropometric muscle-based active appearance model, *IEEE Transactions on Instrumentation and Measurement*, 57(8), 1578–1588, 2008.
36. Q. Chen, M. D. Cordea, E. M. Petriu, A. R. Varkonyi-Koczy, T. E. Whalen, Human–computer interaction for smart environment applications using hand gestures and facial expressions, *International Journal of Advanced Media and Communication*, 3(1–2), 95–109, 2009.
37. D. Vukadinovic, M. Pantic, Fully automatic facial feature point detection using Gabor feature based boosted classifiers, *IEEE International Conference on Systems, Man, and Cybernetics*, Hawaii, USA, 2, 1692–1698, 2005.
38. J. Shi and C. Tomasi, Good features to track, *Proceedings of the IEEE Computer Society Conference on Computer Vision and Pattern Recognition*, Seattle, WA, pp. 593–600, 1994.
39. C. Harris, M. Stephens, A combined corner and edge detector, *Proceedings of the 4th Alvey Vision Conference*, Manchester, UK, pp. 146–151, 1988.
40. J. Friedman, T. Hastie, and R. Tibshirani, Additive logistic regression: A statistical view of boosting, *Annals of Statistics*, 28(2), 337–374, 2000.
41. E. H Mamdani, S. Assilian, An experiment in linguistic synthesis with a fuzzy logic controller, *International Journal of Man–Machine Studies*, 7, 1–13, 1975.
42. Q. Chen, N. D. Georganas, E. M. Petriu, Hand gesture recognition using haar-like features and a stochastic context-free grammar, *IEEE Transactions on Instrumentation and Measurement*, 57(8), 1562–1571, 2008.
43. S. D'Mello, A. Graesser, Multimodal semi-automated affect detection from conversational cues, gross body language, and facial features, *User Modeling and User-Adapted Interaction*, 10147–187, 2010.
44. G. Castellano, L. Kessous, G. Caridakis, Emotion recognition through multiple modalities: Face, body gesture, speech. In C. Peter, R. Beale (Ed.), *Affect and Emotion in Human-Computer Interaction: LNCS*, Vol. 4868. Springer-Verlag, Heidelberg, 2007.
45. F. Parke, Computer generated animation of faces, *ACM National Conference*, New York, USA, pp. 451–457. ACM, 1972.
46. B. deGraf, M. Wahrman, Notes on human facial animation, *ACM SIGGRAPH Course Notes*, 26, 13–15, 1990.
47. F. I. Parke, *A Parametric Model for Human Faces*, PhD thesis, University of Utah, UTEC-CSS-75–074, 1974.
48. S. M. Platt, N. I. Badler, Animating facial expressions, *Computer Graphics*, 15(3), 245, 1981.
49. D. Terzopoulos, K. Waters, Physically-based facial modeling, analysis, and animation, *Visualization and Computer Animation*, 1, 73–80, 1990.
50. D. Terzopolous, A. Witkin, M. Kass, Constraints on deformable models: Recovering 3D shape and nonrigid motion, *Artificial Intelligence*, 36, 91–123, 1988.
51. D. Terzopoulos, J. Platt, K. Fleischer, Heating and melting deformable models, *Proceedings of Graphics Interface '89*, Ontario, Canada, pp. 219–226, June 1989.
52. K. Waters, *Modelling Three-Dimensional Facial Expressions*. Shlumberger Laboratory for Computer Science, Austin, TX in *Processing Images of Faces* by Bruce V. and Burton M. (eds), Ablex, New Jersey, 1992.
53. M. D. Cordea, *Real Time 3D Head Pose Recovery for Model Based Video Coding*, MASc thesis, University of Ottawa, ON, Canada, 1999.
54. M. D. Cordea, *A 3D Anthropometric Muscle-Based Active Appearance Model for Model-Based Video Coding*, PhD thesis, University of Ottawa, ON, Canada, 2007.
55. M. D. Bondy, E. M. Petriu, M. D. Cordea, N. D. Georganas, D. C. Petriu, T. E. Whalen, Model-based face and lip animation for interactive virtual reality applications, *Proceedings of ACM Multimedia 2001*, pp. 559–563, Ottawa, ON, September 2001.

Part II

Methodology, Techniques, and Applications: Coding of Video and Multimedia Content

7

Part Overview: Coding of Video and Multimedia Content

Oscar Au and Bing Zeng

CONTENTS

One major application of multimedia is the vast arrays of consumer electronics, including front-end capturing devices such as digital cameras, camcorders, webcams, camera-phones, surveillance cameras, infrared cameras, and back-end delivery, storage, and display devices such as terrestrial television broadcasting, cable TV, satellite TV, mobile TV, Digital Video Broadcast (DVB), set-top box, peer-to-peer video streaming, YouTube, Facebook, Google, MSN, VCR, Hi-Fi, Digital Video Disk (DVD), Blu-ray, iPad, and so on. With the development of modern networking technologies, particularly high-speed Internet, we can now exchange various kinds of multimedia information with each other in any place, at any time, and using any device. Digital image and video data undoubtedly form the major trunk of the digital multimedia world.

In the raw data format, however, image and video often require a huge amount of bits to represent. For instance, today's high-definition TV (HDTV) images have spatial resolution of 1920×1080 pixels per frame. At 8 bits per pixel per color component and 30 frames per second, this translates into a rate of nearly 1.5×10^9 bps. Obviously, transmitting such images over any of today's networks requires some sort of coding to compress the raw data to fit the available channel capacity. Image and video coding is also needed in many other applications, such as image/video storage, search, and retrieval; multimedia databases/libraries; video streaming over the Internet, geographic image transmission via satellites, video phone and video conferencing, and so on.

By means of image/video compression (or coding), we can eliminate the observable redundancy (both statistical and subjective) as much as possible. This has been an exciting research topic in both the industrial and academic communities for over 40 years. As a result, many efficient coding schemes have been developed and matured to cover different kinds of applications, including differential pulse-coded modulation (DPCM)-based predictive methods, scalar and vector quantization, transform-based coding, and wavelet decompositions [1–8]. In the meantime, a number of international standards have been established, such as JPEG, JPEG2000, MPEG-1/2/4, and H.261/263/H.264 [9–17]. More recently, we have witnessed many advances in distributed video coding (DVC) and multiview video coding (MVC), as well as a big effort on the development of new techniques for the next-generation video coding standard. This part overview is going to focus on the basics of several major image and video coding standards which will serve as a basis for further elaboration of DVC and MVC and next-generation coding in the next few chapters.

In relation to these standards, there are several important international standardization bodies, including the International Organization for Standardization (ISO), the International Electro-technical Commission (IEC), the International Telecommunication Union, and Telecommunication Standardization Sector (ITU-T). ISO is a nongovernment body that sets international standards. IEC is another nonprofit, nongovernment international standard body that establishes standards related to electrical and electronic technologies. ITU-T is an agency under United Nations to deal with information and communication technologies. ISO and IEC work together to establish many joint standards, including ISO/IEC JPEG [10], JPEG 2000 [11], MPEG-1 [12], MPEG-2 [12], MPEG-4 [14], and so on. ITU-T develops many comparable video coding standards including H.261, H.263 [15], and H.264 [16].

The Joint Photographic Experts Group (JPEG) is a workgroup under ISO/IEC mandated to work on the development of the standard for the coding of still pictures, including JPEG, JPEG2000, JBIG, JPEG-LS, motion JPEG, motion JPEG2000, and so on. The Motion Picture Experts Group (MPEG) is another workgroup under ISO/IEC workgroup to work on the development of the MPEG-x series of video coding standards, including MPEG-1, MPEG-2, MPEG-4, MPEG-7, MPEG-21, MPEG-A, MPEG-B, MPEG-C, MPEG-D, MPEG-E, MPEG-H, MPEG-M, MPEG-U, and MPEG-V. The Video Coding Experts Group (VCEG) is the workgroup under ITU-T to develop visual coding standards, including H.261, H.262, H.263, H.264, T.80 to T.89, T.800 to T.812, and so on. Here we focus on the JPEG, MPEG-1, MPEG-2, MPEG-4, H.261, H.263, and H.264 standards.

7.1 The ISO/IEC JPEG Image Coding Standard

The JPEG [10] image coding standard is very popular and is used widely in digital cameras and web pages. It has four modes: sequential mode, progressive mode, hierarchical mode, and lossless mode. Sequential mode is the most common mode used often in digital cameras, camera-phones, web browsers, and so on. In JPEG sequential mode, an image is subdivided into 8×8 nonoverlapping blocks which are processed sequentially in raster scan order. For each block, three steps are applied: 8×8 discrete cosine transform (DCT), quantization, and variable length coding (VLC).

The 8×8 DCT is defined as

$$C_{pq} = \alpha_p \alpha_q \sum_{m=0}^{7} \sum_{n=0}^{7} X_{mn} \cos \frac{\pi(2m+1)p}{16} \cos \frac{\pi(2n+1)q}{16}, p,q = 0,\ldots,7,$$

$$\text{where } \alpha_p = \begin{cases} \frac{1}{\sqrt{8}}, p = 0 \\ \frac{1}{2}, p = 1,\ldots,7 \end{cases}, \alpha_q = \begin{cases} \frac{1}{\sqrt{8}}, q = 0 \\ \frac{1}{2}, q = 1,\ldots,7 \end{cases}.$$

Before the DCT transform, the 256 image pixels within an 8×8 block are 8-bit integers taking on values from 0 to 255. Each pixel value tends to be uniformly distributed in the dynamic range resulting in large statistical entropy and the pixels within the block tend to be highly correlated spatially with similar energy. After the DCT transform, the DCT coefficients are real numbers ranging from -2048 to $+2048$. In a way, DCT performs vector rotation and the coefficients have frequency and perceptual interpretation. Instead of being uniformly distributed within the dynamic range, the coefficients tend to be zero-biased leading to significantly lower entropy, with the only exception of the DC component. They tend to uncorrelate with unequal energy distributions. For natural images, the low-frequency DCT coefficients tend to have a lot of energy while the high-frequency coefficients tend to be very small. It is in this sense that DCT achieves energy compaction and spatial decorrelation. DCT is an approximation of the Karhunen Loeve transform (KLT) for the class of first-order Markov fields. KLT achieves maximum energy compaction and completely decorrelates the spatial pixel values.

Quantization is then applied to the DCT coefficients to convert them into integers. Uniform quantization with an optional dead zone is applied to each coefficient. The quantization step sizes are usually different for different DCT coefficients. The step sizes are stored in an 8×8 quantization matrix. Experiments reveal that the human visual system tends to have different perceptual sensitivity to the DCT coefficients. The human eye tends to be significantly more sensitive to distortion in low frequency than to high frequency. The just-noticeable-difference values of different DCT frequency coefficients, obtained experimentally, are often used to design the quantization matrix. As such, the quantization step size tends to be small for low-frequency DCT coefficients and large for high-frequency ones. For the low-frequency DCT coefficients, they tend to be large for natural images and are subjected to small quantization step sizes and thus their quantized values tend to be nonzero. But for high-frequency DCT coefficients, they tend to be small and are subjected to large quantization step sizes and thus their quantized values tend to be zero.

The quantized DCT coefficients are then scanned in a zigzag order to convert the two-dimensional (2D) data into one-dimensional data. This prepares the data for VLC, which can only be applied to one-dimensional data. The zigzag scanned coefficients tend to have long runs of zero, due to the many quantized high-frequency DCT coefficients being zero. The VLC is a combined runlength coding and Huffman coding. The runlength coding is particularly effective in representing the long runs of zeros. For the Huffman code, shorter code words tend to be given to shorter runlengths.

In the JPEG progressive mode, an image is scanned in multiple scans and encoded with progressively refined details for reconstruction in multiple coarse-to-fine passes. In the JPEG hierarchical mode, an image is encoded at multiple resolution levels so that the lower-resolution image can be accessed without decompressing the higher resolution. In the JPEG lossless mode, an image is encoded with no distortion to guarantee exact recovery of every source image.

7.2 The Video Coding Standards: MPEG-1, MPEG-2, MPEG-4, H.261, and H.263

H.261 [116] was an early standard for video conferencing at p*64 kbit/s where p can be an integer between 1 and 30. MPEG-1 [12] was the standard for encoding VCR-quality video (Common Intermediate Format (CIF) resolution, i.e., 352×288) at 1.5 Mbit/s and was used in the successful Video Compact Disc. MPEG-2 [13] was an improved standard for encoding broadcasting quality video (CCIR-601 resolution, e.g., 720×576) and was used in the hugely popular DVD, DVB, and so on. H.262 and MPEG-2 are the same, as they are developed jointly by MPEG and ITU-T. H.263 [15] is an improved standard over H.261 and is a popular format for video conferencing and video phone. MPEG-4 [14] was an improved standard over MPEG-2 with much additional functionality such as object-based coding. It is mildly successful, and is used for video downloading and streaming.

In these video coding standards, a video sequence is typically organized into Group-of-Picture or GOP structure with essentially three frame types: I-frames, P-frames, and B-frames. One GOP has one I-frame followed by many P-frames and B-frames. Some common GOP structures include IPPP, IBPBP, and IBBP. A GOP may be 2s long in DVB and DVD or 10s long in video conferencing. In these standards, the I-frame coding is very similar to the sequential mode of JPEG. Without predictive coding, I-frames have lower compression efficiency than P-frames and B-frames. But without the dependency on previous frames, an I-frame provides a random access point for a GOP, which is very important for channel switching during broadcasting and for fast-forwarding in DVD. It is also an error resilience tool by stopping error propagation from one GOP to another, and providing resynchronization point for the next error-free GOP.

Both P-frames and B-frames use hybrid coding, with a DPCM feedback loop combining predictive coding and transform coding. In P-frame coding, each current frame is subdivided into 16×16 nonoverlapping macroblocks (MB). For each current MB, a search region is defined in the reference frame, which is typically the previous frame, and unidirectional motion estimation (ME) is performed in search of a best-matched MB in the reference frame. Translational motion is assumed for each MB and the translation, or displacement, is described by a 2D motion vector (MV). The MV precision is integer-pixel for MPEG-1 and H.261, and is increased to half-pixel in MPEG-2 and H.263 and MPEG-4. It can be optionally chosen to be quarter-pixel in MPEG-4. For each candidate MV, the typical cost function used in ME is the sum of absolute difference (SAD), which is

$$\text{SAD} = \sum_{m=1}^{M} \sum_{n=1}^{N} |x_{mn} - p_{mn}|,$$

where M and N are the height and width of one MB, respectively, x_{mn} is the original pixel value at the position (m, n) of the current block, and p_{mn} is the pixel in the reconstructed frame corresponding to the candidate MV.

Half-pixel values are obtained using a simple 2-tap averaging filter. When the best-matched candidate block is found, it is subtracted from the current MB to form a residue block, which is then encoded like JPEG. In other words, a 16×16 residue block is subdivided into four 8×8 blocks, and then 8×8 DCT, scalar quantization, and VLC are applied to each block. The corresponding MV of the MB is encoded and transmitted to the decoder. A simple in-loop deblocking filter is introduced in MPEG-1 and H.261 to suppress the blocking artifacts in the reconstructed frame so as to achieve better motion-compensated prediction, but is removed in MPEG-2 and H.263 as the half-pixel filter can achieve deblocking. Sometimes, the translational motion assumption is badly wrong and ME just does not work. In such cases, an MB can be encoded using intramode, which has no ME and compensation.

B-frame coding is similar to P-frame coding except that there are two reference frames, one in the past and one in the future. For each MB in B-frame, unidirectional ME is performed in the past reference frame (called forward prediction) and also in the future frame (called backward prediction). In addition, bidirectional ME with respect to both reference frames is performed. So for B-frames, there are four possible modes for each MB: intramode with no MV, forward or backward prediction each with one MV, and bidirectional prediction with two MVs. Bidirectional mode is useful when the object appears in both past and future reference frames. Forward mode is useful when the object appears in the past but not the future reference frames, such as in newly covered regions. Backward mode is useful when the object appears in the future but not the past reference frames, such as in newly uncovered regions. When none of the three modes work, the intramode is the backup.

In MPEG-2, frame/field coding is supported in that a 16×16 MB can be subdivided into two 16×8 subblocks, with one containing all the odd rows and the other containing all the even rows. ME is performed on each 16×8 subblock leading to a total of two MVs per MB. Such odd and even subblocks are very effective for interlace video in common broadcasting television systems. In H.263, there are several useful tools. First, a 16×16 MB can be subdivided into four nonoverlapping 8×8 blocks with separate ME such that the MB has four MVs. The smaller blocks tend to achieve better prediction for an MB undergoing rotation and an MB with two objects. Second, the MVs are "unrestricted" in that they are allowed to point outside the reference frame, with the outside pixels generated using mirror reflection. Such unrestricted ME is especially useful for zooming and panning. Third, overlapped block motion compensation (OBMC) is supported. In OBMC, a 24×24 predicted block is copied for each 16×16 MB such that adjacent predicted blocks are overlapping. When weighted averaging is applied to the pixels with multiple predicted pixels, the predicted frame can become less blocky and thus give better prediction.

As the VLC output does not have fixed length but the bandwidth requirement is often fixed, a buffer is needed to smooth out the output bit rate. To prevent buffer overflow or underflow, rate control is supported in both P-frames and B-frames by adjusting a quantization factor QP of each MB which scales the quantization step size for all the DCT coefficients.

There are two main optimization problems associated with video encoding using these standards: fast ME and rate control. While ME is essential for video coding, its computational complexity is extremely high. For example, for a 1080 p video at 30 fps with a modest search window of 64×64, the unidirectional integer pixel ME using exhaustive full search will take 5.0×10^{11} which is excessive even for a 3 GHz quad-core CPU. It is thus very important to reduce the computational complexity of ME. A large number of fast search algorithms were developed in the past two decades. Many algorithms seek to reduce the number of search points for SAD computation. New 3-step search (N3SS) [18] and 4-step search (4SS) [19] assume a local unimodel error surface and perform local hierarchical search, with decreasing step size. Circular [20], diamond [21], and hexagonal [22] zonal search extend such searches by (a) using nonrectangular search pattern, (b) allowing the search center to move, and (c) reducing the step size only when the minimum is at the search center. Another fast strategy is half-stop [23] or early termination [24]. The search is stopped whenever the situation is good enough, such as when the current minimum SAD is less than a certain threshold. While setting the threshold can be tricky, adaptive thresholds [25] can be used to adapt to different video characteristics. Some algorithms seek to simplify the distortion computation, replacing SAD by some simple alternative features such as integral projection [26] or 1-bit feature [27]. Some algorithms seek to subsample the motion field [28], perform more elaborated search for some MB and simplified search for the other. Some seek to calculate partial SAD [29] and skip the search point when it is found to be hopelessly large. Some make use of MV predictors [30] to get close to the global minimum early.

For rate control, it aims to achieve the best visual quality under the storage size or communication bandwidth constraint. There are many rate control methods in the literature, and one good review can be found in [31]. During the rate control design, different issues should be addressed upon its specific applications. For example, in real-time video coding systems, the computational complexity and the end-to-end delay should be reduced in order to avoid the possible delay. For constant bit-rate (CBR) applications, the rate control algorithm should prevent the decoder buffer having overflow or underflow. Although the rate control is a nonnormative part, there is one recommended implementation of rate control in the reference software of each video coding standard, for example, reference model [32] for H.261, adaptive quantization algorithm for MPEG-1 [33], test model (TM) [34] for MPEG-2, test model near-term [35] for H.263, verification model [36] for MPEG-4, and joint model (JM) [37] for H.264. In these standard specific rate control designs, usually two different scenarios: CBR and variable bit rate are considered. The recent research pays more attention to the rate control method in H.264. In JM, the concept of basic unit is proposed in the rate control. One basic unit can be an MB, a slice, or a frame, and the size of one basic unit depends on the coded picture buffer size. All MBs in one basic unit share a common quantization parameter. To solve the chicken and egg dilemma existing in the rate-distortion optimization (RDO) and QP decision [38], a linear model is used in the rate control of H.264 [37]. Afterward, in the literature, many other improved rate control methods have been designed for H.264. Most of them aim to find a more accurate rate-quantization (RQ) model to estimate the relationship between the bit rate and quantization step size, for example, Cauchy distribution-based RQ model [39], Laplacian mixture-based RQ model [40], enhanced RQ model [41], linear RQ model [42], and rho-domain RQ model [43]. Recently, in [44], both the quantization step size and the rounding offset are selected as the tuning parameters to achieve the target bit rate.

7.3 The State-of-the-Art Video Coding Standards: H.264, Audio Video Coding Standard (AVS), and VC-1

After the H.263 standard was finished, a project called H.26L [45] continued in ITU-T VCEG to explore a long-term video coding solution with significantly better coding performance than H.263. In some comparative experiments, H.26L was found to outperform MPEG-4 significantly leading to the formation of the Joint Video Team between MPEG and VCEG to further the work. As such, the resulting standard has two names. On the ITU-T side, it is called H.264 [16]. On the MPEG side, it is called MPEG-4 AVC (Advanced Video Coding). H.264 is currently the state-of-the-art widely deployed video coding standard. It can outperform MPEG-4 by almost two times. It is widely used in Blu-ray, HDTV, satellite TV, cable TV, DVB, and so on.

However, the H.264 licensing terms [46] can be prohibitive in countries with high population density. As a result, China, being one such country, developed its own coding standard called Audio Video coding Standard (AVS) [47,48] which is comparable to H.264 without the burden of the H.264 patent licensing terms. As a few key H.264 coding tools are removed or simplified, the rate-distortion (R-D) performance of AVS tends to be worse than H.264 but still better than MPEG-4.

Meanwhile, Microsoft developed and used its own video coding technology for its Windows Media Video Player software called WMV9 [49]. Based on WMV9, a standard called

VC-1 [50] is developed in the Society of Motion Picture and Television Engineers. It has similar coding performance as H.264. Actually, the tools in the H.264, AVS, and VC-1 standards are similar, with some small differences in details. Thus, we mainly focus on H.264 in this chapter.

Similar to past standards, H.264 also has I-frame, P-frame, and B-frame coding. In I-frame, it performs intraprediction, cosine transform, scanning, and VLC. In P-frame, unidirectional ME and compensation are applied. In B-frame, bidirectional ME and compensation are applied. While they sound the same as past standards, there are several important differences. Innovative tools in H.264 I-frame coding include intraprediction [51], integer cosine transform (ICT) [52], and context-adaptive binary arithmetic coding (CABAC) [53]. Innovative H.264 P-frame and B-frame coding tools include MV of 1/8 pixel precision [54], multiple reference frames [55], marcoblock partitioning [56], and in-loop deblocking filter [57].

For the first time in coding history, intraprediction [58] in H.264 allows predictive coding to be applied also in I-frames. An I-frame can be divided into nonoverlapping 4×4, 8×8, or 16×16 blocks. It is assumed that, a 4×4 subblock is either smooth or contains at most one dominant straight edge. There are nine intraprediction modes, which are shown in Figure 7.1. For each 4×4 or 8×8 subblock, the reconstructed top and left neighboring pixels are extended in eight possible directions to form directional zero-order-hold predictions: horizontal, vertical, diagonal, antidiagonal, and multiples of 22.5°. There is also a 9th mode: the DC mode which calculates the average of the neighboring pixels to form a flat prediction (DC only). The intraprediction mode is encoded by predictive coding. A most-likely mode (MLM) is generated based on two context elements: the intraprediction mode of the left block and the top block. One bit is used to indicate the mode if it happens to be the MLM. If not, 4 bits are used to indicate one of the eight remaining possibilities. For 16×16 blocks, there are only four directional prediction modes: horizontal, vertical, DC, and planar. Planar mode is basically a first-order, instead of zero-order, prediction and is good for smooth regions at the price of higher complexity. The directional prediction in the H.264 I-frames is effective and helps to achieve significantly higher coding efficiency than the nonpredictive I-frames in MPEG-1, MPEG-2, MPEG-4, H.261, and H.263.

As intraprediction is performed on blocks as small as 4×4, it is natural that H.264 uses 4×4 ICT [16]. While DCT tends to achieve very good energy compaction and decorrelation, its definition contains transcendental function making it impossible to guarantee perfect match between the encoder and the decoder. Both ICT and inverse ICT use integers in its definition and guarantees perfect match between the encoder and the decoder. The 4×4

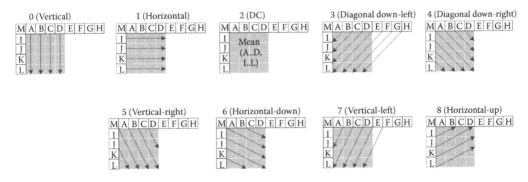

FIGURE 7.1
9-Mode intraprediction for 4×4 blocks.

$$H = \begin{bmatrix} a & a & a & a \\ b & c & -c & -b \\ a & -a & -a & a \\ c & -b & b & -c \end{bmatrix} \cdot \quad H = \begin{bmatrix} 1 & 1 & 1 & 1 \\ 2 & 1 & -1 & -2 \\ 1 & -1 & -1 & 1 \\ 1 & -2 & 2 & -1 \end{bmatrix} \cdot \quad \tilde{H}_{inv} = \begin{bmatrix} 1 & 1 & 1 & 1/2 \\ 1 & 1/2 & -1 & -1 \\ 1 & -1/2 & -1 & 1 \\ 1 & -1 & 1 & -1/2 \end{bmatrix} \cdot$$

FIGURE 7.2
4×4 ICT and inverse ICT matrices in H.264.

ICT and inverse ICT matrices in H.264 are shown in Figure 7.2. Within a 16×16 MB, Hadamard transform is applied to the DC components of the 16 4×4 blocks to achieve additional energy compaction, as they tend to have large magnitude and have remaining correlation. While video denoising [59–66] is considered an important value-added processing, it is not always good. In H.264 experiments, it was found that 4×4 ICT failed to retain a good kind of noise called film grain noise [67]. When movies are made in films, they always carry a kind of film grain noise. Over the years, consumers tend to associate film grain noise with real movie. The problem of the 4×4 ICT is that it tends to remove the film grain noise making the movies not movie-like. Thus, H.264 high profile [68] also supports 8×8 ICT which can retain the film grain noise.

While past standards usually use Huffman coding to encode the DCT coefficients and other mode information, H.264 introduces CABAC [58]. Arithmetic coding is well known to be more efficient than Huffman coding. Binary arithmetic coding helps to keep its complexity low. The context adaptation allows it to adapt to changing statistics so as to achieve maximum possible coding efficiency. CABAC can improve the coding efficiency by about 15%. However, CABAC is very complicated.

Consider P-frame coding of the frame at time t. In past standards, the previous reconstructed frame (frame t-1) is used as the reference frame. In H.264, one of five past reconstructed frames [58] can be chosen as the reference frame. While the most likely reference frame is still frame t-1, sometimes frame t-2, t-3, t-4, and t-5 can give significantly better prediction, such as for blinking eyes. Figure 7.3 illustrates the use of multiple reference frames. Every MB can have a different reference frame. Similarly, in B-frame coding of frame t, past standards would have one previous reconstructed frame and one future reconstructed frame as reference. In H.264, there are two lists of five past or future reconstructed frames. One frame from List0 and one from List1 are used as the reference frames in H.264 B-frame coding. The use of multiple reference frames achieves a coding gain of about 5%.

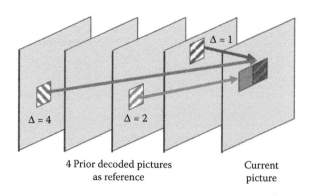

4 Prior decoded pictures
as reference

Current
picture

FIGURE 7.3
Multiple reference frame. (From J. Ostermann et al., Video coding with H.264/AVC: tools, performance, and complexity, *IEEE Circuits and Systems Magazine*, 4(1), 7–28 © (2004) IEEE. With permission.)

While past standards always use the 16×16 MB as a basic unit for ME and compensation, H.264 allows an MB to be partitioned into 8×16, 16×8, 8×8, 8×4, 4×8, and 4×4 subblocks [69]. ME and motion compensation (MC) are performed on each subblock. As ME–MC assume that the whole block contain one rigid body undergoing translational motion, they tend to fail when the block contains more than one object, or when the object motion is rotation or even deformation. With MB partitioning, smaller subblocks can be used and it is much more likely for the assumption to be true in most, if not all, of the subblocks even when it is not true for an MB. MB partitioning can give a bit-rate reduction of about 15%. As subblocks as small as 4×4 are allowed, the 4×4 ICT is matched to the smallest subblock size.

Past standards support integer-pixel and half-pixel precision of MVs. H.264 supports 1/8-pixel precision [16] of MVs making ME more efficient. In addition, while past standards use a simple half-pixel interpolation filter with two coefficients (or two taps), H.264 uses a sophisticated 6-tap half-pixel interpolation filter [70] and a simple 2-tap 1/8 pixel interpolation filter. These help to improve the coding efficiency.

It is well known that past standards tend to incur blocking artifacts in the reconstructed frames, especially when the bit rate is low. Such blocky reconstructed frames would lower the efficiency of ME and MC, because the real-world video contents are almost never blocky. Often deblocking filters are applied as postfilters to improve the visual quality of the reconstructed frames. To this end, H.264 introduces an in-loop deblocking filter [57]. It classifies block boundaries into four classes and applies a spatial varying linear deblocking filter across the boundaries to generate smooth reconstructed frames that can greatly enhance the efficiency of ME and MC and achieve improved coding efficiency. The boundary classification prevents over-smoothing at object boundaries. But the deblocking filter has very high complexity and irregularity, and is usually a major challenge in encoder and decoder designs, in both software and hardware.

7.4 The Future Trends

While H.264 enjoys wide popularity in almost all consumer and professional video systems and services, companies and universities are working hard to optimize the H.264 codec [17] in terms of software [71,72] and hardware [73]. Meanwhile, VCEG and MPEG have been working on the next-generation video coding standard. In July 2005, VCEG started the Key Technology Area (KTA) project [74] with its KTA software to gather new video coding tools. In 2007, MPEG started a similar work called High-Performance Video Coding [75]. In 2010, MPEG and VCEG formed a Joint Collaborative Team on Video Coding (JCT-VC) [76] to develop the next-generation video coding standard, called internally the High-Efficiency Video Coding (HEVC) standard, which is to achieve double the coding efficiency of H.264. A call for proposal was launched as a result of the January 2010 Kyoto Meeting. In the first JCT-VC Meeting held in Dresden in April 2010, a total of 27 high-quality proposals with matching software were submitted and compared. They showed promising results. The top five were significantly better than the rest and were asked to be combined and integrated into a Test Model under Consideration (TMuC) software. TMuC was later tested and simplified into the HEVC test model (HM) software. On the average, the JCT-VC Meetings are held once every 3 months and the HEVC development work is scheduled to end in 2013. Many major companies and universities take part in JCT-VC. As the collaborative and yet competitive work of JCT-VC is ongoing, we give some hints in this section on the research trends.

Over the past three decades, the most popular video coding framework has rooted firmly on a hybrid of prediction and transform, and such a framework will likely be used for many years to come. This framework consists of three fundamental building blocks: (1) prediction (including intraprediction as well as MC), (2) transformation, and (3) quantization. In the following, we highlight some recent advances in each of these three building blocks.

7.4.1 New Prediction Techniques

Different from the traditional DPCM-based prediction techniques, some new intrapredic-tion techniques make use of some well-known features in a typical video frame: many visual patterns exhibit a strong directionality (along horizontal, vertical, diagonal, or oth-ers). By performing a prediction along the dominating direction within each data block, one can anticipate that the residual signal now needs fewer bits to code—leading to an improved R-D performance. For instance, the H.264 standard introduces eight such direc-tional intraprediction modes (for each 4×4 block): vertical (Mode 0), horizontal (Mode 1), diagonal-down-left (Mode 3), diagonal-down-right (Mode 4), vertical-right (Mode 5), horizontal-down (Mode 6), vertical-left (Mode 7), and horizontal-up (Mode 8); along with one traditional DC mode (Mode 2). The mode decision is done for each block according to the SAD [77] or a more sophisticated RDO [77].

The directional intraprediction modes introduced in H.264 have been demonstrated to be a big success. Nevertheless, two minor drawbacks have been identified. First, the mode decision is time consuming. Second, it needs extra bits to represent the mode. A number of works had been carried out during the past several years, which developed some fast-mode decision approaches [78–86]. For the second problem, some very recent works proposed to utilize the "geometric duality" to select the intraprediction mode: a causal patch is selected from the coded region and the mode is decided by utilizing the geometric duality. Since the same mode decision can be done at the decoder side, this scheme does not need any extra bits to code the mode information. More recently, there has been an effort during the development of HEVC to include many more intraprediction modes, for example, 16 directional modes for the 4×4 block, or 33 for the 8×8, 16×16, and 32×32 blocks (in addition to the DC mode) [87].

On the other hand, ME/MC has always been playing a critical role in any video coding scheme. While the full (exhaustive) search always sets up the performance upper bound, we have witnessed a large number of fast search algorithms developed during the past two decades. Good examples include the N3SS [18], 4SS [19], zone-based searches (cir-cular [20], diamond [21], hexagonal [22], etc.) with predictors [23], directional searches [88], early termination [24], and so on. Meanwhile, we have also witnessed that the use of fractional accuracy filter (half-pixel, quarter-pixel, etc.) becomes more and more important. Overall, people have done very extensive research on ME/compensation and therefore we are perhaps very close to the limit of translational motion prediction. How-ever, there are quite a few research issues remaining, including subpixel interpolation filter [89], rotational motion [90], zooming [91], dense motion field [92], chroma prediction [93], and so on. As HEVC is aiming for HDTV and beyond (such as 4000×2000 or even 8000×4000), large block processing is promising to be important. With large blocks [94], block partitioning [94] needs to be further investigated. With all the possible new modes, mode decision [95] tends to require high computational complexity and optimization will be needed.

7.4.2 New Transformations

Since its invention in 1975, the DCT has been dominating at the transformation stage in all transform-based image and video coding schemes. This is mainly because the DCT has been proven to be the best approximation to the optimal KLT under first-order Markov conditions with a strong interpixel correlation. For an image or video frame, the DCT is always applied along each block's vertical and horizontal directions twice so as to form the so-called separable 2D transform. Only very recently, this "Bible-like monster" has been challenged seriously. First, a truly block-based directional transform framework has been proposed, where the first directional transform chooses to follow the dominating direction within each block, while the second transform can be accommodated accordingly. Very similar to the directional intraprediction modes of H.264, these directional transforms take full advantage of directional visual patterns included locally in the source image or video frame. References [96–99] have demonstrated that the directional transforms are extremely efficient for video content with directional features (such as human hair, grass, etc.). In the HEVC development, some new tools are mode-dependent directional transform [100], rotational transform [101], and combined discrete sine transform (DST)/DCT [102]. Much research work would be needed to fully understand how these work.

In the video coding scenario, however, we know that the transform is often applied on some residual signals that are obtained after the MC (in a P-frame) or intraprediction (in an I-frame). In either case, the interpixel correlation [103] may become much weaker or even negative so that the DCT surely will not work as efficiently as one expects. Moreover, after a prediction, the residual signal can hardly follow the first-order Markov conditions so that its covariance matrix may be far away from the Toeplitz-type structure. Some recent works [104,105] have clearly illustrated this feature. For instance, in the intraprediction Mode 0/1 (vertical/horizontal mode), the residual pixel in each column/row will have a bigger energy when it gets farther from the reference pixel. More complicated variations will appear in other directional intramodes (Modes 3–8). In principle, once we know how the intraprediction is done for each intramode, we can always design the most suitable transform—which usually requires some statistical knowledge, for example, the covariance matrix of the 2D block source obtained from an analytical model or experimental tests on practical data. Each of such transforms is bearing the flavor of the so-called KLT and therefore is very costly to be implemented. Nevertheless, research efforts can then be put forward to develop efficient transforms that are much less demanding computationally while offering an improved R-D coding performance over the traditional 2D separate DCT.

7.4.3 Quantization

Quantization plays a key role in any lossy coding scheme: it offers a highly reduced bit rate on the one hand (because many coefficients become zero after quantization) but is also perhaps the sole source to the resulted coding distortion, on the other. So far, quantization has been kept rather simple such that a uniform quantizer (with or without a dead zone) is commonly employed in various scenarios, especially in the block-based transform coding schemes; whereas the optimal (nonuniform) quantizers can be designed once some necessary statistical knowledge of the source is known.

Efforts have been put forward to perform a search on a few nonzero quantized coefficients according to an R-D Lagrangian principle. For instance, a few leading nonzero-quantized coefficients (along the zigzag scanning order) can be adjusted up or down by one step-size so as to produce a better R-D trade-off (Rate Distortion Optimized Quantization (RDOQ)) [106,107]. Clearly, the search space is huge if all nonzero-quantized coefficients are included into such a search process, which unfortunately would become intractable if the number of coefficients becomes large. To overcome this problem, a closed-form solution has recently been developed, which calculates a correction term analytically for each transform coefficient before the quantization is performed. This yields a highly constrained quantization technique [109], which is found to be rather useful in a few coding scenarios such as arbitrarily shaped object coding [108,109], multiple description coding [110], and hierarchical coding [111]. There will also be a need to perform research on scanning order [112], hardware-friendly parallel entropy coding [113], and block processing [114–116].

References

1. M. Yuen and H. R. Wu, A survey of hybrid MC/DPCM/DCT video coding distortions, *Signal Processing*, 70(3), 247–278, 1998.
2. T. Wedi, Motion compensation in H.264/AVC, *IEEE Transactions on Circuits and Systems for Video Technology*, 13, 577–586, 2003.
3. A. Gersho and R. M. Gray, *Vector Quantization and Signal Compression*, Boston, CA: Kluwer, 1992.
4. A. Puri and R. Aravind, Motion-compensated video coding with adaptive perceptual quantization, *IEEE Transactions on Circuits and Systems for Video Technology*, 1(4), 1284–1286, 1991.
5. K. R. Rao and P. Yip, *DCT: Algorithms, Advantages, Applications*, New York: Academic, 1990.
6. Z. Xiong, K. Ramchandran, M. Orchard, and Y. Q. Zhang, A comparative study of DCT- and wavelet-based image coding, *IEEE Transactions on Circuits and Systems for Video Technology*, 9(5), 692–695, 1999.
7. J. M. Shapiro, Embedded image coding using zerotrees of wavelet coefficients, *IEEE Transactions on Signal Processing*, 41, 3445–3463, 1993.
8. A. Said and W. A. Pearlman, A new, fast, and efficient image codec based on set partitioning in hierarchical trees, *IEEE Transactions on Circuits and Systems for Video Technology*, 6, 243–250, 1996.
9. B. G. Haskell, P. G. Howard, Y. A. LeCun, A. Puri, J. Ostermann, M. R. Civanlar, L. Rabiner, L. Bottou, and P. Haffner, Image and video coding—Emerging standards and beyond, *IEEE Transactions on Circuits and Systems for Video Technology*, 8, 814–837, 1998.
10. W. B. Pennebaker and J. L. Mitcell, *JPEG: Still Image Data Compression Standard*, Van Nostrand Reinhold, New York, 1993.
11. *Call for Contributions for JPEG 2000 (JTC 1.29.14, 15444): Image Coding System*, ISO/IEC JTC1/SC29/WG1 N505, March 1997.
12. *Coding of Moving Pictures and Associated Audio for Digital Storage Media at up to about 1.5 Mbit/s—Part 2: Video*, ISO/IEC 11172-2 (MPEG-1 Video), ISO/IEC JTC 1, March 1993.
13. *Generic Coding of Moving Pictures and Associated Audio Information—Part 2: Video*, ITU-T Rec. H.262 and ISO/IEC 13818-2 (MPEG-2 Video), ITU-T and ISO/IEC JTC 1, November 1994.
14. *Coding of Audio–Visual Objects—Part 2: Visual*, ISO/IEC 14 496-2 (MPEG-4 Visual), April 1999.
15. *Video Coding for Low Bit Rate Communication*, ITU-T Rec. H.263, 1995.
16. *Advanced Video Coding for Generic Audiovisual Services*, ITU-T Rec. H.264 and ISO/IEC 14496-10 (MPEG-4 AVC), ITU-T and ISO/IEC JTC 1, May 2003.
17. T. Wiegand, G. J. Sullivan, G.Bjøntegaard, and A. Luthra, Overview of the H.264/AVC video coding standard, *IEEE Transactions on Circuits and Systems for Video Technology*, 13, 560–576, 2003.
18. R. Li, B. Zeng, and M. L. Liou, A new three-step search algorithm for block motion estimation, *IEEE Transactions on Circuits and Systems for Video Technology*, 4, 438–442, 1994.
19. L. M. Po and W. C. Ma, A novel four-step search algorithm for fast block motion estimation, *IEEE Transactions on Circuits and Systems for Video Technology*, 6, 313–317, 1996.

20. A. M. Tourapis, O. C. Au, and M. L. Liou, Fast motion estimation using circular zonal search, *Proceedings of SPIE Symposium on Visual Communications and Image Processing*, Vol. 1, pp. 1496–1504, San Jose, California, January 1999.

21. A. M. Tourapis, O. C. Au, and M. L. Liou, Highly efficient predictive zonal algorithms for fast block-matching motion estimation, *IEEE Transactions on Circuits and Systems for Video Technology*, 12(10), 934–947, 2002.

22. C. Zhu, X. Lin, and L.-P. Chau, Hexagon-based search pattern for fast motion estimation, *IEEE Transactions on Circuits and Systems for Video Technology*, 12(5), 1149–1153, 2002.

23. A. M. Tourapis, O. C. Au, and M. L. Liou, New results on zonal based motion estimation algorithms advanced predictive diamond zonal search, *Proceedings of IEEE International Symposium on Circuits and Systems*, Vol. 5, pp. 183–186, Sydney, Australia, May 2001.

24. L. Yang, K. Yu, J. Li, and S. Li, An effective variable block-size early termination algorithm for H.264 video coding, *IEEE Transactions on Circuits and Systems for Video Technology*, 15, 784–788, 2005.

25. S. Chien, Y. Huang, B. Hsieh, S. Ma, and L. Chen, Fast video segmentation algorithm with shadow cancellation, global motion compensation, and adaptive threshold techniques, *IEEE Transactions on Multimedia*, 6(5), 732–748, 2004.

26. H. W. Wong and O. C. Au, Modified one-bit transform for motion estimation, *IEEE Transactions on Circuits and Systems for Video Technology*, 9(7), 1020–1024, 1999.

27. Y. H. Fok and O. C. Au, An improved fast feature-based block motion estimation, *Proceedings of. IEEE International Conference on Image Processing*, pp. 741–744, Austin, Texas, November 1994.

28. B. Liu and A. Zaccarin New Fast Algorithms for the estimation of block motion vectors, *IEEE Transactions on Circuits and Systems for Video Technology*, 3(2), 148–157, 1993.

29. K. Lengwehasatit and A. Ortega, Probabilistic partial-distance fast matching algorithms for motion estimation, *IEEE Transactions on Circuits and Systems for Video Technology*, 11(2), 139–152, 2001.

30. A. M. Tourapis, O. C. Au, and M. L. Liou, Predictive motion vector field adaptive search Technique (PMVFAST)-enhancing block based motion estimation, *Proceedings of SPIE Visual Communications and Image Processing*, Vol. 4310, pp. 883–892, Sydney, Australia, January 2001.

31. Z. Chen and K. N. Ngan, Recent advances in rate control for video coding, *Signal Processing: Image Communication*, 151, 19–38, 2007.

32. CCITT SG XV WP/1/Q4, *Description of Reference Model 8 (RM8)*, June 1989.

33. E. Viscito and C. Gonzales, A video compression algorithm with adaptive bit allocation and quantization, *Proceedings of SPIE Visual Communications and Image Processing*, Vol. 1605, pp. 58–72, Boston, Massachusetts, 1991.

34. ISO/IEC JTC/SC29/WG11, *MPEG Test Model 5*, April 1993.

35. ITU-T/SG15, *Video Codec Test Model, TMN8*, June 1997.

36. ISO/IEC JTC/SC29/WG11, *MPEG-4 Video Verification Model v18.0*, January 2001.

37. G. Sullivan, T. Wiegand, and K.-P. Kim, Joint model reference encoding methods and decoding concealment methods, *JVT-I049*, September 2003.

38. S. Ma, W. Gao, and Y. Lu, Rate-distortion analysis for H.264/AVC video coding and its application to rate control, *IEEE Transactions on Circuits and Systems for Video Technology*, 15(12), 1533–1544, 2005.

39. N. Kamaci, Y. Altunbasak, and R. M. Mersereau, Frame bit allocation for the H.264/AVC video coder via Cauchy-density-based rate and distortion models, *IEEE Transactions on Circuits and Systems for Video Technology*, 15(8), 994–1006, 2005.

40. C. Pang, O. C. Au, J. Dai, F. Zou, W. Yang, and M. Yang, Laplacian Mixture Model (LMM) based frame-layer rate control method for H.264/AVC high-definition video coding, *Proceedings of IEEE International Conference on Multimedia and Expo*, pp. 25–29, Singapore, July 2010.

41. D.-K. Kwon, M.-Y.Shen, and C.-C. J. Kuo, Rate control for H.264 video with enhanced rate and distortion models, *IEEE Transactions on Circuits and Systems for Video Technology*, 17(5), 517–529, 2007.

42. Y. Liu, Z. G. Li, and Y. C. Soh, A novel rate control scheme for low delay video communication of H.264/AVC standard, *IEEE Transactions on Circuits and Systems for Video Technology*, 17(1), 68–78, 2007.

43. Z. He and D. O. Wu, Linear rate control and optimum statistical multiplexing for H.264 video broadcast, *IEEE Transactions on Multimedia*, 10(7), 1237–1249, 2008.

44. Q. Xu, X. Lu, Y. Liu, and C. Gomila, A fine rate control algorithm with adaptive rounding offset (ARO), *IEEE Transactions on Circuits and Systems for Video Technology*, 19(10), 1424–1435, 2009.

45. H. Schwarz and T. Wiegand, The emerging JVT/H.26L video coding standard, *Proceedings of IBC*, the Netherlands, September 2002.

46. *http://www.mpegla.com/main/programs/avc/Documents/AVC_TermsSummary. pdf*, "SUMMARY OF AVC/H.264 LICENSE TERMS"

47. AVS Video Expert Group, *Information Technology—Advanced Audio Video Coding Standard—Part 2: Video*, AVS-N1063, Audio Video Coding Standard Group of China (AVS), December 2003.

48. L. Fan, S. Ma, and F. Wu, An overview of AVS video standard, *Proceedings of IEEE International Conference on Multimedia and Expo*, Vol. 1, pp. 423–426, Taipei, Taiwan, June 2004.

49. S. Srinivasan, P. Hsu, T. Holcomb, K. Mukerjee, S. L. Regunathan, B. Lin, J. Liang, M.-C. Lee, and J. Ribas-Corbeta, WMV-9: Overview and applications, *Signal Processing: Image Communication*, 231, 851–875, October 2004.

50. H. Kalva and J.-B. Lee, The VC-1 video coding standard, *IEEE Transactions on Multimedia*, 14(4), 88–91, 2007.

51. F. Zou, O. C. Au, W. Yang, C. Pang, and J. Dai, Intra mode dependent quantization error estimation of each DCT coefficient in H.264/AVC, *Proceedings of IEEE International Conference on Multimedia and Expo*, pp. 447–451, Singapore, July 2010.

52. H. S. Malvar, A. Hallapuro, M. Karczewicz, and L. Kerofsky, Low-complexity transform and quantization in H.264/AVC, *IEEE Transactions on Circuits and Systems for Video Technology*, 13(7), 598–603, 2003.

53. D. Marpe, H. Schwarz, and T. Wiegand, Context-based adaptive binary arithmetic coding in the H.264/AVC video compression standard, *IEEE Transactions on Circuits and Systems for Video Technology*, 13(7), 620–636, 2003.

54. D. Quan and Y.-S. Ho, Interpolation scheme using simple multi-directional filters for fractional-pel motion compensation, *IEEE Transactions on Consumer Electronics*, 56(4), 2711–2718, 2010.

55. M. Flierl and B. Girod, Generalized B pictures and the draft H.264/AVC video-compression standard, *IEEE Transactions on Circuits and Systems for Video Technology*, 13(7), 587–597, 2003.

56. M. Wien, Variable block-size transforms for H.264/AVC, *IEEE Transactions on Circuits and Systems for Video Technology*, 13(7), 604–613, 2003.

57. P. List, A. Joch, J. Lainema, G. Bjontegaard, and M. Karczewicz, Adaptive deblocking filter, *IEEE Transactions on Circuits and Systems for Video Technology*, 13(7), 614–619, 2003.

58. *Recommendation and Final Draft International Standard of Joint Video Specification*, ITU-T Rec.H.264 and ISO/IEC 14496–10 AVC, March 2005.

59. M. Mahmoudi and G. Sapiro, Fast image and video denoising via nonlocal means of similar neighborhoods, *IEEE Signal Processing Letters*, 12(12), 839–842, 2005.

60. L. Guo, O. C. Au, M. Ma, and P. H. W. Wong, Integration of recursive temporal LMMSE denoising filtering for video codec, *IEEE Transactions on Circuits and Systems for Video Technology*, 20(2), 236–249, 2009.

61. L. Guo, O. C. Au, M. Ma, and Z. Liang, Fast multi-hypothesis motion compensated filter for video denoising, *Journal of Signal Processing Systems*, 60(3), 273–290, 2009.

62. Y. Chen, O. C. Au, and X. Fan, Simultaneous MAP-based video denoising and rate-distortion optimized video encoding, *IEEE Transactions on Circuits and Systems for Video Technology*, 19(1), 15–26, 2009.

63. L. Guo, O. C. Au, M. Ma, and Z. Liang, Temporal video denoising based on multihypothesis motion, *IEEE Transactions on Circuits and Systems for Video Technology*, 17(10), 1423–1429, 2007.

64. J. Dai, O. C. Au, W. Yang, C. Pang, F. Zou, and X. Wen, Color video denoising based on adaptive color space conversion, *Proceedings of IEEE International Symposium of Circuits and Systems,* Seattle, Washington, May 2010.

65. C. Pang, O. C. Au, J. Dai, W. Yang, and F. Zou, A Fast NL-means method in image denoising based on the similarity of spatially sampled pixels, *Proceedings of IEEE International Workshop on Multimedia Signal Processing,* Rio de Janeiro, Brazil, October 2009.

66. E. Luo, O. C. Au, L. Guo, Y. Wu, and S. F. Tu, Embedded denoising for H.264/ AVC extension— Spatial SVC, *Proceedings of IEEE Pacific Rim Conference on Communications, Computers, Signal Processing,* Victoria, B.C., Canada, August 2009.

67. J. Dai, O. C. Au, C. Pang, W. Yang, and F. Zou, Film grain noise removal and synthesis in video coding, *Proceedings of IEEE International Conference on Acoustics, Speech and Signal Processing,* Prague, Czech Republic, March 2010.

68. G. J. Sullivan, P. N. Topiwala, and A. Luthra, The H.264/AVC advanced video coding standard: Overview and introduction to the fidelity range extensions, *Proceedings of SPIE, Applications of Digital Image Processing* XXVII, Vol. 5558, pp. 454–474, August 2004.

69. J. Ostermann, J. Bormans, P. List, D. Marpe, M. Narroschke, F. Pereira, T. Stockhammer, and T. Wedi, Video coding with H.264/AVC: Tools, performance, and complexity, *IEEE Circuits and Systems Magazine,* 4(1), 7–28, 2004.

70. T. Wedi and H.G. Musmann, Motion- and aliasing-compensated prediction for hybrid video coding, *IEEE Transactions on Circuits and Systems for Video Technology,* 13(7), 577–586, 2003.

71. H. Wang, S. Kwong, and C.-W. Kok, An efficient mode decision algorithm for H.264/AVC encoding optimization, *IEEE Transactions on Multimedia,* 16(4), 882–888, Bangkok, Thailand, 2007.

72. I. Choi, J. Lee, and B. Jeon, Fast coding mode selection with rate-distortion optimization for MPEG-4 part-10 AVC/H.264, *IEEE Transactions on Circuits and Systems for Video Technology,* 16(12), 1557–1561, 2006.

73. J. Zhang, Y. He, S. Yang, and Y. Zhong, Performance and complexity joint optimization for H.264 video coding, *Proceedings of IEEE International Symposium on Circuits and Systems,* Vol. 2, pp. 888–891, Bangkok, Thailand, May 2003.

74. T. Wedi and T. K. Tan, AHG report—Coding Efficiency, *VCEG-AE05,* January 2007.

75. *Draft Call for Evidence on High-Performance Video Coding,* N10117, October 2008.

76. G. J. Sullivan and J.-R. Ohm, Meeting report of the first meeting of the Joint Collaborative Team on Video Coding (JCT-VC), *JCTVC-A200,* Dresden, DE, April 2010.

77. G. J. Sullivan and T. Wiegand, Rate-distortion optimization for video compression, *IEEE Signal Processing Magazine,* 15(6), 74–90, 1998.

78. S.-W. Jung, S.-J. Baek, C.-S. Park, and S.-J. Ko, Fast mode decision using all-zero block detection for fidelity and spatial scalable video coding, *IEEE Transactions on Circuits and Systems for Video Technology,* 20(2), 201–206, 2010.

79. T. Zhao, H. Wang, S. Kwong, and C.-C.J. Kuo, Fast mode decision based on mode adaptation, *IEEE Transactions on Circuits and Systems for Video Technology,* 20(5), 697–705, 2010.

80. H. Zeng, C. Cai, and K.-K. Ma, Fast mode decision for H.264/AVC based on macroblock motion activity, *IEEE Transactions on Circuits and Systems for Video Technology,* 19(4), 491–499, 2009.

81. C.-S. Park, B.-K. Dan, H. Choi, and S.-J. Ko, A statistical approach for fast mode decision in scalable video coding, *IEEE Transactions on Circuits and Systems for Video Technology,* 19(12), 1915–1920, 2009.

82. H. Li, Z. G. Li, and C. Wen, Fast mode decision algorithm for inter-frame coding in fully scalable video coding, *IEEE Transactions on Circuits and Systems for Video Technology,* 16(7), 889–895, 2006.

83. J.-H. Kim, H.-S. Kim, B.-G. Kim, H. Y. Kim, S.-Y. Jeong, and J. S Choi, Fast block mode decision scheme for B-picture coding in H.264/AVC, *Proceedings of IEEE International Conference on Image Processing,* pp. 2873–2876, Hong Kong, China, September 2010.

84. H.-C. Lin, W.-H. Peng, and H.-M. Hang, Fast context-adaptive mode decision algorithm for scalable video coding with combined coarse-grain quality scalability (CGS) and temporal scalability, *IEEE Transactions on Circuits and Systems for Video Technology,* 20(5), 732–748, 2010.

85. H. M. Wong, O. C. Au, A. Chang, S. K. Yip, C. W. Ho, Fast mode decision and motion estimation for H.264 (FMDME), *Proceedings of IEEE International Symposium on Circuits and Systems*, Kobe, Japan, May 2006.

86. B. Meng, O. C. Au, C. W. Wong, and H. K. Lam, Fast mode selection method for 4×4 blocks intra-prediction in H.264, *Proceedings of IEEE International Conference on Image Processing*, Barcelona, Catalonia, Spain, September 2003.

87. Unification of the Directional Intra Prediction Methods in TMuC, *JCTVC-B100*, Geneva, Switzerland, July 2010.

88. H. Jia and L. Zhang, Directional diamond search pattern for fast block motion estimation, *Electronics Letters, 39*, 1581–1583, 2003.

89. B. Triggs, Empirical filter estimation for subpixel interpolation and matching, *Proceedings of IEEE International Conference on Computer Vision*, Kyoto, Japan, 2001.

90. M. Feng and T. R. Reed, Detection and estimation of rotational motions using the 3-D Gabor representation, *Proceedings of IEEE International Conference on Image Processing*, Genoa, Italy, October 2005.

91. A. Smolic, T.Sikora, J.-R. Ohm, Long-term global motion estimation and its application for sprite coding, content description, and segmentation, *IEEE Transactions on Circuits and Systems for Video Technology, 9*(8), 1227–1242, 1999.

92. Yuwen He, Bo Feng, Shiqiang Yang, and Yichuo Zhong, Fast global motion estimation for global motion compensation coding, *Proceedings of IEEE International Symposium on Circuits and Systems*, Rio de Janeiro, Brazil, May 2011.

93. S. H. Lee, J. W. Moon, J. W. Byun, and N. I. Cho, A new intra prediction method using channel correlations for the H.264/AVC intra coding, *Proceedings of Picture Coding Symposium*, Chicago, Illinois, 2009.

94. F. Bossen, V. Drugeon, E. Francois, J. Jung, S. Kanumuri, M. Narroschke, H. Sasai, J. Sole, Y. Suzuki, T. K. Tan, T. Wedi, S. Wittmann, P. Yin, and Y. Zheng, Video coding using a simplified block structure and advanced coding techniques, *IEEE Transactions on Circuits and Systems for Video Technology, 20*(12), 1667–1675, 2010.

95. D. M. Marpe, H. Schwarz, S. Bosse, B. Bross, P. Helle, T. Hinz, H. Kirchhoffer. et al., Video compression using nested quadtree structures, leaf merging, and improved techniques for motion representation and entropy coding, *IEEE Transactions on Circuits and Systems for Video Technology, 20*(12), 1676–1687, 2010.

96. H. Xu, J. Xu, and F. Wu, Lifting-based directional DCT-like transform for image coding, *IEEE Transactions on Circuits and Systems for Video Technology, 17*(10), 1325–1335, 2007.

97. B. Zeng and J. Fu, Directional discrete cosine transforms—A new framework for image coding, *IEEE Transactions on Circuits and Systems for Video Technology, 18*(3), 305–313, 2008.

98. C.-L. Chang and B. Girod, Direction-adaptive partitioned block transform for image coding, *Proceedings of IEEE International Conference on Image Processing*, pp. 145–148, San Diego, California, October 2008.

99. J. Sole, P. Yin, Y. Zheng, and C. Gomila, Joint sparsity-based optimization of a set of orthonormal 2-D separable block transforms, *Proceedings of IEEE International Conference on Image Processing*, pp. 9–12, Cairo, Egypt, November 2009.

100. M. Karczewicz, P. Chen, R. Joshi, X. Wang, W.-J. Chien, and R. Panchal, Video coding technology proposal by Qualcomm, *JCTVC-A121*, Dresden, Germany, April 2010.

101. K. McCann, W.-J. Han, I.-K. Kim, J.-H. Min, E. Alshina, A. Alshin, I. Lee. et al., Video coding technology proposal by Samsung (and BBC), *JCTVC-A121, 79*(16), 2579–2581, 2010.

102. C. Yeo, Y. H. Tan, Z. Li, and S. Rahardja, Choice of transforms in MDDT for unified intra prediction, *JCTVC-C039*, Guangzhou, Connecticut, October 2010.

103. S. L. Sclove, Pattern recognition in image processing using interpixel correlation, pattern analysis and machine intelligence, *IEEE Transactions on Pattern Analysis and Machine Intelligence, PAMI-3*(2), 206–208, 1981.

104. S. Zhu, S.-K. Au Yeung, and B. Zeng, In search of "Better-than-DCT" unitary transforms for encoding of residual signals, *IEEE Signal Processing Letters, 17*(11), 961–964, 2010.

105. S.-C. Lim, D.-Y. Kim, and Y.-L. Lee, Alternative transform based on the correlation of the residual signal, *Proceedings of Congress on Image and Signal Processing*, 1, 389–394, May 2008.
106. M. Karczewicz and Y. Ye, Rate distortion optimized quantization, *VCEG-AH21*, ITU-T VCEG, Antalya, Turkey, 2008.
107. J. Wen, M. Xiao, J. Chen, P. Tao, and C. Wang, Fast rate distortion optimized quantization for H.264/AVC, *Proceedings of Data Compression Conference*, p. 557, Snowbird, Utah, March 2010.
108. G. Shen, B. Zeng, and M. L. Liou, Rate-distortion optimization for arbitrarily-shaped object coding, *Proceedings of IEEE International Conference on Multimedia and Expo*, pp. 1601–1604, New York City, NY, July 2000.
109. S. Zhu and B. Zeng, Constrained quantization in the transform domain with applications in arbitrarily-shaped object coding, *IEEE Transactions on Circuits and Systems for Video Technology*, 20(11), 1385–1394, 2010.
110. M. T. Orchard, Y. Wang, V. Vaishampayan, and A. R. Reibman, Redundancy rate-distortion analysis of multiple description coding using pairwise correlating transforms, *Proceedings of IEEE International Conference on Image Processing*, Vol. 1, pp. 608–611, Hong Kong, China, October 1997.
111. X. Li, P. Amon, A. Hutter, and A. Kaup, Model based analysis for quantization parameter cascading in hierarchical video coding, *Proceedings of IEEE International Conference on Image Processing*, pp. 3765–3768, Cairo, Egypt, November 2009.
112. V. Sze, K. Panusopone, J. Chen, T. Nguyen, and M. Coban (CE coordinators), Description of core experiment 11: Coefficient scanning and coding, *JCTVC-C511*, Guangzhou, October 2010.
113. M. Budagavi and A. Segall (co-chairs), JCT-VC AHG report: Parallel entropy coding, *JCTVC-C008*, Guangzhou, October 2010.
114. W.-J. Chien, P. Chen, and M. Karczewicz (Qualcomm), TE 12: Evaluation of block merging (MRG), *JCTVC-C264*, Guangzhou, October 2010.
115. A. Paul (Hanyang Univ./NCKU), Region of block based dynamic video processing for HEVC, *JCTVC-C283*, Guangzhou, October 2010.
116. *Video Codec for Audio Visual Services at p*64 kbit/s*, ITU-T Recommendation H.261, 1993.

8

Distributed Video Coding

Zixiang Xiong

CONTENTS

8.1 Introduction

Driven by a host of emerging applications, problems of distributed source coding have received increased attention in recent years. Although the underlying information-theoretic results by Slepian and Wolf [1] for distributed lossless coding, Wyner and Ziv [2] for lossy coding with decoder side information, and Berger [3] for multiterminal (MT) source coding have been known for more than 30 years, practical schemes did not appear until recently (e.g., [4,5]). They are now being explored for applications ranging from wireless video [6] to camera arrays (e.g., [7]) to distributed video sensor networks [8]. These networks are becoming increasingly important for a wide range of critical applications such as video surveillance, monitoring of disaster zones and traffic, elderly care, tracking people and vehicles in crowded environments, and providing more realistic images for consumer electronics and entertainment. For example, the growing use of camera arrays allows the viewer to observe a scene from any viewpoint, and the industry (e.g., Sharp and Panasonic) is starting to produce three-dimensional (3D) displays that do not require glasses. The immersive experience provided by these 3D displays are compelling and have the potential to create a growing market for 3D or free viewpoint TV (see [9]) and hence for MT video compression.

After giving the background on Slepian–Wolf (SW) coding and Wyner–Ziv (WZ) coding and reviewing the latest development on MT source coding theory, this chapter surveys recent works on applying distributed source coding principles to video compression, covering three areas of distributed video coding, namely, Witsenhausen–Wyner video coding, layered WZ video coding, and MT video coding. Compared to H.264/AVC [10] and H.26L-FGS video coding [11], the advantage of Witsenhausen–Wyner video coding and layered WZ video coding is error robustness when the compressed video has to be delivered over a noisy channel. The price for this added error robustness is a small loss of compression performance.

In MT video coding for video sensor network applications, since the encoders are not allowed to communicate with each other, the use of distributed source coding is a must. We describe a generic scheme for MT video coding based on SW-coded quantization, where the first view is coded by H.264/AVC and the remaining views are sequentially compressed using WZ coding with decoder side information generated by exploring multiple view geometry in computer vision. We also introduce the latest MT video coding setup where a low-resolution depth camera is employed to assist side information generation at the decoder. We conclude this chapter by highlighting triumphs achieved so far in distributed video coding and pointing out future directions.

8.2   Theoretical Background

8.2.1   SW Coding and WZ Coding

In 1973, Slepian and Wolf showed the surprising result that separates *lossless* encoding of two correlated sources (with joint decoding) suffers no rate loss when compared with joint encoding (and decoding). Their seminal paper [1] laid the theoretical foundation for distributed source coding.

Later, Wyner, and Ziv [2] extended a special case of SW coding to lossy source coding with side information at the decoder, and showed that there is, in general, a rate loss with WZ coding when compared with source coding with side information at both the encoder and the decoder. One special case of WZ coding (with no rate loss) is when the source and side information are jointly Gaussian and the distortion measure is the mean square error (MSE); another no—rate—loss case [12] is when both the source and the side information are binary symmetric and the correlation between them can be modeled as a binary erasure channel and the distortion measure is the Hamming distance. We refer the readers to ref. [5] for overview material on SW and WZ coding.

8.2.2   MT Source Coding

MT source coding deals with separate encoding and joint decoding of multiple correlated sources under distortion constraints. It can be viewed as the lossy version of SW coding [1]. Berger first investigated MT source coding in 1977 [3]. Since then, two classes of MT source coding problems, namely, *direct* MT source coding and *indirect* MT source coding as depicted in Figure 8.1, have been studied. The latter is often referred to as the CEO problem [13], where different terminals observe and separately encode noisy versions of a *single*

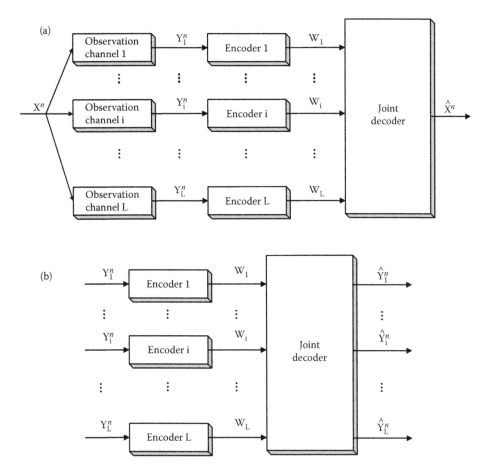

FIGURE 8.1
(a) Direct MT source coding. (b) Indirect MT source coding (the chief executive officer (CEO) problem).

remote source, which is to be reconstructed at the decoder. Recently, the CEO problem has been generalized to the setup with multiple remote sources under a sum-distortion constraint.

Determination of the rate region of MT source coding has been a difficult task ever since the formulation of the problem. Until recently, we only had general (e.g., Berger–Tung [3,14]) inner and outer bounds for both setups. In the quadratic Gaussian case with MSE distortion, Oohama [15] completely solved the indirect/CEO problem for any number of terminals and gave the rate region in 2005. The direct case, on the other hand, has been more challenging due to multiple distortion constraints at the decoders. In addition, unlike SW coding, there is in general a rate loss associated with MT source coding (when compared with joint encoding). For quadratic Gaussian direct two-terminal source coding, Oohama proved side—bound tightness in 1997 [16] and Wagner et al. [17] showed tightness of the Berger–Tung (BT) achievable sum-rate bound [3,14] by proving the converse in 2008. In the same 2008 paper, Wagner et al. also showed that in the positive symmetric case, the BT inner sum-rate bound is tight for any number of terminals and that quantization followed by binning (or SW compression) is optimal. Recently, Yang and Xiong [18] showed tightness of the BT bound for a new class of quadratic Gaussian MT problems. Moreover, sufficient conditions for tightness of the BT sum-rate bound are given by Wang et al. in [19] and Yang et al. in [20]. Albeit these recent progresses, finding the sum-rate bound of quadratic Gaussian MT source coding with more than two terminals is still an open problem.

8.3 Distributed Source Code Design

For classical data compression, there are several well-known entropy coding techniques (e.g., Huffman coding, arithmetic coding, and Lempel–Ziv coding). A less-known approach is channel code-based binning [21,22], but it is the one that can be easily extended to SW (or *lossless* distributed compression). The idea of using linear codes to achieve the SW limit was first illustrated in Wyner's 1974 paper [23] and implemented only recently using turbo or low-density parity-check (LDPC) codes (see [4,5] and references therein).

For WZ and MT source coding, most practical designs are based on the framework of quantization and binning (or SW compression). In the quadratic Gaussian case, state-of-the-art trellis-coded quantization in conjunction with turbo/LDPC code-based SW coding have been shown to be able to approach the WZ rate–distortion limit and the BT sum-rate bound of MT source coding. These recently available code designs, although for ideal Gaussian source(s), provide guidance when we deal with real-world video sources in distributed video coding, which is covered next.

8.4 Monoview Video Compression Based on Distributed Coding Principles

Distributed video coding was first conceptualized by Witsenhausen and Wyner in their 1980 US patent on an "interframe coder for video signals" [24], whose main idea is illustrated in Figure 8.2, which is reproduced from [24]. Assuming that the video signal source is binary, and the current frame X and the previous frame Y are correlated, for example, with i.i.d X and Y satisfying the relationship $X = Y + D$. The syndrome former at the video encoder outputs $S(X_i) = HX_i$, where X_i consists of an n-bit block of X, and H is an $(n - k) \times n$ parity-check matrix of an error correction code (ECC) for the binary channel with X and Y being its input and output, respectively. It thus achieves a compression ratio of $n/(n - k)$.

After receiving the transmitted syndrome $S(X_i)$, the video decoder first computes $S(X_i) - S(Y_i) = H(X_i - Y_i)$, since it already has the previous frame Y_i, before applying hard-decision decoding to recover $D_i = X_i - Y_i$ noiselessly by picking D_i as the leader of the ECC coset indexed by $S(X_i)$. Finally, D_i is added to Y_i to form the decoded X_i.

For general M-ary video sources, the patent suggests the use of M-ary ECC or the above-mentioned binary syndrome-based scheme for the most significant bit-plane in conjunction with conventional coding schemes for the remaining bit-planes.

We now take a closer look at the patent and see how it is related to recent works. The syndrome-based encoding step in the patent is the same as that in [25]. The first decoding step in the patent, forming $HD_i = H(X_i - Y_i)$, could easily be used in [25] and decode for D_i instead of X_i directly. The "noisy" vector in this case would be an all-zeros vector instead of Y_i. The result of the decoding algorithm should then be added to Y_i to form the most

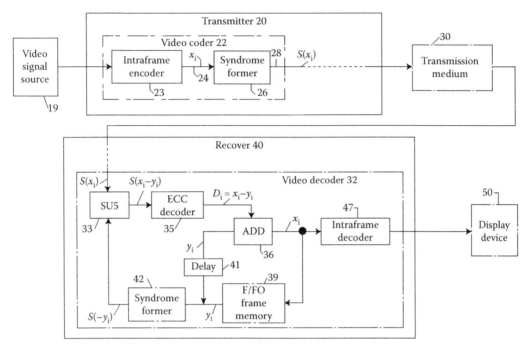

FIGURE 8.2
Block diagram of the interframe video coder proposed by Witsenhausen and Wyner in their 1980 patent.

likely X_i, similar to the patent. The difference in the patent lies in the decoding algorithm. Hard-input decoding is used in the patent, whereas soft-input iterative message-passing decoding is used in [25]. This could be because at the time the patent was written syndromes which were mainly used for decoding of block codes (hard input) while soft-input decoding was only considered possible for convolutional codes through the Viterbi algorithm.

However, the Witsenhausen–Wyner decoder is optimum under the binary symmetric channel (BSC) model, meaning that in the setup of [25] it will give the same performance as the message-passing algorithm. The reason is that because of the BSC model, the soft-input probabilities to the decoder have only two levels (more likely and less likely), thus decoding D_i with the minimum Hamming weight is the maximum-likelihood approach. Of course, this will not be the case when the correlation model is not the BSC and therefore, the soft input to the decoder has more levels, for example, different bits have different cross-over probability or there is an additive white Gaussian noise channel model.

8.4.1 Low-Complexity Encoding?

Despite the flurry of recent research activities on distributed video coding (e.g., [26–28]), Wistenhausen and Wyner's patent expired without actually being implemented. Instead, many existing works on distributed video coding, usually under the name of WZ video coding, focus on performing the computation-intensive motion estimation operation at the decoder, with the hope of moving most of the complexity of standard (e.g., H.264/AVC) video coding from the encoder to the decoder. For example, two representative works [26,27] are highlighted in the 2007 overview paper [28] by the European Union's DISCOVER project team leaders. However, results from the IBM Research team [29] using low-complexity rateless SW encoding for WZ video coding, with no feedback channel, still show a gap of 2.0 dB relative to H.264/AVC motion-compensated interframe coding (and a gain of 3.2 dB over H.264/AVC intraframe coding) for the CIF Cost_Guard sequence at 15 f/s. This performance gap is summarized in [28] to come from the following three aspects:

- Side information construction
- Correlation modeling and estimation
- Rate allocation and coding mode selection

For the first aspect, recent works have been focusing on decoding a low-resolution/quality version of the current frame before iterative refinement of the motion field at the decoder, leading to progressive side information estimation at the decoder (see [30] and references therein). Nevertheless, it still remains elusive in WZ video coding to have low-complexity encoding while achieving performance competitive to H.264/AVC motion-compensated interframe coding, especially at a low rate.

8.4.2 Error Robustness

Then what is the main advantage of applying distributed coding for monoview video? Because of the error-correcting capability of SW-coded bitstreams, WZ video coding has the advantage of increased error-robustness over standard H.264/AVC when transporting video over noisy channels. The price paid for this error-robustness is a small loss in compression performance.

To address the last two aspects of WZ video coding listed above, namely, correlation modeling and rate allocation, one solution is to employ motion estimation at the encoder to find the best possible side information (hence the joint source correlation statistics for rate allocation). This approach does not impose a low-complexity constraint on the encoder. It is still distributed because it codes the video source itself instead of the difference between the source and the side information (or motion-compensated prediction) as in standard H.264/AVC coding. Such an approach was initially pursued in [31,32] and further explored recently in the first implementation of Witsenhausen–Wyner video coding.

8.4.2.1 *Witsenhausen–Wyner Video Coding*

The block diagram of the Witsenhausen–Wyner video codec proposed in [33] is depicted in Figure 8.3, where intercoding in H.264/AVC is replaced by the syndrome-based coding scheme in [24] for the top discrete cosine transform (DCT) bit-planes (the lower bit-planes are entropy coded). The Intra and Skip modes of H.264/AVC coding are kept unchanged since they are inherently error resilient. The side information is first generated from the previously reconstructed frame with motion compensation at the Witsenhausen–Wyner video encoder. However, the encoder does not compute the difference between the original video source and the side information. It only uses the side information for making decisions among syndrome-based Witsenhausen–Wyner coding, Intra, and Skip modes (and classification of Witsenhausen–Wyner coded blocks based on the correlation). The resolution of motion estimation is adaptively selected from full-, half-, and quarter-pixel, based on the available computational resources. The resulting motion vectors are entropy encoded and sent to the decoder to reproduce the side information. Depending on the encoding mode, the decoder either performs Witsenhausen–Wyner decoding or entropy decoding to reconstruct the original video source.

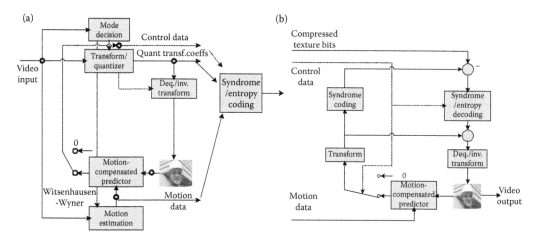

FIGURE 8.3
Witsenhausen–Wyner video coding. (a) Encoding, (b) decoding. (From M. Guo et al., Witsenhausen-Wyner video coding, *IEEE Trans. Circuits Systems Video Tech.*, 21, 1049–1060, 2011. © IEEE. With permission.)

In terms of compression performance, simulations using Football (720 × 486) and Mobile (720 × 576) at 15 f/s show a gap of about 0.5 dB in average peak signal-to-noise ratio (PSNR) with Witsenhausen–Wyner video coding when compared with H.264/AVC in the bit rate range of 2.5–5.5 Mb/s.

For video over noisy channels, simulations are performed on transmitting Witsenhausen–Wyner compressed bitstreams (after protecting them with 25% Reed–Solomon-based forward error correction overhead) over a Qualcomm wireless channel simulator for RTP/IP over 3GPP, which adds random errors to the protocol data unit (PDU) delivered in the radio link control layer. Figure 8.4 compares Witsenhausen–Wyner video coding, H.264/AVC, and H.264/AVC IntraSkip coding (with Intra and Skip modes only in H.264/AVC)—in terms of average PSNR versus PDU loss rate—for Football and Mobile. It

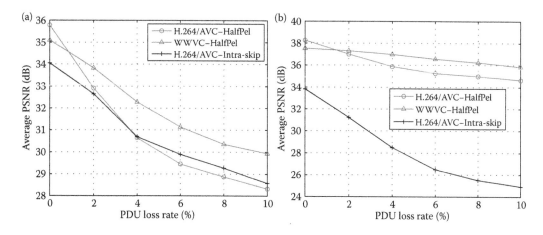

FIGURE 8.4
Witsenhausen–Wyner video coding versus H.264/AVC and H.264/AVC IntraSkip coding when the bitstreams are protected with Reed–Solomon codes and transmitted over a simulated CDMA2000 1X channel. (a) Football with a compression/transmission rate of 3.78/4.725 Mb/s. (b) Mobile with a compression/transmission rate of 4.28/5.163 Mb/s.

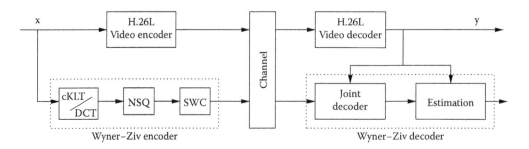

FIGURE 8.5
Block diagram of layered WZ video coding (From Q. Xu and Z. Xiong, Layered Wyner-Ziv video coding, *IEEE Trans. Image Process.*, 15, 3791–3803, 2006. © (2006) IEEE. With permission.)

is seen that Witsenhausen–Wyner video coding is much more error robust against channel errors than H.264/AVC (at the price of a small loss in compression efficiency), while being almost uniformly better than H.264/AVC IntraSkip.

8.4.2.2   Layered WZ Video Coding

In [34], Xu and Xiong proposed a layered video coder based on standard video coding and successive WZ coding—improvements to this work appeared in [35,36]. Treating a standard coded video as the base layer and decoder side information, the layered WZ encoder consists of DCT as an approximation of the conditional Karhunen–Loeve transform (cKLT), nested scalar quantization (NSQ), and SW coding. The WZ decoder performs joint decoding by combining the base layer and the WZ bitstream for enhanced video quality Figure 8.5 shows its block diagram.

WZ coding alleviates the problem of error drifting/propagation associated with encoder–decoder mismatch in standard differential pulse coded modulation (DPCM)-based coders because a corrupted base layer (with errors within a certain range) can still be combined with the enhancement layer for WZ decoding—the loss in compression efficiency is within 0.5 dB in average PSNR. This inherent error resilience in the base layer is the main advantage of the layered WZ coding over standard (e.g., H.26L FGS (fine-grain scalability) [11]) coding.

FIGURE 8.6
Error robustness performance of WZ video coding compared with H.26L FGS for Football. The 10th decoded frame by H.26L FGS (a) and WZ video coding (b) in the 7th simulated transmission (out of a total of 200 runs).

Figure 8.6 compares sample Football frames from layered WZ video coding and H.26L FGS coding when both the base layer and enhancement layer bitstreams are protected with 25% Reed–Solomon-based forward-error protection overhead and transmitted over a simulated CDMA2000 1X channel with 6% PDU loss rate.

8.5 MT Video Coding

Starting from [37], MT video coding for camera arrays and distributed video sensor network has become a very active area of research in recent years (see e.g., [38–44]). In [45], the authors examined MT video coding of two correlated sequences captured by calibrated cameras with known intrinsic (e.g., focal length and pixel width) and extrinsic 3D geometric parameters (e.g., relative positions). Figure 8.7 depicts the camera settings and sample frames.

Two MT video coding schemes were proposed in [45]. In the first scheme, the left sequence of the stereo pair is coded by H.264/AVC and used at the joint decoder to facilitate WZ coding of the right video sequence. The first I-frame of the right sequence is successively coded by H.264/AVC intracoding and WZ coding. An efficient stereo matching algorithm based on loopy belief propagation is then adopted at the decoder to produce pixel-level disparity maps between the corresponding frames of the two decoded video sequences on the fly. Based on the disparity maps, side information for both motion vectors and motion-compensated residual frames of the right sequence are generated at the decoder before WZ encoding. In the second scheme, source splitting was employed on top of classic and WZ coding for compression of both I-frames to allow flexible rate allocation between the two

(a)

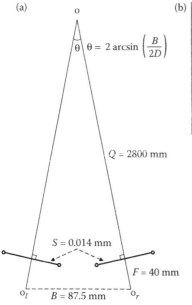

(b)

First frame of the left sequence of "tunnel"

First frame of the right sequence of "tunnel"

FIGURE 8.7
(**See color insert.**) (a) 3D camera settings and (b) first pair of frames from the 720×288 stereo sequence "tunnel."

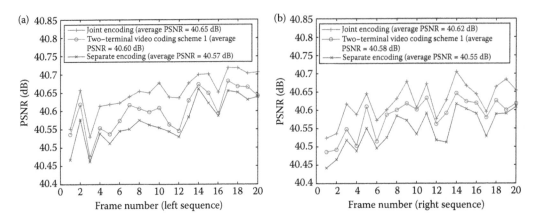

FIGURE 8.8
PSNR versus frame number comparison among separate H.264/AVC coding, two-terminal video coding, and joint encoding at the same sum rate of 6.581 Mbps for (a) the left and (b) the right sequences of the "tunnel."

sequences. Experiments are run on "tunnel" with both separate and joint encoding at the same sum rate of 6.581 Mbps as with two-terminal video coding. Comparsions of these three different schemes are shown in Figure 8.8.

A practical three-terminal video coding scheme that saves the sum-rate over independent coding of three correlated video sequences was presented in [46]. The first video sequence is coded by H.264/AVC, while the other two sequences are compressed sequentially using WZ coding with side information generated from the already decoded sequences. A block diagram of the scheme is depicted in Figure 8.9.

To improve the performance of WZ coding, a depth-map-based view interpolation approach and a novel soft-decision side information generation method are proposed (see Figure 8.10 for an illustration based on the 1024 × 768 Microsoft multiview sequence Ballet).

For the Ballet sequence, the rate used for encoding different types of frames are given in Table 8.1. H.264/AVC is employed for separate coding (and decoding), which is also

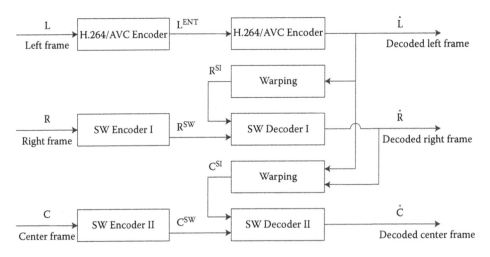

FIGURE 8.9
The general framework proposed in [46] for three-terminal video coding.

FIGURE 8.10
(**See color insert.**) An example of left-and-right-to-center frame warping (based on the first frames of the Ballet sequence). (a) The decoded left frame. (b) The original center frame. (c) The decoded right frame. (d) The left frame warped to the center. (e) The warped center frame. (f) The right frame warped to the center.

TABLE 8.1

Rate (in Bytes) Used for Encoding Different Types of Frames for Ballet

Camera View Positions	H.264/AVC Simulcast		JMVM		MT Video Coding	
	I-Frames	P-Frames	I-Frames	P-Frames	I-Frames	P-Frames
Left	32,714	55,774	32,131	55,998	32,714	55,774
Right	34,646	54,094	–	77,855	34,154	53,953
Center	34,209	52,600	–	66,553	33,365	52,085
Sum-rate	264,037		232,537		262,044	

referred to as *simulcast*, of the three sequences with the IPPP... structure on a total of 10 frames. Quantization parameter (QP) is set to be 22 for both I-frames and P-frames. The joint video coding software JMVM [47] is also based on the H.264/AVC scheme, and the same configurations are used. From Table 8.1, we can see that the sum-rate savings by MT video coding (over separate encoding) is 1993 bytes for the 10-frame sequence (in each view), which is only 1% of the sum-rate of separate encoding. This underscores the difficulty with efficient MT video coding in practice. On the other hand, JMVM joint encoding saves 12% of the sum-rate of separate encoding.

8.5.1 Depth Camera-Assisted MT Video Coding

In [45,46], the depth information was computed from the texture sequences. An alternative is to have a separate camera capture the original depth information and make it available at the decoder (e.g., through a dedicated channel). Not only will 3D rendering and visualization benefit from the original depth information, side information for WZ/MT video coding can also be made much more accurate. Recently, a new type of camera, for example,

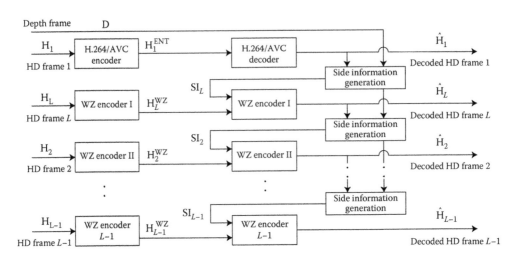

FIGURE 8.11
Depth camera-assisted MT video coding.

SwissRanger series range camera, which can directly generate the depth map of a scene in real time, became available. Although constrained by its low resolution (highest being quarter common intermediate format (QCIF)) and high geometrical distortion, the depth camera can provide more accurate depth information for the background as well as objects that cannot be easily discerned by existing stereo matching algorithms, especially when the number of stereo views is limited (e.g., to no more than three).

Zhang et al. addressed depth camera-assisted MT video coding of high-definition (HD) 1920×1080 video sequences. With depth information compressed and transmitted to the joint decoder, side information is generated more accurately for WZ coding based on refinement of the decompressed low-resolution depth map and texture frame warped from other terminals. Although several approaches have been proposed to compress the depth sequence (see e.g., [48,49]), Zhang et al. used H.264/AVC to compress the depth sequence for the sake of simplicity. Optimal rate allocation between the texture and depth sequences is achieved with equal slope on the operational rate-distortion curves.

Assuming a total of L-terminals, the authors encode the first HD view frames using H.264/AVC before sequentially compressing the remaining $L-1$ views using WZ video coding, where the decoder side information for each WZ coding step is generated based on all previously decoded views in conjunction with the depth frames. The block diagram of the proposed scheme is depicted in Figure 8.11, where cameras 1 through L are placed in a line, with camera 1 being the left most and camera L the right most. The coding order is $1, L, 2, \ldots, L-1$. The side information SI_L and $\text{SI}_2 \sim \text{SI}_{L-1}$ are generated using all previously decoded frames, as well as the depth maps refined by previously decoded frames.

Zhang et al. also described an MT video capturing system with $L = 4$ HD cameras which output colored video frames, and one depth camera which outputs the depth value for every points in the current scene in grayscale format. The four HD cameras are fixed horizontally to a cage, while the depth camera is placed closely above one of the HD cameras. The system is shown in Figure 8.12. To ensure synchronization between different cameras, all cameras are triggered by a single—series rectangular wave at 20 Hz.*

---

* The hardware setup was implemented at AT&T Labs-Research, Florham Park, NJ 07932.

FIGURE 8.12
An MT video capturing system with four HD texture cameras and one low-resolution (QCIF) depth camera.

The depth camera frames are assumed to be available at the decoder for depth map estimation of the HD cameras and side information generation. Due to the large resolution difference between the depth and HD cameras, as well as the calibration inaccuracy induced by this difference, the decompressed low-resolution depth map is first refined before being combined with warped versions of previously decoded texture sequences to generate decoded side information for WZ video coding of the next texture sequence—with the first texture sequence being coded by H.264/AVC. An example of the successive depth refinement result is shown in Figure 8.13. The effectiveness of depth camera assistance can be easily seen by comparing Figure 8.13c and d. We can also see from Figures 8.13b and c that the refined depth map is much closer to the true depth distribution than the original one. In comparison with an original HD texture frame in Figure 8.13a, we can see that the quality of a side information frame with depth camera assistance in Figure 8.13e is much higher in both background and foreground than that without depth camera assistance in Figure 8.13f.

For MT video coding of the four HD texture sequences (partially shown in Figure 8.13), Zhang et al. devided each 100-frame sequence into 10 group of pictures (GOPs), each with 10 frames using an "IPPP..." structure with QP = 22 in the scheme of Figure 8.11. Results, including comparisons with simulcast and JMVM coding, are shown in Table 8.2. It is seen that MT video coding with depth sequence gains 1.43% in sum-rate over that without depth sequence. This means that much needs to be done to improve the performance of MT video coding when the depth information is available. In addition, MT video coding with depth sequence saves 2.59% in sum-rate over simulcast, whereas JMVM does 6.98% better. This underlines the difficulty of significantly outperforming simulcast with both distributed MT video coding and joint JMVM coding.

8.6 What is Next for Distributed Video Coding?

The demand for distributed video coding applications and recent progresses made on SW and WZ code designs have fueled much initial excitements and possibly hype about the topic, leading to some confusion (about what is the right approach to it) or even disappointment (about its performance). This is partially because distributed video coding goes

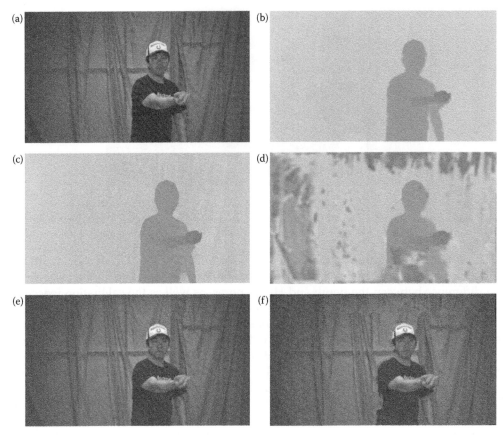

FIGURE 8.13
(**See color insert.**) An example of depth map refinement and side information comparisons. (a) The original HD frame. (b) The preprocessed (warped) depth frame. (c) The refined depth frame. (d) The depth frame generated without the depth camera. (e) Side information with depth camera help. (f) Side information without depth camera help.

above and beyond standard video coding in additionally requiring knowledge on network information theory and expertise on advanced channel code design. After almost a decade, the reality is that research on distributed video coding is still in its infancy and much remains to be done.

In terms of complexity trade-off between the encoder and the decoder, a compromise between approaches (e.g., in [26–28]) with lightweight encoding and those (e.g., in [31–33]) with heavy-duty motion estimation at the decoder is probably the way to go. The approach of multiresolution motion refinement (see e.g., [30]) with low-resolution encoding in conjunction with iterative motion refinement and side information estimation at the decoder holds promise in this regard. Then the immediate research issues are:

Correlation modeling, especially at the encoder: Various existing correlation modeling techniques for WZ video coding are covered in [28]—the most popular one being the Laplacian model for the difference between the video source and the side information. Alternatives include simultaneous encoder and decoder estimation of the side information (e.g., within the framework of [30]) and training-based approaches to come up with a generic correlation model.

TABLE 8.2

Sum-Rates Comparison of Different Schemes, and Percentage of Rate
Savings over H.264/AVC-Based Simulcast

GOP #	Sum-Rate (in Bytes)			
	Simulcast	MT I	MT II	JMVM
1	496,436	489,893	482,118	454,844
2	623,899	616,986	610,627	583,718
3	627,720	620,919	611,322	585,217
4	528,388	521,036	511,645	484,062
5	559,221	552,213	541,295	517,049
6	650,392	643,889	636,377	611,772
7	665,571	658,950	651,919	628,064
8	599,016	591,632	582,869	554,862
9	564,457	556,905	546,755	519,087
10	664,972	658,280	650,261	623,988
Total	5,980,072	5,910,703	5,825,188	5,562,663
Savings	–	1.16%	2.59%	6.98%

Note: MT II is with depth sequence (a rate of 6933 bytes for the depth sequence
is included) and MT I is without. In each case, the same average PSNRs of
46.82, 47.23, 47.14, and 47.21 for the four texture sequences, respectively,
are achieved.

For MT video coding, even though the neighboring views are obviously highly correlated, it remains a challenge to succinctly model this correlation for efficient coding in
terms of achieving sum-rate savings over simulcast. In addition, there are many important issues (e.g., the number of texture/depth cameras and view synthesis/rendering for
camera arrays) outside of MT video coding that have to be addressed before multiview
video-related applications become practical.

Design of short but high-performing channel codes for SW compression: It is well known that
very long codes are needed to asymptotically approach the SW limit. But the delay/latency
requirement in real-time video applications demands short (but good) channel codes for SW
compression. Fortunately, there are recent developments in the channel coding community
on good short-length codes (e.g., quasi-cyclic, protograph, and probably good LDPC codes).
We expect these developments to benefit practical SW compression, hence distributed video
coding.

Besides the above short-term research issues on distributed video coding, two bigger
challenges remain in distributed source coding in general.

Scalable SW coding: Because of the potential applications of scalable/layered coding, successive WZ coding with scalable SW coding is highly desirable. At the end of Slepian and
Wolf's original paper [1], the authors asked (among other things): What is the theory of
variable-length encodings of correlated sources? Unfortunately, there is still no rigorous
theoretical treatment of the important problem of scalable SW coding. On the practical
code design side, various attempts have been made by using rate-adaptive/compatible
LDPC codes or rateless (fountain or Raptor) codes.

In the case of rateless SW coding (e.g., using Raptor codes), modification to the original
random graph-based design technique is needed, because every bit in a rateless bitstream
is created equal. This goes against the requirement of a sequential dependency in a scalable

source-coded bitstream (with the first bit being the most important, and the second bit being less important, etc.). However, we lack a systematic means of modifying the Raptor code design process in terms of making *biased* random connections when forming the sparse-graph of the component Luby-transform code. Consequently, true scalable SW codes (like arithmetic codes in the point-to-point setting) are still not available.

Universal SW coding: When the source correlation is known, SW coding is in general a channel coding problem, as it boils down to designing a channel code for the *correlation channel* between a source pair. When the source correlation is not known *a priori*, it is not clear whether we have a channel coding problem even though Csiszar's 1982 paper [50] showed (among other things) existence of a universal SW decoder. Some recent theoretical works (see e.g., [51]) have started to address the issue of universal SW coding, but we will not have a real breakthrough in practical distributed source/video coding until the invention of a Lempel–Ziv-like universal SW code.

8.7 Concluding Remarks

Even though the theory of distributed source coding dates back to the 1970s, the practice of distributed video coding did not happen until the early 2000s. This is partially due to the fact that its driving applications such as camera arrays and distributed video sensor networks did not arise until recently, and that turbo and LDPC codes were not (re)invented until the mid-1990s. Today, research in distributed video coding has reached a point where the challenges listed in the previous section have to be tackled before real progress can be made.

Before we move forward with distributed video coding, it is helpful to look back at the path we went through with classic video coding, which is underpinned by Shannon's theory (1948 and 1959). It took over 40 years of R&D before popular video coding standards such as MPEG-2 (1993) and H.264 (2003) were developed. And in the process there were major milestones such as Huffman coding (1952), DCT (1974), arithmetic coding (1976), subband coding (1976), and wavelets (1980s).

Compared to classic video coding, distributed video coding is a very young topic and we still have a long way to go. Looking ahead, we expect that milestones such as scalable and universal SW coding be achieved before the possibility of having an international standard on distributed video coding.

Acknowledgment

This work was supported by NSF grants 0729149 and 1017829, the Qatar National Research Fund, and China's 985 Program under grant 293704.

References

1. D. Slepian and J. Wolf, Noiseless coding of correlated information sources, *IEEE Trans. Inform. Theory*, 19, 471–480, 1973.

2. A. Wyner and J. Ziv, The rate–distortion function for source coding with side information at the decoder, *IEEE Trans. Inform. Theory*, 22, 1–10, 1976.
3. T. Berger, Multiterminal source coding, in *The Information Theory Approach to Communications*, G. Longo, ed., New York: Springer-Verlag, 1977.
4. S. Pradhan, J. Kusuma, and K. Ramchandran, Distributed compression in a dense microsensor network, *IEEE Signal Process. Maga.*, 19, 51–60, 2002.
5. Z. Xiong, A. Liveris, and S. Cheng, Distributed source coding for sensor networks, *IEEE Signal Process. Maga.*, 21, 80–94, 2004.
6. B. Girod and N. Farber, Wireless video, in *Compressed Video over Networks*, A. Reibman and M.-T. Sun, eds., Marcel Dekker, New York, 2000.
7. The Stanford multi-camera array. URL: http://graphics.stanford.edu/projects/array/.
8. Workshop on distributed video sensor networks: Research challenges and future directions, DVSN'09, Riverside, CA, May 2009. URL: http://videonetworks2009.cs.ucr.edu/.
9. *IEEE Signal Processing Magazine: Special Issue on Multiview Imaging and 3DTV*, vol. 24, November 2007.
10. T. Wiegand, G. Sullivan, G. Bjøtegaard, and A. Luthra, Overview of the H.264/AVC video coding standard, *IEEE Trans. Circuits Systems Video Tech.*, 13, 560–576, 2003.
11. Y. He, F. Wu, S. Li, Y. Zhong, and S. Yang, H.26L-based fine granularity scalable video coding, in *Proc. ISCAS'02*, Phoenix, AZ, May 2002.
12. S. Verdu and T. Weissman, The information lost in erasures, *IEEE Trans. Inform. Theory*, 54, 5030–5058, 2008.
13. T. Berger, Z. Zhang, and H. Viswanathan, The CEO problem, *IEEE Trans. Inform. Theory*, 42, 887–902, 1996.
14. S. Tung, Multiterminal rate–distortion theory, PhD dissertation, School of Electrical Engineering, Cornell University, Ithaca, NY, 1978.
15. Y. Oohama, Rate–distortion theory for Gaussian multiterminal source coding systems with several side informations at the decoder, *IEEE Trans. Inform. Theory*, 38, 2577–2593, 2005.
16. Y. Oohama, Gaussian multiterminal source coding, *IEEE Trans. Inform. Theory*, 43, 1912–1923, 1997.
17. A. Wagner, S. Tavildar, and P. Viswanath, Rate region of the quadratic Gaussian two-encoder source-coding problem, *IEEE Trans. Inform. Theory*, 54, 1938–1961, 2008.
18. Y. Yang and Z. Xiong, The sum-rate bound for a new class of quadratic Gaussian multiterminal source coding problems, in *Proc. Allerton'09*, Monticello, IL, October 2009.
19. J. Wang, J. Chen, and X. Wu, On the sum rate of Gaussian multiterminal source coding: New proofs and results, *IEEE Trans. Inform. Theory*, 56, 3946–3960, 2010.
20. Y. Yang, Y. Zhang, and Z. Xiong, A new sufficient condition for sum-rate tightness of quadratic Gaussian multiterminal source coding, in *Proc. UCSD ITA'10*, San Diego, CA, February 2010.
21. T. Cover and J. Thomas, *Elements of Information Theory*, 2nd edition, New York: Wiley & Sons, 2006.
22. G. Caire, S. Shamai, and S. Verdu, Lossless data compression with low-density parity-check codes, in *Multiantenna Channels: Capacity, Coding and Signal Processing*, G. Foschini and S. Verdu, eds. Providence, RI: DIMACS, American Mathematical Society 2003.
23. A. Wyner, Recent results in the Shannon theory, *IEEE Trans. Inform. Theory*, 20, 2–10, 1974.
24. H. Witsenhausen and A. Wyner, Interframe coder for video signals, US Patent 4,191,970, March 1980.
25. A. Liveris, Z. Xiong, and C. Georghiades, Compression of binary sources with side information at the decoder using LDPC codes, *IEEE Communi. Lett.*, 6, 440–442, 2002.
26. R. Puri, A. Majumdar, and K. Ramchandran, PRISM: A video coding paradigm with motion estimation at the decoder, *IEEE Trans. Image Process.*, 16, 2436–2448, 2007.
27. B. Girod, A. Aaron, S. Rane, and D. Rebollo-Monedero, Distributed video coding, *Proc. IEEE*, 93, 71–83, 2005.

28. C. Guillemot, F. Pereira, L. Torres, T. Ebrahimi. R. Leonardi, and J. Ostermann, Distributed monoview and multiview video coding: Basics, problems and recent advances, *IEEE Signal Process. Mag.*, 24, 67–76, 2007.

29. D. He, A. Jagmohan, L. Lu, and V. Sheinin, Wyner–Ziv video compression using rateless LDPC codes, in *Proc. VCIP'08*, San Jose, CA, January 2008.

30. W. Liu, L. Dong, and W. Zeng, Motion refinement based progressive side-information estimation for Wyner–Ziv video coding, *IEEE Trans. Circuits Systems Video Tech.*, 20, 1863–1875, 2010.

31. S. Milani, J. Wang, and K. Ramchandran, Achieving H.264-like compression efficency with distributed video coding, in *Proc. VCIP'07*, San Jose, CA, February 2007.

32. S. Milani and G. Calvagno, A distributed video coder based on the H.264/AVC standard, in *Proc. EUSIPCO'07*, Poznan, Poland, September 2007.

33. M. Guo, Z. Xiong, F. Wu, X. Ji, D. Zhao, and W. Gao, Witsenhausen–Wyner video coding, *IEEE Trans. Circuits Systems Video Tech.*, 21, 1049–1060, 2011.

34. Q. Xu and Z. Xiong, Layered Wyner–Ziv video coding, *IEEE Trans. Image Process.*, 15, 3791–3803, 2006.

35. M. Tagliasacchi, A. Majumda, and K. Ramchandran, A distributed source coding based spatial-temporal scalable video codec, in *Proc. PCS'04*, San Francisco, CA, December 2004.

36. H. Wang, N. Cheung, and A. Ortega, A framework for adaptive scalable video coding using Wyner–Ziv techniques, *EURASIP J. Appl. Signal Process.*, 2006, Article ID 60971.

37. X. Zhu, A. Aaron and B. Girod, Distributed compression for large camera arrays, in *Proc. 2003 IEEE Workshop Statistical Signal Processing*, St. Louis, MO, September 2003.

38. M. Flierl and B. Girod, Coding of multi-view image sequences with video sensors, in *Proc. ICIP'06*, pp. 609–612, Atlanta, GA, October 2006.

39. M. Ouaret, F. Dufaux, and T. Ebrahimi, Fusion-based multiview distributed video coding, in *Proc. ACM Workshop on Video Surveillance and Sensor Networks*, Santa Barbara, CA, October 2006.

40. X. Guo, Y. Lu, F. Wu, W. Gao, and S. Li, Distributed multi-view video coding, in *Proc. SPIE VCIP*, San Jose, CA, January 2006.

41. M. Flierl and P. Vandergheynst, Distributed coding of highly correlated image sequences with motion-compensated temporal wavelets, *EURASIP J. Appl. Signal Process.*, Article ID 46747, 2006.

42. B. Song, E. Tuncel, and A. Roy-Chowdhury, Towards a multi-terminal video compression algorithm by integrating distributed source coding with geometrical constraints, *J. Multimed.*, 2, 9–16, June 2007.

43. I. Tosic and P. Frossard, Geometry-based distributed scene representation with omnidirectional vision sensors, *IEEE Trans. Image Process.*, 17, 1033–1046, 2008.

44. N. Gehrig and P. Dragotti, Geometry-driven distributed compression of the plenoptic function: Performance bounds and constructive algorithms, *IEEE Trans. Image Process.*, 18, 457–470, 2009.

45. Y. Yang, V. Stankovic, Z. Xiong, and W. Zhao, Two-terminal video coding, *IEEE Trans. Image Process.*, 18, 534–551, 2009.

46. Y. Zhang, Y. Yang, and Z. Xiong, Three-terminal video coding, in *Proc. IEEE Multimedia Signal Processing Workshop*, Rio de Janeiro, Brazil, October 2009.

47. The JMVM software. URL: http://moscoso.org/pub/video/jsvm/current/jmvm/.

48. H. Oh and Y. Ho, H.264-based depth map sequence coding using motion information of corresponding texture video, *LNCS*, pp. 898–907, December 2006.

49. M. Kang, C. Lee, J. Lee, and Y. Ho, Adaptive geometry-based intra prediction for depth video coding, in *Proc. ICME'10*, pp. 1230–1235, Singapore, July 2010.

50. I. Csiszar, Linear codes for sources and source networks: Error exponents, universal coding, *IEEE Trans. Inform. Theory*, 28, 585–592, July 1982.

51. E. Yang and D. He, Universal data compression with side information at the decoder by using traditional universal lossless compression algorithms, in *Proc. ISIT'07*, pp. 431–435, Nice, France, June 2007.

9

Three-Dimensional Video Coding

Anthony Vetro

CONTENTS

There exist a variety of ways to represent three-dimensional (3D) content, including stereo and multiview video, as well as depth-based video formats. There are also a number of compression architectures and techniques that have been introduced in recent years. This chapter reviews the most relevant 3D representation formats and associated compression techniques.

9.1 Introduction

It has recently become feasible to offer a compelling 3D video (3DV) experience on consumer electronics platforms due to advances in display technology, signal processing, and circuit design. Production of 3D content and consumer interest in 3D has been steadily increasing, and we are now witnessing a global roll-out of services and equipment to support 3DV through packaged media such as Blu-ray Disc and through other broadcast channels such as cable, terrestrial channels, and the Internet.

Despite the recent investments and initial services, there are still a number of challenges to overcome in making 3DV for consumer use at home become a fully practical and sustainable market in the long term. For one, the usability and consumer acceptance of 3D viewing technology will be critical. In particular, mass consumer acceptance of the special eyewear needed to view 3D at home with current display technology is still relatively unknown. In general, content creators, service providers, and display manufacturers need to ensure that the consumer has a high-quality experience and is not burdened with high transition costs or turned off by viewing discomfort or fatigue. The availability of premium 3D content in the home is another major factor to be considered. These are broad issues that will significantly influence the rate of 3D adoption and market size.

Considering the distribution of 3D content, it is clear that the representation format and compression technology that are utilized are central issues that will largely define the interoperability of the system. A number of factors must be considered in the selection of a distribution format. These factors include available storage capacity or bandwidth, player and receiver capabilities, backward compatibility, minimum acceptable quality, and provisioning for future services. Each distribution path to the home also has its own unique requirements that must be considered.

After a brief review of the major applications of 3D and multiview video, this chapter describes the major representation formats for 3DV and their associated encoding techniques. Multiview video coding (MVC) is described with emphasis on the recent extensions to the state-of-the-art H.264/MPEG-4 Advanced Video Coding (AVC) standard. Next, frame-compatible formats for delivery of stereo video through existing infrastructures are introduced. Following this, depth-based 3D formats are presented. The chapter concludes with a summary and discussion on future outlooks.

9.2 3D and Multiview Video Applications

The primary usage scenario for multiview video is to support 3DV applications, where 3D depth perception of a visual scene is provided by a 3D display system. There are many types of 3D display systems [1] including classic stereo systems that require special-purpose glasses to more sophisticated multiview autostereoscopic displays that do not require glasses [2]. The stereo systems only require two views, where a left-eye view is presented to the viewer's left eye, and a right-eye view is presented to the viewer's right eye. The 3D display technology and glasses ensure that the appropriate signals are viewed by the correct eye. This is accomplished with either passive polarization or active shutter techniques. The multiview displays have much greater data throughput requirements relative to conventional stereo displays in order to support a given picture resolution, since 3D is achieved by essentially emitting multiple complete video sample arrays in order to form view-dependent pictures. Such displays can be implemented, for example, using conventional high-resolution displays and parallax barriers; other technologies include lenticular overlay sheets and holographic screens. Each view-dependent video sample can be thought of as emitting a small number of light rays in a set of discrete viewing directions—typically between eight and a few dozen for an autostereoscopic display. Often these directions are distributed in a horizontal plane, such that parallax effects are limited to the horizontal motion of the observer. A comprehensive review of 3D display technologies can be found in [3].

Another goal of multiview video is to enable free-viewpoint video [4,5]. In this scenario, the viewpoint and view direction can be interactively changed. Each output view can either be one of the input views or be a virtual view that was generated from a smaller set of multiview inputs and other data that assist in the view generation process. With such a system, viewers can freely navigate through the different viewpoints of the scene—within a range covered by the acquisition cameras. Such an application of multiview video could be implemented with conventional 2D displays. However, more advanced versions of the free-viewpoint system that work with 3D displays could also be considered. This functionality is already in use in broadcast production environments, for example, to change the viewpoint of a sports scene to show a better angle of a play. Such functionality may also be of interest in surveillance, education, gaming, and sightseeing applications. Finally, this interactive capability may also be provided directly to the home viewer in the foreseeable future, for example, for special events such as concerts.

Another important application of multiview video is to support immersive teleconference applications. Beyond the advantages provided by 3D displays, it has been reported that a teleconference systems could enable a more realistic communication experience when motion parallax is supported. Motion parallax is caused by the change in the appearance of a scene when viewers shift their viewing position, for example, shifting the viewing position to reveal occluded scene content. In an interactive system design, it can be possible for the transmission system to adaptively shift its encoded viewing position to achieve a dynamic perspective change [6–8]. Perspective changes can be controlled explicitly by user intervention through a user interface control component or by a system that senses the observer's viewing position and adjusts the displayed scene accordingly.

Other interesting applications of multiview video have been demonstrated by Wilburn et al. [9]. In this work, a high spatial sampling of a scene through a large multiview video camera array was used for advanced imaging. Among the capabilities shown was an effective increase in bit depth and frame rate, as well as synthetic aperture photography effects. Since then, there have also been other exciting developments in the area of computational imaging that rely on the acquisition of multiview video [10].

The above applications and scenarios are illustrated in Figure 9.1. For all these scenarios, the storage and transmission capacity requirements of the system are significantly increased. Consequently, there is a strong need for efficient 3D and multiview video compression techniques.

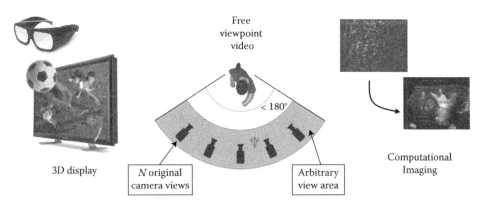

FIGURE 9.1
Applications of 3D and multiview video.

9.3 Multiview Video Coding

This section introduces the basic representation for stereo and multiview video and provides an overview of the recent extensions of the H.264/MPEG-4 AVC standard for MVC. The performance of the MVC standard in terms of both objective and subjective measures is reviewed. Additionally, techniques that have not been included in the MVC standard, but still improve the coding efficiency of multiview video, are also discussed.

9.4 Stereo and Multiview Representations

Stereo and multiview videos are typically acquired at common high-definition (HD) resolutions (e.g., 1920 × 1080 or 1280 × 720) for a distinct set of viewpoints. In the context of this chapter, such video signals are referred to as full-resolution formats. Full-resolution multiview representations can be considered as a reference relative to representation formats that have a reduced spatial or temporal resolution, for example, to satisfy distribution constraints, or representation formats that have a reduced view resolution, for example, due to production constraints. It is noted that there are certain cameras that capture left and right images at half of the typical HD resolutions; such videos would not be considered full resolution. In the case of stereo, the full-resolution representation basically doubles the raw data rate of conventional single-view video. For multiview, there is an N-fold increase in the raw data rate for N-view video. Efficient compression of such data is a key issue and is discussed in the remainder of this section.

9.5 Historical Notes on MVC

One of the earliest studies on coding of multiview images was done by Lukacs [11]; in this work, the concept of disparity-compensated inter-view prediction was introduced. In a later work by Dinstein et al. [12], the predictive coding approach was compared to 3D block transform coding for stereo image compression. In [13], Perkins presented a transform-domain technique for disparity-compensated prediction, as well as a mixed-resolution coding scheme.

The first support for MVC in an international standard was in a 1996 amendment to the H.262/MPEG-2 video coding standard [14,15]. It supported the coding of two views only. In that design, the left view was referred to as the "base view" and its encoding was compatible with that for ordinary single-view decoders. The right view was encoded as an *enhancement view* that used the pictures of the left view as reference pictures for inter-view prediction.

The coding tool features that were used for this scheme were actually the same as what had previously been designed for providing temporal scalability (i.e., frame rate enhancement) [16–19]. For the encoding of the enhancement view, the same basic coding tools were used as in ordinary H.262/MPEG-2 video coding, but the selection of the pictures used as references was altered, so that a reference picture could either be a picture from within the enhancement view or a picture from the base view. A significant benefit of this approach, relative to *simulcast* coding of each view independently, was the ability to use inter-view prediction for the encoding of the first enhancement-view picture in each random-accessible encoded video segment. However, the ability to predict in the reverse-temporal direction, which was enabled for the base view, was not enabled for the enhancement view. This helped to minimize the memory storage capacity requirements for the scheme, but may have reduced the compression capability of the design.

9.6 MVC Extensions of H.264/MPEG-4 AVC

The most recent major extension of the H.264/MPEG-4 AVC standard [20,21] is the MVC design, which was first finalized in July 2009 and later amended in July 2010. The MVC format was selected by the Blu-ray Disc Association as the coding format for 3DV with high-definition resolution. Several key features of MVC are reviewed below; some of which have also been covered in [19,22,23]. Particular aspects of the MVC design such as the adaptation of an MVC bitstream to network and device constraints, as well as an analysis of MVC-decoded picture buffer requirements, could be found in [22].

9.6.1 MVC Design

The basic concept of inter-view prediction, which is employed in all of the described designs for efficient MVC, is to exploit both spatial and temporal redundancy for compression. Since the cameras (or rendered viewpoint perspectives) of a multiview scenario typically capture the same scene from nearby viewpoints, substantial inter-view redundancy is present. A sample prediction structure is shown in Figure 9.2. Pictures are not only predicted from temporal references, but also from inter-view references. The prediction is adaptive, and so the best predictor among temporal and inter-view references can be selected on a block basis in terms of rate-distortion (RD) cost.

Inter-view prediction is a key feature of the MVC design, and it is enabled in a way that makes use of the flexible reference picture management capabilities that had already been designed into H.264/MPEG-4 AVC, by making the decoded pictures from other views available in the reference picture lists for use by the interpicture prediction processing. Specifically, the reference picture lists are maintained for each picture to be decoded in a given view. Each such list is initialized as usual for single-view video, which would include the temporal reference pictures that may be used to predict the current picture. Additionally, inter-view reference pictures are included in the list and are thereby also made available for prediction of the current picture.

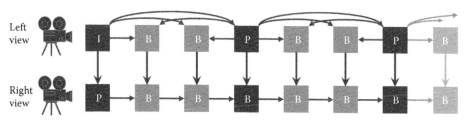

FIGURE 9.2
Illustration of inter-view prediction in MVC.

According to the MVC specification, inter-view reference pictures must be contained within the same access unit as the current picture, where an access unit contains all pictures pertaining to a certain capture or display time instant; in the case of multiview capture, each picture of an access unit has a unique viewpoint. The MVC design does not allow the prediction of a picture in one view at a given time using a picture from another view at a different time. This would involve inter-view prediction across different access units, which would incur additional complexity for limited coding benefits.

It is important to emphasize that the core macroblock-level and lower-level decoding modules of an MVC decoder are the same as that in AVC. In other words, it does not matter to the lower-level decoding modules that a reference picture is a temporal reference or an inter-view reference. The prediction and decoding process is the same at the macroblock-level and this distinction is managed at a higher level of the decoding process.

In terms of syntax, supporting MVC only involves small changes to high-level syntax. Three important pieces of information must be signaled including view identifiers, view dependency information, and information about the resources that are required to decode the stream. The view identifiers are used to associate a particular view to a specific index, while the order of the view identifiers signals the view order index. The view order index is critical to the decoding process as it defines the order in which views are decoded. The view dependency information is composed of a set of signals that indicate the number of inter-view reference pictures for each of the two reference picture lists that are used in the prediction process, as well as the views that may be used for predicting a particular view. The final portion of the high-level syntax is the signaling of a level index to indicate the resource requirements for a decoder; this establishes a bound on the complexity of a decoder, for example, in terms of memory and decoding speed.

A major benefit of not requiring changes to lower levels of the syntax (at the macroblock level and below it) is that MVC is compatible with existing hardware for decoding single-view video with H.264/MPEG-4 AVC. In other words, supporting MVC as part of an existing H.264/MPEG-4 AVC decoder should not require substantial design changes.

Since MVC introduces dependencies between views, random access must also be considered in the view dimension. Specifically, in addition to the views to be accessed (called the target views), any views on which they depend for purposes of inter-view referencing also need to be accessed and decoded, which typically requires some additional decoding time or delay. For applications in which random access or view switching is important, the prediction structure can be designed to minimize access delay, and the MVC design provides a way for an encoder to describe the prediction structure for this purpose.

Another key aspect of the MVC design is that it is mandatory for the compressed multiview stream to include a base-view bitstream, which is coded independently from all other views in a manner compatible with decoders for single-view profile of the standard. This requirement enables a variety of use cases that need a two-dimensional (2D) version of the content to be easily extracted and decoded. For instance, in television broadcast, the base view could be extracted and decoded by legacy receivers, while newer 3D receivers could decode the complete 3D bitstream including nonbase views.

9.6.2 MVC Performance

It has been shown that coding multiview video with inter-view prediction can give significantly better results compared to independent coding [24]. A comprehensive set of results for MVC over a broad range of test material was also presented in [25]. This study used the common test conditions and test sequences specified in [26], which were used throughout

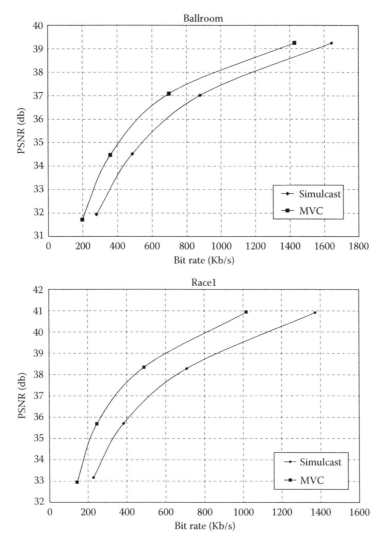

FIGURE 9.3
Sample coding results for Ballroom and Race1 sequences; each sequence includes eight views at video graphics array (VGA) resolution.

the MVC development. For multiview video with up to eight views, an average of 20% reduction in bit rate relative to the total simulcast bit rate was reported with equal quality for each view. All the results were based on the Bjontegaard delta measurements [27]. Figure 9.3 shows sample RD curves comparing the performance of simulcast coding with the performance of the MVC reference software. In other studies [28], an average bit-rate reduction for the second (dependent) view of a typical HD stereo movie content of ~20–30% was reported, with a peak reduction up to 43%. It is noted that the compression gains achieved by MVC using the stereoscopic movie content, which are considered professional HD quality and representative of entertainment quality video, are consistent with gains reported earlier on the MVC test set.

There are many possible variations on the prediction structure considering both temporal and inter-view dependencies. The structure not only affects coding performance, but has notable impact on delay, memory requirements, and random access. It has been confirmed that the majority of gains are obtained using inter-view prediction at random access points, that is, when there is an intracoded picture in the base view. In contrast to independent coding of the views, inter-view prediction is permitted for the encoding of other views at these random access points. An average decrease in bit rate of ~5–15% at equivalent quality could be expected if the inter-view predictions are not used for pictures between the random access points [29]. The benefit of such a prediction structure is that delay and required memory would also be reduced.

Prior studies on asymmetrical coding of stereo video, in which one of the views is encoded with lower quality than the other, suggest that a further substantial savings in bit rate for the nonbase view could be achieved using that technique. In this scheme, one of the views is significantly blurred or more coarsely quantized than the other [30], or is coded with a reduced spatial resolution [31,32], with an impact on the stereo quality that may be imperceptible. With mixed resolution coding, it has been reported that an additional view could be supported with minimal rate overhead, for example, of the order of 25–30% additional rate added to a base view encoding for coding the other views at quarter resolution. Further study is needed to understand how this phenomenon extends to multiview video with more than two views. The currently standardized MVC design provides the encoder with a great deal of freedom to select the encoded fidelity for each view and to perform preprocessing such as blurring if desired; however, the standard currently requires the same resolution for the encoding of all views.

A recent study of subjective picture quality for the MVC Stereo High-Profile targeting full-resolution HD stereo video applications was presented in [33] and also discussed in [34]. For this study, different types of 3DV content were selected with each clip running 25–40 s. In the MVC simulations, the left-eye and right-eye pictures were encoded as the base view and dependent view, respectively. The base view was encoded at 12 and 16 Mbps. The dependent view was coded at a wide range of bit rates, from 5% to 50% of the base-view bit rate. As a result, the combined bit rates range from 12.6 to 24 Mbps. AVC simulcast with symmetric quality was selected as the reference.

Figure 9.4 presents the average mean opinion score (MOS) of all the clips. In the bar charts, each short line segment indicates a 95% confidence interval. The average MOS and 95% confidence intervals show the reliability of the scoring in the evaluation. Overall, when the dependent-view bit rate is no less than 25% of the base-view bit rate, the MVC compression can reproduce the subjective picture quality comparable to that of the AVC simulcast case. In [33,34], a detailed analysis of the performance for each clip is provided, which reveals that the animation clips receive fair or better scores even when the dependent view is encoded at 5% of the base-view bit rate. Also, when the dependent-view bit rate drops below 20% of the base-view bit rate, the MVC-encoded interlaced content starts to receive unsatisfactory scores.

9.7 Block-Level Coding Tools for Multiview Video

During the development of MVC, a number of new block-level coding tools were also explored and proposed. These coding tools were shown to provide additional coding gains beyond the interprediction coding supported by MVC. However, these tools were not adopted to the standard since they would require design changes at the macroblock level. At the time, it was believed that this implementation concern outweighed the coding

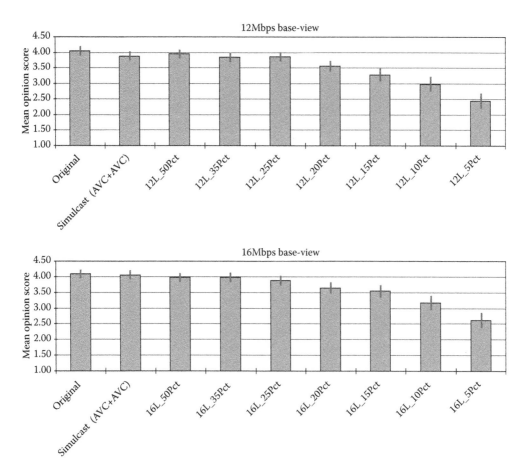

FIGURE 9.4
Subjective picture quality evaluation results given as average MOS with 95% confidence intervals.

gain benefits. However, the benefits of block-level coding tools may be revisited in the specification of future 3DV formats. A brief overview of these coding tools is included below:

Illumination compensation: The objective of this tool is to compensate for illumination differences as part of the inter-view prediction process [34,35]. This tool has shown to be very useful when the illumination or color characteristics vary in different views. This is a likely case since even cameras from the same manufacturer could acquire video with very different color properties. The proposed method determines an offset value that corresponds to the difference in illumination between a current block and its reference. This offset value is calculated as part of the motion estimation process. Rather than computing the typical sum of absolute differences (SAD) between blocks of pixels in the current and reference frame, a mean-removed SAD is computed instead, where there is a mean associated with the current block and a mean associated with a reference block. The difference between these two mean values is the offset that is used for illumination compensation in the decoder. The decoder simply adds this offset value as part of the motion-compensated prediction. The illumination differences between views have been found to be spatially correlated. Therefore, rather than coding the offset value directly, a prediction from neighboring illumination offset values is used to keep rate overhead to a minimum. Coding gains up to 0.6 dB have been reported in comparison with the existing weighted prediction tool of H.264/AVC.

Adaptive reference filtering: It was observed by Lai et al. [36,37] that there are other types of mismatches present in multiview video in addition to illumination differences, which led to the development of an adaptive reference filtering scheme to compensate for focus mismatches between different views. In this approach, the reference frame is filtered by an estimator of the point spread function that models the blur. The filter is depth dependent and designed based on the disparity-compensated difference between the reference frame and the current frame. For sequences with focus mismatch, coding gains up in the range of 0.5–0.75 db have been reported.

Motion skip mode: An effective method to reduce bit rate in video coding is to infer side information used in the decoding process, for example, motion vectors for a particular block, based on other available data, for example, motion vectors from other blocks. This is the basic principle of direct mode prediction in AVC. Extensions to the conventional skip and direct coding modes for MVC were proposed in [38,39]. Specifically, this method infers side information from inter-view references rather than temporal references. A global disparity vector is determined for each neighboring reference view. The motion vector of a corresponding block in the neighboring view may then be used for prediction of the current block in a different view. This signaling is very minimal and this method has the potential to offer notable reduction in bit rate. An analysis of the coding gains offered by both illumination compensation and motion skip mode was reported in [25]. A rate reduction of 10% was reported over a wide range of sequences with a maximum reduction of $\sim$18%.

View synthesis prediction: Another novel macroblock-level coding tool that has been explored for improved coding of multiview video is view synthesis prediction. This coding technique predicts a picture in the current view from synthesized references generated from neighboring views. One approach for view synthesis prediction is to encode depth for each block, which is then used at the decoder to generate the view synthesis data used for prediction, as first described in [40] and fully elaborated on in [41]. Another approach estimates pixel-level disparities at both the encoder and decoder and encodes only disparity correction values [42]. Gains between 0.2 and 1.5 dB compared to inter-view prediction have been reported for views that utilized this type of prediction.

9.8   Frame-Compatible Stereo Formats

To facilitate the introduction of stereoscopic services through the existing infrastructure and equipment, frame-compatible formats have been introduced. With such formats, the stereo signal is essentially a multiplex of the two views into a single frame or sequence of frames. Typically, the left and right views are subsampled and interleaved into a single frame. In the following, a general overview of these formats along with the key benefits and drawbacks are discussed. An architecture that enhances the resolution of these lower-resolution formats is also described.

9.9   Overview

There are a variety of options for the way that the views are subsampled and interleaved. For instance, the two views may be filtered and decimated horizontally or vertically and

FIGURE 9.5
(**See color insert.**) Comparison of full-resolution and frame-compatible formats: (a) full-resolution stereo pair; (b) side-by-side format; (c) top-and-bottom format.

stored in a side-by-side or top-and-bottom format, respectively (see Figure 9.5). Temporal multiplexing is also possible. In this way, the left and right views would be interleaved as alternating frames or fields. These formats are often referred to as frame sequential and field sequential. The frame rate of each view may be reduced so that the amount of data is equivalent to that of a single view. Alternative sampling and packing arrangements are also possible, for example, quincunx filtering and subsampling followed by a rearrangement of resulting pixels into a side-by-side format.

Since frame-compatible formats are able to facilitate the introduction of stereoscopic services through existing infrastructure and equipment, they have received considerable attention from the cable and broadcast industry [44,45]. In this way, the coded video can be processed by encoders and decoders that were not specifically designed to handle stereo video, and only the display subsystem that follows the decoding process needs to be altered to support 3D.

The drawback of representing the stereo signal in this way is that spatial or temporal resolution would be only half of that used for 2D video with the same (total) encoded resolution. The key additional issue with frame-compatible formats is distinguishing the left and right views. To perform the de-interleaving, it is necessary for receivers to be able to parse and interpret some signal that indicates that the frame packing is being used. The signaling for these formats has been standardized as part of the H.264/MPEG-4 AVC standard; further details can be found in [20,23].

The signaling used to indicate the presence of a frame-compatible format in the bitstream and the particular arrangement may not be understood by legacy receivers; therefore, it may not even be possible for such devices to extract, decode, and display a 2D version of the 3D program. However, this may not necessarily be considered so problematic, as it is not always considered desirable to enable 2D video extraction from a 3D stream. The content production practices for 2D and 3D programs may be different, and 2D and 3D versions of a program may be edited differently, for example, using more frequent scene cuts and more global motion for 2D programming than for 3D. Moreover, the firmware on some devices, such as cable set-top boxes, could be upgraded to understand the new signaling that describes the video format, although the same is not necessarily true for broadcast receivers and all types of equipment.

9.10 Resolution Enhancement

Although frame-compatible methods can facilitate easy deployment of 3D services to the home, they still suffer from a reduced resolution, and therefore reduced 3D quality perception. Recently, several methods that can extend frame-compatible signals to full resolution have been proposed. These schemes ensure backward compatibility with already deployed frame-compatible 3D services, while permitting a migration to full-resolution 3D services. A general architecture that illustrates the resolution enhancement of frame-compatible video formats is shown in Figure 9.6.

The Scalable Video Coding (SVC) extension of H.264/AVC [46] is perhaps the most straightforward method to achieve resolution enhancement since spatial scalability is one of the key features of this standard. By leveraging the existing spatial scalability coding tools, it is possible to scale the lower-resolution frame-compatible signal to full resolution. Alternatively, a combination of both spatial and temporal scalability coding tools in SVC

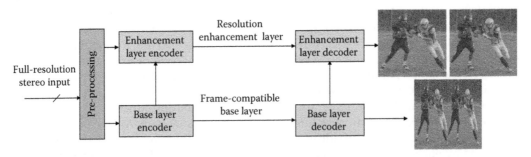

FIGURE 9.6
(**See color insert.**) Illustration of video codec for scalable resolution enhancement of frame-compatible video.

could also be utilized. In this way, rather than applying spatial scalability to the entire frame, only half of the frame relating to a single view is enhanced by applying spatial scalability to a region of interest. Then, the full-resolution second view can be encoded as a temporal enhancement layer.

This second method more closely resembles the coding process used in MVC since the second view is able to exploit both temporal and inter-view redundancy. However, the same view is not able to exploit the redundancies that may exist in the lower-resolution base layer. This method essentially sacrifices exploiting spatial correlation in favor of inter-view correlation. Both of these methods have the limitation that they may not be effective for more complicated frame-compatible formats such as side-by-side formats based on quincunx sampling or checkerboard formats.

With the appropriate preprocessing, the MVC architecture could also be used to enhance a frame-compatible signal to full resolution. In particular, instead of low-pass filtering the two views prior to decimation and then creating a frame-compatible image, one may apply a low-pass filter at a higher cutoff frequency or not apply any filtering at all. Although this may introduce some minor aliasing in the base layer, this provides the ability to enhance the signal to a full or near-full resolution with an enhancement layer consisting of the complementary samples relative to those of the base layer. These samples may have been similarly filtered and are packed using the same frame-compatible packing arrangement as the base layer. The advantage of this method is that one can additionally exploit the spatial redundancies that may now exist between the base and enhancement layer signals, resulting in very high compression efficiency for the enhancement layer coding. Furthermore, existing implementations of MVC hardware could easily be repurposed for this application with minor modifications in the postdecoding stage.

An improvement over this method that tries to further exploit the correlation between the base and enhancement layer was presented in [47]. Instead of directly considering the base layer frame-compatible images as a reference of the enhancement layer, a new process is introduced that first prefilters the base layer picture given additional information that is provided within the bitstream. This process generates a new reference from the base layer that has much higher correlation with the pictures in the enhancement layer.

A final category for the enhancement of frame-compatible signals to full resolution considers filter-bank-like methods [47,48]. Essentially, the base and enhancement layers contain the low- and high-frequency information, respectively. The separation is done using appropriate analysis filters in the encoder, whereas the analogous synthesis filters can be used during reconstruction at the decoder.

All these methods have their own unique benefits and drawbacks. It is not yet clear as to which method will be finally adopted by the industry. There are studies currently underway within the Moving Picture Experts Group (MPEG) committee to evaluate compression efficiency and complexity of various solutions.

9.11 Depth-Based 3DV Formats

Depth-based representations are another important class of 3D formats. As described by several researchers [49–51], depth-based formats enable the generation of virtual views through depth-based image rendering techniques. The depth information may be extracted from a stereo pair by solving for stereo correspondences [52] or obtained directly through special range cameras [53]; it may also be an inherent part of the content, such as with computer-generated imagery. These formats are attractive since the inclusion of depth enables a display-independent solution for 3D that supports generation of an increased number of views, which may be required by different 3D displays. In principle, this format is able to support both stereo and multiview displays, and also allows adjustment of depth perception in stereo displays according to viewing characteristics such as display size and viewing distance.

9.12 Depth Representation

The depth of a 3D scene is expressed relative to the camera position or an origin in the 3D space. The following equation expresses a general relation among a depth value z, the nearest clipping plane Z_{near}, and the farthest clipping plane Z_{far} from the origin in a 3D space or the camera plane:

$$z = \frac{1}{\dfrac{v}{255} \cdot \left(\dfrac{1}{Z_{\text{near}}} - \dfrac{1}{Z_{\text{far}}} \right) + \dfrac{1}{Z_{\text{far}}}},$$

where v is an 8-bit intensity of the depth map value. It is noted that the z, Z_{near}, and Z_{far} values are assumed to be either all positive or all negative.

In much of the literature on multiview video, cameras are assumed to be arranged in one-dimensional parallel configuration and the camera array is almost parallel to the x-axis of a 3D coordinate system. Let f and b be camera focal length and baseline distance between cameras, respectively. Furthermore, let d and Δd be the disparity and camera offset. Then, given the following relations:

$$z = \frac{f \cdot b}{d + \Delta d}; \quad Z_{\text{near}} = \frac{f \cdot b}{d_{\max} + \Delta d}; \quad Z_{\text{far}} = \frac{f \cdot b}{d_{\min} + \Delta d};$$

the intensity v of the depth map with 8-bit precision can be expressed as

$$v = 255 \cdot \frac{d - d_{\min}}{d_{\max} - d_{\min}}.$$

The set of v values for each pixel of an image comprise the depth map or depth image. Note that it is possible to have a depth image that corresponds to each camera viewpoint.

ISO/IEC 23002-3 (also referred to as MPEG-C Part 3) specifies the representation of auxiliary video and supplemental information. In particular, it enables signaling for depth map streams to support 3DV applications. Specifically, the well-known 2D plus depth format as illustrated in Figure 9.7 is specified by this standard. It is noted that this standard does not specify the means by which the depth information is coded, nor does it specify the means by which the 2D video is coded. In this way, backward compatibility to legacy devices can be provided.

The main drawback of the 2D-plus-depth format is that it is only capable of rendering a limited depth range and was not specifically designed to handle occlusions. Also, stereo signals are not easily accessible by this format, that is, receivers would be required to

FIGURE 9.7
(**See color insert.**) Example of 2D-plus-depth representation.

generate the second view to drive a stereo display, which is not the convention in existing displays.

To overcome the drawbacks of the 2D plus depth format, while still maintaining some of its key merits, the MPEG standardization committee is now in the process of exploring alternative representation formats and is considering a new phase of standardization. The targets of this new initiative are discussed in [54]. The objectives are to:

- Enable stereo devices to cope with varying display types and sizes, and different viewing preferences. This includes the ability to vary the baseline distance for stereo video so that the depth perception experienced by the viewer is within a comfortable range. Such a feature could help to avoid fatigue and other viewing discomforts.

- Facilitate support for high-quality autostereoscopic displays. Since directly providing all the necessary views for these displays is not practical due to production and transmission constraints, the new format aims to enable the generation of many high-quality views from a limited amount of input data, for example, stereo and depth.

A key feature of this new 3DV data format is to decouple the content creation from the display requirements, while still working within the constraints imposed by production and transmission. The 3DV format aims to enhance 3D rendering capabilities beyond 2D plus depth. Also, this new format should substantially reduce the rate requirements relative to sending multiple views directly. These requirements are outlined in [55].

9.13 Depth Compression

As discussed earlier, depth information could be used at the receiver to generate additional novel views or used at the encoder to realize more efficient compression with view synthesis prediction schemes. Although the depth data are not directly output to a display and viewed, maintaining the fidelity of depth information is important since the quality of the view synthesis result is highly dependent on the accuracy of the geometric information provided by depth. A depth sample represents a shift value in texture samples from the original views. Thus, coding errors in depth maps result in wrong pixel shifts in synthesized views. This may lead to annoying artifacts, especially along visible object boundaries. Therefore, a depth compression algorithm needs to preserve depth edges much better than traditional coding methods. It is also crucial to strike a good balance between the fidelity of depth data and the overall bandwidth requirement.

A depth signal mainly consists of larger homogeneous areas inside scene objects and sharp transitions along boundaries between objects at different depth values. Therefore, in the frequency spectrum of a depth map, low and very high frequencies are dominant. Video compression algorithms are typically designed to preserve low frequencies and image blurring occurs in the reconstructed video at high compression rates. The need for compression techniques that are adapted to these special characteristics of the depth signal and the requirement of maintaining the fidelity of edge information in the depth maps has motivated research in this area.

As reported in [56], the rate used to code depth video with pixel-level accuracy could be quite high and of the same order as that of the texture video. Experiments were performed to demonstrate how the video synthesis quality varies as a function of bit rate for both the texture and depth videos. It was found that higher bit rates were needed to code the depth data so that the view rendering quality around object boundaries could be maintained.

There have been various approaches considered in the literature to reduce the required rate for coding depth, while maintaining high view synthesis and multiview rendering quality. One approach is to code a reduced resolution version of the depth using conventional compression techniques. This method could provide substantial rate reductions, but the filtering and reconstruction techniques need to be carefully designed to maximize quality. A set of experiments were performed in [57] using simple averaging and interpolation filters. These results demonstrate effective reduction in rate, but artifacts are introduced in the reconstructed images due to the simple filters. Improved down/up-sampling filters were proposed by Oh et al. in [58]. This work does not demonstrate very substantial reductions in the bit rate, whereby bit-rate reductions of >60% were shown relative to full-resolution coding of the depth videos. It also improves the rendering quality around the object boundaries compared to conventional filtering techniques. The impact of the filtering techniques is illustrated in Figure 9.8.

Another approach is to code the depth based on geometric representation of the data. In [59], Morvan et al. model depth images using a piecewise linear function; they referred to this representation as platelets. The image is subdivided using a quadtree decomposition and an appropriate modeling function is selected for each region of the image in order to optimize the overall RD cost. The benefit of this approach for improved rendering quality relative to JPEG 2000 was clearly shown. In subsequent work, comparisons to AVC intracoding were also made and similar benefits have been shown [60].

A drawback of the platelet-based approach is that it appears difficult to extend this scheme to video. An alternative multilayered coding scheme for depth was suggested in [61]. In this approach, it was argued that the quality of depth information around object boundaries needs to be maintained with higher fidelity since it has a notable impact on subjective visual quality. The proposed scheme guarantees a near-lossless bound on the depth values around the edges by adding an extra enhancement layer. This method effectively improves the visual quality of the synthesized images, and is flexible in the sense that it could incorporate any lossy coder as the base layer, thereby making it easily extendible to the coding of depth video.

In [62], conventional coding schemes, such as AVC and MVC, were used to code the depth data; however, in addition to considering the coding quality to the depth data when

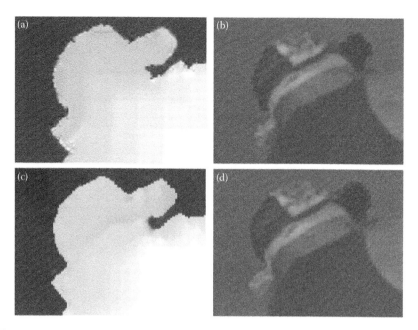

FIGURE 9.8
(**See color insert.**) Effect of down/up sampling filters on depth maps and corresponding synthesis result (a, b) using conventional linear filters; (c, d) using nonlinear filtering as proposed in [58].

applying RD optimization techniques, the quality of the synthesized views at the receiver was also accounted for as part of the optimization. This work also highlights the importance of rate allocation between texture and depth information. Recent studies in the MPEG standardization committee have began to consider the quality of a synthesized view versus bit rate, where optimal combinations of quantization parameter (QP) for texture and depth are determined for a target set of bit rates; a sample plot with various operating points is shown in Figure 9.9. Such coding methods can also be combined with edge-aware synthesis algorithms, which are able to suppress some of the displacement errors caused by depth coding with MVC [63].

Besides the adaptation of compression algorithms to the individual video and depth data, some of the block-level information, such as motion vectors, may be similar for both and thus can be shared. An example compression scheme that makes use of side information between the different data channels is given in [64]. In addition, mechanisms used in SVC can be applied, where a base layer was originally used for a lower-quality version of the 2D video and a number of enhancement layers were used to provide improved quality versions of the video. In the context of multiview coding, a reference view is encoded as the base layer. Adjacent views are first warped onto the position of the reference view and the residual between both is encoded in further enhancement layers.

Other methods for joint video and depth coding with partial data sharing, as well as special coding techniques for depth data, are expected to be available soon in order to provide improved compression in the context of the new 3DV format that is anticipated. The new format will not only require high coding efficiency, but it must also enable good subjective quality for synthesized views that could be used on a wide range of 3D displays.

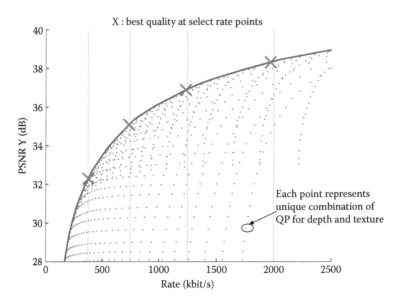

FIGURE 9.9
Sample plot of quality for a synthesized view versus bit rate where optimal combinations of QP for texture and depth are determined for a target set of bit rates.

9.14 Summary

3DV has drawn significant attention recently among industry, standardization forums, and academic researchers. The efficient representation and compression of stereo and multiview video is a central component of any 3D or multiview system since it defines the format to be produced, stored, transmitted, and displayed. This chapter has reviewed a number of important 3D representation formats and the associated compression techniques.

After many false starts, it now appears that 3DV is here to stay. The 3D experience has formed a positive impression on a significant percentage of consumers around the world through high-quality digital theater experiences. The 3D content production is now beginning to ramp up and the infrastructure and equipment have been put in place recently for the distribution of stereoscopic video services to the home. The distribution is now being realized through Blu-ray Discs, which have selected the MVC format as the coding format for full-resolution stereoscopic content. Delivery of stereoscopic 3D services over cable and broadcast networks has also started based on the frame-compatible formats. It is expected that the cable and broadcast services will eventually migrate to full-resolution formats utilizing one of the resolution enhancement schemes that are described in this chapter. It is certain that delivery of premium 3D content through the Internet, including both streaming and download services, will also begin in the very near future.

As the market evolves and new types of displays and services are offered, new technologies and standards will need to be introduced. For example, it is anticipated that a new 3DV format to support the generation of the large number of views required by autostereoscopic displays would be needed. As presented in this chapter, solutions that consider the inclusion of depth map information for this purpose are a significant area of focus for future designs.

References

1. J. Konrad and M. Halle, 3-D displays and signal processing—An answer to 3-D ills? *IEEE Signal Processing Magazine*, 24(6), 97–111, 2007.
2. N. A. Dodgson, Autostereoscopic 3D displays, *IEEE Computer*, 38(8), 31–36, 2005.
3. H. Urey, K. V. Chellephan, E. Erden, and P. Surman, State of the art in stereoscopic and autostereoscopic displays, *Proceedings of the IEEE*, 99(4), 540–555, 2011.
4. A. Smolic and P. Kauff, Interactive 3-D video representation and coding technologies, *Proceedings of the IEEE*, 93(1), 98–110, 2005.
5. M. Tanimoto, M.P. Tehrani, T. Fujii, and T. Yendo, Free-viewpoint TV, *IEEE Signal Processing Magazine*, 28(1), 67–76, 2011.
6. P. Kauff, O. Schreer, and R. Tanger, Virtual team user environments—A mixed reality approach for immersive tele-collaboration, *International Workshop on Immersive Telepresence (ITP 2002)*, pp. 1–4, Juan Les Pins, France, January 2002.
7. I. Feldmann, O. Schreer, P. Kauff, R. Schäfer, Z. Fei, H. J. W. Belt, and Ò. Divorra Escoda, Immersive multi-user 3D video communication, *Proceedings of International Broadcast Conference (IBC 2009)*, Amsterdam, Netherlands, September 2009.
8. C. Zhang, D. Florencio, and Z. Zhang, Improving immersive experiences in telecommunication with motion parallax, *IEEE Signal Processing Magazine*, 28(1), 139–144, 2011.
9. B. Wilburn, N. Joshi, V. Vaish, E.-V. Talvala, E. Antunez, A. Barth, A. Adams, M. Horowitz, and M. Levoy, High performance imaging using large camera arrays, *ACM Transactions on Graphics*, 24(3), 765–776, 2005.
10. R. Raskar and J. Tumblin, Computational photography: Mastering new techniques for lenses, lighting, and sensors, A K Peters, Ltd., ISBN 978–1–56881–313–4, 2010. Available online: http://web.media.mit.edu/~raskar/photo/.
11. M. E. Lukacs, Predictive coding of multi-viewpoint image sets, *Proceedings of IEEE International Conference on Acoustics, Speech and Signal Processing*, Vol. 1, pp. 521–524, Tokyo, Japan, 1986.
12. I. Dinstein, G. Guy, J. Rabany, J. Tzelgov, and A. Henik, On the compression of stereo images: Preliminary results, *Signal Processing: Image Communications*, 17(4), 373–382, 1989.
13. M. G. Perkins, Data compression of stereo pairs, *IEEE Transactions on Communications*, 40(4), 684–696, 1992.
14. ITU-T and ISO/IEC JTC 1, Generic coding of moving pictures and associated audio information—Part 2: Video, ITU-T Recommendation H.262 and ISO/IEC 13818–2 (MPEG-2 Video), 1994.
15. ITU-T and ISO/IEC JTC 1, Final Draft Amendment 3, Amendment 3 to ITU-T Recommendation H.262 and ISO/IEC 13818–2 (MPEG-2 Video), *ISO/IEC JTC 1/SC 29/WG 11 (MPEG) Doc. N1366*, September 1996.
16. A. Puri, R. V. Kollarits, and B. G. Haskell. Stereoscopic video compression using temporal scalability, *Proceedings of SPIE Conference on Visual Communications and Image Processing*, Vol. 2501, pp. 745–756, Taipei, Taiwan, 1995.
17. X. Chen and A. Luthra, MPEG-2 multi-view profile and its application in 3DTV, *Proceedings of SPIE IS&T Multimedia Hardware Architectures*, Vol. 3021, pp. 212–223, San Diego, USA, February 1997.
18. J.-R. Ohm, Stereo/multiview video encoding using the MPEG family of standards, *Proceedings of SPIE Conference on Stereoscopic Displays and Virtual Reality Systems VI*, San Jose, CA, January 1999.
19. G. J. Sullivan, Standards-based approaches to 3D and multiview video coding, *Proceedings of SPIE Conference on Applications of Digital Image Processing XXXII*, San Diego, CA, August 2009.
20. ITU-T and ISO/IEC JTC 1, Advanced video coding for generic audiovisual services, ITU-T Recommendation H.264 and ISO/IEC 14496–10 (MPEG-4 AVC), 2010.
21. T. Wiegand, G. J. Sullivan, G. Bjøntegaard, and A. Luthra, Overview of the H.264/AVC video coding standard, *IEEE Transactions on Circuits and Systems for Video Technology*, 13(7), 560–576, 2003.

22. Y. Chen, Y.-K. Wang, K. Ugur, M. M. Hannuksela, J. Lainema, and M. Gabbouj, The emerging MVC standard for 3D video services, *EURASIP Journal on Advances in Signal Processing*, vol. 2009, Article ID 786015, 13 pages, 2009. doi:10.1155/2009/786015.

23. A. Vetro, T. Wiegand, and G.J. Sullivan, Overview of the stereo and multiview video coding extensions of the H.264/MPEG-4 AVC standard, *Proceedings of the IEEE*, 99(4), 626–642, 2011.

24. P. Merkle, A. Smolic, K. Mueller, and T. Wiegand, Efficient prediction structures for multiview video coding, *IEEE Transactions on Circuits and Systems for Video Technology*, 17(11), 1461–1473, 2007.

25. D. Tian, P. Pandit, P. Yin, and C. Gomila, Study of MVC coding tools, *Joint Video Team (JVT) Doc. JVT-Y044*, Shenzhen, China, October 2007.

26. Y. Su, A. Vetro, and A. Smolic, Common test conditions for multiview video coding, *Joint Video Team (JVT) Doc. JVT-U211*, Hangzhou, China, October 2006.

27. G. Bjøntegaard, Calculation of average PSNR differences between RD-curves, *ITU-T SG16/Q.6, Doc. VCEG-M033*, Austin, TX, April 2001.

28. T. Chen, Y. Kashiwagi, C.S. Lim, and T. Nishi, Coding performance of Stereo High Profile for movie sequences, *Joint Video Team (JVT) Doc. JVT-AE022*, London, UK, July 2009.

29. M. Droese and C. Clemens, Results of CE1-D on multiview video coding, *ISO/IEC JTC 1/SC 29/WG 11 (MPEG) Doc. M13247*, Montreux, Switzerland, April 2006.

30. L. Stelmach, W. J. Tam, D. Meegan, and A. Vincent, Stereo image quality: Effects of mixed spatio-temporal resolution, *IEEE Transactions on Circuits and Systems for Video Technology*, 10(2), 188–193, 2000.

31. C. Fehn, P. Kauff, S. Cho, H. Kwon, N. Hur, and J. Kim, Asymmetric coding of stereoscopic video for transmission over T-DMB, *Proceedings of 3DTV-CON 2007*, pp. 1–4, Kos, Greece, May 2007.

32. H. Brust, A. Smolic, K. Müller, G. Tech, and T. Wiegand, Mixed resolution coding of stereoscopic video for mobile devices, *Proceedings of 3DTV-CON 2009*, pp. 1–4, Potsdam, Germany, May 2009.

33. T. Chen and Y. Kashiwagi, Subjective picture quality evaluation of MVC stereo high profile for full-resolution stereoscopic high-definition 3D video applications, *Proceedings of IASTED Conference on Signal and Image Processing*, Maui, HI, August 2010.

34. Y. L. Lee, J. H. Hur, Y. K. Lee, K. H. Han, S. H. Cho, N. H. Hur, J. W. Kim et al., CE11: Illumination compensation, *Joint Video Team (JVT) Doc. JVT-U052*, Hangzhou, China, 2006.

35. J. H. Hur, S. Cho, and Y. L. Lee, Adaptive local illumination change compensation method for H.264/AVC-based multiview video coding, *IEEE Transactions on Circuits and Systems for Video Technology*, 17(11), 1496–1505, 2007.

36. P. Lai, A. Ortega, P. Pandit, P. Yin, and C. Gomila, Adaptive reference filtering for MVC, *Joint Video Team (JVT) Doc. JVT-W065*, San Jose, CA, April 2007.

37. P. Lai, A. Ortega, P. Pandit, P. Yin, and C. Gomila, Focus mismatches in multiview systems and efficient adaptive reference filtering for multiview video coding, *Proceedings of SPIE Conference on Visual Communications and Image Processing*, San Jose, CA, January 2008.

38. H. S. Koo, Y. J. Jeon, and B. M. Jeon, MVC motion skip mode, *Joint Video Team (JVT) Doc. JVT-W081*, San Jose, CA, April 2007.

39. H. S. Koo, Y. J. Jeon, and B. M. Jeon, Motion information inferring scheme for multi-view video coding. *IEICE Transactions on Communications*, E91-B(4), 1247–1250, 2008.

40. E. Martinian, A. Behrens, J. Xin, A. Vetro, View synthesis for multiview video compression, *Proceedings of Picture Coding Symposium*, Beijing, China, 2006.

41. S. Yea and A. Vetro, View synthesis prediction for multiview video coding, *Image Communication*, 24(1–2), 89–100, 2009.

42. M. Kitahara, H. Kimata, S. Shimizu, K. Kamikura, Y. Yashima, K. Yamamoto, T. Yendo, T. Fujii, and M. Tanimoto, Multi-view video coding using view interpolation and reference picture selection, *Proceedings of IEEE International Conference on Multimedia & Expo*, pp. 97–100, Toronto, Canada, July 2006.

43. A. Vetro, A. M. Tourapis, K. Mueller, and T. Chen, 3D-TV content storage and transmission, *IEEE Transactions on Broadcasting*, 57(2), 384–394, 2011.

44. D. K. Broberg, Infrastructures for home delivery, interfacing, captioning, and viewing of 3D content, *Proceedings of the IEEE*, 99(4), 684–693, 2011.

45. Cable Television Laboratories, Content Encoding Profiles 3.0 Specification OC-SP-CEP3.0-I01–100827, Version I01. Available at http://www.cablelabs.com/specifications/OC-SP-CEP3.0-I01–100827.pdf, August 2010.

46. H. Schwarz, D. Marpe, and T. Wiegand, Overview of the scalable video coding extension of the H.264/AVC standard, *IEEE Transactions on Circuits and Systems for Video Technology, Special Issue on Scalable Video Coding*, 17(9), 1103–1120, 2007.

47. A. M. Tourapis, P. Pahalawatta, A. Leontaris, Y. He, Y. Ye, K. Stec, and W. Husak, A frame compatible system for 3D delivery, *ISO/IEC JTC1/SC29/WG11 Doc. M17925*, Geneva, Switzerland, July 2010.

48. K. Minoo, V. Kung, D. Baylon, K. Panusopone, A. Luthra, and J. H. Kim, On scalable resolution enhancement of frame-compatible stereoscopic 3D video, *ISO/IEC JTC1/SC29/WG11 Doc. M18486*, Guangzhou, China, October 2010.

49. K. Mueller, P. Merkle, and T. Wiegand, 3D video representation using depth maps, *Proceedings of the IEEE*, 99(4), 643–656, 2011.

50. C. Fehn, Depth-image-based rendering (DIBR), compression and transmission for a new approach on 3D-TV, *Proceedings of SPIE Conference on Stereoscopic Displays and Virtual Reality Systems XI*, pp. 93–104, San Jose, CA, USA, January 2004.

51. A. Vetro, S. Yea, and A. Smolic, Towards a 3D video format for auto-stereoscopic displays, *Proceedings of SPIE Conference on Applications of Digital Image Processing XXXI*, San Diego, CA, August 2008.

52. D. Scharstein and R. Szeliski, A taxonomy and evaluation of dense two-frame stereo correspondence algorithms, *International Journal of Computer Vision*, 47(1), 7–42, 2002.

53. E.-K. Lee, Y.-K. Jung, and Y.-S. Ho, 3D video generation using foreground separation and disocclusion detection, *Proceedings of IEEE 3DTV Conference*, Tampere, Finland, June 2010.

54. Video and Requirements Group, Vision on 3D video, *ISO/IEC JTC1/SC29/WG11 Doc. N10357*, Lausanne, Switzerland, February 2009.

55. Video and Requirements Group, Applications and requirements on 3D video coding, *ISO/IEC JTC1/SC29/WG11 Doc. N11829*, Daegu, Korea, January 2011.

56. P. Merkle, K. Mueller, A. Smolic, and T. Wiegand, Experiments on coding of multi-view video plus depth, *ITU-T & ISO/IEC JTC1/SC29/WG11 Doc. JVT-X064*, Geneva, Switzerland, 2007.

57. A. Vetro, S. Yea, and A. Smolic, Towards a 3D video format for auto-stereoscopic displays, *SPIE Conference on Applications of Digital Image Processing XXXI*, Vol. 7073, San Diego, CA, September 2008.

58. K. J. Oh, S. Yea, A. Vetro, and Y. S. Ho, Depth reconstruction filter and down/up sampling for depth coding in 3D video, *IEEE Signal Processing Letters*, 16(9), 747–750, 2009.

59. Y. Morvan, D. Farin, and P. H. N. de With, Depth-image compression based on an R-D optimized quadtree decomposition for the transmission of multiview images, *IEEE International Conference on Image Processing*, pp. V-105–V-108, San Antonio, TX, September 2007.

60. P. Merkle, Y. Morvan, A. Smolic, D. Farin, K. Mueller, P. H. N. de With, and T. Wiegand, The effects of multiview depth video compression on multiview rendering, *Image Communication*, 24(1–2), 73–88, 2009.

61. S. Yea and A. Vetro, Multi-layered coding of depth for virtual view synthesis, *Picture Coding Symposium*, Chicago, IL, May 2009.

62. W.-S. Kim, A. Ortega, P. Lai, D. Tian, and C. Gomila, Depth map coding with distortion estimation of rendered view, *Proceedings of the SPIE on Visual Information Processing and Communication*, 7543, 2010.

63. C. L. Zitnick, S. B. Kang, M. Uyttendaele, S. Winder, and R. Szeliski, High-quality video view interpolation using a layered representation, *ACM SIGGRAPH and ACM Transactions on Graphics*, Vol. 23(3), pp. 600–608, Los Angeles, CA, August 2004.

64. H. Oh and Y.-S. Ho, H.264-based depth map sequence coding using motion information of corresponding texture video, *Advances in Image and Video Technology*, Vol. 4319, Berlin: Springer, 2006.

10

AVS: An Application-Oriented Video Coding Standard

Siwei Ma, Li Zhang, Debin Zhao, and Wen Gao

CONTENTS

10.1 Introduction

In the last few years, the emerging advanced video coding standards or codecs have become the focus of multimedia industries and research institutes, such as H.264/MPEG-4 AVC established by Joint Video Team (JVT) [1], video codec-1 (VC-1) by Microsoft [2], and Audio Video Coding Standard (AVS) [3]. The AVS is developed by the China Audio Video Coding Standard Working Group, which was founded in June 2002. The role of the group is to establish general technical standards for the compression, decoding, processing, and the representation of digital audio–video, thereby enabling digital audio–video equipment and systems with high-efficiency and economical coding/decoding technologies. After 9 years, AVS has published a series of standards. So far, AVS standards are gaining wider acceptance in international standards. The AVS has been accepted as an option by International Telecommunication Union-Telecommunication Standardization Sector Focus Group on Internet Protocol Television (ITU-TFGIPTV) for Internet Protocol Television (IPTV) applications in 2007. This chapter gives an overview of the background and technical features of the AVS.

As with Moving Picture Experts Group (MPEG) standards, the AVS is composed of several parts, such as technical specifications for system multiplexing, video coding, audio coding, supporting specifications for conformance testing, reference software, and so on [4]. For video coding, AVS include two parts: Parts 2 and 7. Part 2, called AVS1-P2 in the AVS Working Group mainly targets to high-definition and high-quality digital broadcasting, digital storage media. Part 7 (AVS1-P7) is targeted to the growing mobile applications, wireless broadband multimedia communication, and internet broadband streaming media. As H.264/AVC, AVS1-P2 defines several profiles, including Jizhun Profile, Jiaqiang Profile, Shenzhan Profile, and Yidong Profile to satisfy different requirements of various video applications. Jizhun Profile, the first defined profile in AVS1-P2, mainly focuses on digital video applications such as commercial broadcasting and storage media, including high-definition applications. It has been approved as the national standard in 2006 in China. After that, an extended work of AVS1-P2 Jizhun Profile, called Zengqiang Profile, was started to further improve the coding efficiency of AVS1-P2. In March 2008, a new profile called Jiaqiang Profile was defined based on the partial work of Zengqiang Profile to fulfill the needs of multimedia entertainment, such as movie compression for high-density storage. Shenzhan Profile in AVS1-P2 focuses on standardizing the solutions for the video surveillance applications, considering the characteristics of surveillance videos, that is, the random noise, relatively lower encoding complexity affordable, and friendliness to events detection and searching required. Yidong Profile was defined in targeting to mobility video applications featured with lower resolution, low computational complexity, and robust error resiliency ability to meet the wireless transport environment.

Table 10.1 shows the history of the AVS video coding standard and the development process of the major video coding tools, such as variable block-size motion compensation (VBMC), multiple reference pictures, quarter-pixel motion interpolation techniques, and so on. AVS provide a good trade-off between the performance and complexity for the applications, because all coding tools in AVS are selected by jointly considering the coding complexity and performance gain for the target applications. This chapter gives a detailed introduction to these coding tools.

The rest of this chapter is organized as follows: Section 10.2 first gives a brief introduction to the framework of AVS video coding standards, and the reason that these coding tools were selected for AVS. In Section 10.3, the performance comparison between AVS standards and H.264/AVC. Finally, Section 10.4 concludes the chapter.

10.2 An Overview of the AVS

As shown in Figure 10.1 the AVS video encoder is based on the traditional transform–prediction hybrid framework. In the coding process, each input macroblock is predicted

TABLE 10.1

History of AVS Standard

Time	Document	Coding Tools and Comments
Jun. 2002	CFP of Jizhun Profile	
Oct. 2002	Jizhun Profile WD	Prediction/transform-based hybrid framework is selected
Jul. 2003	Jizhun Profile CD	8 × 8 Integer transform, 2D VLC entropy coding, improved B frame direct mode and symmetric mode
Oct. 2003	Jizhun Profile FCD	Improved 8 × 8 integer transform, low-complexity loop filter, low-complexity quarter-pixel interpolation, low-complexity intraprediction, motion vector prediction
Dec. 2003	Jizhun Profile DS	AVS1-P2 Jizhun Profile was finished
Feb. 2006	Jizhun Profile NS	AVS1-P2 was approved as China National Standard
Jun. 2007	Start the standardization work for Shenzhan Profile	Video surveillance, etc.
Mar. 2008	Start the standardization work for Jiaqiang Profile	Multimedia entertainment, etc.
Jun. 2008	Shenzhan Profile FCD	Background-predictive picture for video coding, adaptive weighting quantization (AWQ), core frame coding
Jun. 2008	Requirement of Yidong Profile approved	Mobility applications, etc.
Sep. 2008	Jiaqiang Profile FCD	Context binary arithmetic coding (CBAC), AWQ
Mar. 2009	Yidong Profile CD	Adaptive block-size intraprediction and transform, simplified high-precision interpolation, adaptive weighting quantization, flexible slice set, constraint DC prediction, L-slice
Jul. 2009	Yidong Profile FCD	
Jun. 2010	3D video coding	Syntax amendment

Note: CD: committee draft; CFP: call for proposal; FCD: final committee draft; NS: national standard; WD: working draft.

with intraprediction or interprediction. After prediction, the predicted residual is transformed and quantized. Then the quantized coefficients are coded with an entropy coder. At the same time, the quantized coefficients are processed with inverse quantization and inverse transform to produce reconstructed prediction error, and the reconstructed prediction error and prediction sample are added together to obtain the reconstructed picture. The reconstructed picture is filtered and sent to the frame buffer. In the framework, video coding tools are classified as:

- Intraprediction
- VBMC
- Multiple reference picture interprediction
- Quarter/eighth subpixel motion interpolation
- Improved direct mode for B picture

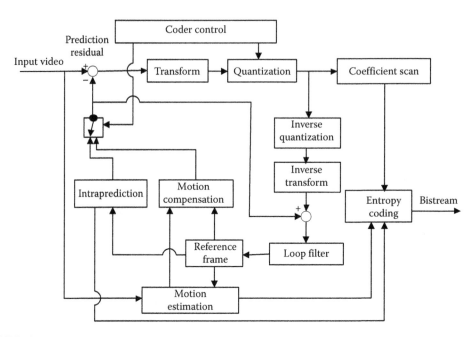

FIGURE 10.1
The block diagram of AVS video encoder.

- Symmetric mode for B picture
- Integer transform
- Quantization, adaptive weighted quantization
- Entropy coding
- Loop filter
- Interlace coding tools
- Error resilience tools
- Three-dimensional (3D) video coding

Table 10.1 has shown the history of the AVS and the development process of major video coding tools. We detail these coding tools and the profile/level definition of AVS1-P2 in the following sections.

10.2.1 Intraprediction

Spatial domain intraprediction has been used both in H.264/AVC and AVS to improve the coding efficiency of intracoding [5]. In AVS, the intraprediction is based on 8 × 8 luma blocks for Jizhun, Jiaqiang, and Shenzhan Profiles and based on adaptive 4 × 4 and 8 × 8 luma blocks for Yidong Profile, as shown in Figure 10.2. Adaptive block prediction was selected for the Yidong Profile since it focuses on the mobility applications, which covers from low-resolution coding to high-resolution coding. On the contrary, other profiles mainly targeted to high resolution, such as standard definition/high-definition coding where the smaller block prediction is not efficient as the bigger block prediction for coding performance improvement.

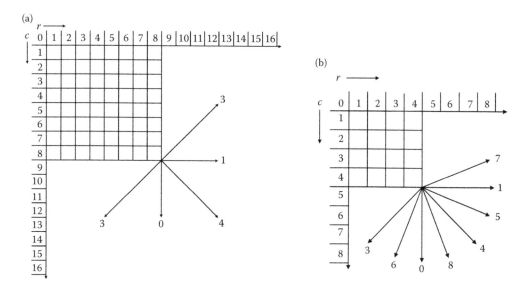

FIGURE 10.2
Neighboring samples used for intraluma prediction. (a): 8 × 8 based. (b): 4 × 4 based.

In Figure 10.2, when the neighbor samples r[8..16] (or c[8..16]) are not available, they are replaced with r[8] (or c[8]). Figure 10.3 shows five intraprediction modes in AVS1-P2. Mode 0 (vertical prediction) and Mode 1 (horizontal prediction) in AVS1-P2 are the same as that in H.264/AVC; for example, the sample at position $[i, j]$ is predicted with $r[i + 1]$ for Mode 0 and $c[j + 1]$ for Mode 1. For the other modes, the filtered neighbor samples in the prediction direction will be used as prediction samples. For example, for Mode 2 (DC prediction), the sample at [0, 0] position is predicted with:

$$((r[0] + r[1] \ll 1 + r[2] + 2) \gg 2 + (c[0] + c[1] \ll 1 + c[2] + 2) \gg 2) \gg 1.$$

For 4 × 4 block-based intraprediction, the same luma prediction directions are used with little difference on samples selection and mode index number, as shown in Figure 10.2. To indicate the intraprediction size, one flag to represent whether there is no intra 4 × 4 in one macroblock is first coded. If this flag is set to be zero, some of the blocks are coded with intra 4 × 4 modes. In this case, another flag will be coded for each 8 × 8 block to indicate whether intra 4 × 4 or intra 8 × 8 mode is chosen after rate-distortion optimization process. Besides, if the current block and its neighboring blocks using different block-sized intraprediction, that is, intra 4 × 4 and intra 8 × 8, mapping is needed between the modes used in intra 4 × 4 and that in intra 8 × 8 before the prediction of most probable mode. The mapping process can be found in Table 10.2.

10.2.2 Variable Block-Size Motion Compensation

VBMC is very efficient for prediction accuracy and coding efficiency improvement; however, it is also the costliest in video encoder due to the high cost of motion estimation (ME) and the rate-distortion optimization mode decision. Macroblock partitions used in AVS standard are shown in Figure 10.4. In AVS1-P2, the minimum macroblock partition is 8 × 8 block. This is because a block size less than 8 × 8 does not give much improvement

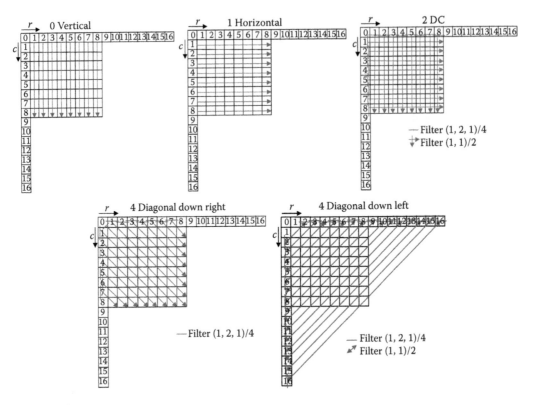

FIGURE 10.3
Five intraluma prediction modes in all profiles in AVS1-P2.

for high-resolution coding. Figure 10.5 shows the VBMC performance for Quarter Common Intermediate Format (QCIF) and 720p HD sequences. Compared with the minimum 4×4 block motion compensation (MC), the minimum 8×8 block MC reduces encoder and decoder complexity significantly.

TABLE 10.2

Mapping between Intra 4×4 and Intra 8×8 Modes in the AVS1-P2 Yidong Profile

Intra 4×4 Modes	Intra 8×8 Modes
0-Intra_4×4_Vertical	0-Intra_8×8_Vertical
6-Intra_4×4_Vertical_left	
8-Intra_4×4_Vertical_right	
1-Intra_4×4_Horizontal	1-Intra_8×8_Horizontal
5-Intra_4×4_Horizontal_down	
7-Intra_4×4_Horizontal_up	
2-Intra_4×4_DC	2-Intra_8×8_DC
3-Intra_4×4_Down_left	3-Intra_8×8_Down_left
4-Intra_4×4_Down_right	4-Intra_8×8_Down_right

Note: This adaptive block-size prediction is only applied for luma component in Yidong Profile. For chroma prediction, AVS1-P2 has also four chroma modes (vertical, horizontal, DC, and plane mode).

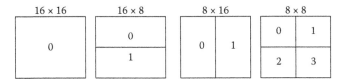

FIGURE 10.4
Macroblock partitions in AVS1-P2.

10.2.3 Multiple Reference Picture MC

Multiple reference picture coding can further improve coding efficiency compared to using one reference [6]. In general, two or three reference pictures give almost the best performance, and more reference pictures will not bring significant performance improvement but increase the complexity greatly, as shown in Figure 10.6, which shows the multireference picture performance comparison. AVS1-P2 restricts the maximum number of reference picture to be 2, and P frame can use two reference frames. Actually, for AVS1-P2 setting the maximum number of reference to be 2 does not increase the reference buffer size relative to MPEG-1 and MPEG-2. In the previous video coding standards, such as MPEG-1 and MPEG-2, although P pictures use only one previous picture to predict the current picture, B pictures use one previous picture and one future picture as references; therefore, the reference buffer size in a decoder has to be twice the picture size. Compared with MPEG-1 and MPEG-2, AVS can improve coding efficiency while using the same reference buffer size.

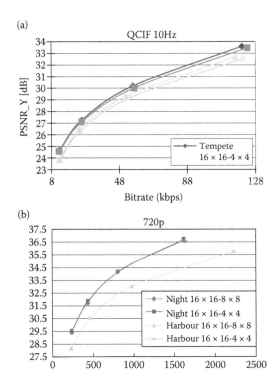

FIGURE 10.5
VBMC performance testing on QCIF and 720p test sequences. (a) QCIF and (b) 1280 × 720 Progressive.

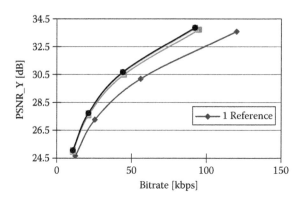

FIGURE 10.6
Multiple reference picture performance testing.

In AVS1-P2 Shenzhan Profile, background picture and background-predictive picture are defined to further exploit the temporal redundancy, suppress background noise, and facilitate video event generation such as object segmentation, and motion detection. The background picture is similar to the long-term reference picture in H.264/MPEG-4 AVC; however, in AVS it is explicitly coded as a special I-picture. The reconstructed background picture is stored in a separate background memory. The background-predictive picture is a special P-picture, which is coded by only being predictable from the reconstructed background picture. The traditional interprediction pictures such as P-pictures or B-pictures are coded by being predictable from the reconstructed background picture besides their traditional reference pictures. Figure 10.7 shows the video codec architecture for AVS1-P2 Shenzhan Profile, where three added modules compared with AVS1-P2 Jizhun Profile are marked.

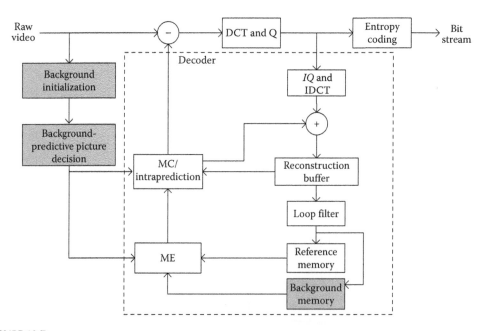

FIGURE 10.7
Video codec architecture for video sequence with static background (AVS1-P2 Shenzhan Profile).

Background picture is initialized by Background initialization. Background picture represents the background of a scene with no occluding foreground objects, which can be precaptured by video surveillance system or constructed from several captured video pictures by any suitable techniques such as median filtering, and so on. Generally, if a pixel shows long-term stability with little variation, it can be assumed to be a background pixel. Once the background picture is obtained, it is encoded and the reconstructed picture is stored into background memory in encoder/decoder and can be updated adaptively during the encoding/decoding process. After that, background-predictive picture can be involved in the encoding process by background-predictive picture decision. The background-predictive picture owns similar utilities as traditional I-picture such as error resilience and random access, and so the pictures that should be coded as traditional I-pictures can be a candidate picture as background-predictive picture, such as the first picture of one group of pictures (GOP), scene change, and so on. ME and MC/intraprediction are performed to obtain the prediction values for each area of a picture. The residues between the prediction values and the raw picture values are calculated, and then the residues are transformed and quantized by discrete cosine transform & quantization (DCT&Q). The quantization results are scanned to one-dimensional values and mapped into bitstream by entropy coding.

10.2.4 Subpixel Interpolation

10.2.4.1 Quarter-Pixel Interpolation

Fractional pixel motion accuracy has already been used since MPEG-1. Since MPEG-4 (Advanced Simple Profile), quarter-pixel interpolation has been developed. For fractional-pixel interpolation, the interpolation filter has significant effects coding efficiency. In [7], the performances of Wiener filter and bilinear filter are studied. In the development process of H.264/AVC, the filter parameter has been discussed in many proposals [8,9,10]. In the final H.264/AVC standard, a 6-tap filter is used for half-pixel interpolation. But in AVS1-P2, a 4-tap filter is used for half-pixel interpolation [11]. This is because the 6-tap filter does better for low-resolution video, such as QCIF and Common Intermediate Format (CIF). But a 4-tap filter can achieve similar performance with a 6-tap filter while with much lower computing and memory access complexity. Figure 10.8 shows the performance comparison between the 6-tap filter of H.264/AVC and 4-tap filter of AVS on several QCIF and 720p sequences.

For the interpolation process in AVS1-P2, as shown in Figure 10.9, the samples at half-sample positions labeled b and h are derived by first calculating intermediate values b' and h', respectively by applying the 4-tap filter $(-1, 5, 5, -1)$ to the neighboring integer samples, as follows:

$$b' = (-C + 5D + 5E - F)$$
$$h' = (-A + 5D + 5H - K)$$

The final prediction values for locations b and h are obtained as follows and clipped to the range of 0 to 255:

$$h = (h' + 4) \gg 3,$$
$$b = (b' + 4) \gg 3.$$

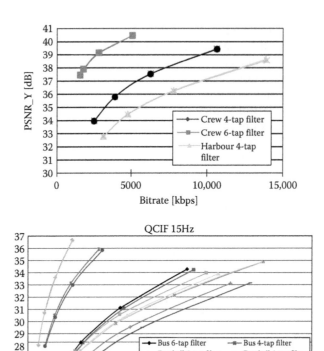

FIGURE 10.8
Interpolation filter performance comparison.

The samples at half-sample positions labeled as j are obtained by applying the 4-tap filter $(-1, 5, 5, -1)$ to the neighboring half-samples as follows:

$$j' = (-bb' + 5h' + 5m' - cc') \text{ or}$$
$$j' = (-aa' + 5b' + 5s' - dd');$$

here, $aa', bb', cc', dd', m',$ and s' are the intermediate values of samples at half-sample positions $aa, bb, cc, dd, m,$ and s, respectively.

The final prediction values for locations j is obtained as follows and clipped to the range of 0–255:

$$j = (j' + 32) \gg 6.$$

The samples at quarter-sample positions labeled as $a, i, d,$ and f are obtained by applying the 4-tap filter $(1, 7, 7, 1)$ to the neighboring integer and half-sample position samples, for example:

$$a' = (ee' + 7D' + 7b' + E'),$$
$$i' = (gg' + 7h'' + 7j' + m''),$$

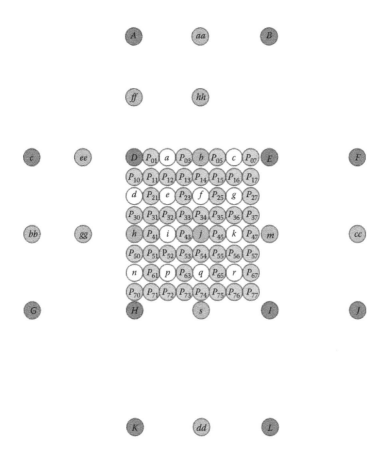

FIGURE 10.9
Filtering for fractional sample accuracy MC. Uppercase letters indicate samples on the full-sample grid, lowercase letters represent samples at half- and quarter-sample positions, and all the rest samples with s integer number subscript are eighth-pixel locations.

$$d' = (ff' + 7D' + 7h' + H'),$$
$$f' = (hh' + 7b'' + 7j' + s'');$$

here, ee', gg', ff', and gg' are the intermediate values of samples at half-sample positions. D', E', H' are the scaled integer samples with 8 times. h'', m'', b'', and s'' are the half-samples scaled with 8 times.

The final prediction values for locations a, i, d, and f are obtained as follows and clipped to the range of 0–255:

$$a = (a' + 64) \gg 7,$$
$$i = (i' + 512) \gg 10,$$
$$d = (d' + 64) \gg 7,$$
$$f = (f' + 512) \gg 10.$$

The samples at quarter-sample positions labeled as e, g, p, and r are obtained by averaging the neighboring integer position samples and j:

$$e = (D'' + j' + 64) \gg 7,$$

$$g = (E'' + j' + 64) \gg 7,$$

$$p = (H +'' j' + 64) \gg 7,$$

$$r = (I'' + j' + 64) \gg 7.$$

D'', E'', H'', and I'' are the integer samples scaled with 8 times.

10.2.4.2 Eighth-Pixel Interpolation

AVS1-P2 Yidong Profile utilizes eighth-pixel interpolation to improve the prediction accuracy for lower-resolution video. The samples at eighth sample positions labeled as P_{01}, P_{41}, P_{10}, and P_{14}, are obtained by applying the 4-tap filter $(-6, 56, 15, -1)$ to the neighboring samples, as follows:

$$P'_{01} = -6ee' + 56D' + 15b' - E',$$

$$P'_{41} = -6gg' + 56h'' + 15j' - m'',$$

$$P'_{10} = -6ff' + 56D' + 15h' - H',$$

$$P'_{14} = -6hh' + 56b'' + 15j' - s'';$$

here, hh', j', ff', h', gg', ee', and b' are the intermediate values of samples at half-sample positions. D', E', H' are the scaled integer samples with 8 times. b'', s'', h'', and m'' are the intermediate half-samples b', s', h', and m' scaled with 8 times.

The final prediction values for locations P_{01}, P_{41}, P_{10}, and P_{14}, are obtained as follows and clipped to the range of 0–255:

$$P_{01} = (P'_{01} + 256) \gg 9,$$

$$P_{41} = (P'_{41} + 256) \gg 9,$$

$$P_{10} = (P'_{10} + 256) \gg 9,$$

$$P_{14} = (P'_{14} + 256) \gg 9,$$

Samples at eighth sample positions labeled as P_{03}, P_{43}, P_{30}, and P_{34} are obtained by applying the 4-tap filter $(-1, 15, 56, -6)$ to the neighboring samples, as follows:

$$P'_{03} = -ee' + 15D' + 56b' + 6E',$$

$$P'_{43} = -gg' + 15h'' + 56j' - m'',$$

$$P'_{30} = -ff' + 15D' + 56h' - H',$$

$$P'_{34} = -hh' + 15b'' + 56j' - s''.$$

The final prediction values for locations P_{03}, P_{43}, P_{30}, and P_{34} are obtained as follows and clipped to the range of 0–255:

$$P_{03} = (P'_{03} + 256) \gg 9,$$
$$P_{43} = (P'_{43} + 256) \gg 9,$$
$$P_{30} = (P'_{30} + 256) \gg 9,$$
$$P_{34} = (P'_{34} + 256) \gg 9.$$

Samples labeled as $P_{st}(1 <= s <= 3$ and $1 <= t <= 3)$ should be obtained by

$$P'_{st} = (4 - m)'(4 - n)'D + m'(4 - n)'b + (4 - m)'n'h + m'n'j,$$
$$P_{st} = (P'_{st} + 512) \gg 10,$$

where $m = s\%4$, $n = t\%4$, j denotes the intermediate values of samples at half-sample position, D' represents the scaled integer samples with 64 times. b'' and h are the intermediate half-samples b' and h' scaled with 8 times. The interpolation process of other samples located at $P_{st}(5 <= s <= 7$ and $1 <= t <= 3)$, $P_{st}(1 <= s <= 3$ and $5 <= t <= 7)$, $P_{st}(5 <= s <= 7$ and $5 <= t <= 7)$ is similar to $P_{st}(1 <= s <= 3$ and $1 <= t <= 3)$.

10.2.5 Improved Direct Mode

The direct mode previously existing in H.263+ and MPEG-4 for global motion is improved in H.264/AVC and AVS [17]. In H.264/AVC, it is classified as temporal and spatial direct mode according to the motion vector derivation scheme. The temporal direct mode and spatial direct mode are independent of each other. However, in AVS1-P2, the temporal direct mode and spatial direct mode are combined together. In the prediction process of direct mode, spatial prediction will be used when the colocated macroblock for temporal prediction is intracoded [18].

Figure 10.10a shows the motion vector derivation process for direct mode in AVS1-P2 frame coding. The motion vectors for the current block in B picture can be derived as follows:

$$mvFw\_x = \text{sign}(mvRef\_x) \left(\left(\left(\frac{16384}{BlockDistanceRef} \right) \right. \right.$$
$$\left. \left. \times \left(1 + \text{abs} \left(\frac{mvRef\_x}{BlockDistanceFw} \right) \right) \right) - 1 \right) \gg 14 \right)$$

$$mvFw\_y = \text{sign}(mvRef\_y) \left(\left(\left(\frac{16384}{BlockDistanceRef} \right) \right. \right.$$
$$\left. \left. \times \left(1 + \text{abs} \left(\frac{mvRef\_y}{BlockDistanceFw} \right) \right) \right) - 1 \right) \gg 14 \right)$$

$$mvBw\_x = -\text{sign}(mvRef\_x) \left(\left(\left(\frac{16384}{BlockDistanceRef} \right) \right. \right.$$
$$\left. \left. \times \left(1 + \text{abs} \left(\frac{mvRef\_x}{BlockDistanceBw} \right) \right) \right) - 1 \right) \gg 14 \right)$$

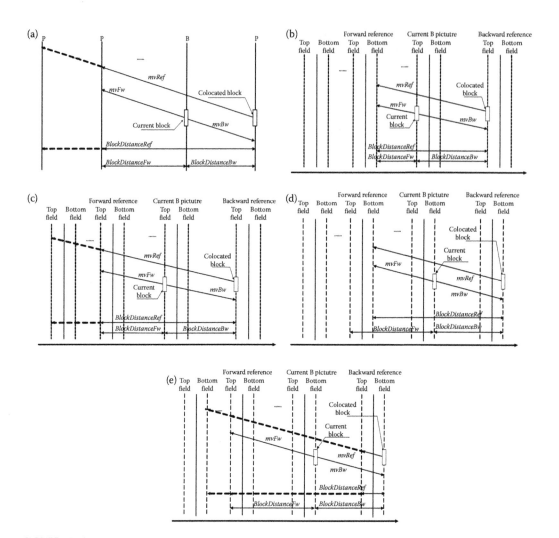

FIGURE 10.10
Temporal direct mode in AVS1-P2. (a) Motion vector derivation for direct mode in frame coding. Colocated block's reference index is 0 (solid line), or 1 (dashed line). (b) Motion vector derivation for direct mode in top field coding. Colocated block's reference index is 0. (c) Motion vector derivation for direct mode in top field coding. Colocated block's reference index is 1 (solid line), 2 (dashed line pointing to bottom field), or 3 (dashed line pointing to top field). (d) Motion vector derivation for direct mode in top field coding. Colocated block's reference index is 1. (e) Motion vector derivation for direct mode in top field coding. Colocated block's reference index is 0 (solid line), 2 (dashed line pointing to bottom field), or 3 (dashed line pointing to top field).

$$
mvBw\_y = -\mathrm{sign}(mvRef\_y)\left(\left(\left(\frac{16384}{BlockDistanceRef}\right)\right.\right.
$$
$$
\left.\left.\times\left(1 + \mathrm{abs}\left(\frac{mvRef\_y}{BlockDistanceBw}\right)\right)\right) - 1\right) \gg 14\Big)
$$

Here, $(mvFw\_x, mvFw\_y)$ and $(mvBw\_x, mvBw\_y)$ are the forward motion vector and backward motion vector of the current block, respectively. $(mvRef\_x, mvRef\_y)$ is the motion vector of the colocated block in the backward reference. *BlockDistanceRef* is the distance between the reference picture to which the colocated block belongs and the colocated block's reference picture. *BlockDistanceFw* is the distance between the current picture and the current block's forward reference picture. *BlockDistanceBw* is the distance between the current picture and the current block's backward reference picture.

For interlace coding, with different field parity of the picture where the current block exists and different reference index of the colocated block, different reference pictures will be selected for the current block, and the corresponding temporal picture distance will be calculated for motion vector derivation, as shown in Figure 10.10b through e.

10.2.6 Symmetric Mode

In AVS1-P2, a new symmetrical mode is used for B picture coding to replace the traditional bidirection coding [19]. For the symmetric mode, only a forward motion vector is coded and the backward motion vector is derived from the forward motion vector. That is to say, at most one motion vector is coded for a block in a B picture in AVS.

As shown in Figure 10.11a, the nearest previous and future reference pictures are selected as the forward and backward reference pictures for the current block in frame coding. The forward motion vector is $mvFW$, and the backward motion vector is derived as

$$mvBW = -\frac{BlockDistanceBw}{BlockDistanceFw}mvFW.$$

Here, *BlockDistanceFw* is the distance between the current picture and the forward reference picture of the current block. *BlockDistanceBw* is the distance between the current picture and the backward reference picture of the current block. For field coding, the forward and backward reference selection is as shown in Figure 10.11b and c.

10.2.7 Transform and Quantization

In H.264/AVC, the traditional float DCT is replaced with integer transform and the mismatch can be removed [12]. In general, a larger-sized transform has better energy compaction property, while a smaller-sized transform has the advantages of reducing ringing artifacts at edges and discontinuities. In the development of H.264/AVC, adaptive block transform (ABT) was proposed and has been adopted into H.264/AVC High Profile [13,14]. As larger transform size is more efficient for high-resolution coding, an 8×8 integer transform is used in AVS1-P2, shown as follows:

$$T_8 = \begin{bmatrix} 8 & 10 & 10 & 9 & 8 & 6 & 4 & 2 \\ 8 & 9 & 4 & -2 & -8 & -10 & -10 & -6 \\ 8 & 6 & -4 & -10 & -8 & 2 & 10 & 9 \\ 8 & 2 & -10 & -6 & 8 & 9 & -4 & -10 \\ 8 & -2 & -10 & 6 & 8 & -9 & -4 & 10 \\ 8 & -6 & -4 & 10 & -8 & -2 & 10 & -9 \\ 8 & -9 & 4 & 2 & -8 & 10 & -10 & 6 \\ 8 & -10 & 10 & -9 & 8 & -6 & 4 & -2 \end{bmatrix}.$$

This transform has the similar feature as that in WMV9, called prescaled integer transform in [16]. As the basis of the transform coefficients is very close, the transform normalization

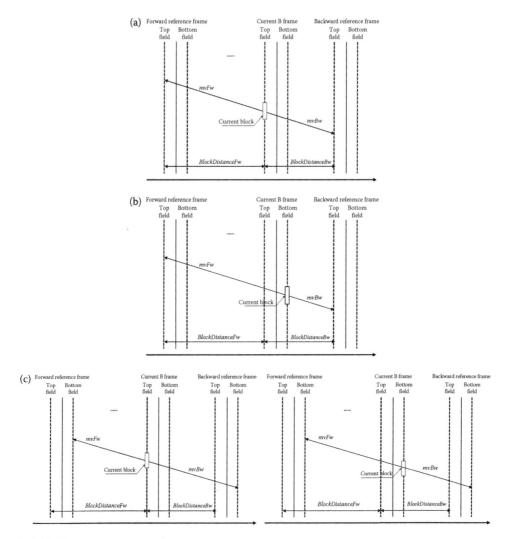

FIGURE 10.11
Motion vector derivation for symmetric mode in AVS1-P2. (a) Frame coding. (b) Field coding, forward reference index is 1, backward reference index is 0. (c) Field coding, forward reference index is 0, backward reference index is 1.

can be accounted for entirely on the encoder side [15,16]. For AVS1-P2 Yidong Profile, one additional 4×4 transform is adopted, which has the same feature as that in AVS1-P2:

$$
T_4 = \begin{bmatrix} 2 & 2 & 2 & 2 \\ 3 & 1 & -1 & -3 \\ 2 & -2 & -2 & 2 \\ 1 & -2 & 3 & -1 \end{bmatrix}.
$$

All these transforms can be implemented within 16 bits. After the transform, the transform coefficient matrix Y is normalized with a scale table *ScaleTbl*:

$$
Y'_{i,j} = (Y_{i,j} \times ScaleTbl[i][j] + 2^{a-1}) \gg a.
$$

For 4×4 transform/quantization case, a is equal to 19, and the scale table is defined as follows:

$$ScaleTbl = \begin{bmatrix} 32768 & 37959 & 36158 & 37958 & 32768 & 37958 & 36158 & 37958 \\ 37958 & 43969 & 41884 & 43969 & 37958 & 43969 & 41884 & 43969 \\ 36158 & 41884 & 39898 & 41884 & 36158 & 41884 & 39898 & 41884 \\ 37958 & 43969 & 41884 & 43969 & 37958 & 43969 & 41884 & 43969 \\ 32768 & 37958 & 36158 & 37958 & 32768 & 37958 & 36158 & 37958 \\ 37958 & 43969 & 41884 & 43969 & 37958 & 43969 & 41884 & 43969 \\ 36158 & 41884 & 39898 & 41884 & 36158 & 41884 & 39898 & 41884 \\ 37958 & 43969 & 41884 & 43969 & 37958 & 43969 & 41884 & 43969 \end{bmatrix}.$$

For 4×4 transform/quantization case, a is equal to 15, and the scale table is defined as

$$ScaleTbl = \begin{bmatrix} 32768 & 26214 & 32768 & 26214 \\ 26214 & 20972 & 26214 & 20972 \\ 32768 & 26214 & 32768 & 26214 \\ 26214 & 20972 & 26214 & 20972 \end{bmatrix}.$$

As H.264/AVC, quantization/dequantization can be implemented with multiply and right shift. All these operations can be completed within 16 bits. After transform normalization, the quantization is done as

$$YQ_{i,j} = (Y'_{i,j} \times Q[QP] + 2^{b-1}) \gg b.$$

$Q[QP]$ is the quantization table that is used by the encoder. The dequantization is as follows:

$$X_{i,j} = (YQ_{i,j} \times IQ[QP] + 2^{s(QP)-1}) \gg s(QP),$$

where QP is the quantization parameter, $IQ(QP)$ is the inverse quantization table, and $s(QP)$ is the varied shift value for inverse quantization. In AVS1-P2, all 8×8 case uses the same quantization and dequantization table as that for 4×4 case, except for a little difference on the value of b. b is 15 for 8×8 transform/quantization and 19 for 4×4 transform/quantization. $Q(QP)$, $IQ(QP)$, and $s(QP)$ are as follows:

QP	0	1	2	3	4
$Q[QP]$	32768	29775	27554	25268	23170
$IQ[QP]$	32768	36061	38968	42495	46341
$s[QP]$	14	14	14	14	14
QP	5	6	...	62	63
$Q[QP]$	21247	19369	...	152	140
$IQ[QP]$	50535	55437	...	55109	60099
$s[QP]$	14	14	...	7	7

In order to improve the overall image quality, AVS-P2 Jiaqiang Profile adopts the technique of AWQ [26]. Similar to the adaptive quantization matrix technique in H.264/AVC, users can define the quantization matrices based on the characteristics of test sequences. To achieve this purpose, three default quantization matrix patterns are defined and corresponding six weighting parameters for each pattern that can be set by users are defined in

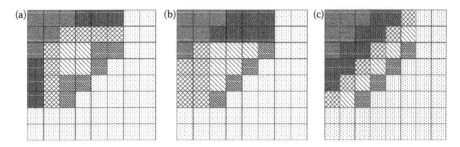

FIGURE 10.12
Quantization matrix patterns in AVS1-P2 Jiaqiang Profile.

AWQ, which are based on the distribution characteristics of transform coefficients as well as the human perception. Figure 10.12 depicts the three patterns defined in AVS1-P2 Jiaqiang Profile. It can be easily observed that in these three patterns, the quantization matrix is partitioned into six different regions. Each region represents similar characteristics of transform coefficients. The quantization weighting value is set to be the same in different frequency subbands of the current region. For each pattern, three predefined weighting parameter sets are given in AVS specification. They represent default weighting quantization case, the case for reserving detail texture information and one for removing the detail information of texture. We take the second pattern depicted in Figure 10.12b for an example, the predefined weighting parameter sets for this pattern are shown in Figure 10.13. Therefore, the encoder only needs to send the matrix pattern together with six offsets of weighting values from the predefined ones to the decoder to form an optimized quantization matrix. The dequantization process can be derived as

$$X_{i,j} = (((((YQ_{i,j} \times WQ_{i,j}) \gg 3) \times IQ[QP]) \gg 4) + 2^{s(QP)-1}) \gg s(QP),$$

where $WQ_{i,j}$ indicates the weighting value for each frequency subband.

In AWQ, the quantization matrix can be optimized frame by frame or macroblock by macroblock. For the macroblock-level AWQ, the encoder decides the best quantization matrix pattern index according to its neighbors' types and modes. In addition, the quantization parameter (QP) for two chrominance components (U and V) in one macroblock can be

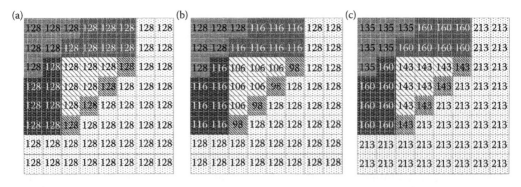

FIGURE 10.13
Predefined quantization weighting parameters in AVS1-P2 Jiaqiang Profile: (a) default parameters, (b) parameters for keeping detail information of texture, and (c) parameters for removing detail information of texture.

adjusted and the difference of QP for U or V blocks with that of QP for luma component is also sent to the decoder.

10.2.8 Entropy Coding

In H.264/AVC, two entropy coding schemes for transformed coefficients are included: Context-adaptive variable length coding (CAVLC) and Context-adaptive binary arithmetic coding (CABAC) [20]. AVS also has two entropy coding schemes to achieve different levels of compression efficiency with different coding complexity. One is context-based two-dimensional (2D–VLC (C2DVLC) entropy coding [21], used in AVS1-P2 Jizhun Profile, and the other one is improved CBAC for AVS1-P2 Jiaqiang Profile [22]. The two schemes have lower complexity compared to CAVLC and CABAC in H.264/AVC.

10.2.8.1 2D Adaptive VLC Coding

The two-dimensional (2D) adaptive VLC coding is used in AVS Jizhun Profile. As in H.264/AVC, kth-order Exp-Golomb code ($k = 0, 1, 2, 3$) is used for entropy coding. Coded block pattern, macroblock coding mode, and motion vectors are coded with 0th-order Exp-Golomp code. The quantized transform coefficients are coded with multiple tables. The difference between AVS and H.264/AVC is that run and level are coded as a pair in AVS and not coded independently as H.264/AVC.

For AVS1-P2 Jizhun Profile, a block of coefficients is coded as follows:

Coefficient scan: As shown in Figure 10.14, two scan patterns are used in AVS1-P2. Zigzag scan is used for progressive coding and alternate scan is used for interlace coding. After coefficient scan, the 2D transformed coefficients are organized into one sequence in the form of (*Level*, *Run*) pairs. To indicate the coding end of current block, a special symbol *EOB* is used, which is one (0, 0) pair.

Table selection: 19 mapping tables are defined in AVS1-P2 to code the codeword after mapping one (*Level*, *Run*) pair into Exp-Golomb code. Seven tables are used for intraluma coefficient coding; seven tables are used for interluma coefficient coding; remaining five tables are used for chroma coding. The (*Level*, *Run*) pairs are coded in reverse scan order, that is to say the last (*Level*, *Run*) pair in scan order will be coded first. Tables are selected

(a)

	0	1	2	3	4	5	6	7
0	0	1	5	6	14	15	27	28
1	2	4	7	13	16	26	29	42
2	3	8	12	17	25	30	41	43
3	9	11	18	24	31	40	44	53
4	10	19	23	32	39	45	52	54
5	20	22	33	38	46	51	55	60
6	21	34	37	47	50	56	59	61
7	35	36	48	49	57	58	62	63

(b)

	0	1	2	3	4	5	6	7
0	0	3	11	16	22	32	38	55
1	1	6	12	20	25	33	42	57
2	2	7	15	21	28	37	43	58
3	4	10	19	27	31	39	47	59
4	5	14	24	30	36	44	50	60
5	8	17	26	35	41	48	52	61
6	9	18	29	40	46	51	54	62
7	13	23	34	45	49	53	56	63

FIGURE 10.14
Coefficient scan in AVS1-P2. (a) Zigzag scan. (b) Alternate scan.

adaptively based on the maximum magnitude of all previous *Levels* in the coding process, as shown in Figure 10.15. First, an initial table is selected for the first nonzero quantized coefficient and its corresponding *Run* value. Afterward, the absolute value of the current coefficient decides which table is used for the next nonzero quantized coefficient following (*Level*, *Run*) pair.

Run-level mapping and coding: The (*Level*, *Run*) pair is converted into a *CodeNum* according to the selected table. An example table is shown in Figure 10.16. In this table, only *CodeNums* for positive *Level* are shown and *CodeNum* + 1 is that for the corresponding negative *Level*. Then the *CodeNum* is mapped into a *Code* according to the mapping function of kth-order Exp-Golomb code. If the coefficient is the last nonzero coefficient, an end of block (EOB) symbol is coded, otherwise go to step 2 to code the next (*Level*, *Run*) pair. When the (*Level*, *Run*) pair is not in the table, it will be coded with an escape code. An escape code is combined with two codes: one code is $59 + Run$; the other one is $abs(Level)$-RefAbsLevel(*Run*). Here $abs()$ denotes the absolute function. RefAbsLevel(*Run*) is decided by the selected table. When the *Run* is not in the table, *RefAbsLevel(Run)* is 1.

Run	EOB	Level > 0						RefAbsLevel
		1	2	3	4	5	6	
	8	–	–	–	–	–	–	
0	–	0	4	15	27	41	55	7
1	–	2	17	35	–	–	–	4
2	–	6	25	53	–	–	–	4
3	–	9	33	–	–	–	–	3
4	–	11	39	–	–	–	–	3
5	–	13	45	–	–	–	–	3
6	–	19	49	–	–	–	–	3
7	–	21	51	–	–	–	–	3
8	–	23	–	–	–	–	–	2
9	–	29	–	–	–	–	–	2

10.2.8.2 Context-Based Binary Arithmetic Coding

To further improve entropy coding efficiency, a new context-based binary arithmetic coding method is developed in the AVS1-P2 enhancement profile. Its complexity is reduced compared with CABAC in H.264/AVC. For CABAC in H.264/AVC, 326 probability models are used for residual data coding. On the other hand, AVS uses only unary binarization and 132 context models. The coding process of a block of coefficients after zigzag scan is shown as follows:

Binarization: The (*Level*, *Run*) pair is in a large range and coding these *Level*/*Run* values directly by an m-ary arithmetic code will have a high computational complexity. So binary arithmetic coding is used in AVS too. Both *Level* and *Run* are binarized using unary binarization scheme. The signed integer *Level* is represented by sign (0/1: +/−) and the unary bits of its magnitude (absLevel: the absolute value of *Level*). *Level* is coded first followed by the *Run*.

Context model selection: In AVS, a two-order context model is used, called as primary context and secondary context. The primary context is defined as the maximum magnitude of all previously coded levels in the current block, denoted with a context variable *Lmax*. *Lmax* will be updated in the coding process. As the dynamic range of the context variable *Lmax* can still be too large, it is reduced by the quantization function defined into five

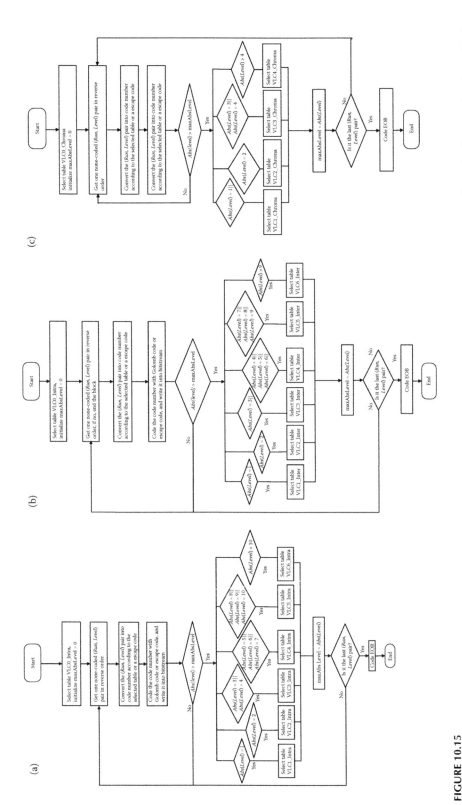

FIGURE 10.15

Coefficient coding process in AVS1-P2 2D VLC entropy coding scheme. (a) Flowchart of coding one intraluma block. (b) Flowchart of coding one interluma block. (c) Flowchart of coding one interchroma block.

Run	EOB	Level > 0						RefAbsLevel
		1	2	3	4	5	6	
	8	–	–	–	–	–	–	
0	–	0	4	15	27	41	55	7
1	–	2	17	35	–	–	–	4
2	–	6	25	53	–	–	–	4
3	–	9	33	–	–	–	–	3
4	–	11	39	–	–	–	–	3
5	–	13	45	–	–	–	–	3
6	–	19	49	–	–	–	–	3
7	–	21	51	–	–	–	–	3
8	–	23	–	–	–	–	–	2
9	–	29	–	–	–	–	–	2

FIGURE 10.16
An example table in AVS1-P2—VLC1_Intra: from (*Run, Level*) to *CodeNum*.

primary contexts. The context quantization function can be defined as

$$\chi_{(Lmax)} = \begin{cases} Lmax & Lmax \in [0,2] \\ 3 & Lmax \in [3,4] \\ 4 & otherwise \end{cases}$$

Under each primary context, seven secondary contexts are defined to code *level* and *run* as follows:

0: first bit of *absLevel* (i.e., the EOB symbol).

1: second bit of *absLevel*, if exist.

2: remaining bits of *absLevel*, if exist.

3: first bit of Run, if *absLevel* = 1.

4: remaining bits of Run when *absLevel* = 1, if exist.

5: first bit of Run when *absLevel* > 1

6: remaining bits of Run when *absLevel* > 1, if exist.

Here, *absLevel* is the absolute value of the *level*.

In order to further improve compression performance, a context weighting technique is used in AVS entropy coding. Another context variable *ReverseP* is introduced into the

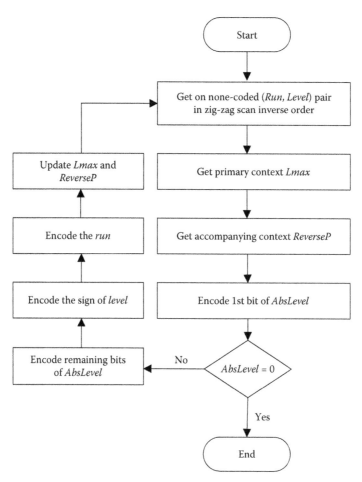

FIGURE 10.17
Coefficient coding process in AVS1-P2 context-adaptive arithmetic coding.

context, called accompanying context, which is the position of current nonzero DCT coefficient in the reverse scanning order. For an 8×8 block the range of *ReverseP* is [0,64], and it is uniformly quantized into 32 accompanying contexts [0,31]. In the binary arithmetic coding of the *run* and *level*, each accompanying context created by *ReverseP* will be combined with the same seven secondary contexts as in the case of primary contexts created. The whole coding process can be seen in Figure 10.17.

10.2.9 Loop Filter

Block-based video coding often produces blocking artifacts especially at low bit rates. As in H.264/AVC [23], AVS also adopts adaptive in-loop deblocking filter to improve the decoded visual quality, which handles 8×8 block boundaries. In AVS1-P2, the boundary may be between two 8×8 luma or chroma blocks, except for picture or slice boundaries. For each edge, boundary strength (*Bs*) is derived based on the macroblock type, motion vector, reference index, and so on. Based on the boundary strength *Bs*, the filter process in AVS1-P2 is shown in Figure 10.18. In the filtering process, two threshold values α and β are used to

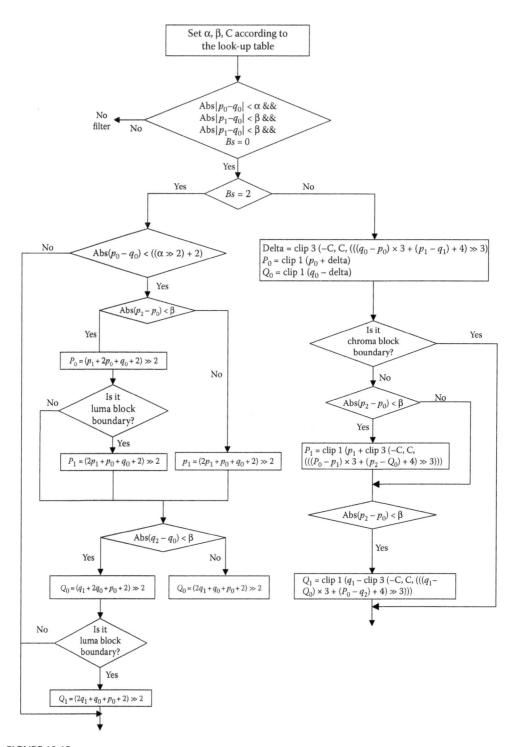

FIGURE 10.18
Deblocking filter process in AVS1-P2.

control the filtering operation on the block edge. α is cross-block gradient boundary, and β is inner block gradient boundary. $|q_0 - p_0| < \alpha \&\& |q_0 - p_1| < \beta \&\& |q_1 - p_0| < \beta$ means the block edge is an artifact and is not an object boundary.

10.2.10 Interlace Coding

H.264/AVC supports both macroblock-level adaptive frame field (MBAFF) and picture-level adaptive frame field coding for interlace coding (PAFF). In general, MBAFF is more complex than PAFF with little performance improvement. However, in AVS1-P2 Jizhun Profiles, only picture-level frame field coding is used to reduce the computational complexity. For PAFF coding, a picture can be coded as either one frame or two fields (a top field and a bottom field).

10.2.11 Error Resilience Tools

Error resilience is very important for video transport, especially when the network is erroneous. H.264/AVC is very promising for low-bit-rate applications such as IP network [24] and wireless communications [25]. To adapt these applications, more error resilience tools are introduced into H.264/AVC, such as parameter sets, flexible macroblock ordering (FMO), redundant slice, and so on. As AVS1-P2 Jizhun Profile, Jiaqiang Profiles mainly target broadcasting, and the network condition usually is error free, and few error resilience tools are defined in AVS1-P2. In AVS1-P2, only slice partition is defined to support error resilience and each slice is composed of one or more rows of macroblocks. However, as Yidong Profile or Shenzhan Profile may be used in wireless network or transport on other erroneous channel, error resilience is very important. So in these two profiles, except for the slice partition, core frame coding, flexible slice set, constrained intraprediction are also introduced.

Core picture [27] is a special interframe. An interframe-coded core picture can only be predicted from another core picture. And the first P-picture afterward can only refer to its nearest previous core picture. If there is a feedback channel from the decoder to the encoder, only the core pictures that have been indicated correctly received at the decoder side should be taken as the reference picture and marked as core pictures. Note that a core picture is different from I-pictures since it allows interframe prediction. Temporal prediction of other coding pictures cannot cross the temporal boundary of a core picture, which will avoid error propagation. Usability of core picture is signaled in the sequence header.

L-slice [28] is another special slice type besides I-slice, P-slice, and B-slice, which is proposed for error resilience coding. In AVS1-P2 Yidong Profile, the coded macroblocks in P-slice can be converted into L-slice losslessly and vice versa. If the prediction picture is correct, the decoded L-slice can produce the same reconstructed picture as P-slice; otherwise, the decoded L-slice can provide a better reconstruction picture than P-slice. The conversion process can be found in Figure 10.19. The input of the conversion from P-slice to L-slice is the quantized residue $Q(E)$ and the prediction y, obtained by decoding the original compressed video stream. The prediction y is transformed by the integer transform and then quantized to get quantized prediction $Q(Y)$. Then the sum of $Q(E)$ and $Q(Y)$ are entropy coded to replace the original entropy-coded residue (E). For the decoder, first y and $Q(Y)$ are obtained by MC, transform and quantization; Second, $Q(Y) + Q(E)$ are entropy decoded and are used to subtract the $Q(Y)$ to obtain $Q(E)$. Finally, with the quantized residue $Q(E)$ and the prediction y, the reconstructed signal is recovered losslessly. In case that the prediction y is not available due to packet loss, we directly dequantize $Q(Y) + Q(E)$ and get an approximate reconstruction by inverse transform.

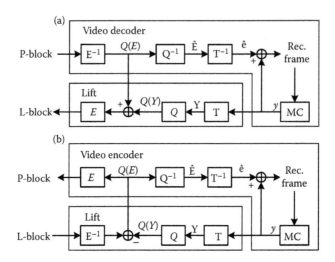

FIGURE 10.19

Slice-type conversion process. E: entropy coding, E^{-1}: entropy decoding, Q: quantization, Q^{-1}: Inverse quantization, T: transform, T^{-1}: inverse transform, MC: motion compensation. (a) Convert P-slice to L-slice. (b) Convert L-slice to P-slice.

Flexible slice set provides robustness to transmission [29]. It allows that slices with the same index of slice group in one picture belong to the same slice group. One slice can refer to other slices in the same slice group. As shown in Figure 10.20, slices B0, B1, and B2 belong to the same slice group and they can refer to each other. A flexible slice set is helpful for both coding efficiency and error robustness.

Constrained DC intraprediction [30] is another error resilience tool defined in AVS1-P2 Shenzhan Profile. One marker bit in the sequence header and one marker bit in the picture header of P and B pictures indicate the use of constrained DC intraprediction mode

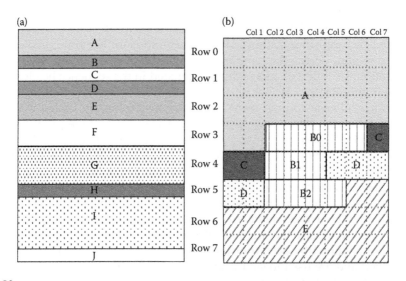

FIGURE 10.20

Slice structure in AVS1-P2. (a) Normal slice structure where the slice can only contain continual lines of macroblocks. (b) Flexible slice set allowing more flexible grouping of macroblocks in slice and slice set.

instead of normal DC mode. Constrained DC intraprediction mode constrains the prediction values of DC mode to be fixed to 128 for luminance 8×8 blocks to avoid possible error accumulation.

10.2.12 3D Video Coding

Nowadays, the demands of 3D video applications are growing rapidly. Driven by the significant improvements of 3D video technologies, including the acquisition, representation, coding, transmission, rendering, and 3D displays, H.264/AVC supports 3D video coding in one bitstream with the indication of packing method in Supplemental Enhancement Information message. To implement stereo video coding without significant modification of AVS existing video coding standard, Multiview Coding with only a single encoder and decoder is adopted. Current AVS1-P2 utilizes the original reserved bits in sequence_display_extension data to support 3D video coding. The syntax amendment of the reserved bits is defined as follows:

Stereo_Packing_Mode	Method of Packing
00	No packing (single view)
01	Side-by-side packing
10	Top-and-bottom packing
11	Reserved bits

Further work on improving the coding efficiency of stereo video and support of multiview video is being under consideration now.

10.2.13 Profile and Level

In AVS1-P2, three additional profiles are defined based on Jizhun Profile, named Jiaqiang Profile, Shenzhan Profile, and Yidong Profile. Jizhun means baseline profile as H.264/AVC in Chinese, which is defined for digital TV broadcasting and high-density storage, and so on. Jiaqiang Profile means an enhancement and is defined for digital movie applications. Shenzhan and Yidong Profiles are defined for surveillance and mobility applications. Different levels are defined in different profiles. Each level specifies the upper limits for the picture size, the maximal video bit rate, the Bitstream Buffer Verifier, buffer size, and so on. The coding tools supported in each profile of AVS1-P2 are listed in Table 10.3. From this table, it can be seen that with the exception of 4:2:2 support, a Jizhun Profile bitstream could be decoded by any other profile decoder.

10.3 Performance Comparisons

The performance comparisons between three new profiles with Jizhun Profile are provided in this section. Moreover, to better illustrate the performance status of AVS1-P2, H.264/AVC Main Profile, and High Profile are also compared.

Tables 10.4 through 10.6 show the coding tools used for the comparison between AVS1-P2 each new profile compared to Jizhun Profile. Table 10.7 lists the main coding tools of H.264/AVC Main and High Profiles tested in our experiments. To give a fair comparison, different test sequences are used based on the application of different profiles. Individual

TABLE 10.3

Profile Definition in AVS1-P2

Coding Tools		Jizhun Profile	Jiaqiang Profile	Shenzhan Profile	Yidong Profile
Color format	4:0:0			✓	
	4:2:0	✓	✓	✓	✓
	4:2:2	✓	✓		
Intraprediction	4 × 4				✓
	8 × 8	✓	✓	✓	✓
VBMC	16 × 16-8 × 8	✓	✓	✓	✓
Multireference prediction	Up to 2	✓	✓	✓	✓
Background-predictive picture				✓	
Entropy coding	C2DVLC	✓	✓	✓	✓
	CBAC			✓	
Loop filter	8 × 8 Based	✓	✓	✓	✓
Interlace coding	Frame/field	✓	✓	✓	✓
B frame		✓	✓	✓	✓
Sample precision	8 Bits	✓	✓	✓	✓
	10/12 Bits			✓	
L-slice					✓
Adaptive quantization			✓	✓	✓

TABLE 10.4

Coding Tools for Comparison between AVS1-P2 Jiaqiang Profile and Jizhun Profile

Coding Tools	Jizhun Profile	Jiaqiang Profile
Intraprediction	All intra 8 × 8 modes	All intra 8 × 8 modes
Multireference frames	2 Reference frames	2 Reference frames
Variable block-size MC	16 × 16-8 × 8	16 × 16-8 × 8
Entropy coding	C2DVLC	CBAC
RDO (Rate Distortion Optimization)	On	On
Loop filter	On	On
ABT	Only 8 × 8	Only 8 × 8
Adaptive quantization	Fixed	Fixed
Subpixel interpolation	Quarter	Quarter
Background-predictive picture	None	None
Background picture generation	None	None

frames of these test sequences that are widely used in video coding standardization groups and research fields are shown in Figure 10.21. The test conditions are listed in Table 10.8.

Figure 10.22 shows the rate–distortion curves of different profiles. It can be observed that for high-definition video application, Jiaqiang Profile in AVS1-P2 outperforms Jizhun Profile at about 0.3–0.5 dB in an average peak signal-to-noise ratio (PSNR) as shown in Figure 10.22a. AVS1-P2 Jizhun Profile shows similar coding performance with H.264/AVC Main Profile, and AVS1-P2 Jiqiang Profile shows comparable coding efficiency with

TABLE 10.5

Coding Tools for Comparison between AVS1-P2 Shenzhan Profile and Jizhun Profile

Coding Tools	Jizhun Profile	Shenzhan Profile
Intraprediction	All intra 8×8 modes	All intra 8×8 modes
Multireference frames	2 Reference frames	2 Reference frames
Variable block-size MC	16×16-8×8	16×16-8×8
Entropy coding	C2DVLC	C2DVLC
RDO	On	On
Loop filter	On	On
ABT	Only 8×8	8×8, 4×4
Adaptive quantization	Fixed	Fixed
Subpixel interpolation	Quarter	Quarter
Background-predictive picture	None	Used
Background picture generation	None	Used

TABLE 10.6

Coding Tools for Comparison between AVS1-P2 Yidong Profile and Jizhun Profile

Coding Tools	Jizhun Profile	Yidong Profile
Intraprediction	All intra 8×8 modes	Adaptive intra 8×8, intra 4×4 modes
Multireference frames	2 Reference frames	2 Reference frames
Variable block-size MC	16×16-8×8	16×16-8×8
Entropy coding	C2DVLC	C2DVLC
RDO	On	On
Loop filter	On	On
ABT	Only 8×8	8×8, 4×4
Adaptive quantization	Fixed	Fixed
Subpixel interpolation	Quarter	Eighth
Background-predictive picture	None	None
Background picture generation	None	None

TABLE 10.7

Coding Tools for Comparisons Used in H.264/AVC Main Profile and High Profile

Coding Tools	Main Profile	High Profile
Intraprediction	All intra 4×4, intra 16×16 modes	All intra 4×4, intra 16×16 and intra 8×8 modes
Multireference frames	2 Reference frames	2 Reference frames
Variable block-size MC	16×16-4×4	16×16-4×4
Entropy coding	CABAC	CABAC
RDO	On	On
Loop filter	On	On
ABT	Only 4×4	8×8, 4×4
Adaptive quantization	Fixed	Fixed
Subpixel interpolation	Quarter	Quarter
Hierarchy B coding	None	No use

FIGURE 10.21
(**See color insert.**) Test sequences: (a) Vidyo 1 (1280 × 720@60 Hz); (b) Kimono1 (1920 × 1080@24 Hz); (c) Crossroad (352 × 288@30 Hz); (d) Snowroad (352 × 288@30 Hz); (e) News; and (f) Paris.

H.264/AVC High Profile. While for typical image resolutions include CIF (352 × 288 format) with bit-rate less than 1.5 Mbps for mobility video applications such as mobile communication, mobile television, and so on, performance comparison between AVS1-P2 Yidong Profile, Jizhun Profile, H.264/AVC Main Profile, and High Profile is illustrated in Figure 10.22c. From this figure, we can see that although Yidong Profile has further improved the coding efficiency of Jizhun Profile for low resolution at about 0.5–1 dB, and the coding

TABLE 10.8

Configurations of Three Test Sets

Platform	AVS1-P2: RM09.06 [31] H.264/AVC: JM17.2 [32]
Coding structure	IBBPBBP...
Intraperiod	1s
Coded frame number	238 frames for Jiaqiang, Yidong profiles; 1198 frames for Shenzhan Profile
ME	Fast ME
Quantization parameters	H.264/AVC: 22, 27, 32, 37 AVS1-P2: 24, 31, 37, 44
Rate distortion optimized quantization (RDOQ)	OFF

gap between Yidong Profile and H.264/AVC Main Profile still exists. For typical scenes in video surveillance, Shenzhan Profile can provide significant coding efficiency gain compared to Jizhun Profile, as illustrated in Figure 10.22b.

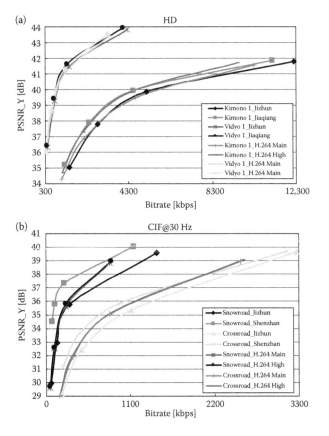

FIGURE 10.22

Rate–distortion curves of different profiles. (a) performance of Jiaqiang Profile, (b) performance of Shenzhan Profile, and (c) performance of Yidong Profile.

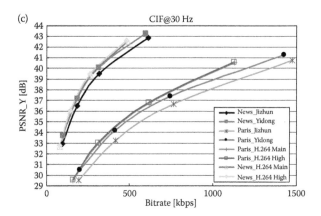

FIGURE 10.22
Continued.

10.4   Conclusions

In this chapter, we have given an overview of AVS including background and technical features. AVS provides a good trade-off between the performance and the complexity of the specific applications, because all coding tools in AVS, including intraprediction, variable block-size MC, multiple reference frames, interpolation filter, loop filter, and entropy coding, are selected by jointly considering the coding complexity and performance gain for the target applications. Experimental results prove that AVS provides similar performance with H.264/AVC while with lower complexity for the specific applications.

References

1. Wiegand T., Sullivan G., Bjøntegaard G., Luthra A. Overview of the H.264/AVC video coding standard. *IEEE Transactions on Circuits and Systems for Video Technology*, 13(7), 560–576 (2003).
2. Srinivasan S., Hsu P., Holcomb T., Mukerjee K., Regunathan S., Lin B., Liang J., Lee M., Corbera J. Windows media video: 9 Overview and applications. *Signal Processing: Image Communication*, 19(9), 851–875 (2004).
3. AVS. Working Group Website: http://www.avs.org.cn.
4. AVS. Document and Software FTP Site: ftp://124.207.250.92.
5. Wiegand T. Version 3 of H.264/AVC. http://wftp3.itu.int/av-arch/jvt-site/2004_07_Redmond/JVT-L012d2wcmRelTod1.doc.
6. Wiegand T., Sullivan G., Bjøntegaard G., Luthra A. Long-term memory motion-compensated prediction. *IEEE Transactions on Circuits and Systems for Video Technology*, 9(1), 70–84 (1999).
7. Wedi T., Musmann H. Motion- and aliasing-compensated prediction for hybrid video coding. *IEEE Transactions on Circuits and Systems for Video Technology*, 13(7), 577–586 (2003).
8. Wedi T. More results on adaptive interpolation filter for H.26L. http://wftp3.itu.int/av-arch/jvt-site/2001_12_Pattaya/VCEG-O28.doc.
9. Boyce J., Gomila C. 4-Tap motion interpolation filter. http://wftp3.itu.int/av-arch/jvt-site/2002_05_Fairfax/JVT-C014.doc.

10. Hallapuro A., Lainema J., Karczewicz M. 4-Tap filter for bi-predicted macroblocks. http://wftp3.itu.int/av-arch/jvt-site/2002_07_Klagenfurt/JVT-D029.doc.

11. Wang R., Huang C., Li J., Shen Y. Sub-pixel motion compensation interpolation filter in AVS. *IEEE International Conference on Multimedia and Expo*, 1, pp. 93–96, Taipei, Taiwan, 2004.

12. Malvar H., Hallapuro A., Karczewicz M., Kerofsky L. Low-complexity transform and quantization in H.264/AVC. *IEEE Transaction on Circuits and Systems for Video Technology*, 13(7), 598–603, 2003.

13. Wien M. Variable block-size transforms for H.264/AVC. *IEEE Transaction on Circuits and Systems for Video Technology*, 13(7), 604–613, 2003.

14. Gordon S., Marpe D., Wiegand T. Simplified use of 8×8 transforms—Updated proposal and results. http://ftp3.itu.ch/av-arch/jvt-site/2004_03_Munich/JVT-K028.doc.

15. Ma S., Gao W., Fan X. Low complexity integer transform and high definition coding. *Proceedings of SPIE 49th Annual meeting*, Denver, USA, 2004.

16. Zhang C., Lou J., Yu L., Dong J., Cham W. The techniques of pre-scaled integer transform. *International Symposium on Circuits and Systems*, 1, pp. 316–319, Kobe, Japan, 2005.

17. Flierl M., Girod B. Generalized B pictures and the draft H.264/AVC video-compression standard. *IEEE Transactions on Circuits and Systems for Video Technology*, 13(7), 587–597, 2003.

18. Ji. X., Zhao D., Gao W., Lu Y., Ma S. New scaling technique for direct mode coding in B pictures. *IEEE International Conference on Image Processing*, 1, pp. 469–472, Singapore, 2004.

19. Ji. X., Zhao D., Gao W. B-picture coding in AVS video compression standard. *Signal Processing: Image Communication*, 23(1), 31–41, 2008.

20. Marpe D., Schwarz H., Wiegand T. Context-based adaptive binary arithmetic coding in the H.264/AVC video compression Standard. *IEEE Transactions on Circuits and Systems for Video Technology*, 13(7), 620–636, 2003.

21. Wang Q., Zhao D., Gao W. Context-based 2D-VLC entropy coder in AVS video coding standard. *Journal of Computer Science and Technology*, 21(3), 315–322, 2006.

22. Zhang L., Wang Q., Zhang N., Zhao D., Wu X., Gao W. Context-based entropy coding in AVS video coding standard. *Signal Processing: Image Communication*, 24(4), 263–276, 2009.

23. List P., Joch A., Lainema J., Bjøntegaard G., Karczewicz M. Adaptive deblocking filter. *IEEE Transactions on Circuits and Systems for Video Technology*, 13(7), 614–619, 2003.

24. Wenger S. H.264/AVC over IP. *IEEE Transactions on Circuits and Systems for Video Technology*, 13(7), 645–656, 2003.

25. Stockhammer T., Hannuksela M., Wiegand T. H.264/AVC in wireless environments. *IEEE Transactions on Circuits and Systems for Video Technology*, 13(7), 657–673, 2003.

26. Zheng J., Zheng X., Lai C. Adaptive weighting technology for AVS Jiaqiang Profile. *AVS Doc. AVS_M2427*, Tianjin, China, September 2008.

27. Rui Chen, Peisong Yu, Xiaohong Huang, Ning Wang. A coding method for error resilience. *AVS Doc. AVS M2192*. 2007.

28. Xiaopeng Fan, Oscar C. Au, Feng Zou, Yannan Wu, Yi Yang. L-slice and lossless error resilient transcoding, *28th AVS Meeting*, AVS-M2533, Hangzhou, China, March 2009.

29. Zhen Mao, Zhongmou Wu, Xiaozhong Xu, Yun He. AVS-S flexible slice group. *AVS Doc. AVS M2305*. 2008.

30. Changcai Lai, Xiaozhen Zheng, Yongbin Lin, Jianhua Zheng, Mingchen Han. Constrained DC intra-prediction in AVS-S. *AVS Doc. AVS M2464*. 2008.

31. RM09.06. ftp://124.207.250.92//incoming/video_codec/AVS1_P2/FourProfiles

32. JM17.2. http://iphome.hhi.de/suehring/tml/download

Part III

Methodology, Techniques, and Applications: Multimedia Search, Retrieval, and Management

11

Multimedia Search and Management

Linjun Yang, Xian-Sheng Hua, and Hong-Jiang Zhang

CONTENTS

11.1 Introduction

With the increasing availability of image and video cameras, the advancement of multimedia communication technologies, and the popularization of media sharing websites, the amount of image and video data on the web and in everyone's personal archive has been growing drastically. According to the report of TechCrunch [1], the amount of images on Flickr* had reached 4 billion till 2008. At the same time, it costs only one year for Facebook[†] to increase from 4.1 billion images in 2007 to 10 billion in 2008.

The incredible proliferation of multimedia documents leads to a strong and emergent demand on effective and efficient multimedia search and management tools for facilitating users' consumption of the large amount of multimedia data. The mainstream search engine companies have released online search engines for images and videos, such as Bing,[‡] Google,[§] and Yahoo![¶] image and video search. Beyond these keyword-based multimedia search engines, query-by-example-based multimedia search has also been taken off the shelf, for example, TinEye[‖] and Google Goggles.[**] In the meantime, the research on multimedia search and management is ushering in a new era with a number of new challenges as well as opportunities.

---

* http://www.flickr.com/
[†] http://www.facebook.com/
[‡] http://www.bing.com/
[§] http://www.google.com/
[¶] http://www.yahoo.com/
[‖] http://www.tineye.com/
[**] http://www.google.com/mobile/goggles/

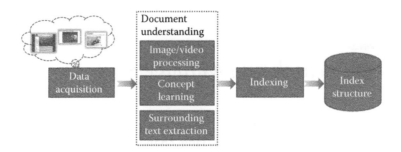

FIGURE 11.1
Overview of the offline processing and indexing process for a typical multimedia search system.

Figures 11.1 and 11.2 illustrate the primary procedures and modules for a general multimedia search engine. In the offline process as shown in Figure 11.1, the multimedia data are first acquired by crawling the web or scanning users' personal archive, and then pass to a *document understanding* component to extract textual and/or visual metadata for indexing. The visual metadata can be generated by extracting low-level visual features from the image or video. The textual metadata can be obtained by analyzing the surrounding text of the image or video in a web page, or converting the visual features into the so-called semantic concepts by concept detection [8]. After that, the metadata, whether they are in text form or in visual feature form, is sent to the *indexing* component, which is typically realized by using an inverted file or Locality-sensitive Hashing (LSH), for efficient search thereafter.

The online procedure is to process the users' queries and to return and present relevant results to users, as shown in Figure 11.2. First, a user submits a query in a *query interface* to express the search intent. The currently supported user interface for multimedia search mainly includes *query by keyword*, *query by example*, and *query by sketch*. Then, the system converts the user's query into a representation which can be recognized by the system, using the *query formulation* component. For example, in query-by-example-based image search, the query image typically is represented by a set of visual words. Then the *ranking* component [7] is invoked to compare the query with the images/videos in the database. The *index structure* [14] is used to speed up the comparison so that the result can be returned quickly. The returned result is finally presented to users to satisfy their search intent. To better understand the user's search intent, *relevance feedback* [17] is often incorporated into the multimedia search system.

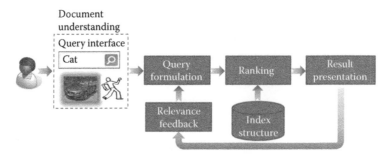

FIGURE 11.2
Overview of the query process for a typical multimedia search system.

Due to the high efficiency of text-based search, the image and video search engines in industry are mostly text based, where the surrounding text associated with the image and video is indexed. However, since the text is not always available and frequently mismatches with the image and video's visual content [26], content-based and concept-based approaches are intensively studied in the research community.

Content-based image and video search/retrieval (CBIR and CBVR for abbreviation) approaches represent a set of methods which describe the image and video using low-level features extracted from their content. The most well-known scenario of CBIR and CBVR is the query-by-example-based image and video search, where the query is an example image or video to express the user's intent. In such a setting, both feature extraction and feature indexing in large-scale search are challenging problems, which will be detailed in Chapters 13 and 14 in this part. Besides, CBIR and CBVR can also be applied to help other multimedia related applications, for example, in [22,26] the content-based features are shown to be able to improve the performance of keyword-based image and video search through reranking.

Since the existence of the well-known *semantic gap*, which means that the state-of-the-art content-based features are still not accordant with the common understanding of the content in image and video by human beings, content-based approaches only achieved limited success. Taking one further step beyond CBIR and CBVR, concept-based approaches intend to bridge the semantic gap by learning concept models from manually labeled training samples. Once the concepts are learned, they can be used to predict the existence of the concepts in a new image or video, and thus multimedia content can be indexed through those predicted concepts [8]. However, this model-based approach, as analyzed in Chapter 13, also has limitations especially in terms of scalability on both the number of concepts and the number of multimedia documents. Consequently, data-driven approaches for scalable image tagging were proposed [10,11], where large-scale text information associated with the images are mined to improve the scalability and accuracy of image concept annotation.

As search is the primary and fundamental functionality of a multimedia search and management system, we focus more on discussing multimedia search techniques in this part. The other aspects of multimedia management, such as browsing and summarization, are also covered. In the rest of this part overview, we first identify a few research trends in multimedia search and management area, which are summarized under different in Section 11.2. Then an overview of three selected topics, that is, the three chapters in this part, is presented in Section 11.3, followed by concluding remarks in Section 11.4.

11.2 Research Trends

Multimedia search and management is a broad, multidisciplinary research topic, built on top of computer vision, natural language processing, information retrieval, and machine learning. In this section, we summarize four important research directions which potentially have big impacts on the development of applicable content-based multimedia search and management systems in the future. However, note that the four directions are not isolated from each other; instead, their boundaries are blurred and they often leverage each other. For example, scalable multimedia search lays the foundation for web multimedia mining and the results of mining can further be utilized as a kind of new modality to bridge the semantic gap.

11.2.1   Scalable Content Understanding and Search

With the ever-increasing amount of multimedia data on the web and in users' personal archives, the scalability requirement of a real-world multimedia search and management system becomes more and more important. To realize a large-scale multimedia search and management system, one must consider the following two requirements. The first is called *offline scalability*, which means the system can process and index a large-scale database within a reasonable time. On the other hand, the *online scalability* requires the system be able to respond quickly to users' search, browsing, and other requests with the large-scale database.

A heavily studied topic under this direction is large-scale image and video similarity search. The basic setting is given a query feature point, which may be the feature vector extracted from an image or video, to search for the k-nearest neighbors of the query in a large database, comprising millions or billions of samples. It is an important foundation for CBIR/CBVR and many data-driven approaches such as image annotation by search [24].

A lot of algorithms have been proposed to tackle these difficulties, among which two of the most well-known approaches are K-d tree [16] and LSH [3]. However, due to the dimensionality curse, the application of those methods on large-scale multimedia data, which is in general of high dimension, achieved limited success. Besides the dimensionality curse, the widely adopted bag-of-words representation of images and videos [18,20] imposes new challenges to large-scale similarity search in multimedia. The reason is that the bag-of-words representation is normally very sparse and of very high dimension (it may be as high as 1 million dimension in near-duplicate image retrieval [15]), and existing methods generally are difficult to achieve a good performance.

With the ever-increasing data scale and the proliferation of multicore and distributed computing systems, effective algorithms which can run on the distributed or parallel computing systems for multimedia search and management are desired. MapReduce [5] is a flexible framework and has been successfully applied in machine learning [2]. Its basic idea is to decompose a complex task into a set of map and reduce operations so that the task can be distributed in many computers. Large-scale multimedia search and management have a strong demand on distributed algorithms that can be realized in MapReduce or similar frameworks and run on cloud computing platforms, for example, Windows Azure* and Amazon EC2.[†]

11.2.2   Bridging the Semantic Gap through Multimodality Integration

While the research on semantic indexing and concept detection has attracted a lot of research efforts for decades, the limited progress makes it still impractical to be applied in real applications due to the aforementioned *semantic gap*. It is admitted that the semantic gap is difficult to be bridged from the visual modality only [4], instead, the information from the other modalities or context, say, Global Positioning System (GPS) signal, is able to help alleviate the problem.

Besides the visual modality, the signals from other sources, including the Exchangeable Image File Format (EXIF), the tags in the social sharing web sites, the surrounding text in the associated web pages, and the GPS information can be regarded as the context to interpret the image or video's content from different perspectives. Yang et al. [25], proposed to utilize the surrounding text and the click-through log associated with the

---

* http://www.microsoft.com/windowsazure/
[†] http://aws.amazon.com/ec2/

query image as "local context" and the entire corpus as "global context" to formulate a more reliable retrieval model, for web image search. It showed that the different modalities have respective advantages and can thereafter complement each other: the visual modality is advantageous to return highly precise (mostly near-duplicate) images while the text modality can be better used to retrieve semantically similar results.

While many modalities such as the location obtained by GPS coordinates and the image-capturing context recorded in EXIF can be integrated to help infer the semantic of an image [12], existing approaches are ad hoc to utilize the contextual modalities in a problem-specific manner, which may not scale up with the number of modalities and applications. One open problem is, thereafter, to develop a unified representation of various modalities and a general framework with new context being easily plugged in.

11.2.3 Mining Rich Information from Web Multimedia

The large-scale multimedia data on the web can be regarded as a huge knowledge base that one can utilize to mine useful information to help tackle existing challenges. Fergus et al. [6] proposed to utilize the image search results from Google to construct training data to learn object category models. Wang et al. [24] further employed a data-driven approach to directly annotate images using image search results, without learning a category model.

Different from general image search, the social sharing web sites provide more useful information for images and videos, for example, tags uploaded by end users. To better utilize such information, Liu et al. [11] and Li et al. [10] developed algorithms to rank the tags in Flickr and showed that a high-precision annotation can then be obtained. Siersdorfer et al. [19] utilized the video duplicates in YouTube for automatic video tagging.

As shown above, the data available for multimedia mining on the web can be obtained from a number of heterogeneous sources. Besides the search results and social tags, the other information including the click-through from search engine logs and the image labeling game proposed in [23] is also available and contains many useful information. However, how to mine information from the many heterogeneous data sources still remains an open challenge.

It is observed that the research on web multimedia mining is overlapped with the multi-modality integration in some respects. Web multimedia mining may handle multimodality data, while multimodality integration may use the result from web multimedia mining as a new kind of modality.

11.2.4 Involving Humans into the Loop

The automatic multimedia search and management approaches cannot yet achieve a satisfactory performance due to the gap between the feature computing and the human's perception. The semantic gap between low-level features and semantic concepts [21] and intention gap between users' desire and query representation [28] are still difficult to be addressed through automatic approaches. A natural solution is to involve humans into the loop of multimedia search and management, so that we can take the respective advantages of humans and computers.

There are two schemes to incorporate humans' intelligence into a search system. The first is to introduce user interaction into the online query processing, so that the system can learn more about the user's intent for example, by relevance feedback. However, since users are often reluctant to provide feedback, this scheme is difficult to be adopted in real search engines [13]. Hence, designing a natural and easy-to-use user interface as well as incentive mechanism to encourage users to interact with the system becomes an important problem.

Visual query suggestion [28] and color-structured image search [9] can be regarded as preliminary attempts along this direction.

The second scheme is to attract and employ people out of the search system to contribute their knowledge (say, data labeling), to continuously improve the index quality and the system performance. Obviously, in this scheme, the first challenge is how to incent people so that they are willing to contribute. Von Ahn and Dabbish [23] proposed an idea to allow people label images using a game so that people can help the multimedia search system during the entertainment. In the human computation framework, such as Amazon Mechanical Turk* and HumanSense proposed by Yang et al. [27], the problems that computers may perform not well can be outsourced to human workers at low costs.

Since the number of problems needed to be solved by humans is very large, the other important problem likes in effectively selecting the most informative problems to ask for solutions from humans and then using computers to address the rest by learning from those that have been already solved. Hua and Qi [8] presented an active annotation system, which can automatically select the most informative videos and queries to ask for labels from people. Since the labeling cost and the system performance are always leading factors to be traded off in the system design and maintenance, they should be modeled jointly into an objective so as to be optimized.

Besides incentive mechanism and effective data selection and usage schemes, labeling quality evaluation and fraud detection are also challenging problems in leveraging large-scale users.

11.3   Overview of the Selected Topics

This part includes three chapters that describe the different aspects of multimedia search and management. In Chapter 12, Zha et al. reviews the recent advances on video modeling and retrieval, which are divided into three parts, including the semantic concept detection, semantic video retrieval, and interactive video search. Semantic concept detection is a fundamental problem in multimedia retrieval, which aims to annotate videos into a set of concepts. In Chapter 12, the basic models of semantic concept detection as well as the recent advances including semisupervised learning, multilabel learning, and cross-domain-learning-based approaches have been reviewed. Once concepts are detected, applying the detected concepts into video retrieval is still a nontrivial problem, which needs mapping query to related concepts and fusing the results from multiple related concepts. The related algorithms are also introduced as key technologies of semantic video retrieval in Chapter 12. Beyond the automatic approaches, Chapter 12 discusses on interactive video search as well.

Chapter 13 is devoted to a comprehensive review of recent advances in image retrieval. The different aspects of an image retrieval system, including visual feature extraction, image annotation, relevance feedback, and large-scale indexing are discussed in sufficient detail. The earlier proposed global features such as color, shape, and texture, as well as the recently proposed local features such as Scale Invariant Feature Transform (SIFT) are all described. Since relevance feedback is an important topic in early CBIR, the various algorithms from the early query point movement approaches to the recent machine-learning-based approaches are introduced. As an important branch of CBIR in recent years, image annotation is treated

---

* https://www.mturk.com/mturk/welcome

in an entire section, for which the typical methods are summarized into two schemes, model based and data driven. Since large-scale indexing is an important problem underlying scalable content understanding and search, the major algorithms are also presented with the analysis.

While Chapters 12 and 13 are about image and video retrieval, Chapter 14 focuses on video management to facilitate users in locating and browsing a large-scale video database. In this chapter, Ngo and Tan proposed a system to summarize a large video database into topic structures. The system is comprised of three components including content structuring, data cleansing and clustering, and summarization and browsing. Content structuring is to decompose videos into temporal units like shots and stories. Then, the full duplicates in the database are removed for efficiency while the near-duplicate videos are utilized to construct a hyperlink of videos. Finally, the summarization and browsing component generates a topic structure for a large video database, which is then combined with Google news for effective browsing of video.

11.4 Conclusion

With the explosion of video and image data available on the Internet, desktops, and mobile devices, multimedia search and management is becoming more and more important. Moreover, mining semantics and other useful information from large-scale multimedia data to facilitate online and local multimedia search, management, and other related applications has also gained more and more attention from both academia and industry. The rapid increase of multimedia data brings us new challenges to multimedia search and management especially in terms of scalability. Both computational costs and accuracy are still far from satisfactory. While on the other hand, large-scale multimedia data also provide us new opportunities to attack these challenges as well as conventional problems in media analysis and computer vision. That is, the massive associated metadata, context, and social information available on the Internet, desktops, and mobile devices, as well as the massive grassroots users, are valuable resources that can be leveraged to solve the aforementioned difficulties. Recently, more and more researchers are realizing both the challenges and the opportunities for multimedia research brought by rapid increases of multimedia data, multimedia users, as well as associated metadata, context, and social information. More and more research and products on aggregating data, model, and users, as well as content, concept, and context, are emerging, which are moving to the right direction of enabling scalable and content-aware multimedia search and management.

References

1. Three billion photos at flickr. http://www.techcrunch.com/2008/11/03/threebillion-photos-at-flickr
2. C.-T. Chu, S. K. Kim, Y.-A. Lin, Y. Yu, G. Bradski, A. Y. Ng, and K. Olukotun. Map-reduce for machine learning on multicore. In B. Schölkopf, J. Platt, and T. Hoffman, editors, *Advances in Neural Information Processing Systems 19*, pp. 281–288, MIT Press, Cambridge, MA, 2007.
3. M. Datar, N. Immorlica, P. Indyk, and V. S. Mirrokni. Locality-sensitive hashing scheme based on p-stable distributions. In *Proceedings of the 20th Annual Symposium on Computational Geometry*, SCG '04, pp. 253–262, ACM, Brooklyn, New York, 2004.

4. R. Datta, D. Joshi, J. Li, and J. Z. Wang. Image retrieval: Ideas, influences, and trends of the new age. *ACM Comput. Surv.*, 40(2):1–60, 2008.

5. J. Dean and S. Ghemawat. MapReduce: Simplified data processing on large clusters. *Communications of the ACM*, 51(1):107–113, 2008.

6. R. Fergus, L. Fei-Fei, P. Perona, and A. Zisserman. Learning object categories from google's image search. In *Proceedings of the 10th IEEE International Conference on Computer Vision—Volume 2*, ICCV '05, pp. 1816–1823, IEEE Computer Society, Washington, DC, 2005.

7. B. Geng, L. Yang, C. Xu, and X.-S. Hua. Content-aware ranking for visual search. In *The 23rd IEEE Conference on Computer Vision and Pattern Recognition*, CVPR '10, pp. 3400–3407, San Francisco, CA, 2010.

8. X.-S. Hua and G.-J. Qi. Online multi-label active annotation: towards large-scale content-based video search. In *Proceeding of the 16th ACM International Conference on Multimedia*, MM '08, pages 141–150, ACM, Vancouver, British Columbia, Canada, 2008.

9. X. Li, C. G. Snoek, and M. Worring. Learning tag relevance by neighbor voting for social image retrieval. In *Proceeding of the 1st ACM International Conference on Multimedia Information Retrieval*, MIR '08, pp. 180–187, ACM, Vancouver, British Columbia, Canada, 2008.

10. D. Liu, X.-S. Hua, L. Yang, M. Wang, and H.-J. Zhang. Tag ranking. In *Proceedings of the 18th International Conference on World Wide Web*, WWW '09, pp. 351–360, ACM, Madrid, Spain, 2009.

11. J. Luo, A. Hanjalic, Q. Tian, and A. Jaimes. Integration of context and content for multimedia management: An introduction to the special issue. *IEEE Transactions on Multimedia*, 11(2):193–195, 2009.

12. C. D. Manning, P. Raghavan, and H. Schtze. *Introduction to Information Retrieval*. Cambridge University Press, New York, NY, 2008.

13. K. Min, L. Yang, J. Wright, L. Wu, X.-S. Hua, and Y. Ma. Compact projection: Simple and efficient near neighbor search with practical memory requirements. In *The 23rd IEEE Conference on Computer Vision and Pattern Recognition*, CVPR '10, pp. 3477–3484, San Francisco, CA, 2010.

14. J. Philbin, O. Chum, M. Isard, J. Sivic, and A. Zisserman. Object retrieval with large vocabularies and fast spatial matching. In *Proceedings of the IEEE Conference on Computer Vision and Pattern Recognition*, CVPR '07, Minneapolis, Minnesota, 2007.

15. J. T. Robinson. The k-d-b-tree: A search structure for large multidimensional dynamic indexes. In *Proceedings of the 1981 ACM SIGMOD International Conference on Management of Data*, SIGMOD '81, pp. 10–18, ACM, Ann Arbor, Michigan, 1981.

16. Y. Rui, T. Huang, M. Ortega, and S. Mehrotra. Relevance feedback: A power tool for interactive content-based image retrieval. *IEEE Transactions on Circuits and Systems for Video Technology*, 8(5):644–655, 1998.

17. L. Shang, L. Yang, F. Wang, K.-P. Chan, and X.-S. Hua. Real-time large scale near-duplicate web video retrieval. In *Proceedings of the International Conference on Multimedia*, MM '10, pp. 531–540, ACM, Firenze, Italy, 2010.

18. S. Siersdorfer, J. San Pedro, and M. Sanderson. Automatic video tagging using content redundancy. In *Proceedings of the 32nd international ACM SIGIR Conference on Research and Development in Information Retrieval*, SIGIR '09, pp. 395–402, ACM, Boston, MA, 2009.

19. J. Sivic and A. Zisserman. Video google: A text retrieval approach to object matching in videos. In *Proceedings of 9th IEEE International Conference on Computer Vision*, volume 2 of ICCV '03, pp. 1470–1477, Nice, France, 2003.

20. A. W. M. Smeulders, M. Worring, S. Santini, A. Gupta, and R. Jain. Content-based image retrieval at the end of the early years. *IEEE Transactions on Pattern Analysis and Machine Intelligence*, 22(12):1349–1380, 2000.

21. X. Tian, L. Yang, J. Wang, Y. Yang, X. Wu, and X.-S. Hua. Bayesian video search reranking. In *Proceeding of the 16th ACM International Conference on Multimedia*, MM '08, pp. 131–140, ACM, Vancouver, British Columbia, Canada, 2008.

22. L. von Ahn and L. Dabbish. Labeling images with a computer game. In *Proceedings of the SIGCHI Conference on Human Factors in Computing Systems*, CHI '04, pp. 319–326, ACM, Vienna, Austria, 2004.

23. J. Wang, X.-S. Hua, and Y. Zhao. Color-structured image search. Technical report, In *Microsoft Research Technical Report*, 2009.

24. X.-J. Wang, L. Zhang, F. Jing, and W.-Y. Ma. Annosearch: Image autoannotation by search. In *Proceedings of the IEEE Conference on Computer Vision and Pattern Recognition*, volume 2 of CVPR '06, pp. 1483–1490, IEEE Computer Society, Los Alamitos, CA, 2006.

25. L. Yang, B. Geng, A. Hanjalic, and X. Hua. A unified context model for web image retrieval. *ACM Transactions on Multimedia Computing, Communications, and Applications (TOMCCAP)*, to appear.

26. L. Yang and A. Hanjalic. Supervised reranking for web image search. In *Proceedings of the International Conference on Multimedia*, MM '10, pp. 183–192, ACM, Firenze, Italy, 2010.

27. Y. Yang, B. B. Zhu, R. Guo, L. Yang, S. Li, and N. Yu. A comprehensive human computation framework: with application to image labeling. In *Proceeding of the 16th ACM International Conference on Multimedia*, MM '08, pp. 479–488, ACM, Vancouver, British Columbia, Canada, 2008.

28. Z.-J. Zha, L. Yang, T. Mei, M. Wang, Z. Wang, T.-S. Chua, and X.-S. Hua. Visual query suggestion: Towards capturing user intent in internet image search. *ACM Trans. Multimedia Comput. Commun. Appl.*, 6:13:1–13:19, 2010.

12

Video Modeling and Retrieval

Zheng-Jun Zha, Jin Yuan, Yan-Tao Zheng, and Tat-Seng Chua

CONTENTS

12.1 Introduction

Recent advancements in processor speed, high-speed network, and the availability of massive digital storages have led to an explosive amount of video data, covering a wide range of topics from broadcast news, to documentaries, meetings, movies, and other videos. There is a compelling need for automatic modeling and efficient retrieval of these data. Most commercial video search engines provide access to videos based on the textual metadata associated with videos, including filename, surrounding text, closed captions, a speech transcript, and so on. However, the problem with text-based video retrieval is that the textual metadata are frequently noisy, incomplete, and inconsistent with the semantics of videos.

    Video modeling and retrieval in academia moves one step ahead, which takes video semantic analysis, especially semantic concept detection, into consideration. It aims to facilitate semantic video retrieval resorting to a set of intermediate semantic concepts, which are defined and utilized to reveal video semantics based on machine learning techniques. Recently, a wealth of techniques for semantic concept detection and semantic video retrieval has been proposed.

In this chapter, we review the active research on video modeling and retrieval in recent years, including semantic concept detection, semantic video retrieval, and interactive video retrieval taking advantage of user interaction. Furthermore, we identify open research issues based on the state-of-the-art research efforts.

12.2   Semantic Concept Detection

Semantic concept detection, also called video semantic annotation or high-level feature extraction in literatures, aims to annotate video clips with semantic concepts. It has shown encouraging progress toward bridging the *"semantic gap"* between data representation and their interpretation by humans [1,13]. Early research efforts focus on the detection of specific semantic concepts with small intraclass and large interclass variability of content, such as *"sunset"* in [43] and *"news anchors"* in [42]. The focus of research in recent years has been shifted to the detection of a large set of generic concepts, including a wide range of categories such as scenes (e.g., *urban, sky, mountain* etc.), objects (e.g., *airplane, bus, face*, etc.), and events (e.g., *people-marching walking–running*, etc.). A wealth of approaches have been proposed for generic semantic concept detection. Most of them are developed based on machine learning techniques, including supervised learning [2,12], semisupervised learning [16,29], multilabel learning [19,20,25], cross-domain learning [3,4,8], and so on. In this section, we first introduce the basic approach based on supervised learning, and then briefly review advanced approaches in semantic concept detection, such as semisupervised learning, multilabel learning, and cross-domain learning methods.

Before going into details, we first introduce certain preliminary knowledge of learning-based semantic concept detection. First, video sequences are divided into workable segments. The most natural segment is video shot, which consists of one or more frames that represent a continuous action in time and space [9]. Then, low-level features are extracted from each segment (e.g., key frames of each segment) to represent its content. More details about video shot detection and content representation could be found in [5–7]. Based on low-level features, semantic concept detection is formalized to learn a set of predefined semantic concepts for each segment. In a typical semantic concept detection framework, a binary classification procedure is conducted for each concept. Given a new segment, it is annotated to be "positive" or "negative" with respect to each concept according to whether it is associated with the concept or not.

12.2.1   Basic Approach

Supervised learning technology has been widely used for semantic concept detection. With a training dataset obtained by manual labeling, it first learns classifiers for the target concepts and then performs concept detection on the unlabeled samples by inferring their labels. The supervised learning approaches for semantic concept detection are required to have the following advantages: (a) it is able to learn robust classifiers from a limited number of training samples since manual labeling is labor intensive and time consuming; (b) it can handle imbalance in the number of positive versus negative training samples; and (c) the resultant classifiers should have good generalization performance such that they can perform well on test samples not used during training. In such heavy demands, the support vector machine (SVM) framework [2,12] has found to be a solid choice and has been widely employed in most semantic concept detection systems [13].

Support Vector Machine Given a training set $\mathcal{L} = \{\mathbf{x}_i, y_i\}_{i=1}^{N}$ for certain concept, where $\mathbf{x}_i \in \mathbb{R}^d$ is the d-dimensional feature vector of ith sample, $y_i \in \{-1, 1\}$ is the label of $\mathbf{x}_i$. y_i indicates that $\mathbf{x}_i$ is a positive ($y_i = 1$) or negative ($y_i = -1$) sample to the concept. SVM aims to learn an optimal hyperplane that can separate the positive samples from the negative ones with a maximal margin. The classification hyperplane is defined as

$$\langle w, \Phi(x) \rangle + b = 0, \tag{12.1}$$

where $\mathbf{w}$ is a weight vector, b is a bias term, $\Phi(\cdot)$ is a mapping from $\mathbb{R}^n$ to a Hilbert Space $\mathcal{H}$, and $\langle \cdot, \cdot \rangle$ is the dot product in $\mathcal{H}$.

Figure 12.1 shows an illustration of SVM. The samples that lie closest to the decision boundary are the support vectors, which are the critical elements of the training set. Here we briefly introduce the soft-margin SVM. Different from the original SVM, soft-margin SVM allows for misclassified samples and chooses a hyperplane that splits the samples as clear as possible. Slack variables ξ_i are introduced to measure the degree of misclassification of the samples $\mathbf{x}_i$. The maximization of the margin and minimization of the training errors can be formulated as the following quadratic optimization problem:

$$\min_{\mathbf{w}, b, \xi} \frac{1}{2} \|w\|^2 + C \sum_{i=1}^{N} \xi_i$$

$$\text{subject to}: \quad y_i(\langle \mathbf{w}, \Phi(x) \rangle + b) \geq 1 - \xi_i; \tag{12.2}$$

$$\xi_i \geq 0, \quad \forall i = 1, 2, \ldots, N,$$

where C is a trade–off parameter.

Equation 12.2 could be transferred into its dual form as follows:

$$\min_{\alpha} \frac{1}{2} \alpha^\mathsf{T} Q \alpha - 1^\mathsf{T} \alpha$$

$$\text{subject to}: \quad 0 \leq \alpha_i \leq C, \quad \forall i = 1, 2, \ldots, N; \tag{12.3}$$

$$\alpha^\mathsf{T} y = 0,$$

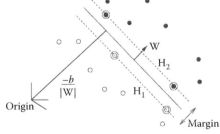

FIGURE 12.1
An illustration of SVM. The support vectors are circled. (Adapted from C. J. C. Burges. *Data Mining and Knowledge Discovery*, 2, 121–167, 1998.)

where $\boldsymbol{\alpha}$ is a vector $[\alpha_1, \alpha_2, \ldots, \alpha_N]^T$ containing Lagrange multipliers α_i, $\mathbf{1}$ is a vector of all ones, and $\mathbf{Q}$ is an $N \times N$ positive semidefinite matrix with $Q(i,j) = y_i y_j K(\mathbf{x}_i, \mathbf{x}_j)$. $K(\cdot)$ is a kernel function and $K(\mathbf{x}_i, \mathbf{x}_j) = \langle \Phi(\mathbf{x}_i), \Phi(\mathbf{x}_j) \rangle$. The matrix $\mathbf{K}$ containing the values of kernel function for all training sample pairs is named Gram matrix. It is worthy to note the importance of the kernel function, since it maps the distance between feature vectors into a higher-dimensional space in which the hyperplane separator and its support vectors are obtained. After getting the optimal $\boldsymbol{\alpha}$, the classification confidence score of a given test sample $\mathbf{x}$ can be obtained following the representation theorem:

$$f(\mathbf{x}) = \sum_{i=1}^{N} \alpha_i K(\mathbf{x}, \mathbf{x}_i) \tag{12.4}$$

Based on $f(\mathbf{x})$, we can compute the posterior probability $p(C|\mathbf{x})$, which denotes the probability of the sample $\mathbf{x}$ associated with the target concept C, if it is required. The sigmoid function is widely employed to transfer $f(\mathbf{x})$ to $p(C|\mathbf{x})$ as

$$P(C|\mathbf{x}) = \frac{1}{1 + exp(Af(\mathbf{x}) + B)}, \tag{12.5}$$

where the parameters A and B are maximum-likelihood estimates based on the training set.

12.2.2 Advanced Approaches

As aforementioned, many recently proposed learning techniques have shown encouraging performance in semantic concept detection. In this section, we briefly review some advanced semantic concept detection approaches based on semi-supervised learning, multilabel learning, and cross-domain learning techniques.

SemiSupervised Learning. In semantic concept detection, the labeled data are usually insufficient due to the fact that manual labeling is highly laborious and time consuming. Recent studies show that labeling 1 h of video with 100 concepts generally takes anywhere between 8 and 15 h [14]. On the other hand, unlabeled data are usually easy to collect. Recent progress on semisupervised learning shows that the unlabeled data used with a small number of labeled samples can improve the learning performance greatly [15]. This has triggered many researches on improving semantic concept detection based on semisupervised learning methods. Among diverse semisupervised learning algorithms, such as self-training [27], cotraining [28], transductive SVM [11], and graph-based algorithms [10,16,29], graph-based semisupervised learning algorithms have been widely employed for semantic concept detection due to their effectiveness and efficiency on this task. The common assumption of graph-based methods is *label consistency* [16,29], that is, the similar samples would be more likely to have the same label. These methods typically model the whole dataset as a graph, where the nodes correspond to labeled and unlabeled samples, and the edges reflect the similarities between samples. Most existing graph-based methods essentially estimate a labeling on a graph, expecting that such a labeling satisfies two properties: (a) it should be as close as possible to the given labeling on the labeled samples and (b) it should be smooth on the whole graph. To this end, a regularized optimization framework is adopted, where a loss function is defined to penalize the deviation of estimated labeling from the given label, and a regularizer is defined for the label consistency over the graph. The typical graph-based methods are similar to each other, and differ slightly in the loss function and the regularizer.

Given a set of N samples $\mathcal{X} = \{\mathbf{x}_i\}_{i=1}^{N}$ in $\mathbb{R}^d$. The first L samples are labeled as $\{y_1, y_2, \ldots, y_L\}$ with $y_i \in \{0, 1\}(1 \leq i \leq L)$. The task is to estimate the labeling on the remaining unlabeled samples $\{\mathbf{x}_{L+1}, \mathbf{x}_{L+2}, \ldots, \mathbf{x}_N\}$. Denote the graph by $\mathcal{G} = (\mathcal{V}, \mathcal{E})$. The node set $\mathcal{V} = \mathcal{L} \cup \mathcal{U}$ corresponds to the labeled samples $\mathcal{L} = \{\mathbf{x}_1, \mathbf{x}_2, \ldots, \mathbf{x}_L\}$ and the unlabeled samples $\mathcal{U} = \{\mathbf{x}_{L+1}, \mathbf{x}_{L+2}, \ldots, \mathbf{x}_N\}$. The edges $\mathcal{E}$ are weighted by an $N \times N$ affinity matrix $\mathbf{W}$ with W_{ij} indicating the similarity between between $\mathbf{x}_i$ and $\mathbf{x}_j$. $\{f_1, \ldots, f_L, f_{L+1}, \ldots, f_N\}$ denoting the estimated labeling on $\mathcal{X}$ can be obtained by minimizing the following objective functions [16,29]:

$$\mu \sum_{i \in \mathcal{L}} (f_i - y_i)^2 + \tfrac{1}{2} \sum_{i \in \mathcal{X}} W_{ij}(f_i - f_j)^2, \tag{12.6}$$

where μ is a trade-off parameter.

As can be seen, the edge weights W_{ij} are crucial for graph-based methods. Most existing methods estimate the weights simply based on the Euclidean distances between samples. In order to derive more effective weights, Tang et al. [30] incorporated the density information of each sample into the weight estimation, while Wang et al. [31] proposed a neighborhood similarity which takes the local structure of each sample and label distribution into consideration.

Multilabel Learning. Multilabel semantic concept detection has attracted increasing attention recently. Different from typical concept detection methods that use binary classification to detect each concept individually, multilabel concept detection models multiple concepts simultaneously and exploits the inherent correlation between concepts. As introduced in recent works [17–19], semantic concepts are usually interacting with each other naturally rather than existing in isolation. For example, *mountain* and *sky* tend to appear simultaneously, while *mountain* typically does not appear with *indoor*. Moreover, an important purpose of modeling concept correlations is its support for inference [13]. The complex concepts can be inferred based on the correlations with other simple concepts that can be well modeled from low-level features. For example, the presence of *people-marching* can be boosted if both *crowd* and *walking–running* occur in a video segment.

Existing multilabel concept detection approaches can be categorized into two classes [19]: (a) context-based concept fusion [21–24] and (b) integrated multilabel concept detection approaches [19,20,25,26]. In context based-concept fusion approaches, each concept is first modeled individually by a classifier, and then a second step is conducted to refine the detection results of the individual classifiers by taking advantages of concept correlations. For example, Smith et al. [22] concatenated the detection scores of individual concept classifiers into a single vector. This vector served as the input for an SVM, which was trained to exploit the concept correlation and refine the individual detection results. Hauptmann et al. [23] proposed to adopt logistic regression to fuse the detections of individual classifiers, while Jiang et al. [24] employed conditional random field to refine the individual detections. Wu et al. [21] and Zha et al. [18] used an ontology-based multiclassification learning for semantic concept detection. A predefined ontology hierarchy is investigated to improve the individual detections. These context-based concept fusion strategies commonly suffer from the error propagation problem, that is, the detection errors of individual classifiers can propagate to the second fusion step. As a result, the final detections will be corrupted. Motivated by this observation, the integrated multilabel concept detection scheme has been proposed to simultaneously model both the concepts and their correlations in an integrated formulation. Qi et al. [19] proposed to model individual detectors and concept correlations in a single graphical model formulation. Wang et al. [25] proposed a transductive approach for multilabel concept detection, while Zha et al. [20] proposed a graph-based semisupervised multilabel concept detection method, which further makes use of unlabeled data to improve detection performance. As shown by these works, the integrated multilabel concept detection is normally more effective than context-based concept fusion strategy.

Cross-Domain Learning. Cross-domain semantic concept detection is a newly emerging technique, which has shown promising performance by leveraging labeled samples from auxiliary domains to learn robust concept classifiers for the target domain with a limited number of labeled samples [3,4,8]. Here a domain can be a video genre, a content provider, or a video program. Compared to the conventional concept detection approaches that build concept classifiers from labeled samples in one or more domains and assume that the test data come from the same domain(s), cross-domain concept detection is superior in performance, computational efficiency, and human resource. Yang et al. [3] proposed an Adaptive Support Vector Machine framework to adapt existing classifiers from one or more auxiliary domains to the target domain. The classifier for the target domain was formulated as an additive combination of a new delta function and the classifiers from auxiliary domains. The new "delta function" was learned optimally based on the labeled samples in the target domain with the assistance of auxiliary classifiers. Duan et al. [4] proposed to make use of both the labeled and unlabeled samples in the target domain. They proposed a Domain Transfer Support Vector Machine to simultaneously learn a kernel function and a robust classifier by minimizing both structural risk function of classifier and distribution mismatch of labeled and unlabeled samples between auxiliary and target domains. Yang and Hauptman [8] pointed out that not all the concepts could be beneficial from the cross-domain concept detection technique. They found that most concept classifiers built using the cross-domain technique are unreliable because these classifiers generalize poorly to the target domains. By examining the properties of the reliable concept classifiers, they reported that the classifiers of frequent concepts, classifiers bloated with a large number of support vectors, and classifiers using a small percentage of positive samples as support vectors tend to be more reliable.

12.3 Semantic Video Retrieval

With the proliferation of video data, the need of video retrieval has drastically increased. Most commercial search engines retrieve videos based on the textual metadata associated with videos. The problem with text-based video retrieval is that the textual metadata are frequently noisy, incomplete, and inconsistent with the semantics of videos. Alternatively, a new search paradigm, named semantic video retrieval or concept-based video retrieval has been extensively studied and found to be a promising direction. Semantic video retrieval utilizes a set of prebuilt concept detectors to predict the presence of semantic concepts in video segments. For a given query, it first maps the query to related concepts and then retrieves answer video segments by fusing individual search results of the related concepts, each of which is generated based on the prediction of the corresponding concept. As illustrated in Figure 12.2, semantic video retrieval typically consists of two main components: (a) concept selection, that is, selecting related concepts for the given query and (b) search result fusion, that is, fusing the individual results of related concepts. In the following sections, we briefly review existing research efforts on these two topics.

FIGURE 12.2
The framework of automatic semantic video search.

12.3.1 Concept Selection

For a given query, semantic video search engine needs to select a set of appropriate concepts to interpret query semantics. The existing concept selection approaches can be categorized into three classes [13]: (a) text-based selection through text matching between query words and concept detector description [32–34,36,37]; (b) visual-based selection based on detection scores of the concept detectors to query examples (e.g., images or video clips) [32,34,44]; and (c) ontology-based selection utilizing external ontology like WordNet [34,37].

Basic text-based concept selection is based on the exact text matching between query and detector description [32–34]. For example, Snoek et al. [34] represented each detector description by a term vector, where each element corresponds to a unique normalized word. The vector space model [40] was then employed to compute the similarity between detector description and query words. Based on the resultant similarity, the concepts were selected automatically. Although exact text matching can achieve highly accurate selection results, its largest limitation is that it is not able to discover the potentially related concepts that do not explicitly appear in the query. To tackle this problem, some sophisticated text-based selection approaches have been developed [36,37]. They performed data-driven query expansion using external text resources and selected the related concepts for the expanded query. One drawback of these approaches is that the query expansion might introduce noisy terms and thus result in inappropriate concept selection.

The concept selection can also be performed based on the concept detection scores to query image or video examples (if any). Visual-based selection of concept detectors might prove a valuable additional strategy when other selection strategies fail [13,37]. For example, it is not possible to relate *mosques* to the query *Helicopters in flight* by text-based selection methods. However, they are really relevant in the videos about *Iraq war*, and this connection can be mined through visual cues [32]. The straightforward visual-based selection approach relies on the posterior probabilities of visual samples with respect to concept detectors [44]. The concepts with the highest posterior probabilities are selected. Snoek et al. [34] argued that the frequently occurring concepts (e.g., person and outdoor) are usually less informative for retrieval. They suggested that it seems better to avoid frequently occurring but nondiscriminative concepts. Li et al. [32] proposed a modified tf-idf measure, named c-tf-idf, to estimate the relevance between the query and concepts. This method treats each concept as *visual term* and each image or key frame as *visual document*. Similar to the typical tf-idf approach, the c-tf-idf formula is represented by the product of two measures: a tf measure corresponding to the confidence score of a query image by a concept detector, and an idf measure which is inversely proportional to the frequency of that concept. Therefore, the c-tf-idf measure considers the relevance between the query and a concept as well as the popularity of that concept.

Ontology-based selection employs semantic ontology for concept selection. The widely used ontology is the WordNet. By utilizing information from WordNet, such as word frequencies and hierarchical structure of words, some ontology reasoning techniques have been developed for estimating linguistic relatedness of words. For a given query, the concepts can be selected based on their relatedness to the query words [47]. To compute word relatedness, Snoek et al. [34] adopted Resnik's measure [41], while Natsev et al. [37] employed Lesk's measure [38]. With the ontology reasoning techniques, a recent work by Wei et al. [46] constructed a vector space, named ontology-enriched semantic space (OSS), by considering the pairwise relatedness of the concepts. In OSS, both query words and concept detectors are represented as vectors, and the relatedness measurement inferred from OSS has the merit of global consistency.

12.3.2 Search Result Fusion

Given a query, the semantic video retrieval engine returns answer video segments by fusing the individual search results of the related concepts. The state-of-the-art search result fusion approach is the linear fusion as

$$r(S_i) = \sum_{k=1}^{K} w_k r_k(S_i) \tag{12.7}$$

where $r(S_i)$ is the relevance score of video segment S_i to the query, $r_k(S_i)$ is the relevance score of S_i to the kth related concept, K is the number of related concepts, and w_k is the weight of concept k with respect to the query. Yan and Hauptmann [35] presented a theoretical framework for monotonic and linear weighted fusion. They studied the upper bounds for fusion functions, and argued that a linear fusion might be sufficient when fusing a small number of query results. Donald and Smeaton [53] compared the performance of linear summation fusion strategy to other strategies, such as multiplication fusion and max score. Their experimental results showed that the linear fusion is the best fusion strategy.

The general problem in linear fusion is the estimation of the weight values. Ideally, the optimal weights could be estimated based on validation set performance, but in practice, the amount of available training samples is usually too small [13]. Alternatively, Chang et al. [33] determined the weights based on a text-match score. The concepts having large text-match scores were assigned with large weights, while Wei et al. [46] proposed to estimate the weight of each concept by relating it to the cosine distance between the concepts and query. Neo et al. [36] incorporated the accuracy of concept detectors into weight estimation. They computed the weight of each concept by the product of its accuracy and its similarity to the query.

Linear fusion does not consider the duplicate information between related concepts. For example, if the query is *vehicle at street* and the relate concepts *car*, *bus*, and *street* are selected, linear fusion computes the relevance of a video segment by summing its prediction scores by these concept detectors. However, the concepts *car* and *bus* are actually duplicate elements for this query. Motivated by this, Wang et al. [39] proposed a query representation to capture the concept structure of the query. An example is shown in Figure 12.3. The query is *multiple people in formation*, and the most salient combination is made up of *demonstration or protest*, *crowd*, and *military*. The other combination is made up of *soldiers*, *crowd*, and *military*. A two-level fusion method was proposed to solve a related concepts combination problem: the bottom level is a AND logic to make sure the selected samples satisfy multiple concepts at the same time, and the upper level uses an OR logic to return the final result from any combination group.

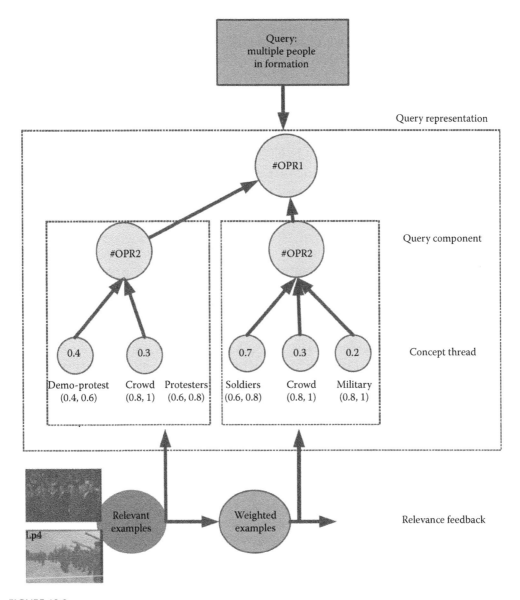

FIGURE 12.3
The query representation as structured concept threads. (Adapted from D. Wang et al., *Journal of Visual Communication and Image Representation*, 20, 104–116, 2009.)

12.4 Interactive Video Search

The skyrocketing amount of videos have thrusted the developments of many fully auto-mated video search systems in the last two decades to cater for precise retrieval. However, with the inability to understand a user's intention, most of these automated video search systems are still far from satisfactory [58]. To enhance searching in large video corpus, research works have recently explored the use of interactive retrieval systems that harness a user's effort in the loop of retrieval to improve search performance [49,67]. In interactive video search, the search session is divided into several continuous rounds. In each round, users provide feedback on the retrieval results by manually annotating returned videos as "relevant" or "irrelevant" to queries. The user feedback then allows the search engine to better understand the queries and video corpus and revert with improved retrieval results in the next round [58].

12.4.1 Relevance Feedback, Active Learning, and User Interface

Recently, many algorithms and systems for interactive retrieval have been proposed. Among which the relevance feedback (RF) techniques are the most effective to improve the performance of content-based information retrieval [50,64,65]. RF was first introduced for information retrieval in text domain [66,70]. Researchers then introduced RF content-based image retrieval (CBIR) in 1990s [59,61,63]. Early RF approaches attempted to find better query points as close as possible to the assumed "ideal query point" and adjust the weights of various features [50]. However, they are not all-purpose methods due to the presence of certain limitations or strong assumptions. Currently, RF in CBIR is an online learning problem since RF can be generally viewed as a particular type of pattern classification that consider positive and negative samples as two different groups. Among the various learning algorithms [55,62,68], the RF based on SVM is the most popular because of its inherent advantages such as fast learning, multichoices of kernel, and reliance on only support vectors. However, SVM-based RF usually faces big challenges including small-sized labeled training set, imbalance between the positive and the negative samples, and high-dimensional features. To tackle these problems, a new asymmetric bagging and random subspace mechanism is designed in [69].

Beyond RF, the active learning technique has been introduced to interactive video retrieval, since it can significantly reduce human cost in sample labeling. By employing active learning techniques which identify and select the most informative unlabeled samples for manual labeling, the interactive video search engine can harvest the most informative samples quickly and thus improve the search performance effectively. The core problem of active learning methods is the sample selection strategy. Existing sample selection strategies explored in interactive video retrieval can be classified into three categories: the risk reduction strategy [73–75], the most uncertainty strategy [76–79], and other strategies considering sample relevance, density, and diversity information [80–83]. Cohen et al. [73] and Roy and McCallu [74] suggested that the optimal active learning method should select samples that minimize the expected risk of current learning model. The reduced risk is estimated with respect to each unlabeled sample, and the most effective samples are selected. However, for most of the existing learning models, it is infeasible to estimate the risk. Therefore, in practice, many active learning methods adopt the most uncertainty sample selection strategy which selects the samples that the current model is most uncertain about. The most popular uncertainty-based active learning method is Support Vector Machine active learning (SVM *active*) [76]. SVM *active* selects samples that are closest to the current SVM boundary to label. In addition, Nguyen et al. [79] chose samples closest to the boundary in an information space based on distances to a number of selected samples. Some other sample selection approaches take sample relevance, density, and diversity information into account. Ayache and Quenot [82] proposed to use relevance criteria in sample selection. The samples that have the highest probabilities to be relevant were selected to label. Density criterion aim to select samples from the dense regions of feature space [80]. The samples in dense regions are usually representative of the underlying sample distribution. Therefore, these samples can add much more information to the learning model as compared to the samples within low-density regions. Brinker [84] incorporated the diversity criterion into sample selection. They selected samples that are diverse to each other.

In addition to refinement using various feedback methods, many researchers find that well-designed user interfaces (UIs) with good visualization are extremely helpful in improving the search performance [58]. Hence, many efficient and novel interfaces are designed for the communication between the users and the system. In fact, interactive systems designed by CMU and MediaMill [51,56] for TRECVID evaluations have demonstrated that efficient UI is crucial for interactions which jointly maximize the performance of both human and computer.

12.4.2 Interactive Video Retrieval Systems

In the last few years, the TRECVID evaluations have provided a testbed to develop various interactive video search systems. This section reviews a few representative systems on TRECVID video corpus with highlights on their distinct characteristics.

VisionGo: The focus of VisionGo system [57,58,60] is adaptive multiple feedback strategies of queries and for different video domains. The motivation is that the multimodal nature of video makes video content understanding a fairly complicated process and the wide range of video domains makes user queries quite diverse. Usually, one feedback method works well only for specific classes of queries or on certain domains. To tackle this issue, the VisionGo system first segregates interactive feedback into three distinct types (recall-driven RF, precision-driven active learning, and locality driven RF). The first emphasizes on analyzing and applying the correlation of general features obtained from labeled positive and negative instances to provide high recall retrieval on the entire corpus. The second uses active learning with a precision-driven sampling strategy to continuously refine the reranking model using a combination of multimodal features. The final strategy exploits high temporal coherence among the neighboring shots within the same story. The advantage is that a generic interaction mechanism with more flexibility can be performed to cover different search queries and different video corpuses. Moreover, to cater to the large number of novice users (nonexpert users), VisionGo provides nonexpert users a recommendation mechanism, which automatically recommends a suitable strategy to perform feedback based on the current situation. Figure 12.4 shows the UI and system framework of VisionGo.

MediaMill: Different from VisionGo, the MediaMill system [52,54] adopts a thread-based navigation by embedding multiple query methods into a single browsing environment. In the system, a query thread is defined to be a shot-based ranking of the video collection according to some aspect of video content, for example, some feature-based similarity measure. Intuitively, the thread connects related videos in certain similarity space. By visualizing a varying number of threads, the system provides flexibility to perform both fast targeted search and exploratory search in a balanced manner.

K-Space: The K-Space system [48,71] focuses on exploring the temporal context in video retrieval. In the interactive search interface, a portion of the result presentation space is allocated to temporal contextual cues about the returned results. Temporal context encompasses semantic relationship among adjacent shots within a video stream. The preceding and

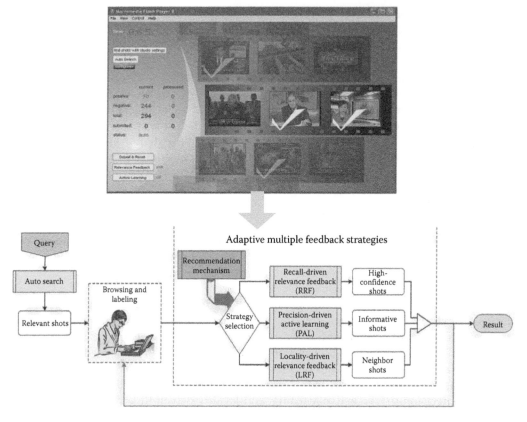

FIGURE 12.4
UI and framework of VisionGo system. (Adapted from H.-B. Luan et al., In *Proceedings of the ACM International Conference on Multimedia*, pp. 293–296, Augsburg, Germany, 2007; H. Luan et al., In *Proceedings of the International Conference on Content-Based Image and Video Retrieval*, pp. 457–464, New York, 2008.)

succeeding shots allow users to better understand the semantic topic or concept as a whole and convey additional relevant information in the temporal progression. The differences between temporal context and other feedback strategies are that temporal context enable users to locate relevant shots by simple exploration, while other feedback schemes rely more on reformulating queries and learning video corpus.

IBM Multi-Query Interactive Retrieval System: The interactive retrieval system developed by IBM [72] emphasizes on optimizing the retrieval performance on multiple query topics. The system classifies user annotation into two types: tagging and browsing. Tagging refers to users annotating images with a chosen set of keywords from a vocabulary, while browsing means users browsing returned videos to annotate their relevance to the given keywords. The proposed system allows users to switch between tagging and browsing interface in a learning-based hybrid approach, so as to maximize user labeling efficiency. Specifically, two formal annotation models are used to track and estimate the retrieval time for each method. Based on these two models, the retrieval system exploits the tagging-based and browsing-based methods adaptively to minimize the overall retrieval time.

12.5 Summary

In this chapter, we have reviewed the active research on video modeling and retrieval, including semantic concept detection, semantic video retrieval, and interactive video retrieval. However, research in this field has achieved many advances in various aspects. However, there are still many open research issues that need to be solved.

12.5.1 Large-Scale Concept Detection

Despite the success of existing semantic concept detection techniques, there are difficulties in handling large-scale semantic concepts with large-scale video data. Due to the complexity of both semantic concepts and video data, huge training datasets and high computation powers are required by existing techniques to achieve acceptable performance. However, it is too difficult, even impossible, to meet this requirement in real-world applications. Therefore, it is worthwhile to develop scalable semantic concept detection technique that can perform the detection of large-scale concepts on large-scale video data effectively and efficiently.

12.5.2 Complex Query Search

Existing semantic retrieval techniques retrieve answer video segments by fusing the individual search resutl of related concepts. This fusion strategy oversimplifies the query semantics into a set of primitive concepts. Working well for simple queries, it is normally ineffective for complex queries that are in the form of phrases or sentences with complex semantic meanings, such as *find the video with a man running through an airport and a woman in a pink dress*. Although a preliminary work has been introduced in [45], complex query search needs to be further studied.

12.5.3 Social Video Analysis

Existing video modeling and retrieval techniques mostly focus on videos in specific domains, such as broadcast news and documentary videos. On the other hand, the increasingly popular video-sharing service, such as YouTube, has attracted millions of users and is perhaps the most heterogeneous and the largest publicly available social video archive. As compared to the videos in specific domains, the social video archive contains more diverse content, richer metadata such as user-generated tags, ratings, categorizations and so on, as well as social information. It brings new opportunities and challenges in video modeling and retrieval.

References

1. A. W. M. Smeulders, M. Worring, S. Santini, A. Gupta, and R. Jain. Content-based image retrieval at the end of the early years. *IEEE Transactions on Pattern Analysis and Machine Intelligence*, 22, 1349–1380, 2000.
2. C. J. C. Burges. A tutorial on support vector machines for pattern recognition. *Data Mining and Knowledge Discovery*, 2, 121–167, 1998.

3. J. Yang, R. Yan, and A. G. Hauptmann. Cross-domain video concept detection using adaptive svms. In *Proceedings of the ACM International Conference on Multimedia*, pp. 188–197, Augsburg, Germany, 2007.

4. L. X. Duan, I. W. Tsang, D. Xu, and S. J. Maybank. Domain transfer SVM for video concept detection. In *Proceedings of the IEEE International Conference on Computer Vision and Pattern Recognition*, pp. 1375–1381, Florida, USA, 2009.

5. A. Hanjalic. Shot-boundary detection: unraveled and resolved. *IEEE Transactions on Circuits and Systems for Video Technology* 12, 90–105, 2002.

6. C. Cotsaces, N. Nikolaidis, and I. Pitas. Video shot detection and condensed representation. a review. *IEEE Singnal Processing Magazine*, 23, 28–37, 2006.

7. P. Aigrain, H.-J. Zhang, and D. Petkovic. Content-based representation and retrieval of visual media: A state-of-the-art review, *Multimedia Tools and Applications*, 3(3), 179–202, 1996.

8. J. Yang and A. G. Hauptmann. (Un)Reliability of video concept detection. In *Proceedings of the ACM International Conference on Image and Video Retrieval*, pp. 85–94, Niagara Falls, Canada, 2008.

9. G. Davenport, T. G. A. Smith, and N. Pincever. Cinematic principles for multimedia. *IEEE Computer Graphics and Applications* 11, 67–74, 1991.

10. M. Belkin, P. Niyogi, and V. Sindhwani. Manifold regularization: a geometric framework for learning from labeled and unlabeled examples. *Journal of Machine Learning Research*, 7, 2399–2434, 2006.

11. T. Joachims. Transductive Inference for text classification using support vector machines. In *Proceedings of the International Conference on Machine Learning*, pp. 200–209, Bled, Slovenia, 1999.

12. V. N. Vapnik. *The Nature of Statistical Learning Theory*. 2nd edition. New York, USA: Springer-Verlag, 2000.

13. C. G. M. Snoek and M. Worring. Concept-based video retrieval. *Foundations and Trends in Information Retrieval*, 2(4), 218–322, 2009.

14. C.-Y. Lin, B. L. Tseng, M. R. Naphade, A. Natsev, and J. R. Smith. MPEG-7 video automatic labeling system. In *Proceedings of the ACM International Conference on Multimedia*, pp. 98–99, California, USA, 2003.

15. O. Chapelle, B. Schölkopf, and A. Zien. *Semi-Supervised Learning*. MIT Press, Cambridge, MA, 2010.

16. D. Zhou, O. Bousquet, T.N. Lal, J. Weston, and B. Schölkopf. Learning with local and global consistency. *Advances in Neural Information Processing Systems* 16, 321–328, 2004.

17. Z.-J. Zha, X.-S. Hua, T. Mei, G.-J. Qi, and Z. Wang. Joint multi-label multi-instance learning for image classification. In *Proceedings of the Internation Conference on Computer Vision and Pattern Recognition*, pp. 1–8, Anchorage, USA, 2008.

18. Z.-J. Zha, T. Mei, Z. Wang, and X.-S. Hua. Building a comprehensive video ontology to refine video concept detection. In *Proceedings of the International Conference on Multimedia Information Retrieval*, pp. 227–236, Augsburg, Germany, 2007.

19. G. Qi, X.-S. Hua, Y. Rui, J. Tang, T. Mei, and H.-J. Zhang. Correlative multilabel video annotation. In *Proceedings of the ACM International Conference on Multimedia*, pp. 17–26, Augsburg, Germany, 2007.

20. Z.-J. Zha, T. Mei, J. Wang, Z. Wang, and X.-S. Hua. Graph-based semi-supervised learning with multiple labels. *Journal of Visual Communication and Image Representation*, 20, 97–103, 2009.

21. Y. Wu, B. L. Tseng, and J. R. Smith. Ontology-based multi-classification learning for video concept detection. In *Proceedings of the IEEE International Conference on Multimedia and Expo*, pp. 1003–1006, Taibei, Taiwan, 2004.

22. J. R. Smith and M. Naphade. Multimedia semantic indexing using model vectors. In *Proceedings of the IEEE International Conference on Multimedia and Expo*, pp. 445–448, Maryland, USA, 2003.

23. A. Hauptmann, M.-Y. Chen, and M. Christel. Confounded expectations: Informedia at TRECVID 2004. In *Proceedings of TREC Video Retrieval Evaluation Online*, 2004. http://www-nlpir.nist.gov/projects/tvpubs/tv.pubs.org.html.

24. W. Jiang, S.-F. Chang, and A. C. Loui. Context-based concept fusion with boosted conditional random fields. In *Proceedings of the IEEE International Conference on Acoustics, Speech and Signal Processing*, pp. 949–952, Hawaii, USA, 2007.

25. J. Wang, Y. Zhao, X. Wu, and X.-S. Hua. Transductive multi-label learning for video concept detection. In *Proceedings of the IEEE International Conference on Multimedia Information Retrieval*, pp. 298–304, Vancouver, Canada, 2008.
26. Y. Li, Y. Tian, L.-Y. Duan, J. Yang, T. Huang, and W. Gao. Sequence multi-labeling: A unified video annotation scheme with spatial and temporal context. *IEEE Transactions on Multimedia* 12, 814–828, 2010.
27. C. Rosenberg, M. Hebert, and H. Schneiderman. Semi-supervised self-training of object detection models. In *Proceedings of the IEEE Workshops Application of Computer Vision*, pp. 25–36, Colorado, USA, 2005.
28. A. Blum and T.M. Mitchell. Combining labeled and unlabeled Data with co-training. In *Proceedings of the IEEE International Conference on Learning Theory*, pp. 92–100, Wisconsin, USA, 1998.
29. X. Zhu, Z. Ghahramani, and J. Lafferty. Semi-supervised learning using gaussian fields and harmonic functions. In *Proceedings of the IEEE International Conference on Machine Learning*, pp. 912–919, Washington, DC, 2003.
30. J. Tang, G.-J. Qi, M. Wang, and X.-S. Hua. Video semantic analysis based on structure-sensitive anisotropic manifold ranking. *IEEE Signal Processing* 89, 2313–2323, 2009
31. M. Wang, X.-S. Hua, J. Tang and R. Hong. Beyond distance measurement: Constructing neighborhood similarity for video annotation. *IEEE Transactions on Multimedia* 11, 465–476, 2009.
32. X. Li, D. Wang, J. Li, and B. Zhang. Video search in concept subspace: A text-like paradigm. In *Proceedings of the ACM International Conference on Image and Video Retrieval*, pp. 603–610, Amsterdam, the Netherlands, 2007.
33. S.-F. Chang, W. Hsu, W. Jiang, L. S. Kennedy, D. Xu, A. Yanagawa, and E. Zavesky. Columbia University TRECVID-2006 video search and high-level feature extraction. In *Proceedings of TREC Video Retrieval Evaluation Online*, http://www-nlpir.nist.gov/projects/tvpubs/tv.pubs.org.html.
34. C. G. M. Snoek, B. Huurnink, L. Hollink, M. de Rijke, G. Schreiber, and M. Worring. Adding semantics to detectors for video retrieval. *IEEE Transactions on Multimedia* 9, 975–986, 2007.
35. R. Yan and A. G. Hauptmann. The combination limit in multimedia retrieval. In *Proceedings of the ACM International Conference on Multimedia*, pp. 339–342, California, USA, 2003.
36. S.-Y. Neo, J. Zhao, M.-Y. Kan, and T.-S. Chua. Video retrieval using high level features: Exploiting query matching and confidence-based weighting. In *Proceedings of the ACM International Conference on Image and Video Retrieval*, pp. 143–152, Arizona, USA, 2006.
37. A.P. Natsev, A. Haubold, J. Tešić, L. Xie, and R. Yan. Semantic concept-based query expansion and re-ranking for multimedia retrieval. In *Proceedings of the ACM International Conference on Multimedia*, pp. 991–1000, Augsburg, Germany, 2007.
38. S. Banerjee and T. Pedersen. Extended gloss overlaps as a measure of semantic relatedness. In *Proceedings of the International Joint Conference on Artificial Intelligence*, pp. 805–810, Acapulco, Mexico, 2003.
39. D. Wang, Z. K. Wang, J. M. Li, B. Zhang, and X. R. Li. Query representation by structured concept threads with application to interactive video retrieval. *Journal of Visual Communication and Image Representation* 20, 104–116, 2009.
40. R.A. Baeza-Yates and B.A. Ribeiro-Neto. *Modern Information Retrieval*. ACM Press/Addison-Wesley, England, 1999.
41. P. Resnik. Using information content to evaluate semantic similarity in a taxonomy. In *Proceedings of the International Joint Conference on Artificial Intelligence*, pp. 448–453, Quebec, Canada, 1995.
42. H.-J. Zhang, S. Y. Tan, S. W. Smoliar, and Y. Gong. Automatic parsing and indexing of news video. *Multimedia Systems*, 2, 256–266, 1995.
43. J. R. Smith and S.-F. Chang. Visually searching the web for content. *IEEE MultiMedia*, 4, 12–20, 1997.
44. J. R. Smith, M. R. Naphade, and A. P. Natsev. Multimedia semantic indexing using model vectors. In *Proceedings of the IEEE International Conference on Multimedia and Expo*, pp. 445–448, Maryland, USA, 2003.

45. J. Yuan, Z.-J. Zha, Z. Zhao, X. Zhou, and T.-S. Chua. Utilizing related samples to learn complex queries in interactive concept-based video search. In *Proceedings of the ACM International Conference on Image and Video Retrieval*, pp. 66–73, Xi'an, China, 2010.

46. X.-Y.Wei, C.-W. Ngo, and Y.-G. Jiang. Selection of concept detectors for video search by ontology-enriched semantic spaces. *IEEE Transactions on Multimedia* 10, 1085–1096, 2008.

47. Y.-G. Jiang, C.-W. Ngo, and S.-F. Chang. Semantic context transfer across heterogeneous sources for domain adaptive video search. In *Proceedings of the ACM International Conference on Multimedia*, pp. 155–164, Beijing, China, 2009.

48. D. Byrne, P. Wilkins, G. J.F. Jones, A. F. Smeaton, and N. E. O'Connor. Measuring the impact of temporal context on video retrieval. In *Proceedings of the International Conference on Content-Based Image and Video Retrieval*, pp. 299–308, Niagara Falls, Canada, 2008.

49. M. G. Christel, C. Huang, and N. Moraveji. Exploiting multiple modalities for interactive video retrieval. In *Proceedings of the IEEE International Conference on Acoustics, Speech, and Signal Processing*, pp. 1032–1035, Quebec, Canada, 2004.

50. X. S. Zhou and T. S. Huang. Relevance feedback in image retrieval: A comprehensive review. *Multimedia Systems* 8, 536–544, 2003.

51. O. de Rooij, C. G. M. Snoek, and M. Worring. Query on demand video browsing. In *Proceedings of the ACM International Conference on Multimedia*, pp. 811–814, Augsburg, Germany, 2007.

52. O. de Rooij, C. G.M. Snoek, and M. Worring. Balancing thread based navigation for targeted video search. In *Proceedings of the International Conference on Content-Based Image and Video Retrieval*, pp. 485–494, New York, Niagara Falls, Canada, 2008.

53. K. McDonald and A. F. Smeaton. A comparison of score, rank and probability-Based fusion methods for video shot retrieval. In *Proceedings of the ACM International Conference on Image and Video Retrieval*, pp. 61–70, Singapore, 2005.

54. O. de Rooij, C. G.M. Snoek, and M. Worring. Mediamill: fast and effective video search using the forkbrowser. In *Proceedings of the International Conference on Content-Based Image and Video Retrieval*, pp. 561–562, Niagara Falls, Canada, 2008.

55. G.-D. Guo, A. K. Jain, W.-Y. Ma, and H.-J. Zhang. Learning similarity measure for natural image retrieval with relevance feedback. *IEEE Transactions on Neural Networks* 13, 811–820, 2002.

56. A. G. Hauptmann, W.-H. Lin, R. Yan, J. Yang, and M.-Y. Chen. Extreme video retrieval: Joint maximization of human and computer performance. In *Proceedings of the ACM International Conference on Multimedia*, pp. 385–394, California, USA, 2006.

57. H.-B. Luan, S.-Y. Neo, H.-K. Goh, Y.-D. Zhang, S.-X. Lin, and T.-S. Chua. Segregated feedback with performance-based adaptive sampling for interactive news video retrieval. In *Proceedings of the ACM International Conference on Multimedia*, pp. 293–296, Augsburg, Germany, 2007.

58. H. Luan, Y. Zheng, S.-Y. Neo, Y. Zhang, S. Lin, and T.-S. Chua. Adaptive multiple feedback strategies for interactive video search. In *Proceedings of the International Conference on Content-Based Image and Video Retrieval*, pp. 457–464, Niagara Falls, Canada, 2008.

59. C. Nastar, M. Mitschke, and C. Meilhac. Efficient query refinement for image retrieval. In *Proceedings of the IEEE Conference on Computer Vision and Pattern Recognition*, pp. 547, Washington, DC, 1998.

60. S.-Y. Neo, H. Luan, Y. Zheng, H.-K. Goh, and T.-S. Chua. Visiongo: bridging users and multimedia video retrieval. In *Proceedings of the International Conference on Content-based Image and Video Retrieval*, pp. 559–560, Niagara Falls, Canada, 2008.

61. Rosalind W. Picard, Thomas P. Minka, and M. Szummer. Modeling user subjectivity in image libraries. In *Proceedings of the IEEE International Conference on Image Processing*, pp. 777–780, Lausanne, Switzerland, 1996.

62. P. H. Qi, Q. Tian, and T. S. Huang. Incorporate support vector machines to content-based image retrieval with relevant feedback. In *Proceedings of the IEEE International Conference on Image Processing*, pp. 750–753, Vancouver, Canada, 2000.

63. Y. Rui, T.S. Huang, M. Ortega, and S. Mehrotra. Relevance feedback: A power tool in interactive content-based image retrieval. *IEEE Transactions on Circuits and Systems for Video Technology*, 8, 644–655, 1998.

64. Y. Rui, T. S. Huang, and S. Mehrotra. Content-based image retrieval with relevance feedback in MARS. In *Proceedings of IEEE International Conference on Image Processing*, pp. 815–818, Washington, DC, USA, 1997.

65. Y. Rui, T. S. Huang, S. Mehrotra, and M. Ortega. A relevance feedback architecture for content-based multimedia information retrieval systems. In *Proceedings of the Workshop on Content-Based Access of Image and Video Libraries*, p. 82, Washington, DC, 1997.

66. G. Salton. *Automatic Information Organization and Retrieval*. McGraw-Hill Text, 1968.

67. C. G. M. Snoek, M. Worring, D. C. Koelma, and A. W. M. Smeulders. A learned lexicon-driven paradigm for interactive video retrieval. *IEEE Transaction on Multimedia*, 9, 280–292, 2007.

68. D. Tao and X. Tang. Random sampling based SVM for relevance feedback image retrieval. In *Proceedings of the IEEE Conference on Computer Vision and Pattern Recognition*, pp. 647–652, Washington, DC, 2004.

69. D. Tao, X. Tang, X. Li, and X. Wu. Asymmetric bagging and random subspace for support vector machines-based relevance feedback in image retrieval. *IEEE Transactions on Pattern Analysis and Machine Intelligence*, 28, 1088–1099, 2006.

70. C. J. van Rijsbergen. *Information Retrieval*. 2nd edition, London: Butterworths, 1979.

71. P. Wilkins, A. F. Smeaton, N. E. O'Connor, and D. Byrne. K-space interactive search. In *Proceedings of the International Conference on Content-based Image and Video Retrieval*, pp. 555–556, Niagara Falls, Canada, 2008.

72. R. Yan, A. Natsev, and M. Campbell. Multi-query interactive image and video retrieval: theory and practice. In *Proceedings of the International Conference on Content-based Image and Video Retrieval*, pp. 475–484, Niagara Falls, Canada, 2008.

73. D. A. Cohn, Z. Ghahramani, and M. I. Jordan. Active learning with statistical models. *Journal of Artificial Intelligence Research*, 4, 129–145, 1996.

74. N. Roy and A. McCallum. Toward optimal active learning through sampling estimation of error reduction. In *Proceedings of the International Conference on Machine Learning*, pp. 441–448, Williamstown, Massachusetts, 2001.

75. R. Yan, J. Yang, and A. Hauptmann. Automatically labeling video data using multi-class active learning. In *Proceedings of the International Conference on Computer Vision*, pp. 516–523, Nice, France, 2003.

76. S. Tong and E. Chang. Support vector machine active learning for image retrieval. In *Proceedings of the ACM International Conference on Multimedia*, pp. 107–118, Ottawa, Canada, 2001.

77. M. Naphade and J. R. Smith. Active learning for simultaneous annotation of multiple binary semantic concepts. In *Proceedings of the International Conference on Image Processing*, pp. 77–80, Singapore, 2004.

78. M.-Y. Chen, M. Christel, A. Hauptmann, and H. Wactlar. Putting active learning into multimedia applications: Dynamic definition and refinement of concept classifiers. In *Proceedings of the ACM International Conference on Multimedia*, pp. 902–911, Singapore, 2005.

79. G. P. Nguyen, M. Worring, and A. W. M. Smeulders. Interactive search by direct manipulation of dissimilarity space. *IEEE Transactions on Multimedia*, 9, 1404–1415, 2007.

80. H. T. Nguyen and A. Smeulders. Active learning using pre-clustering. In *Proceedings of the International Conference on Machine Learning*, pp. 79–85, Alberta, Canada, 2004.

81. C. K. Dagli, S. Rajaram, and T. S. Huang. Leveraging active learning for relevance feedback using an information theoretic diversity measure. In *Proceedings of the International Conference on Image and Video Retrieval*, pp. 123–132, Tempe, Arizona, 2006.

82. S. Ayache and G. Quenot. TRECVID 2007 collaborative annotation using active learning. In *Proceedings of TREC Video Retrieval Evaluation Online*, http://www-nlpir.nist.gov/projects/tvpubs/tv.pubs.org.html

83. S. Ayache and G. Quénot. Evaluation of active learning strategies for video indexing. *Image Communication*, 22, 692–704, 2007.

84. K. Brinker. Incorporating diversity in active learning with support vector machines. In *Proceedings of the International Conference on Machine Learning*, pp. 59–66, Washington, DC, 2003.

13

Image Retrieval

Lei Zhang and Wei-Ying Ma

CONTENTS

13.1 Introduction

Image retrieval, a technique for browsing, searching, and retrieving images from a large database of digital images, has been an active research area since the 1970s. Especially, the explosive growth of digital images due to the advances in the Internet and digital imaging devices greatly necessitates effective and efficient image retrieval techniques.

Basically, there are two image retrieval frameworks: text-based image retrieval and content-based image retrieval (CBIR). Text-based image retrieval can be traced back to the late 1970s. In traditional text-based image retrieval systems, images were first annotated with text and then searched using a text-based approach from traditional database management systems. Through text descriptions, images can be organized by semantic topics to facilitate easy navigation and browsing. However, since automatically generating descriptive texts for a wide spectrum of images is not feasible, most text-based image retrieval systems require manual annotation of images, which is a tedious and expensive task for large image databases, making the traditional text-based methods not scalable. Moreover, as manual annotations are usually subjective, imprecise, and incomplete, it is inevitably for the text-based methods to return inaccurate and mismatched results.

In the early 1990s, because of the dramatically increased number of images produced by various digital devices and commercial applications, the difficulties faced by text-based retrieval became increasingly more severe. To overcome these difficulties, CBIR was then introduced to index and retrieve images based on their visual content, such as color, texture, and shapes. This naturally avoids the problem of manual annotations which are usually subjective and incomplete. This new research direction has attracted researchers from the communities of computer vision, database management, human–computer interface, and information retrieval. As a result, many techniques of visual information extraction, browsing, user query and interaction, and indexing have been developed, and a large number of academic and commercial image retrieval systems have been built. For comprehensive surveys on CBIR, refer to [10,38,71].

Figure 13.1 shows a general framework of a CBIR system. For each database image, the system needs to extract visual features such as color, texture, and shape in an *offline* stage. After all the database images have been processed, the system will build an visual index, which is by design loaded into memory for efficient *online* retrieval. When a user inputs a query image, the system will extract the same visual features from the query image, search for its k-nearest neighbors in the visual feature space, and present the resulting k-nearest

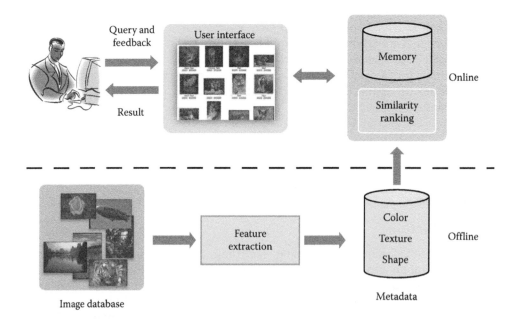

FIGURE 13.1
A general CBIR framework.

neighbors to the user. If the database contains only thousands of images, this framework generally works quite well. However, if there are millions of images in the database, the problem becomes quite challenging.

In its early years, researchers mainly focused on the problem of feature extraction, hoping to find the best visual features that can represent the semantic meaning of images. A remarkable number of different visual features such as color histogram, texture, and edge descriptors have been tried, and great progresses have been made. However, soon after a few years, researchers recognized that the retrieval accuracy is still far from satisfactory. This is because of the well-known semantic gap between low-level image features and high-level semantic concepts, which is also a fundamental problem in computer vision. To bridge the semantic gap, researchers resorted to users' continuous feedback in an iterative way. This direction is called *relevance feedback*. The fundamental idea is to show the user a list of candidate images, ask the user to indicate whether each image is relevant or irrelevant, and learn a good similarity measure to retrieve more relevant images. A comprehensive survey can be found in [107].

Although CBIR (including relevance feedback) has been extensively studied for more than one decade, the semantic gap between low-level image features and high-level semantic concepts is still the key hindrance in the effectiveness of CBIR systems. Meanwhile, since the year 2000, the advances in the Internet and digital imaging devices have greatly increased the number of images on the web. In 2001, Google launched its Image Search, offering access to 250 million images. Although the index is purely based on the keywords extracted from the surrounding text of web images, image search results are often surprisingly good. This is because web images usually have rich metadata, such as filename, URL, and surrounding text, which can be used as semantic descriptions for indexing and searching. In addition, different from the traditional text-based approach, manual annotation is not needed and thus the system is highly scalable. The major problem in keyword-based web image search engines is that the surrounding text of web images are generally noisy and redundant, which usually lead to irrelevant search results.

Motivated by the phenomenal success of web image search engines, researchers started to study the problem of *automatic image annotation*, aiming at automatically generating metadata in the form of captioning or keywords to digital images, and thus facilitating keyword-based image search, browsing, and organization. The problem is also related to object recognition, scene classification, or video concept detection in different research communities. The fundamental problem is to learn the mapping between visual features and textual keywords, and then apply the learned model to analyze and annotate the semantic concepts of new images. The study of this problem is also helpful to clean the noisy surrounding text or user-generated tags of web images, which is also called *image annotation refinement* in the literature. The mainstream of image annotation research is based on statistical approaches to model the relationship between visual features and semantic concepts. In contrast to model-based approaches, recent work shows data-driven approaches to be another promising direction. This is due to the explosive growth of multimedia data and large-scale image datasets readily available on the web. By treating web as a huge repository of weakly labeled images, data-driven approaches complement model-based approaches from a different angle and make image annotation more practical for large-scale image retrieval systems.

In addition to image content analysis, efficient visual index is another key technique to make the CBIR scalable to large-scale image data collections, so as to effectively utilize the huge amount of web images and build large-scale multimedia applications. The research in early years is mainly contributed by the database community, but the advances of this problem in recent years have also attracted more researchers from the communities of computer vision, machine learning, and information retrieval. The capability of utilizing large-scale multimedia data has become a critical factor for image retrieval, analysis, organization, and web search.

In the following sections, we introduce the fundamental techniques for CBIR, including feature extraction, relevance feedback, image annotation, and large-scale visual index. Thereafter, we conclude this chapter and discuss the latest advances in web image search.

13.2   Visual Feature Extraction

Visual feature extraction is the basis of CBIR. The research of this problem is mainly based on the fundamental techniques in computer vision and image processing. Generally speaking, image content may include both semantic and visual content [15]. Semantic content can be obtained by either textual annotation or complex inference procedures based on visual content. Since text-based feature extraction and search have been well studied in the information retrieval community and are quite mature, this section will not touch text-based feature extraction and later sections will discuss automatic image annotation based on visual content. Visual content can be very general (color, texture, shape) or domain specific (human faces, finger prints). Domain-specific feature extraction is application dependent and requires much domain knowledge which is beyond the scope of this chapter. This section concentrates on general visual feature extraction techniques which can be used in a wide range of multimedia applications.

13.2.1   Color

Color is the most widely used visual feature in CBIR. Most color features are based on the statistics of color distribution in an image. Thus, color features are generally quite robust to translation and rotation about the viewing axis and change only slowly with the scale, occlusion, and viewing angle.

Color features are extracted in a certain color space. A color space consists of a color model along with a specific mapping of that model onto an absolute color space. Each pixel of the image can be represented as a point in a three-dimensional (3D) color space. Commonly used color space for image retrieval include *RGB, Munsell, CIE L*a*b*, CIE L*u*v*, HSV* (hue, saturation, and value) (or *HSL* (hue, saturation, and lightness), *HSB* (hue, saturation, and brightness)), and *opponent color* space. Quite a number of studies have been done on color perception and color spaces. Readers interested in learning more can refer to [52,59,93]. In the following sections, we introduce some commonly used color features: color moment, color histogram, and color correlogram.

13.2.1.1   Color Moments

Color moments were first introduced by Stricker and Orengo [80] to describe color distributions of images and compare the similarity of two images. It has been proven in theory that the color distribution, as a probability distribution, can be uniquely characterized by its moments, especially the *first-order* (*mean*), the *second-order* (*variance*), and the *third-order* (*skewness*) color moments. Mathematically, the first three moments are defined as

$$\mu_i = \frac{1}{N} \sum_{j=1}^{N} f_{ij}, \tag{13.1}$$

$$\sigma_i = \left(\frac{1}{N} \sum_{j=1}^{N} (f_{ij} - \mu_i)^2 \right)^{\frac{1}{2}}, \tag{13.2}$$

$$s_i = \left(\frac{1}{N} \sum_{j=1}^{N} (f_{ij} - \mu_i)^3 \right)^{\frac{1}{3}}, \tag{13.3}$$

where f_{ij} is the value of the ith color component of the image pixel j, and N is the number of pixels in the image.

Usually, the color moments defined on $L^*u^*v^*$ and $L^*a^*b^*$ color spaces perform better than those on HSV space [15]. In practice, the first two order moments are more widely used because they are more robust to scene changes than using the first three order moments. Since only six (two moments for each of the three color channels) numbers are used to represent the color content of each image, the discrimination power of color moments is limited due to its compactness, making it more suitable to be used as the first pass to narrow down the search space before other sophisticated features are used or combined with other features to achieve a better retrieval performance.

13.2.1.2 Color Histogram

Color histogram is defined as the distribution of the number of pixels for each quantized bin in a color space. It is the most commonly used color feature representation in image retrieval, because it is easy to compute and effective in characterizing both the global and local distributions of colors in an image.

In 1991, Swain and Ballard proposed the histogram intersection, an L_1 metric, as similarity measure for the color histogram [81]. To improve the matching robustness, Ioka [26] and Niblack et al. [61] introduced an L_2-related metric in comparing two histograms, and Stricker and Orengo proposed using the cumulated color histogram [80].

One problem in computing the color histogram is to determine the color quantization scheme. Clearly, the more bins a color histogram contains, the more discrimination power it has. However, a very fine bin quantization does not necessarily improve the retrieval performance in many applications because a histogram with a large number of bins is usually sparse and sensitive to noise. A simple quantization scheme [32] is to divide the RGB color space into 64 bins ($4*4*4 = 64$), each of the three channels being quantized to four uniform bins. A more complex quantization scheme [79] is to divide the HSV color space into 166 bins ($18*3*3+4 = 166$), with the *hue* channel being quantized to 18 bins, the *saturation* and *value* (*lightness*) channels each being quantized to three bins, and four bins reserved for gray colors. To overcome the problem of inconsistency between human-perceived color distance and computer-calculated color distance, Zhang et al. proposed a nonuniform quantization scheme to divide the HSV color space into 36 bins [37] and obtained improved retrieval performance over the conventional color histogram approach.

13.2.1.3   Color Correlogram

The color correlogram was proposed to characterize not only the color distributions of pixels, but also the spatial correlation of pairs of colors [25]. A color correlogram of an image is a table indexed by color pairs, where the kth entry for (i, j) specifies the probability of finding a pixel of color j at a distance k from a pixel of color i in the image. Let $\mathcal{I}$ represent the entire set of image pixels and $\mathcal{I}_{c_i}$ represent the set of pixels whose colors are c_i. Then, the color correlogram is defined as

$$\gamma_{c_i,c_j}^{(k)}(\mathcal{I}) \triangleq \Pr_{p_1 \in I_{c_i}, p_2 \in I} [p_2 \in \mathcal{I}_{c_j} \mid |p_1 - p_2| = k] \tag{13.4}$$

where $i, j \in \{1, 2, \ldots, N\}$, $k \in \{1, 2, \ldots, d\}$, and $|p_1 - p_2|$ is the distance between pixels p_1 and p_2. If we consider all the possible combinations of color pairs, the size of the color correlogram will be very large ($O(N^2)d$)). Therefore a simplified version called the *color autocorrelogram* is often used to capture spatial correlation between identical colors and reduce the dimension to $O(Nd)$: $\alpha_c^{(k)}(\mathcal{I}) \triangleq \gamma_{c,c}^{(k)}(\mathcal{I})$. Or an even more compact version called the *banded autocorrelogram* [24] is used to further reduce the dimension to $O(N)$: $\alpha_c(\mathcal{I}) \triangleq \sum_{k \in \mathcal{K}} \gamma_{c,c}^{(k)}(\mathcal{I})$, where $\mathcal{K}$ is a specified set of local pair distances.

Comparing to the color histogram, the color autocorrelogram provides the best retrieval result because it captures the spatial correlation of local pairs of colors in an image.

13.2.2   Texture

Texture is one of the most important characteristics used in identifying objects or regions of interests in an image [21]. It is an innate property of virtually all surfaces—the grain of woods, the weave of a fabric, the pattern of crops in a field, and so on. Because of its importance and usefulness in pattern recognition and computer vision, there are rich research results in the past decades. Basically, texture representation methods can be classified into two categories: *structural* and *statistical* [15]. Structural texture representation includes *morphological operator* and *adjacency graph*, describing texture by identifying structural primitives and their placement rules. Statistical texture representation includes *Fourier power spectra, co-occurrence matrices, shift-invariant principal component analysis, Tamura feature, Wold decomposition, Markov random field, fractal model*, and *multiresolution filtering* techniques such as *Gabor and wavelete transform*, characterizing texture by the statistical distribution of the image intensity.

Early studies of texture feature were based on the structural methods, attempting to detect primitive patterns that construct a texture. Since 1970, statistical methods based on the gray-level spatial dependence of texture were widely adopted. In 1973, Haralick et al. proposed the co-occurrence matrix representation of texture features [21]. It first constructed a co-occurrence matrix based on the orientation and distance between image pixels and then extracted meaningful statistics from the matrix as the texture representation.

To make the texture representation meaningful to human visual perception, Tamura et al. developed computational approximations to the visual texture properties found to be important in psychology studies [82]. The six visual texture properties were *coarseness, contrast, directionality, linelikeness, regularity,* and *roughness*. These texture properties are visually meaningful and make the Tamura texture representation very attractive in image retrieval, as it can provide a more user-friendly interface.

Other representative texture features also include Gabor filter-based and wavelet transformation-based features [12,50]. By applying a Gabor filter or wavelet transformation on a texture region, such approaches extract statistical information (e.g., *mean* and *variance*) from a subband filtered by a certain scale and orientation.

Texture representation and feature extraction have been extensively investigated in pattern recognition and computer vision. For comprehensive reviews refer to [63,67,68,98].

13.2.3 Shape

Shape (or contour) plays an important role in human visual perception, and has been widely used as a basic representation for a variety of computer vision tasks [3,76] and CBIK systems [18,20,27,83].

A good shape representation feature for an object should be invariant to translation, rotation, and scaling. The state-of-the-art methods for shape description can be categorized into either boundary-based or region-based methods. The former uses only the outer boundary of the shape while the latter uses the entire shape region [74]. The most successful representatives for these two categories are Fourier descriptor and moment invariants. Fourier descriptor uses the Fourier transformed boundary as the shape feature, whereas moment invariants use region-based moments which are invariant to transformations. For more comprehensive surveys in shape representation, see [53,103].

Compared with color and texture features, shape features are usually described after images have been segmented into regions or objects. Since robust and accurate image segmentation is difficult to achieve, the use of shape features for image retrieval has been limited to special applications where objects or regions are readily available [15]. As a core research problem in computer vision, searching for images to match with a hand-drawn sketch query has become a highly desired feature, especially due to the explosive growth of web images and the popularity of devices with touch screens. Cao et al. [4,5] demonstrated a prototype system called MindFinder for sketch-based image search, enabling many useful applications, such as enhancing traditional keyword-based image search, and enlightening children/designers' drawing. The system shows a promising research direction of shape-based image representation and retrieval.

13.2.4 Local Features

Local features, which are distinctive and invariant to many kinds of geometric and photometric transformations, are widely utilized in a large number of applications, for example object categorization, image retrieval, robust matching, and robot localization.

Local feature detection is the requisite step in obtaining local feature descriptions. It tries to detect salient patches, such as edges, corners, and blobs in an image. These salient patches are considered more important than other patches, such as the regions attracting human attention, which might be more useful for object categorization. Feature detectors can be traced back to Moravec's corner detector [60] in 1981, which looks for the local maximum of minimum intensity changes. The problem of this detector is that its response is anisotropic, noisy, and sensitive to edges. To overcome this problem, the Harris corner detector [22] was developed in 1988. However, it fails to deal with scale changes, which are quite common in images. To address this problem, in 1999, Lowe developed a scale-invariant local feature, namely the scale-invariant feature transform (SIFT) [49]. It consists of a detector and a descriptor. The SIFT detector finds the local maximums of a series of difference of Gaussian images, and the SIFT descriptor builds a histogram for gradient magnitudes and orientations representation. As shown by its name, the SIFT detector is scale and translation invariant but not affine invariant, and the SIFT descriptor achieves a certain property of rotation invariant by normalizing the orientation histogram. Despite such shortcomings, SIFT has shown robust matching across a substantial range of affine distortion, change in 3D viewpoint, addition of noise, and change in illumination. As a result, the SIFT feature is widely used in real applications.

To obtain affine covariant features, several detectors were developed. These improvements include: the "Harris-Affine" detector [54] and the "Hessian-Affine" detector [54] developed by Mikolajczyk and Schmid, an edge-based region detector [88] and an intensity-based region detector [88] developed by Tuytelaars and Van Gool, the "maximally stable extremal region" detector [51] developed by Matas et al., and an entropy-based region detector [31] developed by Kadir et al. We will not introduce these detectors in detail. Interested readers, see [43,55,56,89] for comprehensive reviews and comparisons of these detectors.

To represent points and regions, a large number of different local descriptors have been developed. These descriptors could be computed based on *local derivatives, steerable filters, Gabor filters, wavelet filters, complex filters, Textons*, and *shape context* of the target local region. Among these descriptors, the SIFT descriptor has been proven robust to changes in partial illumination, background clutter, occlusion, and transformations in terms of rotation and scaling. A SIFT keypoint descriptor is created by first computing the gradient magnitude and orientation at each image sample point in a region around the keypoint location. These samples are then accumulated into orientation histograms (eight bins) summarizing the contents over 4×4 subregions, resulting in a 128-dimensional feature vector. An extension to SIFT is the gradient location and orientation histogram, proposed by Mikolajczyk and Schmid [55]. It computes the SIFT descriptor for a log-polar location grid with three bins in radial direction and eight in angular direction, which results in 17 location bins. Note that the central bin is not divided in angular directions. The gradient orientations are quantized in 16 bins. This gives a 272-bin histogram. The size of this descriptor is further reduced to 128 with Principal Component Analysis.

Typically, there are two different ways of utilizing local features in applications [43]: (1) traditional utilization, which involves the following three steps: feature detection, feature description, and feature matching; (2) bag-of-features, which includes the following four steps: feature detection, feature description, feature clustering, and frequency histogram construction for image representation. As bag-of-features can be treated as a document-like representation, the proven effective indexing and ranking schemes developed for web search engines can be used to handle the scale. This has become a common framework for large-scale local feature-based visual search. As by design local features are invariant with viewpoint change, corresponding regions in two images will have similar (ideally identical) vector descriptors [56]. The benefits are that correspondences can be easily established and, since there are multiple regions, the method is robust to partial occlusions.

13.3 Relevance Feedback

As discussed in Section 13.1, CBIR in its early years (roughly speaking, 1990–1998) mainly focused on the problem of feature extraction, attempting to find the best visual features that can represent the semantic meaning of images. However, the retrieval accuracy of today's CBIR algorithms is still limited. The fundamental problem is the gap between low-level image features and semantic concepts of images. This problem arises because visual similarity measures, such as color histograms, do not necessarily match the semantics of images and human subjectivity [15]. Human perception of image similarity is subjective and task dependent. That is, people often have different semantic interpretations of the same image. Furthermore, the same person may perceive the same image differently at different times. In addition, each type of visual feature tends to capture only one aspect of the image property and it is usually hard for a user to specify clearly how different aspects are combined to form an optimal query.

To bridge the semantic gap, relevance feedback techniques have been proposed to incorporate human perception subjectivity into the retrieval process, providing users with the opportunity to evaluate the retrieval results.

Early approaches assume the existence of an ideal query point that, if found, would provide the appropriate answer to the user. These approaches belong to the family of query point movement (QPM) methods, for which the task of the learner consists in finding, at every round, a better query point together with a reweighting of the individual dimensions of the description space.

Recent work often treats relevance feedback as a machine-learning problem in which positive and negative examples are used to learn a new similarity measure to refine the retrieval results. For example, support vector machine (SVM) can be used to learn a classifier to find more relevant images. With SVMs, the data are usually first mapped to a higher-dimensional feature space using a nonlinear transform kernel, and then the images are classified in this mapped feature space using maximum margin strategy [75,104].

13.3.1 General Assumptions

Before developing relevance feedback mechanisms for CBIR, it is necessary to understand some underlying general assumptions. According to [7], the assumptions are:

1. The discrimination between relevant and irrelevant images must be possible with the available image descriptors.
2. There is some relatively simple relation between the topology of the description space and the characteristic shared by the images the user is searching for.
3. Relevant images are a small part of the entire image database.
4. While part of the early work on relevance feedback assumed that the user could (and would be willing to) provide a rather rich feedback, including relevance notes for many images, the current assumption is that this feedback information is scarce: the user will only mark a few relevant images as positive and some very different images as negative.

Figure 13.2 shows the flowchart of a typical CBIR process with relevance feedback [48]. Based on the above general assumptions, a typical scenario for relevance feedback in CBIR is as follows [48,107]:

1. The system provides initial retrieval results through query-by-example, sketch, and so on.
2. User judges the returned results as to whether and to what degree, they are relevant (positive examples)/irrelevant (negative examples) to the query.
3. Machine learning algorithm is applied to learn the user's feedback. Then go back to (2).

Steps (2) and (3) are repeated till the user is satisfied with the results.

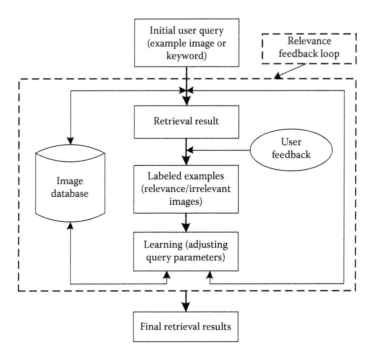

FIGURE 13.2
A typical flowchart of relevance feedback.

Step (3) is comparably the most important step and different approaches can be used to learn the new query. A few generally adopted approaches are introduced in the following.

13.3.2 QPM Approaches

A typical approach in step (3) is called QPM [48]. It improves the estimation of the query point by moving it toward the positive examples and away from the negative examples. A widely adopted query point removing technique is called Rocchio's formula [6]:

$$Q' = \alpha Q + \beta \left(\frac{1}{N_{R'}} \Sigma_{i \in D'_R} D_i \right) - \gamma \left(\frac{1}{N_{N'}} \Sigma_{i \in D'_N} D_i \right) \tag{13.5}$$

where Q and Q' are the original query and the updated query, respectively, and D'_R and D'_N are the sets of the positive and negative images labeled by the user, and $N_{R'}$ and $N_{N'}$ are the set sizes. α, β, and γ are weights. This technique was used in the Multimedia Analysis and Retrieval System (MARS) [73] to replace the document vector with visual feature vectors. Experiments show that retrieval performance can be improved by using these relevance feedback approaches.

13.3.3 Reweighting Approaches

Another approach is to adjust the weights of low-level features to accommodate the users' need, which frees the burden of specifying the feature weights from the user. This reweighting step dynamically updates the weights embedded in the query (not only the weights to different types of low-level features such as color, texture, and shape, but also the weights to different components in the same feature vector) to model the high-level concepts and perception subjectivity [70].

Huang et al. [25] proposed an algorithm called *learning the metric*. Consider the weighted metric defined as

$$D(x^{(1)}, x^{(2)}) = \sum_{i \in [N]} \omega_i \cdot |x_i^{(1)} - x_i^{(2)}| \tag{13.6}$$

For a positive image A_j^+, the feature components that contribute more similarity to the match are considered more important, while the feature components with smaller contribution are considered to be less important. Therefore, the weight for a feature component, ω_i, is updated as

$$\omega_i = \omega_i \cdot (1 + \bar{\delta} - \delta_i), \quad \delta = |f(Q) - f(A_j^+)| \tag{13.7}$$

where $\bar{\delta}$ is the mean of δ. On the other hand, for a negative image A_j^-, the feature components that contribute more to the match should be depressed. That is, the weight is updated as

$$\omega_i = \omega_i \cdot (1 - \bar{\delta} + \delta_i), \quad \delta = |f(Q) - f(A_j^-)| \tag{13.8}$$

If the distance metric is defined as a Mahalanobis distance (or generalized Euclidean distance) $D(x, y) = (x - y)^T M(x - y)$, the reweighting approach can be formulated as to learn the optimal covariance matrix M to minimize the mean distance between the relevant images under the learned distance metric. This idea is illustrated in Figure 13.3, where Figure 13.3a is the original feature space. In the MARS system [73], Rui et al. proposed the standard deviation method to learn the weights. This is equivalent to learn the Mahalanobis distance when the covariance matrix M is diagonal (i.e., Λ), as shown in Figure 13.3b. In the MindReader system, Ishikawa et al. proposed to learn the full covariance matrix, as shown in Figure 13.3c. However, as the number of positive examples specified by users are often much less than the feature dimension, the estimation of the covariance matrix is usually singular and unstable. To overcome this problem, Rui and Huang [72] proposed a hierarchical learning algorithm to support multiple features. The

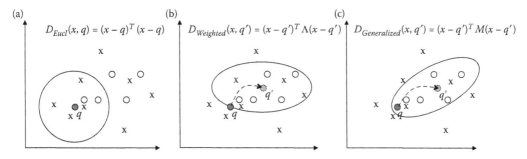

FIGURE 13.3
Three different two-dimensional (2D) distance metrics. The dark gray dot q denotes the initial query point, and the light gray dot q' denotes the learned optimal query point, which is estimated to be the center of all the positive examples. Circles and crosses are positive and negative examples. (a) Euclidean distance; (b) normalized Euclidean distance; and (c) Mahalanobis distance.

proposed method learns a diagonal matrix for each individual feature to avoid the problem of insufficient training data, and learns a full covariance matrix to combine multiple features.

Note that the QPM approaches and the reweighting approaches generally complement each other and can be used together. As shown in Figure 13.3b and c, one can first move the query point q to the center q' of the positive examples, and then apply the reweighting approach at the new query point q'.

13.3.4 Machine Learning Approaches

The key issue in relevance feedback is how to use positive and negative examples to adjust the similarity measure. This is essentially a machine learning problem which targets at predicting unseen data based on the labeled training examples. And it is also a typical *small sample* (the number of training samples are usually small) and *imbalanced* (positive and negative database images are highly imbalanced) learning problem. Since 2000, a remarkable number of different machine learning algorithms have been applied to this problem. A few representative algorithms are introduced subsequently.

Vasconcelos and Lippman [90] treated feature distribution as a Gaussian mixture and used Bayesian inference for learning during feedback iterations in a query session. The richer information captured by the mixture model also made image regional matching possible. The potential problems with this method are computing efficiency and a complex data model that requires too many parameters to be estimated from very limited samples. To handle the small sample problem, Tian et al. [85] formulated image retrieval as a transductive learning problem by combining unlabeled images in supervised learning to achieve better classification, which boosts the classifier learnt from the limited labeled training data.

In [104], SVM was utilized to capture the query concept by applying first the kernel trick to project images onto a hyperspace and then separate the relevant images from irrelevant ones using maximum margin strategy. The advantages of adopting SVM include the following: (1) it has high generalization ability and (2) it works for small training sets. Another step-forward approach is proposed by Tong and Chang [86] called SVM active learning, which selects and presents the most ambiguous samples for users to label, and learns the query concept faster and with better accuracy.

To memorize the learned semantic information to improve the CBIR performance, Lee et al. [36] proposed to use a correlation network to accumulate semantic relevance between image clusters learned from the user's feedback. Through the selection of positive and negative examples based on a given query, the semantic relationships between images are captured and embedded into the system by splitting/merging image clusters and updating the correlation matrix. Image retrieval is then based on the resulting image clusters and the correlation matrix obtained through relevance feedback.

Relevance feedback can also be treated as a manifold propagation problem. He et al. [23] proposed a transductive learning framework named manifold ranking-based image retrieval. Given a query image, it first makes use of a manifold ranking algorithm to explore the relationship among all the data points in the feature space, and then measures relevance between the query and all the images in the database accordingly, which is different from traditional similarity metrics based on pairwise distance.

Relevance feedback has been a very active research topic in CBIR. Almost all the classical algorithms for pattern recognition, statistical learning, and manifold mining have been applied to this problem. For comprehensive surveys, please refer to [15,48,84,108].

13.4 Image Annotation

The research of image annotation was started since the early 2000s. Although there have been great progresses in feature extraction and relevance feedback, it still remains unclear as to how to model the semantic concepts effectively, and as a result the retrieval performance is still not satisfactory. Meanwhile, the phenomenal success of web image search engines has demonstrated a great potential of text-based image retrieval, and thus motivated the study of the problem of *automatic image annotation*, aiming at automatically generating keywords to images, and thus facilitating keyword-based image search, browsing, and organization.

Since 2002 [13], image annotation has become a very active research topic in CBIR. The fundamental problem is to learn the mapping between visual features and textual keywords, and then apply the learned model to analyze and annotate the semantic concepts of new images. The study of this problem is also helpful to clean the noisy surrounding text or user-generated tags of web images, which is also called *image annotation refinement* in the literature.

13.4.1 Model-Based Approaches

Model-based approaches are dominant in early years when large-scale data are not as widely available as in recent years. These approaches generally utilize probabilistic models or classification algorithms to model visual concepts by taking into account the object's appearance, spatial, and structure information. Approaches vary from modeling of a few number of specific concepts [17,39,109] to categorizing a considerable number of visual concepts [40], and from using histograms over the bag-of-words model [8,35] to leveraging visual language models [99]. Particularly, in the past several years, object detection, categorization, and recognition have been a very active research topic in the computer vision community. Such a trend can be measured by the increasing number of publications related to object recognition in Computer Vision and Pattern Recognition and International Conference on Computer Vision in recent years. To encourage researchers to compare their algorithms on a benchmark dataset, since 2005, visual object class (VOC) challenge was designed to provide a benchmark image database every year by enhancing some of the dimensions of variability of visual objects in the database [14]. Through the delicately designed competition tasks, VOC challenge greatly helps advance research in this area.

Comparably, in the multimedia community, image annotation has been receiving increasingly more attentions since 2002 [13]. Typical researches in image annotation attempt to model the joint probability between low-level image features and high-level semantic keywords [2,13,16,29], and then use the learned joint probability to generate keywords for unseen images. Although the image annotation problem usually targets at a larger image database and a larger vocabulary, the boundary between image annotation and object recognition, however, is not that clear. This is because there are many common algorithms and challenges in both problems.

A few representative image annotation approaches are introduced in the following.

Duygulu et al. [13] proposed an approach that treats image annotation as a machine translation problem. A statistical machine translation model was used to "translate" textual keywords to visual keywords, that is image blob tokens obtained by clustering.

Another way of capturing co-occurrence information is to introduce latent variables to model hidden concepts in images. Barnard et al. [2] extended the latent dirichlet allocation (LDA) model to the mix of words and images and proposed a correlation LDA model. This model assumes that there is a hidden layer of topics, which are a set of latent factors and obey the Dirichlet distribution, and words and regions are conditionally independent of the topics, that is, generated by the topics. This work used 7000 Corel photos and a vocabulary of 168 words for annotation.

Inspired by the relevance language models, several relevance models have been proposed. Jeon et al. [29] introduced the cross-media relevance model (CMRM) to model the joint distribution of words and discrete image features. This model was subsequently improved by continuous relevance model (CRM) [28,34], and multiple Bernoulli relevance model (MBRM) [16]. CRM directly associates continuous features with words and MBRM utilizes a multiple Bernoulli model to estimate the word probabilities. The joint distribution can be computed as the expectation over training images, and the annotations for untagged images are obtained by maximizing the expectation as

$$
\begin{aligned}
w^* &= \underset{w \in V}{argmax} P(w|I_u) \\
&= \underset{w \in V}{argmax} P(w, I_u) \\
&= \underset{w \in V}{argmax} \sum_{J \in \mathcal{T}} P(w, I_u) P(J),
\end{aligned}
\tag{13.9}
$$

where J is an image in the training set $\mathcal{T}$, w is a word or a set of words in the annotation set V, and I_u is an untagged image.

With the assumption that the probabilities of observing the word w and the image I_u are mutually independent given an image J, the model can be rewritten as

$$
w^* = \underset{w \in V}{argmax} \sum_{J \in \mathcal{T}} P(w|J) P(I_u|J) P(J)
\tag{13.10}
$$

Jeon and Manmatha [28] extended CRM and built it with 56,000 Yahoo! news images with noisy annotations and a vocabulary of 4073 words. This is the largest vocabulary ever proposed and they discussed noisy annotation filtering and speeding-up schemes.

As a different kind of approach, Pan et al. [64] constructed a two-layered graph whose nodes are images and their associated captions and proposed a random-walk-with-restart (RWR) algorithm to estimate the correlations between new images and the existing captions. Then, in another work [65], they extended the model to a three-layered graph with image regions added.

Li and Wang [41] proposed a 2D multiresolution hidden Markov model to couple images and concepts. They used 60,000 Corel photos with 600 concepts. In one of their recent works, they improved this model and built an interesting real-time annotation system named *Alipr* [42], which attracted a great deal of attention from both academia and industry.

All the aforementioned works require a supervised learning stage. Hence, the generalization capability is a crucial assessment to their effectiveness. Among them, the online demo of *Alipr* shows its robustness to outliers, although the model was trained on Corel images.

Due to space limitation, we are not able to list all of the previous works. Interested readers can refer to a survey [78] written by Smeulders et al. in 2000 and two recent surveys [10,11] after 2005 by Datta et al. for a comprehensive understanding of this area.

There is no doubt that significant progress has been made by model-based approaches in recent years. However, visual concept detection is an extremely difficult computational problem because each physical object can generate an infinite number of different 2D images as the object's position, pose, lighting, and background vary relative to the camera. Due to these challenges, the state-of-the-art research work is only capable of handling a limited number of visual concepts on a small-scale image dataset. How to scale up the number of concepts remains to be very challenging.

13.4.2 Data-Driven Approaches

In contrast to model-based approaches, recent work shows data-driven approaches to be another promising direction. This is due to the explosive growth of multimedia data and large-scale image datasets readily available on the web. By treating the web as a huge repository of weakly labeled images, data-driven approaches complement model-based approaches from a different angle and make image annotation more practical for large-scale image retrieval systems.

In [94], a novel search-based annotation framework was proposed to explore web-based resources. 2.4 million web images were crawled and a system AnnoSearch was built to index the images as a knowledge base for image annotation [14]. Image annotation was formulated as a two-step process: given an input image, first searching for a group of similar images in this large-scale web image database, and then mining key phrases extracted from the descriptions of the images.

The process of [94] is shown in Figure 13.4. It contains three stages: the text-based search stage, the content-based search stage, and the annotation learning stage, which are differentiated using different colors (black, brown, blue) and labels (A, B, C). When a user submits a query image together with a query keyword, the system first uses the keyword to search a large-scale web image database (2.4 million images crawled from several web photo forums), in which images are associated with meaningful but noisy descriptions, as labeled by "A" in Figure 13.4. This step intends to select a semantically relevant image subset from the original pool. Then, visual feature-based search is applied to filter visually similar images from the subset (the path labeled by "B" in Figure 13.4). In this way, a group

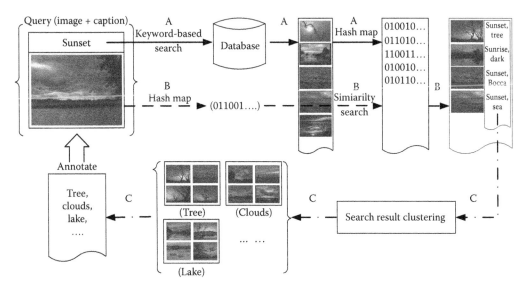

FIGURE 13.4
(**See color insert.**) The framework of search-based annotation.

of resulting images are obtained which are both semantically and visually similar to the query image. To speed up the visual feature-based search procedure, a hash coding algorithm is adopted to map the visual features into hash codes, by which inverted indexing technique can be applied for fast retrieval. Finally, based on the search results, the system collects their associated textual descriptions and applies the search result clustering algorithm [102] to group the images into several clusters. By ranking these clusters and setting a certain threshold, the system selects a group of clusters and merges their names as the final annotations for the query image, which ends the entire process (C in Figure 13.4).

The AnnoSearch system shown in Figure 13.4 requires an initial keyword as a seed to speed up the search by leveraging text-based search technologies. However, the initial keyword might not always be available in real applications. To overcome this problem, Li et al. [44] developed an efficient visual index to remove the dependency to initial keywords. Such a framework effectively leverages large-scale web images and does not need an explicit training stage. It is potentially capable of dealing with an unlimited vocabulary.

Motivated by billions of images freely available online, in 2008, Torralba et al. [87] built a large dataset of 80 million images collected from the web for nonparametric object and scene recognition. The images in the dataset are stored as tiny 32×32 pixel images. Given an input image, the system searches the 80 million image database to find its k-nearest neighbors and uses the associated labels to infer the semantic classes of the input image for object and scene recognition.

In 2010, Wang et al. [95] further extended the search-based image annotation framework to 2 billion web images. Based on simple yet effective near-duplicate detection, the system, *Arista*, is capable of automatically generating accurate tags for popular web images having near-duplicates in the database. The statistics in the 2 billion web image database shows that about 8.1% web images have more than 10 duplicates and this means about 160 million web images can be automatically tagged, which clearly demonstrates the great potential of data-driven approaches when billions of weakly labeled web images are utilized.

However, due to the lack of an efficient visual index, *Arista* only works for popular images which have duplicates in the database. How to use visually highly similar images rather than near-duplicates presents a new challenge to large-scale visual index.

Due to the scalability problem, the mathematical models used in data-driven approaches are usually very simple and it remains to be a problem as how to incorporate the latest developments in model-based approaches to further improve the performance of image annotation.

13.4.3 Image Annotation Refinement

Despite the continuous efforts on image annotation, results of existing image annotation methods are usually unsatisfactory. It would be advantageous if a dedicated approach could refine the inaccurate annotation result. This problem is called *image annotation refinement* [92].

Jin et al. [30] have done pioneering work on annotation refinement using a generic knowledge base WordNet [57]. From a small candidate annotation set obtained from any annotation method, irrelevant annotations can be pruned using WordNet [13]. The basic assumption is that highly correlated annotations should be reserved and noncorrelated annotations should be removed. The major problem is that the annotation refinement process is independent of the original image, which is inappropriate for images with relatively poor initial annotation results.

To overcome this problem, Wang et al. [92] developed an algorithm using RWRs to rerank the candidate annotations. The algorithm not only utilizes the corpus information by defining a co-occurrence-based similarity, but also leverages the ranking and confidence information of original annotations. Wang et al. [91] further improved this framework by taking into account the query image's visual content. A query-biased Markov chain with a query-biased transition matrix is dynamically constructed based on the query image, using both content feature of the query image and corpus information.

Image annotation refinement also applies to tag refinement for social images with user-created tags (like Flickr images). Since amateur tagging is known to be ambiguous, noisy, and personalized, a fundamental problem in social image retrieval is how to reliably learn the relevance of a tag with respect to the visual content it is describing. Li et al. [45] proposed a neighbor voting algorithm for learning tag relevance. The key idea is, by propagating common tags through visual links induced by visual similarity, each tag accumulates its relevance credit by receiving neighbor votes. Wu et al. [100] proposed a multimodality recommendation based on both tag and visual correlation, and formulated the tag recommendation as a learning problem. Each modality is used to generate a ranking feature, and Rankboost algorithm is applied to learn an optimal combination of these ranking features from different modalities. The learned tag relevance can be treated as tag frequency and embedded into current tag-based social image retrieval paradigms.

13.5 Large-Scale Visual Index

Utilizing web image datasets has become very important in both multimedia and computer vision for various multimedia content analysis problems. However, most existing systems suffer from a scalability problem and cannot scale to millions or billions of images because it is difficult to build an effective index for high-dimensional image features [46]. This challenge has motivated a substantial number of research attempts in large-scale image index and retrieval.

13.5.1 Index for Dense Features

Large-scale index for multimedia search is a classical research topic, which has been extensively studied since the beginning of the research on CBIR along with the contribution from the database management community. The major difficulty of image indexing is due to the high dimensionality of image feature vectors, which makes most index techniques developed for database management fail.

This section mainly concentrates on index techniques for dense image features such as color histogram and local feature descriptor. The dimensionality of such features is usually of 10s or a few 100s. Until today, the most widely used index techniques for conducting k-nearest–neighbor search for image retrieval include kd-tree [69] and locality-sensitive hashing (LSH) [9]. The former is partition based and the later is hash based. A kd-tree (short for k-dimensional tree) is a space-partitioning data structure for organizing points in a k-dimensional space. It is a binary tree in which every node is a k-dimensional point. Every nonleaf node can be thought of as implicitly generating a splitting hyperplane that divides the space into two parts, known as subspaces. According to Goodman et al. [19], kd-trees are not suitable for efficiently finding the nearest neighbor in high-dimensional spaces. As a general rule, if the dimensionality is k, then the number of points in the data, N, should be $N \gg 2^k$. Otherwise, when kd-trees are used with high-dimensional data, most of the points in the tree will be evaluated and the efficiency is no better than exhaustive search. Nevertheless, kd-trees are widely used for local feature quantization to speed up the look-up of the nearest cluster center. LSH [9] is a method of performing probabilistic dimension reduction of high-dimensional data. The key idea is to hash data points using several hash functions so as to ensure that, for each function, the probability of collision is much higher for objects which are close to each other than for those which are far apart. Then, one can determine nearest neighbors by hashing the query point and retrieving elements stored in buckets containing that point.

Most of other variants, for example, randomized tree [1], spectral hashing [97], and random projection [58], can be classified to these two categories: space partition based and hash based. Due to space limitation, we will not introduce them with more details. For comprehensive reviews, refer to [19,96] which have a good coverage on nearest–neighbor search techniques for high-dimensional space.

13.5.2 Index for Sparse Bag-of-Words Features

Motivated by the success of web search engines, many researchers have tried to map image retrieval problems to text retrieval problems, hoping that the proven effective indexing and ranking schemes can be used to handle the scale. The basic idea is to map image features to words [66,77]. Typically, images are first represented by local features, and then by clustering, each local feature is mapped to a discrete keyword. Such an image representation is called "bag of features," similar to "bag-of-words" for document representation. With this representation, comparing two images becomes matching words in them, and therefore, text-based search engine technologies can be utilized to reduce the computational and memory cost.

To develop this type of image search engines for a web-scale database, there are still many technical challenges and problems that we need to address [46]:

- *Vocabulary*: What kinds of image features should be used? How to map them to words? The most generally utilized method is clustering. Some researchers also adopted a hierarchical clustering method to generate a vocabulary tree. But it is clear that we need to develop some kinds of visual language models to solve the problem.

- *Long query*: The reason why the text search engine is effective is because text queries usually contain only a few words. So, the query-document matching can be conducted efficiently by an inverted index. Although images can be represented by "bag-of-features," the retrieval problem is still very different from text retrieval because query-by-example is actually equivalent to using a whole document as a query. So, the search is more like document-to-document matching. How can we deal with this kind of "long query" effectively?

- *Content quality*: Web search engine is effective because it can use link analysis to obtain quality and importance measurement (e.g., PageRank) for web pages. Based on PageRank, an efficient cache can be designed to select a small portion of high-quality web pages and keep them into memory. In most cases, top-k documents can be found in this cache and therefore there is no need to go through the disk index. This strategy significantly improves the index scalability. For images, it is hard to obtain a similar kind of measurement because the links are typically not directly associated with images. Without PageRank for images, it will lead to the lack of efficient cache of index, because we do not know how to select high-quality images and keep them into memory to speed up the search process, and therefore we will not be able to take advantage of many top-*k* search techniques typically used in web search.

- *Relevance ranking*: The similarity measure between two images is quite different from text. How are image words weighted in computing the relevance. And how to deal with "word proximity" in images?

- *Distributed computing for web-scale multimedia analysis*: Because of the large volume of image data we need to process and index, the system has to be a distributed system, consisting of hundreds of powerful servers. It is inevitably to confront with the challenges as in text-based search engines, such as fault tolerance, data redundant backup, auto configuration, and so on.

To be able to respond to a query within 1 s, the system has to employ a very efficient index solution, which is probably similar to inverted lists used in text-based search engines. The use of the index solution will make many algorithms depending on sequential scan of the whole database impractical. Therefore, most existing CBIR algorithms need to be reevaluated on a web-scale CBIR system.

To address these technical challenges, much work has been done in recent years, especially since 2006. For example, to generate a larger and more discriminatory *vocabulary*, Nistér et al. [62] developed an efficient scheme to quantize local region descriptors in a *hierarchical vocabulary tree* which leads to a dramatic improvement in retrieval quality. The most significant property of this scheme is that the tree directly defines the quantization. The quantization and the indexing are therefore fully integrated, essentially being one and the same. To generate a more *discriminative vocabulary*, Lian et al. [47] developed a probabilistic model, in which image category information is introduced for generating a dictionary. The obtained vocabulary is a trade-off between minimization of distortions of clusters and maximization of discriminative power of imagewise representations, that is, histogram representations of images. However, due to its high computational complexity, the model is more suitable for generating a small vocabulary for image classification, but not suitable for image retrieval which typically requires a larger vocabulary.

To take into account the spatial information, Zhang et al. [105] developed a scheme to construct a *contextual visual vocabulary*. It considers local features in groups and learns a discriminant distance metric between local feature groups. This group distance is further leveraged to induce visual vocabulary from groups of local features. By introducing spatial information to the vocabulary generation process, this work addresses both the *vocabulary* problem and the *relevance ranking* problem. To explicitly consider geometric relationships among visual words in *relevance ranking*, Wu et al. [101] proposed to bundle image features into local groups. Each group of bundled features becomes much more discriminative than a single feature, and within each group simple and robust geometric constraints can be efficiently enforced. Such algorithms are quite effective in finding partial duplicates from a large-scale image database. However, the restrict geometric constraints make them not suitable for finding similar images.

Another important issue is the *long query* problem. Inverted index structure is particularly efficient for short queries (e.g., 3–5 terms), whereas an image query represented by "bag-of-features" usually contains hundreds of visual terms. This difference makes the inverted index structure inappropriate to index images. For example, given a query with three terms, all images containing them can be obtained by intersecting three inverted lists. However, given a real image query which consists of 1000 visual terms, we have to intersect 1000 inverted lists and will very likely get an empty set. To address this problem, Zhang et al. [106] developed a new index technique to decompose a document-like representation of an image into two components, one for dimension reduction and the other for residual information preservation. The computing of similarity of two images can be transferred to measuring similarities of their components. The good properties of the two components make both index and retrieval very efficient.

In comparison, the problem of *content quality* is less touched. The problem is related to image quality assessment, which is mainly based on the content analysis. Due to the lack of effective feature representation, it is extremely hard to develop a quality measure to match with the human perception about image quality. A straightforward approach is to use the PageRank of a web image's hosting page to approximate the image's static rank. However, in many cases, PageRank does not necessarily reflect the image's static rank. There still remains a large improvement space for this problem.

Furthermore, due to the restrictions imposed by the indexing solution, the user interface for relevance feedback also needs to be restudied. For example, it is not trivial to refine the search result by QPM or distance function modification based on relevance feedbacks in an inverted index-like system. How to leverage users' feedback either explicitly or implicitly will be an interesting research problem in a web-scale content-based image search system. And it is not a trivial task to extract low-level features from every image in a database containing one billion images. We need a flexible platform and infrastructure to provide large-scale data management and data-processing capabilities. The infrastructure should facilitate the extraction and experimentation of various features and similarity measures for image search, so that it can help researchers and engineers to find the best practical solution by carefully evaluate the capabilities and limitations of different features and algorithms.

13.6   Conclusion

In this chapter, we have reviewed the brief history of image retrieval, and introduced a variety of existing techniques for image retrieval, including visual feature extraction, relevance feedback, automatic image annotation, and large-scale visual indexing. Our focus is mainly on CBIR, which has presented huge application opportunities and great technical challenges. Some challenges are fundamental to image understanding. For example, how to extract visual features that are close to semantic concepts and how to generate descriptive annotation keywords are fundamental problems to computer vision. Some challenges require research from multidisciplinary areas. For example, relevance feedback requires the understanding of both visual content and a user's search intent. How to engage users into the feedback loop and how to help users better express their search intents are interesting research topics for both human–computer interaction and information retrieval researchers. Large-scale visual index has become an indispensable technique to scale up an image retrieval system and effectively utilize the huge number of web images. It is clear that the recent advances in this research direction are contributed by multiple communities of computer vision, database, and web search.

The most challenging and fundamental problem in image retrieval is how to bridge the semantic gap between low-level visual features and high-level semantic concepts. The problem is to find more powerful features capable of describing the essential characteristics of semantic concepts. However, low-level visual features such as color histogram, wavelet texture feature, and local regional feature are usually not discriminative enough to differentiate different concepts or objects. A few recent research efforts have started to take into account contextual information to combine local features to increase the discriminative power of visual features. This is a promising direction to generate mid-level descriptors toward high-level semantic concepts.

Applying CBIR techniques to web image search presents a new challenge. Due to space limitation, we did not introduce web image retrieval in detail. Interested readers can refer [33] for a comprehensive survey. The World Wide Web contains a great quantity of image and other visual information such as video and audio. For example, it was reported[*] that as of April 2009, ImageShack[†] had 20 billion images, Facebook[‡] had about 15 billions, and Flickr[§] had over 3.4 billions. Tools and systems for effective retrieval of such information can prove very useful for many applications. The great paradox of the web is that the more information there is available about a given subject, the more difficult it is to locate accurate information. The progress in the past several years was mainly driven by industrial companies, including Google, Microsoft, Yahoo!, and so on. Different from the research of CBIR in academia, commercial image search engines are mainly keyword based. However, we have seen quite a few attempts of incorporating visual search to web image search engines. For example, in 2009, both Google and Microsoft Bing launched the feature of "show similar images" to let users search for visually similar images. The two image search engines also provide a few options to let users filter images based on the content of resulting images, such as face, photo clip art, or line drawing. In 2010, Google launched Google Goggles, a mobile application to recognize objects in an image and return relevant search results. In the same year, Microsoft Bing launched another feature to allow users perform interactive image search by specifying color layout.[¶] Other commercial search systems such as Xcavator,[‖] Like.com,[**] and Gazopa[††] also provide interesting CBIR features on web-scale image databases. Despite various challenges, all these successes show a very promising future of image retrieval.

References

1. Y. Amit and D. Geman. Shape quantization and recognition with randomized trees. *Neural Computation*, 9(7):1545–1588, 1997.
2. K. Barnard, P. Duygulu, D. A. Forsyth, N. de Freitas, D. M. Blei, and M. I. Jordan. Matching words and pictures. *Journal of Machine Learning Research*, 3:1107–1135, 2003.
3. S. Belongie, J. Malik, and J. Puzicha. Shape matching and object recognition using shape contexts. *IEEE Transactions on Pattern Analysis and Machine Intelligence*, 24:509–522, 2002.
4. Yang Cao, Changhu Wang, Liqing Zhang, and Lei Zhang. Edgel index for large-scale sketch-based image search. In *IEEE Conference on Computer Vision and Pattern Recognition (CVPR)*, San Francisco, USA, June 2011.

---

[*] http://techcrunch.com/2009/04/07/who-has-the-most-photos-of-them-all-hint-it-is-not-facebook/
[†] http://imageshack.us/
[‡] http://www.facebook.com
[§] http://www.flickr.com
[¶] http://research.microsoft.com/en-us/um/people/jingdw/searchbycolor/
[‖] http://www.xcavator.net
[**] http://www.like.com
[††] http://www.gazopa.com

5. Yang Cao, Hai Wang, Changhu Wang, Zhiwei Li, Liqing Zhang, and Lei Zhang. Mindfinder: Interactive sketch-based image search on millions of images (demo). In *ACM Multimedia*, Florence, Italy, October 2010.

6. Zhixiang Chen and Binhai Zhu. Some formal analysis of rocchio's similarity-based relevance feedback algorithm. *Information Retrieval*, 5:61–86, 2002.

7. M. Crucianu, M. Ferecatu, and N. Boujemaa. Relevance feedback for image retrieval: A short survey. In *State of the Art in Audiovisual Content-Based Retrieval, Information Universal Access and Interaction Including Datamodels and Languages (DELOS2 Report)*, 2004.

8. G. Csurka, C. R. Dance, L. Fan, J. Willamowski, and C. Bray. Visual categorization with bags of keypoints. In *Workshop on Statistical Learning in Computer Vision, ECCV*, pp. 1–22, Prague, Czech, May 2004.

9. M. Datar, N. Immorlica, P. Indyk, and V. S. Mirrokni. Locality-sensitive hashing scheme based on p-stable distributions. In *Symposium on Computational Geometry*, pp. 253–262, New York, USA, June 2004.

10. R. Datta, D. Joshi, J. Li, and J. Ze Wang. Image retrieval: Ideas, influences, and trends of the new age. *ACM Computing Surveys*, 40(2), 2008.

11. R. Datta, J. Li, and J. Ze Wang. Content-based image retrieval: Approaches and trends of the new age. In *Multimedia Information Retrieval*, pp. 253–262, Singapore, November 2005.

12. J. G. Daugman. Complete discrete 2-D gabor transforms by neural networks for image analysis and compression. *IEEE Transactions on Acoustics, Speech, and Signal Processing*, 36:1169–1179, 1988.

13. P. Duygulu, K. Barnard, J. F. G. de Freitas, and D. A. Forsyth. Object recognition as machine translation: Learning a lexicon for a fixed image vocabulary. In *Proceedings of the 7th European Conference on Computer Vision—Part IV, ECCV '02*, pp. 97–112, London, UK, Springer-Verlag, 2002.

14. M. Everingham, A. Zisserman, C. Williams, L. Van Gool, M. Allan, C. Bishop, O. Chapelle, and N. Dalal. The 2005 pascal visual object classes challenge. In *The First PASCAL Machine Learning Challenges Workshop, MLCW 2005*, Southampton, UK, April 11–13, pp. 117–176, 2005.

15. David Dagan Feng, Wan-Chi Siu, and Hongjiang Zhang. *Multimedia Information Retrieval and Management: Technological Fundamentals and Applications*. Technological Fundamentals and Applications. Springer, Berlin, 2003.

16. Shaolei Feng, Raghavan Manmatha, and Victor Lavrenko. Multiple bernoulli relevance models for image and video annotation. In *IEEE Conference on Computer Vision and Pattern Recognition (CVPR)*, pp. 1002–1009, Washington, DC, USA, June 2004.

17. R. Fergus, P. Perona, and A. Zisserman. Object class recognition by unsupervised scale-invariant learning. In *IEEE Conference on Computer Vision and Pattern Recognition (CVPR)*, pp. 264–271, Madison, WI, USA, June 2003.

18. J. E. Gary and R. Mehrotra. Shape-similarity-based retrieval in image database systems. In *Proceedings of SPIE, Image Storage and Retrieval Systems*, Vol. 1662, pp. 2–8, 1992.

19. J. E. Goodman, J. O'Rourke, and P. Indyk (eds.). Chapter 39: Nearest neighbors in high-dimensional spaces. *Handbook of Discrete and Computational Geometry* (2nd edition). CRC Press, 2004.

20. W. I. Grosky and R. Mehrotra. Index-based object recognition in pictorial data management. *Computer Vision, Graphics Image and Processing*, 52:416–436, 1990.

21. R. M. Haralick, Dinstein, and K. Shanmugam. Textural features for image classification. *IEEE Transactions on Systems, Man, and Cybernetics*, SMC-3:610–621, 1973.

22. C. Harris and M. Stephens. A combined corner and edge detector. In *Proceedings of the 4th Alvey Vision Conference*, pp. 147–151, 1988.

23. Jingrui He, Mingjing Li, Hong-Jiang Zhang, Hanghang Tong, and Changshui Zhang. Manifold-ranking based image retrieval. In *ACM International Conference on Multimedia*, pp. 9–16, New York, NY, USA, October 2004.

24. Jing Huang, S. Ravi Kumar, M. Mitra, and Wei-Jing Zhu. Spatial color indexing and applications. In *Proceedings of the Sixth International Conference on Computer Vision, ICCV '98*, Washington, DC, USA, 1998.

25. Jing Huang, S. Ravi Kumar, M. Mitra, Wei-Jing Zhu, and R. Zabih. Image indexing using color correlograms. In *Proceedings of the 1997 Conference on Computer Vision and Pattern Recognition (CVPR '97)*, Washington, DC, USA, 1997.

26. M. Ioka. A method of defining the similarity of images on the basis of color information. *Technical Report RT-0030, IBM Research*, Tokyo Research Laboratory, November 1989.

27. H. V. Jagadish. A retrieval technique for similar shapes. In *Proceedings of the 1991 ACM SIGMOD international conference on Management of data*, pp. 208–217, New York, NY, 1991.

28. J. Jeon and R. Manmatha. Automatic image annotation of news images with large vocabularies and low quality training data. In *ACM Multimedia*, New York, NY, USA, October 2004.

29. J. Jeon, V. Lavrenko, and R. Manmatha. Automatic image annotation and retrieval using cross-media relevance models. In *ACM SIGIR international conference on Research and development in Information retrieval*, pp. 119–126, Toronto, Canada, July 2003.

30. Yohan Jin, Latifur Khan, Lei Wang, and Mamoun Awad. Image annotations by combining multiple evidence and wordnet. In *ACM Multimedia*, pp. 706–715, Singapore, November 2005.

31. T. Kadir, A. Zisserman, and J. M. Brady. An affine invariant salient region detector. In *European Conference on Computer Vision*. Springer-Verlag, 2004.

32. P. Kerminen and M. Gabbouj. Image retrieval based on color matching. In *Proceedings of the Finnish Signal Processing Symposium (FINSIG-99)*, Oulu, Finland, May 1999.

33. M. L. Kherfi, D. Ziou, and A. Bernardi. Image retrieval from the world wide web: Issues, techniques, and systems. *ACM Computing Surveys*, 36(1):35–67, 2004.

34. Victor Lavrenko, R. Manmatha, and Jiwoon Jeon. A model for learning the semantics of pictures. In *Neural Information Processing Systems (NIPS)*, Vancouver, Canada, December 2003.

35. S. Lazebnik, C. Schmid, and J. Ponce. Beyond bags of features: Spatial pyramid matching for recognizing natural scene categories. In *IEEE Conference on Computer Vision and Pattern Recognition (CVPR)*, pp. 2169–2178, New York, USA, 2006.

36. Catherine S. Lee, HongJiang Zhang, and Wei-Ying Ma. Information embedding based on user's relevance feedback for image retrieval. In *SPIE International Conference on Multimedia Storage and Archiving Systems IV*, pp. 19–22, Boston, USA, 1999.

37. Zhang Lei, Lin Fuzong, and Zhang Bo. A CBIR method based on color-spatial feature. In *IEEE Region 10 Annual International Conference*, Cheju, Korea, 1999.

38. M. S. Lew, N. Sebe, C. Djeraba, and R. Jain. Content-based multimedia information retrieval: State of the art and challenges. *ACM Transactions on Multimedia Computing Communications and Applications*, 2(1):1–19, 2006.

39. F.-F. Li, R. Fergus, and P. Perona. A Bayesian approach to unsupervised one–shot learning of object categories. In *Proceedings of International Conference on Computer Vision (ICCV)*, pp. 1134–1141, Nice, France, 2003.

40. F.-F. Li, R. Fergus, and P. Perona. Learning generative visual models from few training examples: An incremental bayesian approach tested on 101 object categories. *Computer Vision and Image Understanding*, 106(1):59–70, 2007.

41. J. Li and J. Ze Wang. Automatic linguistic indexing of pictures by a statistical modeling approach. *IEEE Transactions on Pattern Analysis and Machine Intelligence*, 25(9):1075–1088, 2003.

42. J. Li and J. Ze Wang. Real-time computerized annotation of pictures. In *ACM Multimedia*, pp. 911–920, Santa Barbara, USA, October 2006.

43. J. Li and N. M. Allinson. A comprehensive review of current local features for computer vision. *Neurocomputing*, 71(10–12):1771–1787, 2008.

44. Xirong Li, Le Chen, Lei Zhang, Fuzong Lin, and Wei-Ying Ma. Image annotation by large-scale content-based image retrieval. In *ACM Multimedia*, pp. 607–610, Santa Barbara, USA, 2006.

45. Xirong Li, Cees G. M. Snoek, and Marcel Worring. Learning tag relevance by neighbor voting for social image retrieval. In *Multimedia Information Retrieval*, pp. 180–187, 2008.

46. Zhiwei Li, Xing Xie, Lei Zhang, and Wei-Ying Ma. Searching one billion web images by content: Challenges and opportunities. In *Multimedia Content Analysis and Mining: International Workshop (MCAM)*, pp. 33–36, Weihai, China, June 30–July 1, 2007.

47. Xiao-Chen Lian, Zhiwei Li, Changhu Wang, Bao-Liang Lu, and Lei Zhang. Probabilistic models for supervised dictionary learning. In *IEEE Conference on Computer Vision and Pattern Recognition (CVPR)*, pp. 2305–2312, San Francisco, USA, June 2010.

48. Ying Liu, Dengsheng Zhang, Guojun Lu, and Wei-Ying Ma. A survey of content-based image retrieval with high-level semantics. *Pattern Recognition*, 40:262–282, 2007.

49. David G. Lowe. Object recognition from local scale-invariant features. In *Proceedings of International Conference on Computer Vision*, pp. 1150–1157, Kerkyra, Greece, 1999.

50. W. Y. Ma and B. S. Manjunath. A comparison of wavelet transform features for texture image annotation. In *Second International Conference on Image Processing (ICIP'95)*, Vol. 2, pp. 256–259, Washington, DC, November 1995.

51. Jiri Matas, Ondrej Chum, Martin Urban, and Tomás Pajdla. Robust wide baseline stereo from maximally stable extremal regions. In *British Machine Vision Conference (BMVC)*, University of Cardiff, UK, September 2002.

52. C. S. McCamy, H. Marcus, and J. G. Davidson. A color-rendition chart. *Journal of Applied Photographic Engineering*, 2(3):95–99, 1976.

53. B. M. Mehtre, M. S. Kankanhalli, and W. F. Lee. Shape measures for content based image retrieval: A comparison. *Information Processing and Management*, 33:319–337, 1997.

54. K. Mikolajczyk and C. Schmid. Scale and affine invariant interest point detectors. *International Journal of Computer Vision*, 60(1):63–86, 2004.

55. K. Mikolajczyk and C. Schmid. A performance evaluation of local descriptors. *IEEE Transactions on Pattern Analysis Machine and Intelligence*, 27(10):1615–1630, 2005.

56. K. Mikolajczyk, T. Tuytelaars, C. Schmid, A. Zisserman, J. Matas, F. Schaffalitzky, T. Kadir, and L. J. Van Gool. A comparison of affine region detectors. *International Journal of Computer Vision*, 65(1–2):43–72, 2005.

57. George A. Miller. Wordnet: A lexical database for English. *Communications of the ACM*, 38(11): 39–41, 1995.

58. K. Min, L. Yang, J. Wright, L. Wu, Xian-Sheng Hua, and Yi Ma. Compact projection: Simple and efficient near neighbor search with practical memory requirements. In *IEEE Conference on Computer Vision and Pattern Recognition (CVPR)*, pp. 3477–3484, San Francisco, USA, 2010.

59. M. Miyahara and Y. Yoshida. Mathematical transform of (r,g,b) color data to munsell (h,s,v) color data. In *SPIE Proceedings: Visual Communications and Image Processing*, Vol. 1001, pp. 650–657, San-Jose, SPIE, 1988.

60. H. P. Moravec. Rover visual obstacle avoidance. In *International Joint Conference on Artificial Intelligence (IJCAI)*, pp. 785–790, Vancouver, Canada, August 1981.

61. W. Niblack, R. Barber, W. Equitz, M. Flickner, E. Glasman, D. Petkovic, P. Yanker, and C. Faloutsos. The qbic project: Querying images by content using color, texture and shape. In *Proceedings of Storage and Retrieval for Image and Video Databases*, Vol. 1908, 1993.

62. D. Nistér and H. Stewénius. Scalable recognition with a vocabulary tree. In *IEEE Conference on Computer Vision and Pattern Recognition (CVPR)*, pp. 2161–2168, New York, USA, 2006.

63. T. Ojala, M. Pietikainen, and D. Harwood. A comparative study of texture measures with classification based on feature distributions. *Pattern Recognition*, 29(1):51–59, 1996.

64. Jia-Yu Pan, Hyung-Jeong Yang, Christos Faloutsos, and Pinar Duygulu. Automatic multimedia cross-modal correlation discovery. In *ACM SIGKDD International Conference on Knowledge Discovery and Data Mining (KDD)*, pp. 653–658, Seattle, USA, 2004.

65. Jia-Yu Pan, Hyung-Jeong Yang, Christos Faloutsos, and Pinar Duygulu. Gcap: Graph-based automatic image captioning. In *International Workshop on Multimedia Data and Document Engineering (MDDE)*, Washington, DC, USA, July 2, 2004.

66. J. Philbin, O. Chum, M. Isard, J. Sivic, and A. Zisserman. Object retrieval with large vocabularies and fast spatial matching. In *IEEE Conference on Computer Vision and Pattern Recognition (CVPR)*, Minneapolis, USA, 2007.

67. T. Randen and J. Hå Husøy. Filtering for texture classification: A comparative study. *IEEE Transactions on Pattern Analysis and Machine Intelligence*, 21:291–310, 1999.

68. T.R. Reed and J.M.H. du Buf. A review of recent texture segmentation and feature extraction techniques. *Graphical Model and Image Processing (CVGIP): Image Understanding*, 57(3):359–372, 1993.

69. J. T. Robinson. The k-d-b-tree: A search structure for large multidimensional dynamic indexes. In *ACM SIGMOD international conference on Management of data*, pp. 10–18, Ann Arbor, Michigan, USA, 1981.

70. Y. Rui, T. S. Huang, M. Ortega, and S. Mehrotra. Relevance feedback: A power tool for interactive content-based image retrieval. *IEEE Transactions on Circuits and Systems for Video Technology*, (5):644–655, 1998.

71. Yong Rui and T. S. Huang. Image retrieval: Current techniques, promising directions and open issues. *Journal of Visual Communication and Image Representation*, 10:39–62, 1999.

72. Yong Rui and Thomas S. Huang. A novel relevance feedback technique in image retrieval. In *ACM International conference on Multimedia*, pp. 67–70, Orlando, Florida, USA, ACM, 1999.

73. Yong Rui, T. S. Huang, and S. Mehrotra. Content-based image retrieval with relevance feedback in MARS. In *IEEE International Conference on Image Processing*, pp. 815–818, Washington, DC, USA, 1997.

74. Yong Rui, A. C. She, and T. S. Huang. Modified fourier descriptors for shape representation—A practical approach. In *Proceedings of First International Workshop on Image Databases and Multi Media Search*, Amsterdam, The Netherlands, 1996.

75. B. Schölkopf and A. J. Smola. *Learning with Kernels: Support Vector Machines, Regularization, Optimization, and Beyond*. MIT Press, 2001.

76. J. Shotton, A. Blake, and R. Cipolla. Multiscale categorical object recognition using contour fragments. *IEEE Transactions on Pattern Analysis and Machine Intelligence*, 30(7):1270–1281, 2008.

77. J. Sivic and A. Zisserman. Video google: A text retrieval approach to object matching in videos. In *Proceedings of International Conference on Computer Vision (ICCV)*, pp. 1470–1477, Nice, France, 2003.

78. A. W. M. Smeulders, M. Worring, S. Santini, A. Gupta, and R. Jain. Content-based image retrieval at the end of the early years. *IEEE Transactions on Pattern Analysis and Machine Intelligence*, 22(12):1349–1380, 2000.

79. J. R. Smith and Shih-Fu Chang. Tools and techniques for color image retrieval. In *Storage and Retrieval for Image and Video Databases (SPIE)*, pp. 426–437, San Jose, USA, 1996.

80. M. Stricker and M. Orengo. Similarity of color images. In *Storage and Retrieval of Image and Video Databases*, pp. 381–392, San Jose, USA, 1995.

81. M. J. Swain and D. H. Ballard. Color indexing. *International Journal of Computer Vision*, 7:11–32, 1991.

82. H. Tamura, S. Mori, and T. Yamawaki. Texture features corresponding to visual perception. *IEEE Transactions on System, Man and Cybernetics*, 6:460–473, 1978.

83. D. Tegolo. Shape analysis for image retrieval. In *Proceedings of SPIE, Storage and Retrieval for Image and Video Databases II*, Vol. 2185, pp. 59–69, San Jose, USA, 1994.

84. B. Thomee and M. S. Lew. Relevance feedback in content-based image retrieval: Promising directions. In *The 13th annual conference of the Advanced School for Computing and Imaging*, pp. 450–456, Heijen, The Netherlands, 2007.

85. Q. Tian, Y. Yu, and T. S. Huang. Incorporate discriminant analysis with em algorithm in image retrieval. In *International Conference on Multimedia and Expo*, pp. 299–302, New York, USA, 2000.

86. S. Tong and E. Y. Chang. Support vector machine active learning for image retrieval. In *ACM Multimedia*, pp. 107–118, Ottawa, Canada, 2001.

87. A. Torralba, R. Fergus, and W. T. Freeman. 80 million tiny images: A large data set for nonparametric object and scene recognition. *IEEE Transactions on Pattern Analysis and Machine Intelligence*, 30(11):1958–1970, 2008.

88. T. Tuytelaars and L. J. Van Gool. Matching widely separated views based on affine invariant regions. *International Journal of Computer Vision*, 59(1):61–85, 2004.

89. T. Tuytelaars and K. Mikolajczyk. Local invariant feature detectors: A survey. *Foundations and Trends in Computer Graphics and Vision*, 3(3):177–280, 2007.

90. N. Vasconcelos and A. Lippman. Learning from user feedback in image retrieval systems. In *Neural Information Processing Systems (NIPS)*, pp. 977–986, Denver, Colorado, USA, 1999.

91. Changhu Wang, Feng Jing, Lei Zhang, and Hong-Jiang Zhang. Content-based image annotation refinement. In *IEEE Conference on Computer Vision and Pattern Recognition*, Minneapolis, USA, 2007.

92. Changhu Wang, Feng Jing, Lei Zhang, and HongJiang Zhang. Image annotation refinement using random walk with restarts. In *ACM Multimedia*, pp. 647–650, Santa Barbara, USA, 2006.

93. Jia Wang, Wen-jann Yang, and Raj Acharya. Color clustering techniques for color-content-based image retrieval from image databases. In *Proceedings of the 1997 International Conference on Multimedia Computing and Systems*, Washington, DC, USA, 1997.

94. Xin-Jing Wang, Lei Zhang, Feng Jing, and Wei-Ying Ma. Annosearch: Image auto-annotation by search. In *IEEE Conference on Computer Vision and Pattern Recognition (CVPR)*, pp. 1483–1490, New York, USA, 2006.

95. Xin-Jing Wang, Lei Zhang, Ming Liu, Yi Li, and Wei-Ying Ma. Arista - image search to annotation on billions of web photos. In *IEEE Conference on Computer Vision and Pattern Recognition (CVPR)*, pp. 2987–2994, San Francisco, USA, 2010.

96. R. Weber, H.-J. Schek, and S. Blott. A quantitative analysis and performance study for similarity-search methods in high-dimensional spaces. In *International Conference on Very Large Databases (VLDB)*, pp. 194–205, New York, USA, 1998.

97. Yair Weiss, Antonio Torralba, and Robert Fergus. Spectral hashing. In *Neural Information Processing Systems (NIPS)*, pp. 1753–1760, Vancouver, Canada, 2008.

98. J. S. Weszka, C. R. Dyer, and A. Rosenfeld. A comparative study of texture measures for terrain classification. *IEEE Transactions on System, Man. and Cybernetics*, 6(4):269–286, 1976.

99. Lei Wu, Xian-Sheng Hua, Nenghai Yu, Wei-Ying Ma, and Shipeng Li. Flickr distance. In *ACM Multimedia*, pp. 31–40, Vancouver, Canada, 2008.

100. Lei Wu, Linjun Yang, Nenghai Yu, and Xian-Sheng Hua. Learning to tag. In *International World Wide Web Conference*, pp. 361–370, Madrid, Spain, 2009.

101. Zhong Wu, Qifa Ke, Michael Isard, and Jian Sun. Bundling features for large scale partial-duplicate web image search. In *IEEE Conference on Computer Vision and Pattern Recognition*, pp. 25–32, 2009.

102. Hua-Jun Zeng, Qi-Cai He, Zheng Chen, Wei-Ying Ma, and Jinwen Ma. Learning to cluster web search results. In *ACM SIGIR International Conference on Research and Development in Information Retrieval*, pp. 210–217, Sheeld, UK, 2004.

103. Dengsheng Zhang and Guojun Lu. Review of shape representation and description techniques. *Pattern Recognition*, 37(1):1–19, 2004.

104. Lei Zhang, Fuzong Lin, and Bo Zhang. Support vector machine learning for image retrieval. *IEEE International Conference on Image Processing*, pp. 721–724, 2001.

105. Shiliang Zhang, Qingming Huang, Gang Hua, Shuqiang Jiang, Wen Gao, and Qi Tian. Building contextual visual vocabulary for large-scale image applications. In *ACM Multimedia*, pp. 501–510, Florence, Italy, 2010.

106. Xiao Zhang, Zhiwei Li, Lei Zhang, Weiying Ma, and Heung yeung Shum. Efficient indexing for large scale visual search. In *Internal Conference on Computer Vision*, Kyoto, Japan, 2009.

107. Xiang S. Zhou and Thomas S. Huang. Relevance feedback in image retrieval: A comprehensive review. *Multimedia Systems*, 8:536–544, 2003.

108. Xiang Sean Zhou and T. S. Huang. Relevance feedback in image retrieval: A comprehensive review. *Multimedia System*, 8(6):536–544, 2003.

109. Long Zhu, Yuanhao Chen, Xingyao Ye, and Alan L. Yuille. Structure-perceptron learning of a hierarchical log-linear model. In *IEEE Conference on Computer Vision and Pattern Recognition (CVPR)*, Anchorage, USA, 2008.

14

Digital Media Archival

Chong-Wah Ngo and Song Tan

CONTENTS

14.1 Introduction

Digital media such as images and videos tell a story by showing millions of pixels in different patterns and at different snapshots of time. However, compared to the textual documents where every word could carry a semantic meaning, a pixel alone indicates intensity value. The absence of semantic structure and order in raw visual data makes the existing file and database systems designed for text indexing not applicable for multimedia data. As a result, the storage and management of digital media has long been an outstanding problem. For many commercial systems, very often the media are not indexed by visual content itself, but rather the surrounding meta data such as filenames, tags, and descriptions associated with media. It is not strange to expect that such systems could only offer very limited search capability, and the performance will depend heavily on the quality and quantity of meta data.

With decades of research efforts in multimedia community, general solutions to the aforementioned problem include the extraction of indexable visual patterns and the generation of short summaries to facilitate efficient browsing of digital media content. For instance, images or frames could be clustered and indexed according to visual words [16] or semantic tags. Traditional data structures such as inverted file indexing [24] developed for text indexing can then be employed to support high-speed search of visual content with text queries. Visual summaries could come in different forms, such as extracting representative keyframes and generating video skims [20], for supporting fast navigation of videos and to locate visual content of interest.

The solutions to media indexing and summarizing are always coupled with knowledge from vertical domain. For instance, news video has a rigid style of narrating events—starting with anchor person shot highlighting news stories, and followed by scenes describing the development of stories from different perspectives. Home video capturing life-logs of personal experiences is full of long-winded shots and a mixing of jerky camera motion. Processing these genre-specific videos requires incorporation of domain knowledge.

The effective management of digital media archival becomes even challenging, with the proliferation of social media websites and the arrival of massive multimedia data in these sites. The issue of scalability comes into picture when majority of web videos for instances are not searchable by any commercial search engines. On the other hand, the advance in media technology makes editing visual content easy and this propagates a huge amount of identical or near-duplicate images and videos in Internet. The search list provided by search engines could be a mixing of many identical or partially identical items. As a result, users often need to painstakingly explore the search list to locate the right items in media archival, which overall degrades user experience and search satisfaction. In addition, abbreviating single video content is no longer enough, and instead, summarizing thousands of videos as a whole become necessarily for large-scale exploratory search.

In short, the challenge of management media archival comes from different aspects including the need for visual indexing beyond pixels, diverse genres of content, and scale of data. This chapter addresses the problems in broad by describing three basic components: content structuring and organization, data cleaning, and summarization, to enable management of large digital media archival. The type of media under discussion in this chapter is mainly about digital video content. Figure 14.1 depicts the flow of different components, where videos are first structured into different units to facilitate the cleaning and clustering of content in subsequent steps. Summaries are then generated for fast navigation of video content.

14.2 Content Structuring and Organization

To manage digital video archival, the first fundamental step is to decompose a video into temporal units. The features in each unit are then extracted and indexed for content-based search. The upper part of Figure 14.1 depicts the procedure of content structuring for video browsing. Basically, a video is decomposed into shots where a shot is an uninterrupted segment of video frame sequence with static or contiguous camera motion. For certain applications involving temporal event or object localization, a shot may be further partitioned into sub-shots such that one or multiple keyframes are extracted from a subshot for content representation. For certain video genres such as instructional videos, each shot could mean a single video segment capturing one presentation slide. Furthermore, each subshot consists of a single unit of action where an action can be a gesture, a pose, or a combination of gesture and pose that highlights a lecturing focus [23]. Such fine-grain decomposition could benefit tasks such as highlight extraction and video editing.

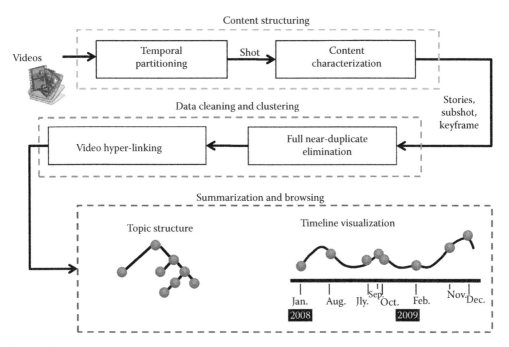

FIGURE 14.1
Large digital video archival management.

Usually, a video may contain thousands of shots; clustering of shots from single or a group of videos is required for efficient browsing. Conventional approaches includes extracting low-level features such as color, texture, motion, and audio from videos for shot clustering. For movie archival, clustering can lead to the formation of scenes where a scene is a series of contiguous shots that are coherent from the narrative point of view and captured in the same place [6]. For news video archival, clustering forms news stories each of which is a series of shots together reporting an event [2].

There have been various studies in the literature on the decomposition of videos into shots [5], and the construction of table of contents for video collection browsing [15]. The focus of research in recent years has indeed been shifted from structuring to redundancy analysis. More specifically, once videos are decomposed into basic units, these units undergo data cleaning such that correlations among videos can be leveraged to better organize video archival. Sections 14.3 and 14.4 further elaborate this issue from the viewpoint of data preprocessing and summarization, respectively.

14.3 Data Cleaning and Clustering

Digital video archival such as web and news videos often contain large sets of duplicate or near-duplicate data. The data redundancy inevitably will result in low efficiency in archival indexing and management. On the other hand, near-duplicate data could highlight key snippets of video content. A common observation in user-generated videos is that

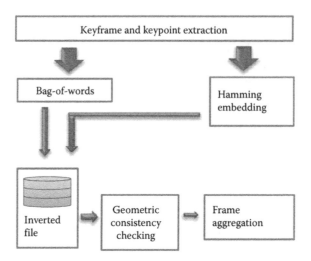

FIGURE 14.2
Near-duplicates detection framework.

the interesting parts of a video will be frequently cut, edited, and then pasted as segments in another video of similar topic. Thus, a general strategy in dealing with data redundancy is to eliminate full duplicate, while utilizing partial duplicate to cluster videos of similar topic.

14.3.1 Near-Duplicate Elimination

To detect near-duplicates and eventually exclude them from further processing in search involves the analysis of visual content. Brute force approaches include the intensive pairwise comparison between shots in two videos. Such approaches are apparently slow. Even for a small video corpus containing only 7000 shots, more than 20 million pairs of comparison are needed. Filtering by rapid retrieval has been a general way of speeding up the process by skipping unnecessary comparisons. Figure 14.2 shows a framework for fast detection of near-duplicates. The videos are preprocessed by performing shot boundary detection and then keyframe extraction. Keypoints are extracted and clustering is carried out to quantize the keypoints into a dictionary composed of visual words. Each keypoint in the keyframes is then encoded with a visual word, and this forms a bag-of-words for each keyframe. Inverted file indexing plus Hamming embedding (HE) [9] are employed to support scalable search of near-duplicate keyframes (NDKs). Geometric consistency checking is further carried out to eliminate false-positive keyframes. Finally, the near-duplicate identity of a video is determined by aggregating the similarity scores of keyframes through weakly considering the temporal consistency with its counterparts.

14.3.1.1 Visual Keywords and Inverted File Indexing

Keypoint descriptors are always used as the features for reliable retrieval. Nonetheless, the number of keypoints in a keyframe can range from hundreds to thousands, while the dimension of descriptor such as scale-invariant feature transform (SIFT) is typically high. Matching keypoints between two keyframes becomes extremely slow. Thus, a clustering approach is adopted to first quantize keypoints into a visual dictionary. Each entry in the dictionary (or centroid of a cluster) corresponds to a word. By mapping each keypoint in a frame to the nearest word, this forms a bag of words which is a histogram, describing the visual content of a keyframe. Each bin in the histogram accumulates the number of words found in the keyframe. Measuring the similarity between two keyframes is then performed by bin-to-bin matching of their histograms. Denote m as the vocabulary size, and $f_k(I_i)$ as the weight of kth bin in keyframe I_i, cosine similarity is used to measure the closeness between keyframes I_i and I_j:

$$sim_{ij} = \frac{\sum_{k=1}^{m} f_k(I_i) \times f_k(I_j)}{\sqrt{\sum_{k=1}^{m} f_k(I_i)^2 \sum_{m}^{k=1} f_k(I_j)^2}} \tag{14.1}$$

To ensure the coverage of dictionary, the number of words is usually large (e.g., 10,000). Directly matching two histograms using Equation 14.1 will not be extremely fast in this case. Nevertheless, since the histogram is normally very sparse, the matching can be efficiently conducted by exploiting structure such as inverted file index [24] which is popularly used in text information retrieval. The index stores the keyword–keyframe relationship, in which each entry (or row) corresponds to a keyword and links to the list of keyframes that contain the word. As a consequence, given a keyframe, the words are hashed into the index and the matched keyframes are retrieved. Cosine similarity is thus only evaluated for a subset of keyframes in the dataset and for those nonzero entries in the histograms.

Two techniques are often used: multilevel vector quantization (VQ) and HE [9] to further speed up the online retrieval time. Multilevel VQ allows efficient encoding of keypoints to keywords without exhaustive search of the nearest words. To reduce the information loss caused by VQ, HE maintains a binary signature (e.g., a 32-bit signature is used in [28]) for each keypoint. The signature is indexed in the inverted file to facilitate the measurement of keypoint distances for keypoints falling into the same visual word. During retrieval, any two matched visual words can be pruned if the Hamming distance between their signatures is large. This results in less words being involved in similarity measuring and also the subsequent steps of geometric checking. The space of inverted file is linear to the number of keywords to be indexed. Basically, each visual word in the inverted file stores the keyframe ID (4 bytes), spatial location of keypoint (4 bytes), scale (2 bytes), dominant orientation (2 bytes), and Hamming signature (4 bytes).

14.3.1.2 Frame-Level Geometric Consistency Checking

Keypoint quantization introduces ambiguity in visual matching. For example, words from the same bin are always matched regardless of their actual distance measured by keypoint descriptors. For words from large clusters, this could cause excessive number of false matches. Thus geometric consistency checking is a postprocessing step aiming to examine the coherency of matches between two sets of visual words. Ideally, by recovering their underlying geometric transformation from the word matches, the dissimilar keyframes can be pruned.

Recovery of transformation is often done by RANdom SAmple Consensus (RANSAC) [13]. However, such estimation is always costly and not appropriate when large number of keyframes are required to be investigated. Weak Geometry Consistency (WGC) [9] is a recently proposed technique which exploits the weak or partial geometric consistency without explicitly estimating the transformation by checking the matches from one keyframe to another. Given two matched visual words $p(x_p, y_p)$ and $q(x_q, y_q)$ from two keyframes respectively, WGC estimates the transformation from p to q as

$$\begin{bmatrix} x_q \\ y_q \end{bmatrix} = s \times \begin{bmatrix} \cos\theta & -\sin\theta \\ \sin\theta & \cos\theta \end{bmatrix} \times \begin{bmatrix} x_p \\ y_p \end{bmatrix} + \begin{bmatrix} T_x \\ T_y \end{bmatrix}, \tag{14.2}$$

where (x_p, y_p) and (x_q, y_q) are the two-dimensional (2D) spatial positions of p and q in the x–y coordinate. In Equation 14.2, there are three parameters to be estimated: the scaling factor s, the rotation parameter θ, and the translation $[T_x, T_y]^t$. In WGC, only the parameters s and θ, are estimated. For efficiency, the scale and rotation can be derived directly from the local patches of p and q without the explicit estimation of Equation 14.2. The scale s is approximated as

$$\tilde{s} = 2^{(s_q - s_p)}, \tag{14.3}$$

where s_p, s_q are the characteristic scales of words p and q, respectively. The scale values of words are known by the time when their corresponding keypoints (or local patches) are detected. For instance, the value s_p indicates the scale level which p resides in the Laplacian of Gaussian (or Difference of Gaussian) (DoG) pyramid [11,13]. Similarly, the orientation θ is approximated as

$$\tilde{\theta} = \theta_q - \theta_p, \tag{14.4}$$

where θ_p and θ_q are the dominant orientations of visual words p and q estimated during keypoint detection [11].

WGC computes $\log(\tilde{s})$ and $\tilde{\theta}$ for each matched visual word between two keyframes. By treating scale and rotation parameters independently, two histograms h^s and h^θ, referring to the scale and orientation consistency, respectively, are produced. Each peak in a histogram means one kind of transformations being performed by a group of words. Ideally, a histogram with one or few peaks hints the consistency of geometry transformation for most visual words in the keyframes. WGC utilizes the consistency clue to adjust the similarity of keyframes computed in Equation 14.1 by

$$sim_{wgc}(i, j) = min(max(h^s), max(h^\theta)) \times sim_{ij}. \tag{14.5}$$

The similarity is boosted, by a factor corresponding to the peak value in scale or orientation histogram, for keyframe pairs which show consistency in geometry transformation.

The merit of WGC lies in its simplicity and thus efficiency in transformation estimation. Nevertheless, such estimation is not always reliable. The main reason for unreliable estimation is due to the fact that the characteristic scale and dominant orientation estimated from keypoint detection are not always discriminative enough. For example, although DoG detector adopts five levels of Gaussian pyramid for keypoint localization, most points are detected at level 1. As a consequence, the scale histogram always has a peak corresponding to level 1.

A variant of WGC, named enhanced-WGC or E-WGC, is proposed in [28] by also including translation information. Combining Equations 14.2 through 14.4, we have the WGC estimation as

$$\begin{bmatrix} \tilde{x}_q \\ \tilde{y}_q \end{bmatrix} = \tilde{s} \times \begin{bmatrix} \cos\tilde{\theta} & -\sin\tilde{\theta} \\ \sin\tilde{\theta} & \cos\tilde{\theta} \end{bmatrix} \times \begin{bmatrix} x_p \\ y_p \end{bmatrix}. \tag{14.6}$$

Deriving from Equations 14.2 and 14.6, the translation τ of the visual word q can be efficiently estimated by

$$\tau = \sqrt{(\tilde{x}_q - x_q)^2 + (\tilde{y}_q - y_q)^2}. \tag{14.7}$$

Ideally, the matched visual words which follow consistent transformation should have similar values of τ, and thus τ can be used to directly adjust the keyframe similarity as in Equation 14.5. There are two advantages with this simple scheme. First, the inclusion of translation information provides another geometric clue in addition to scale and rotation. Second, Equation 14.7 has jointly integrated the clues from scale, rotation, and translation, thus generating one histogram of τ is enough for similarity reranking. In addition, Equation 14.7 can be trivially computed without incurring additional computational cost.

14.3.1.3 Video-Level Similarity Aggregation

Similarity aggregation involves measuring the sequence similarity for videos where their keyframes are fully or partially matched to the keyframes of a query video. Given a set of keyframe pairs from a video V_k and a query Q, the similarity between V_k and Q can be counted by aggregating the number of keyframe matches. Such a measure, nevertheless, does not consider temporal consistency and the noisy matches can be easily included in similarity counting. Hough Transform (HT) is a technique that aimings at aggregating the keyframe matches by weakly considering their temporal consistency [3]. HT is basically a voting scheme which accumulates scores from matches with similar time lags. Given a keyframe pair I_i and I_j with similarity score $sim_{wgc}(i,j)$ as computed in Equation 14.5 and temporally located at time t_1 and t_2 of videos V_k and Q, respectively, the time lag is computed as

$$\delta_{i,j} = t_1 - t_2. \tag{14.8}$$

HT aggregates the similarity score as a result of one keyframe match into a 2D histogram, with one dimension as the video ID and the other dimension as the time lag. In this histogram, video ID is a unique integer assigned to a video, and the range of time lag is quantized into bins by a bandwidth of δ_0. Each keyframe matching pair I_i and I_j contributes a score of $sim_{wgc}(i,j)$ to the bin $[k,b]$, where k is the video ID of V_k and $b = \lfloor \delta_{i,j}/\delta_0 \rfloor$. Consequently, a peak in the histogram, in the form of a triple $[k,b,score_{kb}]$, corresponds to an accumulated score $score_{kb} = \sum_{i,j} sim_{wgc}(i,j)$ of keyframes in V_k which are temporally aligned with the query video Q (i.e., $b = \lfloor \delta_{i,j}/\delta_0 \rfloor$). In other words, peaks in the histogram hint the video segments which are similar to Q. Detecting the peaks is basically equivalent to finding the partial near-duplicates of Q.

Let $\mathcal{H}[k,b]$ be the 2D Hough histogram, where $1 \leq k \leq n$, and n is the number of videos having the matched keyframes with query Q. HT measures the similarity of video V_k to Q by

$$Sim_{ht}(V_k, Q) = max_b(\mathcal{H}[k,b]), \tag{14.9}$$

In other words, the similarity of two videos is determined by the maximum aggregated similarity score of keyframes from both videos which are consistently matched along the temporal dimension.

While HT is efficient to implement, it has the deficiency that the influence of noisy matches is not carefully tackled. The similarity aggregation is often mixed with considerable portion of false-positive matches. The reasons are mainly due to two practical concerns. First, shot boundary detection is not always perfect. False detection can cause excessive number of shots (and keyframes) which are similar to each other within a video. As a consequence, this often results in one keyframe from a video being matched to several keyframes in another video or vice versa. Second, the imprecise matching of visual words due to quantization error, as well as WGC or E-WGC which only weakly considers the geometric transformation, also introduces random false matches. These practical concerns jointly make the similarity aggregation in HT lack of robustness.

To solve this problem, [28] revises Equation 14.9 by taking into account the granularity of matching. The intuitive idea is that a keyframe which matches to multiple keyframes in another video is given less priority when determining video similarity. Thus, the aim is to lower the scores of video segments which include excessive matching. Let the 2D Hough bin $[k, b]$ as the peak which gives rise to the similarity between videos V_k and Q as in Equation 14.9, and assume that the bin corresponds to a segment S_v from V_k and another segment S_q from Q. Let N_q be the number of keyframes in S_q, and η_l be the number of keyframes from S_v which matches with the lth keyframe of S_q. Entropy is employed to measure the associative mapping from S_q to S_v as

$$RE(S_q \rightarrow S_v) = \frac{-1}{\log Z} \left(\sum_{l=1}^{N_q} \frac{\eta_l}{Z} \times \log \frac{\eta_l}{Z} \right), \tag{14.10}$$

where $Z = \sum_{l=1}^{N_q} \eta_l$ is the total number of keyframe matches between videos V_k and Q. The value of entropy ranges within $[0, 1]$. The measure of entropy depends on the granularity of matches from one video to another. Matching which exhibits one-to-one correspondence will receive the highest entropy value of 1. On the contrary, for the cases of one-to-many or many-to-one matching, the entropy value will be low. A special case happens when only one keyframe in S_q has matches, and the keyframe matches to all the keyframes in S_v. In this case, the value of entropy will be 0. Since the definition of entropy here is different from the conventional definition where a value of 1 indicates uncertainty while a value of 0 indicates confident match, the entropy measure in Equation 14.10 is named as *Reverse-Entropy (RE)* measure.

The measure of *RE* is not symmetric, meaning that the matches for $S_q \rightarrow S_v$ will have a different *RE* value from that of $S_v \rightarrow S_q$. Thus the final value of *RE* is defined as

$$RE(V_k, Q) = min(RE(S_q \rightarrow S_v), RE(S_v \rightarrow S_q)). \tag{14.11}$$

The *RE* measure used to estimate the similarity between videos V_k and Q is as follows:

$$Sim_{re}(V_k, Q) = \begin{cases} Sim_{ht}(V_k, Q) \times RE(V_k, Q)^2 & \text{if } RE(V_k, Q) \neq 0 \\ Sim_{ht}(V_k, Q) \times \frac{1}{\sqrt{Z}} & \text{Otherwise.} \end{cases} \tag{14.12}$$

The original similarity is devalued by the square of *RE* to impose a heavier penalize on videos having noisy matching segments. Note that when *RE* equals to 0, it indicates that only one keyframe of a video (either Q or V_k) is matched to keyframes of another video. In this case, the similarity is weighted by $1/\sqrt{Z}$ so as to avoid the similarity score from dropping abruptly to zero.

The elimination of near-duplicates starts from visual word indexing to geometric checking and then frame aggregation. Candidate videos with high similarity scores at the end of HT can be regarded as full duplicates and excluded from further processing.

14.3.2 Video Hyperlinking

In addition to full near-duplicates, partial near-duplicate videos are also popularly found in different genres of video archival, for example, the news videos and web videos. Figure 14.3 depicts how partial near-duplicate segments of web videos are interconnected to form a "media network." Understanding the topology of network offers insight in tracing the manipulation history of multimedia data [10]. Thus, an interesting problem is how to "hyperlink" the videos and utilize the links to facilitate the navigation of digital video archival. The framework presented in Figure 14.2 can be revised to achieve this task. For instance, HT is essentially a voting scheme that facilitates fast discovery of full near-duplicates, but this scheme is not suffice for precise localization of near-duplicate segments. Thus, replacing HT by more sophisticated frame alignment algorithm such as temporal network [17] will enable segment localization.

14.3.2.1 Temporal Network

Conventionally, the alignment of two frame sequences can be considered as an optimization problem to find the set of frame pairs that maximizes the accumulated similarity with temporality as constraints. Temporal network, in contrast, *structurally* embeds the temporal

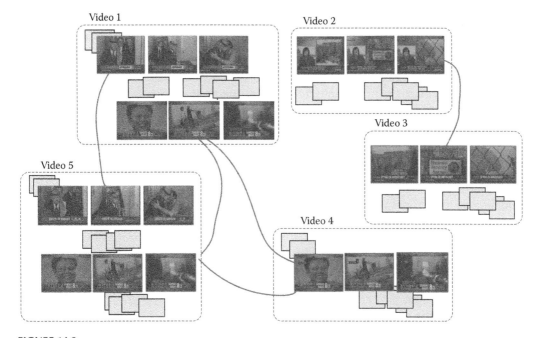

FIGURE 14.3
Partial near-duplicate videos. Given a video corpus, near-duplicate segments create hyperlinks to interrelate different portions of the videos.

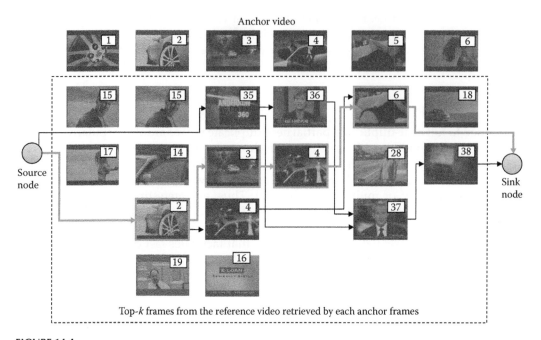

FIGURE 14.4
A temporal network. The columns of the lattice are frames from the reference videos, ordered according to the
k-NN of the query frame sequence. The label on each frame shows its time stamp in the video. The optimal path
is highlighted. For ease of illustration, not all paths and keyframes are shown.

constraint as directed edges in the network. This structural embedding novelly converts
the alignment problem into a transportation problem, or more specifically a network flow
problem, where efficient algorithms are readily available.

Given two videos, a video is designated as the *anchor* video $Q = \{q_1, \ldots, q_{|Q|}\}$ and the
other as the *reference* video $R = \{r_1, \ldots, r_{|R|}\}$, where $|\cdot|$ denotes the number of frames in the
videos. Temporal network is initially formed by querying the top-k similar frames from
R using the query frames q_i. Figure 14.4 illustrates an example where an anchor video
consisting of six frames retrieves six columns of top-k frames from the reference video.
Directed edges are established across the frames in the columns by chronologically linking
frames according to their time stamp values. For example, the frame in a column with
a time stamp value t can link to another frame in a right-hand side column with a time
stamp value larger than t. Two artificial nodes, *source* and *sink* nodes, are included for
modeling so that all paths in the network are originated from the source node and end at
the sink node.

Denote the temporal network as $G = (\mathbf{N}, \mathbf{E})$ where $\mathbf{N} = \{N_1, \ldots, N_{|Q|}\}$ are columns of
frames from R, where each column $N_i = [n_1, \ldots, n_k]$ is the retrieval result using $q_i \in Q$ as
the query, while $\mathbf{E} = \{e_{ij}\}$ is the set of all edges where e_{ij} represents a weighted directed
edge linking any two nodes from column N_i to N_j, respectively. Each edge is characterized
by two terms: weight $w(\cdot)$ and flow $f(\cdot)$. Given an edge e_{ij}, its weight is proportional to
the similarity of the destination node to its query frame in Q. In this network, the weight
signifies the capacity that an edge can carry

$$w(e_{ij}) = Sim(q_j, n_j), \tag{14.13}$$

where n_j is the node in N_j and q_j is the query frame which retrieves n_j. For any edge terminating at the sink node, the weight is assigned to zero. The flow $f(e_{ij})$, under the problem definition, is a binary indicator with a value equal to 1 or 0. A valid solution is an unbroken chain of edges forming a path from the source node to the sink node where the flows at the edges traversed by the path is 1 while for all other edges, the flow value is 0. The network flow which a path can transport is equal to the accumulated weights of its edges from the source to sink nodes. Finding a maximal path with the maximum flow is thus equivalent to searching for a sequence alignment which maximizes the similarity between Q and R in monotonically increasing temporal order.

The optimization is indeed an equivalent of the classical network maximum flow problem in operations research [1]. The objective is to find the optimal values of the flow variables $f(\cdot)$ that maximizes the total accumulated weight. The frame alignment, based on network flow optimization, can be formulated as

$$\text{maximize} \sum_{e_{ij} \in \mathbf{E}} f(e_{ij}) w(e_{ij}), \tag{14.14}$$

where various *flow conservation* constraints can be imposed to control a well-behaved weight transfer from the source to the sink node. The temporal network always has a feasible solution since there always exists at least a valid path from the source node to the sink node and there are no dangling nodes in the structure. Therefore, by the convexity property, Equation 14.14 will always converge into a global solution [1].

14.3.2.2 Potential Applications

One interesting application of partial near-duplicate search is movie tagging [18]. In social media websites, movie fans like to extract and publish interesting scenes from their favorite movies, together with tags, descriptions, and comments about the scenes. These user meta-data can be directly utilized to annotate full-length movies as well as to generate movie highlights. The key enable technology is to partially align the full-length movies with thousands of short video clips of different near-duplicate versions uploaded by users. Figure 14.5 shows an example of how the movie *310 to Yuma* is matched to web videos by temporal network. With the alignment, user supplied data can be easily propagated to tag the partial segments of movie content.

News video archival use to contain large amount of partial near-duplicate clips. These clips are composed of footage about the same news events and are edited by different broadcast stations for various purposes. By detecting these near-duplicate clips, related news events broadcasted from various sources and at different times can be clustered to facilitate the indexing and browsing of news videos. One such early system is mediaWalker [7] which constructs dependency among news stories for large-scale video threading and browsing.

14.4 Summarization and Browsing

The aim of video summarization is to speed up browsing of a large collection of video archival, and achieve efficient access and representation of the video content. By watching

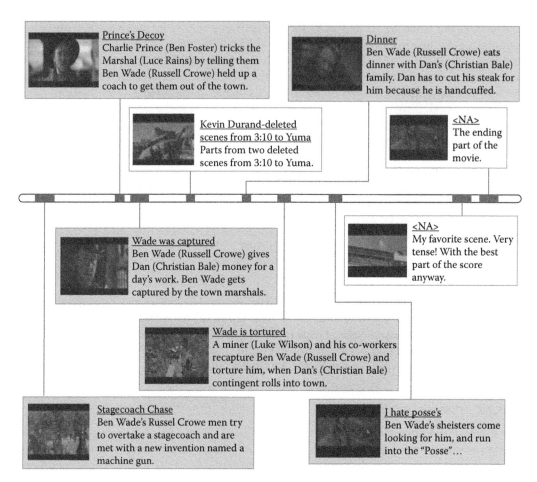

FIGURE 14.5
Automatically tagging the movie *310 to Yuma* using YouTube clips.

the summary, users can make quick decisions on the usefulness of the video. How to decide the importance and usefulness of video segments is the essential component for summary generation. Some approaches numerically measure the importance of video segments using attention model [12]. Attention is a neurobiological conception, implying the concentration of mental powers upon an object by close or careful observing or listening. The attention curve of a video can be numerically computed by audio-visual features. The crests of the curve are more likely to be attractive. Thus, keyframes (or video segmented) can be extracted from these crests to form a static storyboard (or video skim). The different solutions to summarization, nevertheless, often depends on the characteristics of video genres. Here we list three examples.

Music video archival: In music videos, the music plays the dominant role. Thus, summarization is mainly based on music analysis. For instance, the chorus is detected and used as a thumbnail for music content [27]. The analysis of visual content such as shot classification and text (lyrics) recognition are employed for music–visual–text alignment so as to make a meaningful and smooth music video summary.

Sports video archival: When watching sports videos, people are more interested in the exciting moments and great plays. Thus, highlight detection and event classification are usually involved [4]. Based on low-level features, different semantic shots (e.g., Audience, Player Close-up, and Player Following) are classified. Together with domain-specific knowledge, the corresponding interesting objects and events (such as the goal event in a soccer video) are detected with semantic analysis. The interesting events are then selected for summary generation.

Rushes video archival: Rushes are the raw materials captured during video production [14]. Being unedited, rushes contain a lot of redundant and junk information which is intertwined with useful stock footage. Thus, stock footage needed to be located by motion analysis, repetitive shot detection, and shot classification. Sophisticated analysis for object and event understanding is required to select representative clips for composing summary [22].

The aforementioned techniques can condense only the content of single video. There are cases such as news video and web video archival where summarization needs to be performed simultaneously for a large group of videos that are correlated. The remaining two sub-sections will describe this issue, which is sometimes referred to as "multivideo summarization" [21].

14.4.1 Generation of News Topic Structure

News video archival form a major portion of information disseminated in the world everyday, which constitutes an important source for topic tracking and documentation. Most people are interested in keeping abreast of the main story or thread and new events. However, it is becoming very difficult to cope with the huge volume of information that arrives each day. The scenario is more complicated when considering news stories from multiple sources such as CNN, BBC, ABC, and CCTV. Topic threading and autodocumentary appears as one prominent solution to this problem. For example, when searching the topic "Arkansas School Shooting" it is more interesting if the system can provide a concise overview or a fresh development of the topic, rather than just showing a list of items and leaving the viewers alone to find out the dependencies among them.

In news videos, a topic can comprise multiple events, and each event is usually under the umbrella of one theme. In a topic, new themes emerge over time. Some themes evolve slowly while others remain intact throughout the topic. These themes can be described by concepts that include keywords and keyframes that may repeatedly appear, evolve, or change. In news videos, stories are accompanied by shots and speech transcripts that tend to be used repeatedly during the course of a topic. There is a great deal of redundancy in news stories of one topic, especially when they are broadcast from different channels. Generally speaking, the primary interest of viewers is to learn the evolution and highlights of a topic, rather than browsing through every part of the story.

When browsing through a very large-scale corpus, one important task is to construct the dependencies of news stories by threading them under the same topic. The outcome is the semantic organization of news stories that allows viewers to rapidly interpret and analyze a topic. Figure 14.6 shows a framework for topic structure generation and documentary video for summarization [26]. The news videos from different sources are pooled and their keyframes and speech transcripts are coclustered to form news topics. Then stories are rated based on novelty, and a topic structure which chronologically threads the story development is constructed. A main thread can be further extracted to represent the major development of a topic by removing the redundant and peripheral news stories from the topic structure.

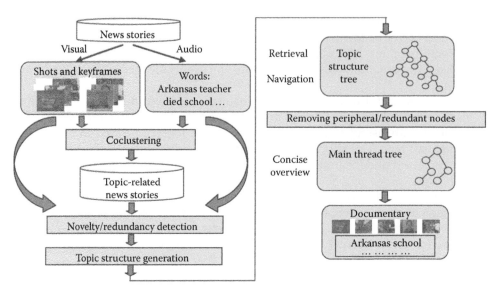

FIGURE 14.6
Topic structure generation and video documentation framework.

14.4.1.1 Topic Threading

Topic threading is the process of identifying dependencies among news stories, and linking them as a graph for navigating story development over time. The main processing part is the utilization of visual (keyframes) and textual (transcripts) cues for distinguishing novel and redundancy stories. Intuitively, named entities that specify the 4W information: who (persons), when (time), where (locations), and what (keywords), would be informative. Novel information is often conveyed through the introduction of new named entities such as the names of people, organizations, and places. On the other hand, visual information, in particular the near-duplicate shots, provides reliable clue to compare the similarity among stories across sources, time, and languages. It is especially useful when the news transcript is very noisy or the transcripts are not available, for instance, if appropriate speech recognition and machine translation tools are absent.

Similar to textual words, news story can be treated as a bag of keyframes in the visual track. Keyframes are classified as NDKs appeared multiple times within or across other news stories and non-near-duplicate keyframes (non-NDKs) that appeared once in the whole corpus. One can view the frequency of NDKs in one news story as term frequency (tf), and the frequency that NDKs appeared in different news stories as the document frequency (df) [25]. As a result, the three difference pieces of information: name entities, textual keywords, and near-duplicate shots, are separately represented as different bag-of-words. Conventional retrieval model such as vector space or language model can be employed directly to measure story novelty.

14.4.1.2 An Example of Topic Structure

Two common structures for threading news stories are chronologically ordered directed graph [8] and binary tree [26]. The basic idea is to chain story evolution, based on content novelty, in a structure such that nonlinear navigation of topic development could be supported.

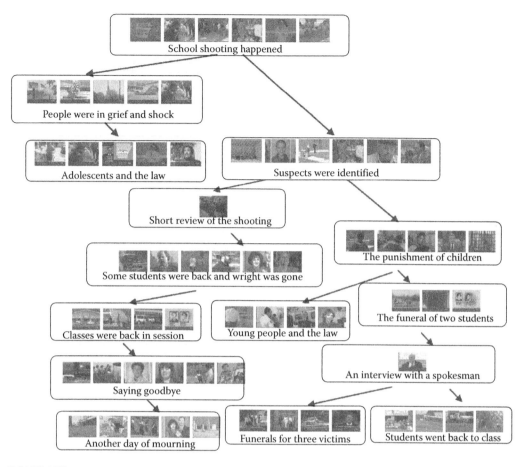

FIGURE 14.7
A graphical view of the topic structure of the news videos about "Arkansas School Shooting."

Figure 14.7 shows an example of topic structure constructed for the topic "Arkansas school shooting." The topic structure is represented by a binary tree $\langle S, E \rangle$, where S is the set of news stories and E is the set of edges. The topic structure is built upon the story dependency pairs. To illustrate the process, assume that we are given a set of news stories $S = \{S_1, S_2, \ldots, S_n\}$ on a given topic T, and these news stories are ordered by their time of publication. Each news story is compared with all previously delivered news stories to find the most similar news story using the novelty/redundancy detection described in Section 14.4.1.1. Next, n story dependency pairs are formed based on their publication orders, resulting in each story having one dependency pair.

Based on the story dependency pair set $D(T)$ of a topic T, a topic structure binary tree can be constructed by using the parent–child relation of dependency pairs. The previously delivered news story S_{parent} that is most similar to the current news story S_{child} is regarded as the parent of S_{child}. S_{child} is linked to the left branch of S_{parent}. If the left branch is occupied, then S_{child} is linked to the right branch of the left child of S_{parent} until the right branch is empty. If a story is novel, its parent is set as the root node. The first story of a topic is the the root node.

Topic Structure

The topic structure offers several unique features to facilitate browsing and autodocumentary. For example, the first story and the nodes in its rightmost path are novel stories. The nodes in the rightmost path without the left child are peripheral (isolated) stories. Stories at the left-hand path of a node represents the evolving development of an event. With reference to Figure 14.7, the topic "Arkansas school shooting" is composed of six main events including "school shooting happened," "suspects were identified," and "the punishment of children" which are linked chronologically on the rightmost path of the tree. The evolving stories of an event, for example, the "suspects were identified," can be navigated by simply tracing the stories on the subtree of left-hand side. In addition, news summary can be generated, for instance, based on the depth-first, breadth-first traversal or time-order traversal of the tree to assemble the autodocumentary.

14.4.2 Browsing Thousands of Video Clips

Commercial search engines are used to return a long sequential list of hundreds or even thousands of video clips in response to a query topic. As a result, even browsing clips under a topic is difficult. Since a clip itself is short and condense, further summarizing clip individually will not gain much advantage. Instead, providing a summary such that users can take a glimpse into the thousands of videos under a topic will greatly facilitate large-scale browsing.

An intuitive solution is to exploit the metadata and speech for clustering clips, as what has been described in Section 14.4.1 for news video archival summarization. However, the tags and descriptions surrounding web videos are always sparse and not discriminative, which makes finding the dependencies between videos challenging. In addition, web videos are diverse where the content and quality vary significantly for different web videos. For example, the videos can be musical videos, video blogs by users, self-made slide shows, and furthermore some videos are speech-free or audio-less where automatic speech recognition is not possible.

To solve this real-world problem, external knowledge is always utilized to supplement the "weak text" in web videos. Figure 14.8 shows a framework of how Google-context information is explored for web video summarization [19]. Given a query, the search trend from *Google Trends* and the upload count of the web videos over time is employed to detect the hot times for the topic. The news articles from *Google News* at the detected "hot time" are then paired up with tags from specific scenes (e.g., near-duplicate threads) in web videos, and to density the original user meta data. The scene–news pairs are presented to the users for timeline-based browsing. Figure 14.9 shows an example of how to browse the topic "US Presidential Election 2008" where important scenes mined from videos are paired up with news article snippets and mapped to timeline for visualization.

The challenge observed in Figure 14.8 is how to extract important video scenes and how to align web videos with news articles. For the former issue, a flexible solution is by assuming partial near-duplicate clips as hot scenes, imagining that these clips are always reused and edited and thus representing the key snapshots of event milestones. Techniques such as temporal network described in Section 14.3.2.1 can be employed to efficiently mined near-duplicate clips from thousands of videos. The later issue can be tackled by matching the texts in news articles and the metadata surrounding near-duplicate clips. To improve precision, the matching can be subjected to additional constraint which preserves the time consistency between the news articles publication date and video upload time. For example, the

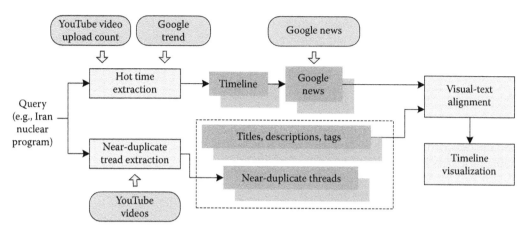

FIGURE 14.8
Google-context video summarization system.

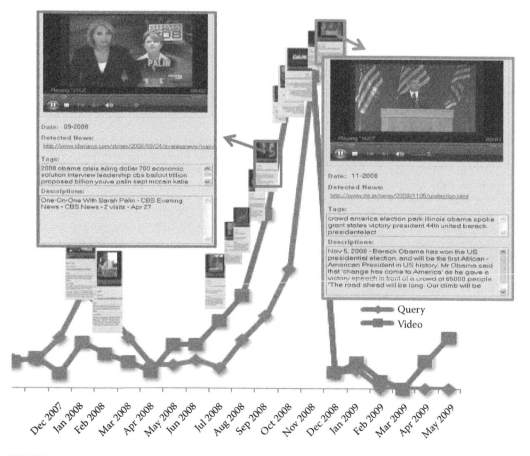

FIGURE 14.9
(**See color insert.**) Timeline-based visualization of videos about the topic "US Presidential Election 2008." Important videos are mined and aligned with news articles, and then attached to a milestone timeline of the topic. When an event is selected, the corresponding scene, tags, and news snippet are presented to users.

similarity measure between a video v_i and a news article n_j can be based on the accumulated tf-idf score of the overlapping words in the metadata of v_i and the terms in n_j as follows

$$sim(v_i, n_j) = \begin{cases} \sum_w \text{tf-idf}(w) & \text{if } T(v_i) = T(n_j) \\ 0 & \text{otherwise,} \end{cases} \tag{14.15}$$

where w is a keyword that appears both in the tags, description, or title of the web video v_i and the news article n_j while $T(\cdot)$ refers to the upload time of a web video or the publication date of the news article.

Compared to the static storyboard summarization such as listing keyframes or thumbnails to represents video clips, timeline-based visualization shown in Figure 14.8 offers a wealth of multimedia information at a glance. The aligned information from web videos and news articles are encrypted within a simple interface to summarize thousands of videos under a user query topic.

References

1. R. K. Ahuja, T. L. Magnanti, and J. B. Orlin. *Network Flows: Theory, Algorithms, and Applications*. Prentice-Hall, Reading, MA, 1993.
2. T. S. Chua, S. F Chang, L. Chaisorn, and W. Hsu. Story boundary detection in large broadcast news video archives—Techniques, experiences, and trends. *ACM Multimedia*, New York, NY, 2004.
3. M. Douze, A. Gaidon, H. Jegou, M. Marszatke, and C. Schmid. Inria-lear's video copy detection system. *TRECVID*, Gaithersburg, MD, 2008.
4. L. Y. Duan, M. Xu, T. S. Chua, Q. Tian, and C. Xu. A mid-level representation framework for semantic sports video analysis. *ACM Multimedia*, Berkeley, CA, 2003.
5. A. Hanjalic. Shot-boundary detection: Unraveled and resolved? *IEEE Transactions on Circuits and Systems for Video Technology*, 12(2):90–105, 2002.
6. A. Hanjalic, R. L. Lagendijk, and J. Biemond Jan. Automated high-level movie segmentation for advanced video-retrieval systems. *IEEE Transactions on Circuits and Systems for Video Technology*, 9(4):580–588, 1999.
7. I. Ide, T. Kinoshita, T. Takahashi, S. Satoh, and H. Murase. Mediawalker: A video archive explorer based on time-series semantic structure. *ACM Multimedia*, Augsburg, Germany, 2007.
8. I. Ide, H. Mo, and N. Katayama. Threading news video topics. *Proceedings of International Workshop on Multimedia Information Retrieval*, Berkeley, CA, 2003.
9. H. Jegou, M. Douze, and C. Schmid. Hamming embedding and weak geometric consistency for large scale image search. *Proceedings of European Conference on Computer Vision*, Marseille, France, 2008.
10. L. Kennedy and S. F. Chang. Internet image archaeology: Automatically tracing the manipulation history of photographs on the web. *ACM Multimedia*, Vancouver, British Columbia, Canada, 2008.
11. D. Lowe. Distinctive image features from scale-invariant keypoints. *International Journal of Computer Vision*, 60(2):90–110, 2004.
12. Y. F. Ma, L. Lu, H. J. Zhang, and M. Li. A user attention model for video summarization. *ACM Multimedia*, San Francisco, CA, 2002.
13. K. Mikolajczyk and C. Schmid. Scale and affine invariant interest point detectors. *International Journal of Computer Vision*, 60(1):63–86, 2004.

14. P. Over, A. F. Smeaton, and P. Kelly. The trecvid 2007 bbc rushes summarization evaluation pilot. *TRECVID 2007 BBC Rushes Summarization Workshop at ACM Multimedia*, Augsburg, Germany, 2007.
15. Y. Rui, T. S. Huang, and S. Mehrotra. Constructing table-of-content for videos. *Multimedia Systems—Special Section on Video Libraries*, 7(5):359–368, 1999.
16. J. Sivic and A. Zisserman. Video google: A text retrieval approach to object matching in videos. *Proceedings of International Conference on Computer Vision*, Nice, France, 2003.
17. H. K. Tan, C. W. Ngo, and T. S. Chua. Efficient mining of multiple partial near-duplicate alignments by temporal network. *IEEE Transactions on Circuits and Systems for Video Technology*, 20(11): 1486–1498, 2010.
18. H. K. Tan, C. W. Ngo, R. Hong, and T. S. Chua. Scalable detection of partial near-duplicate videos by visual-temporal consistency. *ACM Multimedia*, Beijing, China, 2009.
19. S. Tan, H. K. Tan, and C. W. Ngo. Topical summarization of web videos by visual-text time-dependent alignment. *ACM Multimedia*, Firenze, Italy, 2010.
20. B. T. Truong and S. Venkatesh. Video abstraction: A systematic review and classification. *ACM Transactions on Multimedia Computing, Communications, and Applications*, 3(1), 2007.
21. F. Wang and B. Merialdo. Multi-document video summarization. *International Conference on Multimedia and Expo*, New York, NY, 2009.
22. F. Wang and C. W. Ngo. Rushes video summarization by object and event understanding. *TRECVID Workshop on Rushes Summarization in ACM Multimedia Conference*, Augsburg, Germany, 2007.
23. F. Wang, C. W. Ngo, and T. C. Pong. Lecture video enhancement and editing by integrating posture, gesture and text. *IEEE Transactions on Multimedia*, 9(2):397–409, 2007.
24. I. H. Witten, A. Moffat, and T. Bell. *Managing Gigabytes: Compressing and Indexing Documents and Images*. Morgan Kaufmann Publishers, San Francisco, CA, ISBN:1558605703, 1999.
25. X. Wu, A. G. Hauptmann, and C. W. Ngo. Novelty and redundancy detection with multimodalities in cross-lingual broadcast domain. *Computer Vision and Image Understanding*, 110(3):418–431, 2008.
26. X. Wu, C. W. Ngo, and Q. Li. Threading and autodocumenting in news videos. *IEEE Signal Processing Magazine*, 23(2):59–68, 2006.
27. C. Xu, X. Shao, N. C. Maddage, and M. S. Kankanhalli. Automatic music video summarization based on audio–visual–text analysis and alignment. *SIGIR*, Salvador, Brazil, 2005.
28. W. L. Zhao, X. Wu, and C. W. Ngo. On the annotation of web videos by efficient near-duplicate search. *IEEE Transactions on Multimedia*, 12(5):448–461, 2010.

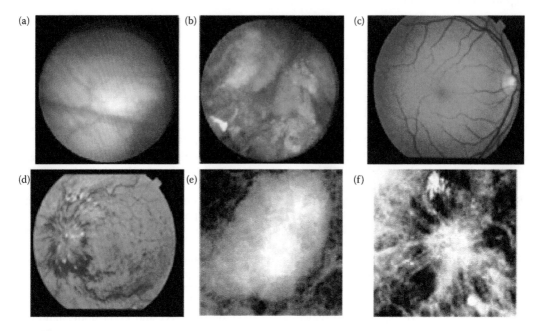

FIGURE 2.33
Medical images exhibiting texture. (a) Normal small bowel, (b) small bowel lymphoma, (c) normal retinal image, (d) central retinal vein occlusion, (e) benign lesion, and (f) malignant lesion. CE was performed on (e) and (f) for visualization purposes.

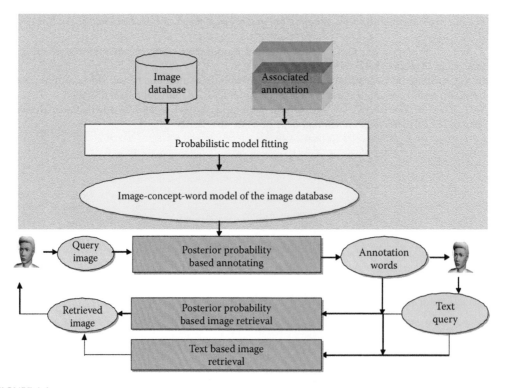

FIGURE 4.4
The architecture of the prototype system.

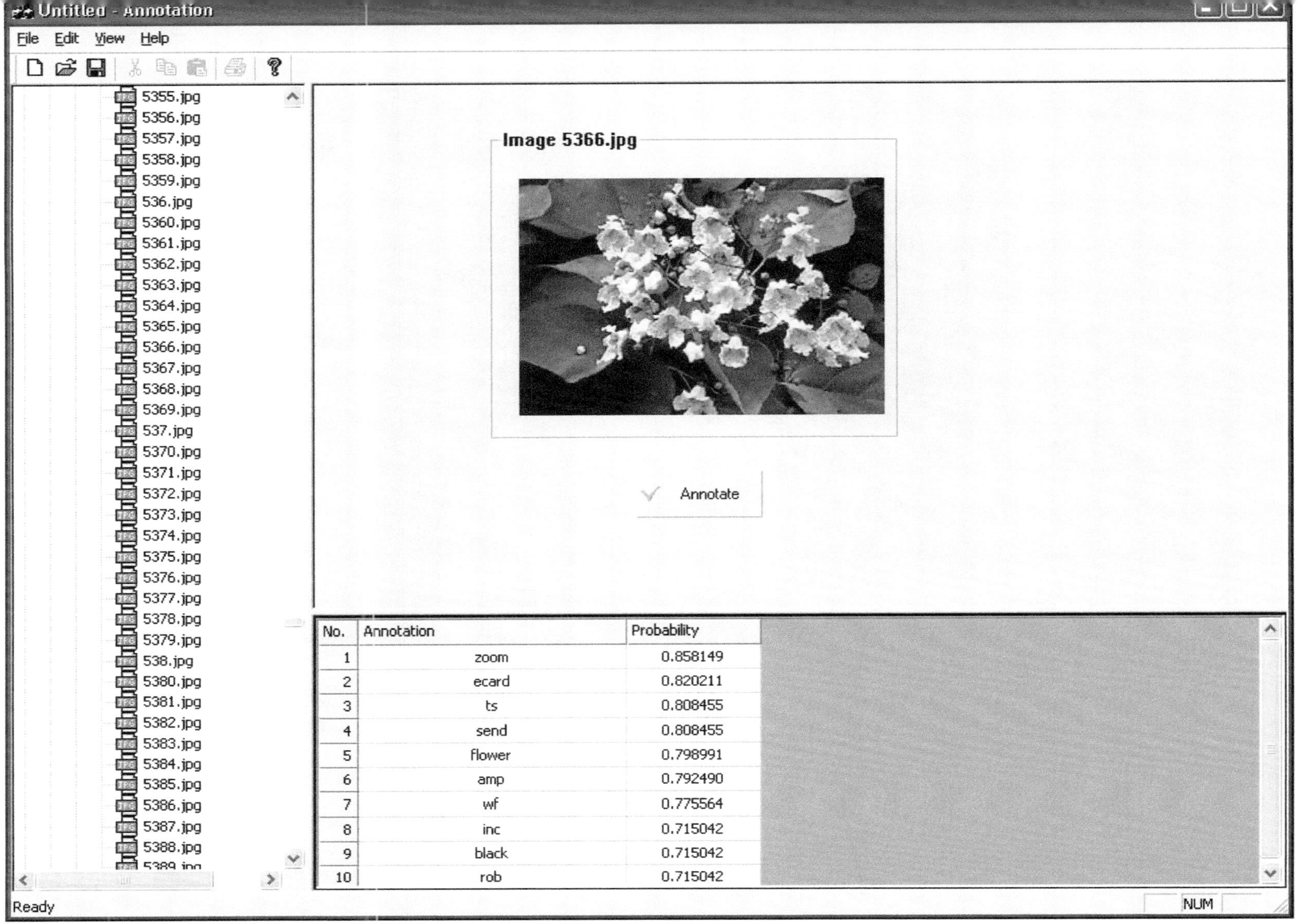

FIGURE 4.6
The interface of the automatic image annotation prototype.

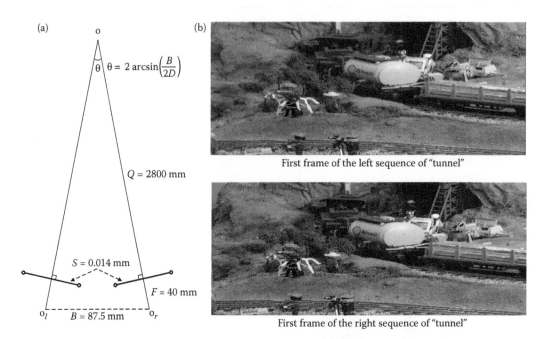

First frame of the left sequence of "tunnel"

First frame of the right sequence of "tunnel"

FIGURE 8.7
(a) 3D camera settings and (b) first pair of frames from the 720 × 288 stereo sequence "tunnel."

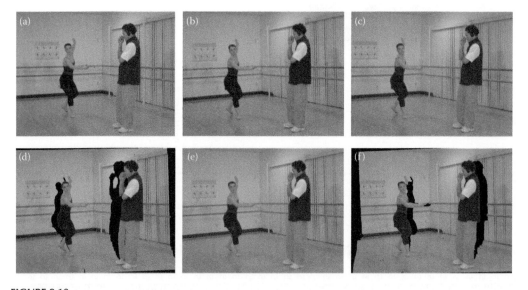

FIGURE 8.10
An example of left-and-right-to-center frame warping (based on the first frames of the Ballet sequence). (a) The decoded left frame. (b) The original center frame. (c) The decoded right frame. (d) The left frame warped to the center. (e) The warped center frame, and (f) The right frame warped to the center.

FIGURE 8.13
An example of depth map refinement and side information comparisons. (a) The original HD frame. (b) The preprocessed (warped) depth frame. (c) The refined depth frame. (d) The depth frame generated without the depth camera. (e) Side information with depth camera help. (f) Side information without depth camera help.

FIGURE 9.5
Comparison of full-resolution and frame-compatible formats: (a) full-resolution stereo pair; (b) side-by-side format; (c) top-and-bottom format.

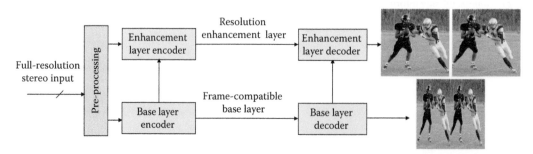

FIGURE 9.6
Illustration of video codec for scalable resolution enhancement of frame-compatible video.

FIGURE 9.7
Example of 2D-plus-depth representation.

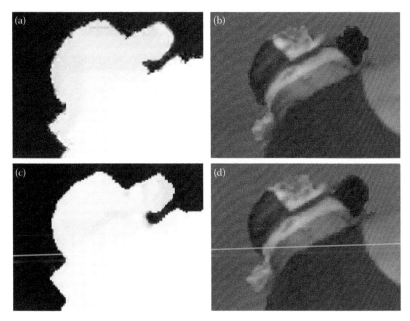

FIGURE 9.8
Effect of down/up sampling filters on depth maps and corresponding synthesis result (a, b) using conventional linear filters; (c, d) using nonlinear filtering as proposed in [58].

FIGURE 10.21
Test sequences: (a) Vidyo 1 (1280 × 720@60 Hz); (b) Kimono 1 (1920 × 1080@24 Hz); (c) Crossroad (352 × 288@30 Hz); (d) Snowroad (352 × 288@30 Hz); and (e) News and (f) Paris.

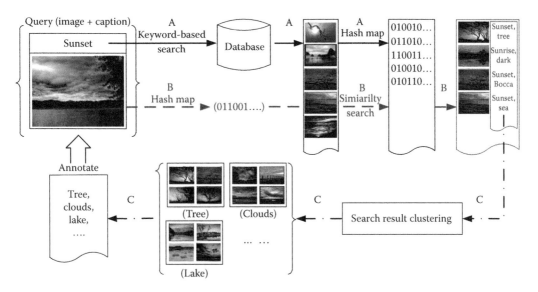

FIGURE 13.4
The framework of search-based annotation.

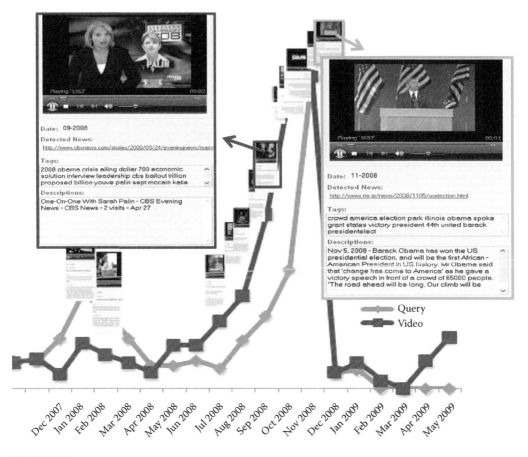

FIGURE 14.9
Timeline-based visualization of videos about the topic "US Presidential Election 2008." Important videos are mined and aligned with news articles, and then attached to a milestone timeline of the topic. When an event is selected, the corresponding scene, tags, and news snippet are presented to users.

Authentic   Forged

FIGURE 15.1
Forgery image examples in comparison with their authentic versions.

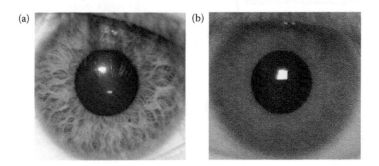

FIGURE 16.5
A colored (a) and a near-infrared (b) version of the same iris.

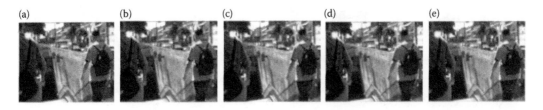

FIGURE 17.9
(a) Original walk pal video. (b) Watermarked pal video by Model 1. (c) Watermarked pal video by Model 2.
(d) Watermarked pal video by Model 3. (e) Watermarked pal video by the combined spatio temporal JND model.

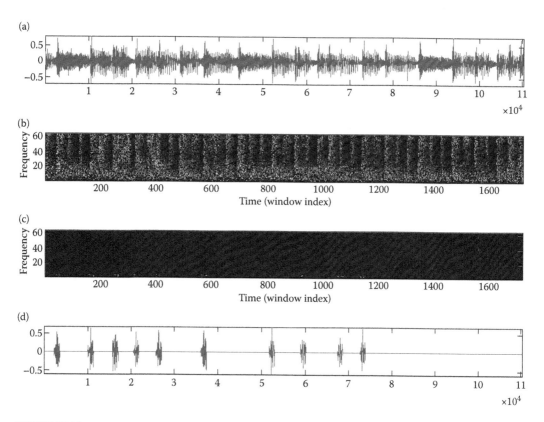

FIGURE 17.16

Example of decomposition with MMP algorithm. (a) The original music signal. (b) The MDCT coefficients of the signal. (c) The molecule atoms after 10 iteration. (d) The reconstructed signal based on the molecule atoms in (c).

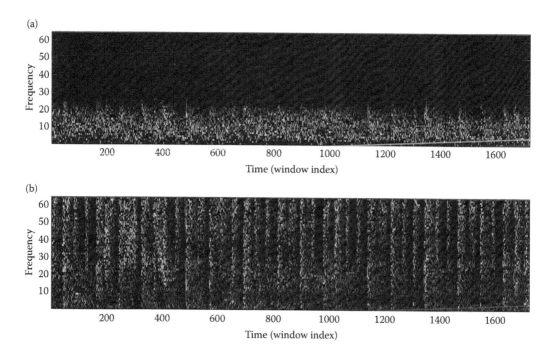

FIGURE 17.19

MDCT coefficients after low-pass filter. (a) MDCT coefficients of the low-pass-filtered signal. (b) MDCT coefficient differences between the original signal and the low-pass-filtered signal.

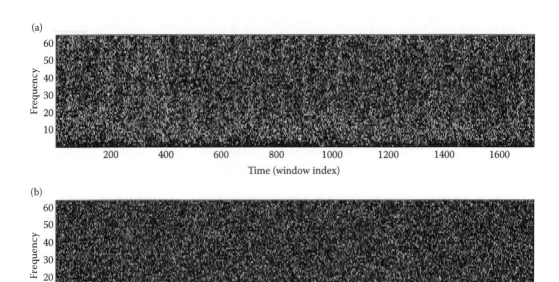

FIGURE 17.20
MDCT coefficients after random noise. (a) MDCT coefficients of the noised signal. (b) MDCT coefficient differences between the original signal and the noised signal.

FIGURE 17.25
Comparison of images before and after fingerprinting. (a) Original Lena. (b) Original Baboon. (c) Original Peppers. (d) Fingerprinted Lena. (e) Fingerprinted Baboon. (f) Fingerprinted Peppers.

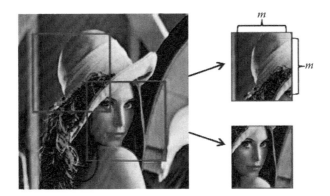

FIGURE 18.3
An example of partitioning an image into overlapping blocks of size $m \times m$. (Adapted from X. Lv and Z. J. Wang, *EURASIP Journal on Information Security*, 1–6, 2009.)

FIGURE 18.5
(a–c) Frames 61, 75, and 90 from a video. (d) A representative frame generated as a result of linearly combining these frames. (M. Malekesmaeili, M. Fatourechi and R. K. Ward, A robust and fast video copy detection system using content-based fingerprinting, *IEEE Transactions on Information Forensics and Security*, 6(1), 213–226, 2011. © (2011) IEEE.)

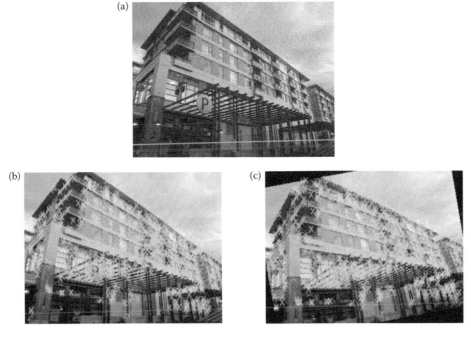

FIGURE 18.6
Example of how SIFT can be used for feature extraction from an image. (a) Original image, (b) SIFT features (original image), and (c) SIFT features (rotated image).

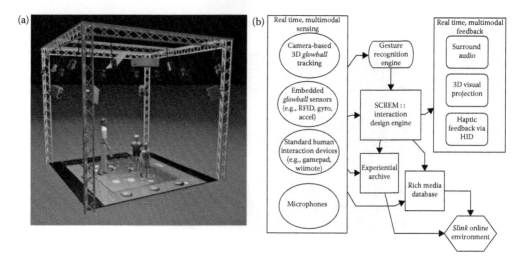

FIGURE 26.2
(a) The *SMALLab* system with cameras, speakers, and project, and (b) *SMALLab* software architecture.

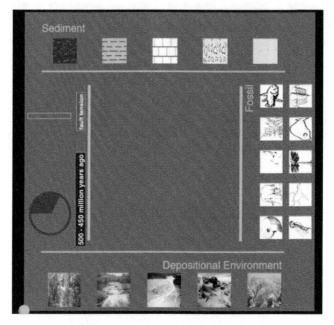

FIGURE 26.4
Screen capture of projected *Layer Cake Builder* scene.

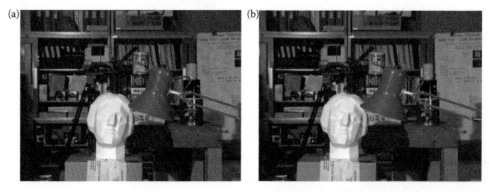

FIGURE 27.1
Tsukuba image pair: left view (a) and right view (b).

FIGURE 27.3
3DTV System by MERL. (a) Array of 16 cameras, (b) array of 16 projectors, (c) rear-projection 3D display with double-lenticular screen (d) front-projection 3D display with single-lenticular screen. (From W. Matusik and H. Pfister, 3D TV: A scalable system for real-time acquisition, transmission, and autostereoscopic display of dynamic scenes. *ACM Trans. on Graphics*, 24, 3, 2004. Available at: http://dl.acm.org/citation.cfm?id=1015805. With permission.)

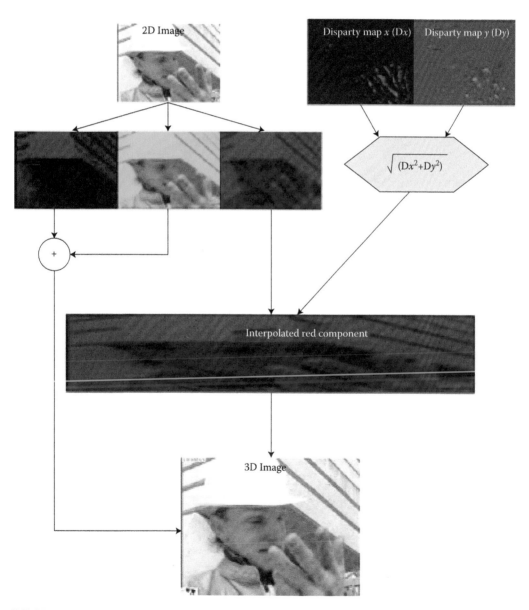

FIGURE 27.5
Flow diagram of the algorithm by Ideses et al. (From I. Ideses, L. P. Yaroslavsky, and B. Fishbain, Real-time 2D to 3D video conversion, *J. Real-Time Image Processing*, 2, 3–9, 2007. With permission.)

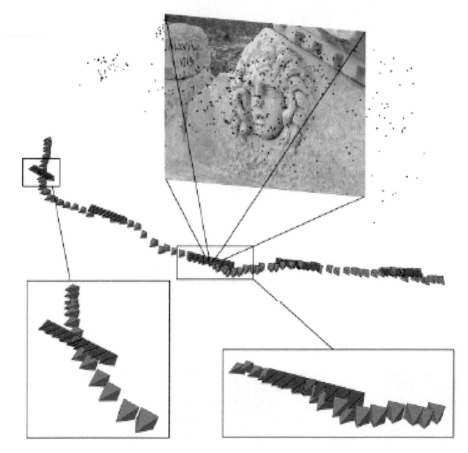

FIGURE 27.9
Multiview synthesis using SfM and DIBR by Knorr et al. Gray: original camera path, red: virtual stereo cameras, blue: original camera of a multiview setup. (S. Knorr, A. Smolic, and T. Sikora, From 2D- to stereo- to multi-view video, *Proc. IEEE 3DTVCON*, pp. 1–4, Kos Island, Greece, 2007. © (2007) IEEE.)

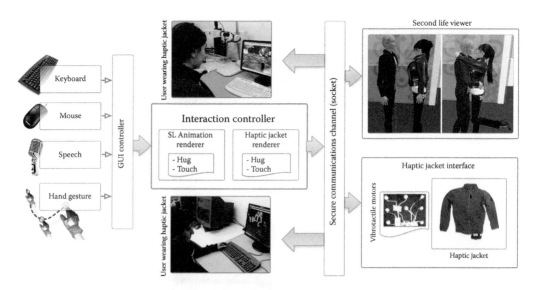

FIGURE 28.1
A basic communication block diagram depicting various components of the SL interpersonal haptic communication system.

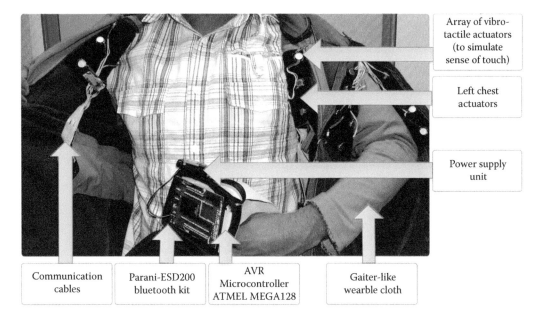

FIGURE 28.2

The Haptic jacket controller and its hardware components. Array of vibro-tactile motors are placed in the gaiter-like wearable cloth in order to wirelessly stimulate haptic interaction.

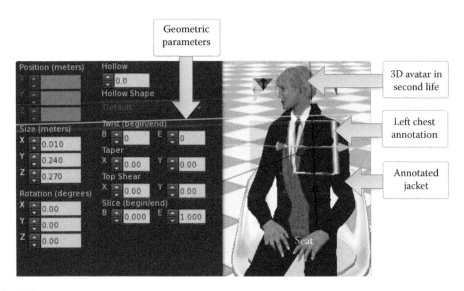

FIGURE 28.3

The flexible avatar annotation scheme allows the user to annotate any part of the virtual avatar body with haptic and animation properties. When interacted by the other party, the user receives those haptic rendering on his/her haptic jacket and views the animation rendering on the screen.

Part IV

Methodology, Techniques, and Applications: Multimedia Security

Part IV

Methodology, Techniques, and
Applications; Multimedia Security

15

Part Review on Multimedia Security

Alex C. Kot, Huijuan Yang, and Hong Cao

CONTENTS

15.1 Information Hiding for Digital Media

The history of secret communication can be traced back to ancient greece [1,2]. Several technologies commonly used to describe the secret communication by hiding information into audio, text, binary images, grayscale and color images, 3D images and video are "Steganography," "Watermarking," and "Data Hiding."

Steganography refers to the techniques used for secret communication by embedding the information into another unsuspected data, which generally relies on the assumption that the existence of the convert communication is unknown to the third parties and the methods are generally not robust [3,4].

Watermarking refers to the techniques to embed a visible or preferably invisible identification code permanently in the cover data and remains present within the cover data after any decryption process [5].

Data hiding refers to the techniques for hiding one into the other information, making the hidden information imperceptible and keeping the existence of the information a secret [5].

Data hiding and watermarking are different from cryptography which protects the content of the messages, as well as steganography which conceals the very existence of the messages [6]. The main purpose of data hiding is to conceal the hidden data in the cover data which could be used for secret data transmission, integrity checking, and authentication. On the other hand, copyright protection of multimedia is the key motivation in developing watermarking technologies. The embedded watermark should be robust against any data manipulations resulting from digital-to-analog format conversion or attack. Hence, watermarking techniques usually have robustness requirements. The existence of the hidden information may not need to be kept a secret.

In this chapter, we mainly review these techniques for images and video. Comprehensive reviews of the watermarking and data hiding techniques for multimedia can be found in [1–9]. In designing a watermarking or data hiding system, some fundamental issues need to be addressed, which are listed below:

Visual quality. The perceptual similarity between the original and watermarked version of the cover work, in other words, the visual distortion incurred by embedding the watermark or secret data. The visual quality is usually assessed using the distortion measures [10,11].

Capacity or payload data. The number of bits that a watermark or datum encodes within a unit of time or within a work, which may vary from a few bits to a large amount of data for different applications, for example, a series number of bits are required to be inserted for identification detection; a large amount of data are required for secret communication; whereas only one single bit ("present" or "absent") may be required for applications such as access control. The capacity is also referred to the "bit rate" of the watermarking or data hiding system.

Blind or informed detector. Whether the original is needed for the watermark extraction and detection. Having the original multimedia source or cover data available for the watermark detection or verification will improve the performance of the detector. However, it is not practical to distribute the original multimedia source or cover data to every customer in most applications.

Robustness. The ability to extract and detect the watermark regardless of whatever signal processing has happened to the host data, which is application dependent. In general, the robustness to known processing such as compression and data format conversion is required for most watermarking schemes.

Security. The ability to resist the hostile attacks such as unauthorized removal, embedding, and detection [5], which is generally achieved by incorporating cryptographical signature or message authentication codes (MACs). A data hiding scheme is truly secure if knowing the exact embedding algorithm does not help an unauthorized party to detect the presence of the embedded data and cannot extract the data in a reasonable amount of time [3].

Watermarking and data hiding have many application areas, for example, broadcast monitoring, ownership identification, transaction tracking, copy control, fingerprinting, authentication, and annotation. The design requirements for a data hiding or watermarking system are different to cater for different applications. For example, high robustness is required for a system designed for ownership protection applications, whereas robustness is not a major concern for a system designed for authentication and tampering detection applications. Since the adversary would rather try every attempt to embed his own watermark and get the forged watermark verified than removing the original watermark. It is desirable to convey as much information as possible for annotation applications even though the robustness requirements are not high. However, the robustness against obliterating and robustness against removal are required for the applications catered for fingerprinting and copy control [8]. The ability of conveying a nontrivial number of bits is also required in these applications. In the following section, we firstly review data hiding techniques for text and binary images, we then review techniques for both images and video, followed by a brief review on techniques for digital fingerprinting.

15.1.1 Data Hiding for Text and Binary Images

Data hiding on binary images can be classified as *fragile/semifragile* and *robust* depending on different applications. From another aspect, it can be classified as *spatial* and *transform* domain techniques based on the domain where data hiding occurs. In addition, it can be classified as *pixel-flipping* and *object*-based techniques depending on whether the watermark or data embedding is done in the pixel level or object level [12]. Specifically, a pixel-flipping technique encodes a data bit by flipping a pixel from white to black or vice versa, whereas an object-based technique encodes a message bit by modulating the spaces or features of characters and words.

It is easy to hide data in gray or color images by slightly perturbing the pixel values, which does not cause obtrusive distortions under normal viewing conditions owing to the wide range of pixel values. Binary images only have one bit-plane, whereby each pixel takes one of the two possible colors, for example, "1" represents black and "0" represents white. Arbitrarily flipping pixels in the nonedge regions in binary images creates dramatic differences in the visual quality of the resultant images. Hence, many papers in the literature address the issues on how to choose the data hiding locations, that is, how to assess the "flippability" of a pixel.

In the past few years, many techniques have been proposed for document watermarking and data hiding such as text line, word or character shifting [13–21]; boundary modifications [22]; fixed partitioning the image into blocks [23]; modification of character features [24]; modification of run-length patterns [25]; and modification of halftone images [26]. Data hiding or watermarking for binary images can be performed either in the spatial domain or in the transform domain. Methods in the transform domain generally involve a pre/post thresholding and binarization process which renders the visual quality of the resultant watermarked image more obtrusive [27,28]. An early investigation on watermarking of binary images in transform domain is made in [27], in which the watermark is embedded in the spatial domain by using the line- and word-shifting method to achieve good visual quality while the watermark extraction and detection are done in the frequency domain to achieve high robustness. A further development in [28] is to embed the watermark on DC components of discrete cosine transform (DCT) with the aid of a preblurring and post biased binarization process. This processing has introduced distortions to the marked images [28]. To address this problem, morphological binary wavelet transform is used in [29–31]. Blind watermark extraction is achieved by using the morphological binary wavelet transform in a redundant way, that is, the interlaced transforms. This method extends the processing of binary images to the smallest unit of 2×2 blocks, which is efficient and a very large capacity is achieved.

Generally, the watermark in the spatial domain can be embedded in binary images by the following methods: (1) incorporating the odd–even features of a group of pixels [32–38]; (2) employing pairs of contour patterns [22]; (3) mapping the block features to the message bit [39] or enforcing the modulo value of the element-wise computation of the image, key and weight matrixes [23,40–42]; (4) enforcing the ratios of black versus white pixels [43] or enforcing the neighborhood pixel ratios [44]; and (5) enforcing the relationship or modulating the white spaces in neighboring lines, words, characters, and segments [14,15,17–21]. Earlier data hiding algorithm enforces the ratio of black versus white pixels in a block to be larger or smaller than one to encode one bit information [43]. Modifications of the pixels are done either on the boundary of pixels for common text documents or distributed throughout the whole block for dithered images. The algorithm is not robust enough to tolerate many distortions/attacks, neither is it secure enough to be directly applied for authentication or other fragile usage [32,33]. Several object-based methods shift the line, word, and character to encode the watermark information in text documents. The line- and word-shifting methods [14,15] have good robustness such that the watermark can survive the printing–photocopying–scanning process. The interword spaces coding [17, 20,21] and inter-character space coding [18,19] methods achieve larger capacity compared to that of the line-shifting method, but the robustness has decreased. The feature-based method [24] encodes the watermark by modifying the length of selected runs of a stroke. Generally speaking, the capacities of these methods are rather low and not targeted for authentication purposes.

Recently, significant efforts have been put in finding good "flippable" locations in binary images for watermark embedding to minimize the visual distortion [16,22,23,32–40,42,45, 46]. These methods include: defining a visual distortion table to assess the "flippability" of a pixel [32–34,36–38]; employing the "connectivity-preserving" patterns to find "flippable" pixels [35,40,46]; defining pairs of contour edge patterns [22] and just using edge locations [23,42]. Employing the random shuffling technique [32,33,36,37] to equalize the uneven embedding capacity of binary images generally requires a large block size which results in a decrease in capacity and an increase in computational load. For the pattern-based approach proposed in [47], flipping the center pixels in some patterns may create distortions such as a protrusion or a hole in a straight line. Employing the denoise patterns [16,39] achieves good visual effects of the watermarked image due to the denoising effects of the patterns. A low capacity can be expected for the images of high resolutions due to the decrease in the number of denoise patterns. Pairs of contour edge patterns that are dual to each other are employed to find suitable locations for data hiding in [22], a minimum length of contour segments is required such that the capacity is not high. A random key matrix and a weight matrix are used to protect the hidden data in [41] to achieve a high embedding rate. However, the randomness in choosing data hiding locations creates poor visual results. Further improvements [23,42] set a constraint such that the edge pixels are chosen to hide the data. Localization of the tampering in binary images is always difficult due to the small capacity and the uneven distribution of the flippable pixels. This issue is addressed in [34,48].

The edge portions are chosen as the data hiding locations in the run-length-based method [25]. Choosing the edge locations alone does not guarantee the good visual quality of the resultant watermarked image. However, the data hiding location is chosen based on the distortion measure as discussed in [49]. Watermarking for binary message under the attack of modulo two addition noise is considered in [50]. Employing the code replacement based on the minimum distance of two codes [51] results in visible distortions for a large number of authentication bits or small number of card holders. The changeable pixels in a block are chosen by computing the distance and weight matrixes in [52]. Errors in one row may be propagated to its subsequent rows due to the row-by-row processing. A watermark of 7 bits is used to represent each alphanumeric character in [53]. The watermark for restoring or detecting tampering for each character is embedded in another character determined by a random or cyclic key. Accurate segmentation and recognition of each character play an important role in implementing the scheme. Embedding information on selected locations of the cover object while keeping the receiver with no information on the selection rule is discussed in [54,55]. The wet paper codes allow a high utilization of pixels with high flippability score, thus improve the embedding capacity significantly. Extensive research has been carried out on data hiding for halftone images [26,56]. In general, the embedding techniques designed for halftone images cannot be directly applied to text document images.

To gain high security, public/private key encryption algorithms are incorporated for binary images authentication [34,36–38,57,58], for example, for binary images [34,36–38] and for halftone images [57,58]. Localization of tampering for binary images is discussed in [34], in which the accuracy is limited to the subimage size, for example, 128 × 128. How to counter against the "parity attack" is a key problem for the algorithms that employ the odd–even enforcement to embed one bit of data [36–38]. Chaining the blocks in the shuffled domain and embedding the image fingerprint computed in one block into the next block [36] help alleviate the attack. However, the last several blocks still suffer the "parity attack." Watermarking on Joint Bi-level Image Experts Group 2 (JBIG2) text images [37] is done by embedding watermark in one of the instances, namely the data-bearing symbol using the pattern-based method proposed in [36]. Recently, a list of 3 × 3 patterns with symmetrical center pixels are employed to choose the data hiding locations in [38]. Flipping the center pixels in many patterns may break the "connectivity" between pixels or create an erosion and protrusion [38]. Hence, the visual quality of the watermarked image is difficult to control. In addition, employing the fixed 3 × 3 block scheme to partition the image leads to small embedding capacity. The random locations that are known both to the embedder and to the receiver are used to carry the MACs for halftone images [57,58]. Recently, the concept of using data hiding and public-key cryptography for secure authentication has been extended to an emerging data type, electronic inks [59]. A point insertion-based lossless embedding scheme is developed to secure the integrity of the electronic writing data and its context.

15.1.2 Watermarking for Images and Video

15.1.2.1 Images Watermarking

Watermarking techniques for digital media can be classified into two major categories: *robust* and *fragile/semifragile* depending on different applications. A robust watermark is designed to survive most signal processings such as rotation, scaling, and translation [60,61]. Fragile and semifragile watermarks, that is, *"hard authentication"* and *"soft authentication"* watermarks, are generally designed for authentication and tampering detection and localization purposes. Any tampering occurred to the marked image may render the authentication fail for **fragile watermarking**, which is usually achieved through the use of the MACs and cryptographic signatures [62–66]. Most of these methods embed the watermark in the LSBs of the image pixel due to the reason that both large capacity and the fragileness of the watermark can be achieved. These schemes are usually used for authentication purposes. Fragile watermarking methods with tampering localization capability are proposed in [67–70]. These methods usually partition the images into blocks and embed block-based signature into each block to provide the localization capability. However, *semifragile watermarking* differentiates the lenient manipulations, for example, compression, from malicious ones, which can be performed on the salient features of the images [71–76]. The salient features can be the block averaging and smoothing [75], and the parent–child relationships in the wavelet domain [72] and so on. These methods generally study how to achieve a good compromise between the fragileness of the watermark to malicious tamperings and its robustness to common signal processings.

Analysis of the security of the watermarking system is discussed in [77,78]. The capacity limits of a watermarking and data hiding system under the specified distortion measures are discussed in [79–82]. Embedding watermark into images unavoidably introduces artifacts and distortions to the host images, which may raise serious concerns in applications such as medical imaging, military imaging, and law enforcement. *Reversible data hiding* methods proposed in [83–87] can be used to recover the original image losslessly once the payload watermark is extracted. A general quantization index modulation-based method [88] embeds the information by modulating the indices with the embedded information and then quantizes the host signal with the associated quantizers. Efficient tradeoffs can be achieved in maximizing information-embedding rate, minimizing distortion between the host signal and watermarked signal, and maximizing the robustness of the embedding.

The robust watermarks are targeted at achieving good robustness when the multimedia data go through the processes such as compression and the printing–photocopying–scanning. To achieve this objective, how to extract robust features are extensively studied. On the other hand, robustness against geometric distortions is heavily investigated in the existing literature. The watermark embedding is usually done in the transform domain such as Fourier Mellin transform, fast Fourier transform, and wavelet transform domain [60]. Error correction and error concealment techniques are incorporated in most methods to resist the random data error occurred during file transmission. Typically, the robust watermark can be embedded by additive embedding using DCT coefficients [89,90] or by quantizing the coefficients in the transform domain such as the wavelet coefficients [67,91–94]. Theoretic analysis of the watermarking from the view point of information theory is presented in [95–97].

15.1.2.2 Video Watermarking

The popular peer-to-peer Internet-based software such as BitTorrent has been used to share copyrighted movie, music, and other materials. These advances have raised serious concerns in digital rights management for digital video [98]. Similar to other media, the watermarking in video signals has several major applications: digital fingerprinting, authentication, and copyright protection. The fingerprinting is used to identify the illegal copy shared in the peer-to-peer system. The authentication of video is basically used in the scenario such as the video created with a surveillance camera. As usual, the copyright protection is to protect the owner of the digital video. Watermarking is one of the effective ways for the content protection of digital video, for example, identifying the owner or recipient of the video owner; access control of video data; content tracking (i.e., fingerprinting). Comprehensive reviews on the watermarking on digital video can be found in [99,100].

The common methods to embed the watermark in video signal can be additive, multiplicative, and quantization based. Embedding the watermark in the raw video format is always time consuming, especially when different watermarks need to be embedded in each of many copies of the video [98]. Many methods are proposed to watermark the uncompressed video signal [101–112]. A hybrid video watermarking scheme based on the scene change analysis and error correction coding is presented in [103]. The watermark can be embedded in wavelet transform domain by replacing some coefficients with the maximum and minimum of the neighboring coefficients [103]. In general, spatial domain watermarking techniques have low robustness. The ridgelet transform is employed in watermark embedding due to its high directional sensitivity and anisotropic [104]. Recently, dual-tree complex wavelet transform (DT-CWT) [105,107] has been used for watermarking [108–111] due to its property of nearly shift-invariance and better direction selectivity. The coefficients of DT-CWT possess six directions, that is, ± 15, ± 45, and ± 75, compared with the three directions exhibited by the normal discrete wavelet transform, that is, 0, 90, and ±45. Embedding a pseudo-random sequence directly to the coefficients of the DT-CWT has always been a difficult problem [108,111]. This is due to the reason that some components of the arbitrary sequence may be lost during inverse transform due to the redundancy of the DT-CWT. To address this problem, Loo and Kingbury suggested to construct the watermark by valid CWT coefficients [108]. Yang et al. proposed two distinctive schemes [94] in which the watermark is embedded by coefficients swapping using low-frequency coefficients and group of coefficients quantization using high-pass complex frequency coefficients of DT-CWT. The mean quantization [112] is adopted in data embedding for higher robustness.

Many compressed domain watermarking techniques for video signal are proposed to reduce the computational cost [113–117]. The watermarks can be embedded in: the quantization indices of intracoded slices [118], the Moving Picture Experts Group (MPEG)-4 or MPEG-2 bitstream by modifying the DCT coefficients [113,114], the coded residuals to make the algorithm robust to intraprediction mode changes [115], the code space in H.264 protocol [116], and the identified segments in the bitstream [117]. These compressed domain watermarking techniques not only reduce the computational load but also make the watermarking scheme more resistant to the common signal processing attacks. Integrating the watermarking process with the compression or other processing is always a good way in designing the watermarking schemes in video.

15.1.3 Digital Fingerprinting

The advancement in the digital technology has made multimedia content widely available and easy to access and process. The ease of duplication, manipulation, and illegal distribution also brings serious concerns. As such, protection of multimedia content has become an important issue. *Digital fingerprinting* is an emerging technology which intends to protect the multimedia content from unauthorized dissemination by embedding a unique ID, known as fingerprints, in every copy of the content distributed to the users. Subsequently, the embedded fingerprints can be extracted to help identify the authorized user and trace the suspicious culprits who use the digital content for unintended use [119–130]. *Collusion attack* is a cost-effective attack against digital fingerprinting, where the colluders combine several copies with the same content but with different fingerprints in an attempt to remove or attenuate the original fingerprints [126].

The digital fingerprinting system should be designed to be robust to collusion attacks and has the capability of identifying the colluders. In order to overcome the complexity issues of fingerprinting at the receiver, decryption, and fingerprinting processes are integrated in [125]. The server encrypts the media using the group key. However, at each receiver a single secret key that is unique for each user is used to decrypt the media to produce a distinct copy for each user. Existing fingerprinting methods in the literature include:

1. Using the orthogonal fingerprinting, which assigns each user a spread spectrum fingerprint sequence and the sequence is typically orthogonal to those used by other users [119,120,123].

2. Using the code modulation to construct each user's fingerprint as a linear combination of the orthogonal noise-like basis signals.

3. Choose appropriate modulation and embedding schemes to design fingerprint codes by exploiting the characteristics of the multimedia signal.

4. Employing coded fingerprinting, designing codes with traceability, and constructing the code using the error correction code.

The performance analysis of a fingerprinting system usually evaluates the probability of at least catching one colluder and the probability of accusing at least one innocent user [123]. Further, the digital fingerprinting system is always evaluated by its robustness against average and nonlinear collusion attacks [126]. Most techniques reported in the literature use binary code, the possibility and feasibility of using other non-binary codes such as Gaussian-like fingerprints are studied in [126]. Detection of the watermarks should be tested under a variety of the assumptions and collusion attacks. The spread spectrum watermark embedding technique has been employed in most of the fingerprinting systems. The investigation of other embedding techniques and the collusion resistance of these techniques may be a good direction. This may give more insight into the design of an efficient and robust digital fingerprinting system. Existing techniques mainly focus on code-level issues with an assumption that the colluders can only change fingerprint symbols, a joint consideration of data embedding and coding techniques would limit the effective ways that the attackers may exploit. For example, the coding and embedding issues are jointly considered for coded fingerprinting systems in [128], where the performances in terms of collusion resistance, detection computational complexity, and distribution efficiency are examined. It is important to study the overall performance across the coding and signal domains by taking into account the coding, embedding, attack, and detection issues jointly. Studying the collusion behavior of a group of users by taking into account the geographical or cultural factors is also an interesting direction. Classifying the users into different categories and designing different fingerprints for each category of the users may help to efficiently trace and identify the colluders. For example, the prior knowledge of the collusion pattern is utilized to design a two-tier group-oriented fingerprinting scheme in [124]. The users who are likely to collude with each other are assigned correlated fingerprints, which can be extended to design a more flexible fingerprinting system.

15.2 Multimedia Forensics

Multimedia acquisition devices such as cameras and camcorders are regarded as trustworthy sensing devices to capture the real-world signals with no bias and the acquired images and videos often represent the truth [131]. The traditional trustworthiness on the multimedia contents largely relies on the notable difficulties to modify their content. For instance, altering a film-based photo in the past had to be made through the so-called darkroom tricks by photo specialists, which is costly, time consuming and moreover the extent of forgery is often very limited. In contrast, today's advances of digital technology have given birth to many powerful devices and digital tools such as high-performance computers, high-resolution digital cameras, high-definition LCD displays, high-fidelity projections, high-speed Internet connections, and highly advanced image editing tools. These commercial tools are made available to a large number of ordinary people. While people enjoy the great conveniences of using these tools, a serious question would be asked like "can we still trust what we see?"

With the popular Photoshop software, almost everything on an image can be deliberately modified to deceive others with regard to the truth. Most of the time, human eyes can hardly differentiate whether an image is genuine or it has been maliciously tampered. Figure 15.1 shows several forgery image examples in comparison with their authentic versions. More

Authentic   Forged

FIGURE 15.1
(**See color insert.**) Forgery image examples in comparison with their authentic versions.

famous image forgery examples can be found in [132–138] and these images were mostly published in the world-renowned newspapers and magazines.

The ability to alter images through electronic manipulation has raised a number of legal, moral, and ethical issues in the publishing industry. The US National Press Photographer Association (NPPA) has also issued a call of guidelines [131] for news photos, which emphasizes on the importance of accurate representation and states that altering the editorial content of photograph, in any degree is a breach of the ethical standards recognized by the NPPA.

While multimedia contents are still commonly used as supporting evidences and historical records in growing number and wide range of applications from journalist reporting, police investigation, law enforcement, insurance claims, medical and dental examinations, military, museum and consumer photography, scientific means that can tell their origin and verify their authenticity and integrity are in urgent needs. This urgency is stimulated by several new trends: (1) Proliferation of digital cameras and other multimedia acquisition devices is fast due to the improving digital photography technology and the reducing cost. Digital cameras equipped on mobile handsets now have become a necessity feature [139] for mobile phone users. (2) Photo sharing on Internet has become increasingly popular. Image and video are the key media that trigger the success of social network giants, Youtube, Flickr, and Facebook, which promote users to upload and share their own multimedia content. These also provide resources for creating the forgery and the common platforms for the forgery media to spread quickly to make great social and political impacts. (3) A significant portion of multimedia content from the public has good news value and is sourced by various news agencies for making journalist reports [140,142]. However, one challenge [142] is how to authenticate the contents from the public, a less trusted source. (4) Following the human stem-cell fake [138], which involves obvious image forgeries, scientific journal editors have been actively seeking efficient forensic tools [141], which detect the deliberate image tweaks.

15.2.1 Multimedia Forgery Categories

We use the term "forgery" to broadly refer to all the fraudulent multimedia contents, which are deliberately created for the purposes of misguidance or deceiving others. Below, we broadly classify multimedia forgery into four categories, altering an existing media, creation from scratch, scenery forgery, and false captioning.

Altering an existing media: Supported with today's high-performance computers and state-of-the-art editing tools, this forgery category has become the easiest and the most convenient means to fake multimedia contents. Depending on the way forgery is made, this category can be further divided into several subcategories, including object based, splicing or photomontage, enhancement, retouching and morphing. The object-based type refers to the object manipulation related operations such as object insertion, deletion, duplication, and changing an object's attributes including size, shape, color, orientation, and so on. The splicing or photomontage forgery refers to compositing several images into one. Enhancement is the broad class of global image manipulation operations such as denoising, sharpening or blurring, adjustment of hue, contrast and brightness, and histogram modification. Retouching, on the other hand, refers to some local small-scale image forgery operations, such as correction of red-eyes, slight modification on an edge curvature, covering up moles, and removing hairs on the skin. Morphing is a common technique to smoothly transform a source entity into a target entity.

Creation from scratch: Artificially creating an image from scratch is generally very difficult but is still made possible with support of start-of-the-art computer graphic and electronic painting tools. With computer graphic tools like Maya and 3ds Max, a forgery maker can first construct some 3D polygonal shape models and then augment them with details including color, texture and illumination. The augmented 3D model is further photorealistically rendered by a virtual camera to create the final syntactic image. Painting, on the other hand, is a traditional way of creating an image from scratch. Supported with high-resolution and high-data rate Tablet LCDs or Tablet Personal Computers (PCs), one can conveniently paint an object electronically, for example a car plate, onto the LCD screen using tools like PaintShop Pro and CorelDraw. The electronic painting enables layered processing and easy correction and this helps skilled people to accomplish a painting task with better quality and less time. However, to paint an image with a feeling of photorealism in general is a difficult task, which requires tremendous skills and photographic knowledge.

Scenery recapturing: In contrast to the direct alteration on an existing media, one can make indirect forgery by capturing on artificially created scenery. The types of artificial scenes can be classified into the physical scenes, artificially displayed scenes, and the mixture of both types of scenes. Similar to what is done in the movie production, one can physically set up a scene even with professional human actors/actresses involved. However, since creating such a physical scene by itself can be a very expensive and even impossible task in many occasions, such forgery is generally very difficult for the ordinary people. On the other hand, the artificially displayed scenes are relatively easy to create. The advances in display technology have brought us the ubiquitous high-resolution display devices, high-quality color printers, and high-fidelity projections. These devices can be easily used to display fake scenes, for example, a manipulated image, so that the forged scene can be recaptured. Note that the recaptured images are still the direct output from a multimedia acquisition device where no obvious tampering traces shall exist. Alternatively, one can fake a scene by mixing artificially displayed objects into the physical environment in order to deceive others.

False captioning: As a special type of forgery, false captioning deliberately misinterprets the multimedia content by modifying its description or metadata tags instead of directly changing the multimedia content. Such kind of forgery is frequently seen on newspapers and magazines.

15.2.2 Current Forensic Issues and Solutions

With an ultimate goal of restoring the traditional trust on the multimedia contents, multimedia forensics investigates the content authentication and data integrity-related issues on different multimedia types and formats. In a broad sense, these issues include:

Source analysis related [143–150]: Which device class or which individual device is used in capturing given media? Are media acquired with a device as claimed? Given a number of media, can we cluster them according to their sources?

Manipulation and tampering discovery [150–193]: Have given media been manipulated and where is the manipulated region? What type and to what degree is the manipulation?

Steganalysis: Have media been concealed with a secret message? What is the message length, content, and embedding method used?

Recovery of processing history [145,146,194]: What processes have media gone through? What are the parameters used?

Recapturing identification [195–199]: Are given media captured on real scenery or on artificially created scenery using the modern display technology?

Computer graphic image identification [199–202]: Are given media acquired using a device or artificially created using photorealistic computer graphic (PRCG) software?

Device temporal forensics [203]: When are given media captured? How can we order a set of media according to their capturing time?

Different from precomputation of the media hash [204] and information hiding approaches, which are described in earlier sections, multimedia forensics is often implicitly associated with the word "passive" indicating that forensic analysis is based on detecting some intrinsic and unintentional evidences. The underlying philosophy is that multimedia content acquired through different acquisition pipelines have their unique characteristics and occupy a small and highly regularized space in the entire high-dimensional space. Highly likely, forgery disturbs some existing regularity, introduces new anomalies, and leads to various forms of inconsistencies, which may not be easily perceived by human beings. If detected, these inconsistencies can be used as tell-tale evidences to expose the forgery. As forensic techniques are nonintrusive in nature and do not require the strict end-to-end protocol needed by information hiding methods, they are expected to have brighter application prospect than the nonpassive approaches.

A large number of papers have been published in the multimedia forensic area, especially in the image forensic area in the recent years. One can refer to several survey papers [205–210] for the variety of ideas explored in the past. In the following section, we categorize and review some works that are directly related to forgery detection. Section 15.2.3 focuses on the image forgery detection techniques. Section 15.2.4 talks about video forgery detection techniques. Section 15.2.5 discusses some tamper hiding techniques.

15.2.3 Image Forgery Detection Techniques

The primary objective in image forgery detection is to reliably reveal the common easy-to-make forgeries based on some underlying forensic evidences. As shown in Figure 15.2, we broadly classify existing techniques into two categories: manipulation and tampering discovery, and detection of unconventional images. The first category includes detection of the forgeries, which directly alter the content of an existing image. This category can be further divided, based on the forensic evidences used, into two subcategories: (1) through detecting the disturbance and the inconsistency of some intrinsic image regularities and (2) through detecting some manipulation anomalies leftover by specific manipulation operations. In the second category, the unconventional images refer to those that are not directly altered from existing images, for example, recaptured images, PRCG images, falsely captioned images and so on.

15.2.3.1 Manipulation and Tampering Discovery

Figure 15.3 shows a typical image acquisition model and the regularities that previous methods have explored for forensic purposes. In general, both scenery regularities and acquisition regularities exist in an output image. Scenery regularities are a collection of high-level cues whose formations are closely associated with the physics law of light transportation and computer vision laws. Using such regularities for tampering detection generally requires humans' interpretation of the picture, common-sense knowledge of the real-world objects, and good facilitating image analysis tools. Acquisition regularities are associated with different parts of an acquisition system and their formations are mainly due to the imperfections of the acquisition modules. As shown in Figure 15.3, they can be further divided into optical component regularities, sensor regularities, processing regularities, and other statistical regularities. Previous works have extensively demonstrated that acquisition regularities are useful for image source identification. Since image forgery creation often involves mixing signals from different sources and some acquisition regularities are also found fragile toward the common image manipulations, the source-related inconsistencies and feature variations can be detected using acquisition regularities.

To analyze the consistency of lighting directions, Johnson and Farid [156] developed a tool to estimate the illumination direction from a point light source on a single image. Based on the occluding edge boundaries, whose surface normal components in the z-axis are zero, the work employs a reduced Lambertian reflectance model to estimate the lighting direction

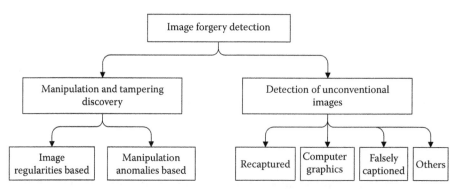

FIGURE 15.2
Categorization of image forgery detection techniques.

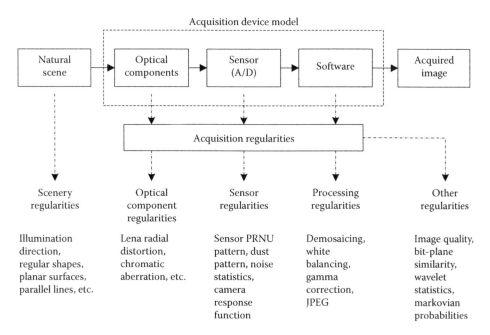

FIGURE 15.3
Image acquisition model and common forensic regularities.

in the XY image plane. With a small set of images including both syntactic and real images, it shows that the inconsistent lighting directions estimated from the different occluding boundaries can be a good tampering indicator in three different lighting scenarios, infinite light source, local light source, and multiple light sources. Along this line, Johnson and Farid have also extended this idea to other forensic scenarios, for example, based on the specular highlights [211] from the 3D human eyes' model and in the complex lighting environment [166]. Another class of useful forensic tools in this category is through projective geometry. Some forgery signs may not be obvious if the planar surface of an object is nonfrontal. To rectify an interested planar surface to be frontoparallel, Johnson and Farid [164] described three methods to estimate the rectification affine transformation based on known planar polygon in the world coordinate system, based on vanishing lines of known angle or known length ratios or based on coplanar circles, respectively. In certain forensic settings, this rectification enables easier understanding of the scene, easier spotting on the forgery cues and possible measurement of the real-world object lengths.

Lateral chromatic aberration (LCA) is caused by the varying refractive indexes of lens materials for different light wavelengths so that light of different colors cannot be perfectly focused. This causes misalignments on an image in its different color channels, which are small near the optical center but become larger at the locations that are further away from the optical center. Johnson and Farid [161] proposed to detect the inconsistent LCA as a tampering indicator. Based on a linear LCA model, the method estimates the global model parameters, for example, coordinates of the optical center and a scalar value, using an image registration technique that maximizes the mutual entropy between different color channels. By also estimating the LCA in localized image patches and comparing the local estimation with the global estimation, the blocks with large deviations in the LCA distortion orientations are deemed as the tampered blocks.

Photo-response nonuniformity (PRNU) noise is the dominant noise of the sensor pattern noise, which is caused by the imperfect manufacturing process. Lukas et al. [144] proposed to extract PRNU noise pattern from normal photos for individual camera identification. The reference PRNU pattern is learned through averaging synchronized training PRNU patterns from a camera, which suppresses the random noise and enhances the pattern noise. With 9 cameras and 300 training photos per camera, the results show that the PRNU patterns can be used as features to identify individual cameras based on a correlation detector with a close-to-zero false rejection rate when the false acceptance rate is fixed at 0.1%. Backed with the good identification results, Lukas et al. [162] further extended the PRNU pattern approach to discover the local tampered region of interest (ROI). By correlating the local PRNU patterns extracted from a test image with the synchronized reference PRNU patterns based on different sliding block shapes and sizes, the forged ROI is automatically determined and it shows relative reliable identification at a JPEG quality factor of 70. Chen et al. [150] extended this PRNU approach by improving the preprocessing techniques, the noise model, and the correlation detector. As a result, fewer training photos are needed to learn the reference PRNU pattern and better results are achieved. Based on a sliding block size of 128×128, the correlation statistics for each pixel is measured and converted into a probabilistic score of tampering, which is subsequently used to determine whether a pixel is tampered.

Demosaicing is a common in-camera software process that interpolates the missing colors due to color filtering by a color filter array (CFA) in a single-sensor commercial digital still camera. As an important process to image fidelity, demosaicing has been researched intensively with many available algorithms [212]. As each demosaicing algorithm tends to introduce some unique and persistent image correlation, its detection is useful for source model identification and tampering discovery. Popescu and Farid [157] proposed an expectation maximization (EM) algorithm for estimating a set of interpolation filter weights, which characterize the pixel correlations in the red, green, and blue channels. As another outcome of this approach, a probability map is generated from pixel prediction errors, which exhibits unique localized peaks in Fourier transformed domain for different demosaicing algorithms. In a block-based manner, these frequency peaks are used as features to achieve good results in determining whether CFA interpolation traces exist in a local image patch. Swaminathan et al. [145] estimate the demosaicing parameters using an intracolor channel demosaicing model. By first detecting the underlying CFA pattern, this work estimates the color interpolation coefficients in three color channels and in three image regions including horizontal edges, vertical edges, and smooth region. The estimated demosaicing parameters are used as features in identifying 19 cameras of different models. In [169], Swaminathan et al. extended this work to manipulation detection based on the fact that manipulation often alters the pixel correlation. By formulating the extrinsic manipulation as a linear time-invariant filtering process, this work estimates the filter parameters through a recursive deconvolution technique. A similarity score is used for detecting different forms of image manipulations. Good results demonstrated that this method works in steganalysis and in detection of PRCGs, scanned images and cut-and-paste forgery and so on. In another work [172], Dirik et al. proposed to compute two demosaicing features, which are related with Bayer CFA pattern number and analysis of the interpolation noise, respectively, to detect image manipulations including blurring, downsizing, upsizing, rotation, and JPEG recompression.

Cao and Kot [146,148,149] proposed another demosaicing detection method based on a partial derivative correlation model. To cater for different demosaicing formulas applied in a demosaicing algorithm, this method employed an EM reverse classification to divide the demosaiced samples into 16 categories through iterative minimization of the model-fitting errors. The underlying demosaicing formula for each category is estimated using regularized least-squares solution in terms of derivative weights. The derivative-based model largely suppressed the detection variations caused by different image contents since the scenery-dependent local DC components are removed. This model also allows extending the detection across the boundary of color channels for characterizing both intracolor channel and interchannel correlation. Comparison results show that the derivative weights can better reestimate demosaiced samples from the sensor samples with less error. Correspondingly, three sets of demosaicing features including derivative weights, error cumulants, and sample distributions are suggested for source model identification. Besides the high-quality photos, Cao and Kot [147,187] also extended the demosaicing features to forensic analysis on the low-end mobile images, where an eigenfeature regularization and extraction (ERE) method [213] is employed to derive a compact set of discriminant features. Through regularizing the unreliable portion of within-class eigenspectrum, this ERE method is demonstrated to work well in the scenario of small training sample size and high feature dimensionality. In blind tamper detection where source information may not be available, Cao and Kot [187] further proposed a metric to quantify the source-related inconsistencies based on cropped blocks at different locations of a mobile image. A new concept of minimizing the within-photo covariance and maximizing between-photo covariance using ERE is introduced to derive the demosaicing features that are insensitive to image content changes but are highly sensitive to image signal mixing from multiple sources.

Bayram et al. [159] combine three sets of other statistical forensic features including image quality metrics, binary similarity, and wavelet statistics for image manipulation detection. These features have also been proposed for camera model identification, steganalysis, and detection of computer graphic images. Through sequential forward floating search feature selection and based on support vector machine classification, this work detects manipulations including scaling, rotation, contrast enhancement, brightness adjustment, blurring/sharpening, and the combinations of these tampering types with different parameters.

Image manipulations often leave some specific anomalies that can be difficult to find in the untampered natural images. Previous works have detected the anomalies associated with different manipulations, such as resampling [158,174], copy move/paste [151,152,163,171, 175,178,183], splicing [154,173,176,184], double JPEG compression [160,165,186,193], sharpening [193], contrast enhancement [182] and so on. Below we briefly describe some selected works.

Resampling, for example, scaling and rotation, refers to the operation of transforming the current discrete image lattice to fit a new lattice. Resampling likely introduces periodical pixel correlations. Popescu and Farid [158] proposed an EM algorithm to detect the presence of periodical pixel correlation on a probability map as an evidence of image resampling. Based on derivatives in a radon-transformed domain, Mahdian and Saic [174] proposed another technique to detect the periodical interpolation traces.

Splicing or photomontage is the operation of combining parts from different images into a single image. To detect the presence of the sharp discontinuities caused by image splicing, Ng et al. [154] and Shi et al. [176] proposed several types of statistical features to detect the presence of sharp image discontinuities caused by the splicing or photomontage forgery. Hsu and Chang [184] proposed an automatic detection method using camera response function (CRF) inconsistencies. The CRF is estimated through using geometric invariants from locally planar irradiance points in each segmented image region.

Copy-and-move forgery refers to copying one part of the image to another image region, which can introduce highly correlated image regions, Fridrich et al. [151] proposed a search algorithm using quantized DCT coefficients as features to find the repeated image regions. Popescu and Farid [152] proposed to find the duplicated image blocks based on principal component analysis coefficients. Pan and Lyu [183] extracted the scale-invariant feature transform (SIFT) keypoints for finding the duplicated regions. These SIFT features are capable of detecting duplicated regions under geometrical and illumination distortions.

Since photos are popularly JPEG compressed by default, double JPEG compression can also be detected as a sign that an image has been potentially altered and resaved in JPEG format. Popescu and Farid [214] observed that the second compression with a different quantization step would lead to the periodical artifacts in histogram of the DCT coefficients, which can be used as a tell-tale sign of double JPEG quantization. Fu et al. [165] further discovered that the first digits of JPEG DCT coefficients for a single JPEG-compressed image closely follow a generalized Benford's law but not for the double JPEG-compressed images. The violation of the generalized Benford's law is detected as an evidence of JPEG compression for more than one time. He et al. [160] proposed to recompress a JPEG photo with a high-quality factor and identify the blocks that do not exhibit double-quantization effect as doctored blocks. Through analyzing the DCT coefficients, the probability for each block being doctored is estimated and this probability map helps a forensics analyst visually identify the tampered image region. Luo et al. [167] and Barni et al. [186] proposed to measure the JPEG blocking artifacts for differentiating single-compressed and double-compressed image blocks.

15.2.3.2 *Detection of Unconventional Images*

For detection of PRCGs, Lyu et al. [200] proposed to use wavelet statistics features to characterize the statistics of natural images. Motivated by the different physical image generation processes for photos and PRCGs, Ng et al. [199] proposed a set of geometry features for classification using a support vector machine (SVD) classifier. To characterize several types of perceptional differences between photos and PRCGs, Chen et al. [201] proposed a set of low-level features related with color, ranked histogram, ranked region size, correlogram and so on.

For detection of recaptured image, Lyu [198] suggested computing 72 wavelet statistics features from the gray image plane to distinguish color photos and the printed-and-scanned photos. Since image recapturing is often performed on a planar surface, Yu et al. [196] proposed to study the specularity distribution for detecting recaptured photos as the specularity is known to be uniform on a planar surface. Based on dichromatic reflectance model, several analyses show that the specularity distribution tends to be Laplacian-alike for natural photos while those for recaptured photos tend to be Rayleigh-alike. The efficacy of this method is demonstrated using the fake recaptured tiger photo in [137]. Cao and Kot [197] proposed to identify camera recaptured images on the liquid crystal display (LCD) screens. Through tuning available settings associated with cameras, LCD screens and environmental settings, their carefully recaptured images show good visual quality. By constructing three sets of features including local binary patterns, multiscale wavelet features and color features, the combination of these features show accurate results in identifying recaptured photos from natural photos based on a large-scale test.

For detection of falsely captioned images, Lee et al. [180] proposed to use common-sense reasoning techniques to discover the ambiguities and anomalies within perceptually meaningful regions (PMR) of an image. The PMRs are obtained through an image segmentation technique.

15.2.4 Video Forgery Detection Techniques

Similar ideas used in image forgery identification have been explored among video forensic works. For MPEG-compressed videos, Wang and Farid [188] proposed techniques to detect the static and temporal artifacts introduced by double MPEG compression. The results show that double JPEG compression is detectable from the *I*-frame and increased motion estimation errors can be observed as a result of frame insertion or deletion. Wang and Farid [189] also proposed to detect frame duplication and intra-frame region duplication. These algorithms show good computation efficiency. Wang and Farid [190] proposed algorithms to expose video forgery for deinterlaced and interlaced videos. For deinterlaced video, the algorithm estimates the correlation between the video and the deinterlacing software. For interlaced video, the work shows that correlation exists between the interfield of a frame and the cross-field of neighboring frames. These correlations can be destroyed by video tampering, and hence are detected as potential tampering indicators. Kobayashi et al. [191] proposed to use noise-level functions to expose the inconsistencies introduced in video splicing forgery. Through analyzing the video noises using a maximum a posteriori model, the work demonstrates good accuracy in localizing the forgery regions. For reprojected video, Wang and Farid [195] proposed to detect a camera skew distortion through using camera calibration method. The skew distortion is formed due to the nonright angle of the video camera relative to the screen during the second recording. The results show that skew distortion normally does not appear in authentic videos and can be detected within only a few frames in a reprojected video.

15.2.5 Tamper Hiding Techniques

Multimedia forensics techniques are also subject to attacks. The recent works in [215–218] have proposed several tamper hiding techniques, each targeted for a well-known forensics methodology. These techniques demonstrated that image forensics through detection of the resampling artifact in [158], the PRNU noise patterns in [162] and the CFA interpolation traces in [157] can all be defeated. Though these attacks are specific to the selected forensic methodologies and each attack likely introduces new tampering artifacts detectable by other forensics methodologies, the works [215–218] have shown that performing image forensics with only a single tool can be easily attacked by a sophisticated attacker. It has also been pointed out in early forensics works like [151] that comprehensive forensics analysis should be based on a suite of forensics tools which examine different image properties. In general, covering up all tampering anomalies and restoring all image regularities simultaneously into a tampered photo is believed to be very difficult than attacking on a single forensics tool.

In the future, there is no doubt that the technology to doctor photos, including those for tamper hiding, will continuously improve and so will be the image forensics techniques. It is hard to assure that the forensic tools can detect every forgery. However, by making forgery a difficult task, the research effort in multimedia forensics will still pay off to restore the traditional trustworthiness on digital multimedia content.

15.3 Multimedia Biometrics

Biometric template protection is an important issue to be resolved so that practical biometric authentication systems can be deployed [219–221]. Personal authentication is to establish the identity of a person. Traditional authentication schemes are primarily based on something that a person knows (Password, PIN) or something that a person possesses (e.g., a smart card or a token) or both. While these techniques are widely used, they have several inherent limitations. For example, neither knowledge-based nor token-based approaches can differentiate an authorized user from a person having unauthorized access to the token or the secret knowledge. Recently, many biometric-based authentication schemes are proposed that can overcome these limitations and have significantly enhanced the performances [220–241]. A typical biometric authentication system consists of the enrollment and authentication processes. In the enrollment stage, samples of a user's biometric measurement are collected and some descriptive features are extracted. Templates are subsequently constructed based on these extracted features for authentication purposes. Both biometric samples and the templates can be encrypted before storing in a database. During authentication, the features extracted from the query sample are matched with the template stored in the database to give the authentication results. However, storing the templates and biometric signal introduces a number of security and privacy risks.

- Biometric signal or template stolen. An attacker may steal the templates or the biometric signal from a database. It is worth noting that with the feature extraction algorithms known to the public, artificial biometric signals can be constructed to pass the authentication process.
- Template regeneration. Biometrics cannot be updated, reissued, or destroyed once it is compromised.
- Privacy issues. The stored biometric signal or template may expose sensitive personal information.
- Cross-matching. The lost biometric signal or template used in one system may be used for cross-matching in the other systems such as Internet banking.

Cancelable biometrics is a technique of transforming a biometric signal or template to another domain using a noninvertible transform or pseudo-random matrix. To ensure security, the biometric template matching should be based on the transformed templates, ideally, such matching should incur small degradation of the matching performance. The template protection schemes can be broadly classified into two categories: the feature transformation-based approach and biometric cryptosystem [219]. In the feature transformation-based approach, a transformation is applied to the biometric template and only the transformed template is stored in the database. The transformation is usually guided by a password or a random key. The helper data-based method [228] consists of two parts: the first part identifies the reliable components with high signal-to-noise ratio and the second part maps the binary representation onto a codeword which is subsequently error–correction encoded to tolerate the noise after quantization. The transform can be further classified as salting and noninvertible transforms depending on the characteristics of the transformation function. The problem with the salting-based approaches lies in that, once the key is compromised, the template is no longer secure. If matching is done in the transform domain, salting should not degrade the performance. For noninvertible transform [220,221], security is better than salting-based approach if the transform is really hard to invert. Diversity and revocability can be achieved by using the application-specific and user-specific functions. Trade-offs between the discriminability and noninvertibility should be better achieved. Some data hiding techniques are proposed to protect the biometric data. Ratha et al. suggested to incorporate one time random string into the fingerprint authentication system to be secured against replay attack [239]; while Jain et al. proposed to embed fingerprint minutiae data or eigen-face coefficients into fingerprint images to achieve added data privacy [240]. Different biometric signals collected are usually stored into different format. For example, fingerprint images are usually stored compressed images and sometimes stored in a binary skeleton images to reduce the storage size. Handwritten signature is usually stored as electronic ink format, which contains X and Y coordinates data as well as data channels like pressure, azimuth, and altitude. Face image is usually stored as a grayscale image. Two recent approaches are proposed to protect the thinned fingerprint images and electrical ink [59,241]. The data are embedded in such a way that the visual quality of the stego-images are maintained.

Some issues exist in the cancelable biometrics. First, how to design a biometric system such that a new biometric template can be reissued if the biometric template in an application is compromised. To address the cross matching problem, different applications should use different biometric templates to secure the biometric signal as well as to preserve the privacy. Second, in the registration of templates, two measurements of a biometric signal should be properly aligned with each other before transformation takes places, for example, corresponding fingerprint minutiae should be matched. Alignment of two fingerprint images based on reference points such as Core or Delta is still a challenging problem to be solved. This, hence, is a challenging issue in generating cancelable templates. Third, how to extract robust features from the biometric measurements play an important role in such applications. Fourth, how to design a non-invertible transform such that the performance is not compromised is another issue. The primary shortcomings of some existing approaches are that small changes in minutiae position may lead to a large change in minutiae position after transformation if the point crosses a sharp boundary, leading to increased intra-user variation [220]. Preferably, the similarity of measurements from a single user and distinctiveness of samples from different users must be preserved, otherwise the matching performance based on transformed templates will be degraded. Methods that can overcome these disadvantages are thus needed to be investigated in future research. Finally, for the noninvertible transform-based approach, the transformation must be noninvertible, that is, it should be computationally impossible for the attacker to inverse the transform to recover the original signal even if the attacker knows the transformation matrix. The transformed signal should not be similar to the original template. Two different transformed signals should not be similar or crossly match in an authentication. A large freedom for the transform is thus required so that the transformed templates can resist the brute-force attack and accommodate multiple applications. Bio-cryptography research has been another challenging topic in recent years [223,224,226,242,243]. The major technical challenges are to derive biometric keys, which are robust against intrauser variations and at the same time, the key provides good entropy.

15.4 Summary

In this chapter, we briefly reviewed the techniques for multimedia security. For the active category, techniques for data hiding and watermarking in binary images, grayscale and color images, video and digital fingerprinting are reviewed. In the forensics category, we review techniques for images and video forgery detection based on detecting various kinds of intrinsic regularities and tampering anomalies. We also review biometric security techniques related to multimedia. These techniques serve for a similar purpose to protect the multimedia. However, different techniques will be chosen based on the application scenarios and constraints. To make the system more robust, secure, and easy to use, several techniques can be combined.

References

1. R. J. Anderson and F. A. P. Petitcolas, On the limits of Steganography, *IEEE Journal on Selected Areas in Communications*, 16(4), 474–481, 1998.
2. F. A. P. Petitcolas, R. J. Anderson, and M. G. Kuhn, Information hiding-a survey, *Proceedings of the IEEE-Special Issue on Identification and Protection of Multimedia Information*, 87(7), pp. 1062–1078, July 1999.
3. M. D. Swanson, M. Kobayashi, and A. H. Tewfik, Multimedia data-embedding and watermarking technologies, *Proceedings of the IEEE*, 86(6), pp. 1064–1087, June 1998.
4. F. Hartung and M. Kutter, Multimedia watermarking techniques, *Proceedings of the IEEE-Special Issue on Identification and Protection of Multimedia Information*, 87(7), pp. 1079–1107, July 1999.
5. I. J. Cox, M. L. Miller, and J. A. Bloom: *Digital Watermarking*, San Mateo, CA: Morgan Kaufmann, 2001.
6. S. Katzenbeisser and F. A. P. Petitcolas, Information hiding techniques for Steganography and digital watermarking, *Artech House*, December 1999.
7. I. J. Cox and M. L. Miller, The first 50 years of electronic watermarking, *EURASIP Journal on Applied Signal Processing*, 2, pp. 126–132, 2002.
8. M. Wu and B. Liu: *Multimedia Data Hiding*, Springer-Verlag, 2002.
9. P. Moulin and R. Koetter, Data-Hiding Codes, *Proceedings of the IEEE*, 93(12), pp. 2083–2126, December 2005.
10. H. Lu, Alex C. Kot, and Yun Q. Shi, Distance-reciprocal distortion measure for binary document images, *IEEE Signal Processing Letters*, 11(2), 228–231, 2004.
11. J. Cheng and Alex C. Kot, Objective distortion measure for binary text image based on edge line segment similarity, *IEEE Transactions on Image Processing*, 16(6), 1691–1695, 2007.
12. H. Yang, PhD thesis, Data hiding for authentication of binary images, *Nanyang Technological University*, 2008.
13. M. Chen, E. K. Wong, N. Memon, and S. Adams, Recent development in document image watermarking and data hiding, *Proceedings of the SPIE Conference 4518: Multimedia Systems and Applications IV*, pp. 166–176, August 2001.
14. J. K. Brassil, S. Low, N. F. Maxemchuk, and L. O'Gorman, Electronic marking and identification techniques to discourage document copying, *IEEE Journal on Selected Areas in Communications*, 13(8), 1495–1504, 1995.
15. J. K. Brassil, S. Low, and N. F. Maxemchuk, Copyright protection for the electronic distribution of text documents, *Proceedings of the IEEE*, 87(7), pp. 1181–1196, July 1999.
16. H. Yang and A. C. Kot, Text document authentication by integrating inter character and word spaces watermarking, *Proceedings of the 2004 IEEE International Conference on Multimedia and Expo. (ICME'2004)*, 2, pp. 955–958, June 27–30, 2004.
17. D. Zou and Y. Q. Shi, Formatted text document data hiding robust to printing, copying and scanning, *Proceedings of the IEEE International Symposium on Circuits and Systems, 2005 (ISCAS'2005)*, 5, pp. 4971–4974, May 2005.
18. N. Chotikakamthorn, Electronic document data hiding technique using inter-character space, *The 1998 IEEE Asia-Pacific Conference on Circuits and Systems*, pp. 419–422, November 1998.
19. N. Chotikakamthorn, Document image data hiding technique using character spacing width sequence coding, *Proceedings of the International Conference on Image Processing*, 2, pp. 250–254, October 1999.

20. D. Huang and H. Yan, Interword distance changes represented by sine waves for watermarking text images, *IEEE Transactions on Circuits and Systems for Video Technology*, 11(12), 1237–1245, 2001.

21. Y.-W. Kim, K.-A. Moon, and I.-S. Oh, A text watermarking algorithm based on word classification and inter-word space statistics, *Proceedings of the Seventh International Conference on Document Analysis and Recognition*, 4, pp. 775–779, August 2003.

22. Q. Mei, E. K. Wong, and N. Memon, Data hiding in binary text document, *Proceedings of the SPIE*, 4314, pp. 369–375, 2001.

23. Y. C. Tseng and H.-K. Pan, Data hiding in 2-color images, *IEEE Transactions on Computers*, 51(7), 873–878, 2002.

24. T. Amamo and D. Misaki, A feature calibration method for watermarking of document images, *Proceedings of the 5th International Conference on Document Analysis and Recognition*, Bangalore, India, pp. 91–94, 1999.

25. K.-F. Hwang and C.-C. Chang, A run-length mechanism for hiding data into binary images, *Proceedings of the Pacific Rim Workshop on Digital Steganography*, Kitakyushu, Japan, pp. 71–74, July 2002.

26. M. S. Fu and O. C. Au, Data hiding watermarking for halftone images, *IEEE Transactions on Image Processing*, 11(4), 477–484, 2002.

27. Y. Liu, J. Mant, E. K. Wong, and S. Low, Marking and detection of text documents using transform-domain techniques, *Proceedings of the SPIE, Electronic Imaging (EI'99) Conference on Security and Watermarking of Multimedia Contents*, San Jose, CA, 3657, pp. 317–328, 1999.

28. H. Lu, X. Shi, Y. Q. Shi, A. C. Kot, and L. Chen, Watermark embedding in DC components of DCT for binary images, *2002 IEEE Workshop on Multimedia Signal Processing*, pp. 300–303, December 2002.

29. H. Yang, A. C. Kot, and S. Rahardja, Orthogonal data embedding for binary images in morphological transform domain-A high capacity approach, *IEEE Transactions on Multimedia*, 10(3), 339–351, 2008.

30. H. Yang and A. C. Kot, Data hiding for binary images authentication by considering a larger neighborhood, *Proceedings of the IEEE International Symposium on Circuits and Systems (ISCAS'2007)*, pp. 1269–1272, 27–30 May 2007.

31. H. Yang and A. C. Kot, A general data hiding framework and multilevel signature for binary images, *Lecture Notes in Computer Science*, 5041, 188–202, 2008.

32. M. Wu, E. Tang, and B. Liu, Data hiding in digital binary image, *IEEE International Conference on Multimedia and Expo.*, 1, pp. 393–396, New York, 30 July-2 August 2000.

33. M. Wu and B. Liu, Data hiding in binary images for authentication and annotation, *IEEE Transactions on Multimedia*, 6(4), 528–538, 2004.

34. H. Y. Kim and R. L. de Queiroz, Alteration-locating authentication watermarking for binary images, *Proceedings of the International Workshop on Digital Watermarking*, pp. 125–136, 2004.

35. H. Yang and A. C. Kot, Data hiding for bi-level documents using smoothing techniques, *The 2004 IEEE International Symposium on Circuits and Systems (ISCAS'2004)*, 5, pp. 692–695, 23–26, May 2004.

36. H. Y. Kim and R. L. de Queiroz, A public-key authentication watermarking for binary images, *Proceedings of the IEEE International Conference on Image Processing (ICIP'2004)*, 5, pp. 3459–3462, October 2004.

37. S. V. D. Pamboukian, H. Y. Kim, and R. L. de Queiroz, Watermarking JBIG2 text region for image authentication, *Proceedings of the IEEE International Conference on Image Processing (ICIP'05)*, 2, pp. 1078–1081, September 2005.

38. H. Y. Kim, A new public-key authentication watermarking for binary document images resistant to parity attacks, *Prof. IEEE International Conference on Image Processing (ICIP'05)*, 2, pp. 1074–1077, September 2005.

39. H. Yang, A. C. Kot, and J. Liu, Semi-fragile watermarking for text document images authentication, *Proceedings of the IEEE International Symposium on Circuits and Systems (ISCAS'05)*, 4, pp. 4002–4005, May 2005.

40. J. Liu, H. Yang, and A. C. Kot, Relationships and unification of binary images data-hiding methods, *Proceedings of the IEEE International Conference on Image Processing (ICIP'05)*, 1, pp. 981–984, September 2005.

41. H.-K. Pan, Y.-Y. Chen, and Y.-C. Tseng, A secure data hiding scheme for two-color images, *Proceedings of the Fifth Symposium on Computers and Communications*, pp. 750–755, July 2000.

42. Y. C. Tseng and H.-K. Pan, Secure and invisible data hiding in 2-color images, *Proceedings of the Twentieth Annual Joint Conference of the IEEE Computer and Communications Societies (INFOCOM'01)*, 2, pp. 887–896, 2001.

43. J. Zhao and E. Koch, Embedding robust labels into images for copyright protection, *Proceedings of the International Congress on Intellectual Property Rights for Specialized Information, Knowledge and New Technologies*, pp. 1–10, Vienna, August 1995.

44. S. Hu, Document image watermarking algorithm based on neighborhood pixel ratio, *Proceedings of the International Conference of Acoustics, Speech, and Signal Processing 2005 (ICASSP'05)*, 2, pp. 841–844, March 2005.

45. H. Yang and A. C. Kot, Data hiding for text document image authentication by connectivity-preserving, *Proceedings of the IEEE International Conference on Acoustics, Speech, and Signal Processing (ICASSP'05)*, 2, pp. 505–508, March 18–23, 2005.

46. H. Yang and A. C. Kot, Pattern-based data hiding for binary image authentication by connectivity-preserving, *IEEE Transactions on Multimedia*, 9(3), 475–486, 2007.

47. G. Pan, Z. Wu, and Y. Pan, A data hiding method for few-color images, *Proceedings of the IEEE International Conference on Acoustics, Speech, and Signal Processing (ICASSP'02)*, 4, pp. 3469–3472, May 2002.

48. H. Yang and A. C. Kot, Binary image authentication with tampering localization by embedding cryptographic signature and block identifier, *IEEE Signal Processing Letters*, 13(12), 741–744, 2006.

49. H. Lu, A. C. Kot, and J. Cheng, Secure data hiding in binary images for authentication, *Proceedings of the 2003 International Symposium on Circuits and Systems (ISCAS'03)*, 3, pp. 806–809, May 2003.

50. V. Korzhik, G. M. Luna, D. Marakov, and I. Marakova, Watermarking of binary messages in conditions of an additive binary noise attack, *IEEE Signal Processing Letters*, 10(9), pp. 277–279, September 2003.

51. C.-H. Tzeng and W.-H. Tsai, A new approach to authentication of binary images for multimedia communication with distortion reduction and security enhancement, *IEEE Communications Letters*, 7(9), 443–445, 2003.

52. T.-H. Liu and L.-W. Chang, An adaptive data hiding technique for binary images, *Proceedings of the 17th International Conference on Pattern Recognition (ICPR'2004)*, 4, pp. 831–834, August 2004.

53. A. Markur, Self-embedding and restoration algorithms for document watermark, *Proceedings IEEE International Conference on Acoustics, Speech, and Signal Processing (ICASSP'05)*, 2, pp. 1133–1136, March 2005.

54. J. Fridrich, M. Goljan, P. Lisonek, and D. Soukal, Writing on wet paper, *IEEE Transactions on Signal Processing*, 53(10), 3923–3935, 2005.

55. M. Wu, J. Fridrich, M. Goljan, and H. Gou, Handling uneven embedding capacity in binary images: a revisit, *SPIE Conference on Security, Watermarking and Stegonography, San Jose, CA, Proceedings of SPIE*, 5681, pp. 194–205, January 2005.

56. S.-C. Pei and J.-M. Guo, Hybrid pixel-based data hiding and block-based watermarking for error-diffused halftone images, *IEEE Transactions on Circuits and Systems for Video Technology*, 13(8), 867–884, 2003.

57. H. Y. Kim and A. Afif, Secure authentication watermarking for binary images, *Proceedings of the XVI Brazilian Symposium on Computer Graphics and Image Processing (SIBGRAPI'03)*, pp. 199–206, October 2003.

58. H. Y. Kim and A. Afif, A secure authentication watermarking for halftone and binary images, *International Journal of Image Systems and Technology*, 14(4), 147–152, 2005.

59. H. Cao and A. C. Kot, Lossless data embedding in electronic inks, *IEEE Transactions on Information Forensics and Security*, 5(2), 314–323, 2010.

60. C.-Y. Lin, M. Wu, J. A. Bloom, I. J. Cox, M. L. Miller, and Y. M. Lui, Rotation, scale and translation resilient watermarking for images, *IEEE Transactions on Image Processing*, 10(5), 767–781, 2001.

61. X. Kang, J. Huang, Y. Q. Shi, and Y. Lin, A DWT-DFT composite watermarking scheme robust to both affine transform and JPEG compression, *IEEE Transactions on Circuits and Systems for Video Technology*, 13(8), 776–786, 2003.

62. P. W. Wong and N. Memon, Secret and public key image watermarking schemes for image authentication and ownership verification, *IEEE Transactions on Image Processing*, 10(10), pp. 1593–1601, 2001.

63. P. S. L. M. Barreto, H. Y. Kim, and V. Rijmen, Toward secure public-key blockwise fragile authentication watermarking, *IEEE Proceedings on Vision, Image and Signal Processing*, 149(2), pp. 57–62, April 2002.

64. A. H. Ouda and M. R. El-Sakka, A practical version of wong's watermarking technique, *Proceedings of the IEEE 2004 International Conference on Image Processing*, 4, pp. 2615–2618, October 2004.

65. Y. Yuan and C. T. Li, Fragile watermarking scheme exploiting non-deterministic block-wise dependency, *Proceedings of the 17th International Conference on Pattern Recognition (ICPR'2004)*, 4, pp. 849–852, August 2004.

66. B. B. Zhu, M. D. Swanson, and A. H. Tewfik, When seeing isn't believing, *IEEE Signal Processing Magazine*, 21(2), 40–49, 2004.

67. D. Kundur and D. Hatzinakos, Digital watermarking for telltale tamper proofing and authentication, *Proceedings of the IEEE*, 87(7), pp. 1167–1179, July 1999.

68. M. U. Celik, G. Sharma, E. Saber, and A. M. Tekalp, Hierarchical watermarking for secure image authentication with localization, *IEEE Transactions on Image Processing*, 11(6), 585–595, 2002.

69. J. Fridrich, Security of fragile authentication watermarks with localization, *Proceedings of the SPIE, Security and Watermarking of Multimedia Contents IV*, 4675, pp. 691–700, 2002.

70. Y. Zhao, P. Campisi, and D. Kundur, Dual domain watermarking for authentication and compression of cultural heritage images, *IEEE Transactions on Image Processing*, 13(3), pp. 430–448, March 2004.

71. C.-Y. Lin and S. F. Chang, A robust image authentication method distinguishing JPEG compression from malicious manipulation, *IEEE Transactions on Circuits and Systems of Video Technology*, 11(2), 153–168, 2001.

72. C.-S. Lu and H.-Y. M. Liao, Structural digital signature for image authentication: an incidental distortion resistant scheme, *IEEE Transactions on Multimedia*, 5(2), 161–173, 2003.

73. Q. Sun and S.-F. Chang, A secure and robust digital signature scheme for JPEG2000 image authentication, *IEEE Transactions on Multimedia*, 7(3), 480–494, 2003.

74. C. W. Wu, On the design of content-based multimedia authentication systems, *IEEE Transactions on Multimedia*, 4(3), 385–393, 2002.

75. L. Xie and G. R. Arce, A class of authentication digital watermarks for secure multimedia communication, *IEEE Transactions on Image Processing*, 10(11), pp. 1754–1764, 2001.

76. L. Xie, G. R. Arce, and R. F. Graveman, Approximate image message authentication codes, *IEEE Transactions on Multimedia*, 2(2), 242–252, 2001.

77. M. Holliman and N. Memon, Counterfeiting attacks on oblivious block-wise independent invisible watermarking schemes, *IEEE Transactions on Image Processing*, 9(3), 432–441, 2000.

78. C. Fei, D. Kundur, and R. H. Kwong, Analysis and design of secure watermark-based authentication systems, *IEEE Transactions on Information Forensics and Security*, 1(1), 43–55, 2006.

79. P. Moulin and M. K. Mihcak, A framework for evaluating the data-hiding capacity of image sources, *IEEE Transactions on Image Processing*, 11(9), Vancouver, Canada, pp. 1029–1042, 2002.

80. D. Kundur, Implications for high capacity data hiding in the presence of lossy compression, *Proceedings of the International Conference on Information Technology: Coding and Computing*, 16, pp. 2127–2129, March 2000.

81. T. Kalker and F. M. J. Willems, Capacity bounds and constructions for reversible data-hiding, In *Security and Watermarking of Multimedia Contents V, Edward J. Delp, III and Ping W. Wong (eds), Proceedings of the SPIE*, 5020, pp. 604–611, 2003.

82. P. H. W. Wong and O. C. Au, A capacity estimation technique for JPEG-to-JPEG image watermarking, *IEEE Transactions on Circuits and Systems for Video Technology*, 13(8), 746–752, 2003.

83. J. Fridrich, M. Goljan, and R. Du, Lossless data embedding for all image formats, *Proceedings of SPIE Electronic Imaging 2002: Security and Watermarking of Multimedia Contents*, 4675, pp. 572–583.

84. J. Tian, Reversible data embedding using a difference expansion, *IEEE Transactions on Circuits and Systems for Video Technology*, 13(8), 890–896, 2003.

85. A. M. Alattar, Reversible watermark using difference expansion of a generalized integer transform, *IEEE Transactions on Image Processing*, 13(8), 1147–1155, 2004.

86. M. U. Celik, G. Sharma, A. M. Tekalp, and E. Saber, Lossless generalized-LSB data embedding, *IEEE Transactions on Image Processing*, 14(2), 253–266, 2005.

87. L. Kamstra and H. J. A. M. Heijmans, Reversible data embedding into images using wavelet techniques and sorting, *IEEE Transactions on Image Processing*, 14(12), 2082–2090, 2005.

88. B. Chen and G. W. Wornell, Quantization index modulation: a class of provably good methods for digital watermarking and information embedding, *IEEE Transactions on Information Theory*, 47(4), 1423–1443, 2001.

89. I. J. Cox, J. Kilian, T. Leighton, and T. Shamoon, Secure spread spectrum watermarking for multimedia, *IEEE Transactions on Image Processing*, 6(12), 1673–1687, 1997.

90. I. J. Cox, M. L. Miller, and A. McKellips, Watermarking as communication with side information, *Proceedings of the IEEE*, 87(7), pp. 1127–1141, 1999.

91. M.-S. Hsieh, D.-C. Tseng, and Y.-H. Huang, Hiding digital watermarks using multiresolution wavelet transform, *IEEE Transactions on Industrial Electronics*, 48(5), 875–882, 2001.

92. M. Wu, Joint security and robustness enhancement for quantization based data hiding, *IEEE Transactions on Circuits and Systems for Video Technology*, 13(8), 831–841, 2003.

93. N. Bi, Q. Sun, D. Huang, Z. Yang, and J. Huang, Robust image watermarking based on multiband wavelets and empirical mode decomposition, *IEEE Transactions on Image Processing*, 16(18), 1956–1966, August 2007.

94. H. Yang, X. Jiang, and A. C. Kot, Image watermarking using dual-tree complex wavelet by coefficients swapping and group of coefficients quantization, *The 2010 IEEE International Conference on Multimedia and Expo. (ICME'10), content protection and forensics workshop*, Singapore, pp. 1673–1678, July 2010.

95. P. Moulin and J. A. O'Sullivan, Information-theoretic analysis of information hiding, *IEEE Transactions on Information Theory*, 49(3), 563–593, 2003.

96. M. Costa, Writing on dirty paper, *IEEE Transactions on Information Theory*, 29(3), 439–441, 1983.

97. J. Fridrich and D. Soukal, Matrix embedding for larger payload, *IEEE Transactions on Information Forensics and Security*, 1(3), 390–395, 2006.

98. E. T. Lin, A. M. Eskicioglu, R. L. Lagendijk, and E. J. Delp, Advances in digital video content protection, *Proceedings of the IEEE*, 93(1), pp. 171–183, January 2005.

99. F. Hartung and B. Girod, Watermarking of uncompressed and compressed video, *Signal Processing*, 66, 1998.

100. G. Doerr and J.-L. Dugelay, A guide tour of video watermarking, *Signal Processing: Image Communication*, 18, 263–282, 2003.

101. M. D. Swanson, B. Zhu, and A. H. Tewfik, Multiresolution scene-based video watermarking using perceptual models, *IEEE Journal on Selected Areas in Communication*, 16(4), 540–550, 1998.

102. W. Zhu, Z. Xiong, and Y.-Q. Zhang, Multiresolution wavelet-based watermaring of images and video, *IEEE Transactions on Circuits and Systems for Video Technology*, 9, 545–550, 1999.

103. P. W. Chan, M. R. Lyu, and R. T. Chin, A novel scheme for hybrid digital video watermarking: approach, evaluation and experimentation, *IEEE Transactions on Circuits Systems Video Technology*, 15(12), 1638–1649, 2005.

104. P. Campisi, D. Kundur, and A. Neri, Robust digital watermarking in the ridgelet domain, *IEEE Signal Processing Letters*, 11(10), 826–830, 2004.

105. N. Kingsbury, A dual-tree complex transform with improved orthogonality and symmetry properties, *Proceedings of the International Conference Image Processing*, 2, pp. 375–378, September 2000.

106. N. Kingbury, Complex wavelets for shift invariant analysis and filtering of signals, *Journal of Applied and Computational Harmonic Analysis*, 10(3), 234–253, 2001.

107. I. W. Selesnick, R. G. Baraniuk, and N. G. Kingsbury, The dual-tree complex Wavelet Transform, *IEEE Signal Process. Magazine*, 22(6), 123–151, 2005.

108. P. Loo and N. Kingbury, Digital watermarking using complex wavlets, *IEEE International Conference on Image Processing*, 3, pp. 29–32, September 2000.

109. N. Terzija and W. Geisselhardt, Digital image watermarking using complex wavelet transform, *Proceedings of the Workshop Multimedia Security*, pp. 193–198, 2004.

110. A. I. Thomson, A. Bouridane, and F. Kurugollu, Spread transform watermarking for digital multimedia using the complex wavelet transform, *ECSIS Symposium on Bio-inspired, Learning, and Intelligent Systems for Security*, pp. 123–132, 2007.

111. L. E. Coria, M. R. Pickering, P. Nasiopoulos, and R. K. Ward, A video watermarking scheme based on the dual-tree complex wavelet transform, *IEEE Transactions on Information Forensics and Security*, 3(3), 466–474, 2008.

112. G.-J. Yu, C.-S. Lu, and H.-Y. M. Liao, Mean-quantization-based fragile watermarking for image authentication, *Optical Engineering*, 40(7), 1396–1408, 2001.

113. A. M. Alattar, E. T. Lin, and M. U. Celik, Digital watermarking of low bit-rate advanced simple profile MPEG-4 compressed video, *IEEE Transactions on Circuits and Systems for Video Technology*, 13(8), pp. 787–800, August 2003.

114. S. Biswas, S. R. Das, and E. M. Petriu, An adaptive compressed MPEG-2 video watermarking scheme, *IEEE Transactions on Instrumentation and Measurement*, 54(5), 1853–1861, 2005.

115. M. Noorkami and R. M. Mersereau, A framework for robust watermarking of H. 264-encoded video with controllable detection performance, *IEEE Transactions on Information Forensics and Security*, 2(1), 14–23, 2007.

116. B. G. Mobasseri and Y. N. Raikar, Authentication of H. 264 streams by direct watermarking of CAVLC blocks, *SPIE Security, Steganography and Watermarking of Multimedia Contents*, 6505, pp. 65051W-1-5, 2007.

117. D. Zou and J. A. Bloom, H.264/AVC stream replacement technique for video watermarking, *Proceedings of the IEEE International Conference on Acoustics, Speech, and Signal Processing (ICASSP'08)*, pp. 1749–1752, 2008.

118. P.-C. Su, M.-L. Li, and I.-F. Chen, A content-adaptive digital watermarking scheme in H.264/AVC compressed videos, *International Conference on Intelligent Information Hiding and Multimedia Signal Process*, pp. 849–852, 2008.

119. F. Ergun, J. Kilian, and R. Kumar, A note on the limits of collusion resistant watermarks, in *Proceedings of the Eurocrypt'09*, pp. 140–149, 1999.

120. J. Kilian, T. Leighton, L. Matheson, T. Shamoon, R. Tarjan, and F. Zane, Resistance of digital watermarks to collusive attacks, in *Proceedings IEEE International Symposium on Information Theory*, pp. 271–293, August 1998.

121. J. Dittmann, Combining digital watermarks and collusion secure fingerprints for customer copy monitoring, *IEE Seminar on Secure Images and Image Authentication*, 6, pp. 1–6, 2000.

122. W. Trappe, M. Wu, Z. J. Wang, and K. J. R. Liu, Anti-collusion fingerprinting for multimedia, *IEEE Transactions on Signal Processing*, 51(4), 1069–1087, 2003.

123. M. Wu, W. Trappe, Z. J. Wang, and K. J. R. Liu, Collusion-resistant fingerprinting for multimedia, *IEEE Signal Processing Magazine*, 21(2), pp. 15–27, 2004.

124. Z. J. Wang, M. Wu, W. Trappe, and K. J. R. Liu, Group-oriented fingerprinting for multimedia forensics, *EURASIP Journal on Applied Signal Processing*, 14, 2153–2173, 2004.

125. D. Kundur and K. Karthik, Video fingerprinting and encryption principles for digital rights management, *Proceedings of the IEEE*, 92(6), pp. 919–932, June 2004.

126. H. V. Zhao, M. Wu, Z. J. Wang, and K. J. R. Liu, Forensic analysis of nonlinear collusion attacks for multimedia fingerprinting, *IEEE Transactions on Image Processing*, 14(5), 646–661, 2005.

127. Z. J. Wang, M. Wu, H. V. Zhao, W. Trappe, and K. J. R. Liu, Anti-collusion forensics of multimedia fingerprinting using orthogonal modulation, *IEEE Transactions on Image Processing*, 14(6), 804–821, 2005.

128. S. He and M. Wu, Joint coding and embedding techniques for multimedia fingerprinting, *IEEE Transactions on Information Forensics and Security*, 1(2), 231–247, 2006.

129. W. S. Lin, S. He, and J. Bloom, Binary forensic code for multimedia signals: resisting minority collusion attacks, In *Proceedings of the SPIE, Media Forensics and Security XI, Edward J. Delp, Jana Dittmann, Nasir D. Memon and Ping Wah Wong (eds)*, 7254, 2009.

130. A. L. Varna, S. He, A. Swaminathan, and M. Wu, Fingerprinting compressed multimedia signals, *IEEE Transactions on Information Forensics and Security*, 4(3), 330–345, 2009.

131. D. A. Brugioni, *Photo Fakery: The History and Techniques of Photographic Deception and Manipulation*, Dulles, Va: Brassey's, 1999.

132. Spanish MP's Photo Used for Osama Bin Laden Poster, *in BBC News*, 16 January, 2009.

133. Israel: Women Photoshopped from Cabinet Picture to Cater to the Ultra-Orthodox, in *The Huffington Post*, 18 July, 2009.

134. H. Farid, Photo Tampering Throughout the History, Available: http://www.cs.dartmouth.edu/farid/research/digitaltampering/.

135. S. Graham, 10 Famous fake photos, in *Bright Hub-Digtal Photography*, 2009.

136. Rare-tiger photo flap makes fur fly in China, *Science*, 318(5852), 893, 2007.

137. Beijing fires officials over tiger photo hoax, in *The Singapore Straits Times*, 30 June 2008.

138. D. Cyranoski, Verdict: Hwang's human stem cells were all fakes, *Nature*, 439, 122–123, 2006.

139. F. Mosleh (Kodak), Cameras in handsets evolving from novelty to DSC performance, despite constraints, *Embedded.com*, 2008.

140. J. Lewis, Don't just stand and stare, shoot it, too, in *The Singapore Straits Times*, 28 April, 2007.

141. H. Pearson, Forensic software traces tweaks to images, *Nature*, 439, 520–521, 2006.

142. T. Douglas, Shaping the media with mobiles, in *BBC News*, 2005.

143. Z. J. Geradts, J. Bijhold, M. Kieft, K. Kurosawa, K. Kuroki, and N. Saitoh, Methods for identification of images acquired with digital cameras, in *Proceedings of the SPIE*, 4232, 505, 2001.

144. J. Lukas, J. Fridrich, and M. Goljan, Digital camera identification from sensor pattern noise, *IEEE Transactions on Information Forensics and Security*, 1(2), 205–214, 2006.

145. A. Swaminathan, M. Wu, and K. J. R. Liu, Nonintrusive component forensics of visual sensors using output images, *IEEE Transactions on Information Forensics and Security*, 2(1), 91–106, 2007.

146. H. Cao and A. C. Kot, Accurate detection of demosaicing regularity for digital image forensics, *IEEE Transactions on Information Forensics and Security*, 4(4-2), 899–910, 2009.

147. H. Cao and A. C. Kot, Mobile camera identification using demosaicing features, in *Proceedings of the IEEE International Symposium on Circuits and Systems (ISCAS'10)*, pp. 1683–1686, 2010.

148. H. Cao and A. C. Kot, A generalized model for detection of demosaicing characteristics, in *Proceedings of the IEEE International Conference on Multimedia Expo (ICME'08)*, pp. 1513–1516, 2008.

149. H. Cao and A. C. Kot, Accurate detection of demosaicing regularity from output images, in *Proceedings of the IEEE International Symposium on Circuits and Systems (ISCAS'09)*, pp. 497–500, 2009.

150. M. Chen, J. Fridrich, M. Goljan, and J. Lukas, Determining image origin and integrity using sensor noise, *IEEE Transactions on Information Forensics and Security*, 3(1), 74–89, 2008.

151. J. Fridrich, D. Soukal, and J. Lukas, Detection of copy-move forgery in digital images, in *Proceedings of the Digital Forensic Research Workshop (DFRWS'03)*, OH, pp. 55–61, August 2003.

152. A. C. Popescu and H. Farid, Exposing digital forgeries by detecting duplicated image regions, *Technical Report*, TR2004-515, Dartmouth College, Computer Science, 2004.

153. I. Avcibas, S. Bayram, N. Memon, M. Ramkumar, and B. Sankur, A classifier design for detecting image manipulations, in *Proceedings of the International Conference on Image Processing (ICIP'04)*, 4, pp. 2645–2648, 2004.

154. T.-T. Ng and S.-F. Chang, A model for image splicing, in *Proceedings International Conference on Image Processing (ICIP'04)*, 2, pp. 1169–1172, 2004.
155. D.-Y. Hsiao and S.-C. Pei, Detecting digital tampering by blur estimation, in *Proceedings of the First International Workshop on Systematic Approaches to Digital Forensic Engineering*, pp. 254–278, 2005.
156. M. K. Johnson and H. Farid, Exposing digital forgeries by detecting inconsistencies in lighting, in *Proceedings of the ACM Multimedia Security Workshop*, pp. 1–10, 2005.
157. A. C. Popescu and H. Farid, Exposing digital forgeries in color filter array interpolated images, *IEEE Transactions on Signal Processing*, 53, 3948–3959, 2005.
158. C. Popescu and H. Farid, Exposing digital forgeries by detecting traces of resampling, *IEEE Transactions on Signal Processing*, 53, 758–767, 2005.
159. S. Bayram, I. Avcibas, B. Sankur, and N. Memon, Image manipulation detection, *Journal of Electronic Imaging*, 15, 041102, 2006.
160. J. He, Z. Lin, L. Wang, and X. Tang, Detecting doctored JPEG images via DCT coefficient analysis, in *Lecture Notes in Computer Science*, 3953: Springer, Berlin, 423–435, 2006.
161. M. K. Johnson and H. Farid, Exposing digital forgeries through chromatic aberration, in *Proceedings of the ACM Multimedia Security Workshop*, 2006.
162. J. Lukas, J. Fridrich, and M. Goljan, Detecting digital image forgeries using sensor pattern noise, in *Proceedings of SPIE*, 6072, p. 60720Y, 2006.
163. W. Luo, J. Huang, and G. Qiu, Robust detection of region-duplication forgery in digital image, in *Proceedings of the International Conference on Pattern Recognition (ICPR'06)*, pp. 746–749, 2006.
164. M. K. Johnson and H. Farid, Metric measurements on a plane from a single image, *Technical Report* TR2006-579, Department of Computer Science, Dartmouth College, 2006.
165. D. Fu, Y. Q. Shi, and W. Su, A generalized benford's law for JPEG coefficients and its applications in image forensics, in *Proceedings of the SPIE*, 6505, p. 65051L, 2007.
166. M. K. Johnson and H. Farid, Exposing digital forgeries in complex lighting environments, *IEEE Transactions on Information Forensics and Security*, 2(3-1), pp. 450–461, 2007.
167. W. Luo, Z. Qu, J. Huang, and G. Qiu, A novel method for detecting cropped and recompressed image block, in *Proceedings of the IEEE International Conference on Acoustics, Speech and Signal Processing (ICASSP'07)*, 2, pp. 217–220, 2007.
168. S. Ye, Q. Sun, and E.-C. Chang, Detecting digital image forgeries by measuring inconsistencies of blocking artifact, in *Proceedings of the IEEE International Conference on Multimedia and Expo (ICME07)*, 2007, pp. 12–15.
169. A. Swaminathan, M. Wu, and K. J. R. Liu, Digital image forensics via intrinsic fingerprints, *IEEE Transactions on Information Forensics and Security*, 3(1), 101–117, 2008.
170. H. Farid, Exposing digital forgeries from JPEG ghosts, *IEEE Transactions on Information Forensics and Security*, 4(1), 154–160, 2009.
171. S. Bayram, H. T. Sencar, and N. Memon, An efficient and robust method for detecting copy-move forgery, in *Proceedings of the IEEE International Conference on Acoustics, Speech and Signal Processing (ICASSP'09)*, pp. 1053–1056, 2009.
172. A. E. Dirik and N. Memon, Image tamper detection based on demosaicing artifacts, in *Proceedings of the International Conference on Image Processing (ICIP'09)*, pp. 1497–1500, 2009.
173. W. Wang, J. Dong, and T. Tan, Effective image splicing detection based on image chroma, in *International Conference on Image Processing (ICIP'09)*, pp. 1257–1260, 2009.
174. B. Mahdian and S. Saic, Blind authentication using periodic properties of interpolation, *IEEE Transactions on Information Forensics and Security*, 3(3), 2008.
175. B. Mahdian and S. Saic, Detection of copy-move forgery using a method based on blur moment invariants, Forensic Science International, 171(2), September 2007.
176. Y.-Q. Shi, C. Chen, and W. Chen, A natural image model approach to splicing detection, in *Proceedings of the 9th workshop on Multimedia and Security (MM&Sec'07)*, pp. 51–62, September 2007.
177. H. Farid, Exposing digital forgeries in scientific images, in *Proceedings of the ACM Multimedia and Security Workshop*, Geneva, Switzerland, pp. 29–36, 2006.

178. S. Bayram, H. T. Sencar, and N. Memon, An efficient and robust method for detecting copy-move forgery, in *Proceedings of the IEEE International Conference on Acoustics, Speech and Signal Processing (ICASSP'09)*, pp. 1053–1056, 2009.

179. N. Khanna, G. T. C. Chiu, J. P. Allebach, and E. J. Delp, Scanner identification with extension to forgery detection, in *Proceedings of the SPIE*, 6819, p. 68190G, January 2008.

180. S. Lee, D. A. Shamma, and B. Gooch, Detecting false captioning using common-sense reasoning, *Digital Investigation*, 3(1), 65–70, 2006.

181. B. Mahdian and S. Saic, Using noise inconsistencies for blind image forensics, *Image and Vision Computing*, 27, 1497–1503, 2009.

182. M. C. Stamm and K. J. R. Liu, Forensic detection of image manipulation using statistical intrinsic fingerprints, *IEEE Transactions on Information Forensics and Security*, 5(3), 492–506, 2010.

183. X. Pan and S. Lyu, Region duplication detection using image feature matching, *IEEE Transactions on Information Forensics and Security*, 5(4), 857–867, 2010.

184. Y.-F. Hsu and S.-F. Chang, Camera response functions for image forensics: an automatic algorithm for splicing detection, *IEEE Transactions on Information Forensics and Security*, 5(4), 816–825, 2010.

185. F. Huang, J. Huang, and Y.-Q. Shi, Detecting double JPEG compression with the same quantization matrix, *IEEE Transactions on Information Forensics and Security*, 5(4), 848–856, 2010.

186. M. Barni, L. Costanzo, and L. Sabatini, Identification of cut & paste tampering by means of double-JPEG detection and image segmentation, in *Proceedings of the IEEE International Symposium on Circuits and Systems (ISCAS'10)*, pp. 1687–1690, 2010.

187. H. Cao and A. C. Kot, Detection of tampering inconsistencies on mobile photos, *Lecture Notes in Computer Science*, 6526, Digital Watermarking, 105–119, 2011.

188. W. Wang and H. Farid, Exposing digital forgeries in video by detecting double MPEG compression, in *Proceedings of the 9th workshop on Multimedia and Security (MM&Sec'06)*, pp. 35–42, 2006.

189. W. Wang and H. Farid, Exposing digital forgeries in video by detecting duplication, in *Proceedings of the ACM Multimedia and Security Workshop*, pp. 35–42, Texas, USA, 2007.

190. W. Wang and H. Farid, Exposing digital forgeries in interlaced and deinterlaced video, *IEEE Transactions on Information Forensics and Security*, 2(3), 438–449, 2007.

191. M. Kobayashi, T. Okabe, and Y. Sato, Detecting forgery from static-scene video based on inconsistency in noise level functions, *IEEE Transactions on Information Forensics and Security*, 5(4), 883–892, 2010.

192. Y. Su, J. Zhang, and J. Liu, Exposing digital video forgery by detecting motion-compensated edge artifact, in *Proceedings of the International Conference on Computational Intelligence and Software Engineering (CiSE'09)*, pp. 1–4, 2009.

193. G. Cao, Y. Zhao, and R. Ni, Detection of image sharpening based on histogram aberration and ringing artifacts, in *Proceedings of the IEEE International Conference on Multimedia Expo (ICME'09)*, pp. 1026–1029, 2009.

194. H. Farid, Blind inverse gamma correction, *IEEE Transactions on Image Processing*, 10(10), 1428–1433, 2001.

195. W. Wang and H. Farid, Detecting re-projected video, *Lecture Notes in Computer Science*, 5284/2008, 72–86, 2008.

196. H. Yu, T.-T. Ng, and Q. Sun, Recaptured photo detection using specularity distribution, in *Proceedings of the ICIP*, pp. 3140–3143, 2008.

197. H. Cao and A. C. Kot, Identification of recaptured photographs on LCD screens, in *Proceedings of the International Conference Acoustics, Speech and Signal Processing (ICASSP'10)*, pp. 1790–1793, 2010.

198. S. Lyu, Natural image statistics for digital image forensics, *PhD thesis*, Dartmouth College, 2005.

199. T.-T. Ng, S.-F. Chang, J. Hsu, L. Xie, and M.-P. Tsui, Physics-motivated features for distinguishing photographic images and computer graphics, in *Proceedings of the ACM International Conference on Multimedia*, pp. 239–248, 2005.

200. S. Lyu and H. Farid, How realistic is photorealistic?, *IEEE Transactions on Signal Processing*, 53(2), 845–850, 2005.
201. Y. Chen, Z. Li, M. Li, and W.-Y. Ma, Automatic classification of photographs and graphics, in *Proceedings of the IEEE International Conference on Multimedia and Expo (ICME'06)*, 9(12), pp. 973–976, 2006.
202. Y. Wang and P. Moulin, On discrimination between photorealistic and photographic images, in *Proceedings of the International Conference Acoustics, Speech and Signal Processing (ICASSP'06)*, 2, pp. 161–164, 2006.
203. J. Mao, O. Bulan, G. Sharma and S. Datta, Device temporal forensics: an information theoretic approach, in *Proceedings of the ICIP*, pp. 1501–1504, 2009.
204. G. L. Friedman, The trustworthy digital camera: restoring credibility to the photographic image, *IEEE Transactions on Consumer Electronics*, 39, 905–910, 1993.
205. T.-T. Ng, S.-F. Chang, C.-Y. Lin, and Q. Sun, Passive-blind image forensics, in *Multimedia Security Technologies for Digital Rights*, Elsevier, 111–137, 2006.
206. W. Luo, Z. Qu, F. Pan, and J. Huang, A survey of passive technology for digital image forensics, *Frontiers of Computer Science in China*, 1(2), 166–179, 2007.
207. H. T. Sencar and N. Memon, Overview of state-of-the-art in digital image forensics, *Part of Indian Statistical Institute Platinum Jubilee Monograph series titled 'Statistical Science and Interdisciplinary Research'*, World Scientific Press, 2008.
208. H. Farid, A survey of image forgery detection, *IEEE Signal Processing Magazine*, 26(2), 16–25, 2009.
209. B. Mahdian and S. Saic, A bibliography on blind methods for identifying image forgery, *Signal Processing: Image Communication*, 25(6), 389–399, 2010.
210. S. Lian and Y. Zhang, Multimedia forensics for detecting forgeries, *Handbook of Information and Communication Security, Part G*, pp. 809–828, 2010.
211. M. K. Johnson and H. Farid, Exposing digital forgeries through specular highlights on the eye, in *Proceedings of the Information Hiding*, pp. 311–325, 2007.
212. X. Li, B. Gunturk, and L. Zhang, Image demosaicing: a systematic survey, in *Proceedings of the SPIE*, 6822, p. 68221J, 2008.
213. X. Jiang, B. Mandal, and A. C. Kot, Eigenfeature regularization and extraction in face recognition, *IEEE Transactions on Pattern Analysis and Machine Intelligence*, 30(3), 383–394, 2008.
214. A. C. Popescu and H. Farid, Statistical tools for digital forensics, in *Proceedings of the 6th International Workshop on Information Hiding*, pp. 128–147, 2004.
215. T. Gloe, M. Kirchner, A. Winkler, and R. Bohme, Can we trust digital image forensics?, in *Proceedings of the International Conference on Multimedia*, pp. 78–86, 2007.
216. M. Kirchner and R. Bohme, Synthesis of color filter array pattern in digital images, in *Proceedings of the SPIE*, 7254, p. 72540K, 2009.
217. M. Kirchner and R. Bohme, Hiding traces of resampling in digital images, *IEEE Transactions on Information Forensics and Security*, 3(4), 582–592, 2008.
218. M. Kirchner and R. Bohme, Tamper hiding: defeat image forensics, *Lecture Notes in Computer Science*, 4567/2008, 326–341, 2008.
219. U. Uludag, S. Pankanti, S. Prabhakar, and A. K. Jain, Biometric cryptosystems: issues and challenges, *Proceedings of the IEEE*, 92(6), pp. 948–960, June 2004.
220. N. K. Ratha, J. H. Connell, and R. M. Bolle, Cancelable biometrics: a case study in fingerprints, *The 18th International Conference on Pattern Recognition*, 4, pp. 370–373, September 2006.
221. N. K. Ratha, S. Chikkerur, J. H. Connell, and R. M. Bolle, Generating cancelable fingerprint templates, *IEEE Transactions on Pattern Analysis and Machine Intelligence*, 29(4), 561–572, 2007.
222. A. Nagar, K. Nandakumar, and A. K. Jain, Securing fingerprint template: fuzzy vault with minutiae descriptors, *Proceedings of the ICPR*, pp. 1–4, 2008.
223. A. Juels and M. Wattenberg, A fuzzy commitment scheme, *Proceedings of the Sixth ACM Conference on Computer and Communication Security*, pp. 28–36, 1999.
224. A. Juels and M. Sudan, A fuzzy vault scheme, *Designs, Codes and Cryptography*, 38(2), 237–257, 2006.

225. T. C. Clancy, N. Kiyavash, and D. J. Lin, Secure smart-card-based fingerprint authentication, *Proceedings of the ACMSIGMM 2003 Multimedia, Biometrics Methods and Applications Workshop*, pp. 45–53, 2003.

226. Y. Sutcu, Q. Li, and N. Memon, Protecting biometric templates with sketch: theory and practice, *IEEE Transactions on Information Forensics and Security*, 2(3), pp. 503–512, 2007.

227. K. Nandakumar, A. K. Jain, and S. Pankanti, Fingerprint-based fuzzy vault: implementation and performance, *IEEE Transactions on Information Forensics and Security*, 2(4), 744–757, 2007.

228. Pim Tuyls, Practical biometric authentication with template protection, In *LNCS*, T. Kanda, A. K. Jain and N. K. Ratha (eds.), 3546, pp. 436–446, 2005.

229. A. T. B. Jin, D. N. C. Ling, and A. Goh, Biohashing: two factor authentication featuring fingerprint data and tokenised random number, *Pattern Recognition 37*, 2245–2255, 2004.

230. Y. C. Feng and P. C. Yuen, Selection of distinguish points for class distribution preserving transform for biometric template protection, *ICB 2007, Lecture Notes in Computer Science 4642*, 636–645, 2007.

231. Y. Sutcu, H. T. Sencar, and N. Memon, A geometric transformation to protect minutiae-based fingerprint templates, *Proceedings of the SPIE*, 6539, pp. 65390E-1-8, 2007.

232. Y. Sutcu, Q. Li, and N. Memon, Secure biometric templates from fingerprint-face feature, *Proceedings of the IEEE Conference on Computer Vision and Pattern Recognition*, pp. 1–6, June 2007.

233. X. D. Jiang and W. Yau, Fingerprint minutiae matching based on the local and global structures, *Proceedings of the 15th International Conference on Pattern Recognition*, 2, pp. 1038–1041, 2000.

234. J. Qi and Y. Wang, A robust fingerprint matching method, *Pattern recognition*, 38, 1665–1671, 2005.

235. S. Chikkerur and N. K. Ratha, Impact of singular point detection on fingerprint matching performance, *Proceedings of the fourth IEEE workshop automatic identification advanced technologies*, pp. 207–212, July 2005.

236. K. Nilsson and J. Bigun, Localization of corresponding points in fingerprints by complex filtering, *Pattern Recognition Letters*, 24, 2135–2144, 2003.

237. T. E. Boult, W. J. Scheirer, and R. Woodworth, Revocable fingerprint biotokens: accuracy and security analysis, *Proceedings of the IEEE conference on computer vision and pattern recognition (CVPR '07)*, pp. 1–8, 2007.

238. H. Yang, X. Jiang, and A. C. Kot, Generating secure cancelable fingerprint templates using local and global features, *The second IEEE International Conference on Computer Science and Information Technology*, pp. 645–649, Beijing, China, August 2009.

239. N. K. Ratha, J. H. Connell, and R. M. Bolle, Secure data hiding in wavelet compressed fingerprint images, in *Proceedings of the ACM Multimedia Workshops 2000*, Los Angeles, CA, pp. 127–130, October 2000.

240. A. K. Jain and U. Uludag, Hiding biometric data, *IEEE Transactions on Pattern Recognition and Machine Intelligene*, 25(11), pp. 1494–1498, November 2003.

241. S. Li and A. C. Kot, Privacy protection of fingerprint database, *IEEE Signal Processing Letters*, 18(2), pp. 115–118, 2011.

242. Y. Dodis, L. Reyzin, and A. Smith, Fuzzy extractors: how to generate strong keys from biometrics and other noisy data, *Advances in Cryptology-EUROCRYPT 2004, Lecture Notes in Computer Science*, 3027, 523–540, 2004.

243. Q. Li, M. Guo, and E.-C. Chang, Fuzzy extractors for asymmetric biometric representations, *IEEE Computer Society Conference on Computer Vision and Pattern Recognition Workshops (CVPRW '08)*, pp. 1–6.

16

Introduction to Biometry

Carmelo Velardo, Jean-Luc Dugelay, Lionel Daniel, Antitza Dantcheva, Nesli Erdogmus, Neslihan Kose, Rui Min, and Xuran Zhao

CONTENTS

The term *biometric* comes from the ancient Greek βιος (bios: life) and μετρον (metron: to measure, to count). Both concepts indicate that there is something related to life (the human nature) that can be measured or counted. Biometry is the science that tries to understand how to measure characteristics which can be used to distinguish individuals. Humans have developed such skills during the evolution: the brain has specialized areas to recognize faces [1] and to link identities with specific patterns (behavioral or physical [2]). Researchers in the biometry field have always tried to automatize such processes making them suitable to be run on a computer or a device.

The study of biometric patterns led to the definition of requirements that have to be respected to make a human trait feasible to be used in a recognition process. A biometric trait can be summarized as a characteristic that each person should have (*universality*), in which any two persons should present some differences (*distinctiveness*), which should not drastically vary over a predefined period of time (*permanence*), and that should be quantitatively measurable (*collectability*).

Furthermore, biometric traits can be divided into the two following classes: *physical* and *behavioral*. To the former class belongs the face appearance, the pattern of the iris and the fingerprint, the structure of the retinal veins as well as the shape of the ear. Each of these traits can be additionally subdivided into *genotypic* and *randotypic*; the former indicates a correlation with some genetic factors (like hereditary similarities in twins), the latter describes traits that develop randomly during the fetal phase. Behavioral biometrics develop as we grow older and are not *a priori* defined. To those traits belong, the gait, or even the keystroke pattern (the way of typing on a keyboard).

In this chapter, we will provide a broad view of what a biometric system is and which techniques are commonly employed to measure and compare systems' performance. An overview of the current biometric traits that are part of the International Civil Aviation Organization (ICAO) standard will be provided. A full presentation and comparison of biometric traits are out of the scope of this work. In Section 16.1, a general discussion defines the components of a biometric system and the tools to measure and compare systems' performance. Section 16.2 concentrates on several biometric traits and their associated recognition techniques. Section 16.3 presents the new trends and challenges that the biometric panorama offers today, and a series of examples of working systems and applications will be presented as well. Finally, the conclusion summarizes the potential of biometrics and the challenges still open.

16.1 Biometric for Person Authentication

Since the beginning of biometry's history the discriminative power of some traits (e.g., fingerprints) was used for the identification and tracking of criminals. Recently, after the attacks of September 11, 2001, a urge of security promoted the use of biometry as a tool that could ease the prevention of such events [3]. International organizations started gathering information about all passengers crossing international borders by using biometrics. Thus, biometric authentication started being used in many airports and train stations. Biometric traits are mainly used to perform identification in order to allow/deny access to restricted areas, to index a set of pictures by person, or to enable decrypting data in biometric-enabled devices. Thus, biometry is now employed in many technologies not always related to security.

Biometric-based authentication can nowadays complement or replace both *knowledge-based* and *token-based* authentications, which require knowledge from an authenticated user to know a secret password and/or to keep a personal token like a physical key. Although a biometric trait cannot be forgotten like a password, or lost like a token, it may be unconsciously disseminated like fingerprints or DNA (e.g., hair); this raises ethical concerns about users' privacy.

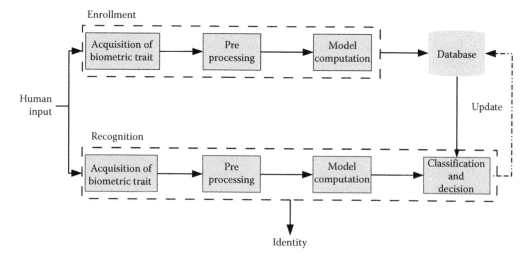

FIGURE 16.1
Scheme of a general biometric system and its modules: enrollment, recognition, and update. Typical interactions among the components are shown.

In Figure 16.1, we present the typical scheme of a general biometric system; three main components can be identified: enrollment, recognition, and update. While the first two components are required, the last step is optional in an automatic application. In this section, a description will be provided for each of the modules, while the last part will focus on techniques that allow performance evaluation of biometric systems.

16.1.1   Enrollment Module

The enrollment module is the first important part of a biometric system. Its first step requires acquiring the biometric trait with a sensor (e.g., fingerprint scanner). Generally, a preprocessing step follows. It extracts important information from the raw representations of the data. The extracted information is then ready to be directly stored as a record in a database (usually associated with the ID of the subject), which constitutes the model of the subject.

16.1.1.1   Acquisition

For capturing biometric traits, specific sensors are needed so that information can be digitized for further processing. It is clear that since each trait is different from the others, the capturing devices need some specificities. Visual patterns like face, iris, and ear generally require cameras to record their image; for other biometrics like fingerprint, specialized scanners are needed.

Moreover, a biometric trait may bring different information, for example, face can be represented with both its texture and shape and also its thermal response; for this reason, a variety of sensors could be used on the same biometric trait. In other cases, the biometric trait can show enhanced characteristics under different acquiring conditions; iris, for example, has shown better performance under infrared lighting, imposing the use of infrared light emitters and near-infrared cameras.

Additionally, appearance variations can cause a temporary *failure to acquire* that forces the user to repeat the acquisition step. The repetition may also be done just to generate a subject model that is more robust to possible variations.

Two classes of sensors exist: *contact* sensor, in which the biometric traits must touch the acquisition device surface (e.g., fingerprint scanners), or *contact-less* sensor if this requirement is not necessary (e.g., camera for face recording). For each of the two kinds of sensors, a variety of technologies can be used to perform the scan. Each one could provide improvements in speed, accuracy, or even overcome adverse conditions (e.g., recording at night using infrared cameras).

The acquisition module is critical as all the following modules depend on the quality of the acquisition. Moreover, the acquisition stability should hold between enrollment and recognition so that the recognition error rate remains low.

16.1.1.2 Preprocessing

After digitizing the biometric trait, several preprocessing techniques might be involved to improve the quality of the recording, to reduce the dimensionality of the captured data, or to extract important features from the biometric trait.

Improving the quality of the recording may include: the restoration of corrupted image areas due to acquisition noise; the compensation of unwanted external elements, like illumination, face expression, and occlusions in face recognition; or the enhancement of some features, like binarizing the ridges in a fingerprint. Many preprocessing techniques aim at extracting some content from the given data. For example, eye detection may be useful for identifying the location of facial features for face registration, or for a successive iris detection. Another useful tool employed in biometric data preprocessing is dimensionality reduction. Mathematical tools like principal component analysis (PCA) or linear discriminant analysis (LDA) allow the extraction of important variations of given biometric traits by discarding redundant information, thus reducing data size.

Features extraction is part of preprocessing; it consists in the extraction of salient information from the given trait. Such saliences should be representative enough to allow a classification of the biometric trait; transforming the raw biometric data to a vector $V = [f_1, f_2, \ldots f_n]$, where f_i represents a single feature. The concept is very close to dimensionality reduction but in this case the information is selected according to given heuristics and algorithms (like Local Binary Patterns LBP [4] or SIFT [5] descriptors for face recognition). More detailed descriptions on how to perform such operations for several biometrics will be given in Section 16.2.

16.1.1.3 Model Computation

After the acquisition and preprocessing steps, data have to be elaborated to create a model (also known as template) to be stored in a secured database. The model is used as reference for the identity of the user. When the user will try to access the system again, his/her scans will be matched against the template.

Directly storing raw scans in the database is a straightforward way to create user models. This has some advantages but also some disadvantages: on one hand the system is not bounded to the use of a single matching strategy that could be changed if needed, and on the other hand the amount of memory demanded to the system could be bigger and the subject privacy lower. Another approach is to save the features, previously extracted in the preprocessing step, so that the required memory is smaller, and the recognition step would not have to compute two models to perform the matching. However, using this paradigm, the system is bounded to the matching algorithm; if this has to be changed, all the systems have to be restructured.

Other approaches are similar to dictionary-like systems; here, a large set of features are analyzed and the discriminant ones are kept as the basis (*words*). Each biometric trait can be described as a collection of such words. Descriptors like SIFT and LBP were both used successfully in dictionary-like methods [6].

16.1.2   Recognition Module

The recognition module utilizes the information stored in the database during the enrollment phase. The biometric modality is sensed again using a similar (but not necessarily the same) device priorly used for the enrollment. However, because of the natural variability of the biometric trait (e.g., face expression, pupil dilatation), or because of the acquisition conditions (e.g., illumination, finger position on the scanner) some differences may arise. For this reason, the elaboration of the model has to consider them and create a template robust to variations.

The recognition module has to verify an identity (*verification*) or to recognize a person in a pool of candidates (*identification*). During *verification*, a subject claims his/her own identity to a system (e.g., through the use of a token) that collects the biometric trait and decides if the extracted features match the ones of the model corresponding to the claimed identity. We can summarize the process as trying to reply to the question: "Is the subject identity the one he/she claims?" In an *identification* paradigm the question changes: "Who is this person?" In this case the system provides an option on the user identity, to be chosen among all the enrolled clients. While in the first case the problem is a *one-to-one* matching (also called *open*), in the second paradigm the match is *one-to-many* (also known as *closed*).

16.1.3   Update Module

As anticipated, each scan of the same biometric may result in a different representation of the same features. Those variations may depend on the variability of conditions at the sensing time, or on the intrinsic nature of biometric traits which do not always appear the same. To compensate for the variability during the enrollment phase one could record many acquisitions of the same trait allowing a generalization of the client's template. Nevertheless, the time can have a critical impact on the biometric trait itself. While some biometric traits are not modified as the subject grows older (e.g., iris), some others are subjected to degradations because of aging effects (e.g., face). Some of those variabilities can be compensated by the use of an *update* module. The purpose of such a step is to tune the enrolled model to make it more robust against natural variations. Thus, when the biometric system recognizes a client within a sufficient confidence range, it can extract features from the current biometric modality, and update the corresponding database entry. Some biometric traits which are more subjected to variations because of their nonpermanent nature (e.g., voice) will particularly benefit of the model update; in these cases such an element should always be part of the system.

16.1.4   Classification and Fusion of Multiple Modalities

Classification is the problem that involves the identification of subpopulations in a set of input data. For biometrics it means finding a transformation that leads from the feature space to a class space. The purpose of a biometric authentication system is mainly to retrieve the identity of a person, or to verify that a person is who he/she claims to be. The verification problem is a binary classification (genuine vs. impostor), whereas the identification problem is an *n*-ary classification, where $n \in \mathbb{N}$ is the number of mutually exclusive identities. A person, represented as a feature set, is classified by measuring its similarity to the template of each class; the person is then said to belong to the class that has the most similar template(s). Classification task is to minimize the *intraclass* variations (i.e., the variations which a biometric trait experiences because of natural conditions) and maximizing the *interclass* variations which occur between different persons. For classification, a similarity measure has to be defined. Such operation measures the distance of a feature projected into the classification space against all the templates.

In a multimodal biometric system, several different feature sets per person are usually available. In order to classify a person according to those data, one could concatenate the features to form a larger array that will be classified by a general-purpose classifier (e.g., support vector machine (SVM), neural network). However, the computational complexity of the classification may exponentially increase in accordance with the bigger dimensionality of the new feature set. An alternative to feature concatenation is *classification fusion*, which merges the results of several low-dimensional classifications.

Combining diverse* biometric systems aim at improving the characteristics of the overall system. For example, the system will be more universal: if a biometric trait is missing because of a failure at acquisition time or because of a handicap, the other modalities could compensate. Also, circumventing the system will become increasingly complex as an impostor will have to deploy several spoofing techniques. Besides being multimodal (more than one modalities, like face, fingerprint, gait), a system can be multisensor (more than one sensor per modality), multisample (more than one acquisition per modality), and multialgorithm (many classifiers per features, and different kind of features extracted per modality).

The information flow of a multimodal biometric can be fused at several levels.

- Fusion at the *sensor* level consolidates raw sensory data before feature extraction; images can be fused at pixel level (stitching and mosaicking), phases of radio or sound waves can be aligned (beam-forming), and data from one sensor (e.g., 2D camera) can help to interpolate the data of another sensor (e.g., 3D camera).

- Fusion at the *feature* level consolidates data either by merging them, or by concatenating them. The former strategy can be used for updating and improving the templates; this requires compatible feature spaces and data normalization. The latter method linearly increases the feature space, and hence exponentially increases the enrollment and authentication computations. A solution to that could be represented by space reduction techniques like PCA and LDA, and assumptions about feature independence.

- Fusion at the *score* level consolidates matcher outputs. Similar to feature fusion, score fusion needs the scores to be compatible, which involves normalization techniques that would be discussed in the following. In case the purpose of the biometric system is to identify people, matchers may provide a ranking of the enrolled people; fusing of ranks has a strong advantage over fusing of scores: ranks are comparable and voting theory provides axiomatic solutions to *rank*-level fusion.

- Fusion at the *decision* level is similar to rank-level fusion, where voting theory provides axiomatic solutions.

---

* Biometrics systems are *diverse* if the cause of an error in one system is unlikely to affect the other systems.

The role of low-level fusions like sensor or feature fusions is to separate the discriminative information from the environmental noise. On the other hand, the role of high-level fusions like score, rank, and decision fusions is to find a consensus among opinions, which are possibly weighted by reliability measures induced from quality measures (e.g., blurriness of the 2D face images).

Score-level fusion is common since an industrial biometric system usually provides scores, and since score is a richer information than rank or decision. However, the scores might be incompatible since they come from different sources. Contrary to ranks and decisions, the scores are usually normalized before being fused. Near-linear normalization techniques, like minmax ($Score_{Normalized} = (Score - min)/(max - min)$), scale and shift the scores to map them onto a common domain. Other normalizations can be applied in order to align score distributions, but such normalizations need a deeper training step during which the score distributions are estimated.

16.1.5 Performance Evaluation: Robustness and Security

A practical way to categorize a biometric system is to analyze it under the three following aspects. The first one considers the *performance* of such a system, like the achievable recognition accuracy, speed, and throughput. User *acceptability* is another important parameter as it gives a measure of how many people are willing to use that biometric system in their life. This could be influenced by benefits, like the access to fast lanes in airports, as well as from cultural factors. A third parameter is the *circumvention* which represents how easily the system can be compromised and spoofed by subjects with malicious intent. Hereafter, we will focus on how to quantitatively measure the system's performance.

16.1.5.1 Evaluation Database

In order to simulate system behavior under normal usage conditions, a set of labeled data are used. These labeled acquisitions allow for measuring the system performance in the presence of simulated clients and impostors. The use of such database guarantees scientific repeatability since data do not vary from one test to the other. Usually, a dataset is associated with one or many protocols, which are defined rules establishing how to perform the experiments. Databases and protocols are used as testbeds for algorithms and as a common platform for the comparison of different systems.

Existing datasets for face recognition are face recognition technology database (FERET), the first one to include big temporal variations, and face recognition grand challenge (FRGC), which includes both 2D and 3D scans of participants' faces. Other databases exist for other biometric traits like iris (UBIRIS and CASIA (Iris Image Database)). Fingerprint databases can be synthesized with the synthetic fingerprint generator (SFINGE) toolbox [7].

Multiple modalities require large databases containing several traits per user. Multiple biometric grand challenge (MBGC), for example, is a multimodal database as it contains both irises and face scans. When multimodal datasets do not exist, it is common practice to mix biometric traits from different databases to create synthetic (*chimeric*) users. Each of these identities will be valid from the experimental point of view, even if it is the result of a mix of different persons' samples.

The research community developed a set of mathematical tools which allow measuring the recognition performance of a biometric system [8]. Here we focus on how to measure

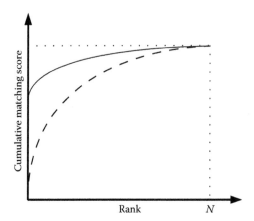

FIGURE 16.2
The lines represent two examples of cumulative matching characteristic curve plots for two different systems. The solid line represents the system that performs better. N is the number of subjects in the database.

recognition performance, while intentionally omit speed and bandwidth measures that can be tuned by simply investing more resources to the system.

The recognition rate is the most used performance measure, but it alone does not provide information on the system's behavior. Indeed, while testing a biometric system, the position of the true positive is also important; we refer to this as the *rank* of the true positive. The higher the test rank, the better the system recognizes that user. Thus, a better biometric system always ranks the person's identity higher. The most common and compact representation of the biometric system's performance is represented by the cumulative matching score (see Figure 16.2), which shows the recognition rate as a function of the rank given to the person's identity by the biometric system.

For what concerns the verification mode, two errors are relevant: *false acceptance* and *false rejection* error. In the first case, the system accepts an impostor as a client, while in the second case, it wrongly rejects a client, considering it as an impostor. In both cases, two error rates can be computed over the entire system. We can refer then to *false acceptance rate* and *false rejection rate*, or *FAR* and *FRR*.

The distributions of the scores for both impostors and clients are represented in Figure 16.3a as well as *FAR* and *FRR*. Intuitively, by modifying the defined threshold, we vary the performance of our system, making it more or less restrictive. Each threshold defines a different *operating point* of the system. Then, testing all the operating points means varying the threshold and recording different values of *FAR* and *FRR*: by plotting those pairs of rates we obtain the *receiver operating characteristic* curve (or simply *ROC* curve). An example can be found in Figure 16.3b. The *equal error rate* corresponds to the point where the FAR is equal to the FRR; it is one operating point particularly relevant as it is used to compare systems' performances. *Security* is another important aspect of a biometric system performance and it should not be confused with *robustness*, which is the recognition capability of the system. Still explored by the research community, security is a key point for the development of commercial applications as it deals with many different aspects and working conditions of the biometric system itself. For this reason, here, we will focus principally on what concerns the security of a system from the biometric point of view: spoofing.

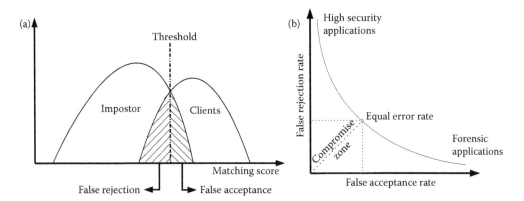

FIGURE 16.3
Typical examples of biometric system graphs. The two distributions (a) represent the client/impostor scores; by varying the threshold, different values of FAR and FRR can be computed. An ROC curve (b) is used to summarize the operating points of a biometric system; for each different application, different performances are required to the system.

16.1.5.2 Spoofing

It has been shown that conventional biometric techniques, like fingerprint or face recognition, are vulnerable to attacks. One of the most important vulnerabilities is *spoofing* attacks, where a person tries to masquerade as someone else by falsifying data and thereby gaining an illegitimate access to the system. Spoofing can be defined as a class of attacks on a biometric security system where a malicious individual attempts to circumvent, at the acquisition phase, the correspondence between the biometric data acquired from an individual and his/her identity. In other words, the malicious individual tries to introduce fake biometric data into a system that does not belong to that individual, either at enrollment and/or recognition [9].

Currently, there is a strong need for efficient and reliable solutions for detecting and circumventing such kind of attacks. The exact techniques for spoofing vary depending on the particular type of biometric trait involved [9]. For example, a prosthetic fake finger can be used for fingerprint spoofing. Early works on the field showed that gelatin and conductive silicon rubber may be used for that purpose [10,11]. On the other hand, for iris spoofing, a high-resolution image of an iris can be used to pass the security check. Also, face appearance-based systems suffer from similar vulnerabilities as a masked fake face; a video of the user or even a photo of the client can be used for spoofing purposes.

The typical countermeasure to a spoofing attack is liveness detection. The aim of liveness testing is to determine if the biometric data are being captured from a live user who is physically present at the point of acquisition [12]. In Ref. [13], liveness detection is grouped in four different ways. The first way is to use available sensors to detect in the signal a pattern characteristic of liveness/spoofing (*software-based*). The second method is to use dedicated sensors to detect an evidence of liveness (*hardware-based*), which is not always possible to deploy. Liveness detection can also exploit *challenge-response* methods by asking the user to interact with the system. Another way is to use recognition methods intrinsically robust against attacks (*recognition-based*). Along those direct methods for liveness detection, multiple modalities could also be exploited (e.g., voice could be jointly used with face recognition in video-based solutions). Ref. [13] presents some examples of countermeasures for face recognition systems that involve, for the first group, skin reflectance/texture/spectroscopy analysis, as well as the use of 3D shape of the head as a way to measure the liveness of a system's user. In the second group, we can mention active lighting, multicamera face analysis, and detection of temperature. The third group involves the challenge–response approach, synchronized lip movement, and speech for liveness detection. For the last group, multispectral scanning of the face may be useful to distinguish live/spoofed face.

16.2 ICAO Biometrics

In this section, we will provide a brief overview of several biometric traits: face, iris, and fingerprint. We selected these because they are included in the open standard of biometric passport created by ICAO. For each biometric we will refer to the scheme in Figure 16.1 using the keywords: *acquisition, preprocessing, model computation,* and *classification*.

16.2.1 Face

The human face is a fundamental element in our social lives because it provides a variety of important signals: for example, it carries information about identity, gender, age, and emotion. For this reason, human face recognition has been a central topic [14,15] in the field of person identification.

Acquisition Face is one of the easiest biometrics to digitize as a normal camera is usually enough. Nevertheless, a camera can only extract the texture information, thus incurring a series of problematics like pose and illumination variations. For this reason, several methods are nowadays explored that make use of innovative sensors like 3D scanners or thermal cameras to extract more information from a face.

Capturing face from a distance makes such a biometric trait nonintrusive, easy to collect, and in general well accepted by the public. However, it is still a very challenging task, as faces of different persons share global shape characteristics, while face images of the same person are subject to considerable variability. This is due to a long list of factors, including facial expressions, illumination conditions, pose, facial hair, occlusions, and aging. Although much progress has been made over the past three decades, automatic face recognition (AFR) is largely considered as an open problem.

In this section, we present two widely adopted approaches to AFR from still intensity images: one deals with face as a whole, the other one is a local feature-based approach. An exhaustive review is out of the scope of this chapter due to the large body of existing work. On the other hand, a brief summary on the recent technologies in AFR and several novel techniques is given at the end of the section.

16.2.1.1 Eigenfaces

Preprocessing Kirby and Sirovich first outlined that the dimensionality of the face space, that is the space of variation between images of human faces, is much smaller than the dimensionality of a single face considered as an arbitrary image [16]; later on, Turk and Pentland applied these considerations into practice to the problem of AFR [17]. As a useful approximation, one may consider an individual face image to be a linear combination of a small number of face components. Such components are called *eigenfaces* and can be derived from a set of reference face images.

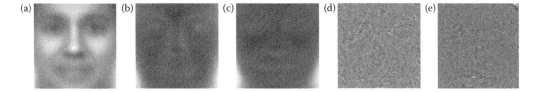

FIGURE 16.4
(a) Average face and (b),(c) eigenfaces 1 to 2, (d),(e) eigenfaces 998-999 as estimated on a subset of 1000 images of the FERET face database.

The name eigenfaces comes directly from the use of eigenvector and eigenvalue decomposition (also known as eigen-decomposition process) used in PCA (cf. [16]) describes how to deduce from a set of data a decreased number of components. Thanks to PCA that for each given set of faces we can deduce a subspace that discards redundant information of the original space. In other words, we obtain from the face space, a set of orthogonal vectors (eigenfaces) that represent the main variations of the original input (Figure 16.4).

To formally describe the PCA process let $\{x_1, \ldots, x_N\}$ be a set of reference or training faces and $\overline{x}$ be the average face. Then we can obtain the centered version of our faces set by computing $d_i = x_i - \overline{x}_i$. Finally, if $\Delta = [d_1, \ldots, d_N]$, the *scatter* matrix S is defined as

$$S = \sum_{i=1}^{N} \delta_i \delta_i^{\mathrm{T}} = \Delta \Delta^{\mathrm{T}}. \tag{16.1}$$

The optimal subspace P_{PCA} is the one that maximizes the scatter of the projected faces:

$$P_{\mathrm{PCA}} = \arg \max |PSP^{\mathrm{T}}|, \tag{16.2}$$

where $|.|$ is the determinant operator. The solution to Equation 16.2 is the subspace spanned by the eigenvectors (also eigenfaces) $[e_1, e_2, \ldots e_K]$ corresponding to the K largest eigenvalues (λ_k) of the scatter matrix S:

$$Se_k = \lambda_k e_k. \tag{16.3}$$

As the number of images in the training set is generally lower than the dimension of the image space, that is the number of pixels in an image, the number of nonzero eigenvalues is $N - 1$ (cf. [14]). We recall that, since the data are normalized to be zero mean, one of the eigenvectors and the corresponding eigenvalue are zero valued.

Due to the size of the scatter matrix S, the direct estimation of its eigenvalues and eigenvectors is difficult. They are generally estimated either through a singular value decomposition (SVD) of the matrix Δ or by computing the eigenvalues and eigenvectors of $\Delta^{\mathrm{T}}\Delta$. It should be underlined that eigenfaces themselves are not usually plausible faces but only directions of variation between face images.

Model computation Each face image x_i is represented by a point w_i in the K-dimensional space: $w_i = [w_i^1, w_i^2, \ldots w_i^K]^{\mathrm{T}} = P_{\mathrm{PCA}} \times \delta_i$. Each coefficient w_i^k is the projection of the face image on the kth eigenface e_k and represents the contribution of e_k in reconstructing the input face image. In other words, by using the eigenfaces we are able to reconstruct the original appearance of the faces that we used to build the space. Additionally, PCA guarantees that, for the set of training images, the mean-square error introduced by truncating the expansion after the Kth eigenvector is minimized.

Classification To find the best match for an image of a person's face in a set of stored facial images, one may calculate the distances between the vector representing the new face and each of the vectors representing the stored faces, and then choose the image yielding the smallest distance. The distance between faces in the face subspace is generally based on simple metrics such as L1 (city-block), L2 (Euclidean), cosine, and Mahalanobis distances (see [18]).

16.2.1.2 Local Binary Patterns

In order to provide a broad view on the two main approaches which are usually used in pattern recognition tasks, we just discussed a *holistic* approach that treats the signal in its entirety. In this part, we will present a typical *local* approach based on the extraction of local features from the original signal.

Preprocessing The LBP [19] operator originally forms labels for the image pixels by thresholding the 3×3 neighborhood of each pixel with the center value and considering the result as a binary number. A histogram of these 2^8 labels is created as the texture descriptor by collecting the occurrences. Due to its computational simplicity and its invariance against monotonic gray-level changes. LBP algorithm rapidly gained popularity among researchers, and numerous extensions have been proposed which prove LBP to be a powerful measure of image texture [20,21]. The LBP method is used in many kinds of applications, including image retrieval, motion analysis, biomedical image analysis, and also face image analysis. The calculation of the LBP codes can easily be done in a single scan through the image. The value of the LBP code of a pixel (x_c, y_c) is given by

$$\text{LBP}_{P,R} = \sum_{p=0}^{P} s(g_p - g_c)2^P \tag{16.4}$$

where g_c corresponds to the gray value of the center pixel (x_c, y_c), g_p refers to gray values of P equally spaced pixels on a circle of radius R, and s defines a Heaviside step function.

Model computation Successively, the histograms that contain data about the distributions of different patterns such as edges, spots, and plain regions are built; the classification is performed by computing their similarities.

Classification Several measures have been proposed for histograms [4] such as Histogram intersection and Log-likelihood statistic. One of the most successful in the case of LBP is chi-square statistic here defined as

$$\chi^2(V', V'') = \sum_{i}^{n} \frac{(V_i' - V_i'')^2}{V_i' + V_i''} \tag{16.5}$$

where V' and V'' are feature histograms and the special case $(0/0) = 0$.

16.2.1.3 New Technologies and Recent Studies

Most algorithms have been proposed to deal with individual images, where usually both the enrollment and testing sets consist of a collection of facial pictures. Image-based recognition strategies have been exploiting only the physiological information of the face, in particular its appearance encoded in the pixel values of the images. However, the recognition performances of these approaches [22] have been severely affected by different kinds of variations, like pose, illumination, and expression changes.

For AFR, various new algorithms and systems are still frequently proposed, targeting one of these different challenges. One of these methods is proposed by Wright et al. [23] for robust face recognition via sparse representation. In this framework, face recognition is casted as penalizing the L1-norm of the coefficients in the linear combination of an over complete face dictionary. *Sparse representation*-based classification has been demonstrated to be superior to the common classifiers such as *nearest neighbor* and *nearest subspace* in various subspaces like Eigenfaces and Fisherfaces.

Some others of these emerging techniques exploit both static and dynamic information from video sequences. There exist approaches that adopt still-image-based techniques to video frames, as well as ones that introduce spatio-temporal representation, in which dynamic cue of the human face contributes to recognition [24,25].

On the other hand, as the 3D capturing process becomes faster and cheaper, 3D face models are also utilized to solve the recognition problem, especially under pose, illumination, and expression variations where 2D face recognition methods still encounter difficulties. Numerous methods have been presented, which treat the 3D facial surface data as 2.5D depth maps, point clouds, or meshes. Even though 3D face recognition is expected to be robust to variations in illumination, pose, and scale, it does not achieve perfect success and additionally introduces some critical problems like 3D mesh alignment [26]. Moreover, intraclass variations related to facial expressions, which cause nonrigid deformations on the facial surface, still need to be dealt with.

16.2.2 Iris

The iris is the colored circular region around the pupil, the small dark hole of the eye through which the light passes to focus on the retina. What makes this part of the eye so peculiar for biometry is the presence of a particular pattern which is determined in a random manner during the fetal life phase. Even though the presence of pigments can increase during childhood, the pattern of the eye does not vary during the life span of a person; along with the strong randomness of the pattern those characteristics make iris a suitable biometric trait. The first automatic iris recognition system is due to the early work of Daugman [27], which first described and introduced algorithms to exploit iris random pattern for people recognition.

An example of iris pattern is shown in Figure 16.5; both a colored version and a near-infrared version are provided. Following the first early works on iris recognition, we can distinguish two main methodologies that are nowadays used to perform iris matching. The former technique refers to the Daugman method, the latter to the work of Wildes.

16.2.2.1 Daugman's Approach

Preprocessing An overview of the methodology can be found in Figure 16.6. The system requires that the eye of the subject is in the field of view of the camera. An automatic mechanism improves the sharpness of the iris image by maximizing the middle- and high-energy bands of the Fourier spectrum, by modifying the focus parameter of the camera, or providing information to the user which will move his/her head accordingly. A deformable template [28] is then used to seek for the position of the eye. The iris can then be described by three parameters: radius r and center position coordinates of the circle, x_0, and y_0. This kind of approximation, at first accepted for iris recognition systems, is no more considered valid nowadays. The latest works [29] deal with the nonuniformity due to deformations

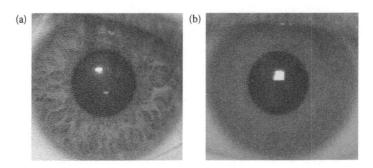

FIGURE 16.5
(**See color insert.**) A colored (a) and a near-infrared (b) version of the same iris.

caused by the inner nature of the iris pattern or due to partial occlusion (caused by eyelashes and eyelids).

After its detection, a normalization of the iris image is needed to compensate for other effects that might influence the scale of the iris. An example could be different acquisition distance or variable light conditions, which make the pupil muscles dilate or shrink the iris pattern. The normalization of Daugman's scheme makes use of polar coordinates to identify each location of the circular pattern. The angular and radial position are then normalized, respectively, between 0° and 360° and 0 and 1. The latter normalization assumes that the iris is modified linearly in its contractions; also, this technique was questioned and explored in a later work [30].

Model computation: After the extraction of the iris boundaries, a technique is needed to encode the information carried by the pattern. Daugman uses convolution of the image with a set of bidimensional Gabor filters.

Classification: The result of such convolution is then quantized and represented as a binary vector which encode the sign of real and imaginary parts of the filter response. A total of 256 bytes is used to represent the signatures. The comparison of two signatures can be made using different distance measures, the one Daugman originally proposed was based on

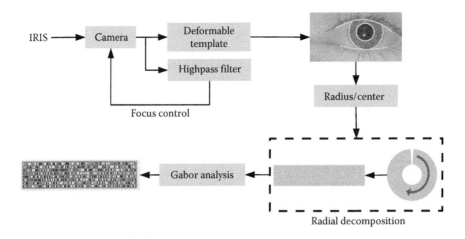

FIGURE 16.6
A scheme that summarizes the steps performed during Daugman approach.

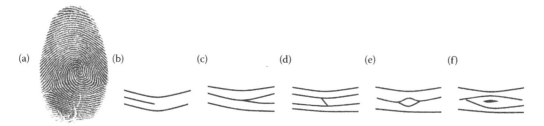

FIGURE 16.7
Example of a fingerprint (a), and of the minutiae: (b) termination, (c) bifurcation, (d) crossover, (e) lake, and (f) point or island.

a XOR operation which simply measures the quantity of different symbols for two given signatures.

16.2.2.2 Wildes' Approach

Preprocessing: Wildes presents a different methodology to the iris recognition task. His approach consists of a binary thresholding of the eye image to perform the localization and segmentation of the iris. In contrast to Daugman's scheme, this solution allows being more robust to noise perturbations while detecting the iris, but in the meanwhile makes the segmentation less sensitive to finer variations.

Classification: The differences between the two approaches continue in the matching of the extracted signatures. Wildes method uses the Laplacian of Gaussian filter at multiple scales to derive a model of the pattern, and a correlation measure to compute the matching score.

The main differences between the two systems can be summarized as follows. Daugman's method is simpler than the latter; on the other hand, Wildes's method allows for a higher information payload as it does not quantize the filter response. This allows better discriminatory power at the cost of a less compact representation of the iris pattern, and a higher computational complexity.

16.2.3 Fingerprint

Every time we touch something, we release information about our identity involuntarily. This information is encoded in the small crests and valleys that draw lines on our fingertips. Those lines are called ridges and together form the complex pattern of the fingerprints. Fingerprints are a random disposition of human skin cells that develops as the fetus grows on the mother's womb during pregnancy. The randomness makes even the fingerprints of mono-zygotic twins completely different.

The fingerprint lines pattern creates several types of configurations which allow differentiating global features; if the fingerprint is considered in its whole appearance, or local features as those lines joint or bifurcate. In Figure 16.7, a fingerprint is shown together with some of the peculiar local features called *minutiae*.

Acquisition The oldest method to acquire a fingerprint is to cover the surface of the fingertips with a layer of ink, and then press the finger against a piece of paper. Nowadays, scanners exist that use several techniques (optic, thermal, electromagnetic, or ultrasounds) to digitize the structure of the fingerprint [31]. The surface of the scanner can be of different size, from very small (as in the case of swiping sensors) to very big (as for full hand fingerprint systems). In the first case, the finger is swiped over the sensor that reconstructs the fingerprint by stitching together the single slices sensed at time, in the second case, all fingerprints of both hands can be digitized together.

Preprocessing: Several methods exist in the literature about evaluation of similarities between two fingerprints. Very few algorithms operate directly on the gray-scale image, in general, each matching algorithm is performed only after a preprocessing of the finger-print image [31]. During this phase, several steps can be performed to enhance the pattern formed by the ridges, or to extract information regarding the global and the local structure of the pattern.

Classification: We can identify three main classes for these matching algorithm: correlation based, minutiae based, and ridge feature based. In the first case, two fingerprint representations are superimposed and a correlation is computed pixel by pixel while varying the rotation and translation of one image over the other; the correlation measures the similarity between the two. The minutiae-based approach is the method applied by human experts, and so far largely popular as an automatic system. It is based on the extraction of minutiae configuration, in both the template and query fingerprint; the matching phase seeks the alignment between the two sets which maximizes the number of corresponding minutiae. For the latter method, the ridge features are extracted from the image; those can be extracted more reliably than minutiae, but in general they are of minor discriminatory power. The first two methods (correlation based and minutiae based) can be considered as subcases of the ridges feature based.

Many different factors may influence the matching process. First of all the position of the finger on the scanner (rotation and translation) may affect the visibility of portions of the fingerprint; also, the humidity of the skin can lead to partial images. Nonetheless, the elasticity of the skin combined with the pressure of the finger on the scanner applies nonlinear transformation to the acquired image. Another source of error is the presence of injuries on the skin surface (voluntary or involuntary) that may lead to temporary or permanent impossibility of correctly acquiring the fingerprint. Moreover, statistics show that for certain population categories (e.g., elderly people), the identification through fingerprint might be inappropriate, and that for 4% of the population, the quality of fingerprint would not suffice for the process [32].

16.3 Biometrics New Trends and Application

Biometrics has been increasingly adopted in security applications, both in the governmental and in the private industry sector. State-of-the-art security systems include at least one biometric trait and this tendency is rising. More and more industries, including e-commerce, cars, and cell phones, are embracing the related benefits. The widespread usage of biometric technology advances associated research, increasing the related performances and innovations.

16.3.1 Soft Biometrics

The latest addition of soft biometrics (also called *semantic* [33]) can increase the reliability of a biometric system and can provide substantial advantages: soft biometric features reveal biometric information, they can be partly derived from hard biometrics, they do not require enrollment, and they can be acquired nonintrusively without the consent and cooperation of an individual.

Soft biometrics are physical, behavioral, or adhered human characteristics classifiable in predefined human-compliant categories. These categories are, unlike in the classical biometric case, established and time proven by humans with the aim of differentiating individuals. In other words, the soft biometric trait instances are created in a natural way, used by humans to distinguish their peers.

Traits accepting this definition include but are not limited to age, gender, weight, height, hair, skin and eye color, ethnicity, facial measurements and shapes, the presence of beard, mustache, and glasses, color of clothes, and so on. An increase in resources (such as an improved resolution of the sensors or an increased computational capability) can lead to expanding of the traits amount and furthermore of the trait instances. We refer to trait instances as the sub-categories that soft biometric traits can be classified into. Example for trait instances of the trait hair color could be blond, red, and black. The nature of soft biometrics features can be binary (e.g., the presence of glasses), continuous (height), or discrete (ethnicity) [34].

Characteristics can be differentiated according to their distinctiveness and permanence, whereby distinctiveness corresponds to the power of a trait to distinguish subjects within a group and permanence relates to the time invariability of a trait. Both these characteristics are mostly in a lower range for soft biometrics than they are for classical biometrics (cf. hair color, presence of beard, presence of glasses, etc.). Furthermore it is of interest with which estimation reliability a trait can be extracted from an image or a video. With respect to these three qualities, namely distinctiveness, permanence, and estimation reliability, the importance of a soft biometric trait can be determined. We note that the classification of soft biometric traits can be expanded and aspects like accuracy and importance can be evaluated or deduced, respectively, depending on the cause for application.

Recently, soft biometric traits have been employed to preliminary narrow down the search in a database, in order to decrease the computational time for the classical biometric trait. A further application approach is the fusion of soft biometric and classical biometric traits to increase the system performance and reliability. Recently, soft biometric systems have also been employed for person recognition and continuous user authentication [35].

Jain et al. first introduced the term *soft biometrics* and performed related studies on using soft biometrics [36,37] for prefiltering and fusion in combination with classical biometric traits. Recent works perform person recognition [34,38] and continuous user authentication [35] using soft biometric traits. Further studies evolve traits extraction algorithms concerning eye color [39], weight [40], clothes color [41], or predictability of human metrology [42].

16.3.2 Applications

Applications that make use of biometrics can generally be divided into three main categories: forensic, government, and commercial. They mainly differ for performance requirements. Here we seek to provide some examples that may clarify some of the practical uses of biometrics.

16.3.2.1 Forensic Applications

The first category is in general devoted to security or control and prevention. This is mainly due to the intrinsic nature of biometric traits that ease the automatic identification task.

Historically, the first application of biometrics was theorized and put into practice by Alphonse Bertillon which invented the "Bertillonage," a system that categorizes and recognizes people according to a biometric signature composed by anthropometric measures. This system was replaced by the more reliable fingerprint recognition system introduced by Francis Galton. The Federal Bureau of Investigation (FBI) followed; assuming responsibility for managing the U.S. national fingerprint collection in 1924 [43], fingerprint matching was performed by human experts. Nowadays, the National Crime Information Center contains up to 39 million criminal records, which are stored electronically and can be accessed by 80,000 law enforcement agencies for data on wanted persons, missing persons, gang members, as well as other information related to other crimes.

16.3.2.2 Government Applications

Many governments started exploring the possibility of using biometrics for identification. Lately, the NEXUS program started as joint collaboration of Canada and the United States. It is designed to facilitate approved, low-risk travelers to cross the United States–Canada border as fast as possible. The clients of the system (only citizens or permanent residents) can use self-check-in gates to speed up the paperwork for crossing the border. The applicant's fingerprints, photographs, and irises are scanned and stored in order to verify his/her identity as needed [44]. A similar project exists between the United States and Mexico under the name of Secure Electronic Network for Travelers Rapid Inspection.

Another promising, though challenging, project is the Multipurpose National Identity Card project, a national Indian project that contemplates the collection of multiple biometric modalities (face appearance, fingerprints, and iris) of a large percentage of the Indian population. According to the specifications of the project [45], the biometric identification profile will be voluntary and for every resident (not only Indians); it will be composed of a random 12-digit number and it will just provide yes/no reply to each authentication query to avoid privacy issues. Many challenges have to be faced, from the acquisition of multiple modalities, to the matching techniques which will involve a very large number of queries performed in parallel. In order to promote each citizen to have a biometric profile, the Indian authorities will include the possibility of availing services provided by the government and private sector (e.g., banks, insurances, and benefits).

An additional use of biometrics to guarantee the identification task is the use of biometric traits inside official documents. For example, the Biometric Passport (or ePassport) contains a microchip which can store fingerprint scans as well as face appearance, and iris images. Those traits can be read from automatic gates at airports, train stations, or state borders. Nowadays, more than 70 countries already adhered to this new identification tool try to standardize biometric identification across nations.

16.3.2.3 Commercial Applications

One of the first commercial systems used for general purposes was the speaker recognition module created for MacOS 9 from Apple [46] which allowed to log in and protect files of a user recognized by his/her voice. Lately, embedded fingerprint systems have seen an increase in popularity and are present in most laptops and computers; they allow to override the password-typing method for both operating system's login and Website's password management. Fingerprint was as well introduced in some portable storing device to be used as keys to decrypt data hidden in the device's memory. VeriSign technology for face

FIGURE 16.8
The two interfaces of Google Picasa (a) and Apple iPhoto (b). Both the systems summarize all the persons present in the photo collection. The two programs give the opportunity to look for a particular face among all the others.

recognition was lately added to Lenovo computers which now allow a full-face recognition system to login into the operating system.

Additionally to access control, biometric for commercial application has seen important uses in daily applications like the one for photo management. Examples in this sector are iPhoto from Apple and Picasa from Google (Figure 16.8). Both the systems implement two different (and proprietary) versions of a face recognition module both allowing face tagging over the entire set of pictures so that the virtual albums can be easily indexed by person. Enabling this function the user is able to divide its multimedia collection by persons easing the usability of such systems. A similar technology was lately announced by Facebook to ease the task of tagging friends' pictures. Some airports already exploit biometric technology for identifying passengers. The Stansted Airport of London, in collaboration with Accenture [47], deployed an automatic border control system which makes use of a face recognition module to speed up the security control of passengers. Paris Roissy-Charles de Gaulle Aiport as well provides security and fast lane access through the use of a fingerprint recognition system. Both the systems exploit biometric passport mechanism previously presented, in order to match the live data with the templates stored either in a database or in the microchip that the passport carries.

16.4 Conclusions

There is no doubt that the exploration of the biometrics domain has reached great heights and its commercial exploitation has just started blooming. Standards are set and regularly improved so that more commercial applications are created by a number of company which operate in this domain (e.g., L1, Safran-Morpho, Thales, etc). Commercial applications make this technology available to an increasing number of people, insomuch as one of the modern biometric challenges is represented by large-scale systems. The UIDAI (Unique Identification Authority of India) Indian project is one of these; it targets 1.2 billion users, the entire Indian population [45]. Such tremendous increase of a system's users yields numerous challenges. As the scale of such systems grows, problems in both speed and accuracy of employed algorithms have to be carefully addressed.

As for many new technologies an ensemble of concerns about privacy and security is also arising for biometrics. In this direction, several aspects we briefly presented in this chapter are currently discussed. One of the aims is establishing quantitative measures of systems security, as well as increasing robustness against spoofing attacks, and guaranteeing privacy for the users of the biometric system.

Security is the main domain of application for biometry but many other applications are finding the discriminative power of the biometric traits useful. Instant login systems exist for personal computers and banking systems which utilize face or fingerprint recognition. The multimedia explosion driven by social networks like Facebook or Flickr is empowered by automatic indexing of pictures by AFR systems.

Furthermore, research boundaries are expanding: shape of ears, stride and gait analysis, soft biometrics, and even *physiological* biometrics (e.g., brain and heart activities) are new experimented traits. This new variety of biometrics creates the ground for new algorithms, theories, and applications. Research still continues and many unsolved challenges exist in this domain. Addressing all these questions will definitely establish biometrics as a leading technology for human identification in the coming years.

References

1. C. A. Nelson, The development and neural bases of face recognition. *Infant and Child Development*, 10(1–2):3–18, 2001.
2. A. J. O'Toole, D. A. Roark, and H. Abdi, Recognizing moving faces: A psychological and neural synthesis, *Trends in Cognitive Sciences*, 6(6):261–266, 2002.
3. US NSTC. National Science and Technology Council. Biometrics in government post 9/11. 2008.
4. T. Ahonen, A. Hadid, and M. Pietikäinen. Face recognition with local binary patterns. In T. Pajdla and J. Matas, eds, *Computer Vision—ECCV 2004*, pp. 469–481. Springer, Berlin, 2004.
5. M. Bicego, A. Lagorio, E. Grosso, and M. Tistarelli. On the use of sift features for face authentication. In *Conference on Computer Vision and Pattern Recognition Workshop, 2006. CVPRW '06.*
6. D. Liu, D. M. Sun, and Z. D. Qiu. Bag-of-words vector quantization based face identification. In *Second International Symposium on Electronic Commerce and Security, 2009. ISECS '09.* vol. 2, pp. 29–33, May 2009.
7. R. Cappelli, D. Maio, D. Maltoni, and A. Erol. Synthetic fingerprint-image generation. In *ICPR*. IEEE Computer Society, Quebec City, Canada, 2000.
8. P. Grother, R. Micheals, and P. Phillips. Face recognition vendor test 2002 performance metrics. In J. Kittler and M. Nixon, eds, *Audio- and Video-Based Biometric Person Authentication*, 2688, chapter Lecture Notes in Computer Science, p. 1057. Springer, Berlin, 2003.
9. M. Pagani. *Encyclopedia of Multimedia Technology and Networking*. London, United Kingdom, Cybertech Publishing, 2005.
10. T. Matsumoto, H. Matsumoto, K. Yamada, and S. Hoshino. Impact of artificial gummy fingers on fingerprint systems. In *Proceedings of SPIE*, vol. 4677, pp. 275–289. Citeseer, 2002.
11. T. Matsumoto. Artificial fingers and irises: importance of vulnerability analysis. In *7th International Biometrics 2004 Conference and Exhibition*, London, UK, 2004.
12. B. Toth. Biometric liveness detection. *Information Security Bulletin*, 10(8), 2005.
13. Tabularasa EU project. http://www.tabularasa-euproject.org/.
14. F. Perronnin. *A probabilistic model of face mapping applied to person recognition*. PhD thesis, Thesis, 11, 2004.
15. W. Zhao, R. Chellappa, P. J. Phillips, and A. Rosenfeld. Face recognition: A literature survey. *ACM Computing Surveys*, 35(4):399–458, 2003.
16. M. Kirby and L. Sirovich. Application of the Karhunen-Loeve procedure for the characterization of human faces. *IEEE Transactions on Pattern Analysis and Machine Intelligence*, 12(1):103–108, 1990.

17. M. A. Turk and A. P. Pentland. Face recognition using eigenfaces. In *IEEE Computer Society Conference on Computer Vision and Pattern Recognition, 1991. Proceedings CVPR '91*, pp. 586–591, June 1991.

18. J. R. Beveridge, K. She, B. A. Draper, and G. H. Givens. A nonparametric statistical comparison of principal component and linear discriminant subspaces for face recognition. In *Proceedings of the 2001 IEEE Computer Society Conference on Computer Vision and Pattern Recognition, 2001. CVPR 2001*, vol. 1, 2001.

19. T. Ojala, M. Pietikäinen, and D. Harwood. A comparative study of texture measures with classification based on featured distributions. *Pattern Recognition*, 29(1):51–59, 1996.

20. X. Wang, T. X. Han, and S. Yan. An hog-lbp human detector with partial occlusion handling. In *2009 IEEE 12th International Conference on Computer Vision*, pp. 32–39, 292009-oct.2 2009.

21. T. Ojala, M. Pietikainen, and T. Maenpaa. Multiresolution gray-scale and rotation invariant texture classification with local binary patterns. *IEEE Transactions on Pattern Analysis and Machine Intelligence*, 24(7):971–987, July 2002.

22. P. J. Phillips, P. Grother, R. J. Micheals, D. M. Blackburn, E. Tabassi, M. Bone, North Fairfax Dr, and United Kingdom. Facial recognition vendor test 2002: evaluation report, 2003.

23. J. Wright, A. Y. Yang, A. Ganesh, S. S. Sastry, and Yi Ma. Robust face recognition via sparse representation. *IEEE Transactions on Pattern Analysis and Machine Intelligence*, 31(2):210–227, 2009.

24. F. Matta and J. Dugelay. A behavioural approach to person recognition. In *2006 IEEE International Conference on Multimedia and Expo*, pp. 1461–1464. IEEE, 2006.

25. M. Paleari, C. Velardo, B. Huet, and J.-L. Dugelay. Face dynamics for biometric people recognition. In *MMSP'09, IEEE International Workshop on Multimedia Signal Processing*, October 5–7, 2009.

26. A. F. Abate, M. Nappi, D. Riccio, and G. Sabatino. 2d and 3d face recognition: A survey. *Pattern Recognition Letters*, 28(14):1885–1906, 2007.

27. J. G. Daugman. High confidence visual recognition of persons by a test of statistical independence. *IEEE Transactions on Pattern Analysis and Machine Intelligence*, 15(11):1148–1161, 2002.

28. A. L. Yuille, P. W. Hallinan, and D. S. Cohen. Feature extraction from faces using deformable templates. *International Journal of Computer Vision*, 8(2):99–111, 1992.

29. J. Daugman. New methods in iris recognition. *IEEE Transactions on Systems, Man, and Cybernetics, Part B: Cybernetics*, 37(5):1167–1175, 2007.

30. Z. Wei, T. Tan, and Z. Sun. Nonlinear iris deformation correction based on gaussian model. *Advances in Biometrics*, 4642:780–789, 2007.

31. D. Maltoni, D. Maio, A. K. Jain, and S. Prabhakar. *Handbook of Fingerprint Recognition*. Springer-Verlag New York, 2009.

32. A. K. Jain, S. Prabhakar, and S. Pankanti. A filterbank-based representation for classification and matching of fingerprints. In *International Joint Conference on Neural Networks, 1999. IJCNN '99*, vol. 5, pp. 3284–3285, 1999.

33. S. Samangooei, M. Nixon, and B. Guo. The use of semantic human description as a soft biometric. In *Proceedings of BTAS*, 2008.

34. A. Dantcheva, C. Velardo, A. D'angelo, and J. L. Dugelay. Bag of soft biometrics for person identification: New trends and challenges. *Multimedia Tools and Applications*, 2:739–777, 2011.

35. K. Niinuma, U. Park, and A. K. Jain. Soft biometric traits for continuous user authentication. *Transactions on Information Forensics and Security*, 5(4):771–780, 2010.

36. A. K. Jain, S. C. Dass, and K. Nandakumar. Soft biometric traits for personal recognition systems. In *Proceedings of ICBA*, pp. 1–40. Springer, 2004.

37. A. K. Jain, S. C. Dass, and K. Nandakumar. Can soft biometric traits assist user recognition? In *Proceedings of SPIE*, 5404, pp. 561–572, 2004.

38. A. Dantcheva, J. L. Dugelay, and P. Elia. Person recognition using a bag of facial soft biometrics (bofsb). In *Proceedings of MMSP*, 2010.

39. C. Boyce, A. Ross, M. Monaco, L. Hornak, and X. Li. Multispectral iris analysis: A preliminary study. In *Proceedings of CVPRW*, 2006.

40. C. Velardo and J. L. Dugelay. Weight estimation from visual body appearance. In *Proceedings of BTAS*, 2010.
41. A. D'Angelo and J. L. Dugelay. People re-identification in camera networks based on probabilistic color histrograms. In *Proceedings of Electronic Imaging*, 2011.
42. D. Adjeroh, D. Cao, M. Piccirilli, and A. Ross. Predictability and correlation in human metrology. In *Proceedings of WIFS*, 2010.
43. Federal Bureau of Investigation. http://www.fbi.gov/about-us/cjis/fingerprints_biometrics.
44. NEXUS. http://www.cbp.gov/xp/cgov/travel/trusted_traveler/nexus_prog/nexus.xml.
45. UIDAI. http://uidai.gov.in/.
46. D. Pogue. *Mac OS X: The Missing Manual*. O'Reilly & Associates, Inc. Sebastopol, CA, USA, 2002.
47. Accenture. http://www.accenture.com/us-en/pages/success-london-stansted-automated-border-clearance-trial-summary.aspx.

17

Watermarking and Fingerprinting Techniques for Multimedia Protection

Sridhar Krishnan, Xiaoli Li, Yaqing Niu, Ngok-Wah Ma, and Qin Zhang

CONTENTS

17.1 Introduction

Digital multimedia is an advancing technology that fundamentally alters our daily life. One critical obstacle is the ease with which digital media content can be accessed and altered. Digital multimedia content can be easily copied and transmitted over networks without any loss of fidelity. This can result in serious loss of revenue for the rightful content owners. Therefore, it is critical for content owners to take steps to safeguard their content from illegal duplication.

Digital rights managements (DRMs) were created to prevent the growing piracy. DRM is the collection of technologies from different disciplines used for copyright, including multimedia signal processing, coding theory, information theory, the human visual/audio system (HVS/HAS), probability theory, and detection and estimation theory. Algorithms derived from multiple disciplines have been incorporated and employed in watermarking and fingerprinting. The objective of this chapter is to highlight some quality research efforts in watermarking and fingerprinting that address the challenges in the emerging area of multimedia protection. Thus, readers will have an overview of the state of the art in this field.

Watermarking provides ways to embed into, and retrieve from, the multimedia data, a secondary signal that is imperceptible to the eye/ear and is well bonded to the original data. For a well-designed watermark, there are many requirements, including imperceptibility, robustness, and capacity. The capacity is the amount of information that is encoded by the watermark. Imperceptibility is another property of the watermark: the distortion, which the watermarking process is bound to introduce, should remain imperceptible to a human observer. Finally, the robustness of a watermarking scheme can be seen as the ability of the detector to extract the hidden watermark from some altered watermarked data. There exists a complex trade-off between these three parameters. In order to maintain the image/audio quality and at the same time increase the probability of the watermark detection, it is necessary to take the HVS/HAS into consideration when engaging in watermarking research. To embed watermark information into the original data, watermarking applies minor modifications to it in a perceptually invisible manner, where the modifications are related to the watermark information, as illustrated in Figure 17.1.

Even though watermarking and fingerprinting have some similar characteristics, for example imperceptibility, robustness, and capacity, *fingerprinting* has an intrinsic difference from watermarking, uniqueness. Traditionally, fingerprinting has been used to generate a short numeric sequence (*fingerprint*) associated with a multimedia signal for the identification of the signal. Recently, another type of fingerprinting, called *watermark fingerprinting*, has been used for a different purpose. Watermark fingerprinting uniquely watermarks (content-related or noncontent-related) each legal copy of a media file so that the copy can be traced to the individual who acquired it [20]. Fingerprinting and watermark fingerprinting can uniquely represent the signal, which is analogous to human fingerprint that uniquely represents an individual person, either by the extracted digital *fingerprint* or by the embedded digital *fingerprint*.

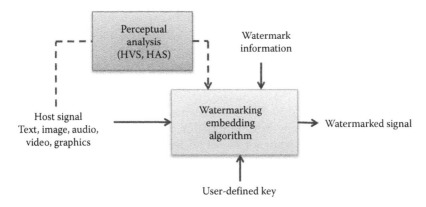

FIGURE 17.1
Generic watermarking process.

Fingerprinting technique (fingerprint extraction) consists of a front-end and a fingerprint modeling block [17]. As Cano et al. [17] studied, the front-end procedure consists of five steps: preprocessing, framing and overlap, transforms, feature extraction, and postprocessing. Among these steps, the pivotal two are the transforms and the feature extraction because they mainly impact on the quality of the fingerprint.

The transforms convert the original signal from one domain (e.g., time, spatial) to the other (e.g., frequency, time-frequency (TF)) for more representable features to be extracted. The most common transformation is the discrete Fourier transform. Some other transforms have been proposed: the discrete cosine transform (DCT), the haar transform (one of the wavelet transforms), or Walsh–Hadamard transform [44].

Since most of the real-world signals are nonstationary, the study and analysis of nonstationary signals is receiving more and more attention in the scientific community. For signal analysis, time series and frequency spectrum contain all the information about the underlying processes of signals. But by themselves, the best representations of nonstationary processes may not be well presented. Due to the time-varying behavior, techniques which give joint TF information are needed to analyze nonstationary signals. The TF distribution is best suited for nonstationary signals that need all the three axes of time, frequency, and energy to represent them efficiently. TF decomposition breaks down a signal into elementary building blocks—TF atoms, to represent the inner structure and the processes. It can better reveal the joint TF relationship and can be useful in determining the nature of the many kinds of nonstationary signals. The success of any TF modeling lies in how well it can model the signal on a TF plane with optimal TF resolution. Different analysis techniques to decompose signals into TF atoms (or basis functions) have been developed. Fourier analysis and wavelet transform are the most common examples of such signal analysis models.

Most researches are interested in the Fourier transform and its derivatives. For example, the fast Fourier transform was used by Brück et al. [15] Seo et al. [43] and Haitsma and Kalker [26]. Ramaligan et al. [40] proposed an audio fingerprinting scheme based on the short-time Fourier transform. On the other hand, Lu [35] used features based on the wavelet coefficients of the continuous wavelet transforms.

Compared to these methods that are based on orthonormal transform, the sparse representation techniques are under development in the field, such as matching pursuit (MP) and its variation molecular matching pursuit (MMP). These techniques aim to find the representation of a signal x as a weighted sum of elements (or atoms) from an over-complete dictionary usually offering better performance with its capacity for efficient signal modeling [28]—using less bases to represent the signal efficiently. On a transformed representation, additional transformation, feature extraction, is applied in order to generate the final feature vector. Even if there is a great diversity of algorithms, the objective is again to reduce the dimensionality and, at the same time, to increase the invariance to distortions. For example, in the case of images, they could be the color, the shape, and changing intensity; in the case of audio, they could be the harmonicity, bandwidth, and loudness presented in a mathematical model. The fingerprint modeling procedure defines the final fingerprint representation, for example a vector, a trace of vectors, a codebook, a sequence of indexes to hidden Markov models sound classes, and so on.

Watermark fingerprinting technique mainly includes fingerprint generation and embedding steps. The technique that generates a unique embeddable fingerprint is required. For this concern, some research groups give their reasonable conclusions. The authors [36] claim that content-based fingerprint generation technique is more favored than the fingerprint generation over codebook because of its obvious shortcomings in the security. First, since the codebook is designed independent of the digital entertainment media, it can be shown that these codes are not appropriately specialized for digital entertainment media. Second, common collusion attacks on uniquely fingerprinted copies using the codebook design do not result in perceptual changes. On the other hand, from the perspective of the correlation between fingerprints, two major classes of fingerprint generation strategies have been proposed [47]. One is called orthogonal fingerprinting and the other is called correlated fingerprinting. Either of them has advantages and disadvantages with respect to generation complexity, detection complexity, resistance to collusion attacks, and supportable individuals. For more details, refer [47]. The fingerprint embedding is the same as watermarking. It is designed to ensure the watermarked fingerprint is secure and imperceptible, as illustrated above.

The first contribution is concerned with spatio-temporal just noticeable distortion model guided video watermarking. HVS makes final evaluations on the quality of videos that are processed and displayed. Perceptual watermarking should take full advantage of the results from HVS studies. Just noticeable distortion (JND), which refers to the maximum distortion that the HVS does not perceive, gives us a way to model the HVS accurately and can serve as a perceptual visibility threshold to guide video watermarking. JND estimation for still images has been relatively well developed. Previous watermarking schemes have only partially used the results of the HVS studies. Many video watermarking algorithms utilize visual models for still images to increase the robustness and transparency. The perceptual adjustment of the watermark is mainly based on the spatial JND model [1,12,13]. An image-adaptive watermarking procedure based on Watson's spatial JND model was proposed in [13]. In [1], the DCT-based watermarking approach uses Watson's spatial JND model in which the threshold consists of spatial frequency sensitivity, luminance sensitivity and contrast masking. An energy-modulated watermarking algorithm based on Watson's spatial JND model was proposed in [12]. During the modulation, Watson's perceptual model is used to restrict the modified magnitude of DCT coefficients. The main drawback of utilizing visual models for still images in video watermarking to increase the robustness and transparency is that it does not satisfactorily take into account the temporal dimension. Thus, the obtained watermark is not optimal in terms of imperceptibility and robustness since it does not consider the temporal sensitivity of the human eye. Motion is a specific feature of video, and temporal HVS properties need to be taken into account to design more efficient watermarking algorithms. An effective spatio-temporal JND model-guided video watermarking scheme in the DCT domain is introduced here. The watermarking scheme

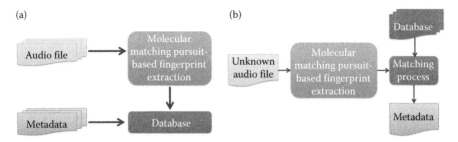

FIGURE 17.2
Fingerprint extraction/registration and identification procedure for legacy content protection. (a) Populating the database and (b) Identifying the new file.

is based on the design of additional accurate JND visual model which incorporates spatial contrast sensitivity function (CSF), temporal modulation factor, retinal velocity, luminance adaptation, and contrast masking. The proposed watermarking scheme, where the JND model is fully used to determine scene-adaptive upper bounds on watermark insertion, allows us to provide the maximum strength transparent watermark.

The second and third contributions focus on MMP-based audio fingerprinting for legacy content protection and fingerprinting for traitor tracing in peer-to-peer (P2P) network. As for the application of identifying the legacy content, the owner can claim the ownership of a piece of multimedia content by registering the content's fingerprint (the compact representation of the content) with the owner's name to a centralized database. If the fingerprint is robust enough to withstand a certain degree of the content alternation, it can be used to prevent anyone else from making a false claim of the ownership of the same piece of multimedia content. The procedure of the fingerprinting scheme is demonstrated in Figure 17.2. For the traitor tracing purpose, the same copies of a file that are distributed to different users should have different fingerprints embedded, so that when the user shares the file with an illegal user, the owner should be able to identify the infringer by sorting the distorted fingerprint in the file. Figure 17.3 depicts the structure of the scheme.

As described in [16], for the content-based fingerprinting, an ideal fingerprint should fulfill several requirements:

- Uniqueness: to have a high discrimination power among all the files of a large database.

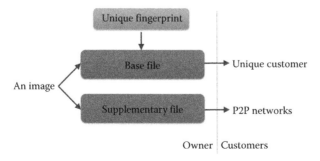

FIGURE 17.3
Structure of the proposed P2P fingerprinting method.

- Compactness: to reduce the high dimensionality of the original file, and to simplify the computation process.
- Robustness to distortions: the system must be able to recognize the file even if it has been severely distorted.

In addition, for a watermark fingerprinting technique such as the traitor tracing application, the imperceptibility needs to be considered.

The fingerprint as the compacted version of the original file, can be successfully obtained by using the following techniques: (1) *Principal component analysis (PCA)*, a data-driven approach (without involving any external basis) to decompose the data which can be further utilized to narrow down the dimension of the fingerprint by only taking into account the most principal component. (2) *Discrete wavelet transform (DWT)*, a very good method for signal decomposition, is generally based on the description of the signal in a given orthogonal wavelet basis. The orthogonal wavelet coefficients support multiresolution of the signal so that the signal with lower resolution consists of sparser supporting coefficients. (3) *Sparse approximation algorithms*. Different from the previous two techniques that decompose the signal into weighted orthogonal bases, sparse approximation algorithms find an optimal decomposition of the signal, including a number of elements higher than the dimension of the signal. The collection of such elements is called a redundant dictionary. Among the existing sparse approximation algorithms, l_1-optimisation principles (Basis Pursuit [19], least absolute shrinkage and selection operator (LASSO)) and greedy algorithms (e.g., MP [37] and its variants) have in particular been extensively studied and proved to have good decomposition performance [41]. The "basis pursuit" approach minimized the norm l_1 of the decomposition resorting to linear programming techniques. The approach is of larger complexity, but the solution obtained yields generally good properties of sparsity, without reaching, however, the optimal solution that would have been obtained by minimizing l_0. The "matching pursuit" approach consists in optimizing incrementally the decomposition of the signal, by searching at each stage the element of the dictionary which has the best correlation with the signal to be decomposed, and then by subtracting from the signal the contribution of this element. This greedy algorithm is suboptimal but it has good properties for what concerns the decrease of the error and the flexibility of its implementation.

Section 17.2 gives a description of a spatio-temporal JND-based watermarking scheme. Section 17.3 elaborates the discussions about the recently proposed fingerprinting scheme based on one of the variants of the MP approach—MMP for the legacy content identification. Section 17.4 introduces the newly designed complete fingerprinting technique for traitor tracing in a P2P file sharing network by utilizing both the advantages of the DWT and PCA techniques. Section 17.5 concludes this chapter.

17.2 A Spatio-Temporal JND Model-Guided Video Watermarking Scheme

In the following, we discuss an effective spatio-temporal JND model-guided video watermarking scheme in the DCT domain. The watermarking scheme, where JND models are fully used to determine scene-adaptive upper bounds on watermark insertion, allows us to provide more strength transparent watermark.

17.2.1 Combined Spatio-Temporal JND Model

Spatio-temporal JND is an efficient model incorporating spatial and temporal sensitivity of the human eye to represent the additional accurate perceptual redundancies for digital videos. Here, we compute the visibility threshold of each DCT coefficient with a combined spatio-temporal JND model which is illustrated in Figure 17.4.

17.2.1.1 Eye Track Analysis

Motion is a specific feature of video imagery. Human eyes tend to track a moving object to keep the retinal image of the object in the fovea. It is necessary to take into account the observers' eye movements to see how well the traced objects can be seen during the presentation of motion imagery for human perceptual visibility analysis.

There are three types of eye movements [2]: natural drift eye movements, smooth pursuit eye movements (SPEM), and saccadic eye movements. The natural drift eye movements are present even when the observer is intentionally fixating on a single position and these movements are responsible for the perception of static imagery during fixation. The saccadic eye movements are responsible for rapidly moving the fixation point from one location to another; thus, the HVS sensitivity is very low. The former are very slow (0.8–$1.5°$/s) and the latter are fast (80–$300°$/s). Fast object motion blurs the HVS perception. The SPEM tend to track the moving object and reduce the retinal velocity, and thus, compensate for the loss of sensitivity due to motion.

Based on eye movements, the retinal velocity v_R is different from the image velocity v_I. The retinal velocity v_R can be expressed as Equation 17.1. Eye movement velocity v_E is determined as Equation 17.2. The image velocity v_I can be obtained with a motion estimation technique as Equation 17.3. The process for eye track analysis to get retinal velocity v_R is shown in Figure 17.5.

$$v_{R\hbar} = v_{I\hbar} - v_{E\hbar} \qquad (\hbar = x, y), \tag{17.1}$$

$$v_{E\hbar} = \min[g_{sp} * v_{I\hbar} + v_{MIN}, v_{MAX}], \tag{17.2}$$

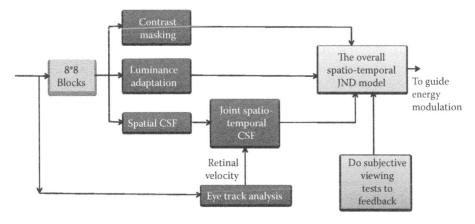

FIGURE 17.4
Overall spatio-temporal JND model.

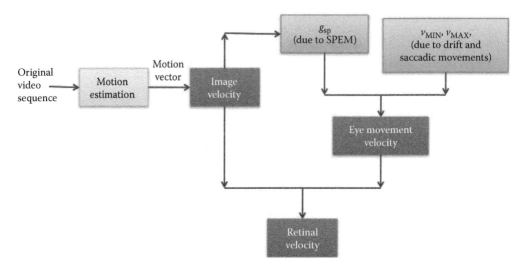

FIGURE 17.5
The process of eye track analysis.

$$v_{I\hbar} = f_{frame} * MV * \theta_{\hbar}, \qquad (17.3)$$

$$\theta_{\hbar} = 2 * \arctan\left(\frac{\Lambda_{\hbar}}{2l}\right), \qquad (17.4)$$

where g_{sp} is the gain of the SPEM, v_{MIN} is the minimum eye velocity due to drift, v_{MAX} is the maximum eye velocity before saccadic movements, f_{frame} is the frame rate of video and MV is the motion vector of each block, $\theta_{\hbar}$ is the visual angle of a pixel obtained by Equation 17.4, an l is the viewing distance and stands for the display width/length of a pixel on the monitor. The average g_{sp} over all observers was $0.956 +/- 0.017$ [3]. The values v_{MIN} and v_{MAX} are set to 0.15 and 80.0°/s.

17.2.1.2 Joint Spatio-Temporal CSF

Human eyes show a band-pass property in the spatial frequency domain. In comparison with various spatial CSF models for still images, the CSF model for videos need to take into account the temporal dimension in addition to the spatial properties. The joint spatio-temporal CSF describes the effect of spatial frequency and temporal frequency on the HVS sensitivity. The spatio-temporal CSF model is the reciprocal of the base distortion threshold which can be tolerated for each DCT coefficient. The literature [4,5] shows that the base threshold for the DCT domain T_{BASE} corresponding to the spatio-temporal CSF model can be expressed by

$$T_{BASE}(k,n,i,j) = T_{BASEs}(k,n,i,j) * F_T(k,n,i,j) \qquad (17.5)$$

where T_{BASEs} is the base threshold corresponding to the spatial CSF model and F_T is the temporal modulation factor; k is the index of the frame in the video sequences, and n is the index of a block in the kth frame; and i and j are the DCT coefficient indices. For still images, the T_{BASE} corresponding to the spatio-temporal CSF is equivalent to T_{BASEs} corresponding to the spatial (static) CSF, and so the JND model is also applicable for guiding image watermarking.

In [5], the base threshold T_{BASEs} is computed by Equation 17.6, where $a = 1.33, b = 0.11, c = 0.18, s = 0.25$, ϕ_i and ϕ_j are DCT normalization factors by Equation 17.7, ω_{ij} is the spatial frequency which can be calculated by Equation 17.8, N is the dimension of the DCT block, θ_x and θ_y are the horizontal and vertical visual angles of a pixel by Equation 17.4, r is set to 0.6, and φ_{ij} stands for the directional angle of the corresponding DCT component by Equation 17.9.

$$T_{BASEs}(n, i, j) = s * \frac{1}{\phi_i \phi_j} * \frac{\exp(c\omega_{ij})/(a + b\omega_{ij})}{r + (1 - r) * \cos^2(\varphi_{ij})}, \tag{17.6}$$

$$\phi_m = \begin{cases} \sqrt{1/N} & m = 0 \\ \sqrt{2/N} & m > 0 \end{cases}, \tag{17.7}$$

$$\omega_{ij} = \frac{1}{2N}\sqrt{(i/\theta_x)^2 + (j/\theta_y)^2}, \tag{17.8}$$

$$\varphi_{ij} = \arcsin(\frac{2\omega_{i,0}\omega_{0,j}}{\omega_{ij}^2}), \tag{17.9}$$

In [4], the temporal modulation factor F_T is computed by Equation 17.10, where the cpd is cycles per degree, the temporal frequency f_t which depends not only on the motion, but also on the spatial frequency of the object is given by Equation 17.11, where f_{sx} and f_{sy} are the horizontal and vertical components of the spatial frequency, which can be calculated by Equation 17.12.

$$F_T(k, n, i, j) = \begin{cases} 1 & f_s < 5\text{cpd and } f_t < 10\,\text{Hz} \\ 1.07^{(f_t - 10)} & f_s < 5\text{cpd and } f_t \geq 10\,\text{Hz} \\ 1.07^{f_t} & f_s \geq 5\text{cpd} \end{cases} \tag{17.10}$$

$$f_t = f_{sx}v_{Rx} + f_{sy}v_{Ry} \tag{17.11}$$

$$f_{sx} = \frac{i}{2N\theta_x} \qquad f_{sy} = \frac{j}{2N\theta_y}, \tag{17.12}$$

As discussed in Section 17.2.1.1., human eyes can automatically move to track an observed object. The retinal velocity v_{Rx} and v_{Ry} can be calculated by Equation 17.1.

17.2.1.3 Luminance Adaptation

HVS is more sensitive to the noise in medium–gray regions, and so the visibility threshold is higher in very dark or very light regions. As our base threshold is detected at the 128 intensity value, for other intensity values, a modification factor needs to be included. This effect is called the luminance adaptation effect. The curve of the luminance adaptation factor is U-shaped, which means that the factor at the lower– and higher–intensity regions is larger than the middle–intensity region. An empirical formula for the luminance adaptation factor a_{Lum} in [5] is shown as Equation 17.13, where $I(k, n)$ is the average intensity value of the nth block in the kth frame.

$$a_{Lum}(k, n) = \begin{cases} (60 - I(k, n))/150 + 1 & I(k, n) \leq 60 \\ 1 & 60 < I(k, n) < 170. \\ (I(k, n) - 170)/425 + 1 & I(k, n) \geq 170 \end{cases} \tag{17.13}$$

17.2.1.4 Contrast Masking

Contrast masking refers to the reduction in the visibility of one visual component in the presence of another one. The masking is strongest when both components are of the same spatial frequency, orientation, and location. To incorporate the contrast masking effect, we employ contrast masking a_{Contrast} [6] measured as

$$a_{\text{Contrast}}(k,n,i,j) = \max(1, (\frac{C(k,n,i,j)}{T_{\text{BASE}}(k,n,i,j) * a_{\text{Lum}}(k,n)})^{\varepsilon}), \qquad (17.14)$$

where $C_{(k,n,i,j)}$ is the (i,j)th DCT coefficient in the nth block of the kth frame, and $\varepsilon = 0.7$.

17.2.1.5 Overall JND Estimator

The overall JND given in Equation 17.15 can be determined by the base threshold T_{BASE}, the luminance adaptation factor a_{Lum} and the contrast masking factor a_{Contrast}.

$$T_{\text{JND}}(k,n,i,j) = T_{\text{BASE}}(k,n,i,j) * a_{\text{Lum}}(k,n) * a_{\text{Contrast}}(k,n,i,j) \qquad (17.15)$$

$T_{\text{JND}}(k,n,i,j)$ is the complete scene-driven spatio-temporal JND estimator which represents the additional accurate perceptual visibility threshold profile to guide watermarking.

17.2.2 The JND Model-Guided Watermarking Scheme

Here we exploit the combined spatio-temporal JND model-guided watermarking scheme to embed and extract watermarking. The scheme first constructs a set of approximate energy subregions using the improved longest processing time (ILPT) algorithm [7], and then enforces an energy difference between every two subregions to embed watermarking bits [8,9] under the control of combined spatio-temporal JND model.

17.2.2.1 The Construction of Approximate Energy Subregions

The watermark bit string is embedded bit by bit in a set of regions (each region is composed of $2n8 \times 8$ DCT blocks) of the original video frame. Each region is divided into two sub-regions (each subregion is composed of $n8 \times 8$ DCT blocks). A single bit is embedded by modifying the energy of two subregions separately. However, for better imperceptibility, approximate energy subregions have to be constructed, so that the original energy of each subregion in one region is approximate. Each bit of the watermark bit string is embedded in its constructed obit-carrying-region. For instance, in Figure 17.6 each bit is embedded in a region of $2n = 16$ 8×8 DCT blocks. The value of the bit is encoded by introducing an energy difference between the low-frequency DCT coefficients of the top half of the region (denoted by subregion A) containing in this case $n = 8$ 8×8 DCT blocks, and the bottom half (denoted by subregion B) also containing $n = 8$ 8×8 DCT blocks. The number of watermark bits that can be embedded is determined by the number of blocks in a region which is used to embed one watermark bit.

17.2.2.2 The Watermark Embedding Procedure

The diagram of combined spatio-temporal JND model-guided watermark embedding is shown in Figure 17.7. The embedding procedure of the scheme is described as the

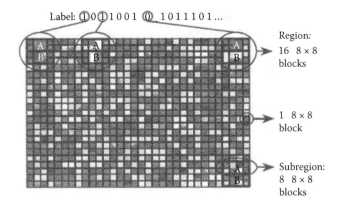

FIGURE 17.6
Watermark bit corresponding to approximate energy subregions.

following steps:

1. Decompose the original video frames into nonoverlapping 8×8 blocks and compute the energy of the low-frequency DCT coefficients in the zigzag sequence.

2. Obtain approximate energy subregions by the ILPT algorithm.

3. Map the index of the DCT blocks in a subregion according to ILPT.

4. Use combined spatio-temporal JND model to calculate the perceptual visibility threshold profile for DCT coefficients.

5. If the watermark to be embedded is 1, the energy of subregion A should be increased (positive modulation) and the energy of subregion B should be decreased (negative modulation). If the watermark to be embedded is 0, the energy of subregion A should be decreased (negative modulation) and the energy of subregion B should be increased (positive modulation). The energy of each subregion is modified by adjusting the low-frequency DCT coefficients according to the combined spatio-temporal JND model as Equation 17.16, where $C(k, n, i, j)^m$ is the modified DCT

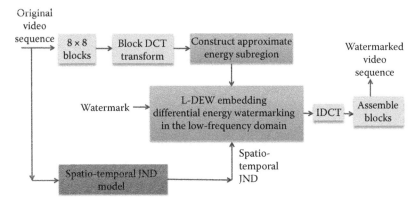

FIGURE 17.7
Diagram of combined spatio-temporal JND model-guided watermark embedding.

coefficient, Sign(.) is the sign function, PM is positive modulation which means increase the energy and NM is negative modulation which means decrease the energy, $T_{JND}(k, n, i, j)$ is the perceptual visibility threshold by our combined spatio-temporal JND model, and $f(.)$ can be expressed by Equation 17.17.

6. Conduct inverse discrete cosine transform (IDCT) to the energy-modified result to obtain the watermark embedded video frames.

$$
C(k, n, i, j)^m
= \begin{cases}
C(k, n, i, j) + \\
\text{sign}(C(k, n, i, j)) * f(C(k, n, i, j), T_{JND}(k, n, i, j)), & \text{PM} \\
\\
C(k, n, i, j) - \\
\text{sign}(C(k, n, i, j)) * f(C(k, n, i, j), T_{JND}(k, n, i, j)), & \text{NM}
\end{cases}
\tag{17.16}
$$

$$
f(C(k, n, i, j), T_{JND}(k, n, i, j))
= \begin{cases}
0, & C(k, n, i, j) < T_{JND}(k, n, i, j) \\
\\
T_{JND}(k, n, i, j), & C(k, n, i, j) \geq T_{JND}(k, n, i, j)
\end{cases}
\tag{17.17}
$$

17.2.2.3 The Watermark Extraction Procedure

The diagram of combined spatio-temporal JND model-guided watermark extraction is shown in Figure 17.8. The extraction procedure is described as follows:

1. Decompose the watermark-embedded video frames into nonoverlapping 8×8 blocks and compute the energy of the low-frequency DCT coefficients in the zigzag sequence.

2. Energy of each subregion is calculated according to the index map.

3. Compare the energy of subregion A with subregion B. If the energy of subregion A is greater than the energy of subregion B, the watermark embedded is 1. If the energy of subregion A is smaller than the energy of subregion B, the watermark embedded is 0. The watermark is extracted accordingly.

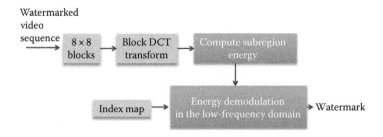

FIGURE 17.8
Diagram of combined spatio-temporal JND model-guided watermark extraction.

17.2.3 Experimental Comparison of Watermarking Methods

Here we performed experiments to evaluate the performance of different JND model-guided watermarking schemes focusing on the watermark's visual quality, capacity, and robustness. Watson's spatial JND model [10] (referred to as Model 1 hereinafter), Zhang's spatial JND model [11] (referred to as Model 2 hereinafter), and Zhenyu Wei's spatio-temporal JND model [4] (referred to as Model 3 hereinafter) were also implemented and compared with the combined spatio-temporal JND estimator. The 720×576 walk pal video sequences are used for experiments described in Section 17.2.3.1, 17.2.3.2, and 17.2.3.3. The suzie video sequences are used for the experiment discussed in Section 17.2.3.4.

17.2.3.1 *Visual Quality*

Figure 17.9a shows the first frame of the walk pal video sequence. Figure 17.9b–e are the first frame of the watermarked video sequence using four JND models. We can see no obvious degradation in Figure 17.9b–e, where the PSNR are 35.5, 47.9, 43.9, and 34.4 dB respectively.

17.2.3.2 *Capacity*

The watermark bit string is embedded bit by bit in a set of regions of the original video frame. Each region is divided into two subregions. A single bit is embedded by modifying the energy of two subregions separately. For each 720×576 video frame, the number of bits that can be embedded is determined by the number of 8×8 DCT blocks in a region. We set the number of blocks at 8 in each region in the following experiments. This means that each 720×576 video frame would embed 810 bits. In Section 17.2.3.3 we focused on testing that for the same number of embedded bits (810 bits) which watermarking scheme is more robust based on the relevant JND model while retaining the watermark transparency.

17.2.3.3 *Robustness*

In practice, the watermarked content may be subject to face a variety of distortions before reaching the detector. We present robustness results with different attacks such as MPEG2 compression, MPEG4 compression, Gaussian noise, and valumetric scaling. Robustness results of the algorithm based on Models 1 to 3 were compared with the results of the algorithm based on the combined spatio-temporal JND model shown in Figures 17.10 through Figure 17.12. For each category of distortion, the watermarked images were modified with a varying magnitude of distortion and the bit error rate (BER) of the extracted watermark was then computed.

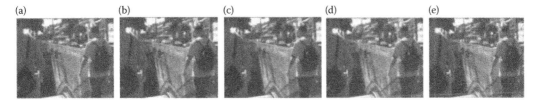

FIGURE 17.9
(See color insert.) (a) Original walk pal video. (b) Watermarked pal video by Model 1. (c) Watermarked pal video by Model 2. (d) Watermarked pal video by Model 3. (e) Watermarked pal video by the combined spatio temporal JND model.

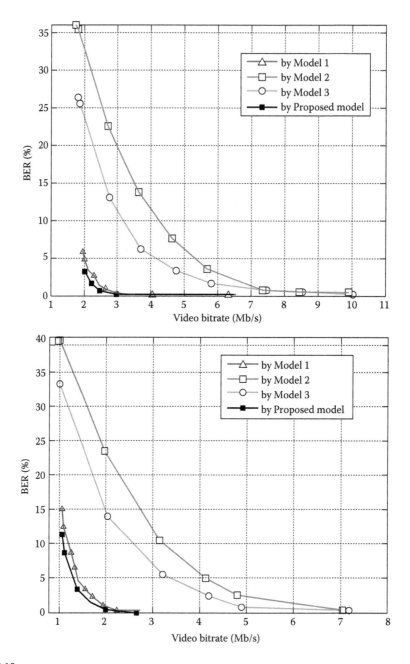

FIGURE 17.10
(a) Robustness versus MPEG2 compression by four models. (b) Robustness versus MPEG4 compression by four models.

1. *MPEG2 and MPEG4 compression*: We test the watermark robustness versus MPEG2 and MPEG4 compression. The MPEG2 and MPEG4 compressions are added to the watermarked video sequences to get a compressed video stream at different bit rates. After decompression and watermark extraction, the watermark bit errors introduced by compression are represented in Figure 17.10a and b. From

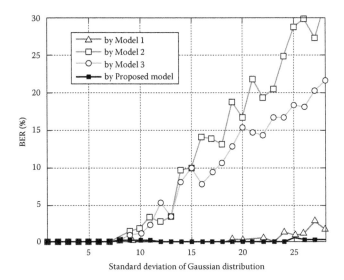

FIGURE 17.11
Robustness versus Gaussian noise.

Figure 17.10, we can see that the watermarking scheme based on the combined spatio-temporal JND model performs better than algorithms based on Models 1–3.

2. *Gaussian noise*: The normal distributed noise with mean 0 and standard deviation σ is added to the watermarked video sequences, where σ varies from 0 to 25. From the experimental results presented in Figure 17.11, the watermarking scheme based on the combined spatio-temporal JND model performs better than algorithms based on Models 1–3.

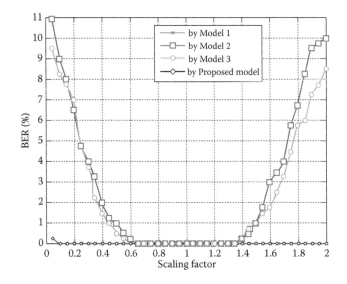

FIGURE 17.12
Robustness versus valumetric scaling.

3. *Valumetric scaling*: The experiment shown in Figure 17.12 reduced the intensities as the scaling factor varied from 1 to 0.1, and increased the intensities as the scaling factor varied from 1 to 2. Figure 17.12 demonstrates that the combined spatio-temporal JND model-guided watermarking scheme's performance against valumetric scaling has an average BER value equal to the algorithm based on Model 1, lower than the algorithm based on Model 2 and 3.

17.2.3.4 Temporal Effect

To test how the JND model-guided video watermarking scheme performs in the presence of motion; we measure the extracted watermark BER results of each frame with different attacks such as MPEG2 compression, Gaussian noise, and valumetric scaling shown in Figures 17.13 through 17.15. As we have discussed, temporal HVS properties have been taken into account to design the spatio-temporal JND model. Thus, when fast object motion occurs, the spatio-temporal JND model-guided video watermarking scheme is capable of yielding higher injected-watermark energy without jeopardizing visual content quality. In suzie video sequences, there is a large increase in average motion energy around frames 50–55; as expected, shows the corresponding drop of BER results of frames 50–55. From evidence below, it can be concluded that the spatio-temporal JND model-guided video watermarking scheme can respond to motion correctly and be more robust with higher motion.

Experimental results confirm the improved performance of the spatio-temporal JND model. The spatio-temporal JND model is capable of yielding higher injected-watermark energy without introducing noticeable distortion to the original video sequences and outperforms the relevant existing visual models. Simulation results show that the proposed spatio-temporal JND model-guided video watermarking scheme is more robust than other algorithms based on the relevant existing perceptual models while retaining the watermark transparency.

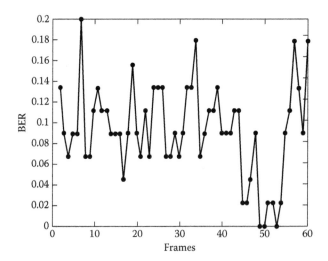

FIGURE 17.13
BER results of each frame versus MPEG2 compression.

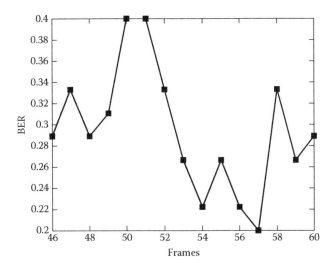

FIGURE 17.14
BER results of each frame versus Gaussian noise.

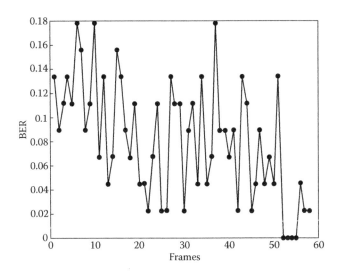

FIGURE 17.15
BER results of each frame versus valumetric scaling.

17.3 Signal Decomposition Technique-Based Audio Fingerprinting for Legacy Content Protection

This section concentrates on describing an audio fingerprinting (fingerprint extraction) scheme, which enables the legacy content identification by comparing the extracted fingerprint with the set of fingerprints associated with metadata in the restored system. Recall that the watermarking technique cannot be applied to identify the legacy content in that watermarks must be embedded in multimedia during the production, and therefore cannot be used to identify multimedia that are already in circulation.

17.3.1 Background

The audio fingerprinting scheme uses one of the suboptimal sparse representation algorithms—MMP algorithm [21]—to obtain the approximate representation atoms of the original signal and further extract the features out of them as the fingerprint of the signal. MMP was first proposed for audio decomposition [21]; Parvaix [39] then adopted the decomposed atoms by means of MMP, combined with the psychoacoustic model, as the watermarking platform. The scheme to be described in this section is the first trial that parametrically analyzes the MMP-derived atoms and explores the most representable features in lower dimension to become the signal fingerprint.

The robustness of the scheme was tested by distorting the original files using the standard attacks. The cross-correlation (CC) approach is then used to extract the fingerprint of the distorted file to identify the source of the file. Linear discriminant analysis (LDA) is also used to improve the computational complexity of the matching algorithm.

Another relative research [38] utilizes a sparse representation algorithm as well. It adopts psychophysiological evidence to build up an over-complete dictionary with 768 atoms for music, and applies the separable atoms, on the promise that there is no modulation error, stemming from the dictionary to linearly present the different music genres. The features were obtained by utilizing dimensionality reduction methods. The classification accuracy is high when the feature dimension goes up to a certain large number. Since the classification does not analyze the atoms of each genre, its accuracy just relies on the number of features. The identification accuracy comparison between this research and the following scheme is presented in Section 17.3.5.

17.3.2 Principle of Audio Fingerprinting

In audio fingerprinting, the main objective is to create a mechanism that is able to detect similarities between two audio samples by comparing the associated fingerprints. An audio fingerprinting scheme consists of two steps:

1. Construction of the database: in this stage, the audio signals to be recognized are presented to the system. The system processes them by extracting a unique fingerprint of each of them based on the characteristics of the content. This fingerprint is then associated with its corresponding metadata (e.g., artist, song name, and information about the copyright) and they are finally stored together in the database. An overview of this step according to the scheme has been illustrated in Figure 17.2a.

2. Identification of a new audio file: when a new unknown file is presented to the system, it is processed to extract its fingerprint. This fingerprint is then compared to those of the database. If a match is found, the corresponding metadata is obtained from the database. Figure 17.2b presents the procedure of this stage.

17.3.3 Fingerprint Extraction

The drawbacks of traditional sparse representation techniques are the computational speed and the control of dictionaries. MMP improves the techniques by reducing the MP times and setting explicit dictionary structure. For readers to better understand the MMP technique, the classic sparse representation technique, MP is introduced, followed by MMP.

17.3.3.1 Matching Pursuit

The MP algorithm was developed in 1993 by Mallat and Zhang [37]. This is a decomposition algorithm based on a dictionary. Let u_λ be the elements of a dictionary D and α_λ be the coefficient of the original file x projection on the element u_λ. The original signal x can be written in the form

$$x = \sum_\lambda \alpha_\lambda u_\lambda. \tag{17.18}$$

However, in most of the cases, it is sufficient to approximate most of the energy of the signal with a small subset of elements such that x can be expressed by N elements and the residual (error) R_N:

$$x = \sum_{i=0}^{N-1} \alpha_\lambda u_\lambda + R_N. \tag{17.19}$$

Each element of the dictionary is also called an *atom*. The MP algorithm is used to find a set of atoms and their corresponding coefficients to minimize the residual R_N. The algorithm is given as follows:

1. Initialization step: $R_0 = x$ and $i = 0$. Computation of each coefficient $\alpha_\lambda =\ <R_0, u_\lambda>$ for all the elements of D.
2. Find the maximum among all the coefficients: $\alpha_{\lambda_i} = max|\alpha_\lambda|$.
3. New residual calculation: $R_{i+1} = R_i - \alpha_{\lambda_i} u_{\lambda_i}$.
4. Coefficients updating: $\alpha_\lambda =\ <R_{i+1}, u_\lambda>$.
5. Stop criterion: stop if $\alpha_{\lambda_i} < \epsilon$ otherwise $i \Leftarrow i + 1$ and return to Step 2.

Note that other stopping criteria can be used. For example, the criterion can be based on a certain amount of the energy of the initial signal.

Nevertheless, the major problem with this algorithm is its high computational cost. Indeed, in the case of a large dictionary, Steps 2 and 4 can be very long because these two steps need to go through each element of D one by one. The extensions of the MP algorithm, such as MMP, were proposed to reduce the computational cost.

17.3.3.2 Molecular Matching Pursuit

As mentioned before, the MMP algorithm is one of the most recent techniques implemented for audio decomposition. The general idea of this algorithm is to group atoms with similar TF properties to form molecules so that several atoms can be subtracted from the residual at the same time for each iteration.

In fact, there exist other extensions of the MP than MMP such as harmonic matching pursuit. Moreover, the idea of selecting groups of atoms at each iteration was already developed in the harmonic matching pursuit algorithm [25], which looks for harmonic structures made of (quasi) harmonically related Gabor atoms. One of the main drawbacks is its unsuitability for a large number of musical sounds that are not harmonic (e.g., percussive sounds). Daudet

[21] showed that MMP provides better results than harmonic matching pursuit. The MMP approach [21] claims that the local TF/timescale grouping is a stronger and more robust assumption about the structure of real audio signals than the harmonicity, because MMP considers dictionaries made by concatenation of a small number of orthonormal bases that are sufficiently incoherent. This is a major difference from the harmonic matching pursuit, where the dictionary is made of a large number of very coherent atoms that make it more difficult to consider local TF/timescale structures.

In order to obtain most of the energy of the initial signal concentrated in a small number of elements, the choice of an appropriate dictionary is very important. Most of the audio signals can be modeled as the sum of two elementary components: the tonal part (sum of sinusoids) and the transient part (sum of Diracs). That is why modified discrete cosine transform (MDCT) with atoms μ_λ and DWT with atoms v_λ are used to construct a 2-times redundant dictionary $\mathcal{D}$ for MMP, with $\mathcal{D} = \mathcal{C} \cup \mathcal{W}$, where $\mathcal{C}$ is an orthogonal basis of lapped cosines (also called an MDCT basis), and $\mathcal{W}$ is an orthogonal basis of discrete wavelets. MDCT atoms are used to present the tonal part, and DWT atoms are used to present the transient part.

The last step (Step 4 mentioned below) of each iteration is to group atoms within a certain window around the significant molecule found. The MMP algorithm could be summarized as follows:

1. Initialization: $R_0 = x$, and $i = 0$. Compute each MDCT coefficient $c_\lambda = <R_0, \mu_\lambda>$ for all the elements of $\mathcal{C}$ and each DWT coefficient $w_\lambda = <R_0, v_\lambda>$ for all the elements of $\mathcal{W}$.

2. Compute the molecule index $\mathcal{T}$ that a set of μ_λ defines and the molecule index $\mathcal{K}$ that a set of v_λ defines; find $K = \max\mathcal{K}$ and $T = \max\mathcal{T}$.

3. Identify the most significant structure. If $T \geq K$, then the most significant structure is of type "tonal molecule"; otherwise, $K > T$, and it is of type "transient molecule."

4. For a tonal molecule, identify atoms that define the most significant tonal molecule: $M_i = \sum_{\lambda=1...m_i} c_\lambda$. Update the residual: $R_{i+1} = R_i - \sum_{\lambda=1...m_i} c_\lambda\mu_\lambda$. Update the coefficients from the new residual: $c_\lambda = <R_{i+1}, \mu_\lambda>$.
   For a transient molecule, identify atoms that define the most significant transient molecule: $M_i = \sum_{\lambda=1...m_i} w_\lambda$. Update the residual: $R_{i+1} = R_i - \sum_{\lambda=1...m_i} w_\lambda v_\lambda$. Update the coefficients from the new residual: $w_\lambda = <R_{i+1}, v_\lambda>$.

5. Stop criterion: stop if $\max(K, T) < \epsilon$, otherwise $i \Leftarrow i + 1$ and return to Step 2.

By comparing with the MP algorithm in Section 17.3.3.1, it can be observed that only one step was added to this algorithm. This significantly reduces the number of times of Step 4. The main advantages of this algorithm for fingerprinting are evidently the reduction of computational cost and the capability of the extraction of very descriptive features so that the fingerprint fulfills the uniqueness requirement.

An example of a decomposition of an audio signal is shown in Figure 17.16, where a 5 s duration signal sampled at 44.1 kHz is demonstrated. The reconstructed signal is obtained after 10 iterations and contains more than 18.5% of the energy of the initial signal.

17.3.3.3 Features Extracted

Here, the scheme extracts features from MDCT coefficients only because they contain more information about the audio file than DWT coefficients. The MDCT function is

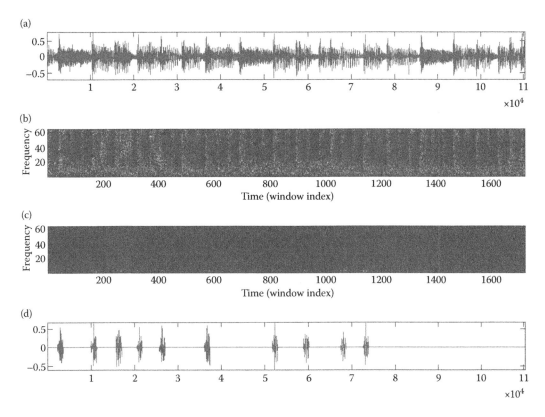

FIGURE 17.16
(See color insert.) Example of decomposition with MMP algorithm. (a) The original music signal. (b) The MDCT coefficients of the signal. (c) The molecule atoms after 10 iteration. (d) The reconstructed signal based on the molecule atoms in (c).

given by

$$X(m) = \sum_{k=0}^{n-1} x(k)\cos(\frac{\pi}{2n}(2k + 1 + \frac{n}{2})(2m + 1)), \quad \text{for } m = 0 \sim \frac{n}{2} - 1, \quad (17.20)$$

where $x(k)$, the overlap-segmented frame with length n derived from the original signal, is obtained by Kaiser–Bessel–Derived (KBD) windows. A TF matrix A is then derived by the MDCT transform. Thus, for each iteration i, MMP provides an MDCT matrix A_i. An example of these MDCT matrices is given in Figure 17.17. The coordinates of each coefficient in the matrix gives information about time and frequency. As explained in Section 17.3.3.2, the matrix M contains the coefficients forming the molecule that contributes the maximum amount of energy to the signal.

The higher the iterations process, the more precise the presentation of the initial signal. In fact, the features that are found do not require a very large number of iterations. Only the first N iterations were taken into account for simplicity's sake and because they contain the most relevant information about the file. Let N be the number of iterations and L be the number of extracted features. In the ith iteration, the feature vector $\mathcal{F}_i = [f_j]_{j=1,\dots,L}, i = 1, \dots, N$ is derived. Then the scheme takes the mean of these vectors to get $\mathcal{F}_{\text{avg}} = \frac{1}{N} \sum_{i=1}^{N} \mathcal{F}_i$. It then

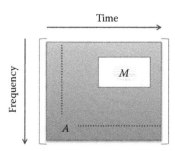

FIGURE 17.17
Example of decomposition with MMP algorithm.

proceeds in the same manner for the variance to obtain $\mathcal{F}_{\text{var}} = \frac{1}{N} \sum_{i=1}^{N} (\mathcal{F}_i - \mathcal{F}_{\text{avg}})^2$. So at the end, the complete fingerprint $\mathcal{F} = \mathcal{F}_{\text{avg}} \cup \mathcal{F}_{\text{var}}$ forms a vector with $2L$ elements. In this experiment, for each iteration, 14 feature elements ($L = 14$) are generated. Therefore, the file fingerprint is a 28-dimensional (28D) vector. The small size makes it easy to compute, which meets the compactness requirement.

The 14 features from the ith iteration were extracted as follows:

- Maximum value of the MDCT coefficients $\max(c_{\lambda_i})$ and its logarithm $\log \max(c_{\lambda_i})$.
- Ratio between previous and current MDCT coefficients $\frac{\max(c_{\lambda_i})}{\max(c_{\lambda_{i-1}})}$ and its logarithm $\log \frac{\max(c_{\lambda_i})}{\max(c_{\lambda_{i-1}})}$.
- Corresponding mean frequency of the molecule M_i: $\overline{f(p_{M_i}, q_{M_i})}$ where p_{M_i} and q_{M_i} are the coordinates of the ith molecule matrix M_i, and $f(p_{M_i}, q_{M_i})$ indicates a function to calculate the frequency expression of the two parameters p_{M_i} and q_{M_i}.
- The dominant principal component, that is, the eigenvector that corresponds to the highest eigenvalue of the frequency matrix windowed by $W1_{4 \times 21}$: $E(W1_{4 \times 21})$ with the four elements, where $E(\cdot)$ stands for deriving a principal eigenvector; this window centers on the frequency that matches the highest MDCT coefficient.
- The elements in the autocorrelation of the frequency matrix windowed by $W2_{2 \times 21}$, which centers on the frequency that matches the highest MDCT coefficient; the autocorrelation matrix which is derived by $W2_{2 \times 21} \times W2_{2 \times 21}^T$ has four elements as well.
- The position of the most significant MDCT coefficient $s(\max(c_{\lambda_i}))$, where $s(\cdot)$ represents deriving the segment index of the coefficient.

Therefore, the set of the features for the ith iteration forms the vector $[\max(c_{\lambda_i}), \log \max(c_{\lambda_i}), \frac{\max(c_{\lambda_i})}{\max(c_{\lambda_{i-1}})}, \log \frac{\max(c_{\lambda_i})}{\max(c_{\lambda_{i-1}})}, \overline{f(p_{M_i}, q_{M_i})}, E(W1_{4 \times 21}), W2_{2 \times 21} \times W2_{2 \times 21}^T, s(\max(c_{\lambda_i}))]$.

The following points will further explain each item in the feature vector:

- The first two items in the vector are two types of computational methods of the feature—maximum value of the MDCT coefficient, and they aim at enhancing the feature strength.
- The third and fourth items actually describe the time relationship between two significant atoms in the previous iteration and the current iteration. Since the first

iteration captures the significant atoms, the more stable these atoms are, the more robust these features will be.

- The fifth item maps the molecular coordinates to an average frequency value in order to measure the stability of the molecular coordinates.

- The sixth item presents the major shape of the matrix of the frequencies corresponding to the selected molecule by calculating its principal eigenvector.

- The seventh item emphasizes the dominant frequency values in each iteration by calculating their autocorrelation matrix.

- The last item in the feature vector precisely presents which segment of the signal the significant MDCT coefficient belongs to by extracting the column index of the coefficient, because the column index matches the segment of the signal.

Note that the scheme does not use the time information. Indeed, because of the uncertainty principle of Heisenberg explained in [48,49], it is impossible to obtain a high resolution for time and frequency at the same time. Actually, the frequency coordinate gives a precise information while the time coordinate provides only the number of temporal windows used by the MDCT.

The experiment shows that one particular molecule is quite possibly selected in the ith iteration from the source signal but selected in the $(i + 1)$th iteration from the attacked signal, especially when two molecules make a similar contribution to the signal. Therefore, the features correspondingly have the same issue, that is, the order of their values appearing over the number of iterations from the source signal and the attacked signals are not exactly the same. However, after the summation and average operations, the values approximately remain the same before and after the attacks.

17.3.4   Fingerprint Matching

In audio fingerprinting, the main problem for the matching part is to find a matching algorithm with low computational cost. The scheme uses the combination of CC for its efficiency and LDA algorithm for its speed.

The CC algorithm is well adapted to compare fingerprints as it gives a measure of similarity. Between two real signals x and y, the CC coefficient is computed with the following equation:

$$C_{xy}(n) = \sum_j x(j)y(n+j) \tag{17.21}$$

The more the two signals are correlated, the higher the final result is, so that we just need to find the strongest likelihood by comparing the new fingerprint with all the fingerprints of the database.

However, it can take a very long time to go through each element of the database with the CC algorithm, especially when this database is large. In order to reduce this computational cost, LDA is chosen for genre (or interclass) classification. LDA is chosen is under the assumption that the features are linearly separable. The results prove this point as well.

However, if the features of the fingerprints provided are not uncorrelated enough to distinguish the number of classes requested, the LDA classifier sometimes fails and gives a wrong file belonging to a wrong class. Thus, to fix this problem, an empirical confidence-level threshold of the final match can be given so that if the LDA algorithm fails, the system returns to the step of the global matching until it finds a sufficient correlation.

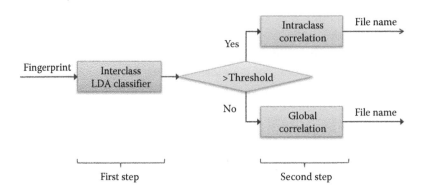

FIGURE 17.18
Fingerprint matching.

In other words, LDA plays the role of a classifier for the fingerprints of the database. As there are three types of music (classical, pop, and rock) in this scheme, the matching search is made in two steps:

1. Find the class of the file by looking for which side of the boundary gives the degree of confidence greater than the confidence-level threshold, and go to (a) of the next step; otherwise go to 2(b).
2. Find the matched file:
   (a) Within this class, search for the file by seeking the fingerprint which provides the highest likelihood and stop.
   (b) Among the classes, search for the file by seeking the fingerprint which provides the highest likelihood and stop.

An overview of the entire system is shown in Figure 17.18.

The advantage of using LDA is not only that it classifies the musical genre, but also that the computational time is much less compared to the CC-based global matching search. More precisely, the search time can be reduced by a factor of n, where n is the number of classes.

17.3.5 Robustness Results

In this section, we study the robustness of the approach to the distortion. For that purpose, the scheme is tested on the benchmark database-GTZAN genre collection [50]. Twenty music files were taken from each classical, pop, and rock collection, respectively. As a fingerprint is a 28D vector, the database becomes a 60×28 matrix. Even though the data set is not high, the leave-one-out cross-validation technique has been used to make the accuracy result more generally reliable. Three types of attacks were tested:

- Low-pass filtering: a filter with an order 23 and a normalized cut-off frequency 0.3.
- Additive noise: the signal-to-noise ratio is always set to ~15 dB.

- Compression: the original files with 352 kbps are compressed to a bit rate of 16 kbps via MP3 processing (compression ratio ≈ 20:1).
- Compression: the original files with 352 kbps are compressed to a bit rate of 128 kbps via MP3 processing (compression ratio ≈ 3:1).

Figures 17.19 through 17.21 illustrate examples of comparisons between MDCT coefficients before and after attacks of the signal.

The results are shown in Tables 17.1 and 17.2. Both tables indicate that the most problematic attack is MP3 compression with bit rate at 16 kbps. Compared to the original bit rate at 352 kbps, this compression ratio is extremely high, which is very unlikely to happen. Instead, the MP3 compression at the popular bit rate of 128 kbps is tested.

The results shown in Table 17.1 are obtained by utilizing the LDA classifier combined with correlation coefficients upon the features mentioned in Section 17.3.4. The confidence-level threshold used in LDA for genre classification among classical, pop, and rock is 0.95. The overall interclass classification accuracy of the approach is 87.9% under distortion.

The results shown in Table 17.2 are obtained without utilizing the LDA classifier, which means that the identification is processed using correlation coefficients upon the features only. It is the same as the global correlation shown in Figure 17.18. The drawback is longer computational time but compensated with higher accuracy. The overall accuracy is 92.5%. To make the result more generalized and comparable to other schemes, the leave-one-out cross-validation method is applied. The identification accuracy becomes 78.3%. Note that the result is under distortion and is still higher than the accuracy 75% without distortion in [38], when they both use the same dimension of the features. The results prove the robustness of the approach even under a certain distortion.

(a)
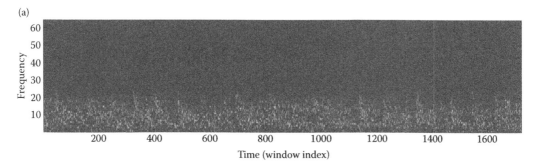

(b)
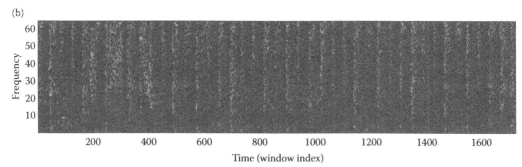

FIGURE 17.19
(**See color insert.**) MDCT coefficients after low-pass filter. (a) MDCT coefficients of the low-pass-filtered signal. (b) MDCT coefficient differences between the original signal and the low-pass-filtered signal.

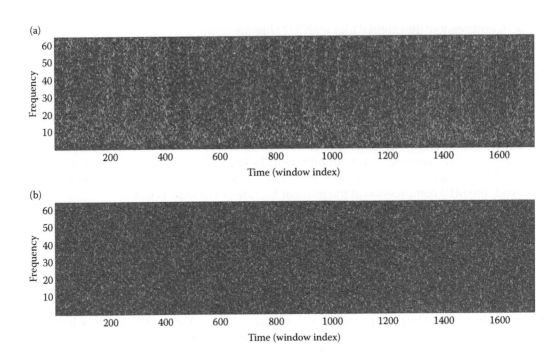

FIGURE 17.20
(**See color insert.**) MDCT coefficients after random noise. (a) MDCT coefficients of the noised signal. (b) MDCT coefficient differences between the original signal and the noised signal.

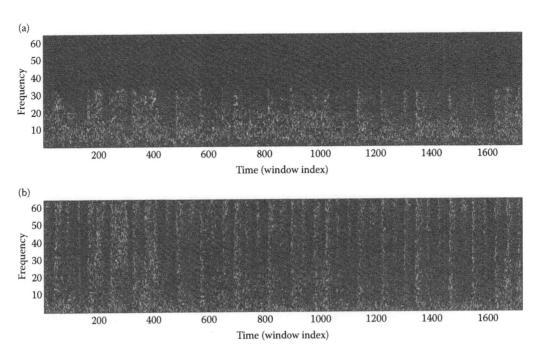

FIGURE 17.21
MDCT coefficients after MP3 compression. (a) MDCT coefficients of MP3 signal with bit rate 16 kbps. (b) MDCT coefficient differences between the original signal and the MP3 signal.

TABLE 17.1

LDA-CC-Based Fingerprint Identification Accuracy on GTZAN after Attacks—Low Pass Filter, Additive Noise, and MP3 Compression

Attacks	Classical Music (%)	Pop Music (%)	Rock Music (%)	Average Accuracy (%)
Low-pass filter [23 0.3]	75	95	90	86.7
Additive noise (SNR = 15 dB)	90	95	90	91.6
MP3 compression at 16 kbps	80	85	70	78.3
MP3 compression at 128 kbps	95	95	95	95

TABLE 17.2

CC-Based Fingerprint Identification Accuracy on GTZAN after Attacks—Low Pass Filter, Additive Noise, and MP3 Compression

Attacks	Classical Music (%)	Pop Music (%)	Rock Music (%)	Average Accuracy (%)	Leave-One-Out Accuracy (%)
Low-pass filter [23 0.3]	80	100	95	91.6	81
Additive noise (SNR = 15 dB)	95	100	95	96.7	80.1
MP3 compression at 16 kbps	85	90	70	81.6	70
MP3 compression at 128 kbps	100	100	100	100	82

The proposed algorithm can be used for the first step of the content registration process. When a content is submitted for registration, its fingerprint is extracted and will be compared with the fingerprints in the database. The proposed algorithm can identify the candidate in the database that has the similar fingerprint. After that, experts may be required to judge if the submitted content was derived from the registered contents or it is an original. This part is outside the scope of the scheme.

17.4 Fingerprinting for Traitor Tracing in P2P Network

This section describes an effective digital watermark fingerprinting technique for a video file based on PCA and wavelet for P2P networks. This work aims to benefit the multimedia producers who like to share their valuable multimedia with the subscribed customers privately within the public P2P networks.

17.4.1 Background

Literature review shows that very few researchers have worked on fingerprinting for P2P applications so far. One research group [46] proposed their watermarking scheme lately for P2P application. There are multiple keys as the watermark is cast into the image by the pseudorandom sequence of a Gaussian distribution generator. However, the paper did not mention the robustness of the scheme against common attacks. Many researchers have proposed algorithms mainly for watermarking, and among those watermarking schemes, there are two mainstreams: the ones which embed a watermark directly in the spatial domain and the others which implement it in a frequency domain. It is found that the transform domain watermarking schemes are typically much more robust to image manipulation as compared to the spatial domain schemes [23].

Among the schemes applying wavelet techniques, one [29] proposes an algorithm in the PCA/wavelet-transform domain. They first apply PCA to produce eigenimages and then decompose them into multiresolution images. Correspondingly, the watermark image is also decomposed into a multiresolution image in the same scale. Finally, the HVS as the strength parameter, is adopted for watermark embedding. The scheme is applicable for embedding one mark because of the uniqueness of the strength parameter. Liu et al. [33] proposed their algorithm based on singular value decomposition (SVD). The host image is originally presented as USV^{-1} where the matrix S contains the singular values and U, V are the singular vectors. The algorithm adds the watermark to the singular values S; thus, the modified S is presented by $U_w S_w V_w^{-1}$. Then the newly generated singular values S_w will replace the original S to generate the watermarked image. The singular vectors U_w and V_w are kept by the owner just for watermark detection. Since S_w is approximately equal to S, the visual quality of the image is preserved. To extract the watermark, the watermarked image will be decomposed again using SVD. The corrupted singular values S'_w and the singular vectors U_w, V_w will recover the watermark. The main issue of this method is that the attacker can also claim his/her watermark easily by providing another set of singular vectors such as U_a, V_a. In other words, the recovered watermark depends more on the selected singular vectors. It proves that embedding a watermark (or fingerprint) only on singular values is unreliable. Hien et al. [27] also proposed a PCA method. The difference is that they embed the watermark into the eigenvectors. First, the PCA process decomposes the image into eigenvectors and eigenvalues. Then the image is projected onto each eigenvector and becomes a coefficient matrix. The watermark is embedded into the coefficient matrix based on the selected components. Finally, the watermarked image is obtained by applying the inverse PCA process. The robustness becomes the issue of this method. As the eigenvectors are normalized, the numerical value of each component of the eigenvector is very small and can be easily corrupted by distortion methods.

In the proposed method, the fingerprint causes changes in both eigenvalues and eigenvectors. Since any attack that corrupts the image will change both the eigenvalues and eigenvectors, if the attack is not strong enough, the fingerprint remains relatively intact. Hence, the proposed method is more robust to common attacks compared to the other two methods.

Unlike the traditional server–client mode networks, the node in P2P networks is not only the client but also the server. This feature makes P2P networks maximize resource utilization of the networks. On the other hand, this feature raises a question. Since each peer finally has the same copy of the shared file, how does the fingerprint uniquely identify each peer and prevent the peer from sharing it with other peers? To resolve this issue, the source file is decomposed into two parts: *base file* and *supplementary file*. The base file will then carry the embedded unique fingerprint for each peer and be distributed using the traditional server–client mode, while the supplementary file will be freely distributed in P2P networks. Thus, it resolves the conflict of traitor tracing and free sharing. However, this solution requires the fingerprint not only to be small enough to alleviate the load of the server but also to keep two other trade-offs: robustness and invisibility. The structure of the fingerprint distribution has been shown in Figure 17.3.

17.4.2 P2P Fingerprint Generation and Embedding

To simplify the description of the P2P fingerprint generation and embedding method, we only discuss how it works on images as well as the generation and embedding of a unique fingerprint.

Since the base file will be distributed from the central server to all the clients, it should be designed to have small size but contain the most important information. Thus, the load of the server can be alleviated to some extent, while the supplementary file can be larger but contains less important information. By doing this, the peer who has the supplementary file has no commercial motivation to leak the supplementary file alone because of its low quality without the base file. One possible approach to derive a small-sized base file is to decompose the file into two parts—the base pixel matrix and the detail pixel matrix. The base pixel matrix can give us a rough outline of the image. Since the base pixel matrix has higher correlation information, its entropy value is small so that it can be compressed into a very small size with no quality loss.

The P2P fingerprinting method employs wavelet transform to model the low-frequency feature of the image and PCA to further decompose it into eigenvectors. After the pre-processing, any one vector can be adopted to generate one fingerprint by following a rule. Literature review shows that wavelet [18] and PCA [42] techniques as features were utilized to detect the image information. This is the first time that wavelet and PCA techniques are employed for fingerprint generation and embedding.

In wavelet transform, an image is split into one approximation (also called approximation coefficient) w_a and three details in horizontal, vertical, and diagonal directions which are named w_h (or horizontal coefficient), w_v (vertical coefficient), and w_d (diagonal coefficient) respectively. The approximation is then itself split into second-level approximation and details, and the process is repeated. For a J-level decomposition, the approximation and the details are described in Equation 17.22,

$$w_{aJ} = \langle I \cdot A_J \rangle, \tag{17.22}$$

$$w_{hj} = \langle I \cdot H_j \rangle, \quad j = J, ..., 1$$

$$w_{vj} = \langle I \cdot V_j \rangle, \quad j = J, ..., 1$$

$$w_{dj} = \langle I \cdot D_j \rangle, \quad j = J, ..., 1$$

where I denotes image and A_J, H_j, V_j, and D_j are wavelet bases. For image decomposition, even the size of the coefficients in different levels is different, but the coefficients are still a two-dimensional (2D) matrix. Equation 17.23 indicates how the image is recovered:

$$I = a_J + \sum_{j=1}^{J} d_j = w_{aJ}A_J + \sum_{j=1}^{J} w_{hj}H_j + \sum_{j=1}^{J} w_{vj}V_j + \sum_{j=1}^{J} w_{dj}D_j. \quad j = J, ..., 1 \tag{17.23}$$

The original definition can be found in [24].

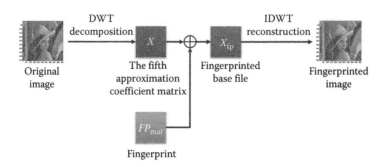

FIGURE 17.22
Fingerprint embedding flowchart.

In this method, the fingerprint is small but strong and robust compared to the multimedia file. Figure 17.22 shows the brief procedure of the fingerprint generation and embedding. For simplicity reason, this research only investigates the fingerprint method on color channel of the image. In reality, it can be extended to other color channels. Even the message in the blue channel is less sensitive to the human eyes [31,32], the channel selection is not very important at the current stage for showing the feasibility and the advantages of the method. First, the host image with size 512×512 was decomposed into 5th level by DWT. For five-level decomposition, the coefficient set is $W = [w_{a5}, w_{h5}, w_{v5}, w_{d5}, w_{h4}, w_{v4}, w_{d4}, \ldots, w_{h1}, w_{v1}, w_{d1}]$. At the 5th level, the size of the approximate coefficient w_{a5} is significantly reduced from the original to 16×16. This coefficient, called X from now on, is defined as the base file as mentioned before. Correspondingly, the other coefficients are defined as supplementary file. The base file was then used to calculate its principals. It goes through three steps according to Equations 17.24 through 17.26.

$$\text{Step 1.} \quad X' = X - \overline{X}, \tag{17.24}$$

$$\text{Step 2.} \quad X'' = Cov(X', X'^T), \tag{17.25}$$

$$\text{Step 3.} \quad X''P = P\Lambda, \tag{17.26}$$

where $\overline{X}$ denotes the mean of X. P and Λ present the set of eigenvectors and the set of eigenvalues of X'' (or $X'X'^T$). Since X'' is a reversible matrix in this case, there are a total of 16 eigenvectors which make up the columns of P. It is presented as

$$P = [\vec{p}_{16}|\vec{p}_{15}|\ldots|\vec{p}_1], \tag{17.27}$$

where the eigenvectors are arranged in descending order according to their principal components and each eigenvector is a 16×1 vector. Equation 17.28 illustrates how the preparatory fingerprint matrix $FP^{\text{pre}}_{\text{mat}}$ is derived:

$$FP^{\text{pre}}_{\text{mat}} = Y \times (\vec{S} \times \vec{p}_m^T)_{\text{full}}, \tag{17.28}$$

where a scale vector defined as $\vec{S}$, which is a 16×1 vector, is multiplied with Y and one of the eigenvectors, for instance, $\vec{p}_m$ ($m = 1, 2, \ldots, 16$). Y is a visually meaningful full matrix with all positive elements, and only known by the source owner. Thus, it can be a company's logo, another low resolution of a portion of the original host image, or simply a portion of the host image. It is utilized to prove the right ownership fingerprint. The elements of $\vec{S}$ can be viewed as the coefficient of the fingerprint amplitude. The bigger the values of $\vec{S}$, the more the visible distortion the fingerprint creates; the smaller the values of $\vec{S}$, the weaker the fingerprint energy will be. The value s_i ($i = 1, 2, \ldots, 16$) in the scale vector is chosen on the basis of empirical optimization. T indicates the transpose operation. *full* denotes the transformation operation of the nonreversible matrix to reversible matrix by fine tuning the singular value matrix of the nonreversible matrix. After the multiplication, the matrix size is 16×16 which is the same as Y. Also, even the data items in each column have different magnitudes but they have the same sign, and this sign matches with the corresponding entries of the eigenvector $\vec{p}_m^T$. Equations 17.29 through 17.31 provide the explanation:

$$\because y_{ij} > 0, \quad (i, j = 1, \ldots, 16; \quad y_{ij} \in Y) \tag{17.29}$$

$$s_j \geqslant 0, \quad (j = 1, \ldots, 16; \quad s_j \in \vec{S}) \tag{17.30}$$

$$\therefore \text{for } i, j = 1 \text{ to } 16,$$

$$p_{m_i} \times fp^{\text{pre}}_{\text{mat}ji} \geqslant 0. \quad \left(i, j = 1, \ldots, 16; \quad p_{m_i} \in \vec{p}_m, \right.$$

$$\left. fp^{\text{pre}}_{\text{mat}ji} \in FP^{\text{pre}}_{\text{mat}} \right) \tag{17.31}$$

According to Equations 17.29 through 17.30, $Y \times \vec{S}$ is a 16×1 vector with all elements positive. Hence, the ith component p_{m_i} of $\vec{p}_m$ determines the sign of the ith column of the matrix $FP^{\text{pre}}_{\text{mat}} = Y \times \vec{S} \times \vec{p}_m^T$. This feature implies that the preparatory fingerprint in the horizontal direction follows the trend of the applied eigenvector. The only problem is that it makes the distortion visible after the image is reconstructed. The reason is that data items in each column of matrix Y have a similar stretch scale because of the matrix ($\vec{S} \times \vec{p}_m^T$). The solution can be done by fine adjustment on each element so that the obvious boundary between the columns is invisible. This procedure can be modeled as Equation 17.32. We call the procedure *Column Unify*, because the fine adjustment coefficients from c_{1j} to c_{nj} are created to adjust all the elements in the jth column of the matrix $FP^{\text{pre}}_{\text{mat}}$ to have the same value. The rule of unification not only ensures that the value keeps the sign as before but also maintains a certain difference between two columns. To prevent the previous steps of implementation from creating visual distortion, the variation of the whole matrix was limited by an empirical-based perceptually lossless threshold. Thus, the scale vector and the fine adjustment coefficients should be adjusted accordingly. The generated fingerprint matrix is named FP_{mat}:

$$FP_{\text{mat}} = \text{ColumnUnify}(FP^{\text{pre}}_{\text{mat}}) \tag{17.32}$$

$$= \text{diag}[c_{11}, c_{21}, \ldots, c_{n1}][FP^{\text{pre}}_{\text{mat}}]_{COL1} + \text{diag}[c_{12}, c_{22}, \ldots, c_{n2}][FP^{\text{pre}}_{\text{mat}}]_{COL2}$$

$$+ \ldots + \text{diag}[c_{1n}, c_{2n}, \ldots, c_{nn}][FP^{\text{pre}}_{\text{mat}}]_{COLn},$$

where n is equal to 16 in this case. The term $[FP^{\text{pre}}_{\text{mat}}]_{COLi}$ presents the ith column of the matrix $FP^{\text{pre}}_{\text{mat}}$.

FP_{mat} as well as $\overline{X}$ are then added to X' and the fingerprinted image can be reconstructed based on the fingerprinted 5th approximation matrix X_{fp} (alternatively called the fingerprinted base file) using an inverse DWT as shown in Figure 17.22 along with Equation 17.23.

Liu and Tan [33] mentions that the singular values of an image have very good stability; that is, when a small perturbation is added to an image, its singular values do not change significantly. In this case, the singular values are eigenvalues. Also, the same concept applies to the eigenvectors, which also means that the differences of each term in SVD of X and X_{fp} are similar. For example, the SVD of X is defined as

$$X = P\lambda Q^{\mathrm{T}}, \tag{17.33}$$

where P as mentioned in Equation 17.26 is the set of eigenvectors of $X'X'^{\mathrm{T}}$. The eigenvectors of $X'^{T}X'$ make up the columns of Q. The singular values in λ are square roots of the eigenvalues from $X'X'^{T}$ or $X'^{T}X'$. It can be defined as

$$\Lambda = \lambda\lambda^{\mathrm{T}}, \tag{17.34}$$

since X' is a square matrix, λ equals λ^{T}. while the SVD of X_{fp} can be similarly defined as

$$X_{fp} = \tilde{P}\tilde{\lambda}\tilde{Q}^{\mathrm{T}}. \tag{17.35}$$

Since the elements of FP_{mat} are small enough compared to the base file X, by adding the fingerprint, the vectors between P and $\tilde{P}$, and Q and $\tilde{Q}$ are very close based on their correlation coefficients. It can be proved by calculating the overall correlation coefficient of X and X_{fp} as well. For images such as Lena, Baboon, and Peppers, their corresponding correlation coefficients between before and after fingerprinting are 0.9988, 0.9997, and 0.9995, respectively. Therefore, the visual aspect of the image is preserved after the fingerprint is embedded.

The fingerprint embedding method not only involves the wavelet but also uses the PCA technique. The reason the wavelet technique is used is because of the advantages that it can provide a scalable approximation matrix associated with the scalable precision, and also because the approximation matrix contains the most important low-frequency information in small size. The PCA technique, on the other hand, finds out the orthogonal eigenvectors that a pattern can project into.

Under the unification operation, the robustness of the fingerprint is enhanced. Because the absolute magnitude of elements in each column is replaced by one value, the sign of each column is kept which means that the discrimination feature between columns is maintained. The result is presented in Section 17.4.5.

Ideally, X" has 16 eigenvectors and the same number of fingerprints according to the host image with size 512 × 512. Such a small number of fingerprints is evidently not enough to represent the large number of users around the world. Fortunately, the multimedia file does not have one frame only; it has many frames in sequence. Any number of frames, for example, 20 out of a total number of 1000 frames can be chosen as the target images and their 16 local eigenvectors, assuming each frame has the size of 512 × 512, can be determined. If one out of 16 different eigenvectors is chosen from one image, and also other eigenvectors are chosen respectively from the other 19 images to label each customer, there will be 16^{20} different combinations for labeling. The mapping of the labeling and the customer will only be known by the owner.

17.4.3 Fingerprint Detection

Since only the owner, for example, the media producer, keeps the mapping between the fingerprint and the customer, as long as the producer successfully tracks back the fingerprint for a suspect video, for example, the pirate customer can be revealed. The suspected video is defined as a video which is freely distributed out of the scope of owners' authorized P2P networks.

In this case, to identify the embedded fingerprint, the multimedia producer needs to decompose the fingerprinted image into level 5 using the wavelet technique so that a 16×16 approximate matrix X'_{fp} is obtained. By deducting X_{fp}, the difference is the fingerprint matrix FP^*_{mat}. Then the signs (based on the majority rule) of the columns in this matrix will be compared with the signs of each eigenvector using the Hamming distance. The eigenvector that has the minimum Hamming distance to the matrix will be claimed as the embedded fingerprint.

To prove that the ownership of the fingerprint is issued by the right source owner, the matrix Y^*, approximately equal to Y is first extracted. Then, the correlation coefficient is used to decide whether the matrix Y exists. Equations 17.36 through 17.37 denote the operations:

$$Y^* = FP^*_{mat} \times (\vec{S} \times \vec{p}_m^T)^{-1}_{full},\tag{17.36}$$

$$C_{y,y^*} = \frac{Y \cdot Y^*}{\| Y \|}.\tag{17.37}$$

17.4.4 P2P Fingerprint Distribution

The fingerprint generation method described in the previous section can be used to generate a unique fingerprint or a sharable fingerprint. The only difference is that the unique fingerprint will be individually distributed with the base file, while the sharable fingerprint will be distributed with the image in which it is embedded. The sharable fingerprinting, as a security enhancement method, is necessary and will be explained shortly.

As mentioned above, the central server, which is the owner's server, will only distribute the unique fingerprint-embedded base files to the customer, while the P2P networks deliver the supplementary files, sharable fingerprint embedded files, and regular files. Figure 17.23 depicts the relationship between unique fingerprints and sharable fingerprints in a video with 9 I frames. To simplify the illustration, P frames and B frames are neglected in the figure. These frames are included into the scope of regular files. The reason of choosing I frame is that these types of frames are encoded independently, while P frames and B frames are encoded on the prediction from I frames. The 1st, 4th, and 7th images are embedded with unique fingerprints in their base files and the 2nd and 6th images are embedded with sharable fingerprints. The files in red will be distributed by the central server only and the files in blue will be delivered in P2P networks.

In P2P networks, if a new multimedia source file is available to its customers, the peers that join the P2P networks at the very beginning most possibly download the whole video file from the central server, but the peers who joined after may have the file from different peers. For example, a partial video file is downloaded from Peer A with A's sharable fingerprint; the rest is downloaded from Peer B with B's sharable fingerprint or from the owner's server directly. The procedure is illustrated in Figure 17.24.

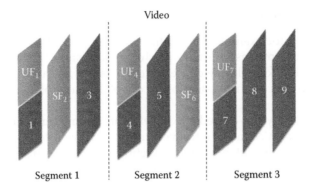

FIGURE 17.23
Two kinds of fingerprints in a video. UF denotes that a unique fingerprint is embedded and SF denotes that a sharable fingerprint is embedded.

The purpose of involving a sharable fingerprint in the P2P sharing networks is to enhance the efficiency of traitor tracing. The sharable fingerprint functions as an assistance to provide the source owner a hint about from which peers this video was downloaded. This is especially useful if the attacker successfully removed or attacked the unique fingerprint. It is because a multimedia file consists of many segments and is usually obtained through the P2P networks. Different segments may come from different peers and have different sharable fingerprints. These sharable fingerprints together can be treated as a fingerprint,

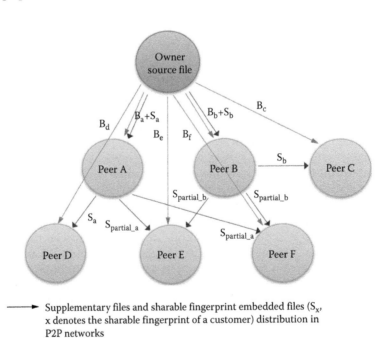

Supplementary files and sharable fingerprint embedded files (S_x, x denotes the sharable fingerprint of a customer) distribution in P2P networks

Unique fingerprint embedded base files (B_{id}, id denotes the unique fingerprint of a customer) distribution

FIGURE 17.24
The topology of base file and supplementary file distribution.

which may not be unique, but can still help to narrow down the suspected traitors and to provide further evidence to support the result derived from the unique fingerprint.

17.4.5 Attacks

The fingerprint method is studied on a series of common image attack processes that include Gaussian noise, median filter, lossy compression, and geometric distortion. The test images—Lena, Baboon, and Peppers—are 512 × 512, and the fingerprinted images are obtained as the example shown in Figure 17.25. Under the same subject, visual quality of the fingerprinted image, the results show that this method is far more resistant to many common attacks than other methods—Liu and Tan [33] and Hien et al. [27]—which also use the concepts of eigenvalues and eigenvectors. The measurement of the resistance for each image among Liu, Hien, and the proposed methods is detection rate. For Liu and Hien methods, the detection rate is defined as the number of correctly detected embedded eigenvalues (or eigenvectors) divided by 16. For the proposed method, the detection rate is the number of successfully detected embedded fingerprints divided by 16.

The experiments suggested that the wrong claims are usually derived from those less principal eigenvectors. It implies that the most principal eigenvectors which have a larger Hamming distance gap should be chosen as the vectors to generate the fingerprints for better robustness performance. The eigenvectors which are utilized for fingerprints in this study are selected according to their principals and their Hamming distances.

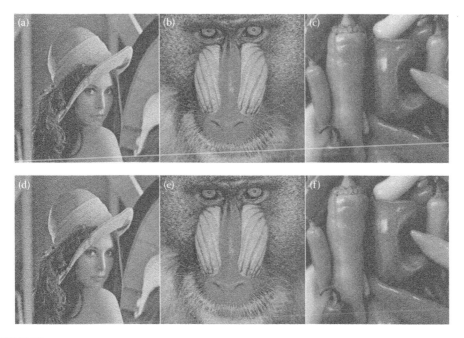

FIGURE 17.25
(See color insert.) Comparison of images before and after fingerprinting. (a) Original Lena. (b) Original Baboon. (c) Original Peppers. (d) Fingerprinted Lena. (e) Fingerprinted Baboon. (f) Fingerprinted Peppers.

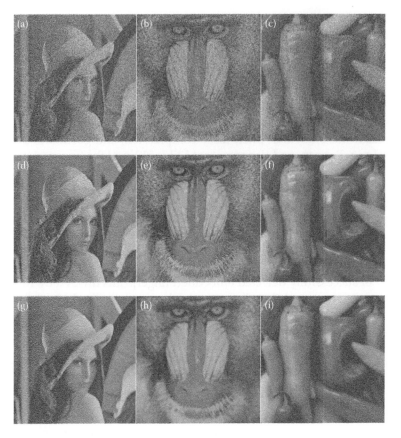

FIGURE 17.26
Images after Gaussian white noise, compression, and median filter. (a) Lena with noise power at 7000. (b) Baboon with noise power at 7000. (c) Peppers with noise power at 7000. (d) Lena at quality 5 of JPEG (joint photographic experts group) compression. (e) Baboon at quality 5 of JPEG compression. (f) Peppers at quality 5 of JPEG compression. (g) Lena with median filter [9 9]. (h) Baboon with median filter [9 9]. (i) Peppers with median filter [9 9].

Figure 17.25a–c and d–f presents the comparison between original images and finger-printed images, respectively, where the fingerprints are included in the red channel in this case.

Anticollusion attacks are another concern in fingerprint design. Some research groups [14,22,45] proposed their schemes against collusion attack. They use the binary codes based on marking assumption or linear combination of a set of orthogonal vectors. Although the proposed scheme does not design against the anticollusion attack, the sharable fingerprint described above does provide some form of anticollusion function.

17.4.6 Robustness Results

Due to space limitation, all the attacks are not described individually. Instead, the results are summarized in Table 17.3 and the images after attacks are demonstrated in Figure 17.26. It indicates that the method is strongly not vulnerable to common attacks such as Gaussian white noise, lossy compression, median filter, and border cropping. It also proves that how to choose the eigenvectors to generate the fingerprints is very important. That is,

TABLE 17.3

Fingerprint Method Average Robustness on Lena, Baboon, and Peppers at Size 512 × 512

	Accuracy Rate of Detection		
Attacks	Proposed Method (%)	Liu Method (%)	Hien Method (%)
AWGN power at 7000	92	12.5	0
Compress to quality 10	87.5	8	4
Median filter at size [7 7]	91.7	21%	0
Rotate 1 degree	75	8	4
Shift 2 lines	79.2	12.5	0
Border crop 5 lines	100	12.5	4

the fingerprints derived from the chosen eigenvectors (based on their principals and their Hamming distance) outperform the ones derived from all the available eigenvectors.

17.5 Conclusion

Perceptual watermarking should take full advantage of the results from HVS studies. JND, which refers to the maximum distortion that the HVS does not perceive, gives us a way to model the HVS accurately. An effective spatio-temporal JND model-guided video watermarking scheme in a DCT domain is designed based on the additional accurate JND visual model which incorporates spatial CSF, temporal modulation factor, retinal velocity, luminance adaptation, and contrast masking. The proposed watermarking scheme, where JND models are fully used to determine scene-adaptive upper bounds on watermark insertion, allows us to provide the maximum strength transparent watermark. Experimental results confirm the improved performance of our spatio-temporal JND model. Our spatio-temporal JND model is capable of yielding higher injected-watermark energy without introducing noticeable distortion to the original video sequences and outperforms the relevant existing visual models. Simulation results show that the proposed spatio-temporal JND model-guided video watermarking scheme is more robust than other algorithms based on the relevant existing perceptual models while retaining the watermark transparency.

The MMP-based audio fingerprinting for legacy content protection scheme shows that the MMP algorithm, a form of the sparse approximation technique, is very promising for audio fingerprinting. The final results demonstrated the efficiency of this algorithm for some popular distortions, particularly the addition of noise which is one of the most destructive attacks. With 10 iterations or less, the fingerprints with 28 dimensions of audio signals were still quite well matched among three music genres—classical, pop, and rock.

In the MMP algorithm, the atoms derived from the MDCT transform are only used as the atoms because they contain more information about the audio signal. Nevertheless, it should be interesting to optimize the MMP algorithm by adding wavelet elements in the dictionary without significantly affecting its computational time as it can provide more interesting features to have a better characterization of the original signal, because the signal which has a lot of transient parts needs wavelet elements to sparsely represent their main features.

The proposed P2P fingerprinting scheme is specific for P2P networks and will benefit those multimedia producers who want to share their big file, such as video file, utilizing the convenience of P2P networks. The videos that involve content-complicated frames (except for medical applications where original contents are needed without any distortion) can apply this fingerprinting scheme.

The fingerprint embedding scheme involves the wavelet and PCA techniques. Unlike other conventional fingerprinting techniques which suffer from poor scalability mentioned in [34], this scheme is scalable, not only because the scheme reduces the burden of the media owner's server by only sending the small-size base file and making use of the P2P network infrastructure to support the majority of the file transfer process [34], but also because it provides a large number of unique fingerprints. Moreover, the wavelet technique makes the base file into necessary information for the customer and the fingerprint robust enough to common attacks. The PCA technique, on the other hand, determines the orthogonal eigenvectors, which makes it possible to maximally distinguish the different fingerprint while maintaining low computation complexity. This scheme has shown that the unique fingerprint has strong robustness against most common attacks such as Gaussian white noise, lossy compression, median filter, and border cropping. The sharable fingerprint, on the other hand, enhances the invulnerability to the collusion attack of the scheme to some extent.

References

1. R. B. Wolfgang, C. I. Podilchuk, and E. J. Delp, Perceptual watermarks for digital images and video. In *Proceedings of IEEE, Special Issue on Identification and Protection of Multimedia Information*, Vol. 87, pp. 1108–1126, 1999.
2. S. Daly, Engineering observations from spatio velocity and spatiotemporal visual models. In *Proceedings of SPIE*, Vol. 3299, pp. 180–191, 1998.
3. S. Tourancheau, P. L. Callet, and D. Barba, Influence of motion on contrast perception: Suprathreshold spatio-velocity measurements. In *Proceedings of SPIE*, 6492, 64921M, 2007.
4. Z. Wei and K.N. Ngan, A temporal just-noticeble distortion profile for video in DCT domain. In *Proceedings of the 15th IEEE International Conference on Image Processing*, Hanover, Germany, pp. 1336–1339, 2008.
5. Z. Wei and K.N. Ngan, Spatial just noticeable distortion profile for image in DCT domain. In *Proceedings of IEEE International Conference on Multimedia and Expo*, pp. 925–928, 2008.
6. G. E. Legge, A power law for contrast discrimination, *Vision Research*, Vol. 21, pp. 457–467, 1981.
7. Zou, Fu hao, Research of robust video watermarking algorithms and related techniques, A Ph.D. dissertation in Engineering, Hua Zhong University of Science and Technology, 2006.
8. G. C. Langelaar, and R. L. Lagendijk, Optimal differential energy watermarking of DCT encoded images and video. *IEEE Transactions on Image Processing*, 10(1), 148–158, 2001.
9. H. F. Ling, Z. D. Lu, and F. H. Zou, Improved differential energy watermarking (IDEW) algorithm for DCT-encoded imaged and video. In *Proceedings of the Seventh International Conference on Signal Processing (ICSP2004)*, Beijing, China, pp. 2326–2329, 2004.
10. A. B. Watson, DCTune: A technique for visual optimization of DCT quantization matrices for individual images, *Society for Information Display Digest of Technical Papers XXIV*, 946–949, 1993.
11. X. K. Zhang, W. S. Lin, and P. Xue, Improved estimation for justnoticeble visual distortion, *Signal Processing*, 85(4), 795–808, 2005.
12. H. F. Ling, Z. D. Lu, F. H. Zou, and R. X. Li, An energy modulated watermarking algorithm based o Watson perceptual model, *Journal of Software*, 17(5), 1124–1132, 2006.

13. C. I. Podilchuk, and W. Zeng, Image-adaptive watermarking using visual models. In *Proceedings of the IEEE*, 16, 525–539, 1998.

14. D. Boneh and J. Shaw, Collusion-secure fingerprinting for digital data, *IEEE Transactions on Information Theory*, 44, 1897–1905, 1998.

15. J. M. Brück, S. Bres, and D. Pellerin, Construction d'une signature audio pour l'indexatioin de documents audiovisuels, CORESA, 2004.

16. P. Cano, E. Battle, T. Kalker, and J. Haitsma, A review of algorithm for audio fingerprinting, *IEEE Workshop on Multimedia Signal Processing*, December 2002.

17. P. Cano, E. Batlle, T. Kalker, and J. Haitsma, A review of audio fingerprinting, *Journal of VLSI Signal Processing Systems*, 41(3), 271–284, 2005.

18. E. Chang, J. Wang, C. Li, and G. Wiederhold, RIME: A replicated image detector for the World Wide Web, *SPIE Multimedia Storage and Archiving Systems III*, pp. 58–67, November 1998.

19. S. S. Chen, D. L. Donoho, and M. A. Saunders, Atomic decomposition by basis pursuit, *SIAM Journal on Scientific Computing*, 20(1), 33–61, 1998.

20. S. Craver, M. Wu, and B. Liu, What Can We Reasonably Expect from Watermarks? In *Proceedings of the IEEE Workshop on Applications of Signal Processing to Audio and Acoustics*, New Paltz, NY, pp. 223–226, October 2001.

21. L. Daudet, Sparse and structured decompositions of signals with the molecular matching pursuit, *IEEE Transaction on Speech and Audio Processing*, 14(5), 1808–1816, September 2006.

22. J. Dittmann, Combining digital watermarks and collusion secure fingerprints for customer copy monitoring, *Proceeding of IEEE Seminars on Secure Images and Image Authentication*, Savoy Place, London, pp. 128–132, March 2000.

23. R. Dugad, K. Ratakonda, and N. Ahuja, A new wavelet-based scheme for watermarking images, *Image Processing, 1998. ICIP 98. Proceedings. 1998 International Conference*, Chicago, Illinois, Vol. 2, pp. 419–423, October 1998.

24. R. C. Gonzalez and R. E. Woods, *Digital Image Processing* New York: Addison-Wesley, 1992.

25. R. Gribonval and E. Bacry, Harmonic decompostions of audio signals with matching pursuit, *IEEE Transactions on Signal Processing*, 51(1), 101–111, 2003.

26. J. Haitsma and T. Kalker, A highly robust audio fingerprinting system, *Third International Symposium on Music Information Retrieval*, Paris, France, pp. 107–115, October 2002.

27. T. D. Hien, Z. Nakao, K. Miyara, Y. Nagata, and Y. W. Chen, A new chromatic color image watermarking and its PCA-based implementation, *ICAISC 2006*, LNAI 4029, Zakopane, Poland, pp. 787–795, 2006.

28. K. Huang and S. Aviyente, Sparse representation for signal classification. In *Advances on NIPS*, Vancouver, BC, Canada, pp. 609–616, 2006.

29. A. Kaarna and P. Toivanen, Digital watermarking of spectral images in PCA/wavelet-transform domain, *Geoscience and Remote Sensing Symposium, 2003. IGARSS '03. Proceedings of the 2003 IEEE International*, Toulouse, France, Vol. 6, pp. 3564–3567, July 21–25, 2003.

30. D. Kundur and K. Karthik, Video fingerprinting and encryption principles for digital rights management, *Proceedings of the IEEE Special Issue on Enabling Security Technologies for Digital Rights Management*, 92(6), 918–932, 2004.

31. M. Kutter, S. K. Bhattacharjee, and T. Ebrahimi, Towards second generation watermarking schemes, *Proceedings of the IEEE International Conference on Image Processing, ICIP 99*, Kobe, Japan, Vol. 1, pp. 320–323, 1999.

32. M. Kutter, F. Jordan, and F. Bossen, Digital signature of color image using amplitude modulation, in I. K. Sethi, R. Jain (Eds.) *Storage and Retrieval for Image and Video Databases V*, Vol. 3022, SPIE, San Jose, CA, 1997, pp. 518–526.

33. R. Liu and T. Tan, An SVD-based watermarking scheme for protecting rightful ownership, *IEEE Transactions on Multimedia*, 4(1), 121–128, 2002.

34. D. G. Lowe, Object recognition from local scale-invariant features. In *Proceedings ICCV*, Kerkyra, Greece, pp. 1150–1157, 1999.

35. C. S. Lu, Audio fingerprinting based on analyzing time-frequency localization of signals, *IEEE Workshop on Multimedia Signal Processing*, St. Thomas, Virgin Islands, pp. 174–177, December 2002.

36. W. Luh and D. Kundur, New paradigms for effective multicasting and fingerprinting of entertainment media, *IEEE Communication Magazine*, 43(6), 77–84, 2005.
37. S. G. Mallat and Z. Zhang, Matching pursuits with time-frequency dictionaries, *IEEE Transaction on Signal Processing*, 41(12), 3397–3415, December 1993.
38. Y. Panagakis, C. Kotropoulos, and G. R. Arce, Music genre classfication via sparse representations of auditory temporal modulations. In *17th European Signal Processing Conference (EUSIPCO 2009)*, Glasgow, Scotland, August 2009.
39. M. Parvaix, An audio watermarking method based on molecular matching pursuit, *IEEE International Conference on Acoustics, Speech, and Signal Processing*, Las Vegas, Nevada, pp. 1721–1724, 2008.
40. A. Ramaligan and S. Krishnan, Gaussian mixture modeling of short-time fourier transform features for audio fingerprinting, *IEEE Transaction on Information Forensics and Security*, 1(4), 457–463, December 2006.
41. METISS (a joint research group between CNRS, INRIA, Rennes 1 University and INSA), Modélisation et Expérimentation pour le Traitement des Informations et des Signaux Sonores, http://ralyx.inria.fr/2006/Raweb/metiss/metiss.pdf
42. J. M. Sanchez, X. Binefa, J. Vitria, and P. Radeva, Local color analysis for scene break detection applied to TV commercials recognition, *International Conference on Visual Information System*, Amsterdam, the Netherlands, pp. 237–244, 1999.
43. J. S. Seo, M. Jin, S. Lee, D. Jing, S. Lee, and C. D. Yoo, Audio fingerprinting based on normalized spectral subband moments, *IEEE International Conference on Acoustics, Speech, and Signal Processing*, Philadelphia, Pennsylvania, pp. 209–212, November 2005.
44. S. Subramanya, R. Simha, B. Narahari, and A. Youssef, Transform-based indexing of audio data for multimedia database. In *Proceedings of International Conference on Multimedia Computing and Systems*, Ottawa, Ontario, Canada, pp. 211–218, June 1997.
45. W. Trappe, M. Wu, Z. J. Wang, and K. J. R. Liu, Anti-collusion fingerprinting for multimedia, *IEEE Transactions on Signal Processing*, 51, 1069–1087, 2003.
46. D. Tsolis, S. Sioutas, and T. Papatheodorou, Digital watermarking in peer to peer networks. In *16th International Conference on Digital Signal Processing*, Santorini, Greece, pp. 1–5, July 2009.
47. M. Wu, W. Trappe, Z. J. Wang, and K. J. R. Liu, Collusion-resistant fingerprinting for multimedia, *Signal Processing Magazine, IEEE*, 21(2), 15–27, 2004.
48. http://www.techno-science.net/?onglet=glossaire&definition=8040
49. http://plato.stanford.edu/entries/qt-uncertainty/
50. Music analysis, retrieval and synthesis for audio signals. http://marsyas.info/download/data_sets.

18

Image and Video Copy Detection Using Content-Based Fingerprinting

Mehrdad Fatourechi, Xudong Lv, Mani Malek Esmaeili, Z. Jane Wang, and Rabab K. Ward

CONTENTS

18.1   Background

18.1.1   Introduction

The past decade has witnessed a significant growth in the volume of multimedia assets that are shared online by users around the world. This phenomenal growth in the online multimedia industry has resulted in great opportunities as well as significant challenges. As an example of the new opportunities, advertisers have found a new source of revenue by displaying ads next to online videos or images. Leading high-tech companies such as YouTube, Facebook, and Flickr have greatly benefited from this new source of revenue. On the other hand, the huge growth in the volume of online multimedia assets has raised serious concerns about the protection of the rights of copyright holders, as manual monitoring of content has become impossible.

Several research approaches have been proposed to find whether a certain multimedia asset is a copied version of an existing one:

Text-based search is perhaps the most basic approach, where based on specific keyword(s) a database of multimedia assets is searched for the relevant copies. This approach assumes that the textual information is accurate and rich. Unfortunately, in many occasions, users who distribute the assets might intentionally or unintentionally change the textual information. As an example, when a video is uploaded on YouTube, the names of artists or their albums might be mistyped or the wrong keywords are used. This results in failure of text-based search tools that rely on textual information or metadata to find the copies of assets of interest. As a result, more elegant content identification and discovery algorithms have been proposed.

Watermarking is a popular approach, which has been used extensively over the past decade for copy detection purposes. In this approach, the content owner first embeds a hidden watermark signal into the original media. Examples of such a hidden message include the logo, the name of the copyright holder or the time, the date, or the location where the digital media had been generated. Later when the ownership is questioned, if the watermark is successfully extracted, then it is determined that the copyright belongs to the original content owner [1]. Unfortunately, watermarking algorithms suffer from two main shortcomings: first, a message needs to be embedded in the digital media asset. As a result, the content (and subsequently the quality) of the signal is altered to some extent. Second, the watermark must be embedded in the original asset *before* its distribution. Otherwise, the watermark detection process is of no use.

An alternative solution for detecting copies of multimedia assets that addresses the above concerns is the *content-based fingerprinting* (CF) *approach*. The main idea is to generate a *unique fingerprint* (also called a *signature* or a *robust hash*) from the input signal. This fingerprint should be generated such that it can uniquely identify the multimedia asset it is generated from.

Figure 18.1 shows the overall structure of how a content-based fingerprint is generated. To extract the fingerprint, the signal is preprocessed before unique features are extracted from it. Postprocessing might be applied to further improve the performance. These features are then converted into a unique sequence of numbers or bits using a *fingerprint generation* algorithm. The resulting sequence is called the *fingerprint*. Examples include compressing the fingerprints to reduce the size of the storage space. The fingerprint generation process is usually performed using a secret key. This secret key is only known to the

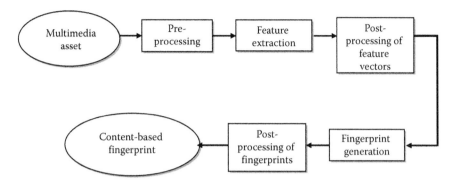

FIGURE 18.1
The building blocks of a CF algorithm.

original content-owner or an authorized representative. The inclusion of a secret key greatly increases the security of the system to adversary attacks.

Figure 18.2 shows how CF can be applied for detection of copies of multimedia assets. First, the fingerprints of the original multimedia content (which we are interested in detecting) should be generated and stored in a table (called the *hash table* or the *fingerprint database*). When a digital multimedia file is shared over the Internet (e.g., by the original content-owner or perhaps a malicious user), it usually undergoes certain distortions such as compression. This process introduces changes to the digital asset, resulting in a slightly different copy that is different from the original one. In order to determine if the query is a copy of the original multimedia asset, its fingerprint should be extracted. To extract the fingerprint of the query,

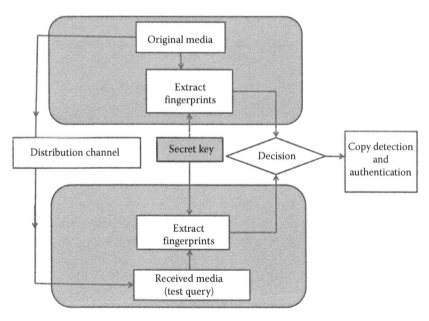

FIGURE 18.2
Overall scheme for finding copies of an original digital media using CF.

the features are first extracted from it, based on the same secret key used to generate the database of fingerprints. A fingerprint is then generated from these features. The *Decision module* then compares the fingerprint extracted from the test query and the fingerprints of the original multimedia files. The query is determined to be a copy of the original one, if its fingerprint is sufficiently close to the fingerprints in the database, according to a certain similarity (matching) criterion.

As seen above, CF is able to find copies of a given multimedia content without altering its content. These algorithms have therefore become very popular in recent years for multimedia copy detection purposes [2,3]. As a side note, we should state that various other terms have also been used in the literature to refer to the above procedure including *multimedia hashing,** *perceptual hashing, content-based copy detection, content-based copy identification, content-based digital fingerprints*, among others. However, we believe that the term CF is more general and intuitive and thus we use it throughout this chapter.

18.1.2 Mathematical Formulation

From the mathematical point of view, the CF problem can be formulated as follows: Let M_1 and M_2 denote two perceptually different digital media signals. Let us also assume that N_1 is a perceptually similar version of M_1 and N_2 is a perceptually similar version of M_2. In other words, a human being perceives M_1 and N_1 and also M_2 and N_2 to have the same content. Furthermore, let $H_k(\cdot)$ be a fingerprint generation function that generates the fingerprint for an input signal based on a secret key k. Given the above assumptions, a CF algorithm should satisfy the following properties:

1. *Compactness*: The size of the fingerprint should be much smaller than the original size of the media signal, that is,

$$Size(H_k(M)) << Size(M) \quad \text{for any input query } M \tag{18.1}$$

2. *Robustness*: Perceptually similar signals should have similar (ideally, identical) fingerprints based on the same secret key, that is,

$$p(H_K(M_1) = H_K(N_1)) \geq 1 - \varepsilon_1, \quad 0 \leq \varepsilon_1 < 1, \tag{18.2}$$

where p is the probability operator and ε_1 should ideally be very close to 0.

3. *Discriminancy*: Perceptually distinct signals should have different fingerprints, that is,

$$p(H_K(M_1) \neq H_K(M_2)) \geq 1 - \varepsilon_2, \quad 0 \leq \varepsilon_2 < 1, \tag{18.3}$$

where ε_2 should ideally be very close to 0.

4. *Unpredictability*: Without having the secret key, the fingerprint should be very difficult to compute. In other words, given two different keys k and k', we shall have

$$p(H_k(M) \neq H_{k'}(M)) \geq 1 - \varepsilon_3, \quad 0 \leq \varepsilon_3 < 1, \forall k \text{ and } k', \tag{18.4}$$

where ε_3 should ideally be very close to 0.

---

* The term "hash" used here is different from traditional hash functions used in cryptography such as message-digest algorithm 5 (MD5) that are used to check the integrity of files.

5. *Low Computational Complexity*: The computational complexity of generating the fingerprints and searching a fingerprint database should be low.

6. *Temporal Granularity*: The fingerprinting algorithm should be able to identify the digital media input such as an audio or a video signal, even if its duration is short. Temporal granularity is especially important in the case of *mash-ups* where small segments of different audio (or video) inputs are mixed together sequentially.

7. *Scalability*: The fingerprinting algorithm should be designed such that it can be applied even to very large online databases, where millions of titles exist.

As mentioned above, ideally $\varepsilon_1, \varepsilon_2, \varepsilon_3$ in Equations 18.2 through 18.4 are very close to 0. Unfortunately, achieving near-zero values for the above parameters is very challenging in practical scenarios. Therefore, some application-dependent trade-offs need to be considered. As an example, for multimedia forensic purposes, the fingerprints of the modified (attacked) digital media should be as close as possible to fingerprints of the original media, and so $\varepsilon_1 \approx 0$. To achieve this goal, a value of ε_2, which is slightly higher than 0, might be tolerated. By using this strategy, we ensure that we can detect almost all copies of the original digital signal but this is at the expense of misclassifying some perceptually different test queries as copies of the digital media of interest.

18.1.3 Outline of this Chapter

In this chapter, we review and discuss the state of the art in the design of CF algorithms. We also address current important issues and discuss areas that need further attention for future research. Our focus will particularly be on reviewing CF algorithms that are applied to images and videos. For a discussion on audio signals and the associated fingerprinting algorithms, the reader can refer to [4,5].

This chapter is organized as follows: In Sections 18.2 through 18.5, we discuss and review several key components of a CF system, including popular preprocessing of input signals (Section 18.2), feature extraction from input signals (Section 18.3), postprocessing of features (Section 18.4), as well as fingerprint generation and associated postprocessing techniques (Section 18.5). In each section, we cite the literature related to the application of CF to images and video signals. Later, in Sections 18.6 and 18.7, we discuss how the fingerprints should be compared against each other and how the performance of CF algorithms should be evaluated. In Section 18.8, we review two case studies from the literature (one related to content-based image fingerprinting and the other related to content-based video fingerprinting). Finally, Section 18.9 is dedicated to conclusions, discussion, and providing suggestions for future research.

18.2 Preprocessing of Input Queries

As shown in Figure 18.1, the first step in extracting fingerprints is to preprocess the input signal. This is done for several reasons such as decreasing the computational complexity and ensuring that the input signals are "standardized" (i.e., they all have the same size, file format, sampling rate, etc.). In this section, we discuss the common preprocessing algorithms applied to input images and video signals.

18.2.1 Preprocessing of Images

The common preprocessing steps applied in dealing with images or individual video frames are as follows:

a. *Conversion of color images to grayscale images*: The aim of this operation is to reduce the computational complexity involved in color images (three-dimensional signals) compared to two-dimensional grayscale images. Another application is to extend a CF algorithm to binary images. For some examples of this preprocessing technique, see [6–10].

b. *Conversion of the color space*: Converting color images to grayscale images reduces the dimensionality of the input signals. This operation, however, results in some information loss. Some researchers have therefore decided to keep the color information while extracting the fingerprints. Sometimes it is recommended that the image in one color space is transformed into another color space (e.g., from RGB (red, green, and blue) to HSV (hue, saturation, and value), where the former color space explicitly separates chromaticity and luminosity [11]).

c. *Resizing*: In order to extract features from an image, the input usually needs to be in a "standard" size. Images are usually resized to a predetermined (generally smaller) size [7,10,12–14]. This results in consistency in the performance when processing images of different sizes. Reducing the size also decreases the associated computational complexity and therefore, improves the computational speed.

d. *Filtering*: To reduce the effect of noise, some researchers filter the input images in the spatial domain to get a *smoother* image before extracting the features [12,15,16].

e. *Block partitioning*: To improve the detection performance as well as the security of a CF algorithm, images are sometimes partitioned into overlapping or nonoverlapping subimages or blocks, as shown in Figure 18.3 [7,9,10,15]. This image partitioning process is usually based on a secret key. Features are then extracted from these blocks instead of the whole image.

f. *Histogram equalization/normalization of luminance*: To deal with attacks which change the luminance distribution of an image (such as gamma correction), the brightness of the image is sometimes adjusted or normalized [16,17].

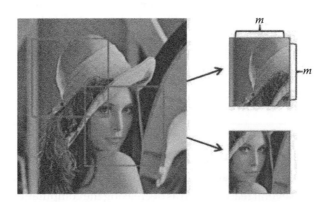

FIGURE 18.3
(**See color insert.**) An example of partitioning an image into overlapping blocks of size $m \times m$. (Adapted from X. Lv and Z. J. Wang, *EURASIP Journal on Information Security*, 1–6, 2009.)

18.2.2 Preprocessing of Video Signals

Preprocessing of a video signal might involve the processing of three components: single frames, the audio, and the temporal information. Preprocessing of a single frame has already been addressed in Sections 18.2.1. For preprocessing the audio component, refer to [4,5]. In this section, we focus on methods for temporal preprocessing of videos:

a. *Reducing the frame rate*: This procedure involves reducing the frame rate by selecting 1 out of every N consecutive frames and then dropping the rest of the frames [12]. An alternative way is to resample the video at a fixed rate of S frames per second, as shown in Figure 18.4 [9,10,18].

b. *Dividing the video into smaller segments:* A video can be too long and this affects the detection speed and the temporal granularity of a CF algorithm. As a result, it is usually necessary to divide each video into segments with shorter durations and extract the fingerprint for each segment as shown in Figure 18.4 [18–20].

c. *Calculating the difference between successive frames:* This step finds any moving objects between successive video frames [21].

d. *Temporal filtering of video frames:* The temporal filtering of video spreads the changes (such as the movement of an object) over several consecutive frames. Therefore, it smoothens the video frames over time [12]. Temporal filtering is also recommended when the fingerprint is obtained using difference of features extracted from successive frames [15].

e. *Selecting a representative frame*: This procedure involves selecting a representative frame for a sequence of frames according to a specific criterion. For example, *key frames* are frequently used to specify the start and the end points of any smooth transition in the video. As a preprocessing step, key frames are extracted first and then fingerprints are generated from them [19,22–24].

f. *Generating a representative frame*: This procedure involves *generating* a representative frame for a sequence of M frames, for example, by combining some selected frames using a weighted function [18,20]. Figure 18.5a–c shows three selected frames from a video sequence and Figure 18.5d shows the results of generating a representative

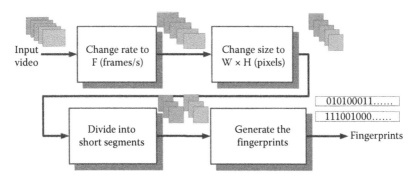

FIGURE 18.4
Some of the common preprocessing algorithms for content-based video fingerprinting. (M. Malekesmaeili, M. Fatourechi and R. K. Ward, A robust and fast video copy detection system using content-based fingerprinting, *IEEE Transactions on Information Forensics and Security*, 6(1), 213–226, 2011. © (2011) IEEE.)

FIGURE 18.5
(**See color insert.**) (a–c) Frames 61, 75, and 90 from a video. (d) A representative frame generated as a result of linearly combining these frames. (M. Malekesmaeili, M. Fatourechi and R. K. Ward, A robust and fast video copy detection system using content-based fingerprinting, *IEEE Transactions on Information Forensics and Security*, 6(1), 213–226, 2011. © (2011) IEEE.)

frame by linearly combining these frames. As seen, the representative frame captures, to some extent, the motion information of the video [18,20].

18.3   Feature Extraction

After applying preprocessing, features are extracted from the preprocessed media signal. In real-world applications, we deal with large media databases. For this reason, a large number of features are usually needed to uniquely identify the media signals. These features form a *feature vector* (FV). Ideally, these FVs should be very close to each other when the inputs are perceptually similar. On the other hand, for inputs that are perceptually different, the corresponding FVs should be as distant as possible from each other.

Similar to the previous section, we present feature extraction algorithms commonly used in the CF literature according to the type of the input signal.

18.3.1   Feature Extraction from Digital Images

Some of the widely used feature extraction algorithms for images are:

a. *Original (raw) pixel values*: Some researchers have used raw pixel values directly as the initial set of features [6,7]. While the use of exact original values is simple, it has the disadvantage of dealing with a large volume of data. In particular, the FV extracted from a grayscale image with size $W \times H$, where W and H are the width and the height of the image, respectively, will have a length of $W \times H$. Obviously, such a large size would quickly become a major processing issue due to the curse of dimensionality.

b. *Image statistics*: As mentioned above, the disadvantage of using raw pixel values as features is the very large length of the FV. To reduce the dimension of the feature space, image statistics such as the mean, variance, and higher moments such as skewness and kurtosis have been employed [15,25,26]. These statistics may be extracted in the spatial or in the frequency domain. They may be extracted from the whole image or only a portion of it. In the same category, image histograms show the distribution of the intensity or the different color spaces and thus have also been explored as useful features in the CF literature [23,27]. Another feature extraction tool in this category is the *centroid of gradient orientations* based on the difference in pixel values. These features are closely related to the distribution of edges and thus can provide useful information regarding the content of a video frame [9,28].

The main advantage of using image statistics is obviously the reduction in the size of the FV. On the other hand, these features are generally not very robust to major manipulations and distortions in an image. As an example, it is very easy to change the color information in an image, while the mean intensity or the color histogram remains intact [2].

c. *Coefficients of transformed images*: One useful feature extraction technique is to transform the image to another domain and then extract the features from the transformed image. These features might include the coefficients of the transformed image or some combinations of them. Examples of such transformations are the Discrete Cosine transform (DCT) [29,30], Discrete Wavelet transform [31,32], Gabor Filter [33], Discrete Fourier transform [31,34], Fourier–Mellin transform (FMT) [7,23]), and Radon transform [35,36].

d. *Local feature points*: The image statistics and the coefficients of the transformed image generally provide information regarding the global properties of the image. For this reason, they are usually referred to as *global* features. On the other hand, *local* features give local information such as edges and corner points of an image [37]. Examples of such features that have been used in the CF literature include Scale-Invariant Feature Transform (SIFT) [23,38,39] and Harris Corners [40]. However, many other local feature descriptors such as Speeded Up Robust Feature (SURF) [41], Gradient Location and Orientation Histogram (GLOH) [42], Histogram of Oriented Gradients (HOG) [43], and Features from Accelerated Segment Test (FAST) [44] can also be used as local features. Local features are usually more robust to geometric transformations (see Figure 18.6). At the same time, their sensitivity to some classical signal processing distortions such as Gaussian noise addition is generally higher than global features [2].

18.3.2 Feature Extraction from Digital Video

A video is a sequence of frames displayed over the course of time. It is usually accompanied by an audio component. As a result, the number of features that can be extracted from a video could potentially be much higher than those of images or audio inputs.

Video features can be divided into four general categories, as discussed below.

a. *Spatial features*: These features are only extracted from individual frames, while the temporal and audio information is ignored. In the case of spatial features, the feature extraction algorithms used are the same as those applied to images (Section 18.3.1).

b. *Audio features*: These features are extracted from the audio component of the video, while the temporal and the visual components are ignored. For more information, please refer to [5].

c. *Temporal features*: Temporal features extract useful and unique relationships between frames. Popular temporal feature extraction techniques include:

- *Shot duration*: Each video can be segmented into different shots, where each shot consists of a single camera action [45]. The duration of a sequence of shots can then be used as a feature that describes the video [46]. In order

(a)

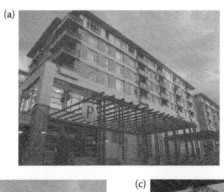

(b)

(c)

FIGURE 18.6
(**See color insert.**) Example of how SIFT can be used for feature extraction from an image. (a) Original image,
(b) SIFT features (original image), and (c) SIFT features (rotated image).

to extract good features that can distinguish videos from each other in a
large database, the size of the fingerprint and subsequently the length of
the video should be relatively large. As a result, this technique cannot be
applied to detect small segments of videos.

- *Location of key frames*: The locations of key frames can be used as a temporal
  fingerprint. This feature extraction algorithm is similar to the duration of
  shots. Therefore, it suffers from the same issues of using shot durations as
  features.

d. *Hybrid features*: These features are extracted from a combination of features in different modalities (e.g., visual, audio or temporal information). As an example,
spatio-temporal features are extracted from spatial information gathered from a
sequence of frames. Some popular hybrid features are:

- *Motion vectors*: Motion vectors can be estimated from image blocks and
  then used as features [29,47].

- *Temporal path of spatial features*: Spatial features vary slightly between two
  consecutive frames. As a result, a useful spatio-temporal fingerprint is the
  one that can track the temporal path of a specific feature over a number of
  successive frames [48].

- *Coefficients extracted from spatio-temporal transformations*: Transformations
  such as the DCT can be applied to extract features from the whole (3D)
  video sequence. Some of the examples reported in the literature include
  applying 3D DCT [12,49] and 3D random bases transform [12] to video
  signals. One main shortcoming of 3D transformation techniques is the

computational complexity associated with computing the coefficients from a 3D input signal [20].

18.4   Postprocessing of Feature Vectors

Depending on the required performance and the available processing units, postprocessing operations are sometimes performed on the extracted FVs. Typical postprocessing operations include:

a. *Combination of features in the same modality*: Different feature extraction techniques are robust in the presence of different types of attacks. Generally, it is very difficult to find a feature extraction algorithm that performs universally well in the presence of various manipulations and distortions. For this reason, in order to increase the detection performance, two or more feature extraction techniques from a specific modality (such as the visual modality or the audio modality) might be combined. The aim is to benefit from strong attributions of individual feature extraction techniques, while overcoming their shortcomings. As an example, Lu et al. [40] have proposed a robust image hashing algorithm by detecting the Harris corner points from the subbands of a wavelet transform.

   Unfortunately, combining feature extraction algorithms is more complicated than it might appear at first glance. First, a proper feature combination algorithm that can overcome the vulnerabilities of the individual feature extraction algorithms while keeping their strengths must be developed. Developing such an information fusion algorithm is dependent on the feature extraction techniques employed and the types of manipulations on the signals. Furthermore, the computational complexity will increase compared to those using individual feature extraction algorithms. As a result, the proper means for controlling the overall computational complexity need to be considered [7].

b. *Dimension-reduction techniques*: Extracted FVs usually have a large size that might even reach several thousands of entries for each input signal. To reduce the size of the FVs and to speed up the search process, the FV needs to be shortened using dimensionality reduction algorithms such as principal component analysis (PCA) [50].

   One of the most common dimensionality reduction techniques is random projection, where a high-dimension FV is projected into a lower dimension using a predefined random transformation. The inclusion of randomness increases the security of the algorithm. Examples of random projection techniques include the Gaussian random projection [6] and the Fast Johnson–Lindenstrauss Transform (FJLT) [7,14].

c. *Matrix decomposition*: In this approach, a matrix of features is decomposed and some invariant features are extracted. Some of the popular matrix decomposition algorithms are singular value decomposition (SVD) [8,51] and nonnegative matrix factorization (NMF) [6,24].

d. *Difference of features*: Calculating the difference between features in the spatial domain for images and in the spatio-temporal domain for videos have been suggested as a useful postprocessing step. This operation increases the robustness with respect to some variations in the input query such as changes in global luminance [15,21,52].

e. *Normalization*: Normalizing the extracted features to have zero means and the unit standard deviation ensures that all the features are within a specific range [14].

f. *Ordinal ranking*: Extracted features might be ranked according to some metric and their ranks are then used as final features. As an example, an image can be divided into different blocks, and then features such as the average pixel value can be extracted from each block (see Figure 18.3). However, instead of using the averages of the actual feature values directly, they are ranked globally and their ranks are used as features [23,52,53]. Note that ordinal ranking is not limited to the spatial domain and it can also be used in the spatio-temporal domain as well. For example, in [54] image blocks were ranked in the same spatial position over a number of successive frames and the temporal ordinal ranks of those blocks were used as features. A major deficiency of using ordinal postprocessing is its vulnerability to global ranking. Even a slight change in the image such as adding a layover text or logo can affect the values of some features and subsequently affect the ranking of features [3].

18.5 Generating Fingerprints from Feature Vectors

The conversion of feature vectors (FVs) to fingerprints is called *signature generation* or *fingerprint generation* and usually involves converting *real-valued* FVs to either *quantized-values* or *binary-valued* sequences. However, it should be noted that in some approaches, the real-valued features are directly used as fingerprints.

The key factor in creating a fingerprint from an FV is that the discriminative property and the robustness of the fingerprints should be preserved. In the case of video signals, if a fingerprint is generated for the whole video, then it would not be possible to identify short clips within the video [15]. For this reason, a short sliding window over time is usually chosen (e.g., the length of the window is 1 s). The video fingerprint for that particular time frame (called *fingerprint block* [15]) is extracted and stored in the fingerprint database. Next, the window is slid over the video with some temporal overlap. The fingerprint block from the next window is extracted and stored in the database. This process is repeated for all future video segments. For a query video, the same process is applied, and then its fingerprint block is matched against the fingerprint blocks stored in the fingerprint database.

Some of the popular approaches for generating fingerprints from FVs are as follows:

a. *No quantization*: In this approach, the real-valued FVs are directly used as fingerprints [6,7]. The main advantage is that no further processing is needed to convert the real-valued FVs to binary strings. The disadvantage is the storage space needed to store all the real-valued numbers.

b. *Quantization and binarizarion*: Interval quantization is a popular solution to reduce the space needed for storing an FV and make it more robust to minor changes in the signal [37,55].

Another popular approach for increasing the robustness of a CF algorithm during the fingerprint generation process is *abstraction*. Here, each feature value is compared to a specific threshold. Feature values, which are equal to or higher than the threshold, will be converted into the binary value "1." Feature values which are lower than the threshold, are converted to the binary value "0" [15,21]. A popular choice for the threshold is the median or the mean of the FV [18].

c. *Traditional cryptography*: Traditional cryptography methods such as Rivest, Shamir and Adleman algorithm [32] or message-digest algorithm 5 (MD-5) [56] have also been applied for generating secure digital fingerprints.

The generated fingerprints using one of the approaches discussed above, may also be further postprocessed. The most common postprocessing technique applied to fingerprints is compression. Some of the compression algorithms applied in the literature include Wynder-Ziv coding [57], JPEG2000 [58], Slepian-Wlof coding [59], and variable-length coding [60].

18.6 Fingerprint Matching

As stated in the introduction, a CF algorithm usually involves two steps. In the first step, a database of fingerprints is generated for the original multimedia assets of interest. In the second step, the fingerprint generated for each test query needs to be compared with the original fingerprints in the database in order to determine whether a copyright has been infringed (see Figure 18.2). Two factors need to be considered during the fingerprint matching process: (1) Defining the similarity between two fingerprints and (2) once the similarity between the two fingerprints is measured, making a proper decision regarding possible copyright infringement. These topics are discussed next in this section.

18.6.1 Similarity Metrics

The comparison of two fingerprints is based on some *similarity* criterion. Different similarity metrics can be employed depending on whether the generated fingerprints are binary or real-valued. For nonbinary fingerprints, the Euclidean distance [6,7,9] and the L_1–Norm [33,61] have been frequently used. For binary strings, the Hamming distance is the most widely used metric [12–14]. A sample normalized Hamming distance for a video fingerprinting algorithm is shown in Figure 18.7 [18]. As seen, the normalized hamming distance is expected to be small when comparing two perceptually similar videos and large for perceptually different videos.

18.6.2 Classification

The *comparison procedure* between two fingerprints can be carried out by means of a classifier. The most widely used classifier in the CF literature is *threshold-based*. This classifier considers two fingerprints to belong to perceptually different inputs if the following condition holds:

$$dist(H_K(M), H_K(M')) \geq Thr, \tag{18.5}$$

where $H_K(M)$ and $H_K(M')$ are the fingerprints extracted from the input signals M and M' based on the secret key K. In Equation 18.5, *dist* is a similarity measurement function (Section 18.6.1), and *Thr* is a predetermined threshold value. The value of *Thr* is usually

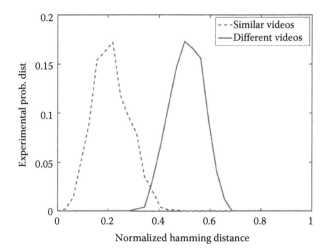

FIGURE 18.7
Normalized Hamming distance. (M. Malekesmaeili, M. Fatourechi, and R. K. Ward, A robust and fast video copy detection system using content-based fingerprinting, *IEEE Transactions on Information Forensics and Security*, 6(1), 213–226, 2011. © (2011) IEEE.)

determined during the training of the CF algorithm (e.g., using the receiver operating characteristic (ROC) curve) [6,9].

The reason for having such a simple structure for a classifier in a CF system is the presence of an infinite number of classes, as each unique input signal can be regarded as a unique class. Training a sophisticated classifier to learn so many classes is therefore a very challenging task. Having said this, examples of more complicated classifiers such as Support Vector Machines (SVMs) have been recently reported in the literature [62].

18.7 Performance Evaluation of CF Algorithms

In this section, we discuss the performance evaluation of CF algorithms. We start by reviewing the manipulations that usually occur to images and videos. We then discuss various metrics needed to summarize different aspects of the performance including the detection performance, the size of the storage space, the computational complexity, and the security of the algorithm.

18.7.1 Content-Preserving versus Content-Changing Manipulations

Manipulations applied to a multimedia asset can be categorized as either *content-preserving* or *content-changing* (or content-altering). In the case of content-preserving attacks, the attacked version remains perceptually the same.

Popular examples of content preserving attacks for digital images or individual frames include *additive noise (Gaussian, salt and pepper, speckle,* and *uniform), filtering (Gaussian, median, average, Wiener, circular, motion filtering,* and *image sharpening), geometric attacks (cropping, rotation, scaling, shearing, changing the aspect ratio, affine transforms, horizontal or vertical shifts of the input image, bending,* and *warping), color/brightness/contrast manipulation (Gamma*

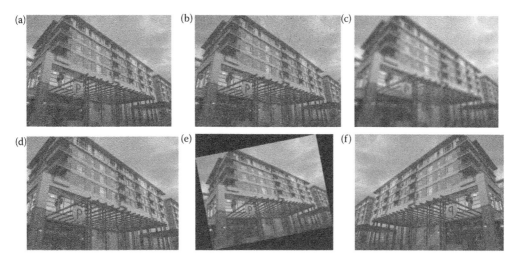

FIGURE 18.8
(a) An original image and (b–f) sample content-preserving attacks.

correction, changing the parameter values of a color space such as hue, saturation and value), Joint Photographic Experts Group (JPEG) compression, image mirroring, watermark embedding, logo insertion, text overlay, and *hybrid attacks* (combination of at least two of the above attacks) [2,7,18]. See Figure 18.8 for examples of content-preserving attacks.

The addition of temporal information creates a more complicated scenario for videos. Content-preserving attacks applied to a video can be categorized as follows:

- *Spatial attacks*: Here, only the spatial information in each video is changed. For example, the videos are compressed frame by frame. However, the temporal and audio information remains unaffected. These attacks are similar to attacks for images discussed above.

- *Audio-based attacks*: Attacks such as time-domain filtering only affect the audio quality. However, the temporal and the visual information remain unchanged. Examples include audio compression and filtering. For more information regarding these types of attacks, refer to [4,5].

- *Temporal attacks*: Content-preserving temporal attacks are also called *storyline-preserving* attacks. The main property of these attacks is that they preserve the storyline of the video even when the temporal content of the video is slightly altered (e.g., a specific number of frames have been dropped). Some of the more popular storyline-preserving attacks are random frame dropping, frame rate change, and time shift.

- *Hybrid attacks*: Hybrid attacks target at least two information channels of videos. Combining attacks in the temporal, audio, and spatial domains result in a wide range of hybrid attacks for a given video. Some popular hybrid attacks are *compression, transcoding, transmission over a lossy channel,* and *fade over.*

Content-changing attacks, on the other hand, introduce small, but perceptually significant changes in the signal. Therefore, to an observer the signal becomes perceptually different than the original one [2]. Studying content-changing attacks is especially relevant for *content-integrity verification*, where slight but intentional changes in the perceptual quality of the input multimedia might have occurred. As an example, the face of a person is replaced with the face of another person in an input image.

Some of the more popular content-altering attacks for images are *adding/removing/ manipulating objects*, *changing the image background*, and *hybrid (composite) attacks*, where at least two of the attacks discussed above are applied to an image. As for the videos, similar to content-preserving attacks, the attacks can occur in four domains. Attacks in the spatial domain are similar to content-changing attacks for images. Examples of content-changing attacks for audio include replacing a song with another one or creating a mash-up with another audio input [4,5]. Examples of temporal attacks (also called *storyline-changing* attacks) are *mash-up* or *substitution attack*, where at least one segment of the video is replaced by a segment of another video. Another example is *segment deletion* where a segment of the video is completely removed. As for hybrid attacks, any combination of above attacks such as compressing the video while its audio is replaced by a song, falls into this category.

It should be noted that the content-preserving and content-changing attacks might be applied concurrently. For example, an object is removed from the image, while it has also been compressed. Overall, this operation results in a *hybrid content-changing attack*.

Please note that content-preserving attacks have been studied more vigorously in the literature, because they can model distortions such as noise addition or compression that frequently happen during transmission and sharing of multimedia content [7,9].

We should also state that while it is relatively easier to analyze the performance of a CF algorithm, when attacks are considered separately, composite attacks usually occur in real-life situations. For example, millions of compressed mash-up videos or mash-up videos with text layover currently exist on video sharing web sites such as YouTube. As a result, studying hybrid attacks is also of great importance.

18.7.2 Detection Performance

To determine the detection performance of a CF algorithm, two metrics are commonly used: *false rejection rate* (FRR) and *false acceptance rate* (FAR). FRR shows the percentage of the query inputs that are falsely identified as not being copies, while FAR represents the percentage of original queries that are incorrectly classified as copies of another query. A fitness function then combines these two metrics in a single value. Some of the popular fitness functions in the CF literature are *classification accuracy*, *bit error rate* [15,40], *F*-score [18], and using ROC curve for demonstration of trade-off between FRR and FAR [9,52].

18.7.3 Computational Cost

In the design of a CF algorithm, a balance between the detection performance and the computational complexity of the algorithm needs to be considered. While the goal of designing an algorithm with a very high detection performance may be accomplished by using sophisticated information retrieval algorithms, the computational cost of the algorithm might also be considerably high, limiting its practicality. In the CF literature, the *execution time* and the *computational complexity* have been more widely used to measure the computational efficiency of CF algorithms [14,20,37]. While the execution time is more intuitive to consider, the hardware used as the test platform as well as the efficiency of the algorithm implementation can greatly impact the run time of the algorithm. On the other hand, although computational complexity is less intuitive, it is hardware/software independent, making it a good candidate for evaluating the computational cost of CF algorithms.

18.7.4 Length of Fingerprints

The length of a fingerprint can greatly affect the size of the fingerprint database, the storage size, and subsequently the speed of the search process. As a result, it becomes very important for a CF algorithm to generate compact fingerprints. For this reason, the lengths of fingerprints reported in the literature are typically below 1000 bits per input image or video block [15,18].

18.7.5 Security of a CF Algorithm

For multimedia forensic purposes, the security of fingerprints is very important: a malicious user should not be able to easily generate the same fingerprint(s) employed by the original owner. The basic idea is to incorporate *pseudorandomness* into the process of generating fingerprints based on a *secret key* [2]. This key is only known to the original content owner or his/her authorized representatives. Users should then be able to generate the fingerprint of the input signal if they have the secret key and it should be impossible for them to generate the fingerprints without having access to the secret key. The unpredictability of the fingerprint based on the secret key could be incorporated in any step in the generation of a fingerprint as discussed below:

a. *Preprocessing*: The security of a CF algorithm can be increased if a secret key is used in the preprocessing stage. A popular approach when dealing with images or video frames is *randomized tiling*. Here, the input image or video frame is randomly divided into overlapped or nonoverlapped blocks and features are extracted from each block (see Figure 18.3) [6–8,14].

b. *Feature extraction*: Secret keys can be used in the feature extraction phase as well. A popular example is to apply a *randomized transform*. In this approach, the extracted features are transformed into another *pseudorandomized domain* determined using a secret key [63].

c. *Fingerprint generation*: Fingerprinting security can also be increased by including a randomization process during the fingerprint generation process. As an example, *random projection* techniques can be used to map robust features into a lower dimension using pseudorandomized variables determined by a secret key [7,14].

Of course, the above randomization techniques can be applied together to further increase the security. Security is rarely discussed in the design of CF algorithms, perhaps because of the computational complexity associated with estimating it. However, *differential entropy* is a metric that has been recommended for measuring security [16]. Han and Chu [2] have used this metric to estimate the security of some of the state-of-the-art CF algorithms.

18.7.6 Database Search

A fast database search algorithm is needed to quickly search a large fingerprint database. As the fingerprints generated for two perceptually similar multimedia signals might not be exactly identical, a simple lookup table will not necessarily yield the correct match. As a result, a *similarity matching* algorithm needs to be developed. This algorithm compares the similarity between the fingerprints of a test query with those extracted from the original assets. Depending on the application, the closest match or a sorted list of fingerprints should then be returned.

As the size of the fingerprint database can easily become very large for a database of multimedia assets, *exhaustive search* is not a proper solution. Alternatively, fast approximate search algorithms that increase the search speed by several orders of magnitude have been developed. The main idea is to design a nearest-neighbor (NN) search algorithm that returns the search result much faster than the exhaustive search approach [40]. This will be at the expense of having a slightly higher error rate [3]. A popular approach is to use the search in two stages: in the first stage, a fast and computationally efficient search is performed to find the most relevant candidates. In the second stage, a more refined and more costly search is performed to find the best match [64].

Among the approximate search algorithms, the *Locality Sensitive Hashing* (LSH) [65–67] is perhaps the most popular one. The main idea is to design hash functions such that a *similar set of fingerprints* are mapped to the *same bucket* with high probability. The number of buckets is much smaller than the size of the original search space. In other words, LSH performs NN search in high dimensions. Once a number of suitable buckets are found for an input fingerprint, an exhaustive search can be performed to find the closest match(es). Other possible search algorithms include the extensions of inverted file techniques for CF algorithms [15,18,28].

18.8 Case Studies

Now that we have reviewed different design aspects of CF algorithms and to better understand how a CF algorithm works, in this section, we examine in detail two state-of-the-art CF algorithms that have been recently designed.

18.8.1 An Image Fingerprinting Algorithm Based on Random Projection

Inspired by the state-of-the-art image hashing schemes that are based on dimensionality reduction such as SVD and NMF, we have proposed a new global image fingerprinting algorithm based on the FJLT random projection algorithm [7,68].

The algorithm works as follows: After resizing the input image to a standard size and converting it into grayscale, it is divided into randomly selected blocks (with overlaps) based on a secret key. The pixel values from each block are concatenated and based on another secret key, the resulting matrix is transformed to a lower dimension using the FJLT. The reduced dimension matrix forms the feature matrix. This matrix is then multiplied by an *ordered random weighting vector* to yield the final FV set (see Figure 18.9) [7].

The experimental results tested on an image database have shown that the proposed algorithm is robust to a large class of content-preserving attacks such as additive noising, blurring, and compression. However, the results have shown that the proposed method is still vulnerable to *rotation* attacks. Therefore, the rotation-invariant Fourier-Mellin Transform (FMT) was added as a preprocessing step. As the input features are first transformed to a rotation-invariant domain, the robustness of the modified FJLT fingerprinting algorithm against rotation attacks is greatly improved at the cost of slight degradation of robustness against other attacks. As a result, in [7] a simple joint decision scheme was proposed to combine these two schemes and form a hybrid content-based fingerprinting (HCF) algorithm (see Figure 18.9).

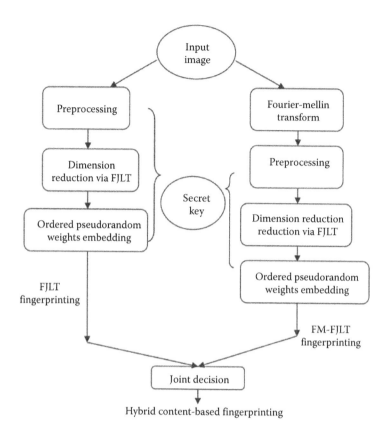

FIGURE 18.9
The overall structure of FJLT, FMT-FJLT, and HCF algorithms. (From X. Lv and Z. J. Wang, *EURASIP Journal on Information Security*, 1–6, 2009. With permission.)

We compared the detection performance of CF algorithms based on NMF [6], FJLT, and the HCF algorithm using an image database with 1000 original images and 99,000 distorted copies. The detailed robustness performance results are shown in Table 18.1. The corresponding ROC curves, which show the relation between the probability of true detection and the probability of the false alarms, are shown in Figure 18.10. The results clearly show that the HCF outperforms both FJLT and NMF.

18.8.2 A Complete Video Fingerprinting Scheme that Considers Detection Performance, Fingerprint Security, and Computational Speed

In this section, we summarize the complete video fingerprinting algorithm proposed in [18] (see Figure 18.11). Here, each video is preprocessed first by changing its frame size and frame rate to a fixed frame size $W \times H$ and a fixed frame rate F, respectively. A window with the length T s is then slid over the video with an overlap of $T/2$ s and the features are extracted from each window. In other words, there is 50% overlap between the video segments used to generate the fingerprint database. After this segmentation, each segment is transformed into a single image that contains both the spatial and the temporal information about the frames of the video segment. This is done by applying a weighted average on the segment's frames (see Figure 18.5). The weighting factors are chosen such that some

TABLE 18.1

Comparison of the Robustness of NMF, FJLT, and HCF Algorithms

	Identification Performance (%)		
Manipulations	**NMF**	**FJLT**	**CBF**
Gaussian noise	59.4	69.5	62.4
Salt and pepper noise	81.9	96.9	97.7
Speckle noise	78.3	99.8	99.8
Gaussian blurring	98.3	99.5	99.0
Circular blurring	98.4	99.5	99.1
Motion blurring	98.9	99.8	99.7
Rotation	16.4	36.9	86.5
Cropping	16.7	96.6	96.1
Scaling	98.5	100	100
JPEG compression	99.7	100	100
Gamma correction	5.2	86.6	74.3

perceptible motion information is contained in the resulting representative image. The representative image called temporally informative representative image (TIRI) is then divided into overlapping blocks of size $M \times N$. Low-frequency DCT coefficients are extracted as features from each block based on a secret key. Finally, the chosen coefficients from all blocks are concatenated to form a feature vector and the feature vector is then converted into a vector of binary numbers through abstraction. The binary fingerprints are then stored in the fingerprint database along with some metadata to precisely localize the video segment that the fingerprint refers to.

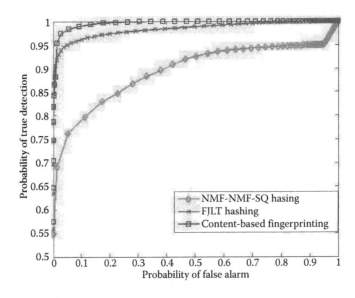

FIGURE 18.10
The ROC curves for NMF, FJLT, and HCF fingerprinting algorithms when tested on a wide range of attacks. (From X. Lv and Z. J. Wang, *EURASIP Journal on Information Security*, 1–6, 2009. With permission.)

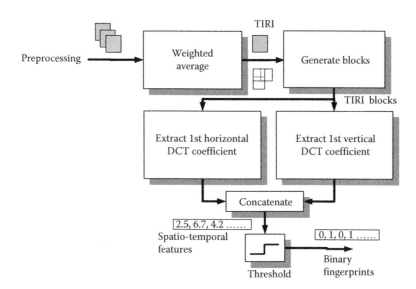

FIGURE 18.11
A nonsecure version of the proposed content-based video fingerprinting algorithm.

For a certain query video, the system follows the same process as above to extract the fingerprints of the query (i.e., with the same parameter values and the same secret key). The query fingerprints are then fed to a matching block. The matching block finds the fingerprint in the database that is closest to the query's fingerprint, using an NN approach. It determines this entry in the fingerprint database to be a match if the NN's distance does not exceed a certain threshold.

There is a trade-off between security and the detection performance. Choosing coefficients based on a secure key increases the security and at the same time decreases the detection performance of the algorithm. Figure 18.12 compares the performance of a secure version to the nonsecure one in the presence of noise and time shifts. The F-score is used as the detection

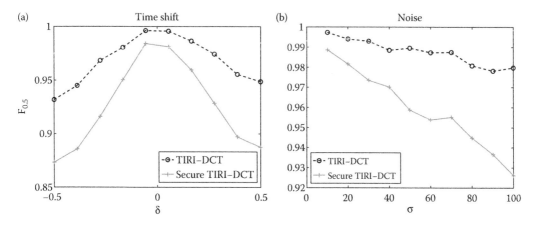

FIGURE 18.12
Comparison of the secure and nonsecure version in presence of (a) time shift from -0.5 s to $+0.5$ s and (b) noise with variance σ^2.

performance metric. While both algorithms show high robustness, the secure version of the proposed algorithm (called the secure TIRI-DCT) has a lower detection performance compared to the nonsecure version (called TIRI-DCT).

To decrease the search time, a clustering-based approach has been proposed in [18], where the fingerprints of the database are first clustered to K clusters, and then for each query a sorted list of potential clusters are generated. Each query is only compared to the fingerprints of these potential clusters one after another. The results in [18] show that the proposed cluster-based approach outperforms exhaustive search and inverted file search approaches.

18.9 Discussion, Conclusions, and Future Research Directions

Research over the past years has shown that content-based fingerprinting (CF) algorithms are reliable alternatives to traditional watermarking algorithms for copy detection purposes. Currently, most state-of-the-art CF algorithms can achieve a high detection performance in the presence of a wide range of attacks. As an example, [12] reported very high robustness (true positive rate or *TPR* ≈ 100%) for their proposed video fingerprinting algorithm in the presence of various content-preserving attacks such as blurring, noise addition, frame drop, fade-over, and so on. These results were achieved at a relatively low FAR (1%). Also refer to Section 18.8 for two relevant case studies from the literature.

Over the past decade, CF approaches have also managed to find a respectable place in the industry as an efficient tool for multimedia copy detection. *YouTube* (http://www.youtube.com) is perhaps the most popular example of companies that use audio and video fingerprinting algorithms to find copies of videos. Some of the other companies that design and use CF include *Audible Magic* (http://audiblemagic.com/, audio fingerprinting), *BroadbandTV* (http://www.broadbandtvcorp.com, audio/video fingerprinting), *Civolution* (http://www.civolution.com/,audio/video fingerprinting), *iPharro Media* (http://www.ipharro.com/, video fingerprinting), *Vercury* (http://www.vercury.com/, audio and video fingerprinting), *Vobile* (http://www.vobile.com/, audio and video fingerprinting), *YUVsoft* (http://www.yuvsoft.com/, video fingerprinting), and *Zeitera* (http://www.zeitera.com/, audio and video fingerprinting).

Despite these significant academic and industrial achievements, there are several research areas where further improvements can be made. These areas are summarized below:

- *Improving the detection performance*: Current CF algorithms mostly report low FRR and low FAR values. It should be noted that these values need to be further improved for real-world multimedia databases such as YouTube and Flickr, which contain millions of original files as well as their pirated copies. Special attention needs to be paid to hybrid attacks, as they are usually ignored in research papers, yet they constitute a large portion of attacks in online digital media databases.

- *Evaluation of detection performance*: So far little work has been carried out to find a proper evaluation metric that combines FRR and FAR and can summarize the detection performance. Classification accuracy, as mentioned earlier, is an obvious choice. However, it can be a misleading metric in the case of imbalanced number of test samples which is the case for CF algorithms [69,70]. This is because in a CF algorithm, the total number of comparisons between two original signals in Equation

18.5 is much higher than the total number of comparisons between original signals and their attacked versions. Therefore, we are dealing with a classification problem with imbalanced number of test samples. In this case, metrics derived from an ROC curve such as the *area under the ROC* (AUC) are perhaps more suitable. Unfortunately, the calculation of ROC-based metrics is usually costly. Recently, the pattern recognition community has paid more attention to metrics such as *F-score* and *Kappa coefficient* [71]. Nevertheless, their efficiency for summarizing the performance of CF algorithms still needs to be investigated.

- *Exploring audio information from video*: So far only a limited number of papers have explored the audio and visual information together for CF purposes. As an example [29], detects video and audio fingerprints separately and then applies a decision fusion algorithm to combine the results. The results show improvements on the performance as a result of combining the outcomes of audio and video fingerprinting algorithms. More research, however, needs to be conducted to design efficient decision/feature fusion algorithms. Also, the performance trade-offs need to be investigated in more details.

- *Measuring the security*: The security of CF algorithms is usually ignored in the performance reports. Given the importance of security in preventing malicious manipulations of a CF system, it is necessary that more effort is taken in increasing the security of a specific CF algorithm as well as measuring and comparing it against the security of other CF algorithms.

- *Database search*: Once the fingerprints of the multimedia assets in an online multimedia database have been extracted, the computational overhead will later be mostly associated with searching the database to match a query video with an existing one in this database. As a result, fast approximate search methods that are tailored to a particular CF algorithm need to be further investigated.

- *Performance trade-off*: Unfortunately, it is not easy to develop feature extraction techniques that have a very high detection performance under a wide range of content-preserving attacks which usually occur in practice. These attacks include noise addition and rotation in the case of images, audio compression in the case of audio inputs and frame dropping in the case of videos. Furthermore, achieving a higher performance usually necessities designing more complicated algorithms, and in turn, increasing the computational complexity, and the size of the fingerprint database. For this reason, a trade-off between the detection performance and the computational complexity is required. To the best of our knowledge, no research paper has quantitatively analyzed the trade-off between the computational complexity and the detection metrics yet. Similarly, the trade-offs between the above performance metrics and security, the size of the fingerprint, and the speed of database search are interesting topics that can be further investigated.

- *Automation*: CF algorithms typically have many parameters, which need to be determined. Exhaustive search and manual tuning are inefficient search methods, when dealing with a large number of parameters. For this reason, more efficient automatic search methods such as *gradient descent* or *evolutionary algorithms* [72] need to be investigated [18].

Acknowledgment

This work was supported in part by NSERC Grant STPGP 365164-08.

References

1. S. J. Lee and S. H. Jung, A survey of watermarking techniques applied to multimedia, In *the Proceedings of IEEE International Symposium on Industrial Electronics (ISIE)*, Pusan, South Korea, pp. 272–277, 2002.
2. S. H. Han and C. H. Chu, Content-based image authentication: Current status, issues, and challenges, *International Journal of Information Security*, 9, 19–32, 2010.
3. J. Lu, Video fingerprinting for copy identification: From research to industry applications, In *the Proceedings of SPIE Media Forensics and Security*, 725402-725402-15, 2009.
4. P. Cano, E. Batlle, E. Gómez, L. de CT Gomes, and M. Bonnet, Audio fingerprinting: Concepts and applications, *Computational Intelligence for Modelling and Prediction*, 2, 233–245, 2005.
5. P. Cano, E. Batlle, T. Kalker, and J. Haitsma, A review of audio fingerprinting, *The Journal of VLSI Signal Processing*, 41, 271–284, 2005.
6. V. Monga and M. K. Mhcak, Robust and secure image hashing via non-negative matrix factorizations, *IEEE Transactions on Information Forensics and Security*, 2, 376–390, 2007.
7. X. Lv and Z. J. Wang, An extended image hashing concept: Content-based fingerprinting using FJLT, *EURASIP Journal on Information Security*, 1–6, 2009.
8. S. S. Kozat, R. Venkatesan and M. K. Mihçak, Robust perceptual image hashing via matrix invariants, In *the Proceedings of IEEE. International Conference on Image Processing (ICIP'04)*, Singapore, pp. 3443–3446, 2005.
9. S. Lee and C. D. Yoo, Robust video fingerprinting for content-based video identification, *IEEE Transactions on Circuits and Systems for Video Technology*, 18, 983–988, 2008.
10. M. Puri and J. Lubin, Robust efficient video fingerprinting, In *the Proceedings. of SPIE*, 7254, 2009.
11. R. Cucchiara, C. Grana, M. Piccardi, and A. Prati, Detecting moving objects, ghosts, and shadows in video streams, *IEEE Transactions on Pattern Analysis and Machine Intelligence*, 28(10), 1337–1342, 2003.
12. B. Coskun, B. Sankur, and N. Memon, Spatio–temporal transform based video hashing, *IEEE Transactions on Multimedia*, 8, 1190–1208, 2006.
13. X. Su, T. Huang, and W. Gao, Robust video fingerprinting based on visual attention regions, In *the Proceedings of IEEE. International Conference on Acoustics, Speech and Signal Processing (ICASSP 2009)*, Taipei, Taiwan, pp. 1525–1528, 2009.
14. M. Fatourechi, X. Lv, Z. J. Wang, and R. K. Ward, Towards automated image hashing based on the fast Johnson–Lindenstrauss transform (FJLT), In *the Proceedings of 1st IEEE. International Workshop on Information Forensics and Security (WIFS 2009)*, London, UK, pp. 121–125, 2010.
15. J. Oostveen, T. Kalker, and J. Haitsma, Feature extraction and a database strategy for video fingerprinting, *Recent Advances in Visual Information Systems*, 2314, 67–81, 2002.
16. A. Swaminathan, Y. Mao, and M. Wu, Robust and secure image hashing, *IEEE Transactions on Information Forensics and Security*, 1, 215–230, 2006.
17. J. Fridrich, Visual hash for oblivious watermarking, In *the Proceedings of SPIE International Society for Optical Engineering*, 286–294, 2000.
18. M. Malekesmaeili, M. Fatourechi, and R. K. Ward, A robust and fast video copy detection system using content-based fingerprinting, *IEEE Transactions on Information Forensics and Security*, 6(1), 213–226, 2011.

19. A. Massoudi, F. Lefebvre, C. H. Demarty, L. Oisel, and B. Chupeau, A video fingerprint based on visual digest and local fingerprints, In *the Proceedings of IEEE. International Conference on Image Processing. (ICIP'06)*, pp. 2297–2300, 2006.

20. M. Malekesmaeili, M. Fatourechi, and R. K. Ward, Video copy detection using temporally informative representative images, In *the Proceedings of International Conference on Machine Learning and Applications (ICMLA'09)*, Miami Beach, Florida, pp. 69–74, 2009.

21. R. Radhakrishnan and C. Bauer, Content-based video signatures based on projections of difference images, In *the Proceedings 9th IEEE Workshop on Multimedia Signal Processing*, Crete, Greece, pp. 341–344, 2007.

22. S. Lee and C. D. Yoo, Robust video fingerprinting based on 2D-OPCA of affine covariant regions, In *The Proceedings of IEEE. International Conference on Image Processing (ICIP'08)*, San Diego, California, pp. 2156–2159, 2008.

23. A. Sarkar, P. Ghosh, E. Moxley, and B. S. Manjunath, Video fingerprinting: Features for duplicate and similar video detection and query-based video retrieval, In *The Proceedings of SPIE Multimedia Content Access: Algorithms and Systems II*, San Jose, California, 2008.

24. O. Gursoy, B. Gunsel, and N. Sengor, Transform invariant video fingerprinting by NMF, *Computer Analysis of Images and Patterns*, 5702, 452–459, 2009.

25. M. Abdel-Mottaleb, G. Vaithilingam, and S. Krishnamachari, Signature-based image identification, In *The Proceedings of SPIE*, 22, 1999.

26. C. Kailasanathan and R. S. Naini, Image authentication surviving acceptable modifications using statistical measures and k-mean segmentation, In *the Proceedings of IEEE-EURASIP Workshop on Nonlinear Signal and Image Processing*, Baltimore, Maryland, 1, 2001.

27. A. M. Ferman, A. M. Tekalp, and R. Mehrotra, Robust color histogram descriptors for video segment retrieval and identification, *IEEE Transactions on Image Processing*, 11, 497–508, 2002.

28. A. Hampapur and R. Bolle, VideoGREP: Video copy detection using inverted file indices, *IBM Research*, 3, 6, 2001.

29. A. Saracoglu, E. Esen, T. K. Ates, B. O. Acar, U. Zubari, E. C. Ozan, E. Ozalp, A. A. Alatan, and T. Ciloglu, Content based copy detection with coarse audio-visual fingerprints, In *the Proceedings of 7th International Workshop on Content-Based Multimedia Indexing (CBMI'09)*, Crete, Greece, pp. 213–218, 2009.

30. C. Y. Lin and S. F. Chang, A robust image authentication method distinguishing JPEG compression from malicious manipulation, *IEEE Transactions on Circuits and Systems for Video Technology*, 11, 153–168, 2002.

31. M. Mihcak and R. Venkatesan, New iterative geometric methods for robust perceptual image hashing, *Security and Privacy in Digital Rights Management*, 2320, 13–21, 2002.

32. C. S. Lu and H. Y. M. Liao, Structural digital signature for image authentication: An incidental distortion resistant scheme, *IEEE Transactions on Multimedia*, 5, 161–173, 2003.

33. Y. Meng and E. Y. Chang, Image copy detection using dynamic partial function, In *the Proceedings of SPIE*, Santa Clara, California, Vol. 5021, pp. 176–186, 2003.

34. P. Brasnett and M. Bober, Fast and robust image identification, In *the Proceedings of 19th International Conference on Pattern Recognition (ICPR 2008)*, Tampa, Florida, pp. 1–5, 2008.

35. F. Lefebvre, J. Czyz, and B. Macq, A robust soft hash algorithm for digital image signature, In *the Proceedings of IEEE. International Conference on Image Processing (ICIP'03)*, Barcelona, Spain, 2003.

36. J. S. Seo, J. Haitsma, T. Kalker, and C. D. Yoo, A robust image fingerprinting system using the Radon transform, *Signal Processing: Image Communication*, 19, 325–339, 2004.

37. V. Monga and B. L. Evans, Perceptual image hashing via feature points: Performance evaluation and tradeoffs, *IEEE Transactions on Image Processing*, 15, 3452, 2006.

38. Z. Xu, H. Ling, F. Zou, Z. Lu, and P. Li, A novel image copy detection scheme based on the local multi-resolution histogram descriptor, *Multimedia Tools and Applications*, 52(2–3), 445–463, 2010.

39. D. G. Lowe, Distinctive image features from scale-invariant keypoints, *International Journal of Computer Vision*, 60, 91–110, 2004.

40. C. S. Lu, C. Y. Hsu, S. W. Sun, and P. C. Chang, Robust mesh-based hashing for copy detection and tracing of images, In *the Proceedings of IEEE International Conference on Multimedia and Expo*, Taipei, Taiwan, pp. 731–734, 2005.

41. H. Bay, T. Tuytelaars, and L. Van Gool, Surf: Speeded up robust features, In *the Proceedings of 9th European Conference on Computer Vision (ECCV 2006)*, Graz, Austria, pp. 404–417, 2006.

42. K. Mikolajczyk and C. Schmid, A performance evaluation of local descriptors, *IEEE Transactions on Pattern Analysis and Machine Intelligence*, 27(10), 1615–1630, 2005.

43. N. Dalal and B. Triggs, Histograms of oriented gradients for human detection, In *the Proceedings of IEEE Computer Society Conference on Computer Vision and Pattern Recognition (CVPR'05)*, San Diego, California, 1, pp. 886–893, 2005.

44. E. Rosten and T. Drummond, Machine learning for high-speed corner detection, In *the Proceedings of 9th European Conference on Computer Vision (ECCV 2006)*, Graz, Austria, pp. 430–443, 2006.

45. D. Brezeale and D. J. Cook, Automatic video classification: A survey of the literature, *IEEE Transactions on Systems, Man, and Cybernetics, Part C: Applications and Reviews*, 38, 416–430, 2008.

46. P. Indyk, G. Iyengar, and N. Shivakumar, Finding pirated video sequences on the internet, *Technical Report*, Stanford University, Palo Alto, California, 1999.

47. A. Hampapur, K. Hyun, and R. Bolle, Comparison of sequence matching techniques for video copy detection, In *the Proceedings of Conference on Storage and Retrieval for Media Databases*, San Jose, California, pp. 194–201, 2002.

48. J. Law-To, O. Buisson, V. Gouet-Brunet, and N. Boujemaa, Robust voting algorithm based on labels of behavior for video copy detection, In *Proceedings of the 14th ACM International Conference on Multimedia*, New York, pp. 835–844, 2006.

49. B. Coskun and B. Sankur, Robust video hash extraction, In *the Proceedings of 12th IEEE Signal Processing, Communication and Applications Conference*, Turkey, pp. 292–295, 2004.

50. N. Nikolaidis and I. Pitas, Image and video fingerprinting for digital rights management of multimedia data, In *the Proceedings of International Symposium on Intelligent Signal Processing and Communications (ISPACS'06)*, Yonago, Japan, pp. 801–807, 2007.

51. R. Radhakrishnan, W. Jiang, and C. Bauer, A review of video fingerprints invariant to geometric attacks, In *the Proceedings of SPIE Conference on Media Forensics and Security*, San Jose, California, pp. 725407–725407-10, 2009.

52. C. Kim and B. Vasudev, Spatiotemporal sequence matching for efficient video copy detection, *IEEE Transactions on Circuits and Systems for Video Technology*, 15, 127–132, 2005.

53. D. N. Bhat and S. K. Nayar, Ordinal measures for image correspondence, *IEEE Transactions on Pattern Analysis and Machine Intelligence*, 20, 415–423, 2002.

54. L. Chen and F. W. M. Stentiford, Video sequence matching based on temporal ordinal measurement, *Pattern Recognition Letters*, 29, 1824–1831, 2008.

55. X. Guo and D. Hatzinakos, Content based image hashing via wavelet and radon transform, In *the Proceedings of Advances in Multimedia Information Processing Comference (PCM 2007)*, Hong Kong, China, pp. 755–764, 2007.

56. C. Skrepth and A. Uhl, Robust hash functions for visual data: An experimental comparison, *Pattern Recognition and Image Analysis*, 2652, 986–993, 2003.

57. M. Tagliasacchi, G. Valenzise, and S. Tubaro, Hash-based identification of sparse image tampering, *IEEE Transactions on Image Processing*, 18, 2491–2504, 2009.

58. R. Norcen and A. Uhl, Robust visual hashing using JPEG 2000, In *Proceedings of the 8th IFIP TC6/TC11 Conference on Communications and Multimedia Security (CMS '04)*, Lake Windermere, UK, vol. 175, pp. 223–235, 2005.

59. Y. C. Lin, D. Varodayan, and B. Girod, Image authentication based on distributed source coding, In *the Proceedings of IEEE. International Conference on Image Processing (ICIP'07)*, San Antonio, Texas, pp. III-5–III-8, 2007.

60. J. Dittmann, A. Steinmetz, and R. Steinmetz, Content-based digital signature for motion pictures authentication and content-fragile watermarking, In *the Proceedings of IEEE. International Conference on Multimedia Computing and Systems*, Florence, Italy, pp. 209–213, 2002.

61. C. Kim, Content-based image copy detection, *Signal Processing: Image Communication*, 18, 169–184, 2003.
62. J. H. Hsiao, C. S. Chen, L. F. Chien, and M. S. Chen, A new approach to image copy detection based on extended feature sets, *IEEE Transactions on Image Processing*, 16, 2069–2079, 2007.
63. A. Meixner and A. Uhl, Security enhancement of visual hashes through key dependent wavelet transformations, In *the Proceedings of Image Analysis and Processing Conference (ICIAP)*, Cagliari, Italy, pp. 543–550, 2005.
64. X. Lin, Comparative study of content-base image retrieval and video fingerprinting, In *the Proceedings of SPIE*, 7540, 754012–754012-8, 2010.
65. A. Gionis, P. Indyk, and R. Motwani, Similarity search in high dimensions via hashing, In *the Proceedings of the 25th International Conference on Very Large Data Bases*, Edinburgh, Scotland, pp. 518–529, 1999.
66. P. Indyk and R. Motwani, Approximate nearest neighbors: Towards removing the curse of dimensionality, In *the Proceedings of 30th Annual ACM Symposium on Theory of Computing*, Dallas, Texas, pp. 604–613, 1998.
67. A. Andoni and P. Indyk, Near-optimal hashing algorithms for approximate nearest neighbor in high dimensions, In *the Proceedings of 47th Annual IEEE. Symposium on Foundations of Computer Science (FOCS'06)*, Berkeley, California, pp. 459–468, 2006.
68. X. Lv and Z. J. Wang, Fast Johnson-Lindenstrauss transform for robust and secure image hashing, In *the Proceedings of 10th IEEE Workshop on Multimedia Signal Processing*, Cairns, Australia, pp. 725–729, 2008.
69. J. Huang and C. X. Ling, Using AUC and accuracy in evaluating learning algorithms, *IEEE Transactions on Knowledge and Data Engineering*, 17, 299–310, 2005.
70. M. Fatourechi, R. K. Ward, S. G. Mason, J. Huggins, A. Schlogl, and G. E. Birch, Comparison of evaluation metrics in classification applications with imbalanced datasets, In *the Proceedings of 7th International Conference Machine Learning and Applications (ICMLA'08)*, San Diego, California, pp. 777–782, 2008.
71. J. Cohen, A coefficient of agreement for nominal scales, *Educational and Psychological Measurement*, 20, 37–46, 1960.
72. R. O. Duda, P. E. Hart, and D. G. Stork, *Pattern Classification*. 2nd edition, Wiley-Interscience, New York, 2000.

Part V

Methodology, Techniques, and Applications: Multimedia Communications and Networking

19

Emerging Technologies in Multimedia Communications and Networking: Challenges and Research Opportunities

Chang Wen Chen

CONTENTS

19.1   Introduction

With tremendous advancement in several emerging technologies in digital media, communications, and networking, multimedia has become an indispensable aspect in contemporary daily life. Multimedia data in a variety of popular formats such as animation, audio, and video clips have become the mainstream contents on the Internet and are being accessed by various mobile devices through wireless links. The proliferation of diverse distributed multimedia applications has been driving the field of multimedia communications and networking into a new era of true ubiquitous media consumption at any time and from anywhere. The demand for widespread deployment of ubiquitous media services is the motivating force for the research and development of numerous paradigm-shifting technologies.

Among all types of digital media, digital video has been considered as the dominating media form because of its volumetric spatial–temporal presentation nature. Over the past few years, digital video has already become the main traffic payload for Internet and major wireless networks. Every minute there are over 26 h of video uploading to YouTube and over 1.5 million of video clips downloading from YouTube [1]. These constitute only about 40% of the exponentially growing Internet video streaming [2]. Among them, a significant portion is accessed via mobile devices. By 2013, video streaming will be 90% of all consumer IP traffic and 64% of mobile traffic [3]. Such a massive traffic of video content has reached over a diverse population of media consumers through increasingly heterogeneous networks with vastly different operating conditions in data rate, transport mechanism, media access protocol, and physical layer link hardware.

Unlike conventional data communications and networking, contemporary multimedia communications and networking have been severely constrained by several intrinsic characteristics of multimedia data in terms of necessary compression at the source and real-time presentation at the destination. The necessary compression at the source produces a digital media bitstream whose data have become closely coupled with each other. The real-time representation at the destination usually requires more stringent latency and loss constraints for the communication and networking links. As a result, new generations of communication and networking technologies need to be continuously developed for emerging multimedia applications.

Over the past decade, innovative communication and networking technologies have become powerful enablers for the deployment of numerous multimedia applications, creating new distribution channels and business opportunities. While the Internet is still undergoing significant development and expansion, new communication and networking paradigms continue to emerge with both wires and wireless links. Among them, recently emerging peer-to-peer networking paradigm has prompted the revisit of many technical issues related to multimedia communications and promised significant performance improvement of multimedia applications beyond what is supported by the current Internet. The increasingly widespread use of low-power wireless mobile devices and sensors also demands a paradigm shift in the design of the next-generation wireless and mobile multimedia systems. Among them, recent widespread deployment of wireless local area networks (WLAN) and the continued advancement of cellular networks from 3G to 4G and beyond have created numerous emerging multimedia applications for all ranges of mobile wireless users. In particular, the recent innovations in multi-input multi-output (MIMO) antenna technologies promise multifold performance enhancement over conventional signal antenna technologies to meet the continuous demand in bandwidth and diversity for multimedia applications. Unprecedented progress in mobile wireless communication and networking is enabling true pervasive multimedia communications, reaching the media consumers anywhere and at any time.

19.2 Challenges in Contemporary Multimedia Communications and Networking

A new array of multimedia communications and networking applications naturally demands a new class of advanced technologies to meet the technical challenges in these paradigm-shifting developments. These challenges are either originated from the inherent nature of compressed media or constrained by the intrinsic characteristics of wired and wireless communication and networking links. To overcome these barriers, comprehensive understanding of compressed media and communication and networking links as well as their inevitable mutual interdependencies will become necessary to develop a new class of advanced technologies able to meet the quality of service (QoS) expected by the contemporary media consumers.

First, digital media are usually compressed to reduce the amount of data to meet both storage and transport needs for a broad range of communication and networking applications. The compression of digital media presents unique challenges for multimedia communication and networking unseen in conventional data communication and networking. Most of these media compression schemes are based on exploiting various redundancies within the digital media. Take video compression as an example: video coding usually exploits spatial and temporal redundancies as well as psychovisual redundancies of video signals and often generates variable bit rate (VBR) compressed video bitstream for delivery to digital video consumers. Over the past three decades, video compression has achieved remarkable progress toward redundancy reduction and has undoubtedly been one of the key factors in unprecedented success of digital media in contemporary information technology era. However, the underlining principles of video coding based on a variety of correlation and prediction strategies have fundamentally changed the characteristics of compressed video bitstream so that each bit within such a stream is no longer independent of each other. Rather, compressed digital video exhibits strong sequential dependence in that some received bits cannot be reconstructed without the successful reception of some other bits within the same stream. Such fundamental nature of compressed digital media demands that the design of digital media transmission be completely different from that of data transmission to ensure that the entire segment of digital media can be successfully reconstructed at the receiving end.

Second, most digital media applications are of real time in nature. Such real-time characteristics in digital media present another set of challenges in multimedia communication and networking in terms of time-critical transport of certain media data along the path of communication and networking to guarantee the required QoS for end user consumption. The real-time requirement for digital media is in sharp contrast to the best effort nature of the packet-switch-based Internet, which has been the major networking infrastructure for numerous digital media applications via both wired and wireless links. To overcome the media transport problems caused by real-time requirement, elegant algorithms and strategies need to be developed to guarantee the seamless delivery of time-critical digital media to their consumers. For many digital media applications, the path from source to destination may need to travel through numerous networking nodes across heterogeneous regional and local networks. The task to ensure seamless transport of real-time digital media over such networks may become extremely challenging, especially at the interface between two different types of networks such as the gateway between wired and wireless networks or the access point of WLAN.

Third, digital media are usually disseminated to a group of distributed consumers. For many contemporary digital media applications, the user group can be very large in the order of millions. To serve such a large group of users distributed in diverse geographical areas, the traditional client–server model or content distribution network (CDN) where distributed servers serve individual participants directly by simple unicast cannot be applied [4]. This is because the conventional model clearly is not scalable to large groups due to high requirements on server processing power, maintenance cost, and network bandwidth. To overcome the challenges imposed by the conventional media delivery model, a new media delivery strategy has been developed. This new strategy is called peer-to-peer (P2P) streaming and has shown to be effective to serve large groups for many live applications [5–8]. Since such a scheme does not rely on dedicated servers, each peer can contribute its own resource for media streaming by uploading the media content already received [9,10]. In P2P streaming, contents are distributed among end-hosts mainly using the upload bandwidth of the peers, while content servers only deliver streams to a small number of the peers. As a result, servers are no longer a performance bottleneck and the scalability of digital media distribution can be achieved. Despite its huge potential in media content distribution to a large number of media consumers, the management of practical P2P system still faces tremendous challenges, especially in the allocation of resources among these distributed peers that serve as both server and client. The resource allocation will need to consider a variety of limitations in flow conservation, in upload and download capacities, in buffer and storage capacities, and in specific application-layer requirement in QoS, in a distributed fashion without centralized control.

Finally, numerous contemporary digital media applications are being designed to penetrate the worldwide mass of mobile wireless consumers. For mobile media applications, additional technical challenges beyond those we encountered in media distribution over wired networks will need to be overcome to ensure a robust quality of experience while the mobile receiving terminal is constantly on the move. These additional challenges are resulting from the inherent nature of open-air wireless links and the portable requirement for mobile terminals and devices and are common among a variety of wireless networks [11], including cellular phone networks, wireless personal area networks (PAN), WLAN, wireless metropolitan area networks (MAN), and wireless sensor networks. One major challenge in these wireless networks is how to effectively cope with the time-varying wireless channel status that usually causes substantial fluctuations in transport capacity and channel error behavior, two important factors for robust media communication and networking. Another major challenge in wireless networks is how to efficiently manage the limited resources that often are the performance bottleneck of wireless multimedia applications in terms of battery power, computational capability, memory and storage capability, and display size of the mobile devices by the media consumers. Over the past two decades, tremendous technical advances have pushed the field of multimedia over wireless networks to flourishing through innovations in all layers of network protocols, ranging from application layer, to transport, network, access, and physical layers. In particular, recent advances in cross-layer strategies have been driving some unprecedented performance enhancement in various wireless networks. However, practical implementations of these cross-layer strategies are facing significant barriers because of the system-level complexity of networking layers in contemporary and future-generation wireless networks.

19.3 Pressing Needs for Emerging Technologies

To successfully resolve the technical challenges discussed in the previous section, we shall first analyze the needs for developing advanced multimedia communications and networking technologies that are capable of exploiting unique and compound characteristics in delivering compressed multimedia data over heterogeneous network links and terminal devices. In particular, we shall consider the following topics: P2P streaming for distributed media content distribution, resource allocation for distributed video communication, HTTP streaming for video distributions, video streaming over WLAN, video adaptation for heterogeneous mobile devices, video over multiple antenna wireless MIMO systems, and media authentication for wireless delivery. We shall identify the needs for each of these research topics and present associated technical challenges as well as research opportunities for these emerging topics.

19.3.1 The Needs for P2P Technologies

As we indicated earlier, P2P technologies have been developed to address the challenges in meeting the demand of simultaneous access to multimedia contents by an extremely large group of end users. It is clear that the traditional server–client model or CDN model based on unicast principles is unable to scale up to millions of users. Despite the desired scalability for P2P that solves the problem of server bottleneck, simple P2P technologies are experiencing a new set of problems in practical systems. For example, peer churns will happen when the leaving of certain peers during a live streaming session causes undesired packet loss of their descendants. Locality problem will occur for many existing P2P protocols that do not carefully consider cross-ISP (Internet service provider) streaming traffic. The locality problem will lead to overly long connection and unnecessary large volume of data transfer between different ISPs.

Recently, a two-tier proxy-based P2P network has been proposed to address some of the new problems of P2P. In essence, this scheme places reliable proxies in different ISPs close to the P2P user pools as the first tier and the peers under such same proxy will form the second tier. The P2P protocol for stream distribution for the first tier of proxies may be different from the P2P protocol for streaming distribution for the second tier of P2P users. The content server first distributes video to these proxies and then the end users under the same proxy receive the contents from this proxy. Such two-tier proxy-P2P provides several benefits and solves the problems from existing P2P schemes. First, proxy-P2P is more resilient to peer churn and flash crowd because the stable proxies provide better resource availability at the beginning of any streaming session. They can also be used to recover streaming errors when peers leave or fail. Second, proxy-P2P is more ISP friendly because it is able to overcome network address translation (NAT) and firewall problems. It is also more locality aware, which avoids the undesired long-distance connections. Chapter 20 presents the details of this new scheme with design, implementation, and experimental results to confirm the effectiveness of the proposed new proxy-P2P. This new scheme is an integrated scheme based on two main protocols they have developed early: *FastMesh* and *SIM*.

FastMesh has been designed for an overlay network that is suitable for efficient proxy-level communication [12] while *SIM*, or scalable island multicast, is a fully distributed protocol that has been designed to effectively integrate IP multicast and overlay multicast with multiple substreams [13].

19.3.2 The Needs for Optimal Resource Allocation

Recently, more and more contemporary multimedia applications have been developed based on communication over distributed systems. P2P streaming introduced above is one such special class of distributed multimedia communication. Other such systems include multimedia over ad hoc networks and wireless visual sensor networks. Because of the lack of centralized control, one key challenge in such systems is how to allocate the resource to each node of the network to achieve optimal overall system performance subject to the resource constraints at each node. The resource constraints include the flow conservation, the limitation of upload and download capacities, the limitation of buffer and storage capacities, the limitation of power supply, and the application-layer requirement (e.g., the distortion requirement). Since there is no centralized controller in the distributed systems, a distributed algorithm is the desired solution in terms of the scalability and the communication overhead.

Chapter 23 presents an effective scheme for resource allocation for distributed multimedia communication systems. Chapter 23 begins with an excellent overview of recent advances in optimal video communication over distributed systems. The overviews in P2P video streaming, P2P video-on-demand (VoD) systems, and video over wireless multihop networks not only provide a nice summary of recent innovations in resource allocation for multimedia distributed systems but also present an array of challenges and technical barriers for accomplishing the grand objective of optimal resource allocation in distributed systems. Such a summary will be very useful in guiding the design of resource allocation algorithm when a specific scenario of distributed multimedia communication is considered. To illustrate the design principle in resource allocation, the authors have described in great detail a special case in wireless visual sensor networks in which resource allocation has been designed to maximize the network lifetime for such a distributed system. In this system, each video sensor has a camera component to capture the video, and a processing component to compress the video. The video sensors construct a mesh network topology, and they communicate with each other within a limited transmission range. A distributed algorithm has been developed to maximize the network lifetime by jointly optimizing the source rates, the encoding powers, and the routing scheme. It has been shown that a fully distributed solution can be designed to solve such an optimization problem with convex minimization through a dual-based approach. Real video transmission simulations have been carried out to verify the effectiveness of the proposed distributed algorithm. The details of the approach are given in Chapter 23.

19.3.3 The Needs for New-Generation HTTP Streaming Protocols

In general, P2P distributed streaming requires significant resources from individual nodes, for example, CPU time, memory, and network bandwidth. When the end users are using mobile devices as their terminals for video streaming via wireless networking, users' playback devices are usually battery powered, with a less powerful processor and less memory. As a result, P2P video streaming is much less popular in wireless video streaming than it is in wired networks.

Because of limitations for wireless P2P streaming, transmission control protocol (TCP)-based HTTP video streaming has also been exploited for video streaming. As TCP was originally designed for wired networks, the TCP-based HTTP video streaming encounters considerable challenges when it is extended to wireless networks. However, recent engineering practices have found strong evidences that the advantages for wireless TCP streaming outweigh its drawbacks. Compared with user datagram protocol (UDP)-based P2P video streaming, the TCP-based approach has major advantages of (1) being able to penetrate firewalls without adding explicit rules; (2) requiring no additional user end communication software beyond a regular web browser; and (3) guaranteeing fairness among concurrent TCP flows. However, in last-hop WiFi access networks, which is the most widely used network configuration nowadays for home, campus, coffee shop, airport, and so on, fading and interfering make the underlying TCP's additive increase multiplicative decrease (AIMD) congestion control algorithm too conservative in reducing the congestion window. This results in increased traffic dynamics and decreased average throughput. In contrast, high throughput with low dynamics is very much desired for video streaming. Balancing these conflicting factors in TCP video streaming constitutes the major challenge in contemporary mobile media applications.

The mismatch between TCP's congestion control algorithm and wireless channel's error-prone characteristics is a well-known technical challenge. Among existing approaches, the most natural wisdom is divide and conquer, that is, separating the long TCP connection into a cascade of wired section and wireless section (split-TCP). Existing research results from both theoretical analyses and practical measurements confirm that split-TCP is able to improve long TCP connection's throughput [14,15]. However, for wireless TCP video streaming, the solution of simply adopting split-TCP has not been working well. A completely new TCP proxy design to overcome the drawbacks of split-TCP becomes necessary. Some of the recent approaches to HTTP video streaming will be presented in Section 19.4.1, in which the new proxy not only transparently splits a long TCP connection from video source to a user into two cascade TCPs but also performs video adaptation last-hop TCP enhancement at the proxy node for wireless transmission.

19.3.4 The Needs for Scalable Video Streaming Scheme for WLAN

Among contemporary wireless networks, WLAN has been the most widely adopted means for mobile terminals to access multimedia data via Internet. In particular, IEEE 802.11-based WLAN [16] have been massively deployed for home use as well as for public and institutional use. Initially, IEEE 802.11 has been designed for mostly web access without significant traffic in real-time services. Its wireless and mobile access convenience was the main driving force for the widespread deployment of 802.11-based WLAN. As the demand for real-time applications over WLAN grew over the past decade, the 802.11 working group has been working on new protocols to facilitate the needs for WLAN to provide real-time services such as streaming over 802.11 WLAN. The 802.11e is indeed such a new standard to offer the necessary QoS for real-time applications such as video streaming and voice over IP (VoIP) services [17]. In particular, 802.11e defines an enhanced distributed channel access (EDCA) for the contention-based channel access by introducing multiple queues (called access categories) to provide the MAC layer with per-class service differentiation. With 802.11e, the base station of WLAN will be able to offer different services to different classes of data traffic so that QoS-optimized schemes may be designed for real-time applications that demand more stringent delay and loss tolerance during the contention and transmission stages.

Video streaming over 802.11 WLAN has received tremendous attention in the past few years owing to the popularity of WLAN and the growing demand for Internet video services. A variety of schemes have been developed to take advantage of 802.11e EDCA protocol in its capability to differentiate service classes to meet the stringent QoS constraints [18–21]. However, the mission to develop an optimal scheme for video over 802.11e still remains very challenging because 802.11e only defines the MAC layer protocol while successful delivery of compressed video over WLAN is dependent on multiple layers of wireless networking.

Chapter 22 presents one such effective scheme, particularly suitable for the transmission of scalable video bitstreams over 802.11e WLAN, by taking advantage of cross-layer design principles. Chapter 22 is based on its authors' pioneering experience in the development of video streaming over WLAN and their successful implementation of optimal cross-layer approach to such a challenging task [22,23]. In particular, the authors describe how the scalability of scalable video coding (SVC) and the QoS provided by the IEEE 802.11e EDCA can be exploited to achieve an optimal performance for video streaming over WLAN. Two scenarios are considered in their research. The first one is to transmit scalable video over one EDCA queue. By adaptively adjusting the retry limit setting of EDCA, a strong loss protection for critical video traffic can be maintained. In the second scenario, two queues are used for scalable video streaming and a cross-layer framework is designed to adaptively map video packets of different classes into the two different queues, which preemptively drops some less important video packets so as to maximize the transmission protection for the more important video packets. Such a cross-layer approach has been proven to produce enhanced service performance for SVC over 802.11e in terms of both quality of the received video and the number of simultaneous active video consumers.

More recent advances in video streaming over 802.11 WLAN have also examined several technical issues beyond just service class differentiation as defined in 802.11e. Among them, deadline award approaches have investigated approaches to take the deadline of the video packets into consideration for scheduling the transmission of video packets to make sure that a video packet is still valid when it arrives at the receiver end [24]. The deadline-aware strategy has also been integrated with variable importance of video packets to further enhance the video streaming performance [25] and with virtual contention-free scheme to ensure that the QoS of real-time application such as video streaming is guaranteed while non-real-time application can also be properly maintained [26].

19.3.5 The Needs for Intelligent Video Adaptation for Mobile Devices

Recent advances in contemporary communication and networking are able to provide various multimedia applications adequate infrastructures for the transport of bandwidth-hungry multimedia data to end users. However, there still exist significant challenges to deliver appropriate media content over heterogeneous links and to a large array of different terminals. Such challenges have recently been intensified due to continued enhancement in spatial resolution of digital images and videos at the media source and increasing demand from mobile devices to access networked high-resolution media contents. It is under such contemporary application environments that an emerging research topic in video adaptation has received much attention recently. Video adaptation refers to the operation applied to the original video source to generate appropriately recomposed video data for transmission to the target users whose terminal devices are unable to accommodate the video source in its original format.

Traditionally, the research in video adaptation has been focused on investigating scalable video in terms of spatial, temporal, and signal–noise ratio (SNR) adaptation to suit for the bandwidth or resource limitation of the terminal devices [27–29]. The standard scalable video solution often cannot meet the requirement of mobile devices when the display size of the mobile devices is small compared to the resolutions of the original video source. Because of the significant mismatch between the original video resolution and the mobile device display, simple reduction of spatial resolution will lead to unrecognizable region of interests within small video frames. This is because the small display size of these mobile devices demands even more adaptive solutions for the region of interest to be clearly displayed. An intelligent video adaptation is therefore needed to extract the video contents and transmit only the region of interest to the mobile device for a better and more pleasant display.

There are several major technical challenges in meeting the requirement for video delivery to mobile devices with limited display size. In practice, real-time video adaptation by simple transcoding may produce significantly reduced resolution from the original video so that the objects of the interest in the scene may lose necessary details for mobile device viewer. To maximize the quality of experience, preserving the details of the regions of interest is very much desired for mobile device users and the major challenge in designing an acceptable video adaptation scheme. Even under the display size constraint, users would generally expect the video adaptation system to provide an intelligent solution based on either user feedback or user attention model that is capable of selecting regions of interest for adaptation and transmission. The technical barriers for video adaptation will then rest on how well the mobile users are able to provide semantically important feedback to the system for proper identification of desired region of interests. It is certainly a challenging task to inject appropriate intelligence in the process of content understanding and extraction, before the size-reduced video data are transported to the mobile users for pleasant viewing.

19.3.6 The Needs for New Technologies for Video over Wireless MIMO Systems

As wireless communication technologies evolve to the next level, it is expected that the future generation of mobile devices will be equipped with multiple antennas to enhance the throughput or signal reception. MIMO systems have recently emerged as one of the most prominent techniques [30–33]. There are two successful approaches in MIMO: space–time coding [30,31] and spatial multiplexing [32,33], to achieve high data rates for wireless systems. The core idea of space–time coding is not to directly increase data rate but rather to maximize diversity using multiple spatially distributed antennas so as to improve the SNR of the mobile receivers. However, to directly and fully utilize the advantage of MIMO systems, it is desired to develop spatial-multiplexing technique so as to increase data rates by turning multipath propagation, usually considered as a drawback for wireless multimedia communication, into an essential part of enhancement without the cost of extra bandwidth.

To successfully implement spatial multiplexing for transmitting media data such as video, the first technical challenge is to appropriately decompose the original media source data into multiple bitstreams for simultaneous transmission over MIMO systems. For digital media that has already been compressed, it is not straightforward to decompose the media stream into multiple streams for simultaneous transmission over multiple antennas. Even when the source data can be decomposed into a number of bitstreams matching the number of antennas in spatial-multiplexing MIMO transmission, variable importance of the decomposed media substreams may not correspond well with the status of MIMO subchannels. It is a greater challenge to design a spatial-multiplexing system that can really exploit the inherent characteristics of media source data and dynamic behavior of the MIMO subchannels. The prioritized spatial-multiplexing scheme to be introduced in the next section is indeed able to achieve significant performance gains over existing schemes when the channel state information (CSI) can be obtained [34]. This scheme can also achieve desired performance gain even when the receiving terminal is in fast motion and the CSI cannot be timely estimated [35].

19.3.7 The Needs for Reliable and Secure Media over Wireless Networks

Recently, the need for secure media delivery, such as confidentiality and authentication, over mobile wireless links has become one of the major concerns in pervasive multimedia communication applications [36,37]. Conventional data authentication cannot be directly applied for streaming media when an unreliable channel is employed and packet loss may be inevitable. This is because digital signature as a natural authentication solution requires strict verification such that even a single bit of flipping would fail the verification. However, transmission errors such as packet loss and bit errors are inevitable in wireless networks due to ambient interferences and open-air operation mode.

Media security is also fundamentally different from the generic data communication security in that media content integrity, rather than the data stream integrity, needs to be preserved during the transmission. For authenticating media for wireless networks, the semantic meaning of the media data, instead of the entirety of the data, needs to be verified. It is this semantic attribute of media data that makes the lossy encoding and transmission of media data acceptable in almost all contemporary networked multimedia streaming and transmission applications. In addition, the real-time constraint of the media stream often requires that the media transmission be low latency without the adoption of the retransmission strategy.

With these fundamental differences, it is necessary to consider information from both transmission channel and media in designing desired, efficient, reliable, and secure media transmission scheme. There are several major technical challenges to be overcome in order to develop a scheme for authenticating compressed media over unreliable wireless channels. The first major challenge comes from the limited channel capacity for wireless networks. To combat against channel error, robust authentication performance is achieved by appending redundant authentication information. Therefore, low authentication overhead and high authentication performance become competing requirements. Both low authentication performance and the reduction of source coding rate due to the introduction of authentication overhead could cause nonnegligible quality decrease at the receiver side. The second challenge comes from the inherent coding dependency between media slices and the inequality of their importance once the digital media is compressed. The complexity of such dependency relation constitutes the major barrier to develop a joint reliable and secure media transmission scheme over wireless networks and hence demands paradigm-shifting solution. One recently developed scheme has tried to answer these challenges by designing a layered joint media error and authentication protection (JMEAP) approach that is able to achieve significant performance gain over competing algorithms that did not exploit adequately the complex interdependency between source coding, channel coding, and media authentication.

The rest of this chapter is organized as follows. Section 19.4 presents in more detail several recent advances to address these technical challenges as outlined above. Along with the description of these recent advances, we also identify research opportunities associated with some technical challenges, anticipating some new innovations and developments to meet the technical challenges in the near future. Section 19.5 will summarize the potential impact of these research activities on the next-generation multimedia communications and networking, especially those applications that are intended for seamless delivering of multimedia signals to mobile devices via wireless links.

19.4 Some Recent Advances and Future Research Opportunities

In this section, we shall present in more detail some recent research advances in the design of novel schemes to meet the technical challenges as applied to the design of next-generation multimedia communications and networking. These recent research advances have been developed to meet some specific pressing needs as we indicated in Section 19.3. Several chapters of this book provide technical details in (1) proxy-based P2P live streaming network (Chapter 20), (2) optimal resource allocation in distributed multimedia systems (Chapter 23), and (3) scalable video streaming over 802.11e WLAN (Chapter 22). In this chapter, we discuss the following research topics in more detail: (1) new HTTP streaming proxy design for wireless video streaming, (2) intelligent video adaptation for mobile devices, (3) scalable video streaming over MIMO wireless systems, and (4) joint reliable and secure media transmission over wireless networks. In particular, technical challenges associated with these advances will be identified and future research opportunities to further advance these emerging areas in multimedia communications and networking will be highlighted.

19.4.1 New HTTP Streaming Proxy Design for Video Services

TCP-based HTTP streaming originally designed for wired networks cannot be directly extended to wireless video streaming because of transmission error due to channel fading that occurs frequently in wireless networks. Novel solutions have been sought to apply HTTP streaming to wireless networking. A recent paper has provided a nice summary of solutions for wireless TCP [38]. Among these solutions, we believe that split-TCP is the most suitable basis for wireless TCP video streaming.

To design a new TCP proxy that is capable of robust video streaming, we consider a wired-cum-wireless network as shown in Figure 19.1. With such a network, H.264/AVC-encoded

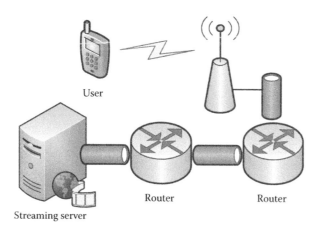

FIGURE 19.1
Illustration of the wired-cum-wireless networking scenario.

video data are sent from the source using TCP connection to a wireless user via a WiFi access point. This new TCP proxy design not only transparently splits a long TCP connection between video source and a user into two cascade TCPs but also performs video adaptation and last-hop TCP enhancement at the proxy node for wireless transmission.

First, we design a new reliable relay protocol for splitting a long TCP at the TCP proxy. Rather than using vanilla TCP in both component TCPs like split-TCP, we propose raw-TCP for wireless connection. Raw-TCP conforms to TCP's syntax, while offloads its congestion control function to a flow-based fair wireless packet scheduler in the newly designed proxy. Second, we wedge a video adapter in the TCP streaming proxy. This is necessary because we need to reorder the video bitstream in such a way that the truncation of video bitstream can be implemented according to rate-distortion criteria. The new proxy does not require modification of the protocol stack of either the video source or the users.

19.4.1.1 Protocol Design

To ease media processing, contemporary HTTP/TCP streaming protocols attach each transmission unit with a header containing time stamp and other necessary information, for example, Adobe's real-time messaging protocol (RTMP). In this research, we adopt a simple solution of wrapping each H.264 network abstraction layer units (NALU) with an RTP header. This solution makes it possible for TCP proxy to process video data simply and easily.

In our scheme, the innovation lies in the novel strategy in optimizing the last-hop wireless transmission. The proxy located at the wireless AP splits a long TCP into wired and wireless parts. The wired part uses vanilla TCP while the wireless part uses a simplified variant of TCP. We call it raw-TCP. Raw-TCP inherits TCP's syntax and disables TCP's AIMD congestion control algorithm. As discussed above, AIMD is too conservative in wireless networks. Disabling it can better explore the wireless channel's capacity. However, unilaterally disabling congestion control can exhaust limited wireless bandwidth. So, we propose a fair scheduler at the proxy to enforce fair bandwidth allocation.

This new proxy design approach is convenient, without modification of the protocol stack of both the video source and users. Furthermore, the proxy is easy to implement. Both the relay protocol and video adaptation algorithms consume very little system resources. These features make this solution promising for practical deployment. The block diagram of the proxy is illustrated in Figure 19.2.

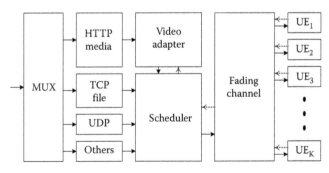

FIGURE 19.2
Illustration of the proposed HTTP streaming proxy.

19.4.1.2 Video Adaptation

Video adaptation is a key innovation in the new HTTP streaming proxy design. To accommodate the mismatch between wired and wireless networks, video source needs to be properly processed for the proxy scheduler to transmit the video data. Video data are first partitioned into fixed time length segments. Segments are encoded independently using H.264/AVC video coding standard. Segmentation of compressed video data facilitates the potential to improve the response time of random access operations. Similar mechanism can be found from Apple's *HTTP live streaming* and Adobe's *HTTP dynamic streaming* solution.

Owing to wireless link dynamics, the network cannot always guarantee to deliver the whole segment before its decoding deadline. We reorder video stream in a segment according to rate-distortion criteria to minimize the effect of data truncation. Important data are placed at the front of the buffer, seeking for early delivery. The adapter adaptively truncates the video stream according to decoding deadline, buffer fullness, and wireless link condition. The truncation algorithm trades off video quality with playback jitter.

In this research, we assume frame granularity reordering, which can be fast processed by the wireless access point by parsing NALU headers. Note that by enabling data partition function in H.264/AVC encoder, it is possible to support fine granularity reordering, but which in turn requires more computation and memory resources.

Without loss of generality, we assume that video is encoded with hierarchical B structure, as illustrated in Figure 19.3. N_B B frames are inserted into the IPPP... structure. Let the frame rate of the basic IPPP... structure be f_{IP}, then B frame's frame rate is approximately

$$f_B = f_{IP} \cdot N_B. \tag{19.1}$$

Let the frame rate of layer i of B frames be f_{B_i}. Then, as illustrated in Figure 19.3, we have

$$\begin{cases} f_{B_{i+1}} = 2f_{B_i} \\ f_B = \sum_{i=1}^{L} f_{B_i} \end{cases} \tag{19.2}$$

where L is the number of layers in the B frame hierarchy. For example, in Figure 19.3, $L = 3$. Assume that IPPP... forms layer 0. Reordering is performed on segment basis. NALUs within one segment are ordered according to the following rules:

1. Non-VCL (video coding layer) NALUs always precede VCL NALUs. This is because non-VCL NALU usually contains important information for the decoder and these NALUs are usually small.

2. Lower-layer NALUs are in front of higher-layer NALUs. Within the same layer,

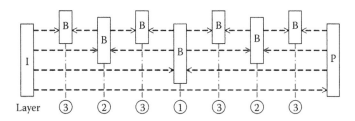

FIGURE 19.3
Example of B frame hierarchy.

a. Layer 0 NALUs are ordered chromatically.

b. Assume that layer $i > 0$ frames are denoted by $f_{0i}, f_{1i}, f_{2i}, \ldots$. We group it into i-ordered sets.

$$S_j^i = \left(f_{ji}, f_{(j+2^{i-1})i}, f_{(j+2\cdot2^{i-1})i}, f_{(j+3\cdot2^{i-1})i}, \ldots \right), \quad 0 \le j < 2^{i-1} \tag{19.3}$$

NALUs in layer i are ordered as

$$\left(S_0^i, S_1^i, \ldots, S_{i-1}^i \right) \tag{19.4}$$

The above order guarantees that decoded video quality deteriorates gradually after truncation from tail. It is easy to tell that our frame reordering policy is also directly applicable to H.264/SVC extensions with temporal scalability. Note that if a higher-priority NALU has not arrived at the proxy, the proxy selects the most important NALU currently available and sends it.

19.4.1.3 Adaptive Truncation

Assume that the beginning of transmission time of the ith segment is denoted by t_i. Its transmission duration is T_i. Segment i is scheduled to begin to play at time $\Delta_0 + iI$, where Δ_0 is initial playback delay. There are two requirements in designing the truncation algorithm: (a) no data from segment i should be transmitted from the proxy after $\Delta_0 + iI$ and (b) neighboring decoded frames' visual quality should change slowly. These requirements can be formulated as

$$\begin{cases} t_i + T_i \le \Delta_0 + iI \\ T_i \le \dfrac{\Delta_0 + (i-1)I - (t_{i-1} + T_{i-1})}{2} + I \end{cases} \tag{19.5}$$

The second equation in Equation 19.5 restricts neighboring frames' quality fluctuation. In fact, $\Delta_0 + (i-1)I - (t_{i-1} + T_{i-1})$ can be seen as available time pool for segment i. It is the accumulated saved time from previous segments. We force segment i to use at most half of this time pool to prevent the pool against exhaustion by a single segment whose transmission time period is with very bad link condition. It helps to ensure that video quality deteriorates gradually under bad link conditions.

This new TCP proxy solution for video streaming over wireless networks provides transparent proxy to both video source in wired networks and wireless downstream user. It performs video adaptation operations based on the video's rate-distortion criteria to resolve short-term throughput dynamic. It splits long TCP and performs last-hop TCP enhancement. It does not require modification to either the source server or end users' protocol stack. Simulation results show that this new design significantly improves the visual quality of decoded video and reduces playback jitters. Furthermore, this new TCP proxy is flexible enough to integrate new video processing functions, such as video transcoding, content-aware adaptation, and so on.

There are ample research opportunities in HTTP streaming for video delivery over wireless networks, especially with the emergence of cloud computing and data center development. One potential high-impact research is in the development of more robust scheme for dynamic adaptive streaming over HTTP (DASH). There are at least two major technical issues to be resolved in order to develop the next-generation HTTP streaming for digital video over wireless networks. One of the research opportunities is to design a scheme that can adaptively match the dynamics of video data burst with the dynamics of the TCP bandwidth fluctuations. Another research opportunity is to design a new HTTP streaming scheme that can take full advantage of the recent development in cloud computing and data center in which the digital media are stored separately in multiple content distribution servers that can be accessed via DASH by the media consumers.

19.4.2 Intelligent Video Adaptation for Content Delivery to Mobile Devices

As the proliferation of mobile devices progresses at an unprecedented pace, mobile IPTV may be considered as one of the killer applications by the mobile communication operators because of the potential profit sharing of the multibillion dollar pay-TV market. However, the broadcast nature of the traditional TV programs may not be the most desired format for multimedia-capable mobile users. Furthermore, both small display size of most mobile devices and limited available bandwidth of the mobile link are hindering the direct adoption of TV programs for delivery to mobile devices. This is because limited resource of the mobile devices usually requires that the original TV signals be subsampled in both spatial and temporal domain to fit for the small size of mobile devices. The subsampling is also necessary for many other applications, including video streaming and remote video surveillance, when the mobile devices are used to access such digital video signals.

To facilitate the delivery of reduced-size video to mobile devices, real-time video adaptation is necessary to convert the original high-resolution video to appropriate size for display in mobile devices. Therefore, video adaptation plays crucial roles in numerous mobile digital video applications. As we indicated earlier, conventional real-time video adaptation has been focusing on transcoding, or simply changing the spatial, temporal, and SNR resolutions to meet either bandwidth constraints or receiver limitations. In the case of non-real-time video adaptation, content-based video summarization can usually be carried out to maximize the information rate for resource-constrained access. This is another emerging area of research and the readers are referred to an excellent summary in Ref. [39] for more details.

We illustrate two representative approaches to intelligent video adaptation: user feedback-based and user attention model-based. In the case of user feedback-based approach [40], the video adaptation is carried out to deliver only the user-selected portion of the video to guarantee maximum quality for this portion of video. The user selection is implemented through a feedback channel. The scheme is shown in Figure 19.4 [40].

Note that there are two channels of feedback for this type of video adaptation scheme. In addition to user preference feedback that specifies the user interest region for adaptation, a second channel of feedback is implemented for QoS feedback to ensure that appropriate video quality of the specified user interest region is guaranteed. The underlining assumption for this type of video adaptation is that the access network offers the user the appropriate feedback channel for both user interest region and QoS level of the user.

However, the access network does not always offer feedback channel to mobile device users. This is particularly true for some wireless broadcast networks in which only downlink streaming is provided to mobile users. In the absence of user feedback, the video adaptation will need to rely on some user attention model to identify the information-carrying portion of the video for intelligent adaptation and transmission.

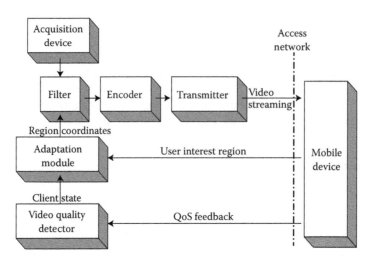

FIGURE 19.4
User feedback-based video adaptation.

This user attention model-based approach to video adaptation has been developed recently [41,42] and has shown much improved video quality on delivering the regions of interest streams to the mobile users. This scheme not only generates the information-carrying portion of the video but also transcodes this portion with H.264-compatible bitstream for delivery to the mobile devices. The key technologies for this user attention model-based approach are in the design of the user attention model-based identification; appropriate cropping of informative region of interest; and finally generation of smooth video sequence-based regions. Each of these steps is crucial to the overall success of user attention model-based video adaptation.

The extraction of attention regions is based on an algorithm to combine several apparent attributes of visual attention, including motion, face, text, and saliency. The generation of smooth video sequence is based on an algorithm of virtual camera control [41]. Figure 19.5 illustrates this user attention model-based video adaptation for mobile devices that also conform to the H.264 standard [41].

The approach presented in Ref. [41] assumes that a video server prestores high-quality videos and serves various mobile terminals, including PCs, smart phone, and PDA. When a mobile user client requests for a service, the server sends a video to the client. We assume that this system is placed on a server or a proxy and will adapt the high-resolution original video to generate a low-resolution adapted video suitable for the display size of the user's mobile device and the bandwidth of the mobile link. For different users, the reduction of the video resolution can be different and will be decided by the real display sizes of mobile devices. The system consists of three main modules: decoder, attention area extractor, and transcoder as shown in Figure 19.5.

Although this region-of-interest (ROI)-based transcoding is able to perform video adaptation for mobile devices with small display and limited bandwidth, this system has two critical shortcomings that need to be overcome to maximize the quality of experience of the mobile video receivers. The first shortcoming of ROI-based transcoding is the need to perform the detection of four types of attention objects separately to obtain a robust ROI within a given video. The computational operations to perform these detections and to combine the detection results will become a significant burden for either proxy server. The

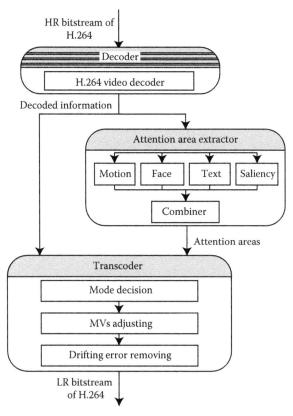

FIGURE 19.5
User attention-based video adaptation scheme.

second shortcoming of ROI-based transcoding is the need to perform ROI detection for different users every time these users request the video streaming service. Such repeated operations may sometimes overwhelm the proxy server.

However, these shortcomings can be overcome if the compressed video at the original server can be augmented with ROI information for the proxy server to access. If we are able to embed the ROI information into the bitstream of compressed video and extract with the proxy server, the burden of the computational complexity can be shifted from the transcoding to encoding. This analysis leads to the design of an attention-information-based spatial adaptation framework for accessing video via mobile devices.

This intelligent design of spatial adaptation is based on the assumption that the videos stored in the video server are usually offline generated and the computational complexity is not an issue with offline operations. Furthermore, we also assume that the attention objects in each video frame may remain the same even for different mobile users. This is because the attention model we adopted is quite generic for a wide variety of users. If we are able to move the attention detection operation from the transcoding process to the encoding process, then, we will be able to shift the complexity from proxy to the video server. The principle of shifting complexity from proxy to video server has been successfully implemented for browsing video using mobile devices and was reported in a recent paper [42].

Numerous challenges still remain to implement this intelligent real-time video adaptation at the consumer scale. First, the model for user attention needs to robustly capture the true region of interest in the video frame for extraction. This is not a trivial task for a wide variety of consumer video with significantly different contents, ranging from sports scenes to financial news. Second, the algorithm for region extraction needs to be of low complexity for real-time implementations. We assume that the region extraction can be simultaneously carried out with the original video encoding. Third, the transcoding needs to conform to the H.264 video coding standard for widespread use. Without conformation, the video contents can only be delivered to special mobile devices that are able to decode the nonconformed video streams. In addition to these implementation challenges, several cross-layer networking transport issues also need to be resolved for video adaptation over heterogeneous mobile wireless networks.

The research opportunities to address these tough challenges are also abundant. The video adaptation with user attention model provides the opportunity to integrate image understanding with video coding. These two fields have been working on different technical problems and have not seen significant collaborations over the past few decades. The development of reduced-complexity video adaptation for mobile devices offers an opportunity for hardware–software codesign and may lead to new hardware architecture that is most suitable for mobile media applications. The reliable and intelligent transport of video signal over heterogeneous mobile wireless networks offers many unique opportunities in cross-layer design ranging from the physical layer, to the routing layer and network layer, and to the application layer.

19.4.3 Scalable Video over Next-Generation MIMO Wireless Systems

To achieve high-data-rate wireless multimedia communication, spatial-multiplexing technique [32,43], in which multitransmit antennas are used to simultaneously transmit independent data, has recently emerged as one of the most prominent techniques for MIMO systems. If the perfect CSI is available at the transmitter [44], we can maximize the channel capacity through the well-known water-filling solution after singular value decomposition. However, from a practical point of view, the perfect CSI is not attainable and the delay of feedback CSI is inevitable. Therefore, the performance of a MIMO system is heavily dependent on how to cope with inaccurate CSI estimations. Furthermore, when MIMO wireless systems are used to transmit video data, the unique characteristics of video different from generic data need to be fully considered to develop the next-generation MIMO systems suitable for multimedia communications and networking.

Some of the early efforts in exploiting video characteristics, especially scalable video coded bitstream, with spatial-multiplexing MIMO systems has generated excellent performance gains over conventional schemes [45–47]. However, these early efforts involve sophisticated power reallocation at the transmitter end, which may not be practical in reality. More recently, a new scheme is available that is capable of prioritizing the transmit antennas for a much improved performance gain for video streaming over spatial-multiplexing MIMO systems.

Precoding is a generalized beamforming technique to support multilayer transmission in MIMO wireless systems. Beamforming techniques can be applied to allow multimedia data transmitted from each of the transmit antennas to be appropriately weighted to achieve maximum signal power at the receiver end. When the receiver has multiple antennas [33], conventional single-layer beamforming cannot simultaneously maximize the signal level at all the receive antennas. In this case, precoding techniques, as a generalized beamforming, can be applied to maximize the throughput performance of a multiple receive antenna system. In precoding, the multiple streams of the signals are transmitted from the transmit antennas with independent and appropriate weighting per each antenna such that the link throughput can be maximized at the receiving end.

We consider a spatial-multiplexing scheme with N_t transmit antennas and N_r receive antennas for closed-loop MIMO system in a flat fading channel. At the transmitter, four-layer scalable video bitstreams are generated by H.264 SVC standard codec. They consist of a base layer and three enhancement layers. The base layer contains all the necessary information of a video sequence, while the enhancement layers, once received, will improve the quality of video stream after combining with the bitstream from the base layer.

At the receiver side, the received signals are detected by ZF-VBLAST detector and finally reconstructed as a video sequence. The received signal vector can be formulated as

$$\mathbf{y} = \mathbf{HFs} + \mathbf{z} \tag{19.6}$$

where $\mathbf{y} \in \mathbb{C}^{N_r \times 1}$, the transmit signal vector $\mathbf{s} \in \mathbb{C}^{N_t \times 1}$ is precoded by a precoding matrix $\mathbf{F} \in \mathbb{C}^{N_t \times N_t}$, $\mathbf{H} \in \mathbb{C}^{N_r \times N_t}$ is the channel matrix whose elements have independent and identical complex Gaussian distribution with unit variance, and $\mathbf{z} \in \mathbb{C}^{N_r \times 1}$ is the additive complex Gaussian noise with zero mean and the covariance matrix $\sigma_z^2 \mathbf{I}_{N_r}$ where $\mathbf{I}_{N_r}$ denotes an identity matrix with dimension N_r. The autocorrelation matrix of $\mathbf{s}$ is assumed to be $E\{\mathbf{ss}^H\} = \sigma_s^2 \mathbf{I}_{N_t}$ where $E\{\cdot\}$ represents the expectation operation, $(\cdot)^H$ denotes the conjugate transpose operation. The SNR of the received signal can be written as

$$\rho = \frac{E\left\{\mathbf{s}^H \mathbf{F}^H \mathbf{Fs}\right\}}{\sigma_z^2} = \frac{\sigma_s^2}{\sigma_z^2} \mathrm{Tr}\{\mathbf{F}^H \mathbf{F}\} = \frac{1}{\alpha} \mathrm{Tr}\{\mathbf{F}^H \mathbf{F}\} \tag{19.7}$$

where $\mathrm{Tr}\{\cdot\}$ indicates the trace of a matrix, and $\alpha \triangleq \sigma_z^2 / \sigma_s^2$.

19.4.3.1 *Conventional SVD-Based Precoding*

SVD precoding has been proven to achieve the desired channel capacity of MIMO systems at the cost of feeding back signaling of CSI from the receiver to the transmitter. We assume that MIMO channel matrix $\mathbf{H}$ can be decomposed by the SVD scheme as

$$\mathbf{H} = \mathbf{U\Sigma V}^H \tag{19.8}$$

where $\mathbf{U} \in \mathbb{C}^{N_r \times K}$ and $\mathbf{V} \in \mathbb{C}^{N_t \times K}$ are semiunitary matrices, is a diagonal matrix whose diagonal elements $\lambda_1 \geq \lambda_2 \geq \cdots \geq \lambda_K > 0$ are singular values of $\mathbf{H}$, and K is the rank of $\mathbf{H}$. The capacity of the MIMO channel is

$$C = \log_2 \left| \frac{\sigma_z^2 \mathbf{I} + \mathbf{HF}\sigma_s^2 \mathbf{F}^H \mathbf{H}^H}{\sigma_z^2 \mathbf{I}} \right| = \log_2 \left| \mathbf{I} + \alpha^{-1} \mathbf{U\Sigma}^2 \mathbf{\Phi U}^H \right|. \tag{19.9}$$

To achieve the maximum capacity, the precoding matrix $\mathbf{F}_{\mathrm{SVD}}$ is designed as [44]:

$$\mathbf{F}_{\mathrm{SVD}} = \mathbf{V\Phi}^{\frac{1}{2}} \tag{19.10}$$

where $\boldsymbol{\Phi} \in \mathbb{C}^{K \times K}$ is a diagonal matrix obtained by the water-filling method

$$\phi_i = \left(\mu - \frac{\alpha}{\lambda_i^2} \right)^+, \quad i = 1, \ldots, K, \tag{19.11}$$

$$s.t. \sum_{i=1}^{K} \phi_i = P_T.$$

P_T is the total transmit power constraint. The water-filling threshold μ in Equation 19.11 can be calculated by

$$\mu = \frac{1}{K} \left(P_T + \alpha \left(\frac{1}{\lambda_1^2} + \frac{1}{\lambda_2^2} + \cdots + \frac{1}{\lambda_K^2} \right) \right). \tag{19.12}$$

Substituting the derived formula (19.8) and (19.10) into (19.6), the received signal can be expressed as

$$\mathbf{y} = \mathbf{U}\boldsymbol{\Sigma}\boldsymbol{\Phi}^{\frac{1}{2}}\mathbf{s} + \mathbf{z}. \tag{19.13}$$

At the receiver end, the filter $\mathbf{U}^H$ is multiplied to the received signal vector

$$\tilde{\mathbf{y}} = \mathbf{U}^H \mathbf{y}. \tag{19.14}$$

Let $\boldsymbol{\Lambda} \triangleq \boldsymbol{\Sigma}\boldsymbol{\Phi}^{1/2} = \mathrm{diag}\{\delta_{11}, \ldots, \delta_{KK}\}$ and $\tilde{\mathbf{z}} \triangleq \mathbf{U}^H \mathbf{z}$, then the filtered signal (19.14) is rewritten as

$$\tilde{\mathbf{y}} = \boldsymbol{\Lambda}\mathbf{s} + \tilde{\mathbf{z}}. \tag{19.15}$$

Hence, we can obtain K parallel scalar subchannels by using a VBLAST detector

$$\tilde{y}_i = \delta_{ii}s_i + \tilde{z}_i, \quad i = 1, 2, \ldots, K. \tag{19.16}$$

19.4.3.2 Prioritized Spatial Multiplexing

The power allocation of the conventional SVD scheme only considers the quality of subchannels according to CSI. Such SNR information may be significantly different from the desired SNRs of each layer of multimedia video bitstreams. In particular, the base layer of the video bitstream is of paramount importance and requires adequate SNR to guarantee its successful transmission. It is therefore very much desired to be able to incorporate the information of necessary SNR for base layer video transmission into the design of a novel precoding to achieve prioritized spatial multiplexing. We will illustrate in the following a novel precoding approach that is able to accomplish such a desired design goal.

Let $\tilde{\mathbf{s}} \triangleq \boldsymbol{\Lambda}\mathbf{s}$ in Equation 19.13; we get $\tilde{\mathbf{y}} = \tilde{\mathbf{s}} + \tilde{\mathbf{z}}$. The received SNR diagonal matrix has a form of

$$\boldsymbol{\Gamma} = \frac{E\left\{ \tilde{\mathbf{s}}\tilde{\mathbf{s}}^H \right\}}{E\left\{ \tilde{\mathbf{z}}\tilde{\mathbf{z}}^H \right\}} = \frac{\boldsymbol{\Sigma}^2 \boldsymbol{\Phi}\sigma_s^2}{\sigma_z^2 \mathbf{I}_K} = \begin{bmatrix} \gamma_1 & & & \\ & \gamma_2 & & \\ & & \ddots & \\ & & & \gamma_K \end{bmatrix} \tag{19.17}$$

where $\gamma_i = \lambda_i^2 \phi_i / \alpha$ is the resulting SNR for the ith subchannel. Hence, the diagonal values of Φ can be expressed as

$$\phi_i = \gamma_i \alpha / \lambda_i^2, \quad i = 1, \ldots, K \tag{19.18}$$

and ϕ_i satisfies the power constraint $\sum_{i=1}^{K} \phi_i = P_T$.

According to Equations 19.12 and 19.18, we can get

$$\frac{\gamma_1 \alpha}{\lambda_1^2} + \frac{\gamma_2 \alpha}{\lambda_2^2} + \cdots + \frac{\gamma_K \alpha}{\lambda_K^2} = K\mu - \left(\frac{\alpha}{\lambda_1^2} + \frac{\alpha}{\lambda_2^2} + \cdots + \frac{\alpha}{\lambda_K^2} \right) = P_T \tag{19.19}$$

Then the water-filling threshold μ can be recalculated as

$$\mu = \frac{\alpha}{K} \left(\frac{1 + \gamma_1}{\lambda_1^2} + \frac{1 + \gamma_2}{\lambda_2^2} + \cdots + \frac{1 + \gamma_K}{\lambda_K^2} \right). \tag{19.20}$$

According to the property of the SVC video, the base layer has the highest priority to be allocated more resource so as to establish a high-quality subchannel. Therefore, we use the subchannel corresponding to the largest singular value λ_1 to transmit the information of the base layer, and the subchannels corresponding to the singular values $\lambda_2, \ldots, \lambda_K$ to transmit the 1st, $\ldots$, $(K-1)$th enhancement layer, respectively. This assignment is apparently based on the quality of subchannels according to CSI. Since Equation 19.20 also contains the resulting SNR γ_i of the ith subchannel, we have effectively made use of the desired SNR information and have found an elegant way to realize the prioritized spatial multiplexing. This scheme is designed not only according to the CSI but also the desired SNRs of the base layer and the enhancement layers of the video bitstreams.

The closed form of probability of bit error for M-QAM, where $M = 2^k$ and k is even, is given by [48]

$$P_e \approx \frac{2(1 - M^{-1})}{\log_2 M} Q\left[\sqrt{\left(\frac{3 \log_2 M}{M^2 - 1} \right) \frac{2E_b}{N_0}} \right]. \tag{19.21}$$

Based on the different tolerances of BER for each layer, the requirements of SNRs of each layer can be calculated and denoted as $\gamma_1, \gamma_2, \ldots, \gamma_K$. These desired SNRs will be adopted in Equation 19.20 to allocate the power to each subchannel correspondingly. We first relax the power constraint and obtain as new water-filling threshold the equation

$$\mu_D = \mu + \frac{P_T}{K} = \frac{1}{K} \left(P_T + \alpha \left(\frac{1 + \gamma_1}{\lambda_1^2} + \frac{1 + \gamma_2}{\lambda_2^2} + \cdots + \frac{1 + \gamma_K}{\lambda_K^2} \right) \right). \tag{19.22}$$

Interestingly, Equation 19.22 can also be obtained through the water-filling method when each element of diagonal matrix Φ_D is written as

$$\phi_{D,i} = \left(\mu_D - \frac{\alpha(1 + \gamma_i)}{\lambda_i^2} \right)^+ \tag{19.23}$$

where $\phi_{D,i}$ satisfies the power constraint condition $\sum_{i=1}^{K} \phi_{D,i} = P_T$. Finally, this novel precoding matrix $\mathbf{F}_{\text{proposed}}$ for spatial multiplexing can be represented as

$$\mathbf{F}_{\text{proposed}} = \mathbf{V} \Phi_D^{\frac{1}{2}}. \tag{19.24}$$

There exist numerous research opportunities in video over MIMO systems, especially for spatial-multiplexing wireless MIMO systems. As we indicated early, when the MIMO channel information cannot be estimated due to practical system constraints such as fast-moving terminals, the system will need to resort to different approaches, such as blind equalization, to facilitate the matching between the multiple media streams and the properly selected antennas [35]. Furthermore, when a MIMO base station is serving multiple users, an effective user scheduling algorithm should be developed to make the compound QoS for all users maximized with deadline-aware scheduling and fairness between media consumer and data users.

19.4.4 Joint Reliable and Secure Media Transmission

There are two major categories for end-to-end media authentication, namely stream-based and content-based techniques. One key feature in designing media authentication for delivery over error-prone transport links is to develop authentication schemes that exploit the unequal importance of different packets for compressed media stream. By applying conventional cryptographic hashes and digital signatures to the media packets, the system security is similar to that achievable in conventional data security. Instead of optimizing packet verification probability, we optimize the quality of the authenticated media, which is determined by the packets that are received which can also be decoded and authenticated. The quality of the authenticated media is optimized by allocating the authentication resources unequally across streamed packets based on their relative importance, thereby providing unequal authenticity protection. The media authentication schemes discussed in the following will show the effectiveness of such approach for different types of wired and wireless channels.

We describe in this section an elegant design of joint layered coding to embed authentication into appropriated locations with desired unequal error protection (UEP). This joint design is able to address inevitable quality degradation in many existing schemes that usually separate authentication from source and channel coding.

19.4.4.1 Convert Source Coding Dependency into Utility-Size Ratio Inequality

As a scalable coding standard, JPEG-2000-encoded media has the important property that low-layer media slice does not have coding dependence on any higher-layer media slices. As a result, a low-quality bitstream could be obtained by simply truncating the higher-quality bitstream. Furthermore, because higher-layer media slices require lower-layer media slices in decoding, we can safely assume that lower-layer media slices have larger utility value than high-layer media slices.

Let the encoded media stream have the format of a sequence of slices $M_1, \ldots, M_T$ with utility $U_1, \ldots, U_T$ and size $S_1, \ldots, S_T$. Here, the utility is defined as the amount by which the overall distortion is reduced if slice M_i is consumed.

The proposed design to exploit media source coding dependency is the major innovations for a joint layered coding scheme to authenticate media over wireless networks. Coding dependency between media slices could introduce possible undesired degradation into end-to-end quality when channel noise exists. For example, if slice M_a is coding dependent on another slice M_b, it is possible that M_a is received and verifiable but not decodable if M_b is not received or verifiable. It is challenging to design an authentication graph that aligns with the coding dependency graph in the graph-based authentication approach. In this research, we convert the coding dependency relationship into the utility–size ratio inequality, that is, the utility–size ratio of a slice is no less than that ratio of any other slices with coding dependent on it. If the calculation of utility has taken the coding dependency into consideration, this property is achieved. Otherwise, we revise the utility–size ratio of a slice to be the maximum of that of all other slices coding dependent on it, including the slice itself. Later, we will see that the error protection priority is determined by the utility–size ratio so that after the preprocessing, a media slice M_i will be protected at a higher level than any other slices dependent on it. In other words, if a media slice dependent on M_i is recoverable from possible channel impairments, then M_i must also be available. Hence, no additional quality degradation will be introduced by the coding dependency.

19.4.4.2 Unequal Error Protection/Packetization of Media Slices

The compressed media slices have different weights or utilities. Given a constraint on the overall transmission rate, it is natural to allocate more channel protection bits and more authentication bits to protect more important media slices. Layered coding-based media error protection is pursued in our system in the channel coding component to reduce quality degradation by channel noise.

Basically there are two packetization strategies for media error protection: (i) parallel packetization in which each media slice is contained in one or several transmission packets and (ii) orthogonal packetization in which any media slice is distributed into many transmission packets. In parallel packetization, verifying media slice is equivalent to verifying transmission packets and it could fit into graph-based schemes perfectly. However, it also inherits the aforementioned limitations of the graph-based stream authentication schemes. Note that a joint source-authentication-channel (JSCA) system using parallel packetization and graph-based unequal authenticity protection has been developed in Ref. [49]. In orthogonal packetization, verifying media slice is different from verifying transmission packets so that it provides the possibility to verify the media slice when some transmission packets are lost. Furthermore, it is convenient to apply UEP on media slices using orthogonal packetization, such as in priority encoding transform (PET) [50]. Based on these reasons, orthogonal packetization is used in our system.

19.4.4.3 Integration of UEP and Authentication

In Ref. [51], we proposed an integration of UEP and authentication scheme, namely Joint Error Control Coding (JECC)-based media error and authentication protection scheme, as illustrated in Figure 19.6a. In this scheme, the hash of all media slices are concatenated and signed. All the authentication information is then protected at the same level as the most important media slice by the UEP scheme. As long as any media slice is successfully constructed, its authentication information too is successfully constructed. Hence, any reconstructed media slice could be verified at the receiver side. Such a scheme achieves good authentication performance while maintaining the authentication overhead at low level. However, we can see that if some media slices could not be reconstructed due to out-of-protection-range number of packet loss, their authentication information is still possibly available at the receiver side. In other words, their authentication information is overprotected because such authentication information is useless if we cannot reconstruct the corresponding media slice at the receiver end.

To eliminate such an overprotection on authentication information, we could further modify the integration of UEP and authentication as shown in Figure 19.6b. The hash of

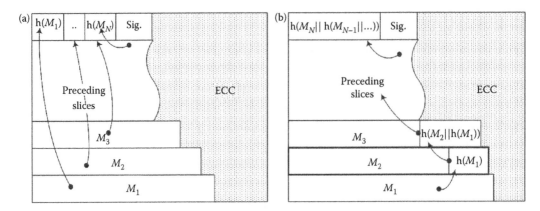

FIGURE 19.6
Integration of UEP and authentication. (a) Joint ECC-based scheme. (b) Joint media error and authentication protection.

slice M_i is only appended to a media slice being protected at the next higher level. Since the hash data are placed at the immediate next level, the authentication overhead is minimized. By the principle of UEP, if a media slice is decodable, then its hash is also decodable since the hash is protected at a higher level. Hence, the media slice is also verifiable. Therefore, the *effective verification probability* reaches 100%, that is, any decodable media slice is also verifiable. In fact, the authentication overhead can be further reduced by generating only one hash for all the media slices being protected at the same level and append this hash to media slices at the immediate higher level.

The advantages of the layered coding-based approach are as follows: (i) There is no need for the layered coding-based media authentication scheme to produce multiple copies of hash; hence, the authentication overhead is reduced. (ii) It is straightforward for the layered coding-based media authentication to adopt UEP strategy to match the varying importance of the compressed media slices. (iii) It is natural for the layered coding-based authentication protection to adapt to time-varying wireless links by allocating an alterative coding rate to achieve optimal end-to-end media quality.

19.4.4.4 Joint Media Error and Authentication Protection

Based on the discussion of the individual components in the media transmission system, we propose the JMEAP system as illustrated in Figure 19.7, and the detailed structure of the transmission packet is illustrated in Figure 19.8. Based on the PET [50], orthogonal packetization is adopted to provide UEP on media slices and authentication information, where a media slice is distributed to multiple transmission packets. The proposed authentication scheme could be easily embedded into existing PET-based media delivery systems.

At the server site, the compressed media slices are protected by channel coding at L different levels using $(N, k_1), \ldots, (N, k_L)$ erasure codes, respectively, where k_i refers to the number of packets required to recover the ith slice. Next, a hash is generated for all the media slices at each level and appended to the media slices at higher level, until level 1 is reached. In other words, the media slices being protected at the same level are considered as a uniform group for both error protection and authenticity protection. Compared with the transmission packet-based scheme, the number of hash required is greatly reduced. Finally,

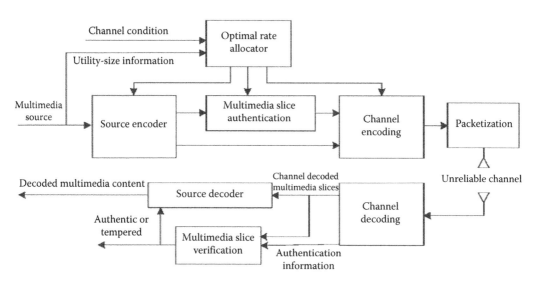

FIGURE 19.7
Block diagram of the JMEAP system.

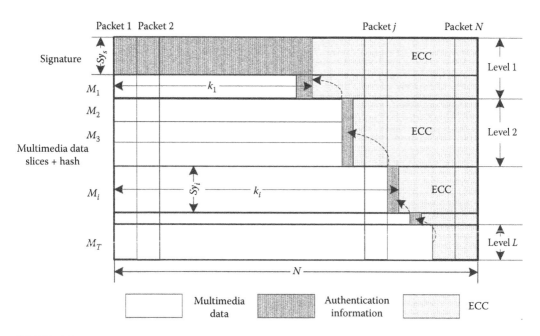

FIGURE 19.8
Structure of transmission packets. The dashed arrows represent hash appending.

from the media slices at level 1 and the appended hash from level 2, a hash is generated and signed. The signature would also be protected at level 1, same as the most important media slices. After all the media slices are authenticated, the error controlling codes-based channel coding is applied uniformly on the media slices and on the authentication information within each protection level. This process is followed by the packetization structure as shown in Figure 19.8.

At the receiver site, if the number of the received packets is more than k_i, all the slices and authentication information at ith level and higher levels could be successfully channel decoded. Since the information to authenticate slices at the ith level is protected at a higher level, these slices are believed to be verifiable as long as they can be correctly reconstructed in the proposed system. Hence, by the nature of the proposed scheme, all the decodable media slices are verifiable. Finally, the authenticated media slices are source decoded and assembled to produce media content.

Under this framework, several media quality degradation issues in the existing schemes can be simultaneously resolved. We elaborate the details of these benefits in the following:

1. *Coding dependency impact*: In this joint protection scheme, the authentication dependency has been made consistent with the coding dependency. Therefore, no additional quality degradation will be introduced by the mismatch of verifiable media slices and decodable media slices.

2. *Channel impairment impact*: The design of UEP for the layered coding-based media authentication scheme will enable an optimal rate allocation (described in the next section) so as to minimize the channel impairment impact.

3. *Guaranteed authentication verification*: The smart placement of hash data under layered coding-based scheme is able to achieve 100% verification probability. Hence, the potential additional distortion introduced by unverifiable media slices can be eliminated.

4. *Minimum authentication overhead*: In the proposed scheme, only one hash is generated for all media slices at any given protection level. As the number of protection levels is generally much less than the number of the transmission packets or the number of media slices, the overhead introduced by authentication is minimized. As we discussed earlier, traditionally, two or more copies of each hash per packet/slice are needed to maintain high verification probability.

Since the proposed scheme does not change the coding dependency of the JPEG-2000 slices, arbitrary truncation in transcoding does not affect the verification of the transcoded codestream. Hence, the proposed scheme fits well with the wireless transmission in terms of compatibility with transcoding.

In summary, the proposed joint error and authentication protection scheme is able to simultaneously address several issues in media quality degradation in wireless networks. An optimal rate allocation scheme is developed to achieve the optimal end-to-end quality under the proposed scheme. The readers should check Ref. [52] for more details.

One major technical challenge is to extend the current joint error and authentication protection scheme to authenticating video data over wireless networks. There are several technical barriers, and therefore research opportunities, in the development of an optimal and robust video authentication scheme for wireless networking applications. First, the temporal dependency of the compressed digital video makes the design of the joint scheme much more difficult. For example, H.264-encoded video may exhibit both intraprediction dependency and interprediction dependency. In addition to considering the channel error-induced distortion in the wireless networking environment, appropriate exploitation of rate-induced distortion for compressed video becomes necessary for the purpose of optimal rate allocation among source coding, channel coding, and media authentication. Second, the estimation of channel information may be inaccurate and with delay. Such departure from true channel information will cause the optimization of the joint scheme to deviate from true optimal solutions, including the crucial optimal rate allocation among source coding, channel coding, and media authentication. Finally, for video over heterogeneous networks and terminal devices, video adaptation that selects only the region of interest from video frames may be adopted. It is certainly challenging to design an authentication scheme that is able to authenticate appropriate regions from original video frames for transmission over wireless networks. This is quite unconventional since a good portion of a video frame will be discarded after video adaptation and the original streaming server may not know where the regions of interest are before encoding.

19.5 Summary and Discussion

In this chapter, we have discussed several emerging technical challenges as well as research opportunities to develop next-generation networked mobile video communication systems. We begin with the recognition of a broad spectrum of technical challenges in contemporary multimedia communications and networking that are inherent from the characteristics of both compressed media and the error-prone networks. We then present a variety of needs for developing emerging technologies that are able to meet the challenges we have identified. In particular, we focus on the following needs: (1) needs for robust P2P media streaming technologies, (2) needs for optimal resource allocation, (3) needs for new-generation HTTP streaming protocols, (4) needs for scalable video streaming over WLAN, (5) needs for intelligent video adaptation for mobile devices, (6) needs for video streaming over MIMO wireless systems, and (7) needs for new strategy in joint reliable and secure media transmission. To report some recent emerging technologies in related research areas, we have devoted a significant portion of this chapter to introduce in great detail several emerging technologies and identify potential future research opportunities related to these technologies.

The impact of these technologies will be far reaching on several relevant research disciplines beyond multimedia communications and networking. These disciplines include signal processing, communications, networking, and mobile device development. We demonstrated several prime examples in multimedia communications and networking to illustrate the complexity and challenging nature of these technologies. In particular, these emerging technology development examples show that end-to-end quality-of-experience optimization for multimedia communication and networking will drive the research in these research communities to expand their horizon from simple bandwidth reduction for video coding and capacity enhancement in communication technology to an overall systematic optimization. We need to address inherent characteristics of compressed digital media as well as communication and networking links so as to design an optimal solution for next-generation media system under the constraints that limited resource along the communication path and at the receiving terminal can be appropriately exploited.

Acknowledgment

The research presented in this chapter has been supported by US NSF Grants 0915842 and 0964797, a grant from Kodak Research Laboratories, and a grant from Huawei Technologies,

Inc. I would like to thank members of my research group for their contributions to several research projects in multimedia communications and networking. In particular, I would like to thank Wei Pu, Xinglei Zhu, Wenyuan Yin, and Qian Liu for their contributions.

References

1. http://www.youtube.com/t/fact_sheet.
2. http://www.comscore.com/Press_Events/Press_Releases/2009/1/US_Online_Video_Viewing.
3. http://www.cisco.com/en/US/solutions/collateral/ns341/ns525/ns537/ns705/ns827/white_paper_c11–481360_ns827_Networking_Solutions_White_Paper.html.
4. X. Hei, C. Liang, J. Liang, Y. Liu, and K. W. Ross, Insights into PPLive: A measurement study of a large-scale P2P IPTV system, In *Proceedings of IPTV Workshop, International World Wide Web Conference*, Edinburgh, Scotland, 2006.
5. S. Banerjee, B. Bhattacharjee, and C. Kommareddy, Scalable application layer multicast, In *Proceedings of ACM SIGCOMM'02*, Pittsburgh, PA, August 2002, pp. 205–217.
6. X. Zhang, J. Liu, B. Li, and T.-S. P. Yum, CoolStreaming/DONet: A data-driven overlay network for live media streaming, In *Proceedings of IEEE INFOCOM*, Miami, FL, March 2005, pp. 2102–2111.
7. X. Zhang, J. Liu, and B. Li, On large-scale peer-to-peer live video distribution: Coolstreaming and its preliminary experimental results, In *IEEE Multimedia Signal Processing Workshop*, Shanghai, China, October 2005.
8. N. Magharei and R. Rejaie, Prime: Peer-to-peer receiver-driven mesh-based streaming, *IEEE/ACM Trans. Netw.*, 17, 1052–1065, 2009.
9. X. Hei and C. Liang, A measurement study of a large-scale P2P IPTV system, *IEEE Transac. Multimedia*, 9(8), 2007.
10. T. Silverston and O. Fourmaux, Measuring P2P IPTV systems, In *Proceedings of the 17th ACM International Workshop on Network and Operating Systems Support for Digital Audio and Video (NOSSDAV07)*, Urbana-Champaign, IL, June 2007.
11. http://www.wireless-nets.com/resources/downloads/wireless_industry_report_2007.pdf.
12. D. Ren, H. Li, and G. S.-H. Chan, Fast-mesh: A low-delay high-bandwidth mesh for peer-to-peer live streaming, *IEEE Trans. Multimedia*, 11, 1446–1456, 2009.
13. X. Jin, K.-L. Cheng, and S.-H. G. Chan, SIM: Scalable island multicast for peer-to-peer media streaming, In *Proceedings of the IEEE International Conference on Multimedia Expo (ICME)*, Toronto, Canada, July 2006, pp. 913–916.
14. W. Wei, C. Zhang, H. Zang, J. Kurose, and D. Towsley, Inference and evaluation of split-connection approaches in cellular networks, In *Proceedings of ACM PAM*, Adelaide, Australia, 2006.
15. F. Baccelli, G. Carofiglio, and S. Foss, Proxy caching in split TCP: Dynamics, stability and tail asymptotics, In *Proceedings of IEEE INFOCOM'08*, Phoenix, AZ, April 15–17, 2008.
16. Wireless LAN Medium Access Control (MAC) and Physical Layer (PHY), IEEE 802.11 Std., 1999.
17. Wireless LAN Medium Access Control (MAC) and Physical Layer (PHY) Specifications Amendment 8: Medium Access Control (MAC) Quality of Service Enhancements, IEEE Std. 802.11e-2005, 2005.
18. A. Ksentini, M. Naimi, and A. Gueroui, Toward an improvement of H.264 video transmission over IEEE 802.11e through a cross-layer architecture, *IEEE Commun. Mag.*, 107–114, January 2006.
19. H. Liu and Y. Zhao, Adaptive EDCA algorithm using video prediction for multimedia IEEE 802.11e WLAN, *IEEE International Conference on Wireless and Mobile Communications*, 2006, p. 10.
20. H.-C. Jang and Y.-T. Su, A hybrid design framework for video streaming in IEEE 802.11e wireless network, In *Proceedings of the 22nd International Conference on Advanced Information Networking and Applications*, Okinawa, Japan, March 25–28, 2008, pp. 560–567.

21. R. MacKenzie, D. Hands, and T. O' Farrell, QoS of video delivery over 802.11e WLANs, In *IEEE ICC. 2009*, Dresden, Germany, June 2009.
22. Y. Zhang, Z. Ni, C. H. Foh, and J. Cai, Retry limit based ULP for scalable video transmission over the IEEE 802.11eWLANs, *IEEE Commun. Lett.*, 498–500, June 2007.
23. C. H. Foh, Y. Zhang, Z. Ni, J. Cai, and K. N. Ngan, Optimized cross layer design for scalable video transmission over the IEEE 802.11e networks, *IEEE Trans. Circuits Syst. Video Technol.*, 17(12), 1665–1678, 2007.
24. M.-H. Lu, P. Steenkiste, and T. Chen, A time-based adaptive retry strategy for video streaming in 802.11 WLANs, *Wireless Commun. Mob. Comput.*, 7, 187–203, 2007.
25. J. Du and C. W. Chen, A deadline-aware transmission framework for H.264/AVC video over IEEE 802.11e EDCA wireless networks, In *Proceedings of VCIP2010*, Huangshan, China, 2010.
26. Q. Liu, Z. Zou, and C. W. Chen, A deadline-aware virtual contention free EDCA scheme for H.264 video over IEEE 802.11e wireless networks, In *Proceedings of ISCAS2011*, Rio de Janeiro, Brazil, 2011.
27. M.-T. Sun and A. Reibman, *Compressed Video over Networks*, Marcel Dekker, Inc. New York, NY, 1997.
28. J. G. Kim, Y. Wang, and S.-F. Chang, Content adaptive utility based video adaptation, In *Proceedings of IEEE ICME*, Baltimore, MD, 2003.
29. Z. Lei and N. D. Georganas, Rate adaptation transcoding for video streaming over wireless channels, In *Proceedings of IEEE ICME*, Baltimore, MD, 2003.
30. V. Tarokh, H. Jafarkhani, and A. R. Calderbank, Space-time block coding for wireless communications: Performance results, *IEEE J. Select. Areas Commun.*, 17, 451–460, 1990.
31. S. M. Alamouti, A simple transmit diversity scheme for wireless communications, *IEEE J. Select. Areas Commun.*, 16, 1451–1458, 1998.
32. G. J. Foschini, Layered space-time architecture for wireless communication in a fading environment when using multielement antennas, *Bell Labs Tech. J.*, 41–59, Autumn 1996.
33. G. J. Foschini and M. J. Gans, On limits of wireless communications in a fading environment when using multiple antennas, *Wireless. Pers. Commun.*, 6, 311–335, 1998.
34. Q. Liu, S. Liu, and C. W. Chen, A novel prioritized spatial multiplexing for MIMO wireless system with application to H.264 SVC video, In *Proceedings of IEEE ICME 2010*, Singapore, July 2010.
35. Q. Liu and C. W. Chen, Blind channel equalization for fast moving terminals in prioritized spatial multiplexing MIMO systems, In *Proceedings of IEEE Globecom 2010*, Miami, FL, December 2010.
36. R. H. Deng and Y. Yang, A study of content authentication in proxy-enabled multimedia delivery systems: Model, techniques and applications, *ACM Trans. Multimedia Comput. Commun. Appl.*, 5(4), 28:1–28:20, 2009.
37. M. Hefeeda and K. Mokhtarian, Authentication schemes for multimedia streams: Quantitative analysis and comparison, *ACM Trans. Multimedia Comput. Commun. Appl.*, 6(1), 1–24, 2010.
38. B. Sardar and D. Saha, A survey of TCP enhancements for last-hop wireless networks, *IEEE Commun. Survey Tutorials, 3rd Quarter 2006*, 8(3), 20–34, 2006.
39. S.-F. Chang, Content-based video summarization and adaptation for ubiquitous media access, In *Proceedings of ICIAP*, Mantova, Italy, 2003.
40. D. Cotroneo, G. Paolillo, C. Pirro, and S. Russo, A user-driven adaptation strategy mobile video streaming applications, In *Proceedings of the 25th IEEE International Conference on Distributed Computer Systems Workshops*, Columbus, OH, 2005.
41. Y. Wang, X. Fan, H. Li, and C. W. Chen, An attention based spatial adaptation scheme for H.264 videos over mobiles. *IJPRAI special issue on Intelligent Mobile and Embedded Systems*, 2006.
42. H. Li, Y. Wang, and C. W. Chen, An attention information based spatial adaptation framework for browsing videos via mobile devices, *EURASIP J. Adv. Signal Process.*, 1–2, 2007.
43. P. W. Wolniansky, G. J. Foschini, G. D. Golden, and R. A. Valenzuela, V-BLAST; An architecture for realizing very high data rates over the rich-scattering wireless channel, In *Proceedings of IEEE ISSSE*, Pisa, Italy, September 1998.
44. I. E. Telatar, Capacity of multi-antenna Gaussian channels, *Eur. Trans. Telecommun.*, 10, 585–595, 1999.

45. D. Song and C. W. Chen, Novel layered scalable video coding transmission over MIMO wireless systems with partial CSI and adaptive channel selection, Preprint, In *Proceedings of the SPIE Conference on Multimedia on Mobile Devices*, San Jose, CA, January 2007.
46. D. Song and C. W. Chen, QoS guaranteed SVC-based video transmission over MIMO wireless systems with channel state information, In *Proceedings of IEEE ICIP2006*, Atlanta, GA, October 2006.
47. D. Song and C. W. Chen, Scalable H.264/AVC video transmission over MIMO wireless systems with adaptive channel selection based on partial channel information, *IEEE Trans. Circuits Syst. Video Technol.*, 17(9), 1218–1226, 2007.
48. B. Sklar, *Digital Communications – Fundamentals and Applications*, Prentice Hall, Upper Saddle River, NJ, 2001.
49. Z. Li, Q. Sun, Y. Lian, and C. W. Chen, Joint source-channel-authentication resource allocation and unequal authenticity protection for multimedia over wireless networks, *IEEE Trans. Multimedia*, 9(4), 837–850, 2007.
50. A. Albanese, J. Blomer, J. Edmonds, M. Luby, and M. Sudan, Priority encoding transmission, *IEEE Trans. Inf. Theory*, 42(6), 1737–1744, 1996.
51. X. Zhu, Z. Zhang, Q. Sun, and C. W. Chen, A joint ECC based media error and authentication protection scheme, In *Proceedings of IEEE International Conference on Multimedia and Expo (ICME'08)*, Hannover, Germany, 2008, pp. 13–16.
52. X. Zhu, Z. Zhang, and C. W. Chen, A joint layered coding scheme for unified reliable and secure media transmission with implementation on JPEG2000 images, *Proc. IEEE International Conference on Multimedia and Expo (ICME'09)*, New York, NY, July 2009.

20

A Proxy-Based P2P Live Streaming Network: Design, Implementation, and Experiments

Dongni Ren, S.-H. Gary Chan, and Bin Wei

CONTENTS

20.1 Introduction

Rapid penetration of residential broadband networks has enabled video streaming over the Internet. In recent years, we have witnessed the success of many live streaming applications over the Internet, such as Internet TV, distance learning, movie on demand, press or teleconferencing, and so on [1–3]. By the end of the year 2010, Internet video had exceeded half of the total Internet traffic consumption in the United States [4]. According to the Cisco visual networking index forecast, by 2013, the various forms of video (TV, video on demand (VoD), Internet video, and peer to peer (P2P)) will exceed 90% of global consumer traffic [5]. In order to provide such services to a group of distributed users, traditionally, the client–server model or content distribution network is used where distributed servers serve individual participants directly by simple unicast [6]. This model clearly is not scalable to large groups due to high requirements on server processing power, maintenance cost, and network bandwidth.

P2P streaming has shown to be effective to serve large groups for many live applications [7–10]. It does not rely on dedicated servers; instead, each peer contributes its own resource for streaming [11,12]. In P2P streaming, contents are distributed among end hosts mainly using the upload bandwidth of the peers, while content servers only deliver streams to a small number of the peers. As a result, servers are no longer a performance bottleneck.

Despite its scalability, P2P streaming nowadays still suffers from the following problems:

- *Peer churns:* Participating peers may leave the network anytime during a streaming session, which may cause their children to suffer from unpredictable packet loss. Such loss may propagate downstream, adversely affecting video continuity and system robustness. Addressing peer churns is particularly important to a highly dynamic network [13].
- *High delay:* If the P2P structure is not carefully constructed, the delay from the source to the end hosts will be high due to delay accumulation. Many current P2P protocols are based on *ad hoc* random connections, leading to unsatisfactory delay performance for large groups.
- *Firewall/network address translation (NAT) problem:* In a streaming network, many peers may be behind the firewall or NAT. It is often challenging for these peers to upload data to other peers to share the resources.
- *Locality problem:* Many existing P2P protocols have not taken sufficient advantage of locality (due to obvious reasons of unpredictable peer dynamics and costly network measurement techniques), and hence are not very Internet service provider (ISP)-friendly. This leads to a network with overly long connections and a large volume of cross-ISP traffics.

To address the above problems, we propose a two-tier proxy-based P2P network. In a proxy-P2P streaming network, reliable proxies (which are supernodes or lightweight content servers) are placed in different ISPs close to user pools [14,15]. The proxies are usually stable, with high sustainable bandwidth and high fanout so that they can form a robust and high-bandwidth "backbone" for streaming. To further scale up the system and to leverage locality, peers that are under the same proxy use another P2P protocol to distribute streams among themselves. Figure 20.1 shows an example of a proxy-based P2P (the so-called proxy-P2P) streaming network, where the content server first distributes video to proxies, and then end users form another P2P network rooted at each proxy to receive contents. Compared to pure P2P networks, such a network provides the following advantages:

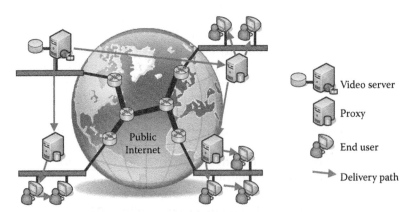

FIGURE 20.1
A proxy-based P2P streaming network.

- *Better system resilience:* Proxy-P2P has better resilience against peer churn and flash crowd, that is, a drastic increase of users at the beginning of a popular program. The stable proxies provide better resource availability at the beginning of a streaming session. They can also be used as reliable nodes for data recovery to confine error propagation upon peer failure/leave.

- *ISP-friendly:* Proxy-P2P overcomes many NAT and firewall problems and is more ISP-friendly. It has better locality awareness than pure P2P. Peers can join the closest proxies or another peer in their network, and hence there are significantly less long-distance connections. Because peers tend to connect to proxy or peer networks within the same local network, peers can often communicate with each other without NAT and firewall problems. The inter-ISP traffic is also greatly reduced.

- *Locality and backward compatibility:* Proxy-P2P enhances the performance of streaming unpopular channels. In proxy-P2P, as compared with the pure P2P approach, proxies serve as resource amplifiers for unpopular channels, hence achieving better performance. It is also backward-compatible with existing client–server architecture, where streams are directly delivered from proxies to peers.

We have designed and developed a proxy-P2P streaming network, where the proxies use a distributed protocol called FastMesh [16] while peers use another distributed P2P protocol called scalable island multicast, or SIM [17]. The objective of our FastMesh–SIM network is to construct ISP-friendly and high-bandwidth P2P streaming with the help of proxies and IP multicast. Compared to other proxy-P2P networks [18,19], FastMesh and SIM are specially designed and optimized given the special characteristics of the proxy and peer networks.

In our proxy–P2P network, proxies may span different public networks from different ISPs and proxies may be deployed over large geographical regions. Therefore, the content is distributed in the proxy backbone mainly based on unicast in the absence of global IP multicast support. TCP (Transmission Control Protocol) is used in stream transmission between proxies for better error control. These proxies can be considered reliable since the introduction or removal happens infrequently. Therefore, a tree-like mesh structure is formed among proxies to achieve low delay and high bandwidth. Our protocol, called FastMesh, meets a certain bandwidth requirement with heterogeneous uplink capacity of proxies.

On the other hand, peers are often considered as unreliable and dynamic, which means that they may join, leave, or fail unexpectedly. The peers in the same local network may have IP-multicast capability. Utilizing IP multicast can achieve low delay, high bandwidth, network-efficient streaming, and lower join latency. It also reduces peer dependency, which leads to better mitigation against churns and better stream continuity. This peer-level protocol is called SIM.

20.2 FastMesh–SIM Protocol

20.2.1 FastMesh: A Low-Delay High-Bandwidth Mesh for P2P Live Streaming

The goal of FastMesh is to design an overlay network that is suitable for efficient proxy-level communications. The overlay, which consists of supernodes, proxies, and content distribution servers, is mildly dynamic. FastMesh achieves the following:

- *Low delay:* An overlay which offers low source-to-proxy delay is desirable for live streaming. FastMesh seeks to minimize the maximum delay from the source to the proxy.
- *Meeting streaming bandwidth requirement in the presence of heterogeneous uplink bandwidths:* Proxies in the network may have diverse uplink bandwidth depending on their access network (such as Assymetric Digital Subscriber Line (ADSL), broadband Ethernet, cable, etc.). The overlay should achieve a certain streaming rate at each proxy despite this bandwidth heterogeneity or asymmetry.
- *Accommodation of proxy churns:* Proxy network can be dynamic, that is, a proxy may be deployed or removed. FastMesh accommodates this dynamic and is adaptive to the change to achieve high performance.
- *Distributed, simple, and self-improving:* FastMesh is distributed so that its performance is scalable to a large number of proxies. The protocol should be self-improving in the sense that it continuously improves and adapts to the overlay based on the existing proxy location and heterogeneous characteristics. Simplicity is also important to reduce overheads.

In FastMesh, the video stream is divided into multiple substreams (say 4–5) of similar bandwidth. Each of the substreams spans all the proxies in the network as a spanning tree, so that the aggregation of the trees is a mesh. Ideally, a proxy should choose nodes that are close to the source and with high bandwidth as its parents. FastMesh hence seeks to balance between delay and bandwidth of nodes in parent selections, and makes use of a concept similar to *power* to achieve that. Traditionally, in networking, *power* is defined as the throughput divided by delay. FastMesh similarly defines the *power* between a node i and its candidate parent j, denoted as $P_i(j)$, as the rate that j can serve i divided by the source-to-end delay of i from j. The larger the $P_i(j)$, the better the node j is as a parent of the node i.

A new proxy k contacts a *Rendezvous Point* (RP) which caches a list of recently arrived proxies. The RP returns a few nodes to i. Proxy i checks its delay from these nodes to the source and requests their residual bandwidths. It then evaluates its power to each of them, and chooses parents in a greedy manner, that is, selecting the node with the highest power as its parent. If this parent cannot fully serve it, it then connects to the second one, and so on. This process repeats until the proxy is fully served. If the proxies returned by the RP cannot fully serve the newcomer, the newcomer requests the neighbor of these proxies and the above process is repeated till it is fully served.

FastMesh adapts to the current network environment by continuously improving the mesh. This means that some nodes may need to shift in position in the streaming mesh to achieve lower overall delay. An adaptation algorithm is proposed to periodically optimize the mesh by moving nodes with low delay and high bandwidth closer to the source.

20.2.2 SIM: Leveraging IP Multicast in Local Networks

For peer-level networks, we have designed a fully distributed protocol called scalable island multicast, which effectively integrates IP multicast and overlay multicast with multiple substreams. Hosts in SIM first form an overlay tree using a distributed, and hence scalable protocol. They then detect IP multicast islands and utilize IP multicast whenever possible. SIM is push-based, achieving lower end-to-end delay, lower link stress, and lower resource usage as compared with traditional overlay protocols. The following is an overview of the protocol steps:

1. *Construction of the overlay tree:* A new host first contacts a public RP to obtain a list of current hosts in the system. It then iteratively pings other hosts and selects a close peer with enough forwarding bandwidth as the parent. Each overlay tree corresponds to a substream and is constructed independently. The overlay branches are used for streaming *across* islands, as within an island, IP multicast is used.

   If a host leaves, its children need to rejoin the tree and find new parents. A rejoining host starts the process from its grandparent and then follows the joining procedure.

2. *Island management:* After a host joins the overlay tree, it detects its island and joins the island, if any. Each streaming session has two unique class-D IP addresses for IP multicast management. One is used for multicasting control messages, and the other is used for multicasting streaming data. We call the groups corresponding to these two IP addresses as CONTROL group and DATA group, respectively.

   Each island has a unique ingress host, which is responsible for accepting streaming data from outside the island and multicasting them within the DATA group. Other hosts within the island accept streaming data from IP multicast instead of overlay unicast. We call a host within the island a border host if its overlay parent is not within the island. A border node is detected if it is not on the same multicast island as its parent. Clearly, the ingress must be a border host. In SIM, border hosts (including the ingress) join both the CONTROL group and the DATA group, and nonborder hosts join only the DATA group.

   An ingress host periodically multicasts KeepAlive messages in the CONTROL group, which contains its source distance. It also multicasts streaming data within the DATA group. Initially, the ingress of an island is the island's first joining host. The noningress border host with the smallest source-to-end delay in the island becomes the new ingress if the current ingress leaves or fails, or the border host has a lower source distance than the current ingress.

3. *Island detection:* The two class-D IP addresses are maintained by the RP. When a new host joins the session, it first obtains the class-D addresses and a list of already joined hosts from the RP. The new host then joins the overlay tree as described above. Afterward, it joins both the DATA and CONTROL groups. If an island exists, the host will receive KeepAlive messages from the ingress in the CONTROL group. The host then detects whether it is a border host by checking if its parent is in the same island ID. If it is, it remains in the CONTROL group and further joins the DATA group. Otherwise, it exits the CONTROL group. If the host does not find any island to join, it forms an island (with a new island ID) consisting of only itself and becomes the island ingress.

4. *System resilience:* A single delivery tree may not provide good streaming quality, especially in a dynamic P2P system. We address this with multiple trees for data delivery, and using a quick recovery scheme to recover temporary packet loss. In the recovery scheme, each host selects a few recovery neighbors. Whenever a packet loss is detected, retransmission request is sent to these neighbors to recover loss.

20.3   System Implementation

We have designed and developed a novel push-based proxy-P2P streaming system called FastMesh–SIM [20] using the two protocols proposed above, which consists of a distributed proxy network augmented with P2P streaming. FastMesh–SIM is a government-funded industry-supported project by Hong Kong Innovation Technology Fund (ITF). Figure 20.2 shows the overview of the FastMesh–SIM architecture, where proxies form a low-delay, high-bandwidth content distribution backbone using the FastMesh protocol. SIM protocol is then applied to distribute streams among end users within the domain of a proxy [21–23]. Peers belonging to the same IP-multicast network form a multicast island to achieve better bandwidth experience and robustness.

20.3.1   FastMesh Node

Figure 20.3a shows the structure of FastMesh program. The proxy program consists of three major components, receiver buffer, socket manager, and message scheduler. The receiver buffer receives video segments from network modules, and constructs the node's buffermap (even though FastMesh is push based, a buffermap is still necessary for error recovery and connection maintenance). The message scheduler maintains a queue of control and data messages for the node to process. The socket manager constructs and manages connections with other proxies or clients, which are used in sending/receiving messages and data segments.

Besides the core module, there are several concurrent threads handling all the connections and streaming logic of the proxy:

- *Mesh join thread:* It runs the FastMesh join algorithm to join the node to the existing proxy network.
- *Adaptation thread:* FastMesh adaptation is implemented and processed in this thread. It sends out requests to the parent nodes to find out more optimized

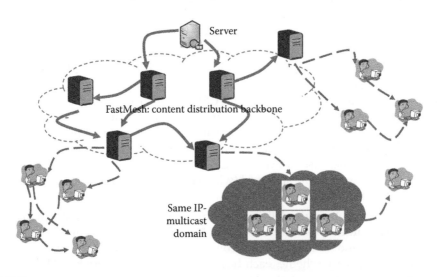

FIGURE 20.2
Overview of FastMesh–SIM architecture.

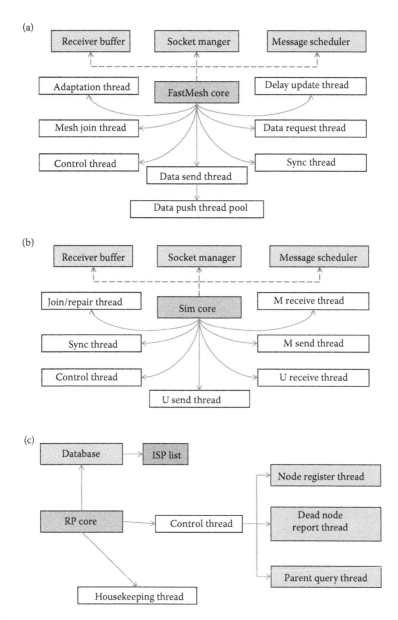

FIGURE 20.3
Software design. (a) FastMesh architecture; (b) SIM architecture; (c) RP architecture.

position to move in the mesh every few seconds (4 or 5 s). When grant messages (confirmation in FastMesh protocol to accept new children) are received, the proxy adapts to the new position in the mesh.

- *Control thread:* Multiple control requests are handled here, for example, ping request, child query and forwarding, join request, root synchronization, and so on.
- *Delay update thread:* Delay query is processed and replied here. This query is used to calculate the current source-to-end delay of the node, which is used in peer join and adaptation.

- *Data request thread:* This thread is used to send out data request of substreams to the corresponding parents. It also manages the sockets used in data receiving.
- *Data send thread:* It organizes a pool of data push threads, each of which is responsible to deliver a substream to one child of the proxy. When a node joins the proxy, the proxy constructs a new data push thread to serve the newcomer. Since end-to-end bandwidth between two nodes is limited by the TCP throughput, in FastMesh, each individual data push thread uses multiple TCP sockets to deliver streams. In this way the parallel connections can boost the bandwidth between proxies.
- *Synchronization thread:* This thread keeps the clock, source-to-end delay, and substream information of a node synchronized with its parents and children.

20.3.2   SIM Node

The code structure for the SIM client is shown in Figure 20.3b. It shares some similar modules with the FastMesh, such as control thread, synchronization thread, receiver buffer, socket manager, and message scheduler. The unique modules of SIM are its join/repair mechanism and threads handling data delivery, which are elaborated in the following:

- *Join/Repair Thread:* This thread runs the SIM algorithm for peers to join the SIM network. When a peer's parent leaves, the thread first conducts the fast repair process by requesting recovery from the peer's grandparents, current proxy, and then backup proxies. After this fast recovery, it rejoins the peer to the local SIM network.
- *Unicast Send Thread:* This thread handles unicast delivery for data and control messages. For peers between different islands, *Unicast Send Thread* transmits the video stream via unicast. We have implemented two different delivery approaches, push and pull modes which can be configurably activated.
- *Unicast Receive Thread:* Packets are received and stored in the receiver buffer within this thread. Different receive mechanisms are applied for the push or pull mode.
- *Multicast Send Thread:* This thread performs multicast sending. It conducts ingress selection among border nodes in a distributed manner. The ingress node then broadcasts the stream to other peers within the same IP-multicast island.
- *Multicast Receive Thread:* For noningress nodes in an IP-multicast island, *Multicast Receive Thread* is used to receive streams delivered by IP multicast.

20.3.3   Rendezvous Point

There is an RP to handle joins, leaves, and repair processes. Figure 20.3c shows the code structure of the RP. The RP keeps track of a (partial) list of proxies and peers in the system. ISP information is stored in database for proxy assignment upon a peer request. *Housekeeping Thread* periodically updates and backs up the proxy list, and provides information about public and NAT peers. *Node Register Thread* adds a new joining peer or proxy to the list in the RP. *Dead Node Report Thread* processes dead node reports received from peers. It then sends out PING messages to the reported nodes. If no PONG (PING reply) is received, then the node is considered dead and removed. *Parent Query Thread* handles the parent list request from the proxies and peers.

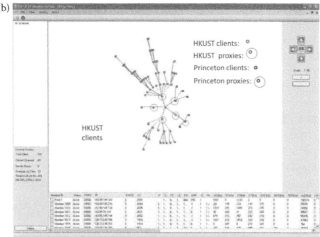

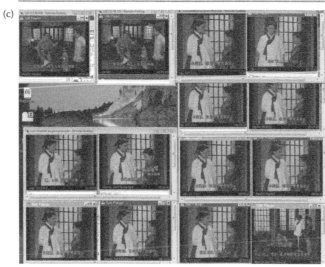

FIGURE 20.4
HKUST-Princeton trials. (a) A lab snapshot; (b) a topology snapshot; (c) screen capture.

20.4   Experiments

We have performed measurement studies on FastMesh–SIM together with Professor Mung Chiang, Professor Jennifer Rexford, Joe Wenjie Jiang from Princeton Unversity, and other collaborating universities and companies from different countries (e.g., Caltech (USA), AT&T (USA), Imperial College (UK), KAIST (Korea), Sun Yat-Sen University (China), etc.) [24]. We have also conducted experiments on PlanetLab at more than 40 sites from five different continents. In our measurement studies, we carefully planned a series of trials that quantify the impact of different parameters on system performance, for example, pull- or push-based data delivery, effect of peer churns, advantages of IP multicast, segment size variation, different streaming rates, and so on. Each trial lasts many hours involving peers from various sites. Figure 20.4a shows our trial conducted in spring 2010 with 140 peers (desktops and laptops) deployed in Hong Kong and the United States. Figure 20.4b shows an example of FastMesh–SIM topology, captured by our monitoring system. Red circles in the figure represent peers and proxies located in the Hong Kong University of Science and Technology (HKUST), whereas black circles represent machines in the United States (at Princeton). We see that FastMesh–SIM system can successfully construct an efficient live streaming network by arranging the joining peers into optimal positions in the topology. Figure 20.4c shows the screens of remote clients in the United States. The video playback is smooth and source-to-end delay is well constrained for the oversea nodes.

Figure 20.5 shows the delay distribution of peers in the field trial. Most peers in Asia achieve very low delay (over 90% of peers have playback delay of less than 300 ms). Since the streaming server is placed in Hong Kong, peers in the United States inevitably have a longer delay. It is clear that the delay is kept low by our distributed protocols and only a small percentage (5%) of the U.S. peers have playback delay longer than 6 s.

Figure 20.6 shows the delay distribution with and without the use of IP multicast. With IP multicast, very few peers suffer delay larger than 700 ms. Clearly, with the use of IP multicast, the delay in the network is significantly reduced.

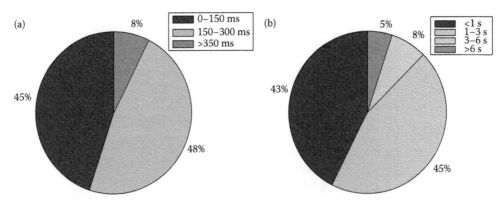

FIGURE 20.5
Peer delay distribution. (a) Asian Peers; (b) US peers.

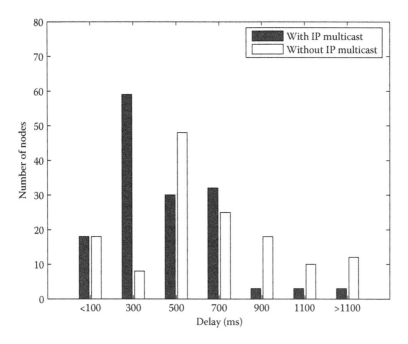

FIGURE 20.6
Delay reduction by IP multicast.

20.5 Conclusion

In this chapter, we discussed the design and implementation of a two-tier proxy-based P2P live streaming network, which consists of a low-delay high-bandwidth proxy backbone and a peer-level network leveraging IP multicast capability. The protocols we designed and developed for the proxy network and peer network are called FastMesh and SIM, respectively. From the field trials and measurement studies conducted with collaborating universities and research labs across the world, we show that indeed our proxy-based P2P streaming is effective in reducing peer delay and providing high-quality video streaming services.

References

1. YouTube, http://www.youtube.com
2. Hulu, http://www.hulu.com
3. Youku, http://youku.com
4. Limelight Networks, http://www.limelightnetworks.com
5. Cisco Visual Networking Index Forecast, http://www.cisco.com/en/US/solutions/collateral
6. X. Hei, C. Liang, J. Liang, Y. Liu, and K. W. Ross, Insights into PPLive: A measurement study of a large-scale P2P IPTV system, in *In Proceedings of IPTV Workshop, International World Wide Web Conference*, Edinburgh, Scotland, 2006.

7. S. Banerjee, B. Bhattacharjee, and C. Kommareddy, Scalable application layer multicast, in *Proceedings ACM SIGCOMM'02*, New York, NY, August 2002, pp. 205–217.

8. X. Zhang, J. Liu, B. Li, and T.-S. P. Yum, CoolStreaming/DONet: A data-driven overlay network for live media streaming, in *Proceedings of IEEE INFOCOM*, Miami, FL, March 2005, pp. 2102–2111.

9. X. Zhang, J. Liu, and B. Li, On large-scale peer-to-peer live video distribution: Coolstreaming and its preliminary experimental results, in *IEEE Multimedia Signal Processing Workshop*, Shanghai, China, October 2005, invited paper.

10. N. Magharei and R. Rejaie, Prime: Peer-to-peer receiver-driven mesh-based streaming, *IEEE/ACM Transactions Networks*, 17, 1052–1065, 2009.

11. X. Hei and C. Liang, A measurement study of a large-scale P2P IPTV system, *IEEE Transactions on Multimedia*, 9(8), 1672–1687, 2007.

12. T. Silverston and O. Fourmaux, Measuring P2P IPTV systems, in *Proceedings of 17th International Workshop on Network and Operating Systems Support for Digital Audio and Video (NOSSDAV07)*, Urbana-Champaign, IL, June 2007.

13. N. Magharei, R. Rejaie, and Y. Guo, Mesh or multiple-tree, a comparative study of live P2P streaming approaches, in *IEEE International Conference on Computer Communications*. Anchorage, AK: IEEE, May 2007, pp. 1424–1432.

14. C. Wu, B. Li, and S. Zhao, Multi-channel live P2P streaming: Refocusing on servers, in *IEEE INFOCOM*. Phoenix, AZ, IEEE, April 2008, pp. 1355–1363.

15. F. Wang, J. Liu, and Y. Xiong, Stable peers: Existence, importance, and application in peer-to-peer live video streaming, in *IEEE International Conference on Computer Communications*. Phoenix, AZ, IEEE, April 2008, pp. 1364–1372.

16. D. Ren, H. Li, and G. S.-H. Chan, Fast-mesh: A low-delay high-bandwidth mesh for peer-to-peer live streaming, *IEEE Transactions on Multimedia*, 11, 1446–1456, 2009.

17. X. Jin, K.-L. Cheng, and S.-H. G. Chan, SIM: Scalable island multicast for peer-to-peer media streaming, in *Proceedings of IEEE International Conference on Multimedia Expo*, Toronto, Canada, July 2006, pp. 913–916.

18. H. Yin, X. Liu, T. Zhan, V. Sekar, F. Qiu, C. Lin, H. Zhang, and B. Li, Livesky: Enhancing CDN with P2P, *Transactions on Multimedia Computing, Communications, and Applications*, 6(3), 16-1–16-19, 2010.

19. H. Yin, X. Liu, F. Qiu, N. Xia, C. Lin, H. Zhang, V. Sekar, and G. Min, Inside the bird's nest: measurements of large-scale live VOD from the 2008 Olympics, in *Internet Measurement Conference*, A. Feldmann and L. Mathy, eds., New York, NY, ACM, 2009, pp. 442–455.

20. Fastmesh-SIM, http://mwnet.cse.ust.hk/fastmesh-sim.

21. X. Jin, K.-L. Cheng, and S.-H. G. Chan, Scalable island multicast for peer-to-peer streaming, *Advances in MultiMedia*, 2007, 10–10, 2007. Available at http://dx.doi.org/10.1155/2007/78913.

22. X. Jin, K. leung Cheng, and S.-H. G. Chan, Island multicast: Combining ip multicast with overlay data distribution, *IEEE Transactions on Multimedia*, 11(5), 1024–1036.

23. X. Jin, H.-S. Tang, S. H. G. Chan, and K.-L. Cheng, Deployment issues in scalable island multicast for peer-to-peer streaming, *IEEE MultiMedia*, 16, 72–80, 2009. Available at http://portal.acm.org/citation.cfm?id=1515606.1515651.

24. J. W. Jiang, M. Chiang, J. Rexford, S. h. Gary Chan, K. f. Simon Wong, and C. h. Philip Yuen, Proxy-P2P streaming under the microscope: Fine-grain measurement of a configurable platform, 2009, invited paper.

21

Scalable Video Streaming over the IEEE 802.11e WLANs

Chuan Heng Foh, Jianfei Cai, Yu Zhang, and Zefeng Ni

CONTENTS

21.1 Introduction

As a result of the high performance-to-price ratio, the IEEE 802.11-based *wireless local area networks* (WLANs) have been massively deployed in public and residential places for various wireless applications. Given the growing popularity of real-time services and multimedia-based applications, it has become more and more critical to tailor IEEE 802.11 *medium access control* (MAC) protocol to meet the stringent requirements of such services. The IEEE 802.11 working group has developed a new standard known as the IEEE 802.11e [1] to provide the *quality of service* (QoS) support. The IEEE 802.11e defines a single coordination function, called the *hybrid coordination function*, which includes two medium access mechanisms: contention-based channel access and controlled channel access. In particular, the contention-based channel access is referred as *enhanced distributed channel access* (EDCA), which extends the legacy *distributed coordination function* (DCF) [2] by introducing multiple queues (called *access categories* (ACs)) to provide the MAC layer with per-class service differentiation.

Among various applications, video streaming is one of the most attractive applications for WLANs. However, due to the characteristics of wireless networks such as high error rate, limited bandwidth, time-varying channel conditions, limited battery power of wireless devices, and dynamic network users, wireless video streaming faces many challenges. From the application-layer coding point of view, wireless streaming requires video coding to be robust to channel impairments and adaptable to the network and diverse scenarios. The need for video adaptation becomes obvious. In general, video adaptation can be implemented in many ways such as bitstream switching [3] and transcoding [4]. *Scalable video coding* (SVC) [5] is an advanced video coding technique designed for video adaptation. SVC provides great flexibility in video adaptation since it only needs to encode a video once and the resulted bitstream can be decoded at multiple reduced rates and resolutions.

In the past few years, we have seen extensive studies [6–9] on streaming scalable video over lossy channels. The common idea is to use *unequal error/loss protection* (UEP/ULP), that is giving the more important information more protection, to explore the fine granularity scalability provided by SVC. Such a ULP idea has been implemented differently in different network protocol layers. For example, in the application layer or the MAC layer, the ULP is often provided through using different FEC (Forward Error Correction) codes [7] or different ARQ (Automatic Repeat reQuest) strategies [8]. In the network layer, the DiffServ is often used to provide the ULP [9]. There are also quite a few physical layer approaches such as using OFDM (Orthogonal Frequency Division Multiplexing) to provide different physical channels with different priorities [10], unequally distributing transmission powers [11], and using different modulations. Most of these existing ULP approaches only consider a single end-to-end connection and focus on optimally distributing network resource among different priorities under the constraint of a fixed total network resource. However, from the entire network point of view, the resource distribution in one connection is not independent of other competing connections. In other words, the ULP adjustment at one user will affect other competing users. Thus, the ULP strategy for one user should aim at not only maximizing its own video quality but also minimizing the harmful effect to other users.

In this chapter, we summarize our recent studies on exploring the scalability of SVC and the QoS provided by the IEEE 802.11e EDCA to achieve an optimal performance for video streaming over WLANs. In particular, we consider two scenarios. The first one is to transmit scalable video over one EDCA queue. By adaptively adjusting the retry limit setting of EDCA, we are able to maintain a strong loss protection for critical video traffic. In the second scenario, where we use two queues for scalable video streaming, we design a cross-layer framework to adaptively map video packets of different classes into the two different queues, which preemptively drops less important video packets so as to maximize the transmission protection to the important video packets.

The rest of the chapter is organized as follows. Section 21.2 gives some background information. Sections 21.3 and 21.4 present our designs for the one-queue and two-queue scenarios, respectively. Finally, Section 21.5 concludes the chapter.

21.2 Background Review

In this section, we briefly review the two contention-based media access mechanisms, the legacy DCF in 802.11 and the EDCA in 802.11e, and also describe the scalable video codec we developed.

21.2.1 Overview of IEEE 802.11 DCF

DCF is based on carrier sense multiple access/collision avoidance where stations listen to the medium to determine when it is free. If a station has frames to send and senses the medium is busy, it will defer its transmission and initiate a backoff counter. The backoff counter is a uniformly distributed random number between zero and *contention window* (CW). Once the station detects that the medium has been free for a duration of *DCF Interframe Space*, it starts a backoff procedure, that is decrementing its backoff counter as long as the channel is idle. If the backoff counter has reduced to zero and the medium is still free, the station begins to transmit. If the medium becomes busy in the middle of the decrement, the station freezes its backoff counter, and resumes the countdown after deferring for a period of time, which is indicated by the *network allocation vector* stored in the winning station's frame header.

It is possible that two or more stations begin to transmit at the same time. In such a case, a collision occurs. Collisions are inferred by no *acknowledgment* (ACK) from the receiver. After a collision occurs, all the involved stations double their CWs (up to a maximum value, CW_{max}) and compete the medium again. If a station succeeds in channel access (inferred by the reception of an ACK), the station resets its CW to CW_{min}.

We can see that DCF does not provide QoS supports since all stations operate with the same channel access parameters and have the same medium access priority. There is no mechanism to differentiate different stations and different traffic.

21.2.2 Overview of IEEE 802.11e EDCA

In the IEEE 802.11e standard, the EDCA mechanism extends the DCF access mechanism to enhance the QoS support in the MAC layer through introducing multiple ACs to serve different types of traffic. In particular, a node implementing IEEE 802.11e MAC protocol provides four ACs that have independent transmission queues as shown in Figure 21.1. Each AC, basically an enhanced variant of the DCF, contends for *transmission opportunity* (TXOP) using one set of the EDCA channel access parameters including

- CW_{min}: minimal CW value for a given AC. Assigning a smaller value to CW_{min} gives a higher TXOP. Each AC is given a particular CW_{min}.
- CW_{max}: maximal CW value for a given AC. Similar to CW_{min}, CW_{max} is also assigned on a per AC basis.
- AIFSN: *arbitration interframe space number*. Each AC starts its backoff procedure after the channel is idle for a period according to AIFSN setting.
- TXOPlimit: the limit of consecutive transmission. During a TXOP, a node is allowed to transmit multiple data frames but limited by TXOPlimit.

Note that if the backoff counters of two or more ACs colocated in the same node elapse at the same time, a scheduler within the node treats the event as a *virtual collision*. The TXOP is given to the AC with the highest priority among the colliding ACs, and the other colliding ACs defer and try again later as if the collision occurred in the medium. Details of the IEEE 802.11e MAC protocol operation is described in Ref. [1]. In short, through differentiating the services among multiclass traffic, EDCA provides a certain level of QoS in WLANs for multimedia applications.

21.2.3 Scalable Video Codec

Various SVC schemes have been proposed in the literature. Some of them utilize wavelet transform, either entirely based on 3D wavelet or combining wavelet transform with motion

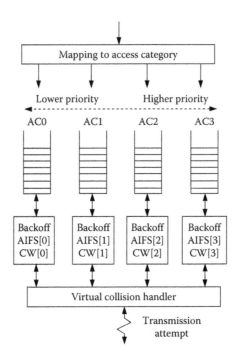

FIGURE 21.1
The four ACs in an EDCA node.

compensated predictive coding [12,13]. Many others are based on the traditional hybrid video coding scheme. The current SVC standard is based on H.264 extensions [14], which offers spatial, temporal, and SNR (Signal-to-Noise Ratio) scalabilities at bitstream level. The SNR scalability also include *coarse-granularity scalability* and *medium-granularity quality scalability*.

Considering that the main focus of the research is on the adapting of video scalability and the MAC QoS, for our study we develop a simple scalable video codec that is easy to perform optimal bit allocation and quantitatively analyze the importance of individual SVC packets. We would like to point out that the transmission frameworks of SVC over WLANs which we are going to describe in this chapter are general and they can be applied for other scalable video codecs as well, including the sophistic and comprehensive H.264 SVC.

Our simple scalable video codec is based on the integration of the *motion compensated temporal filtering* (MCTF) [15] and JPEG2000 [16]. The lifting scheme of MCTF [15] is an effective technique for temporal decomposition since it can efficiently explore multiple-frame redundancies which is hardly achievable by conventional frame-to-frame (MPEG-4) or multiframe prediction (H.264) methods. On the other hand, the latest image coding standard, JPEG2000 [16], provides a highly scalable (including component, quality, spatial, and positional scalability) and also highly efficient (in terms of *rate-distortion* or R-D performance) image codec. By exploring the unique features of MCTF and JPEG2000, the developed codec achieves not only a competitive R-D performance and the property of high scalability but also the features of easy to be analyzed and robust to packet loss.

Figure 21.2 shows the structure of the developed encoder, and the decoding process simply follows an inverse procedure. The entire encoding process consists of three main components: MCTF, spatial coding, and optimal bit truncation. In particular, each color

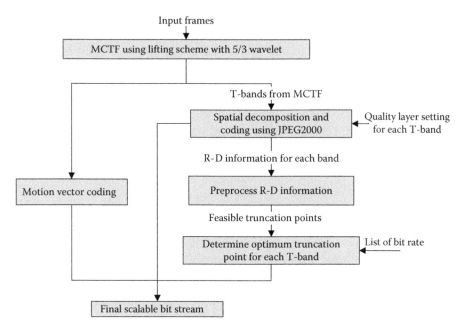

FIGURE 21.2
The encoding structure.

component (YUV) of the original frames F_k is first filtered using MCTF with 5/3 wavelet. MCTF is applied iteratively to the set of low-pass bands in order to provide multiple frame rates in the final scalable bitstream. Through MCTF, we generate the *motion vectors* (MVs) and many *temporal bands* (T-bands). Each T-band can be treated as an individual image. Then, JPEG2000 is used to encode these T-bands into multiple quality layers, each of which has an R-D value. After removing those nonfeasible truncation points, optimal bit truncation is performed to reach the given target bit rate. The final video bitstream consists of the MV information generated by MCTF and the JPEG2000 bitstream for each T-band. Because all the code blocks of all the T-bands are independently encoded, a feature of JPEG2000, the final bit stream is highly error-resilient.

Optimal Bit Allocation. In the developed codec, after encoding each T-bands using JPEG2000, the next step is to truncate each T-band JPEG2000 stream (for each component) to achieve a target bit rate R. The problem of optimal bit allocation can be summarized as: *given a target bit rate R, how to optimally truncate each T-band JPEG2000 stream so that the overall distortion can be minimized?* For our developed fully SVC, there are many ways to allocate bits among T-bands by making tradeoff among frame rates, spatial resolution, and SNR of individual frames. However, an optimal tradeoff among the three scalabilities is still an open question. In this research, for simplicity, we only consider using the SNR scalability to match the estimated network bandwidth. In particular, considering each T-band JPEG2000 stream contains a number of quality layers whose corresponding rates and distortions are known, we directly use these quality layers of JPEG2000 streams as the available truncation points although JPEG2000 provides even finer truncation points. The bit allocation algorithm is similar to the optimal truncation in JPEG2000 [17].

SVC Packetization with Relative Priority Index. For the packetized video transmission, T-band streams and MV data need to be assembled into individual network packets. In this research, we simply put different color component into different packets and place

MVs into the first packet of their associated temporal high bands. The size of a packet is limited by a predefined maximum packet size P_{size} (by default 500 bytes).

For QoS-enabled networks, it is highly desired for the application layer to provide the relative priorities of video packets. In this research, we apply the concept of relative priority index (RPI) proposed in Ref. [9] to categorize different video packets. In particular, we use the loss impact of a packet to calculate its RPI. The loss impact of a packet is defined as the corresponding distortion increase in reconstructed video in the case that the packet is lost while all other packets are correctly received. Mathematically, the loss impact of the i-th packet in the T-band B of the color component C is calculated as

$$W_C \times G_B \times (D_{i-1,B,C} - D_{i,B,C}), \tag{21.1}$$

where $D_{i,B,C}$ is the distortion up to the i-th packet, G_B is the energy gain associated for the T-band B and W_C is the weight ($W_C = 1$ by default) for each color component if we want to assign the YUV components with different priorities.

Moreover, in order to enable smooth adjustment of the QoS mapping in the network sublayer, we map the loss impact values obtained in Equation 21.1 into integer RPI values (e.g., 0–63 with 8 bits representation), and also uniformly distribute the integer RPI values into different packets. The detailed procedure of generating RPI values is summarized as follows. First, we sort all the packets according to their calculated loss impact values in a decreasing order. Then, for each packet i, we identify its position Pos_i in the sorted list. Assuming the RPI values ranging from 0 to $M - 1$, we define that the packets with smaller RPI value are more important. For a total number of N packets, we calculate RPI as

$$RPI_i = \lfloor Pos_i/N \times (M - 1) \rfloor. \tag{21.2}$$

Figure 21.3 shows an example of the loss impact results in MSE (Mean Squared Error) for different packets.* From the figure, it can be seen that some packets have extremely large loss impact. These packets are the first packets of each T-band stream, and their loss will

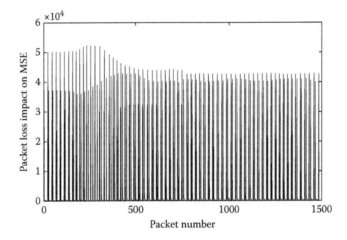

FIGURE 21.3
An example of the loss impact results.

---

* Sequence: Stefan CIF, with two levels of MCTF, bit rate 600 kbps, packet size 500 bytes.

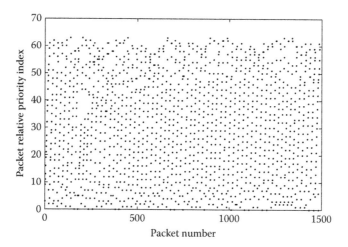

FIGURE 21.4
An example of the RPI value for each packet.

cause the entire T-band undecodable. Figure 21.4 shows the corresponding RPI for each packet with 8-bit quantization, which illustrates the uniform distribution of RPI values.

21.3 Scalable Videos over WLANs Using Single Access Category

In this section, we discuss the transmission of scalable videos over the IEEE 802.11e WLANs using one AC (say, AC2) as recommended in IEEE 802.11e. Our basic idea is to use adaptive retry limit settings to provide an effective ULP for scalable video traffic delivery. It is commonly known that video traffic transmission is sensitive to delay but tolerable to loss. The retry limit setting is one of the main parameters in EDCA governing the packet loss while others such as CW and AIFS merely affect the queuing delay. Controlling the retry limit setting adaptively helps scale the video traffic to match the network available bandwidth. This can be achieved by preemptively dropping low prioritized scalable video traffic locally when necessary via adjusting the retry limit setting.

21.3.1 Retry Limit-Based Unequal Error Protection

Our developed ULP scheme [18] for scalable video traffic delivery over WLANs is based on the control of the retry limit setting in EDCA. In such a design, scalable video packets are divided into a number of groups based on their RPI values (see Section 21.2.3). To avoid implementation complexity, we recommend two groups, namely G_1 and G_2, where G_1 carries higher prioritized video packets than that of G_2. Each group uses a particular retry limit setting for its packet transmissions. Differentiating the retry limit settings ensures ULP among various groups.

We first study the packet loss behavior in EDCA. Based on our earlier performance study on EDCA in Ref. [19], the packet loss probability, $P(r)$ as a function of the retry limit, r, can be expressed as

$$P(r) = p^r, \tag{21.3}$$

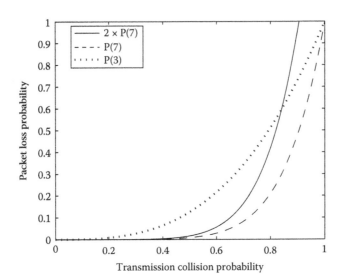

FIGURE 21.5
Relationship between packet loss probability, retry limit, and transmission collision probability.

where p is the *collision probability* of a packet transmission. The quantity p is also commonly used as an indication of network load conditions since networks with higher loads lead to higher collision probabilities during packet transmission.

In Figure 21.5, we plot $P(r)$ versus p for several r values. Consider a simple ULP scheme implementing fixed retry limit setting where traffic of G_1 (resp. G_2) uses the retry limit setting of $r_1 = 7$ (resp. $r_2 = 3$) for packet transmission. The packet loss probabilities for the two traffic streams of G_1 and G_2 under various network load conditions are plotted as $P(7)$ and $P(3)$ respectively in Figure 21.5. A comparison between $P(7)$ and $P(3)$ shows that $P(7) \leq P(3)$ for any p. While this indicates ULP for the two traffic streams, the ULP differentiation between the two traffic streams is, however, uneven across p. Moreover, we see that under a heavy load condition where the collision probability is high, their levels of loss protection become less differentiated.

To overcome the described uneven property of retry limit-based ULP, a practical solution is to vary the retry limit setting based on the network load condition. The solid line presented in Figure 21.5 shows an example of an even ULP that satisfies

$$P_2(r_2) = \min(\alpha P_1(r_1), 1), \tag{21.4}$$

where $P_i(r_i)$ is the packet loss probability for group G_i implementing retry limit of r_i, and α is a constant specifying the differentiation of loss protection between the two groups. To maximize protection for critical salable video traffic transmission, using the typical retry limit setting [20], our design chooses $r_1 = 7$. Consequently, to meet the condition in Equation 21.4, r_2 must vary according to p.

In our design, we consider $\alpha = 2$ which ensures that traffic of G_1 experiences only half the packet loss probability compared to that of the traffic of G_2.* Since r_2 must be a positive

---

* Under very heavy network load conditions, inevitability, traffic of G_2 will experience 100% loss. When this
  upper bound reaches, the packet loss probability of G_1's traffic will purely depend on the network congestion,
  and the target that $\alpha = 2$ cannot be maintained.

integer, our objective is to compute r_2 such that it satisfies Equation 21.4 given a particular p with $r_1 = 7$ and $\alpha = 2$. This involves finding the solution for the following problem

$$\begin{cases} \text{Maximize} & r_2 \\ \text{subject to} & p^{r_2} \le \min(\alpha p^{r_1}, 1), \\ & r_2 \in \mathbb{Z}^+, 0 < p < 1 \end{cases} \qquad (21.5)$$

where $\mathbb{Z}^+ = \{0, 1, \ldots\}$. Given $r_1 = 7$ and $\alpha = 2$, with default values of $r_2 = 7$ (resp. $r_2 = 0$) for $p = 0$ (resp. $p = 1$), solving for r_2 in Equation 21.5 yields

$$r_2 = \begin{cases} 7, & 0 \le p \le 2^{-1} \\ x, & 2^{-\frac{1}{7-x}} < p \le 2^{-\frac{1}{8-x}}, x = 1, 2, \ldots, 6. \\ 0, & 2^{-\frac{1}{7}} < p \le 1 \end{cases} \qquad (21.6)$$

The very last result forms the basis of our design. The retry limit r_2 is adjusted based on the collision probability, p, of the packet transmission of G_1's traffic. Given p, based on Equation 21.6, we determine the retry limit value for G_2's traffic that meets the condition of Equation 21.4.

The value p can be easily obtained by monitoring the statistics of the past packet transmission experiences. According to the definition, p is the ratio of the total number of unsuccessful packet transmission to the total number of processed packets. For each event of either successful packet transmission or packet loss, the value p is updated, leading to the appropriate adjustment of r_2. The adjustment directly follows Equation 21.6. To ensure stability, we apply a simple smoothing strategy that

$$p_u = \beta p_o + (1 - \beta) p_n, \qquad (21.7)$$

where p_u, p_o, and p_n are the updated value, old value, and new value for p, respectively. In our design, we set $\beta = 0.2$ which ensures a quick reaction to the network load condition.

21.3.2 Simulation Results

We use the simple scalable video codec we developed and packetize video data into packets. The size of a packet is limited by a predefined maximum packet size (500 bytes) and each packet is given an RPI value (0–63 with 8 bits representation), as described in Section 21.2.3. We implement our ULP scheme in ns2 simulator for performance evaluation. In Figure 21.6, we illustrate PSNR (Peak Signal-to-Noise Ratio) performance comparison of scalable video delivery over EDCA with (i) no ULP, (ii) fixed retry limit-based ULP, and (iii) adaptive retry limit-based ULP. Details of our simulation setup are described as follows. We consider a single-hop WLAN where all senders and receivers are placed in the same Basic Service Set which operates 11 Mbps of the channel data rate. All generated video traffic is carried by AC2 in EDCA as recommended by the IEEE 802.11e standard. The experiment begins with one sender. For every 4 s of operation, an additional sender is activated in the WLAN, until the number of senders reaches 15. Each sender repeatedly transmits the first 256 frames of the "Table-tennis" CIF sequence with 800 kbps source rate.

Since EDCA is unaware of the scaling property of SVC, the results presented in Figure 21.6 indicate a low support for the number of senders. A simple calculation shows that EDCA supports only five senders before a congestion occurs. This is equivalent to 4-Mbps of

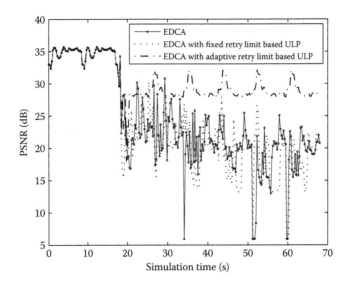

FIGURE 21.6
PSNR performance of scalable video traffic delivery over EDCA and EDCA with various ULP schemes.

goodput. A closer investigation by evaluating the packet loss rates of G_1's and G_2's traffic presented in Figure 21.7 reveals that uniform packet loss appears among video packets of the two groups. This shows that scalable video delivery over EDCA is ineffective without additional handling.

Our second experiment uses fixed retry limit-based ULP on EDCA whereby retry limit settings of $r_1 = 7$ and $r_2 = 3$ apply to the transmission of G_1's and G_2's transmissions, respectively. Interestingly, as reported in Figure 21.6, this setup does not have perceptible PSNR advantage over that of EDCA without ULP. Investigation on packet loss rates plotted in Figure 21.8 also shows similar packet loss behavior with that reported in Figure 21.7. This is because when congestion occurs, p becomes high which causes less differentiation in the

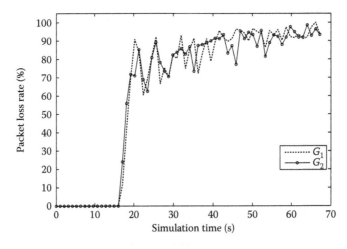

FIGURE 21.7
Packet loss rate of scalable video traffic delivery over EDCA.

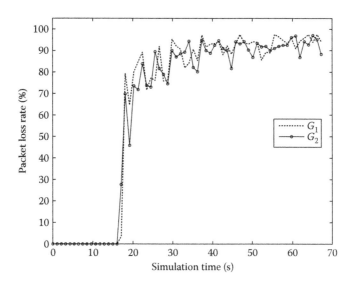

FIGURE 21.8
Packet loss rate of scalable video traffic delivery over EDCA with fixed retry limit-based ULP.

loss protection between the two groups as illustrated in Figure 21.5. This makes the fixed retry limit-based ULP ineffective.

In the third experiment, we implement our ULP scheme on EDCA. G_1's traffic accounts for one third of the total traffic in terms of bytes.* As can be seen in Figure 21.6, a clear PSNR performance benefit is reported. The packet loss rates of the two groups are depicted in Figure 21.9 showing the effectiveness of ULP. The sacrifice of G_2's traffic allows the network

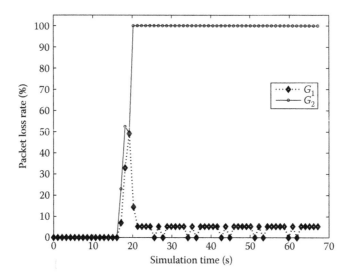

FIGURE 21.9
Packet loss rate of scalable video traffic delivery over EDCA with adaptive retry limit-based ULP.

---

* This selection is based on the observation that the full protection of G_1's traffic gives just above 28 dB of PSNR which is our predefined target for QoS requirement.

to continue to support more senders. A spike in the G_1's packet loss rate indicates a network congestion event and the ability for our scheme to react quickly to the congestion. Since $r_2 \in \mathbb{Z}^+$, for a particular p that pushes r_2 to a value below 1, r_2 will be set to 0 according to Equation 21.6, leading to the dropping of all G_1's packets. Consequently, the packet loss rate of G_2 becomes 100%, which immediately eases the packet loss rate of G_1 creating a spike in G_1's packet loss rate shown in Figure 21.9.

21.4 Scalable Videos over WLANs Using Multiple Access Categories

In this section, we propose a practical cross-layer design for transmitting scalable video over IEEE 802.11e WLANs. Our basic idea is to use two different ACs (AC2 and AC1) for delivering video packets with different priorities. In particular, we propose a macro and a micro rate control schemes at the application layer and the network sublayer, respectively. At the application layer, the macro rate control minimizes the distortion of the video quality given the bandwidth constraint by the optimal bit allocation in our developed SVC. At the network sublayer, the micro rate control performs further rate cut by packet drops when network experiences congestion before the application can react with the macro rate control. Through an adaptive QoS mapping, the micro rate control enforces ULP that preemptively sacrifices video packet with low importance to protect the transmission of those with high importance. This combination ensures optimization of video streaming over the IEEE 802.11e WLAN.

21.4.1 Cross-Layer Design

Our proposed cross-layer design [21] is presented in Figure 21.10. The design uses rate control to optimize quality performance for transmitting the scalable video over IEEE 802.11e EDCA. The application layer and network sublayer cooperate to decide the optimal transmission strategy for the scalable video stream. The common knowledge of the two layers is the available bandwidth which is the main factor dictating the employed transmission strategy at each layer. In brief, based on the detected network available bandwidth at the network sublayer, the application layer decides the encoding strategy and produces video streams that fit into the available bandwidth with the best possible video quality. However, the slower timescales at the application layer makes it difficult to respond to rapid bandwidth variation. To cope with this bandwidth variation at a smaller timescale, especially a sudden downside change in the available bandwidth, using the traffic information passed down from the application layer, the network sublayer reacts with a further transmission rate adjustment subjected to minimum distortion. This design of rate controls, where the application layer and the network sublayer perform *macro* and *micro* rate controls, respectively, provides all timescale rate adaptation for video delivery over EDCA. As shown in Figure 21.10, the macro rate control is achieved by the coordination between the bandwidth estimation and optimal bit allocation components while the micro rate control is achieved by the coordination between the SVC packetization with RPI and adaptive QoS mapping components. The optimal bit allocation component and the SVC packetization components have been described in Section 21.2.3. In the following, we introduce the other two components in the network sublayer.

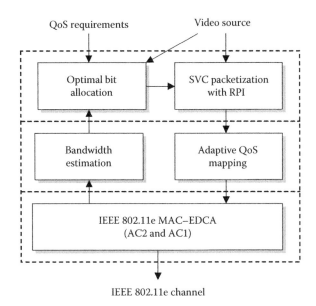

FIGURE 21.10
Block diagram of the proposed cross-layer QoS design.

Bandwidth Estimation in EDCA. Bandwidth estimation is an important component in our cross-layer design. The application layer and network sublayer depend on this estimation to achieve rate control and optimize QoS. The main role of the bandwidth estimation is to measure the network condition during a macro interval and provide estimation of available capacity that a node can access in the next macro interval.

Our design adopts IdleGap [22] to estimate the network bandwidth availability. However, the estimated available bandwidth will be shared among all nodes, thus the bandwidth accessible by each node is usually lower. To ensure that each node only utilizes its share of the available bandwidth, we introduce a simple method described as follows. We first notice that multimedia transmissions usually occurs in a form of continuous streams. The detection of a video packet from a node indicates the participation of multimedia transmissions. Hence we propose that during a macro interval, not only a node estimates the network available bandwidth, $\tilde{R}$, using IdleGap, it also detects and counts the number of different nodes, $\tilde{N}$, that transmit AC2 packets. Using the two quantities, also including itself as a transmitting source, $\tilde{R}/(\tilde{N}+1)$ represents the estimated available bandwidth for a node in the next interval. If a node also transmitted $\tilde{T}$ video traffic during the current macro interval, then by the end of the current interval, the node estimates its available bandwidth, R, for the next interval simply by

$$R = \tilde{T} + \frac{\tilde{R}}{\tilde{N}+1}. \tag{21.8}$$

Adaptive QoS Mapping. Scalable video traffic consists of packets carrying video information of different importance indicated by the RPI. These packet transmissions must receive different error and loss protection to exploit the scalable video benefit. Using the characteristics of the EDCA queues, we design a stream mapping strategy that adaptively maps packets of scalable video traffic onto two EDCA queues with ULP. We choose AC2 as one of the queue since its default purpose is to carry video traffic. AC1, which is used for best effort according to the IEEE 802.11e standard, is another chosen queue. It is used here to carry less important scalable video traffic. Packets mapped onto AC1 are prepared to make sacrifices in forms of packet drops under heavy network load conditions. These sacrifices make room for more important video packets to be transmitted successfully, and hence achieving higher QoS of video transmissions.

The studies of IEEE 802.11e EDCA characteristics in Ref. [21] revealed the service rate ratio of AC2 and AC1 and proposed the service ratio for the mapping to be set to two for ULP. As result, 2/3 of the top prioritized SVC traffic needs to be transmitted via AC2 and the remaining via AC1. In the macro rate control, the application layer controls the source rate based on the bandwidth R estimated by the network layer within an macro interval. Furthermore, the RPI values of video packets is uniformed distributed. With this advantage of knowing the volume characteristics of generated video traffic, it is hence easy for the adaptive QoS mapping scheme to maintain a two-to-one service ratio of AC2 to AC1 for the mapping based on the remaining traffic volume rather than the fixed total traffic volume. The QoS mapping algorithm, according to Ref. [21], is detailed in Algorithm 1.

Algorithm 1 Adaptive QoS Mapping Algorithm at Network Sublayer

// RPI ranges from 0 to $M - 1$.
// v_i: remaining traffic volume of each $RPI = i$ (in bits).
// c_i: expected traffic volume of each $RPI = i$ during a macro interval.
// Λ: Mapping point, packets with $RPI < \Lambda$ is inserted to AC2, otherwise to AC1.

for Each macro interval $T_{interval}$ **do**
　for each i in [0, M-1] **do**
　　$v_i = c_i$
　end for
　$\Lambda = max(p)$ such that $\sum_{0 \le i \le p} v_i = \frac{2}{3} \sum_{0 \le i \le M-1} v_i$

　for Each packet (with $RPI = i$ and size s in bits) departing the network
　　(either successfully transmitted or lost in the network) **do**
　　$v_i = v_i - s$
　　$\Lambda = max(p)$ such that $\sum_{0 \le i \le p} v_i = \frac{2}{3} \sum_{0 \le i \le M-1} v_i$
　end for
end for

21.4.2 Simulation Results

In this section, we present the performance of transmitting the scalable video over the IEEE 802.11e WLANs. In our experiments, we consider the first 256 frames of the "Table-tennis" CIF sequence as our video source. The sequence is repeatedly transmitted for simulation time requirement. Two levels of MCTF decomposition is used during encoding of the video. The maximum video packet size is set to 500 bytes. The RPI ranges between 0 and 63. The macro interval is set to 32/30 s. We take the Y component PSNR as our video quality measurement.

The experiments are conducted using ns2 [23]. The EDCA parameter settings follow the IEEE 802.11e standard. The performance is also compared with DCF and EDCA. The

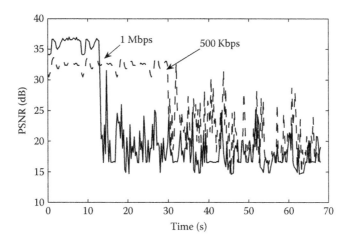

FIGURE 21.11
PSNR of received video for DCF.

former refers to the legacy DCF with setting following the IEEE 802.11b standard [2], and the latter refers to the EDCA using AC2 for video transmission with no cross-layer consideration. Our considered network is a single-hop private infrastructure WLAN. This setup allows us to focus solely on the WLAN. Each experiment starts with one node, labeled as node-0, transmitting a certain rate of video traffic. For every four macro interval duration, a new video node is added to the WLAN, until the number of nodes reaches 15 in the WLAN.

Figures 21.11 and 21.12 plot the PSNR performance measured at node-0 under DCF and EDCA, respectively. Two fixed source rates, which are 500 kbps and 1 Mbps, are used. As can be seen, DCF supports merely three (seven) nodes for the case of 1 Mbps (500 kbps) of video source rate before the PSNR plunges quickly due to network congestion. A simple calculation reveals that the maximum achievable network throughput is no more than 4 Mbps. Similar observation is made for EDCA as indicated in Figure 21.12. The results also

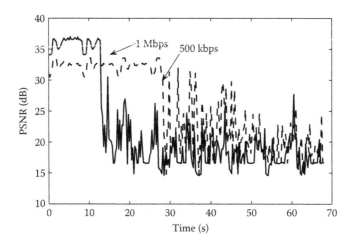

FIGURE 21.12
PSNR of received video for EDCA.

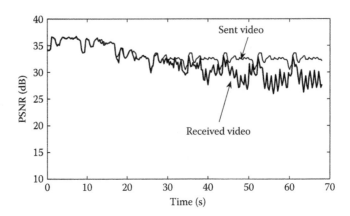

FIGURE 21.13
PSNR for our cross-layer design.

show the maximum achievable network throughput of no more than 4 Mbps. This result reveals that when source rate is left uncontrolled, the network may be pushed to saturation where the protocol operates at a lower throughput level.

For our scheme, since the macro rate control is in place, each node varies its source rate according to the estimated available bandwidth from no less than 500 kbps to no more than 1 Mbps. A node always starts from the maximum specified rate of 1 Mbps. The performance result is presented in Figure 21.13. An immediate comparison among Figures 21.11 and 21.12 show the performance advantage of our cross-layer design. The soft capacity property of our scheme is also clearly illustrated. As can be seen from Figure 21.13, when the network consists of a small number of nodes, each node receives high-quality video streams. As the simulation progresses, the number of nodes increases, all nodes adapt themselves to the changed network condition to accommodate more nodes by reducing the sending rate. Our theoretical study of EDCA capacity indicates that EDCA may only support 11 nodes transmitting 500 kbps each. However, for our proposed scheme, even though we specify our minimum macro rate to 500 kbps, the micro rate control preemptively drops video packets of low importance via AC1 so that the quality of the video streams maintained even at 15 nodes.

21.5 Summary

In this chapter, we have shown two cases of studies that exploit the adaptivity of scalable video and the QoS differentiation provided in the IEEE802.11e MAC layer to achieve optimal video streaming performance with low implementation complexity. In particular, in the first case, we illustrated that the packet loss behavior of EDCA under various retry limit settings, and suggested an adaptive retry limit-based ULP scheme to improve the effectiveness of ULP for scalable video traffic transmissions over the IEEE 802.11e WLANs. In the second case, we have presented the cross-layer design that achieves all timescales rate control for optimizing scalable video transmission over IEEE 802.11e WLANs. Our studies shed light on the issue of developing optimization problems for multimedia networking. With more flexibilities added into media source and more diverse services introduced in different network layers, cross-layer design will play an even significant role.

References

1. *Wireless LAN Medium Access Control (MAC) and Physical Layer (PHY) specifications Amendment 8: Medium Access Control (MAC) Quality of Service Enhancements*, IEEE Std. 802.11e-2005, 2005.
2. *Wireless LAN Medium Access Control (MAC) and Physical Layer (PHY)*, IEEE 802.11 Std., 1999.
3. M. Karczewisz and R. Kurceren, The SP- and SI- frames design for H.264/AVC, *IEEE Trans. Circuits Syst. Video Technol.*, 13(3), 637–644, 2003.
4. A. Vetro, C. Christopoulos, and H. Sun, Video transcoding architectures and techniques: An overview, *IEEE Signal Process. Mag.*, 20(2), 18–29, 2003.
5. J.-R. Ohm, Advances in scalable video coding, in *Proceedings of the IEEE*, 93(1), pp. 42–56, January 2005.
6. P. A. Chou and Z. Miao, Rate-distortion optimized streaming of packetized media, in *Microsoft Research Technical Report*, February 2001.
7. X. K. Yang, C. Zhu, Z. G. Li, G. N. Feng, S. Wu, and N. Ling, A degressive error protection algorithm for MPEG-4 FGS video streaming, in *Proceedings of IEEE ICIP*, pp. 737–740, Rochester, NY, September 2002.
8. C. Liu and S. S. Chen, Providing unequal reliability for transmitting layered videostreams over wireless networks by multi-ARQ schemes, in *Proceedings of IEEE ICIP*, pp. 100–104, Rochester, NY, 1999.
9. J. Shin, J. Kim, and C.-C. J. Kuo, Quality-of-service mapping mechanism for packet video in differentiated services network, *IEEE Trans. Multimedia*, 3(2), 219–231, 2001.
10. H. Zhang, X.-G. Xia, Q. Zhang, and W. Zhu, Precoded OFDM with adaptive vector channel allocation for scalable video transmission over frequency-selective fading channels, *IEEE Trans. Mob. Comput.*, 1, 132–141, 2002.
11. Q. Zhang, Z. Ji, W. Zhu, and Y.-Q. Zhang, Power-minimized bit allocation for video communication over wireless channels, *IEEE Trans. Circuits Syst. Video Technol.*, 12(6), 398–410, 2002.
12. C. I. Podilchuk, N. Jayant, and N. Farvardin, Three-dimensional subband coding of video, *IEEE Trans. Image Process.*, 4(2), 125–139, 1995.
13. J.-R. Ohm, Three-dimensional subband coding with motion compensation, *IEEE Trans. Image Process.*, 3(5), 559–571, 1994.
14. H. Schwarz, D. Marpe, and T. Wiegand, Overview of the scalable video coding extension of the H.264/AVC standard, *IEEE Trans. Circuits Syst. Video Technol.*, 17(9), 1103–1120, 2007.
15. A. Secker and D. Taubman, Motion-compensated highly scalable video compression using an adaptive 3D wavelet transform based on lifting, in *Proceedings International Conference Image Processing*, pp. 1029–1032, 2001.
16. D. S. Taubman and M. W. Marcellin, JPEG2000: Standard for interactive imaging, in *Proceedings of the IEEE*, 90, pp. 1336–1357, August 2002.
17. D. Taubman, High performance scalable image compression with EBCOT, in *Proceedings IEEE International Conference Image Processing*, 9, July 2000.
18. Y. Zhang, Z. Ni, C. H. Foh, and J. Cai, Retry limit based ULP for scalable video transmission over the IEEE 802.11e WLANs, *IEEE Commun. Lett.*, 11(6), 498–500, 2007.
19. J. W. Tantra, C. H. Foh, and A. B. Mnaouer, Throughput and delay analysis of the IEEE 802.11e EDCA saturation, in *Proceedings IEEE International Conference on Communications (ICC'05)*, pp. 3450–3454, 2005.
20. A. Ksentini, M. Naimi, and A. Gueroui, Toward an improvement of H.264 video transmission over IEEE 802.11e through a cross-layer architecture, *IEEE Commun. Mag.*, 44(1), 107–114, 2006.

21. C. H. Foh, Y. Zhang, Z. Ni, J. Cai, and K. N. Ngan, Optimized cross-layer design for scalable video transmission over the ieee 802.11e networks, *IEEE Trans. Circuits Syst. Video Technol.*, 17(12), pp. 1665–1678, 2007.

22. H. K. Lee, V. Hall, K. H. Yum, K. I. Kim, and E. J. Kim, Bandwidth estimation in wireless LANs for multimedia streaming services, in *Proceedings IEEE International Conference on Multimedia and Expo (ICME'06)*, pp. 1181–1184, 2006.

23. *NS*-2 network simulator. Online available: http://www.isi.edu/nsnam/ns/

22

Resource Optimization for Distributed Video Communications

Yifeng He and Ling Guan

CONTENTS

22.1 Introduction

In recent years, we have witnessed a dramatic growth in multimedia applications. Many multimedia applications involve real-time video communications over distributed systems, in which there is no centralized controller. Examples of such distributed systems include peer-to-peer (P2P) networks, wireless *ad hoc* networks, and wireless sensor networks (WSNs).

Internet video streaming applications have become extremely popular. Typically, the video streaming applications are deployed via a client–server architecture. However, the server may be overrun by a large number of users; thus it is nonscalable. It is appealing to apply P2P technology into video streaming applications to relieve the server upload burden by taking advantage of the uplink capability of the peers. P2P live streaming and P2P video-on-demand (VoD) systems are promising in proving real-time video to a large number of users.

Wireless *ad hoc* networks are multihop wireless networks without a preinstalled infrastructure. They can be deployed quickly at conventions, disaster recovery areas, and battlefields. When deployed, nodes cooperate with each other to find routes and relay packets for each other. There is a compelling need for video streaming over wireless *ad hoc* networks. For example, a group of visitors in a museum would like to share their captured video in real time. They can set up a wireless *ad hoc* network using their personal digital assistants, and then multicast the video to each group member.

A wireless sensor network is a system consisting of geographically distributed sensor nodes that communicate with each other over wireless channels. Without the need for a communication infrastructure, the WSN is self-organized and highly dynamic, with each node sensing and forwarding the data [1]. A wireless visual sensor network (WVSN) is a special kind of wireless sensor network with video capture and processing capabilities. A WVSN captures digital visual information about target events or situations and delivers the video data to a remote control unit for further information analysis and decision making. Because of its unique features of rapid deployment, flexibility, low maintenance cost, and robustness, WVSNs have been used in a wide range of important applications, including security monitoring, emergence response, environmental tracking, and health monitoring.

Although each type of distributed system has its own features, they share two common characteristics as follows: (1) Each node only knows about its neighbors, and does not have global knowledge. (2) There is typically no centralized controller, which can coordinate the behaviors of all the nodes. Therefore, a centralized algorithm is not practical for distributed systems. Instead, distributed algorithms are desired.

There are many challenges for video communications over distributed systems. In P2P video applications, high video quality and low startup delay are two major goals. However, it is difficult to achieve these two goals. First, the access bandwidth of the peers is often limited, especially the upload bandwidth. Second, the peers may leave the network at any time which creates a highly unreliable and dynamic network fabric. Third, each user requests the video at a different time, and the video content is delivered from the heterogenous peers. Fourth, it has been observed that users seek frequently to a different position rather than watch the video sequentially in VoD applications, which imposes a great challenge on playback continuity.

In wireless *ad hoc* networks, a wireless link usually has a high transmission error rate because of shadowing, fading, and interferences from other transmitting users. An end-to-end path in *ad hoc* networks has a higher error rate since it is the concatenation of multiple wireless links.

In WVSNs, each video sensor operates under a set of unique resource constraints, including limited energy supply, limited onboard computational capability, and low transmission bandwidth. In conventional WSNs, the power for signal processing at each sensor is very small. In contrast, the video sensor in WVSNs compresses the video before transmission. The compression takes a large amount of power, and raises a greater challenge for maintaining a long network lifetime.

Optimal resource allocation provides an efficient solution to the problem for video communications over distributed systems. The problem in the distributed systems can be formulated into a resource allocation problem with an objective to maximize (or minimize) a performance metric, subject to the resource constraints at each node. The resource constraints include the flow conservation, the limitation of upload and download capacities, the limitation of buffer and storage capacities, the limitation of power supply, and the application-layer requirement (e.g., the distortion requirement). Since there is no centralized controller in the distributed systems, a distributed algorithm is the desired solution in terms of the scalability and the communication overhead.

In this chapter, we first provide a review of recent advances on optimal resource allocation for video communications over some major distributed systems including P2P streaming systems, wireless *ad hoc* networks, and WVSNs. We then provide a case study on network lifetime maximization for WVSNs, in which we present the system models, the problem formulation, the distributed solution, and the simulation results.

The remainder of this chapter is organized as follows. The state of the art of optimal resource allocation for video communications over distributed systems is given in Section 22.2. A case study on network lifetime maximization for WVSNs is presented in Section 22.3. Finally, we discuss the future research directions in Section 22.4 and give the summary in Section 22.5.

22.2 Recent Advances on Optimal Resource Allocation for Video Communications over Distributed Systems

In this section, we present the recent advances on optimal resource allocation for video communications over P2P streaming systems and multihop wireless networks.

22.2.1 Optimal Resource Allocation in P2P Streaming Systems

P2P streaming is one of the major applications on P2P overlay networks. In P2P streaming systems, each peer contributes its resources (e.g., bandwidth, buffer, and storage) to the community. Meanwhile, it retrieves the requested streams from other peers. The collaboration among peers enables each user to receive a high video quality. Depending on the applications, P2P streaming systems can be classified into P2P live streaming systems and P2P VoD streaming systems. In a P2P live streaming system, all the users watch the same segment of the video at the same time, such that they can share the content with each other. In a P2P VoD streaming system, each user requests the video at a different time, and therefore a peer can only request the content from the peer with an earlier playback progress.

22.2.1.1 P2P Live Streaming Systems

The optimization of the scheduling problem in data-driven P2P live streaming systems is studied in Ref. [2]. In data-driven P2P live streaming systems, the video is divided into blocks. There is an exchanging window at each peer containing the blocks that the peer is requesting. Every peer periodically notifies each of its neighbors the availability of the blocks. Then, peers will request their absent blocks from neighbors. Different blocks have different priorities. For instance, the blocks that have fewer suppliers should be requested preemptively such that they can be spread more quickly. Two factors have been considered in the priority definition: rarity factor and emergency factor. The rarity factor is considered first to guarantee the rarest block should be requested in priority, while the emergency factor is used to reduce the probability that the requested blocks miss the playback deadline. The objective of the optimal scheduling problem in Ref. [2] is to maximize the sum of priorities of all requested blocks in the overlay under the bandwidth constraints. To solve the optimal scheduling problem, a heuristic algorithm is developed, which is fully distributed and asynchronous with only local information exchange.

The end-to-end latency in P2P live streaming systems is investigated in Ref. [3], in which the optimization problem is formulated to minimize the average end-to-end streaming latency, subject to the constraints of the peer upload and download bandwidth. A distributed solution to the optimization problem is designed using Lagrangian decomposition and subgradient method. In the distributed algorithm, each peer carries out distributed steps with only local information, and such distributed execution achieves the global optimal objective.

In Ref. [4], Setton et al. propose that distortion-optimized retransmission requests are issued by receiving peers in a tree-based P2P live system to recover the most important missing packets while limiting the induced congestion. After detecting a parent disconnection, a peer can determine a list of missing packets and iteratively select the most important ones to request. This choice should depend on the time at which packets are due, and on the contribution of each packet to the overall video quality.

In the P2P application-layer overlay, multiple content distribution or media streaming sessions are expected to be running concurrently. It is important to provide differentiated services among the sessions. For example, a live streaming session should be handled with a higher priority than a content distribution session of bulk data. The problem of service differentiation across P2P communication sessions is examined in Ref. [5]. In order to select better paths and guarantee the rate for high-priority sessions, a utility function at link l for session s is given by $U_l^s = C_s \log(1 + Q_l x_l^s)$, where Q_l denotes the quality of link l, C_s is the priority of session s, and x_l^s denotes the bandwidth allocation to session s at link l. The optimization problem of service differentiation is formulated to maximize the summation of the utilities among all the sessions with the constraints of the heterogeneous upload and download capacities at each peer. A fully distributed algorithm is developed to solve the optimization problem using the subgradient method and Lagrangian relaxation.

Due to bandwidth constraint, most of the current P2P live streaming systems provide the video at a low bit rate. Users would like to watch the video at a higher quality. However, the P2P network may not be able to deliver the video at a high rate. Therefore, what is the upper bound of the streaming rate in a P2P system becomes an attractive topic. In Ref. [6], Sengupta et al. define the *streaming capacity* as the maximum supported streaming rate that can be received by every receiver, and compute the streaming capacity in a multitree P2P live streaming system. The streaming capacity problem is formulated to maximize the streaming rate r, given by $r = \sum_{t \in \mathbf{T}} y_t$ where $\mathbf{T}$ is the set of all allowed trees for the live session, and y_t is the rate of the substream supported by tree t. The constraint in the streaming capacity problem is the upload capacity at each peer. An iterative combinatorial algorithm is designed to solve the streaming capacity problem approximately.

Most P2P live video systems offer a large number of channels, with users switching frequently between the channels. Wu et al. [7,8] study the performance of multichannel P2P live video systems by using infinite-server queueing network models. In multichannel P2P streaming systems, optimal utilization of cross-channel resources can improve the system performance. In Ref. [9], a view-upload decoupling scheme is prosed to decouple peer downloading from uploading, bringing stability to multichannel systems and enabling cross-channel resource sharing. In Ref. [10], Wang et al. formulate linear programming problems to maximize the sum of the bandwidth satisfaction ratios of all channels for three bandwidth allocation schemes, namely naive bandwidth allocation approach, passive channel-aware bandwidth allocation approach, and active channel-aware bandwidth allocation approach.

22.2.1.2 *P2P VoD Streaming Systems*

In P2P VoD streaming systems, the peers watching the same video can be organized into an overlay based on the playback progress, such that the peer with an earlier playback time can supply streams to the peer with a later playback time. The min-cost flow routing problem in P2P VoD systems is studied in Ref. [11]. The flow routing problem is formulated as a linear program with an objective of minimizing the aggregated link cost, subject to the inequality constraints of the peer upload and download capacities and the equality constraint that each peer has the same playback rate. A distributed auction algorithm is proposed to solve the min-cost flow routing problem.

Most of the existing P2P VoD systems only adopt single-layer video coding [11,12]. If the source rate is high, the peers with a limited or low download bandwidth may not be able to accommodate it. On the other hand, if the source rate is too low, the peers with a higher download bandwidth may underutilize their download bandwidth. Therefore, a scalable source coding is a good solution for P2P VoD applications with heterogeneous bandwidth. In a scalable P2P VoD system, one of the goals is to maximize the aggregate throughput over all the peers. In Ref. [13], the throughput maximization problem in the scalable P2P VoD system is formulated to maximize the aggregate throughput by optimally allocating the link rates, subject to the *source rate constraint* representing that the received rate at any peer is no larger than the maximal source rate, the *download bandwidth constraint*, the *upload bandwidth constraint*, and the *link-forwarding constraint* representing that each outgoing link from peer *i* carries a rate no larger than the total incoming rate into the peer. A distributed algorithm is developed to solve the throughput maximization problem using Lagrangian decomposition and subgradient method.

Depending on the forwarding approach, the existing P2P VoD systems can be classified into two categories: buffer-forwarding architecture [11,14] and storage-forwarding architecture [12,15]. In buffer-forwarding architecture, each peer buffers the recently received content, and forwards it to the child peers. In storage-forwarding architecture, the video content is distributed over the storage of peers. When a peer wants to watch a video, it first looks for the serving peers who are storing the content, and then requests the content from them in parallel. In order to fully utilize the resources in the P2P VoD systems, a hybrid architecture integrating both the buffer-forwarding approach and storage-forwarding approach is proposed [16,17]. The throughput maximization problem in the hybrid P2P VoD architecture is formulated, and a distributed algorithm is designed to maximize the throughput by optimizing the link rates.

Unlike the users in P2P live streaming systems who watch the broadcast video passively, the users in P2P VoD streaming systems may seek to any position that he or she is interested in, as demonstrated in Ref. [18]. The behaviors of random seek place a great challenge to the playback continuity. Zheng et al. propose a prefetching scheme to improve the playback continuity [18]. The user seeking pattern is obtained from the previous seeking statistics. Based on the seeking pattern, the segments that will be prefetched are determined optimally to minimize the expected seeking distance, the deviation between the desired seeking position and the scheduled position. Lloyd algorithm is used to solve the prefetching optimization problem.

The work in Refs. [18] represents the seeking pattern using one-dimensional probability density function (PDF) $P(y)$ where y is the destination segment of a seek. In order to capture the seeking behaviors in a more accurate way, the work in Ref. [19] uses two-dimensional PDF, denoted as $P(x, y)$, representing the probability that a user performs a seek from the start segment x to the destination segment y. In Ref. [20], the concept of *guided seek* is introduced. With the guidance of the two-dimensional PDF, users can perform efficient seeks to the desired positions. The guidance can be obtained from collective seeking statistics of other peers who have watched the same title in the previous and/or concurrent sessions. *Hybrid sketches* are designed to capture the seeking statistics at significantly reduced space and time complexity. Furthermore, an optimal prefetching scheme and an optimal cache replacement scheme are proposed to minimize the expected seeking delay by optimally determining the segments to be prefetched.

In P2P VoD streaming systems, a peer can only request the desired segments from the peer with an earlier playback progress. Therefore, the streaming capacity in P2P VoD streaming systems is different from that in P2P live streaming systems. The streaming capacity in a P2P VoD streaming system is formulated to maximize the streaming rate that can be received by every peer, subject to the limitations of upload and download capacities at each peer [21,22]. In Ref. [23], helpers are introduced in the P2P VoD system and the helper resources are optimized to improve the streaming capacity. The streaming capacity for multichannel P2P VoD systems is studied in Ref. [24], in which the cross-channel resource sharing schemes are proposed to maximize the average streaming capacity.

22.2.2   Optimal Resource Allocation for Video Streaming over Wireless *Ad Hoc* Networks

Depending on the number of simultaneous receivers, video streaming over wireless *ad hoc* networks can be classified into two classes: unicast and multicast [25].

In unicast video streaming, path diversity is very attractive since it provides an effective means to combat transmission errors in wireless *ad hoc* networks. In Ref. [26], a path selection scheme is proposed for multipath video streaming over wireless *ad hoc* networks. The optimization problem is formulated to minimize the concurrent packet drop probability by selecting two optimal paths.

The approach of selecting paths by minimizing packet loss [26] does not guarantee the minimization of the expected video distortion. A distortion-minimized scheme is proposed in Ref. [27] for unicast video streaming over wireless *ad hoc* networks. The received video distortion consists of the encoding distortion and the transmission distortion, which is given by $D_{dec} = D_{enc} + D_{tran}$. The encoding distortion D_{enc} is introduced by the quantization at the encoder, and it can be calculated by $D_{enc} = D_0 + \theta/(R - R_0)$ where R is the rate of the video and (D_0, θ, R_0) are the parameters relative to video encoding. The transmission distortion is caused by the packet loss, and it can be calculated by $D_{tran} = \kappa(P_r + (1 - P_r)e^{-(C'-R)T/L'})$ where P_r is the random packet loss due to transmission errors, and (κ, C', T, L') are positive parameters which are discussed in Ref. [27]. At lower rates, reconstructed video quality is limited by coarse quantization, whereas at high rates, the video stream will cause more network congestion and therefore will lead to longer packet delays. These, in turn, translate into higher loss rates, hence reduced video quality. Therefore, the optimal video rate can be obtained by $R^* = \arg \min D_{dec}$.

The combination of multiple description coding (MDC) and multipath transport has shown the superiority in terms of error resilience [28]. Optimization of the routing scheme for a single unicast video session in wireless *ad hoc* networks is examined in Ref. [29]. The video is encoded into two MDC descriptions. The optimization problem is formulated to minimize the expected video distortion by finding the optimal path for each description. A genetic algorithm (GA)-based solution is provided to solve the optimization problem.

The work in Refs. [26,27,29] studies the unicast video streaming from a single sender to a single receiver. The optimization of multipath routing scheme for a video streaming session from two senders to a single receiver is presented in Ref. [30]. The optimization problem is formulated to minimize the expected distortion of the MDC video by selecting the optimal sender and the optimal path from the sender to the receiver for each MDC description.

Joint optimization of the source rate and the routing scheme for unicast video streaming is studied in Ref. [31]. The received video distortion is given by $D_{\text{dec}} = D_0 + \theta_0/((s_r - \sum_{l \in L} p_l x_l) + \phi_0)$, where s_r is the source rate, x_l is the link rate at link l, p_l is the packet loss rate at link l, and (D_0, θ_0, ϕ_0) are the predetermined parameters. The optimization problem is stated as follows. Given a wireless *ad hoc* network and a unicast session from a sender to a receiver, to minimize the received video distortion D_{dec} by optimally determining the source rate s_r and the link rate x_l at link l, subject to the network flow constraint and the link capacity constraint. A distributed algorithm using dual decomposition is developed to solve the optimization problem.

Different from the single video session, multiple concurrent video sessions in a wireless *ad hoc* network have to compete for limited network resources. Such interactions make the performance of an individual flow coupled with that of other flows. Joint optimization of the source rate and the routing scheme for multiple concurrent video sessions in wireless *ad hoc* networks is investigated in Refs. [32,33]. In Ref. [32], the optimization problem is formulated to minimize the sum of the average distortions of all concurrent video sessions by jointly optimizing the source rate and the routing path for each video session. A greedy heuristic algorithm is proposed to find the near-optimal solution to the optimization problem. In Ref. [33], the objective of the optimization problem is to minimize the Lagrangian sum of the total video distortion and the overall network congestion. A distributed algorithm is proposed to solve the optimization problem.

Multicast video streaming over wireless *ad hoc* networks is bandwidth efficient compared to multiple unicast sessions. Multiple-tree routing algorithms are proposed to explore the path diversity for each receiver. Two typical multiple-tree video multicast schemes in wireless *ad hoc* networks are presented in Refs. [34,35], respectively.

In Ref. [34], Zakhor and Wei proposed a multiple-tree construction protocol that builds two nearly disjoint trees simultaneously in a distributed manner. However, the trees are built based on the network-layer metrics, and the application performance has not yet been optimized.

In Ref. [35], two multicast trees are constructed to deliver two MDC descriptions. Each description is layered encoded to meet the heterogeneous capacity of the receivers. The optimization problem is formulated to minimize the total video distortion of all receivers by constructing two optimal trees. A GA-based heuristic is proposed to solve the optimization problem.

A joint optimization of the source rate and the routing scheme for multicast video streaming over wireless *ad hoc* networks is presented in Ref. [36]. A joint optimization of the source rate, the routing scheme, and the power allocation for multicast video streaming is studied in Ref. [37]. In Ref. [37], a prioritized coding scheme, a combination of the layered source coding and the network coding [38], is employed to enable the heterogeneous receivers to reconstruct the video at different quality levels. The network coding eliminates the delivery redundancy. Therefore, a larger throughput at a receiver can lead to a smaller distortion. The optimization problem is stated as follows. Given a wireless *ad hoc* network and a multicast session from a sender to multiple receivers, to maximize the aggregate throughput by optimally determining the source rate, the routing scheme and the power allocation, subject to the constraint of flow conservation, the link capacity constraint, and the power constraint. A distributed algorithm using hierarchical dual decompositions is developed to solve the optimization problem.

22.2.3 Optimal Resource Allocation for WVSNs

Sensor nodes are typically battery powered, and battery replacement is infrequent or even impossible in many sensing applications. Hence, tremendous research efforts in WSNs have been focused on energy conservation. One aspect of this research is to maximize the network lifetime, which is typically defined as the time from the start of the network until the death of the first node [39,41]. In Ref. [39], the network lifetime maximization problem is formulated into a convex optimization problem [40]. A distributed algorithm using Lagrangian decomposition is developed to maximize the network lifetime by optimizing the routing scheme.

The performance of wireless sensor network applications is typically a function of the amount of data collected by the individual sensors. There is an inherent trade-off in simultaneously maximizing the network lifetime and the application performance. Such trade-off is investigated in Refs. [41,42]. In Ref. [41], the application performance is characterized by a network utility function, which is strictly concave and increasing with respect to the data rate. The network lifetime is the minimum of the node lifetime, which is characterized by a lifetime-penalty function for each node. The optimization problem is formulated to maximize the Lagrangian sum of the network utility and the network lifetime by jointly optimizing the source rates and the routing scheme.

The algorithms [39] that maximize the network lifetime in conventional wireless *ad hoc* networks cannot be applied directly to the WVSNs, since they omit the encoding power consumption at the sensor nodes. The network lifetime maximization for WVSNs is studied in Refs. [43,44]. In Ref. [44], the relationship among the encoding power, the source rate, and the encoding distortion is characterized by $d_{sh} = \sigma^2 e^{-\gamma \cdot R_h \cdot P_{sh}^{2/3}}$ [45], where d_{sh} is the encoding distortion of the video encoded at sensor h, R_h is the source rate generated at sensor h, P_{sh} is the encoding power consumption at sensor h, σ^2 is the average input variance, and γ is the encoding efficiency coefficient. The problem of maximum network lifetime is stated as follows. Given the topology of a static WVSN and the initial energy at each node, to maximize the network lifetime by jointly optimizing the source rate and the encoding power at each video sensor, and the link rates for each session, subject to the constraint of flow conservation and the requirement of the collected video quality. A fully distributed algorithm using the properties of Lagrangian duality is developed to solve the network lifetime maximization problem.

The ultimate goal of the WVSN is to utilize its limited resources to collect as much visual information as possible. A metric, called *accumulative visual information* (AVI), is introduced in Ref. [46] to measure the amount of visual information collected by the video sensor. The AVI is a function of the bit rate R and the encoding power P, which is given by $I = f(R, P)$. The optimal resource allocation problem in WVSN can be formulated to maximize the AVI by optimizing the allocation of the bit rate and encoding power for each frame of the video.

The power consumption P_0 at a video sensor mainly consists of the encoding power P_s and the transmission power P_t [47], which is given by $P_0 = P_s + P_t$. If the encoding power P_s is decreased, the distortion d of the video is increased, that is, $P_s \downarrow \Rightarrow d \uparrow$. On the other hand, since the total power consumption P_0 is fixed, the transmission power P_t will be decreased if the encoding power P_s is increased. This implies that less bits can be transmitted because the transmission energy is proportional to the number of bits to be transmitted. Therefore, $P_s \uparrow \Rightarrow d \uparrow$. In other words, when the encoding power P_s goes too low or too high, the distortion d will become large. The distortion can be minimized by optimally allocating the encoding power P_s and the transmission power P_t [47].

22.3 Case Study: Network Lifetime Maximization for WVSNs

In this section, we present a distributed algorithm to optimize the resources at each sensor to maximize the network lifetime of the wireless sensor network [44].

The topology of the WVSN is illustrated in Figure 22.1. Each video sensor has a camera component to capture the video, and a processing component to compress the video. The video sensors construct a mesh network topology, and they communicate with each other within a limited transmission range. The video captured and encoded at each sensor is transmitted to a sink for further analysis and decision making.

We next present a distributed algorithm to maximize the network lifetime by jointly optimizing the source rates, the encoding powers, and the routing scheme.

22.3.1 System Models

A static WVSN can be modeled as a directed graph $\mathbf{G} = (\mathbf{N}, \mathbf{L})$, where $\mathbf{N}$ is the set of nodes and $\mathbf{L}$ is the set of directed wireless links. Among the node set $\mathbf{N}$, one node belongs to the sink set $\mathbf{T}$, while the other nodes belong to video sensor set $\mathbf{V}$. Thus, $\mathbf{N} = \mathbf{V} \bigcup \mathbf{T}$.

The relationship between a WVSN node and its connected links is represented with a node-link incidence matrix $\mathbf{A}$, whose elements are defined as

$$a_{il} = \begin{cases} 1, & \text{if link } l \text{ is an outgoing link from node } i, \\ -1, & \text{if link } l \text{ is an incoming link into node } i, \\ 0, & \text{otherwise.} \end{cases} \quad (22.1)$$

The relationship between a WVSN node and its outgoing links is represented with a matrix $\mathbf{A}^+$, whose elements are given by

$$a_{il}^+ = \begin{cases} 1, & \text{if link } l \text{ is an outgoing link from node } i, \\ 0, & \text{otherwise.} \end{cases} \quad (22.2)$$

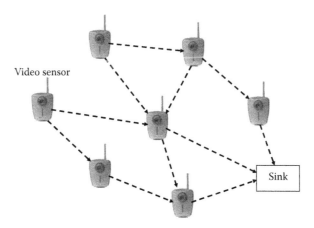

FIGURE 22.1

Illustration of a WVSN. (From Y. He, I. Lee, and L. Guan, Distributed algorithms for network lifetime maximization in wireless visual sensor networks, *IEEE Transactions on Circuits and systems for Video Technology*, 19(5), 704–718, 2009. © (2009) IEEE. With permission.)

The relationship between a WVSN node and its incoming links is represented with a matrix $\mathbf{A}^-$, whose elements are given by

$$a_{il}^- = \begin{cases} 1, & \text{if link } l \text{ is an incoming link into node } i, \\ 0, & \text{otherwise.} \end{cases} \tag{22.3}$$

Hence, $\mathbf{A} = \mathbf{A}^+ - \mathbf{A}^-$.

We assume that a standard medium access control protocol is applied to resolve the link interference problem. Sensor node $h (\forall h \in \mathbf{V})$ can capture and encode the video, and then generate data traffic with a source rate R_h ($R_h = 0$ if sensor h is not on the capture and encoding mode). We define session h as the traffic flow originating from the sensor node h to the sink. For each session, the flow conservation law holds at each node:

$$\sum_{l \in \mathbf{L}} a_{il} x_{hl} = \eta_{hi}, \quad \forall h \in \mathbf{V}, \forall i \in \mathbf{N}, \tag{22.4}$$

where x_{hl} is the flow rate at link l for session h, and η_{hi} is defined as

$$\eta_{hi} = \begin{cases} R_h, & \text{if } i \text{ is the source node of session } h, \\ -R_h, & \text{if } i \text{ is the sink of session } h, \\ 0, & \text{otherwise.} \end{cases} \tag{22.5}$$

Video sensor h captures and encodes the video before it transmits the traffic to its downstream node. The distortion of the compressed video depends on the source rate R_h and the *encoding power consumption* P_{sh}. According to the power-rate-distortion analytical model in Ref. [45], the encoding distortion is computed by

$$d_{sh} = \sigma^2 e^{-\gamma \cdot R_h \cdot P_{sh}^{2/3}}, \tag{22.6}$$

where σ^2 is the average input variance and γ is the encoding efficiency coefficient.

Based on a power consumption model in WSNs [48], the *transmission power consumption* at link l can be formulated as

$$P_{tl} = c_l^s y_l \text{ and } c_l^s = \alpha + \beta d_l^{n_p}, \tag{22.7}$$

where y_l is the aggregate rate transmitted through link l, c_l^s is the transmission energy consumption cost of link l, α is the energy cost of the transmit electronics, β is a coefficient term corresponding to the energy cost of the transmit amplifier, d_l is the distance between the transmitter and the receiver along link l, and n_p is the path-loss exponent [49].

The *reception power consumption* at a node i can be formulated as

$$P_{ri} = c^r \sum_{l \in \mathbf{L}} a_{il}^- y_l, \tag{22.8}$$

where c^r is the energy consumption cost of the radio receiver, and $\sum_{l \in \mathbf{L}} a_{il}^- y_l$ represents the aggregate rate received at node i.

In general, the total power dissipation at node $i (i \in \mathbf{N})$ consists of *the encoding power consumption, the transmission power consumption,* and *the reception power consumption.* It is given by

$$P_i = P_{si} + P_{ti} + P_{ri} = P_{si} + \sum_{l \in \mathbf{L}} a_{il}^+ (c_l^s y_l) + c^r \sum_{l \in \mathbf{L}} a_{il}^- y_l, \tag{22.9}$$

where $P_{si} = 0$, if i is not in the video sensor set $\mathbf{V}$.

The network lifetime is defined as the *minimum node lifetime* [39,42]. In a WVSN, sensor node i has an initial energy B_i, and the lifetime of node i is given by $T_i = B_i/P_i, \forall i \in \mathbf{N}$. Then the network lifetime is given by $T_{\text{net}} = \min_{i \in \mathbf{N}}\{T_i\} = \min_{i \in \mathbf{N}}\{B_i/P_i\}$. In the following, we will use the minimum node lifetime as the network lifetime.

22.3.2   Problem Formulation

We assume that the transmission power is properly chosen such that the bit error rate at the receiver can be neglected, and the transmission loss is 0. In a WVSN without transmission error, the total distortion is equal to the encoding distortion since the transmission distortion is 0. In this case, the received video at the sink is measured by the encoding distortion in mean squared error (MSE).

We state the network lifetime maximization problem as follows. Given the topology of a static WVSN, and the initial energy at each node, the network lifetime maximization problem is to maximize the network lifetime by jointly optimizing the source rate and the encoding power at each video sensor, and the link rates of each session, subject to the requirement of the collected video quality. Mathematically, the problem can be formulated as follows:

$$
\begin{aligned}
\text{maximize}_{(\mathbf{R},\mathbf{x},\mathbf{P_s})} \quad & T_{\text{net}} \\
\text{subject to} \quad & \sum_{l \in \mathbf{L}} a_{il} x_{hl} = \eta_{hi}, \forall h \in \mathbf{V}, \forall i \in \mathbf{N}, \\
& \sum_{h \in \mathbf{V}} x_{hl} = y_l, \forall l \in \mathbf{L}, \\
& \sigma^2 e^{-\gamma \cdot R_h \cdot P_{sh}^{2/3}} \le D_h, \forall h \in \mathbf{V}, \\
& T_{\text{net}} = \min_{i \in \mathbf{N}}\{T_i\} \\
& = \min_{i \in \mathbf{N}} \left\{ \frac{B_i}{P_{si} + \sum_{l \in \mathbf{L}} a_{il}^+ (c_l^s y_l) + c^r \sum_{l \in \mathbf{L}} a_{il}^- y_l} \right\}, \\
& x_{hl} \ge 0, \quad \forall h \in \mathbf{V}, \forall l \in \mathbf{L}, \\
& R_h \ge 0, \quad \forall h \in \mathbf{V}, \\
& P_{sh} > 0, \quad \forall h \in \mathbf{V},
\end{aligned}
\tag{22.10}
$$

where B_i is the initial energy at node i, x_{hl} is the link rate at link l for session h, y_l is the aggregate flow rate at link l, R_h is the source rate of session h, P_{sh} is the encoding power at the source node of session h, and D_h is the upper bound of the encoding distortion in MSE for session h. The first constraint $\sum_{l \in \mathbf{L}} a_{il} x_{hl} = \eta_{hi}$ represents the flow conservation at each node for each session, the second constraint $\sum_{h \in \mathbf{V}} x_{hl} = y_l$ represents that the aggregate flow rate y_l at a link is the summation of the link rates of all the sessions at this link, and the third constraint $\sigma^2 e^{-\gamma \cdot R_h \cdot P_{sh}^{2/3}} \leq D_h$ represents that the encoding distortion for session h is required to be no larger than the corresponding upper bound D_h.

We replace the variable T_{net} using $q = 1/T_{\text{net}}$. Since $T_{\text{net}} \leq B_i/(P_{si} + \sum_{l \in \mathbf{L}} a_{il}^+ (c_l^s y_l) + c^r \sum_{l \in \mathbf{L}} a_{il}^- y_l), \forall i \in \mathbf{N}$, we have $q B_i \geq P_{si} + \sum_{l \in \mathbf{L}} a_{il}^+ (c_l^s y_l) + c^r \sum_{l \in \mathbf{L}} a_{il}^- y_l, \forall i \in \mathbf{N}$. Then the problem (22.10) is converted to an equivalent formulation as follows:

$$
\begin{aligned}
\text{minimize}_{(\mathbf{R},\mathbf{x},\mathbf{P_s})} \quad & q \\
\text{subject to} \quad & \sum_{l \in \mathbf{L}} a_{il} x_{hl} = \eta_{hi}, \forall h \in \mathbf{V}, \forall i \in \mathbf{N}, \\
& \sum_{h \in \mathbf{V}} x_{hl} = y_l, \forall l \in \mathbf{L}, \\
& \log(\sigma^2/D_h)/(\gamma P_{sh}^{2/3}) \leq R_h, \forall h \in \mathbf{V}, \\
& P_{si} + \sum_{l \in \mathbf{L}} a_{il}^+ (c_l^s y_l) + c^r \sum_{l \in \mathbf{L}} a_{il}^- y_l \\
& \qquad \leq q B_i, \quad \forall i \in \mathbf{N}, \\
& x_{hl} \geq 0, \quad \forall h \in \mathbf{V}, \forall l \in \mathbf{L}, \\
& R_h \geq 0, \quad \forall h \in \mathbf{V}, \\
& P_{sh} > 0, \quad \forall h \in \mathbf{V}.
\end{aligned}
\tag{22.11}
$$

In problem (22.11), we minimize the variable q by jointly optimizing the source rate and the encoding power at each video sensor, and the link rate at each link for each session. However, the optimization problem (22.11) cannot be solved in a fully distributed manner, because the value of q needs to be broadcast to each node. In order to develop a fully distributed algorithm, we introduce an auxiliary variable $q_i(\forall i \in \mathbf{N})$ for node i. In problem (22.11), each node maintains a common q, which is equivalent to the case that node i maintains an individual q_i while $q_i = q_j(\forall i, j \in \mathbf{N})$. The equality constraint $q_i = q_j(\forall i, j \in \mathbf{N})$ can be represented in another way $\sum_{i \in \mathbf{N}} a_{il} q_i = 0(\forall l \in \mathbf{L})$.

Since $q = (1/T_{\text{net}}) > 0$, the objective that minimizes q is equivalent to the one that minimizes $|\mathbf{N}|q^2$, where $|\mathbf{N}|$ is the number of nodes in the WVSN. By using auxiliary variable $q_i(\forall i \in \mathbf{N})$ to replace the common q, the objective function $|\mathbf{N}|q^2$ is equal to $\sum_{i \in \mathbf{N}} q_i^2$ under the equality constraint $q_i = q_j(\forall i, j \in \mathbf{N})$, which can be expressed in another way $\sum_{i \in \mathbf{N}} a_{il} q_i = 0(\forall l \in \mathbf{L})$. Therefore, the optimization problem (22.11) is converted to the following equivalent formulation:

$$
\begin{aligned}
\text{minimize}_{(\mathbf{R},\mathbf{x},\mathbf{P_s},\mathbf{q})} \quad & \sum_{i \in \mathbf{N}} q_i^2 \\
\text{subject to} \quad & \sum_{l \in \mathbf{L}} a_{il} x_{hl} = \eta_{hi}, \forall h \in \mathbf{V}, \forall i \in \mathbf{N}, \\
& \log(\sigma^2/D_h)/(\gamma P_{sh}^{2/3}) \leq R_h, \forall h \in \mathbf{V}, \\
& P_{si} + \sum_{l \in \mathbf{L}} a_{il}^+ \left(c_l^s \sum_{h \in \mathbf{V}} x_{hl} \right) \\
& \qquad + c^r \sum_{l \in \mathbf{L}} a_{il}^- \sum_{h \in \mathbf{V}} x_{hl} \leq q_i B_i, \forall i \in \mathbf{N},
\end{aligned}
\tag{22.12}
$$

$$\sum_{i \in \mathbf{N}} a_{il} q_i = 0, \quad \forall l \in \mathbf{L},$$

$$x_{hl} \geq 0, \forall h \in \mathbf{V}, \quad \forall l \in \mathbf{L},$$

$$q_i > 0, \quad \forall i \in \mathbf{N},$$

$$R_h \geq 0, \quad \forall h \in \mathbf{V},$$

$$P_{sh} > 0, \quad \forall h \in \mathbf{V}.$$

We will use the primal-dual method [50] to develop a distributed algorithm for the optimization problem (22.12). However, the objective function in problem (22.12) is not strictly convex with respect to variables $(\mathbf{R}, \mathbf{x})$. Therefore, the corresponding dual function is nondifferentiable, and the optimal values of $(\mathbf{R}, \mathbf{x}, \mathbf{P_s}, \mathbf{q})$ are not immediately available. We add a quadratic regularization term for each link rate variable and each source rate variable to make the objective function strictly convex. Then the optimization problem (22.12) is approximated to the following:

$$\text{minimize}_{(\mathbf{R},\mathbf{x},\mathbf{P_s},\mathbf{q})} \sum_{i \in \mathbf{N}} q_i^2 + \sum_{h \in \mathbf{V}} \sum_{l \in \mathbf{L}} \delta x_{hl}^2 + \sum_{h \in \mathbf{V}} \delta R_h^2$$

$$\text{subject to} \qquad \text{the same constraints as in (22.12)}$$

(22.13)

where $\delta(\delta > 0)$ is the regularization factor. When the regularization factor δ is close to 0, the objective value in problem (22.13) will be close to the objective value in problem (22.12).

22.3.3 Distributed Solution

In problem (22.13), the objective function is strictly convex, the inequality constraint functions are convex, and the equality constraint functions are linear. Therefore, it is a convex optimization problem [40]. In addition, there exists a strictly feasible solution that satisfies all the inequality constraints in the problem (22.13). In other words, the Slater's condition is satisfied, and the strong duality holds [40]. Thus, we can obtain the optimal solution indirectly by first solving the corresponding dual problem [40,51]. The dual-based approach leads to an efficient distributed algorithm [50].

We introduce dual variables $u_{hi}(\forall h \in \mathbf{V}, \forall i \in \mathbf{N})$, $v_h(\forall h \in \mathbf{V})$, $\lambda_i(\forall i \in \mathbf{N})$, and $w_l(\forall l \in \mathbf{L})$ to formulate the Lagrangian corresponding to the primal problem (22.13) as below:

$$L(\mathbf{R}, \mathbf{x}, \mathbf{P_s}, \mathbf{q}, \mathbf{u}, \mathbf{v}, \mathbf{l}, \mathbf{w})$$

$$= \sum_{i \in \mathbf{N}} q_i^2 + \sum_{h \in \mathbf{V}} \sum_{l \in \mathbf{L}} \delta x_{hl}^2 + \sum_{h \in \mathbf{V}} \delta R_h^2 + \sum_{h \in \mathbf{V}} \sum_{i \in \mathbf{N}} u_{hi} \left(\sum_{l \in \mathbf{L}} a_{il} x_{hl} - \eta_{hi} \right)$$

$$+ \sum_{h \in \mathbf{V}} v_h (\log(\sigma^2/D_h)/(\gamma P_{sh}^{2/3}) - R_h) + \sum_{i \in \mathbf{N}} \lambda_i \left(P_{si} + \sum_{l \in \mathbf{L}} a_{il}^+ (c_l^s \sum_{h \in \mathbf{V}} x_{hl}) \right)$$

$$+ c^r \sum_{l \in \mathbf{L}} a_{il}^- \sum_{h \in \mathbf{V}} x_{hl} - q_i B_i) + \sum_{l \in \mathbf{L}} w_l \sum_{i \in \mathbf{N}} a_{il} q_i$$

(22.14)

$$
= \sum_{i \in \mathbf{N}} \left(q_i^2 + q_i \left(\sum_{l \in \mathbf{L}} a_{il} w_l - \lambda_i B_i \right) \right) + \sum_{h \in \mathbf{V}} (v_h \log(\sigma^2/D_h)/(\gamma P_{sh}^{2/3}) + \lambda_h P_{sh})
$$

$$
+ \sum_{h \in \mathbf{V}} \sum_{l \in \mathbf{L}} \left(\delta x_{hl}^2 + x_{hl} \left(c_l^s \sum_{i \in \mathbf{N}} \lambda_i a_{il}^+ c^r \sum_{i \in \mathbf{N}} \lambda_i a_{il}^- + \sum_{i \in \mathbf{N}} u_{hi} a_{il} \right) \right)
$$

$$
+ \sum_{h \in \mathbf{V}} \left(\delta R_h^2 - v_h R_h - \sum_{i \in \mathbf{N}} u_{hi} \eta_{hi} \right).
$$

The Lagrange dual function $G(\mathbf{u}, \mathbf{v}, \boldsymbol{\lambda}, \mathbf{w})$ is the minimum value of the Lagrangian over primal variables $(\mathbf{R}, \mathbf{x}, \mathbf{P_s}, \mathbf{q})$, and it is given by

$$
G(\mathbf{u}, \mathbf{v}, \boldsymbol{\lambda}, \mathbf{w}) = \min_{(\mathbf{R}, \mathbf{x}, \mathbf{P_s}, \mathbf{q})} \{ L(\mathbf{R}, \mathbf{x}, \mathbf{P_s}, \mathbf{q}, \mathbf{u}, \mathbf{v}, \boldsymbol{\lambda}, \mathbf{w}) \}. \tag{22.15}
$$

The Lagrange dual problem corresponding to the primal problem (22.13) is then given by

$$
\begin{aligned}
\text{maximize}_{(\mathbf{u}, \mathbf{v}, \boldsymbol{\lambda}, \mathbf{w})} \quad & G(\mathbf{u}, \mathbf{v}, \boldsymbol{\lambda}, \mathbf{w}) \\
\text{subject to} \quad & v_h \geq 0, && \forall h \in \mathbf{V}, \\
& \lambda_i \geq 0, && \forall i \in \mathbf{N}.
\end{aligned} \tag{22.16}
$$

The objective function in the Lagrange dual problem is a concave and differentiable function. Therefore, we can use the subgradient method [52] to find the maximum of the objective function. If the step size $\theta^{(k)}(\theta^{(k)} > 0)$ at the kth iteration follows a nonsummable diminishing rule:

$$
\lim_{k \to \infty} \theta^{(k)} = 0, \quad \sum_{k=1}^{\infty} \theta^{(k)} = \infty, \tag{22.17}
$$

the subgradient method is guaranteed to converge to the optimal value [52].

With the subgradient method, the dual variable at the $(k+1)$th iteration is updated by

$$
u_{hi}^{(k+1)} = u_{hi}^{(k)} - \theta^{(k)} \left(\eta_{hi}^{(k)} - \sum_{l \in \mathbf{L}} a_{il} x_{hl}^{(k)} \right), \quad \forall h \in \mathbf{V}, \quad \forall i \in \mathbf{N}, \tag{22.18}
$$

$$
v_h^{(k+1)} = \max \left\{ 0, v_h^{(k)} - \theta^{(k)} \left(R_h^{(k)} - \frac{\log(\sigma^2/D_h)}{\gamma(P_{sh}^{(k)})^{2/3}} \right) \right\}, \quad \forall h \in \mathbf{V}, \tag{22.19}
$$

$$
\lambda_i^{(k+1)} = \max \Bigg\{ 0, \lambda_i^{(k)} - \theta^{(k)} \Bigg(q_i^{(k)} B_i - \sum_{l \in \mathbf{L}} a_{il}^+ c_l^s \sum_{h \in \mathbf{V}} x_{hl}^{(k)}
$$

$$
- c^r \sum_{l \in \mathbf{L}} a_{il}^- \sum_{h \in \mathbf{V}} x_{hl}^{(k)} - P_{si}^{(k)} \Bigg) \Bigg\}, \quad \forall i \in \mathbf{N}, \tag{22.20}
$$

$$
w_l^{(k+1)} = w_l^{(k)} + \theta^{(k)} \sum_{i \in \mathbf{N}} a_{il} q_i^{(k)}, \quad \forall l \in \mathbf{L}. \tag{22.21}
$$

The step size we use in our algorithm is $\theta^{(k)} = \omega/\sqrt{k}$, where $\omega > 0$. It follows non-summable diminishing rule.

Given the dual variables at the kth iteration, we calculate the primal variables as follows:

1. q_i at node i:

$$q_i^{(k)} = \arg\min_{(q_i>0)} \left(q_i^2 + q_i \left(\sum_{l\in\mathbf{L}} a_{il} w_l^{(k)} - \lambda_i^{(k)} B_i \right) \right), \quad \forall i \in \mathbf{N}. \tag{22.22}$$

2. The encoding power P_{sh} at video sensor h:

$$P_{sh}^{(k)} = \arg\min_{(P_{sh}>0)} \left(v_h^{(k)} \frac{\log(\sigma^2/D_h)}{\gamma P_{sh}^{2/3}} + \lambda_h^{(k)} P_{sh} \right), \quad \forall h \in \mathbf{V}. \tag{22.23}$$

3. The source rate R_h at video sensor h:

$$R_h^{(k)} = \arg\min_{(R_h\geq 0)} \left(\delta R_h^2 - v_h^{(k)} R_h - \sum_{i\in\mathbf{N}} u_{hi}^{(k)} \eta_{hi} \right), \quad \forall h \in \mathbf{V}. \tag{22.24}$$

4. The link rate x_{hl} at link l for session h:

$$x_{hl}^{(k)} = \arg\min_{(x_{hl}\geq 0)} \left(\delta x_{hl}^2 + x_{hl} \left(c_l^s \sum_{i\in\mathbf{N}} \lambda_i^{(k)} a_{il}^+ \right. \right.$$
$$\left. \left. + c^r \sum_{i\in\mathbf{N}} \lambda_i^{(k)} a_{il}^- + \sum_{i\in\mathbf{N}} u_{hi}^{(k)} a_{il} \right) \right), \quad \forall h \in \mathbf{V}, \forall l \in \mathbf{L}. \tag{22.25}$$

The above algorithm is fully distributed. Each node computes the primal variables: (1) the auxiliary variable q_i, (2) the encoding power P_{sh}, (3) the source rate R_h, and (4) the outgoing link rate x_{hl} from this node, using the dual variables of itself and its neighboring nodes. When the dual variables converge, the primal variables also converge to their optimal values. The message exchange is limited within the one-hop neighbors; thus, the communication overhead is greatly reduced.

22.3.4 Simulation Results

We consider a static WVSN with 10 nodes randomly located in a square region of 50 m × 50 m. Node 10 is the sink, and the other nodes are video sensors. Each node has a maximum transmission range of 30 m. Video sequence "Foreman" in common intermediate format (CIF) is used in the simulations. The values of the model parameters are chosen as follows. Average input variance of the video in MSE is 3500, the encoding efficiency coefficient is 55.54 $W^{3/2}$/Mbps, the energy cost of the transmit electronics is 0.5 J/Mb, the coefficient term of the transmit amplifier is 1.3×10^{-8} J/Mb/m^4, the path-loss exponent is 4, the energy consumption cost of the radio receiver is 0.5 J/Mb, the initial energy at node i is 5.0 MJ, the regularization factor is 0.2, the step size parameter is 0.15, and the upper bound of the encoding distortion D_h in MSE is set to 100 by default, corresponding to a peak signal-to-noise ratio (PSNR) of 28.13 dB.

The proposed algorithm jointly optimizes both the source (e.g., the source rate and the encoding power) and the routing scheme. We compare the proposed algorithm to two other schemes: (1) the routing-optimized scheme (ROS) proposed in Ref. [39], in which the routing scheme is optimized, while the encoding power at each sensor node is fixed at the same value; and (2) video-based distributed activation based on predetermined routes (V-DAPR) presented in Ref. [53], in which a single route is predetermined for each session based on the cost metric. In order for fair comparison, the total encoding power of all the sensor nodes in ROS or V-DAPR is equal to that in the proposed scheme. The comparison of the power consumption at each sensor node is illustrated in Figure 22.2. In the proposed scheme, the power consumption including the encoding power and the transmission/reception power at each sensor node (represented in bar C) converges to the same level, meaning that each sensor node will exhaust its energy almost at the same time. In ROS (represented in bar B) or V-DAPR (represented in bar A), the power consumption at different sensor nodes is uneven; thus, some nodes will die before other nodes. The network lifetime is determined by the lifetime of the node which has the highest power consumption. The highest power consumption in the proposed algorithm is 0.53 W, smaller by 0.06 W compared to that in ROS, and smaller by 0.16 W compared to that in V-DAPR, respectively.

There is a trade-off between the collected video quality (represented in PSNR) and the achievable maximum network lifetime. If a visual sensor network desires a high-quality video, it will have to sacrifice its network lifetime. On the other hand, a sensor network expecting a longer lifetime has to lower the quality of the collected video. As shown in Figure 22.3, the proposed algorithm supports a longer network lifetime for different quality requirements compared to ROS or V-DAPR, because the proposed algorithm optimizes

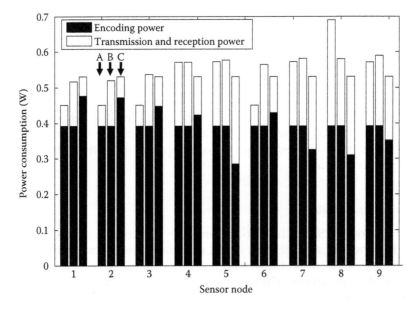

FIGURE 22.2
Comparison of power consumption at each sensor node. (From Y. He, I. Lee, and L. Guan, Distributed algorithms for network lifetime maximization in wireless visual sensor networks, *IEEE Transactions on Circuits and Systems for Video Technology*, 19(5), 704–718, 2009. © (2009) IEEE. With permission.)

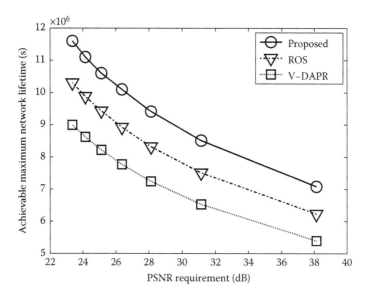

FIGURE 22.3
Trade-off between the PSNR requirement and the achievable maximum network lifetime in lossless transmission. (From Y. He, I. Lee, and L. Guan, Distributed algorithms for network lifetime maximization in wireless visual sensor networks, *IEEE Transactions on Circuits and Systems for Video Technology*, 19(5), 704–718, 2009. © (2009) IEEE. With permission.)

the allocation of the encoding power, the transmission power, and the reception power at each node.

We show the reconstructed picture of frame 1 in Foreman CIF sequence in Figure 22.4. If all the video sensors are required with an upper bound of encoding distortion $D_h = 300.0(\forall h \in \mathbf{V})$, the reconstructed frame 1 has a PSNR of 24.55 dB, as shown in Figure 22.4a. By sacrificing the quality of the collected videos, the WVSN can operate for a long network lifetime, 1.16×10^7 s. If D_h is reduced to 100.0, the sink can receive frame 1 at a PSNR of 28.68 dB, as shown in Figure 22.4b. If D_h is further reduced to 10.0, the PSNR of frame 1 is increased to 38.38 dB as shown in Figure 22.4c, indicating that the sink collects the videos at an excellent quality. To obtain such excellent quality, each video sensor has to

FIGURE 22.4
Comparison of the visual quality at frame 1 in Foreman CIF sequence with different distortion requirement $D_h, \forall h \in \mathbf{V}$: (a) $D_h = 300.0$, (b) $D_h = 100.0$, and (c) $D_h = 10.0$. (From Y. He, I. Lee, and L. Guan, Distributed algorithms for network lifetime maximization in wireless sensor networks, *IEEE Transactions on Circuits and Systems for Video Technology*, 19(5), 704–718, 2009. © (2009) IEEE. With permission.)

consume more power on encoding and transmitting the bit streams with a high rate, thus shortening the network lifetime to 7.08×10^6 s.

22.4 Future Research Directions

Despite the variety of progress made in this research area, there remain a number of important open issues as follows.

The dynamic behaviors in distributed systems place a great challenge to the optimal resource allocation. The dynamic behaviors include the source dynamics, the channel dynamics, and the topology dynamics. The source dynamics are represented by the time-varying rate-distortion characteristics of the encoded multimedia content. The channel dynamics are caused by the time-varying channel conditions. The topology dynamics are due to the dynamic clients, which may join or leave the system at any time. It is challenging to learn the dynamic behaviors online and then make a quick decision on resource allocation to maximize the performance.

Although the streaming capacities of P2P live streaming systems and P2P VoD streaming systems on structured overlays have been examined in Refs. [6,21] respectively, the streaming capacity of a P2P streaming system on unstructured overlay (e.g., data-driven P2P live streaming system) is an open problem. In a data-driven P2P live streaming system, each peer exchanges the video blocks in the sliding window with the randomly chosen neighbors. The order of the block requests at a peer depends on the scheduling algorithm. For example, the scheduling algorithm that places a higher priority to the dissemination of the blocks will request the rarest block in the neighborhood first, while the scheduling algorithm that places a higher priority to the playback continuity will request the block closest to the playback deadline first. The streaming capacity in a data-driven P2P live streaming system is dependent on the peer bandwidth, the constructed overlay, and the scheduling algorithm.

Most P2P streaming systems offer a lot of channels. There are resource imbalances among channels. The cross-channel resource allocation is expected to improve the system performance. However, it is challenging to optimally utilize the cross-channel resources. We may consider setting up collaborations in both the channel level and the peer level. In the channel level, we can establish channel partnership, via which the resource-rich channel helps the resource-poor channel. In the peer level, we can establish peer partnership, via which the resource-rich peer helps the resource-poor peers.

Cloud computing is an emerging technology aiming to provide a variety of computing and storage services over the Internet [54]. Cloud computing technology has offered great opportunities for multimedia applications. The cloud computers, the clients, and the networks connecting them constitute a client–cloud multimedia system, in which the cloud computers handle the data-intensive computing tasks, and the clients become much more lightweight and mobile. The performance of the client–cloud multimedia system can be improved by jointly optimizing the resources in the cloud, at each client, and along the network path.

The emerging cognitive radio technology [55] enables unlicensed users to sense and access the underutilized spectrum bands dynamically as long as the communications among licensed users are not affected. In a multihop cognitive radio network, each node may have a different set of frequency bands with different bandwidths. Efficient utilization of these frequency bands is expected to improve the performances for video streaming applications.

22.5 Summary

Optimal resource allocation can greatly improve the performance for video communications over distributed systems. In general, the resource allocation problem in a distributed system can be formulated into a constrained optimization problem, with an objective to maximize (or minimize) a performance metric, subject to the resource constraints at each node. Since there is no centralized controller in the distributed system, a distributed algorithm is efficient in terms of the scalability and the communication overhead.

In this chapter, therefore, we first provide a review of recent advances on optimal resource allocation for video communications over distributed systems. We next present a case study on network lifetime maximization for WVSNs. Finally, we provide future research directions in the area of video communications over distributed systems.

References

1. I. Akyildiz, W. Su, Y. Sankarasubramaniam, and E. Cayirci, A survey on sensor networks, *IEEE Communication Magazine*, 8, 102–114, 2002.
2. M. Zhang, Y. Xiong, Q. Zhang, and S. Yang, On the optimal scheduling for media streaming in data-driven overlay networks, in *Proceedings of the IEEE GLOBECOM*, pp. 1–5, San Francisco, USA, November 2006.
3. C. Wu and B. Li, Optimal peer selection for minimum-delay peer-to-peer streaming with rateless codes, in *Proceedings of the ACM MM*, pp. 69–78, Singapore, November 2005.
4. E. Setton, J. Noh, and B. Girod, Rate-distortion optimized video peer-to-peer multicast streaming, in *Proceedings of the ACM MM*, pp. 39–45, Singapore, November 2005.
5. C. Wu and B. Li, Diverse: Application-layer service differentiation in peer-to-peer communications, *IEEE Journal on Selected Areas in Communications*, 25(1), 222–234, 2007.
6. S. Sengupta, S. Liu, M. Chen, M. Chiang, J. Li, and P. A. Chou, Streaming capacity in peer-to-peer networks with topology constraints, *Microsoft Research Technical Report*, 2008.
7. D. Wu, Y. Liu, and K. W. Ross, Queuing network models for multichannel P2P live streaming systems, in *Proceedings of the IEEE INFOCOM*, pp. 73–81, Orlando, USA, April 2009.
8. D. Wu, Y. Liu, and K. W. Ross, Modeling and analysis of multichannel P2P live video systems, *IEEE/ACM Transactions on Networking*, 18(4), 1248–1260, 2010.
9. D. Wu, C. Liang, Y. Liu, and K.W. Ross, View-upload decoupling: A redesign of multi-channel P2P video systems, in *Proceedings of the IEEE INFOCOM*, pp. 2726–2730, Orlando, USA, April 2009.
10. M. Wang, L. Xu and B. Ramamurthy, Linear programming models for multi-channel P2P streaming systems, in *Proceedings of the IEEE INFOCOM*, pp. 1–5, San Diego, USA, March 2010.
11. Z. Li and A. Mahanti, A progressive flow auction approach for low-cost on-demand P2P media streaming, in *Proceedings of the ACM QShine*, p. 42, Waterloo, Canada, August 2006.
12. W. P. Yiu, X. Jin, and S. H. Chan, Distributed storage to support user interactivity in peer-to-peer video streaming, in *Proceedings of the IEEE ICC*, 1, pp. 55–60, Istanbul, Turkey, June 2006.
13. Y. He, I. Lee, and L. Guan, Distributed rate allocation in P2P streaming, in *Proceedings of the ICME*, pp. 388–391, Beijing, China, July 2007.

14. C. Huang, J. Li, and K. W. Ross, Peer-assisted VoD: making internet video distribution cheap, in *Proceedings of the IPTPS*, Bellevue, USA, February 2007.
15. Y. Shen, Z. Liu, S. S. Panwar, K. W. Ross, and Y. Wang, Streaming layered encoded video using peers, in *Proceedings of the ICME*, pp. 966–969, Amsterdam, the Netherlands, July 2005.
16. Y. He, I. Lee, and L. Guan, Distributed throughput maximization in hybrid-forwarding P2P VoD applications, in *Proceedings of the ICASSP*, pp. 2165–2168, Las Vegas, USA, April 2008.
17. Y. He, I. Lee, and L. Guan, Distributed throughput optimization in P2P VoD applications, *IEEE Transactions on Multimedia*, 11(3), 509–522, 2009.
18. C. Zheng, G. Shen, and S. Li, Distributed prefetching scheme for random seek support in peer-to-peer streaming applications, in *Proceedings of the ACM MM*, pp. 29–38, Singapore, November 2005.
19. Y. He, G. Shen, Y. Xiong, and L. Guan, Probabilistic prefetching scheme for P2P VoD applications with frequent seeks, in *Proceedings of the ISCAS*, pp. 2054–2057, Seattle, USA, May 2008.
20. Y. He, G. Shen, Y. Xiong, and L. Guan Optimal prefetching scheme in P2P VoD applications with guided seeks, *IEEE Transactions on Multimedia*, 11(1), 138–151, 2009.
21. Y. He and L. Guan, Streaming capacity in P2P VoD systems, in *Proceedings of the ISCAS*, pp. 742–745, Taipei, Taiwan, May 2009.
22. Y. He and L. Guan, Solving streaming capacity problems in P2P VoD systems, *IEEE Transactions on Circuits and Systems for Video Technology*, 20(11), 1638–1642, 2010.
23. Y. He and L. Guan, Improving the streaming capacity in P2P VoD systems with helpers, in *Proceedings of the ICME*, pp. 790–793, New York, USA, July 2009.
24. Y. He and L. Guan, Streaming capacity in multi-channel P2P VoD systems, in *Proceedings of the ISCAS*, pp. 1819–1822, Paris, France, May 2010.
25. W. Wei and A. Zakhor, Multipath unicast and multicast video communication over wireless *ad hoc* networks, in *Proceedings of the IEEE BroadNets*, pp. 496–505, San Jose, USA, October 2004.
26. W. Wei and A. Zakhor, Path selection for multi-path streaming in wireless *ad hoc* networks, in *Proceedings of the IEEE ICIP*, pp. 3045–3048, Atlanta, USA, October 2006.
27. X. Zhu, E. Setton, and B. Girod, Congestion-distortion optimized video transmission over *ad hoc* networks, *Journal of Signal Processing: Image Communications*, 20, 773–783, 2005.
28. S. Mao, S. Lin, S. Panwar, Y. Wang, and E. Celebi, Video transport over *ad hoc* networks: Multi-stream coding with multipath transport, *IEEE Journal on Selected Areas in Communications*, 21(10), 1721–1737, 2003.
29. S. Mao, Y. T. Hou, X. Cheng, H. D. Sherali, and S. F. Midkiff, Multipath routing for multiple description video over wireless *ad hoc* networks, in *Proceedings of the IEEE INFOCOM*, pp. 740–750, Miami, USA, March 2005.
30. S. Mao, X. Cheng, Y. T. Hou, H. D. Sherali, and J. H. Reed, Joint routing and server selection for multiple description video streaming in *ad hoc* networks, in *Proceedings of the IEEE ICC*, 5, pp. 2993–2999, Seoul, Korea, May 2005.
31. Y. He, I. Lee, and L. Guan, Optimized multi-path routing using dual decomposition for wireless video streaming, in *Proceedings of the ISCAS*, pp. 977–980, New Orleans, USA, May 2007.
32. S. Mao, S. Kompella, Y. T. Hou, H. D. Sherali, and S. F. Midkiff, Routing for concurrent video sessions in *ad hoc* networks, *IEEE Transactions on Vehicular Technology*, 55(1), 317–327, 2006.
33. X. Zhu, J. P. Singh, and B. Girod, Joint routing and rate allocation for multiple video streams in *ad hoc* wireless networks, *Journal of Zhejiang University, Science A*, 7(5), pp. 727–736, 2006.
34. A. Zakhor and W. Wei, Multiple tree video multicast over wireless *ad hoc* networks, in *Proceedings of the ICIP*, pp. 1665–1668, Atlanta, USA, September 2006.
35. S. Mao, X. Cheng, Y. T. Hou, and H. D. Sherali, Multiple description video multicast in wireless *ad hoc* networks, in *Proceedings of the IEEE BROADNETS*, pp. 671–680, San Jose, USA, October 2004.
36. Y. He, I. Lee, and L. Guan, Video multicast over wireless *ad hoc* networks using distributed optimization, in *Proceedings of the IEEE PCM*, pp. 296–305, Hong Kong, China, December 2007.

37. Y. He, I. Lee, and L. Guan, Optimized video multicasting over wireless *ad hoc* networks using distributed algorithm, *IEEE Transactions on Circuits and Systems for Video Technology,* 19(6), 796–807, 2009.
38. R. Ahlswede, N. Cai, S.-Y. R. Li, and R. W. Yeung, Network information flow, *IEEE Transactions on Information Theory,* 46, 1204–1216, 2000.
39. R. Madan and S. Lall, Distributed algorithms for maximum lifetime routing in wireless sensor networks, *IEEE Transactions on Wireless Communications,* 5(8), 2185–2193, 2006.
40. S. Boyd and L. Vandenberghe, *Convex Optimization,* Cambridge University Press, Cambridge, UK, 2004.
41. H. Nama, M. Chiang, and N. Mandayam, Utility-lifetime trade-off in self-regulating wireless sensor networks: A cross-layer design approach, in *Proceedings of the IEEE ICC,* 8, pp. 3511–3516, Istanbul, Turkey, June 2006.
42. J. Zhu, K. Hung, B. Bensaou, and F. Abdesselam, Tradeoff between network lifetime and fair rate allocation in wireless sensor networks with multi-path routing, in *Proceedings of the ACM MSWiM,* pp. 301–308, Torremolinos, Malaga, Spain, October 2006.
43. Y. He, I. Lee, and L. Guan, Network lifetime maximization in wireless visual sensor networks using a distributed algorithm, in *Proceedings of the ICME,* pp. 2174–2177, Beijing, China, July 2007.
44. Y. He, I. Lee, and L. Guan, Distributed algorithms for network lifetime maximization in wireless visual sensor networks, *IEEE Transactions on Circuits and Systems for Video Technology,* 19(5), 704–718, 2009.
45. Z. He, Y. Liang, L. Chen, I. Ahmad, and D.Wu, Power-rate-distortion analysis for wireless video communication under energy constraint, *IEEE Transactions on Circuits and Systems for Video Technology,* 15(5), 645–658, 2005.
46. Z. He and D. Wu, Accumulative visual information in wireless video sensor network: Definition and analysis, in *Proceedings of the IEEE ICC,* 2, pp. 1205–1208, Seoul, Korea, May 2005.
47. Z. He and D. Wu, Resource allocation and performance analysis of wireless video sensors, *IEEE Transactions on Circuits and Systems for Video Technology,* 16(5), pp. 590–599, May 2006.
48. W. Heinzelman, A. Chandrakasan, and H. Balakrishnan, An application-specific protocol architecture for wireless microsensor networks, *IEEE Transactions on Wireless Communications,* 1(4), pp. 660–670, October 2002.
49. T. S. Rappaport, *Wireless Communications: Principles and Practice,* 2nd edition, Prentice Hall, New Jersey, USA, 2002.
50. D. Palomar and M. Chiang, A tutorial on decomposition methods and distributed network resource allocation, *IEEE Journal on Selected Areas in Communications,* 24(8), 1439–1451, 2006.
51. M. Chiang, S. H. Low, A. R. Calderbank, and J. C. Doyle, Layering as optimization decomposition: A mathematical theory of network architectures, *Proceedings of the IEEE,* 95(1), pp. 255–312, January 2007.
52. D. P. Bertsekas, A. Nedic, and A. E. Ozdaglar, *Convex Analysis and Optimization,* Athena Scientific, Nashua, NH, USA, 2003.
53. S. Soro and W. Heinzelman, On the coverage problem in video-based wireless sensor networks, in *Proceedings of the International Conference on Broadband Networks,* 2, pp. 932–939, Boston, USA, October 2005.
54. M. Armbrust, A. Fox, R. Griffith, A. D. Joseph, R. Katz, A. Konwinski, G. Lee, D. Patterson, A. Rabkin, I. Stoica, and M. Zaharia, Above the clouds: A berkeley view of cloud computing, *Technical Report No. UCB/EECS-2009-28,* Feb. 2009.
55. S. Haykin, Cognitive radio: Brain-empowered wireless communications, *IEEE Journal on Selected Areas in Communications,* 23(2), 201–220, 2005.

Part VI

Methodology, Techniques, and Applications: Architecture Design and Implementation for Multimedia Image and Video Processing

23

Algorithm/Architecture Coexploration

Gwo Giun (Chris) Lee, He Yuan Lin, and Sun Yuan Kung

CONTENTS

23.1 Introduction

Niklaus Emil Wirth introduced the innovative idea that *Programming = Algorithm + Data Structure*. Inspired by this paradigm, we advance this to the next level by stating that *Design = Algorithm + Architecture* in this chapter.

Traditional design methodologies are usually based on the execution of a series of sequential stages: the theoretical study of a fully specified algorithm, the mapping of the algorithm to a selected architecture, the evaluation of the performance, and the final implementation. However, these straightforward design procedures are no longer adequate to cope with the increasing demands of video design challenges. Conventional sequential design flow yields the independent design and development of the algorithm from the architecture. However, with the ever-increasing complexity of both algorithm and system platforms in each successive generation, such sequential steps in traditional designs will inevitably lead to the scenario that designers may either develop highly efficient but highly complex algorithms that cannot be implemented or else may offer platforms that are impractical for real-world applications because the processing capabilities cannot be efficiently exploited by the newly developed algorithms. Hence, seamless weaving of the two previously autonomous algorithmic and architecture development will unavoidably be observed.

As the algorithms in forthcoming multimedia image and video systems become more complex, many applications in digital image and video technologies must be deployed with different profiles having different levels of performance. Figure 23.1 ranks the spectrum of visual signal processing algorithms based on qualitative complexity analysis.

Extrapolating from a high-level description, as illustrated in Figure 23.1, future visual signal processing algorithms will have better content adaptivity, extended use of temporal information, and further increases the physical size and resolution of the image sequences as they continuously deliver better visual quality. Recent and future video coding standards such as Moving Picture Experts Group (MPEG) and Video Coding Experts Group (VCEG) have been and will continue to focus on video coding tools that are better adapted to the content and that are more refined in motion estimation for more accurate motion compensation models that yield greater complexity. Furthermore, the increase in image-sequence sizes,

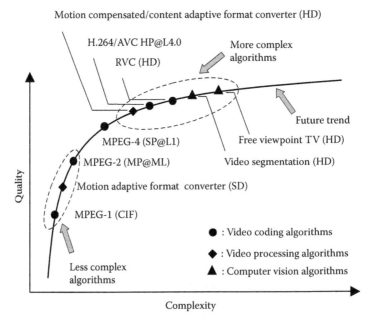

FIGURE 23.1
Complexity spectrum for advanced visual computing algorithms.

from standard definition to high definition (HD) toward ultrahigh definition and beyond, is also within the already defined roadmaps of video coding standards development.

Similarly, the complexity of video processing algorithms, such as motion-adaptive deinterlacers [1,2], scan-rate converters, and other format converters, is also characterized by studying the three qualitative complexity features discussed above. Content-adaptive algorithms for HD video processing, such as those in scalers, deinterlacers, and so on, which are based on the texture contents from surrounding neighbors, have also been documented in the literature [3,4] and used in many applications. Motion-compensated and content-aware algorithms for increasingly higher-definition video processing technologies will also be seen in the future.

In addition to being content adaptive, computer vision algorithms are even more complex due to their content awareness as the required cognitive capabilities are added. Computer graphics algorithms are highly computationally intensive: some graphics algorithms, such as real-time rendering, demand a huge amount of processing power, and visualizing graphic content is possible when using many tremendously complex graphics algorithms. The recent evolution of the processing power of graphics processing units is a good indication that the graphics algorithms require ever-increasing computing power to provide better user visual experiences.

Rapid and continuous improvements in semiconductor and architecture design technologies have yielded innovations in system architectures and platforms that offer advanced multimedia image and video algorithms that target different applications. Each application has versatile requirements for trading off the performance per unit of silicon area (performance/silicon area), flexibility of usage, algorithm changes, and power consumption. Conventional implementations of algorithms were usually placed at two architectural extremes: pure hardware or pure software. Although application-specific integrated circuit (ASIC) implementation of algorithms provides the highest speed or best performance, this is however achieved via trading off platform flexibility. Pure software implementations on single-chip processors or CPUs are the most flexible, but require a higher power-overhead and yield a slower processing speed. Hence, several other classes of architecture, such as Instruction Set Digital Signal Processors (DSPs) and Application-Specific Instruction Set Processors (ASIPs) have also been introduced (Figure 23.2). Recently, embedded multicore processors and reconfigurable architectures have become the leading trend in architectures for the realization of image and video algorithms in the design of versatile visual systems.

As multimedia image and video processing algorithms are becoming ever more complex, successfully mapping them onto platforms optimal for versatile applications is a key consideration for forthcoming design methodologies. The aforementioned sequential design flow may provide either excellent visual algorithms that are highly complex and, therefore, cannot be implemented, or that can be used only on system platforms with limited applications because of their poor visual quality. In this chapter, we introduce the concept of Algorithm/Architecture Co-Exploration (AAC), which is now a leading paradigm.

The concurrent exploration of algorithm and architecture optimizations consists of extracting complexity measures of algorithms featuring architectural characteristics that, together with the dataflow model, yield the best mapping onto a platform for targeted applications. It is important to note that the traditional order of growth used in the complexity analysis of computer algorithms is insufficient because it is based on the ideal random access machine, which is merely a single point or platform within the broad spectrum of platforms (Figure 23.2). As contemporary and future platforms are beyond the nanometer-scale range, the complexity measures we discuss in the subsequent sections in this chapter

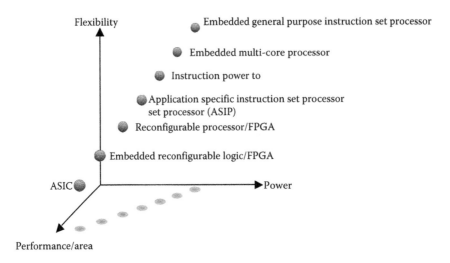

FIGURE 23.2
Spectrum of platforms.

provide quantitative measures of the intrinsic complexity of algorithms for envisioning even the scope or spectrum of future system architectures and platforms.

Furthermore, many complex algorithms such as those for computer vision that was traditionally hard to implement can now be better modeled by dataflow representation discussed in this chapter and hence making realization feasible.

The chapter is divided as follows. Section 23.2 conceptually introduces the advanced AAC methodology based on dataflow modeling and complexity analysis. Fundamental dataflow modeling techniques and four complexity metrics are also discussed herein. Section 23.3 presents three case studies for illustrating the novelty of AAC. Section 23.4 concludes this chapter.

23.2 Algorithm/Architecture Coexploration

23.2.1 Overview of AAC Methodology

This section briefly introduces the overview of the concurrent optimization of both algorithms and architectures, which is referred to as the AAC, with dataflow modeling and complexity analysis.

23.2.1.1 Levels of Abstraction

As signal processing applications such as visual computation and communication are becoming increasingly more complicated, the design exploration in electronic system level (ESL) is facing more problematic challenges as well. For a specific application, its design space is a set composed of all the feasible very-large-scale integration (VLSI) implementation instances. The design space is therefore spanned by different VLSI implementation attributes.

In general, a design starts from the development of an algorithm to VLSI implementation. Abstracting unnecessary design details and separating the design flow into several hierarchies of abstraction level as shown in Figure 23.3 can efficiently enhance the design capability. For a specific application, the levels of abstraction include algorithmic level, architectural level, register transfer level (RTL), gate level, and physical design level. As the design goes into a lower abstraction level, more and more design details are added and the space of the lower abstraction level consequently gets larger as shown in Figure 23.3.

Figure 23.4 shows the features of design details in each level of abstraction. In the algorithmic level, the features considered are the functionalities, and the time unit used is second. For example, the real-time processing is a common constraint for visual applications that the temporal domain precision is measured in terms of frames per second (FPS).

In the architectural level, the remarkable features include hardware/software partition and codesign, memory configuration, bus protocol, and the transaction between each module composing the system. The time unit is measured in terms of number of cycles.

For silicon intellectual property (IP) at the macrolevel, the corresponding microarchitecture including datapath and control logic also has cycle accuracy in timing. In the module level, the features could be the various arithmetic units composing datapath. The longest delay path between two registers should be studied, since the detailed circuit behaviors with parasitical capacitance and inductance are abstracted by time delay. In the gate level, the feature is logic operation for digital circuit. In the circuit level, voltage and current are notable. In the device level, the electrons are taken care of.

From the discussions above, it is easy to find that the timing and physical scales are coarser in the higher levels of abstraction and finer in the lower abstraction levels. In the traditional cell-based design flow, most design efforts are spent in the RTL, thanks to abstracting the unnecessary details together with the assistance of logic synthesis, auto-place and routing, and backend manufacturing. In the ESL design, the level of abstraction is further raised. The dataflow modeling and transaction-level modeling (TLM) play a very important role in this level of abstraction, as described in Section 23.2.2.

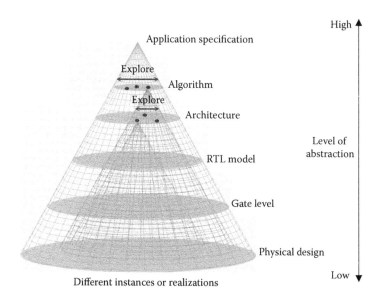

FIGURE 23.3
Levels of abstraction.

Levels	Symbols	Features	Time units
Algorithm		System functionality	Second
Architecture	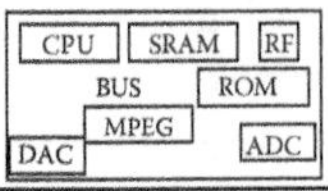 CPU SRAM RF BUS ROM MPEG DAC ADC	System architecture	Number of cycles
IP (Macro)	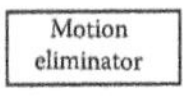 Motion eliminator	IP functionality and micro-architecture	Number of cycles
Module	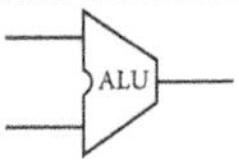 ALU	Arithmetic operation	Cycle
Gate	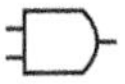	Logic operation	ns
Circuit	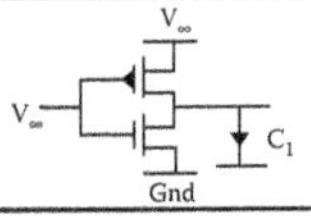 V_{cc} V_{in} C_1 Gnd	Voltage, current	ps
Device	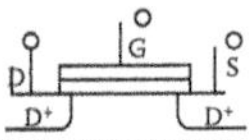 G D S D^+ D^+	Electron	ps

FIGURE 23.4
Features in various levels of abstraction.

23.2.1.2 Joint Exploration of Algorithm and Architecture

Traditionally, designers gradually explore and optimize each abstraction level with a top-down design manner. The realization choices of lower levels are, however, restricted by the instances of higher levels. Once a decision at higher abstraction is made, the size of the space at lower abstraction level is shrunk. Consequently, algorithm/architecture co-design exploration methodology jointly optimizes the realization of algorithm and architecture by charactering the complexity of algorithm together with back-annotating design details from lower abstraction level, as shown in Figure 23.5.

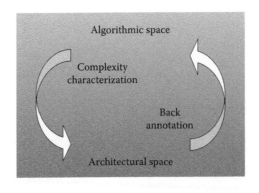

FIGURE 23.5
Concept of AAC.

In order to efficiently design emerging platforms for advanced visual computing applications, this chapter introduces the concept of AAC so as to increase the values of the visual computing systems including flexibility, scalability, gate count, power consumption, maintainability, design turn-around time, the risk of failure, and so on. The AAC approach is composed of three levels of design abstractions including algorithm exploration, AAC and architecture exploration, as shown in Figure 23.6.

Following this advanced methodology, an architect starts to design the algorithm according to the application specification. An appropriate algorithm is then developed to match the application requirements in functionality and performance. As long as an algorithm is developed, many architectural attributes of its VLSI implementation have already been determined. So, the architect can extract the complexity of the algorithm with high-level dataflow modeling so as to reveal the system-level architectural attributes.

If the corresponding system-level architecture does not meet the application specification such as area, power, or memory constraints, the architect needs to go back to modify the developed algorithm. Consequently, the coexploration of algorithm/architecture sometimes needs several iterations to achieve optimization subject to the design constraints. With the early feedback of architectural attributes, the algorithm/architecture codesign space exploration can be well done by this design methodology.

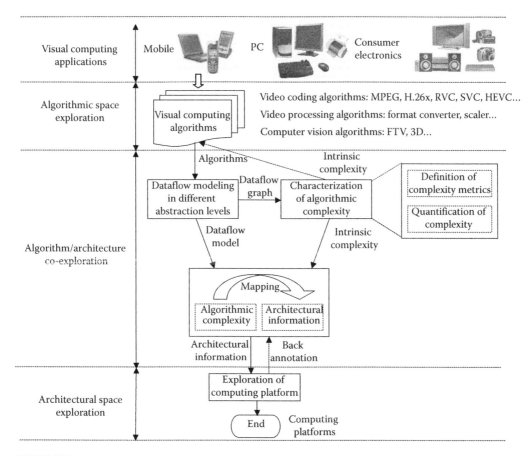

FIGURE 23.6
Advanced visual system design methodology.

23.2.2 Dataflow Model for the Representation and Codesign of Algorithms and Architectures

Convertible from mathematical representations and signal flow graphs in digital signal processing [5], the dataflow model contains both algorithmic and architecture information [6], and the corresponding dataflow model can be easily obtained. The dataflow model is capable of depicting the interrelationships between computations and communications. Consequently, the graph generated by the dataflow model can clearly reveal the data dependencies between each operation via vertexes and directed edges, where the vertexes denote the operations and the directed edges represent the sources and destinations of the data, that is, the flow of data. In addition, the dataflow model can reveal the concurrency of algorithms via the concept of data driven.

In the initial stage of developing an algorithm to satisfy requirements of an application, the algorithmic functionality and performance are relatively important. So, high-level programming languages used in this stage focus on the ease of describing the functional behaviors but not on the dataflow and transaction. Therefore, some programming techniques such as pointer, data structure, functional calls, and recursive coding style are used in these high-level programming languages. Although these techniques increase the flexibility of programming and significantly reduce the coding effort, the dataflow information including the source of the data and destination of the data is abstracted and hence directly porting such programs onto embedded and multicore system-on-chip (SoC) platforms is rather inefficient.

As compared to traditional high-level programming models, dataflow models inherited from the mathematical expression can not only contain relatively intrinsic complexity information but also reveal the architectural information, which is transparent to implementation details. Consequently, in the AAC, the dataflow model is definitely the essential bridge between algorithms and architectures, which are originally two different worlds in traditional design methodologies. In the new emerging reconfigurable video coding (RVC) standard specified by MPEG, a dataflow language namely, Caltrop Actor Language (CAL), is employed to implement RVC reference software to contain the dataflow information [7–9].

Figure 23.7 illustrates the dataflow model of a 4-tap finite-impulse response (FIR) filter whose difference equation is expressed as

$$y[n] = ax[n] + bx[n-1] + cx[n-2] + dx[n-3]. \tag{23.1}$$

From Figure 23.7, one can easily characterize the complexity of the above difference equation representing an algorithm or system. The number of operations needed to generate

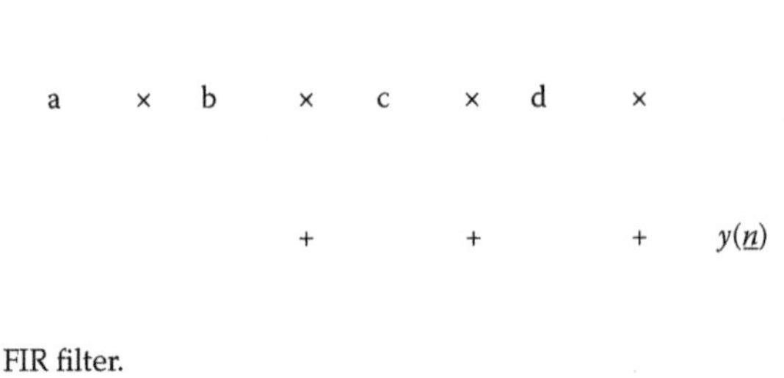

FIGURE 23.7
Dataflow model of a 4-tap FIR filter.

an output is three additions and four multiplications. Based on the dependency displayed by the dataflow graph (DFG), the four multiplications can be executed concurrently and hence reveal the parallelism. In addition, this algorithm needs memories to store at least three samples. Eventually, the amount of data transfer including read/write can be known from the number of edges in the DFG. Obviously, the four preliminary complexity analyses of algorithms can be easily obtained from DFGs.

Figure 23.8 shows the pipeline view of dataflow for the filter under the assumption that one sample is inputted per clock cycle and shifted from the registers D_0, D_1 to D_2 and the output data rate is also one sample per cycle. Based on Figure 23.8, the transaction level or RTL architectural information, such as the types of datapath, memory configuration, longest delay path, clock rate, bandwidth, and so on can be revealed as well as in the early design phases. Hence, in the AAC, the dataflow model is definitely the essential bridge between algorithms and architectures, which are originally two isolated worlds in traditional design methodologies.

Therefore, we intend to write the dataflow models of the to-be-implemented algorithms in different levels of abstraction as described in the following: In the high-level dataflow modeling of algorithms, the functionalities and the data dependencies of algorithms are described without any timing information. Such a dataflow model can generate causation traces for complexity analysis in terms of number of operations and amount of data transfer as described in Section 23.16.1. The key of doing such a model in the high level of abstraction is to implement all the communications between functions in a message-passing manner so as to explicitly monitor the dataflow.

In the middle-level dataflow modeling of algorithms, we describe not only the functionalities and data dependencies of algorithms but also the processing order with approximated timing information, since doing so can facilitate the qualification of memory size and configuration and average bandwidth, which will be discussed in detail in Section 23.2.3. The concept of this middle-level dataflow modeling is similar to that of TLM that ignores unnecessary design details.

In the low-level dataflow modeling of algorithms, the cycle-accurate timing information will be involved in the models with detailed processing schedules. Such a dataflow model can reveal the cycle count for computation and hence provide the lower bound the clock rate for real-time applications. Again, the intention of dataflow modeling in the low level of abstraction is to extract the approximated cycle count of operation and therefore the detailed control information and handshaking between modules must be abstracted. In addition, the width of the data should be ignored as well. So the level of abstraction of such a model is much higher than that of RTL.

Clock cycles	Input	D_0	D_1	D_2	Output
0	$x[0]$				$y[0] = ax[0]$
1	$x[1]$	$x[0]$			$y[1] = ax[1] + bx[0]$
2	$x[2]$	$x[1]$	$x[0]$		$y[2] = ax[2] + bx[1] + cx[0]$
3	$x[3]$	$x[2]$	$x[1]$	$x[0]$	$y[3] = ax[3] + bx[2] + cx[1] + dx[0]$
4	$x[4]$	$x[3]$	$x[2]$	$x[1]$	$y[4] = ax[4] + bx[3] + cx[2] + dx[1]$
...	...	...	...	...	...

FIGURE 23.8
Pipeline view of dataflow in a 4-tap FIR filter.

As following the top-down design flow, we need to add the incremental information between two levels of abstraction and therefore the extra design effort in refining dataflow models from high level to low level is limited. However, with the assistance of dataflow modeling in different levels of abstraction, the design exploration and system integration and verification do not become problematic anymore.

23.2.2.1 Directed Acyclic Graph

Directed acyclic graphs (DAGs) are acyclic graphs composed of vertexes and directed edges, where vertexes indicate operation activities and directed edges between vertexes represent the dependency of operations. Arrows indicate the directions in which the dataflow through the operation vertexes. DAG is one of the direct methods to represent DFGs. The granularity of the vertexes can be at the level of logic operations, arithmetic operations, tasks, or even packet processing [10–12].

23.2.2.2 Synchronous Dataflow Graph

Synchronous dataflow graph (SDFs) are graphs extended from DAGs, where the edges are accompanied with numbers representing fixed input/output data rates of nodes, which are specified a priori. If the input/output data rate of the nodes are not specified or known a priori, the DFG is asynchronous.

23.2.2.3 Kahn Process Networks

Kahn process networks (KPNs) are directed graphs consisting of vertexes and edges [13]. The vertexes indicate processes having some specific functionality and the directed edges represent unbounded and infinite first-in-first out (FIFO) queues between processes. The execution of KPN is based on data driven mechanism via blocking read and nonblocking write. Assuming that two processes are connected via a FIFO Channel in KPN, the first process or the process at the earlier stage is capable of writing or outputting data into the channel FIFO freely.

However, the reading or input of data from the channel FIFO into the second or the later stage process requires that certain conditions hold and hence providing conditional control or the blocking read mechanism. A process is fired or activated when its input data exit and the corresponding output data are produced by the process and then sent out via an unbounded FIFO. Benefiting from the data driven mechanism, any process can be fired as long as it has the input data. Consequently, the KPNs are capable of modeling concurrent processing. In general, the KPNs can be applied to model the dataflow of deterministic systems.

23.2.2.4 Y-chart Application Programming Interface

Y-chart application programming interface (YAPI) [14] is an extension of KPNs. KPNs model deterministic (synchronous) dataflow, but YAPI is capable of modeling nondeterministic (asynchronous) dataflow by allowing dynamic decision on selecting which channel to communicate. The selection of port/channel depends on the amount of data: Let $c(port_k)$ be the number of data or tokens existing in the kth port namely $port_k$ and N_k be the number of data or tokens required to be consumed from $port_k$ at a time. A port is selected and activated as long as the compounding channel has more than N_k data or tokens. Compared to KPNs, YAPI provides additional controlling capability.

23.2.2.5 *Control Dataflow Graph*

Control dataflow graph (CDFGs) In CDFGs [12], the synchronous or asynchronous DFGs are encapsulated as vertexes and connected by control edges that represent control procedures. Obviously, the CDFG is capable of more explicitly describing controlling behaviors of an algorithm as compared to KPNs or YAPI.

23.2.2.6 *Transaction-Level Modeling*

Based on the dataflow concept, Transaction-level modeling (TLM) is capable of efficiently describing a system by abstracting unnecessary details in computation and communication [15,16]. In TLM, computations are executed by modules, and data transactions are done via channels between modules at several levels of abstractions. This feature makes TLM capable of specifying not only the untimed functional model of an algorithm, but also the approximated-time or cycle-accurate model of the architecture composed of both software and hardware. TLM is a good example of applying dataflow to ESL design.

23.2.3 Algorithmic Intrinsic Complexity Characterization

Intrinsic complexity metrics of algorithms that provide important architectural information is critical in AAC, since they are capable of being feedback- or back-annotated in early design stages so as to facilitate concurrent optimizations of both algorithm and architectures. The complexity metrics have to be intrinsic to the algorithm and hence are not biased toward either hardware or software. In other words, they should be platform independent so as to reveal the anticipated architectural features and electronic ingredients in the early design stages. In order to characterize the complexity of algorithms, this chapter introduces the essential algorithmic intrinsic complexity and the corresponding quantification methods based on the metrics.

23.2.3.1 *Number of Operations*

The number of arithmetic and logic operations is one of the most intuitive metrics that can measure the intrinsic complexity of an algorithm during computation. An algorithm possessing more operations requires more computational power in either the software on processor-based platforms or the hardware on other system platforms. Consequently, the number of operations in terms of the four basic arithmetic, including addition, subtraction, multiplication, and division and logic, operations can be used to characterize the complexity of the algorithm and hence to provide insight into architectures such as number of processing elements (PE) needed and the corresponding operating clock rate for real-time applications.

Estimating the number of operations of an algorithm can provide designers with the intrinsic complexity that is independent of whether implementation is in software or hardware. The number of operations can exhibit the gate count estimation if implementation is intended in ASICs. Furthermore, extracting the common operations and the number of operations in an algorithm can help engineers figure out feasible field programmable gate array (FPGA) configurations. On the contrary, if an algorithm is mapped into software, one can know what kind of instruction set architecture is required in the general-purpose CPU or DSP coprocessors. Since this metrics can give designers insight into either software or hardware implementation in early design stages, it can effectively facilitate software/hardware partition and codesign.

To make this metric more accurate, the types of computational operations have to be particularly distinguished, since various operations have different costs in implementation. Among the four basic arithmetic operations, the complexity of addition and subtraction are similar and simplest, multiplication is so complex that it can be executed by a series of additions and shifts based on Booth's algorithm [17], and division is the most complicated, since it can be performed by shifts, subtractions, and comparisons. In CPU profiling, different types of operations spend distinct CPU cycles according to the instruction set architecture. In ASIC and FPGA designs, each basic mathematical operation and logic operation has different gate counts and number of configurable logic blocks (CLBs), respectively. Furthermore, other than gate count and the number of CLBs, one can estimate the average power consumption at algorithmic level according to the numbers of operation per second.

In addition to the types of operation, the precision of operand in terms of bit depth and type of operand (fixed point or floating point) can significantly influence the implementation cost and hence need to be especially specified. In general, the gate count of PE increases as the precision grows higher. Besides, the hardware propagation delay is affected by the precision as well. Hence, the precision is an important factor in determining the critical path length, maximum clock speed, and hence the throughput of electronic systems. If an algorithm is implemented on the processor-orientated platforms composed of general-purpose processors, single-instruction multiple data (SIMD) machines, or application-specific processors, the precision of operand will directly determine the number of instructions needed to complete an operation. Consequently, the operand precision is also a very important parameter as measuring the number of operations.

Furthermore, whether the input of an operator is variable or constant has to be differentiated, since a complicated constant-input operation can be executed via a few simple operations. For example, a constant-input multiplication can be implemented by fewer additions and shifts, where the shifts can be efficiently implemented by just wiring in hardware. In software, the constant operations can be executed by immediate-type instructions that need less access to registers. Hence, the variable or constant-input operant is also a significant factor that should be considered.

The number of different types of operations can be easily quantified according to the algorithm descriptions. Horowitz et al. [18] introduced a complexity analysis methodology based on calculating the number of fundamental operations needed by each subfunction together with the function call frequency in statistics for different video contents. The worst-case and average-case computational complexity can then be estimated according to the experimental results. This method can efficiently estimate the number of operations for content-adaptive visual computing algorithms. Besides, Ravasi and Mattavelli presented a software instrumentation tool capable of automatically analyzing the high-level algorithmic complexity without rewriting program codes [19,20]. This can be done by instrumentation of all the operations that take place as executing the program. These two techniques can dynamically quantify the relatively intrinsic algorithmic complexity on number of operations for ESL design.

23.2.3.2 *Degree of Parallelism*

The degree of parallelism is another metric characterizing the complexity of algorithms. Some partial operations within an algorithm are independent. These independent operations can be executed simultaneously and hence reveal the degree of parallelism. An algorithm whose degree of parallelism is higher has larger flexibility and scalability in architecture exploration. On the contrary, greater data dependence results in less parallelism, thereby giving a more complex algorithm. The degree of parallelism embedded within algorithms is one of the most essential complexity metrics capable of conveying architectural information for parallel and distributed systems at design stages as early as the algorithm development phase. This complexity metric is again transparent to either software or hardware. If an algorithmic function is implemented in hardware, this metric is capable of exhibiting the upper bound on the number of parallel PEs in datapath. If the function is intended in software, the degree of parallelism can provide insight and hence reveal information pertaining to parallel instruction set architecture in the processor. Furthermore, it can also facilitate the design and configurations of multicore platforms.

Amdahl's law introduced a theoretical maximum speedup for parallelizing a software program [21]. The theoretical upper bound is determined by the ratio of sequential part within the program, since the sequential part cannot be paralleled due to the high data dependencies. Amdahl's law provided an initial idea in characterizing parallelism. In a similar manner, the instruction-level parallelism (ILP) that is more specific for processor-oriented platforms is quantified at a coarser data granularity based on the graph theory [22]. The parallelization potential defined based on the ratio between the computational complexity and the critical path length is also capable of estimating the degree of parallelism [23]. The computational complexity is measured by means of the total number of operations, and the critical path length is then defined as the largest number of operations that have to be sequentially performed. The parallelization potential based on the number of operations reveals more intrinsic parallelism measurements at a finer data granularity as compared to Amdahl's law and the ILP method.

Kung's array processor design methodology [24] employed the dependency graph (DG) to lay out all basic operation to the finest details in one single step based on single assignment codes. Hence, DG is capable of explicitly exhibiting data dependencies between detailed operations of dataflow at the finest granularity. This design methodology provides more insight into the exploitation of algorithmic intrinsic parallelism. For instance, the systolic arrays architecture can efficiently implement algorithms possessing regular dependency DFGs, such as the full search motion estimation. As considering algorithms having irregular data dependencies, the outlines of causation trace graphs [25] generated by dataflow models were used by Janneck et al. in rendering a comparative characterization of parallelism. Similar to Parhi's folding and unfolding techniques [26], the thinner portion of a causation trace graph contains more sequential operations, while the wider portion of has relatively higher degree of parallelism.

One of the versatile parallelisms embedded within algorithms can be revealed as the independent operation sets that are independent of each other and hence can be executed in parallel without synchronization. However, the independent operation sets are composed of dependent operations that have to be sequentially performed. Hence, in a strict manner, the degree of parallelism embedded in an algorithm is equal to the number of the fully independent operation sets. To efficiently explore and quantify such parallelism, Lee et al. [27] proposed to represent the algorithm by a high-level dataflow model and analyze the corresponding DFG. The high-level dataflow model is capable of well depicting the interrelationships between computations and communications. The generated DFG can clearly reveal the data dependencies between the operations by vertexes and directed edges, where the vertexes denote the operations and the directed edges represent the sources and destinations of the data, which is similar to the DG used in Kung's array processor design methodology [24] and the causation trace graphs proposed by Janneck et al. [25].

Inspired by the principal component analysis in the information theory, Lee et al. [27] further employed the spectral graph theory [28] for systematically quantifying and analyzing the DFGs via eigen-decomposition, since that the spectral graph theory can facilitate the analysis of data dependency and connectivity of the DFGs simplistically by means of linear algebra. Consequently, it is capable of quantifying the parallelism of the algorithm with robust mathematically and theoretical analysis applicable to a broad range of real-world scenarios.

Given a DFG G of an algorithm composed of n vertexes that represent operations and m edges that denote data dependency and flow of data, in which the vertex set of G is $V(G) = \{v_1, v_2, \ldots, v_n\}$ and the edge set of G is $E(G) = \{e_1, e_2, \ldots, e_m\}$. The spectral graph theory can study the properties of G such as connectivity by the analysis of the spectrum or eigenvalues and eigenvectors of the Laplacian matrix $\mathbf{L}$ representing G, which is defined as [28,29]

$$\mathbf{L}(i,j) = \begin{cases} \text{degree}(v_i) & \text{if } i = j, \\ -1 & \text{if } v_i \text{ and } v_j \text{ are adjacent,} \\ 0 & \text{others.} \end{cases} \tag{23.2}$$

where degree(v_i) is the number of edges connected to the ith vertex v_i. In the Laplacian matrix, the ith diagonal element shows the number of operations that are connected to the ith operation and the off-diagonal element denotes whether two operations are connected. Hence, the Laplacian matrix can clearly express the DFG by a compact linear algebraic form.

Based on the following well-known properties of the spectral graph theory: (1) the smallest Laplacian eigenvalue of a connected graph equals 0 and the corresponding eigenvector $= [1, 1, \ldots, 1]^T$, (2) there exists exactly one eigenvalue $= 0$ for the Laplacian matrix of a connected graph, and (3) The number of connected components in the graph equals the number of eigenvalue $= 0$ of the Laplacian matrix, it is obvious that in a strict sense, the degree of the parallelism embedded within the algorithm is equal to the number of the eigenvalue $= 0$ of the Laplacian matrix of the DFG. Besides, based on the spectral graph theory, the independent operation sets can be identified according to the eigenvectors associated with the eigenvalues $= 0$. Furthermore, by comparing the eigenvalues and eigenvectors of each independent operation set, one can know whether the parallelism is homogeneous or heterogeneous, which is critical in selecting or designing the instruction set architecture.

This method can be easily extended to the analysis of versatile parallelisms at various data granularities, namely multigrain parallelism. These multigrain parallelisms will eventually be used for the exploration of multicore platforms and reconfigurable architectures or Instruction Set Architecture (ISA) with coarse and fine granularities, respectively. If the parallelism is homogeneous at fine data granularity, the SIMD architecture is preferable, since the instructions are identical. On the contrary, the very long instruction word (VLIW) architecture is favored for dealing with the heterogeneous parallelism composed of different types of operations. As the granularity goes coarser, the types of parallelism can help design the homogeneous or heterogeneous multicore platforms accordingly. In summary, this method can efficiently and exhaustively explore the possible parallelism embedded in algorithms with various granularities. The multigrain parallelism extracted can then facilitate the design space exploration for the advanced AAC.

By directly setting eigenvalues of $\mathbf{L} = 0$, it is easy to prove that the degree of parallelism is equal to the dimension of the null space of $\mathbf{L}$ and the eigenvectors are the basis spanning the null space. In general, the number of operations needed to derive the null space of a Laplacian matrix is proportional to the number of edges. Hence, this method provides an efficient approach to quantify the degree of parallelism and the independent operation sets. This method is applicable to large-scale problems by avoiding the computation-intensive procedures of solving traditional eigen-decomposition problem. In addition, since the Laplacian matrix is sparse and symmetrical, it can be efficiently implemented and processed by linking list or compressed row storage (CRS) format.

Figure 23.9 displays a simple example to illustrate the quantification of the algorithmic intrinsic parallelism. The DFG composed of six operations represented by vertexes labeled with different numbers. The corresponding Laplacian matrix $\mathbf{L}$ of the DFG with the arbitrary label is

$$\mathbf{L} = \begin{bmatrix} 1 & 0 & 0 & 0 & 0 & -1 \\ 0 & 1 & 0 & 0 & 0 & -1 \\ 0 & 0 & 1 & 0 & -1 & 0 \\ 0 & 0 & 0 & 1 & -1 & 0 \\ 0 & 0 & -1 & -1 & 2 & 0 \\ -1 & -1 & 0 & 0 & 0 & 2 \end{bmatrix}. \tag{23.3}$$

The eigenvalues and the corresponding eigenvectors of $\mathbf{L}$ are

$$\lambda = \quad 0 \quad 0 \quad 1 \quad 1 \quad 3 \quad 3,$$

$$\mathbf{x} = \begin{bmatrix} 1 \\ 1 \\ 0 \\ 0 \\ 0 \\ 1 \end{bmatrix} \begin{bmatrix} 0 \\ 0 \\ 1 \\ 1 \\ 1 \\ 0 \end{bmatrix} \begin{bmatrix} 0 \\ 0 \\ -1 \\ 1 \\ 0 \\ 0 \end{bmatrix} \begin{bmatrix} 1 \\ -1 \\ 0 \\ 0 \\ 0 \\ 0 \end{bmatrix} \begin{bmatrix} 0 \\ 0 \\ 1 \\ 1 \\ -2 \\ 0 \end{bmatrix} \begin{bmatrix} -1 \\ -1 \\ 0 \\ 0 \\ 0 \\ 2 \end{bmatrix}, \tag{23.4}$$

where λ and x are the eigenvalues and eigenvectors of $\mathbf{L}$, respectively. From the above result, we can know that the DFG is composed of two independent operation sets, since it has two Laplacian eigenvalues $= 0$. So, the degree of parallelism in this algorithm is two. Subsequently, by observing the first eigenvector associated with $\lambda = 0$, we can find that the values corresponding to v_1, v_2, and v_6 are nonzero, indicating that the three operations form a connected dataflow subgraph. In a similar manner, the other eigenvectors associated with $\lambda = 0$ can reveal the rest connected dataflow subgraph. Besides, one can find that the two independent operation sets should be isomorphic, since their eigenvalues and eigenvectors are identical. Hence, the parallelism in this algorithm is homogeneous. This example precisely explains the parallelism extraction and analysis method based on the spectral graph theory.

The spectral parallelism quantification method has several advantages. First of all, it provides a theoretically robust method in quantifying the parallelism of algorithms, whereas

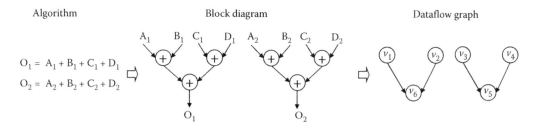

FIGURE 23.9
An example for an illustration of quantifying the algorithmic degree of parallelism.

the causation trace [24] provided only comparative information for the potentials of parallelisms. Besides, benefiting from dataflow modeling, this method is also applicable for characterizing algorithms with irregular data dependencies. In addition, as compared to the analysis based on the high-level programming model in [23] and the quantification of ILP in [22], the parallelism metric is more intrinsic and hence will not be specific only to processor-oriented platforms and is capable of mapping algorithms onto generic platforms and even those for distributed systems. However, the quantification of ILP [22] is used primarily for software implementations. Furthermore, the data structures in instruction-level programming models could influence the parallelism extracted in [23].

In traditional graph theory, connected components can be identified by the depth first search (DFS) or breadth first search (BFS). In general, the algorithmic complexity of the DFS and BFS in terms of the number of operations is linearly proportional to the number of edges plus the number of vertexes. However, the number of operations required by the spectral framework is just proportional to the number of edges when solving the null space of the Laplacian matrix. In addition, the multigrain spectral analysis is capable of systematically decomposing DFGs in a top-down manner, since the eigenvalues and eigenvectors of a graph is the union of those of its individual components. Besides, the spectral method can effectively tell whether the parallelism is either homogeneous or heterogeneous. Furthermore, the spectrum of Laplacian matrix is invariant of the graph matrix regardless of orders in which the vertices are labeled and the Laplacian matrix can be efficiently implemented in CRS format. These features make the handling of matrices representing DFGs efficient in computers and hence preferable for very efficient design automation.

23.2.3.3 *Storage Configuration*

In the theory of signals and systems, a system is said to be memoryless if its output depends on only the input signals at the same time. However, in visual computing applications such as video coding and processing, some intermediate data have to be stored in memory depending on the dataflow of algorithms in higher abstraction levels. Consequently, in order to perform the appropriate algorithmic processing, data storage must be properly configured based on the dataflow scheduling of the intermediate data. Hence, the algorithmic storage configuration is another essential intrinsic complexity metric in AAC design methodology, which is transparent to either software or hardware designs. For software applications, the algorithmic storage configuration helps design the memory modules such as cache or scratch-pad and the corresponding data arrangement schemes for the embedded CPU. In hardware design, the immediate data can be stored in local memory to satisfy the algorithmic scheduling based on this complexity metric.

The minimum storage requirement of an algorithm is determined by the maximum amount of data that needed to be stored at a same time instance, which of course depends on the data lifetime [26]. However, the exact data lifetime analysis can be performed only after scheduling and allocation for completing hardware or software design [30]. Balasa et al. [31] introduced a mathematical method to estimate the background memory size for multidimensional signal processing based on its dataflow with affine indexes. This method is capable of characterizing the relatively intrinsic storage size of algorithms, since seldom scheduling information is involved. Based on the similar idea, a storage requirement estimation method with partially fixed execution ordering is proposed in [32]. This technique can provide designers upper and lower bounds of storage requirement in early design stages. As designs go into lower levels of abstraction, more and more executing orders are fixed and the uncertainty of memory size is gradually diminished.

By analyzing the lifetime of the to-be-stored data based on high-level dataflow scheduling, the algorithmic intrinsic storage size can then be systematically measured. For visual computing applications, the storage requirements result in a specific configuration depending on the data lifetime for different processing in horizontal, vertical, and temporal directions. The size of the algorithmic storage can reveal the potential architectural memory configuration and hierarchy early in the algorithmic exploration design stage for AAC.

In general, video signals can be coordinated in three-dimensional (3D) space spanned by horizontal, vertical, and temporal axes. The input and output order of a typical visual computing system scans pictures one after another. Each individual picture is scanned in a raster scan order line by line. In order to shorten the lifetime of required data and hence minimize the storage requirement, the executing order of a visual system tends to be also raster scan that slides from the top-left corner of a picture to the bottom-right corner. To facilitate measuring the lifetime of input data for visual applications, we express a 3D video signal $f(x, y, n)$ in terms of a one-dimensional (1D) signal $f'(n')$, where x, y, and n are the horizontal, vertical, and temporal indexes, respectively and $n' = x + y \times \text{width} + n \times \text{width} \times \text{height}$. Without loss of generality, assume that the input data list S of an algorithm is $\{f(x, y, n), f(x-1, y, n), f(x, y-1, n), f(x, y, n-1) \dots\}$ and the corresponding 1D input data list $S' = \{f'(n_1), f'(n_2), f'(n_3), f'(n_4), \dots\}$. Benefiting from the regular and well-structured coordinate of video signals, the lifetime of the required data can be estimated according to the range of the index of the input data list. The lifetime of data can then be calculated by the following equation:

$$LT = \max\{n_1, n_2, n_3, \dots\} - \min\{n_1, n_2, n_3, \dots\} + 1, \qquad (23.5)$$

where LT is the lifetime of input data, max{} and min{} are the functions returning the maximum value and minimum value in the input list, respectively. This is because if a datum is slid by the filter window initially at the time stamp, min $\{n_1, n_2, n_3, \dots\}$, the datum will still be required until the time stamp, max $\{n_1, n_2, n_3, \dots\}$.

Figure 23.10 further illustrates the lifetime analysis of input data for typical visual computing systems, in which a datum is inputted at each time instance. Assuming that the lifetime of each datum is equal to N, it is easy to derive that the maximum number of the data required will also be equal to N. Consequently, the storage requirement can also be estimated according to the range of the index of the input data.

Time	$f(0)$	$f(1)$	$f(2)$	$f(3)$	$\cdots$	$f(n-1)$	$f(n)$	$f(n+1)$	$\cdots$	$\cdots$	# of data
0											1
1											2
2											3
3											4
$\cdots$											$\cdots$
$\cdots$											$\cdots$
$N-1$											N
N											N
$N+1$											N
$N+2$											N
$N+3$											N
$\cdots$											N
$\cdots$											N

Lifetime (vertical axis label on left)

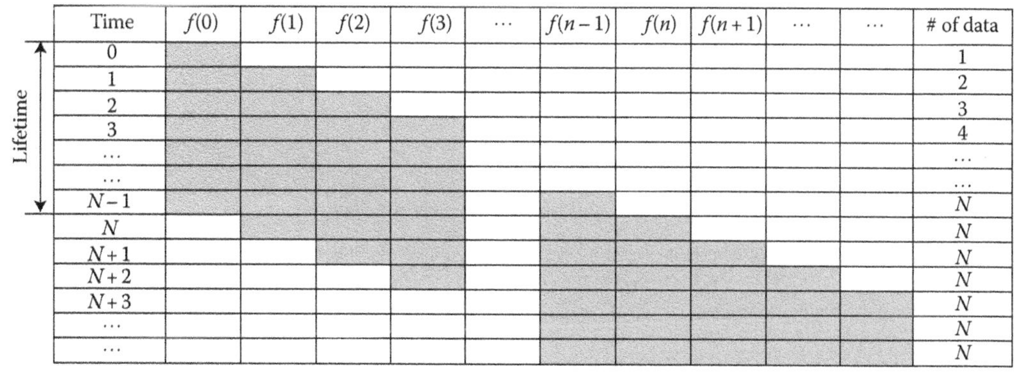

FIGURE 23.10
Lifetime analysis of input data for typical visual computing systems.

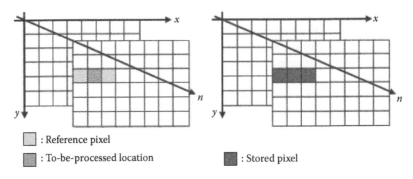

FIGURE 23.11
Filter support of a 3-tap horizontal filter.

Depending on the types of filter windows and the size of storage requirements, the storage requirement of a visual computing algorithm can be categorized into three classes including pixel storage, line storage, and picture storage. Figure 23.11 illustrates the filter support of a three-tap horizontal FIR filter. Based on Equation 23.5, it is easy to derive that the storage requirement of the filter is $\{(x + 1) + y \times W + n \times W \times H\} - \{(x - 1) + y \times W + n \times W \times H\} + 1 = 3$ pixels, where W and H represent picture width and height, respectively. It is clear that a vertical filter can introduce some pixel storages depending on its number of taps.

In a similar manner, the algorithmic storage size of a 3-tap vertical FIR filter shown in Figure 23.12 can be estimated, which is two lines and 1 pixel. It can be found that some line storages are required by a vertical FIR filter according to the analysis of data lifetime.

Figures 23.13 and 23.14 show the storage requirements of a 3-tap temporal FIR filter and a spatial–temporal filter having a $3 \times 3 \times 3$ support, respectively. The 3-tap temporal FIR filter needs two pictures and 1 pixel and the spatial–temporal filter requires storage to store two pictures, two lines, and 3 pixels.

So far, the storage requirements of three major types of filter in visual computing applications have been studied. In summary, filters requiring input data in various domains introduce different classes of storage: horizontal filters need pixel storage, vertical filters demand for line storage, and temporal filter requires picture storages. The storage sizes and configurations discussed in this chapter are the theoretical minimum requirements according to the high-level input/output dataflow of visual computing applications, which are platform independent. As the design stage goes into lower levels of abstraction, the

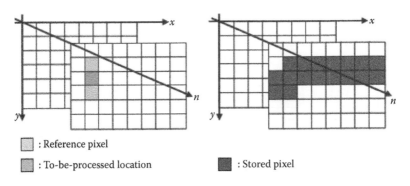

FIGURE 23.12
Filter support of a 3-tap vertical filter.

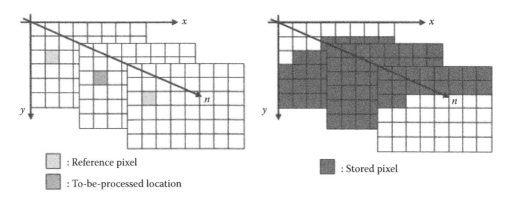

: Reference pixel

: To-be-processed location

: Stored pixel

FIGURE 23.13
Filter support of a 3-tap temporal filter.

architectural memory configurations will be gradually refined according to the low-level dataflow with the considerations of design constraints. In addition, based on the low-level dataflow at fine data granularity, some data are reused thereby resulting extra local storage requirement. Studying different classes of storage requirement of an algorithm can facilitate arranging data onto various physical memories such as dynamic random access memory (DRAM), static random access memory (SRAM), register file, which is a part of architectural space exploration.

23.2.3.4 Data Transfer Rate

In addition to the number of operations and data storages, the amount of data transfer is also an intrinsic complexity metric as executing an algorithm. In real-time applications, visual computing algorithms need to transfer a large amount of data within a specified time. Consequently, the average data transfer rate is a measure of the algorithmic complexity by providing an estimate of the amount of data transferred in 1 s. The average data transfer directly estimated from the algorithm is intrinsic to the algorithm itself and is independent of software or hardware implementations. As the design goes into the architectural level, the corresponding average bus or memory bandwidth can be estimated based on the algorithmic average data transfer rate. However, the peak bandwidth requirements of

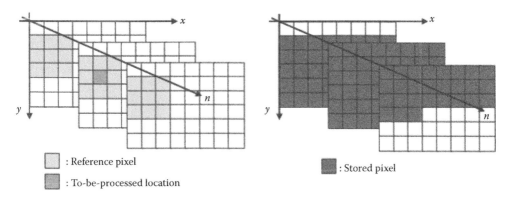

: Reference pixel

: To-be-processed location

: Stored pixel

FIGURE 23.14
Filter support of a $3 \times 3 \times 3$ spatial–temporal filter.

algorithms are significantly influenced by several architectural design details such as the memory hierarchy, data alignment in memory, data transaction type, and datapath. Therefore, the peak bandwidth is not intrinsic to algorithms but is dependent on architectures. Nevertheless, the peak bandwidth provides more insight into bus bit widths and clock speed in the ESL design.

To execute the functions of an algorithm, input data are read and consumed by operators to produce output data. Consequently, the amount of input data transfer can be easily calculated according to the input data list S of an algorithm for processing a unit data granularity. If input data reuse is not considered, some data will be transferred several times. This introduces the maximum amount of data transfer that is an intrinsic complexity metric measuring the theoretical upper bound of amount of data transfer without data reuse. On the contrary, as considering data reuse, the amount of data transfer can be quantified according to the nonoverlapped input data needed by two consecutively processed granularities. The overlapped input data of the previous and current processed granularities are stored and reused. In order to fetch the sufficient input data from outside for the currently processed granularity, only the nonoverlapped portion should be read. Hence, the amount of data transfer needed by the noninitial processed granularities can be estimated based on the nonoverlapped portion of input data. After taking throughput requirements in real-time applications into consideration, the corresponding average data transfer rate can be calculated accordingly.

In VLSI systems, data are read either from external memory or internal memory such as cache, scratchpad, embedded SRAM, and register file. This leaves freedom for trade-off between external bandwidth and internal memory size during architectural space exploration. However, the total amount of data read/written per second is always conserved and is equal to the average algorithmic maximum data transfer rate, which is referred to here as conservation of data transfer, since data are read from either internal memory or external memory and the total data read is conserved. Furthermore, the external bandwidth and internal memory size can be estimated based on the data transfer rate with data reuse. Consequently, these algorithmic complexity metrics can facilitate the exploration of bandwidth and memory configuration in AAC.

This section takes the motion estimation as an example for the quantification of data transfer rate. Figure 23.15 shows the search windows of the motion estimation whose search range is $[-S_H, S_H)$ in the horizontal direction and $[-S_V, S_V)$ in the vertical direction and the block size is $B \times B$. In the case that no datum in the search window is reused, the maximum

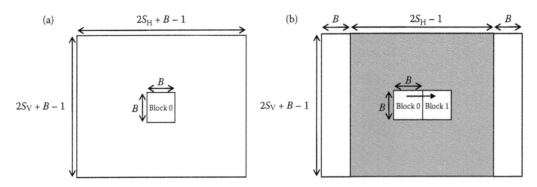

FIGURE 23.15
Search windows for motion estimation: (a) Search window of a single block. (b) Search window reuse of two consecutive blocks, where the gray region is the overlapped region.

amount of data transfer for finding a motion vector (MV) is equal to $N_S \times B \times B$ plus $B \times B$, where N_S is the number of search location depending on the search strategy. For full search (FS), the amount of data transfer needed is $4 \times S_H \times S_V \times B \times B$. As the data in the search windows are reused, the amount of data transfer needed to figure out the MV of a block is equal to $(2S_H + B - 1) \times (2S_V + B - 1)$ plus $B \times B$ and the local storage size for search window is equal to $(2S_H + B - 1) \times (2S_V + B - 1)$ as well. The average data transfer rate can then be calculated according to the number of blocks per second for real-time applications.

The horizontal search window data reuse scheme is proposed in [33] to reduce the external data transfer rate with little overhead on local storage size. As shown in Figure 23.15, only the nonoverlapped region of the search windows of two consecutive blocks in the horizontal direction has to be read. Consequently, the amount of data transfer for the second block is only $B \times (2S_V + B - 1)$. Furthermore, the local storage size becomes $(2S_H + 2B - 1) \times (2S_V + B - 1)$. In other words, the overhead of the local storage size is equal to the amount of the nonoverlapped input data. This example illustrates the data transfer rates and the associated local storage sizes for different levels of data reuse.

One intelligent way to efficiently increase the level of data reuse, thereby reducing external data transfer rate is to employ a coarser data granularity for processing so as to benefit from local spatial correlation of video. In Figure 23.15b, the data reuse scheme utilizes the search window overlap of two adjacent blocks only in the horizontal direction. On the contrary, the data reuse scheme shown in Figure 23.16 uses not only the horizontal overlap but also the vertical overlap at coarse data granularity, where G_V vertically adjacent blocks are encapsulated into a big one [34]. Figure 23.16a displays the search windows of a single big block. Clearly, the size of the search windows at coarse data granularity is $(2S_H + B - 1) \times (2S_V + G_V \times B - 1)$. Furthermore, Figure 23.16b shows the union of the search windows needed by two horizontally adjacent big blocks.

According to the search window overlap shown in Figure 23.16, the amount of input data for the search window update needed by a big block is $B \times (2S_V + G_V \times B - 1)$. The corresponding local storage size of the overall search window is $(2S_H + 2B - 1) \times (2S_V + G_V \times B - 1)$. The data reuse scheme shown in Figure 23.15 is a special case, in which the value of G_V is 1.

Table 23.1 tabulates the comparisons of average data transfer rate per block and local storage size at various levels of data reuse. This table compactly reveals the leverage between average data transfer rates and local storage size at various data granularities. Obviously, using a coarser data granularity for processing is capable of effectively raising the level of

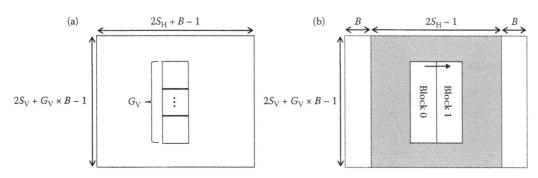

FIGURE 23.16
Search windows for motion estimation at coarser data granularity. (a) Search window of a single big block. (b) Search window reuse of two consecutive big blocks, where the gray region is the overlapped region.

TABLE 23.1

Comparison of Different Levels of Data Reuse

Reuse Scheme	Average Data Transfer Rate per Block (Pixels/Block)	Local Storage Size of Search Window (Pixels)
No data reuse	$4 \times S_H \times S_V \times B \times B$	$B \times B$
Horizontal search window reuse	$B \times (2S_V + B - 1)$	$(2S_H + 2B - 1) \times (2S_V + B - 1)$
Horizontal and vertical search window reuse	$[B \times (2S_V + G_V \times B - 1)]/G_V$	$(2S_H + 2B - 1) \times (2S_V + G_V \times B - 1)$

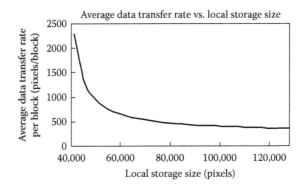

FIGURE 23.17
Average external data transfer rates versus local storage at various data granularities.

data reuse so as to reduce the average data transfer rate with overhead on local storage size. Figure 23.17 further illustrates the data rate versus storage size curve for the horizontal and vertical search window reuse scheme, in which the horizontal search range, $S_H = 128$, the vertical search range, $S_V = 64$, and the block size, $B = 16$. As the data granularity increases, the average data transfer rate decreases approximately in a reciprocal order. As a result, one can significantly reduce the external memory bandwidth by paying little overhead on the local storage size. For example, the average data transfer rate is reduced as 55.6% with 11% increase in the local storage size if the value of G_V is set to 2. This curve showing the average data transfer rate and storage size can then facilitate the exploration of architectural space in terms of the embedded memory size and external average bandwidth.

23.3 Dataflow Modeling and Complexity Characterization for Systems Design

In multicore platforms, the algorithmic complexity analysis, especially of the degree of parallelism helps map applications onto homogeneous or multigrain heterogeneous architectures. In addition, the complexity analysis also provides essential information to develop retargetable compilers for multicore platforms. Furthermore, it is capable of even facilitating porting operating systems onto the platforms, since designers are aware of the algorithmic intrinsic complexity, thereby understanding how to appropriately schedule the task.

As the data granularity of the dataflow studied is fine enough, the algorithmic complexity analysis can be used to extract features common to different algorithms and formats that are adaptive to versatile video contents. The commonality extracted can, of course, help in designing datapath and controllers from the hardware perspective, thereby resulting in highly efficient and flexible reconfigurable architectures in visual computing applications. For instance, the definition of functional units in MPEG RVC is done based on such a concept.

Consequently, building a dataflow model at a proper data granularity followed by thoroughly quantifying the complexity characterizing the algorithms reveals system architecture information and hence provides a systematic top-down design methodology for mapping visual applications onto the broad spectrum of platforms at different levels of granularity and performance. In addition, early understanding and if necessary feedback or back-annotation of architectural information or electronic ingredients enables optimization of algorithms. This section then shows three case studies for illustrating a reconfigurable discrete cosine transform (DCT) based on fine-grain dataflow, an H.264/AVC implementation on multicore platforms at various data granularities and an algorithm/architecture codesign of high-efficient motion estimator.

23.3.1 Reconfigurable DCT

The two-dimensional (2D) DCT is widely adopted in various video coding standards, such as MPEG-1/2/4 and H.264/AVC. The type-II DCT is expressed as follows:

$$y(k,l) = \frac{c_k c_l}{4} \sum_{i=0}^{N-1} \sum_{j=0}^{N-1} x(i,j) \cos\left(\frac{(2i+1)k\pi}{2N}\right) \cos\left(\frac{(2j+1)l\pi}{2N}\right),$$

$$c_k = \begin{cases} \dfrac{1}{\sqrt{2}} & \text{if } k = 0 \\ 0 & \text{otherwise} \end{cases}, \tag{23.6}$$

where $x(i,j)$ is the 2D input signal and $y(k,l)$ is the transformed DCT coefficients. Since the 2D DCT is separable, we can rewrite (23.6) as the form shown in (23.7):

$$y(k,l) = \frac{c_k}{2} \sum_{i=0}^{N-1} \left[\frac{c_l}{2} \sum_{j=0}^{N-1} x(i,j) \cos\left(\frac{(2j+1)l\pi}{2N}\right) \right] \cos\left(\frac{(2i+1)k\pi}{2N}\right) \tag{23.7}$$

The original DCT needs N^4 multiplications and $N^2(N^2-1)$ additions but the later one requires only $2N^3$ multiplications and $2N^2(N-1)$ additions. In other words, the 2D-DCT can be implemented by performing 1D-DCT twice with reduction in numbers of operations. However, this requires an additional transposed memory to store the intermediate data generated by the first 1D-DCT. The size of the transposed memory is $N \times N$. Both the original 2D-DCT and the separated one require the same amount of input data, which is $N \times N$. The above revision of DCT shows an example of trade-off between the number of operations and internal memory. In MPEG-1/2/4, the 8×8 type-II DCT is used for compression. In H.264/AVC, a low-complexity integer DCT is adopted by joint consideration of the DCT coefficients together with the quantization matrix. Furthermore, the H.264/AVC supports two DCT sizes that are 8×8 and 4×4 to adapt to versatile textures at various scales. Hence, it is critical to develop a reconfigurable DCT architecture to support multiple standards.

Sun et al. [35] proposed a low-complexity multipurpose reconfigurable DCT architecture that supports transitional 8×8 DCT, 4×4 integer DCT, and 8×8 integer DCT based on Loeffler DCT algorithm. In order to extract the commonality, the coordinate rotation digit compute (CORDIC) algorithm [36] is applied to the Loeffler DCT. In addition, the CORDIC algorithm has another advantage that can reduce the computational complexity by replacing multiplication by simple shifts and additions subject to the sufficient precision. Consequently, this reconfigurable DCT architecture is a representative example of AAC.

Based on the separated form (23.7), the following section mainly focuses on the complexity analysis of 1D-DCTs in terms of number of operation and degree of parallelism. Figure 23.18 shows the DFG of the 8-point Loeffler DCT. Benefiting from the symmetry of the DCT basis, the common operations within the DCT algorithm are executed only once and therefore the number of multiplications and additions are reused. In addition, the cosine-based coefficient can be calculated by vector rotation. These features result in the butterfly and lattice structure. After extracting the rotation coefficient and putting them to the outputs, the 8-point Loeffler DCT needs only 11 multiplications and 29 additions. Although the Loeffler DCT needs fewer operations, its degree of parallelism is degraded, since some common operations are shared, thereby increasing the dataflow dependency. As compared to the original 8-point DCT composed of eight independent operation sets, the Loeffler DCT has only two.

On the basis of the CORDIC algorithm [36], the multiplications in the dataflow of the Loeffler DCT can be substituted by shifts and additions as shown in Figure 23.19a. In addition, Figure 23.19b and c display the DFGs of the 8-point DCT and 4-point DCT, respectively. In video coding applications, the scale operation at the last stage can be combined with the quantization matrix so as to diminish the requirement of operation. According to the DFGs of various DCTs, the common operations are extracted. Obviously, the 4-point integer DCT is totally equivalent to the calculations of the even outputs of the 8-point DCT. In addition, the operations circled by the light-gray rectangles, the white rectangles with solid

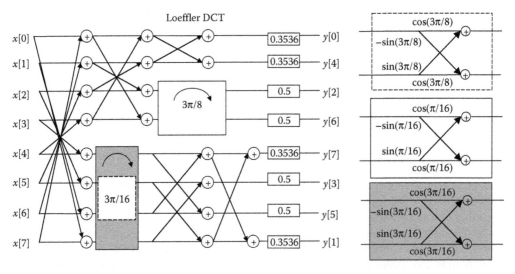

FIGURE 23.18
Dataflow graph of Loeffler DCT.

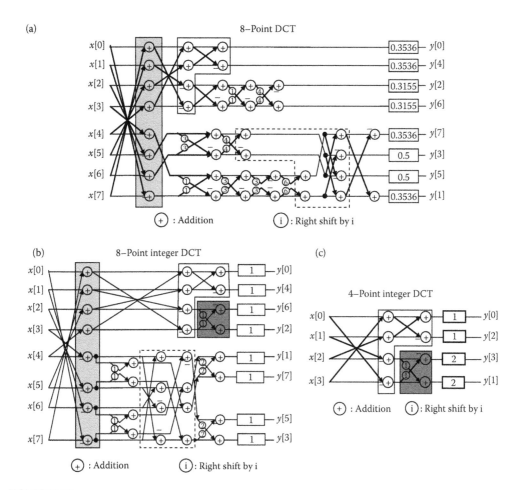

FIGURE 23.19
Dataflow graphs of various DCT: (a) 8-point CORDIC-based Loeffler DCT, (b) 8-point integer DCT, and (c) 4-point integer DCT.

and dotted lines (Figure 23.19a and b) further reveal the commonality between the traditional type-II 8-point DCT and 8-point integer DCT. Based on the commonality extracted, a reconfigurable dataflow for the three DCTs is shown in Figure 23.20.

In the reconfigurable DCT dataflow, six reconfigurable modules are employed for the different DCT algorithms. To serve the functionality of 8-point DCT, the reconfigurable modules are configured by (A). In addition, the input and output orders are accordingly changed as shown in Figure 23.19. Furthermore, the scale coefficients at the last stage are set correspondingly. In a similar manner, the 8-point integer DCT can be implemented by configuring the modules as type (B). It is worth noting that, as performing the 4-point integer DCT, the operations circled by the dotted line is bypass, thereby this reconfigurable dataflow can perform two 4-point integer DCT simultaneously.

Table 23.2 compares the number of operations of three DCTs and the reconfigurable one. Only additions and shifts are compared but the multiplication for the scale operation at the last stage is not involved. According to Table 23.2, it is obvious that the number of 8-point type-II DCT dominates the overall reconfigurable dataflow. The reconfigurable dataflow needs additional two additions and two shifts as compared to the type-II DCT because of

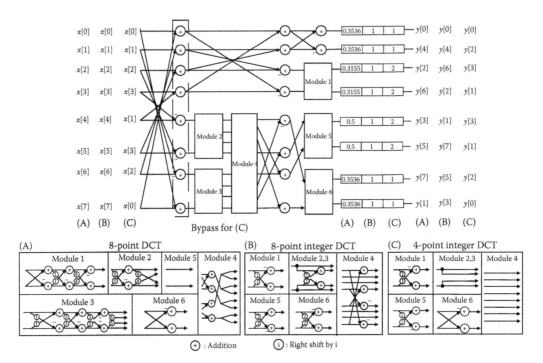

FIGURE 23.20
Reconfigurable dataflow of the 8-point type-II DCT, 8-point integer DCT, and 4-point DCT.

the module 5 needed by the integer DCT. If the three DCT algorithms are implemented separately, in total, 78 additions and 30 shifts are required. However, the reconfigurable dataflow needs only 40 additions and 18 shifts, which benefits extraction of common operations. The overhead of the reconfigurable dataflow is the complexity in control and multiplex. Obviously, this case study shows that via analyzing the algorithmic complexity together with dataflow modeling, the reconfigurable datapath possessing less silicon gate count and higher flexibility can then be efficiently designed at a fine granularity.

23.3.2 H.264/AVC on Multicore Platform at Various Data Granularities

Recently, H.264/MPEG-4 AVC is one of the most popular video coding standards, since it is capable of a high compression ratio. However, its complexity is higher than the other traditional coding standards, such as MPEG-2 and MPEG-4 part 2. Figure 23.21 shows the macroblock-level DFG of the H.264/MPEG-4 AVC decoder under the assumption that the

TABLE 23.2

Number of Operation Comparison for DCT

DCT	Number of Additions	Number of Shifts
8-Point type-II DCT	38	16
8-Point integer DCT	32	10
4-Point integer DCT	8	4
Reconfigurable DCT	40	18

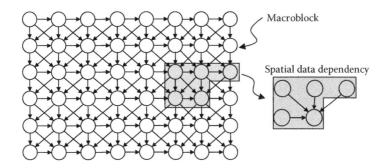

FIGURE 23.21
Dataflow graph of H.264/AVC.

entropy decoding and parsing processes have been performed before the inverse quantization (IQ), the inverse IDCT, the intra-/interprediction, and the deblocking (DB) filter processes, since the entropy decoder is composed of a series of operations [37].

This section introduces a case study of mapping H.264/MPEG-4 AVC onto multicore platforms based on the data partition at different granularities. As opposed to the functional partition used in the design of hardware video pipe and traditional single core, the data partition approach has several advantages. First, this approach provides high scalability. In addition, it inherently results in a higher degree of data locality, thereby significantly reducing the cross-core communication and bandwidth. Besides, the workload of equally partitioned data granularities is extremely balanced and therefore eases the synchronization among multiple cores so as to achieve a better speedup of parallelization. Furthermore, by using different sizes of data granularities, this section focuses on the trade-off between data transfer rate and local storage size for design space exploration as introduced in Section 23.2.3. Hence, the detailed number of operations is not analyzed.

Figure 23.22 displays the dataflow schedule of H.264/MPEG-4 AVC at the granularity of the macroblock level. The dataflow table explicitly tabulates the execution schedule for multicore platforms. The spatial data dependency of the intraprediction and DB filter makes

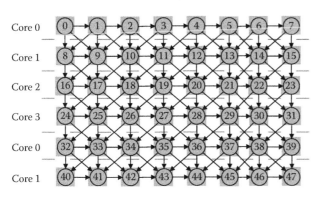

Time	Core 0	Core 1	Core 2	Core 3
0	G(0)			
1	G(1)			
2	G(2)	G(8)		
3	G(3)	G(9)		
4	G(4)	G(10)	G(16)	
5	G(5)	G(11)	G(17)	
6	G(6)	G(12)	G(18)	G(24)
7	G(7)	G(13)	G(19)	G(25)
8	G(32)	G(14)	G(20)	G(26)
9	G(33)	G(15)	G(21)	G(27)
10	G(34)	G(40)	G(22)	G(28)
11	G(35)	G(41)	G(23)	G(29)
12	G(36)	G(42)		G(30)
13	G(37)	G(43)		G(31)
14	G(38)	G(44)		
15	G(39)	G(45)		
16		G(76)		
17		G(47)		

FIGURE 23.22
Dataflow graph schedule of H.264/AVC at a fine granularity.

the dataflow schedule of the H.264/MPEG-4 AVC skewed. According to this dataflow, the maximum degree of parallelism of for H.264/MPEG-4 AVC at such a granularity is only four due to the high spatial data dependency. Based on the parallelism analysis above, one can intuitively obtain the dataflow for H.264/MPEG-4 AVC. If designers intend to implement the H.264/AVC on a multicore platform after considering design constraints and applications, they can employ the platform containing up to four cores to achieve maximum speedup. It takes at least 18 time units to complete the decamping processes if a picture is portioned into four cores for parallel process. Limited by the degree of algorithmic parallelism, the speedup is constrained even if more parallel cores are available. In this dataflow schedule, the reference data from the left macroblock are transferred locally within each core. However, the reference data from the upper macroblock are between cores and therefore introduce bandwidth requirements. The amount of data transfer can be quantified by estimating the reference data passed through the dashed lines shown in Figure 23.22.

In [38], the staircase shape of partition shown in Figure 23.23 is proposed to encapsulate spatially adjacent macroblocks into a coarse granularity without violating the data dependency constraint. In the coarse data granularity, the top-left macroblock has to be decoded prior to the top-right and bottom-left ones. The decoding order of the top-right and bottom-left macroblocks can be freely scheduled. However, the bottom-right macroblocks can be decoded only after all the other three are done. As compared to the fine-grain dataflow schedule, the coarse one needs less cross-core data transaction, since every two macroblock rows are locally decoded within a single core. This can significantly reduce the average data transfer rate. In this case, only two dashed lines are shown in Figure 23.23 but Figure 23.22 contains five dashed lines. In other words, one can easily know that the average data transfer rate of the coarser data granularity is only two-fifths of that of fine-grained one. However, the local storage size of using coarser granularity is also larger. Besides, the dataflow employing the coarser data granularity has the lower degree of parallelism.

On the basis of this concept, Tol et al. [38] presented a mapping of H.264 decoding on multiprocessor architecture. The target multicore platform template consisted of eight VLIW multimedia processors (e.g., TriMedia VLIW [39]) with a four-level memory hierarchy. Each core possesses a Level-1 (L1) cache and a Level-2 (L2) cache. The L1 cache stores data and instructions separately, while the L2 cache is a combined 128-kB data-instruction cache. In addition, a 12-MB Level-3 (L3) cache is used for coherency between each core. Eventually, a globally external shred memory is employed.

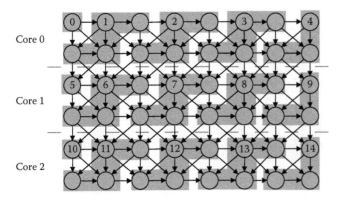

Time	Core 0	Core 1	Core 2
0	G(0)		
1	G(1)		
2	G(2)	G(5)	
3	G(3)	G(6)	
4	G(4)	G(7)	G(10)
5		G(8)	G(11)
6		G(9)	G(12)
7			G(13)
8			G(14)

FIGURE 23.23
Dataflow graph schedule of H.264/AVC at a coarse granularity.

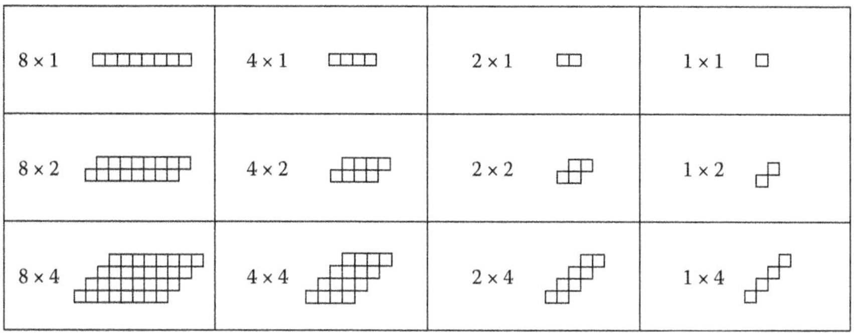

8 × 1	4 × 1	2 × 1	1 × 1
8 × 2	4 × 2	2 × 2	1 × 2
8 × 4	4 × 4	2 × 4	1 × 4

FIGURE 23.24
Data granularities possessing various shapes and sizes.

Tol et al. explored several data granularities with various shapes and sizes for the following considerations. Coarser granularities result in less data transition but need larger local cache sizes. However, finer data granularities imply better load balanced for multi-core platforms. For instance, the number of macroblocks in each partition varies from one to four as shown in Figure 23.23. The regular partition contains four macroblocks, but the corner case partitions have only one or three macroblocks.

Figure 23.24 illustrates 11 data granularities for mapping H.264/MPEG-4 AVC onto the target multicore platform. After taking the architecture design parameters such as the number of parallel cores, the size of the L2 cache, the cache line size and the data arrangement in memory into consideration, the data granularity of 8 × 4 macroblocks is selected, in which every 16 × 8 pixels are mapped onto each cache line. The experimental results conducted with several HD (1920 × 1088 at 30 Hz) sequences show that average bandwidth requirements of the 8 × 4 granularity for I-, P-, and B-picture are 208, 383, and 479 MB/s, respectively. In addition, the traditional functional partition at the data granularity of macroblock-level needs 1336 MB/s for B-pictures. The experiment result shows that using coarser data granularity together with load-balanced partition can reduce up to 65% bandwidth requirement. This case study clearly reveals the effectiveness of mapping complicated applications on to multicore platforms based on exploration of dataflow at multiple data granularities.

23.3.3 Algorithm/Architecture Codesign of Spatial–Temporal Recursive Motion Estimation

The motion estimation that can exploit motion trajectories of objects plays an important role in many video processing and coding applications. Block matching is a sort of algorithm for motion estimation. In general, the complexity of the motion estimation is relatively high due to the intensive computation. For example, the full search motion estimation exhaustively exams every search location within the search range and therefore needs a great number of operations. As a result, several heuristic search methods are proposed in the literature to reduce the complexity, and meanwhile maintain the performance in terms of visual quality as much as possible [40–45].

The 2D logarithmic search [45] is one of the heuristic search examples. Given the search range $[-S, S)$, this method, in the first step, calculates only nine search locations at $(0, 0)$, $(0, r_1)$, $(0, -r_1)$, $(r_1, 0), \ldots,$ and $(-r_1, -r_1)$, where $r_1 = 2d - 1$ and $d = \log_2 S$. In the second step, this method searches the eight locations around the best matching location of the first step with distance $r_2 = 2d - 2$. This method continues to search until the search distance equals 1. As compared to the HS, the 2D logarithmic search significantly reduces the number of search locations in a logarithmic order.

As the search range is $[-8, 8)$, the 2D logarithmic method is referred to as the three-step search (TSS) algorithm, since $\log_2 8 = 3$. In general, the number of search locations of L-step motion estimation is $1 + 8\,L$, since the first step tests nine locations and the rest of the steps examine eight locations. As the search range is larger, the number of search steps increase accordingly. Hence, the number of operations of the overall motion estimation grows as well. The exact number of operations can be estimated according to the matching criterion as discussed in the following paragraphs. The degree of coarse-grain parallelism is equal to the number of blocks. In other words, the searching of motion vector for each block is independent and can be done in parallel. As considering the median-grain parallelism, it is easy to find that the maximum parallelism for calculating the motion vector of a block is nine, which is equal to the number of search locations in the first step. In addition, the eight search locations in the second and third steps can be performed in parallel. Moreover, the parallelism embedded within searching each location depends on the matching criterion used.

In the literature, many matching criteria are investigated, such as sum of absolute difference (SAD) and mean square error. Calculating the SAD of a 16×16 block needs 256 subtractions and 255 additions. The degree of coarse-grain parallelism of the SAD criterion is only one. By incorporating the calculation of SAD into the TSS motion estimation, the overall number of additions and subtractions for searching a motion vector can then be quantified. This requires 6400 additions and 6375 subtractions. Once the image resolution and frame rate are taken into account, the number of operations per second can then be measured correspondingly.

In high-resolution applications, the search range increases as well. The number of search locations of the 2D logarithmic search is 57 for the search range $[-128, 128)$ for HD video. This is quite a larger number as compared to the TSS that needs only 22 search location. Consequently, Lee et al. [46] presented an algorithm/architecture codesign of spatio-temporal motion estimation (STME) for HD real-time applications based on the feature of motion trajectories in video sequences [46]. Via tracking motion trajectories, the STME is capable of significantly reducing the algorithmic complexity and finding relatively true motion vectors.

For video data of which the frame rate is 30 Hz or 60 Hz, the time interval between two consecutive pictures is 1/30 or 1/60 s. This is regarded as a short interval in which the trajectory of moving objects are to be a straight line and the velocity is to be constant as shown in Figure 23.25. Consequently, the STME adopts the backward and forward motion vectors of the previous picture that pass through the current block via linear extrapolation or interpolation

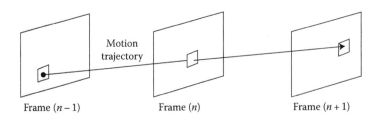

FIGURE 23.25
Linear motion trajectory in spatio-temporal domain.

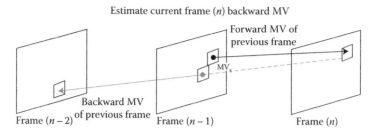

FIGURE 23.26
Spatio-temporal motion search strategy for backward motion estimation.

as the initial motion vector candidates so as to track motion trajectories as shown in Figure 23.26. Subsequently, local search is performed around the initial motion vector candidates to find the best match motion vector so as to deal with the small disturbance or minor error accompanied with linear motion trajectories.

This idea is innovated based on the physical features and properties of moving objects so as to estimate the true motion vectors. In addition, from the viewpoint of optimization theories, starting from proper initial points can not only speed up the convergence of finding optimal solutions but also significantly reduce the complexity. As compared to the logarithmic search methods presented in the previous paragraphs, the recursive motion estimation has lower degree of parallelism since its initial search locations are dependent on the motion vectors of the neighboring blocks in the spatial and temporal directions. Hence, the adjacent block cannot be processed in parallel. However, the number of operations of this recursive motion estimation is much less than that of the logarithmic ones, since the motion search of the STME is based on the contextual information and historical experience. This case shows a good example for the leverage between the degree of parallelism and the number of operations.

From the aspect of AAC, the number of search locations in STME can be viewed as an arbitrary parameter, which is independent of search range and video resolution for trade-off between algorithmic performance in terms of visual quality and the algorithmic complexity. It is a practical parameter for algorithm/architecture codesign of STME for HD real-time applications. As considering the algorithmic complexity, the number of operations is directly proportional to the number of search locations. Moreover, for the case that the reference data of a search location are stored in internal buffers and then reused by the next search location, the input data rate can be estimated by the following equation:

$$DR = (256 + 256 \times SL \times \text{miss\_ratio}) \times \frac{\text{blocks}}{\text{frame}} \times \frac{\text{frames}}{\text{second}}, \tag{23.8}$$

where DR is the input data rate, SL is the number of search locations, and miss_ratio is the percentage of nonoverlapping reference data needed by two consecutive search locations. Figure 23.27 shows the data rates needed by various number of search locations from 10 to 40 points for HD applications. The results are obtained by the simulation of algorithmic C codes of STME. The simulation results show that the data rate is almost linearly dependent on SL. Clearly, the algorithmic complexity of STME can be easily controlled by the number of search locations.

Regarding visual quality analysis, this STME is applied to H.264/AVC. Figure 23.28 shows the peak signal-to-noise ratio (PSNR) curve for different numbers of search locations from 10 to 40 points. Under the assumption that 16 PEs calculating absolute difference of 2 pixels

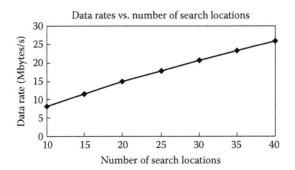

FIGURE 23.27
Data rate comparison of the STME for various number of search locations.

are used, it is easy to find that the possible range of minimum clock rate for real time are from 54 to 216 MHz for 10–40 search locations if the datapaths are employed at 75% utilization. However, as considering the PSNR trend, one can find that using 15 search locations could be a practical decision for low-complexity and low-cost application while providing satisfactory visual quality, since its PSNR is only 0.02 dB lower than that of the 40 search locations.

Figure 23.29 shows the PSNR comparisons of STME with FS, adaptive irregular pattern search (AIPS) [40], enhanced hexagonal search (EHS) [41], and modified hierarchical search (MHS) [42]. Obviously, STME has higher PSNR as compared to other motion estimation (ME) algorithms, except for the FS. However, for the search range $[-128\ 128)$ in the horizontal direction and $[-64\ 64)$ in the vertical direction, the number of search locations of FS is 16,384, which is around 1092-fold as compared to STME. Such a high number of operations need an extremely large number of PEs and consumes more power. Hence, the FS algorithm is not practical for HD applications. On the contrary, STME examining only 15 search locations for satisfactory visual quality is much more practical for VLSI implementation.

Figure 23.30 displays the block diagram of the STME architecture consisting of memory interface, controller, cache table, return data manager, three cache modules, cross path for rotation/windowing, and a parallel SAD calculator. Input to this architecture is a 128-bit data bus along with estimated MVs as the output port. The 16×16 blocks in the same row are allocated in DRAMs with the same row address. One strip of 16 horizontal pixels of a block maps to one DRAM address. Such a configuration efficiently reduces the

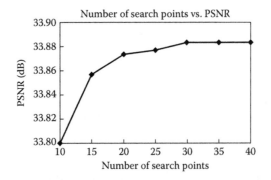

FIGURE 23.28
PSNR comparison of the STME for various number of search locations.

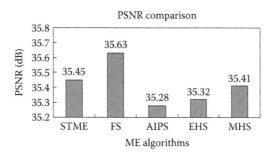

FIGURE 23.29
PSNR comparison of ME algorithms.

DRAM row address strobe latency. All blocks in a reference picture are partitioned into two groups, that is, odd blocks and even blocks, differentiated by the horizontal block index. Odd and even blocks are stored in the odd reference block cache and even reference block cache, respectively. A reference block at an arbitrary position is fetched by simultaneously accessing the odd and even caches with proper rotation and windowing. SAD calculation for an MV is separated into 16 horizontal strips. Each strip goes through cache hit detection, cache data read, and partial SAD calculation. Once cache miss occurs, reading the missing strip from DRAMs is issued. After the reference strip is ready in the caches, the SAD of the strip is obtained by the 16 PEs performing absolute difference in the parallel SAD calculation module. Eventually, the final SAD of a 16×16 bock is summed up so as to determine the best matching MV amount in the search locations.

Table 23.3 lists the architectural comparison of STME with an FS implementation in [47]. The gate count of STME is slightly smaller than that of the FS implementation, since both architectures contain 16 PEs that calculate SADs. However, the throughput of STME is about 15 times that of the FS ME. This is mainly because of the early back-annotation of architectural information and smart development of algorithm by tracking motion trajectories in STME. Hence, STME is definitely one of the representative examples of AAC. More details can be found in [46].

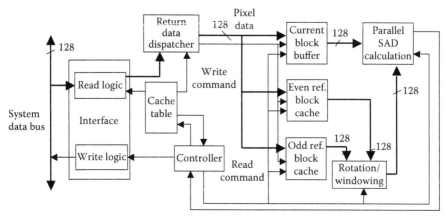

FIGURE 23.30
Block diagram of the STME architecture.

TABLE 23.3

Architectural Comparison of Motion Estimators

Terms	STME [46]	FS/Partial SAD [47]
Application	1920 × 1080 at 30 FPS	720 × 480 at 10 FPS
Technology	0.18 mm	0.18 mm
Clock rate	81 MHz	110.8 MHz
Total cell area	51.2 K gates	66.6 K gates

23.4 Conclusions

To cope with the increasing complexity of future visual computer algorithms and with expectations in rendering the best visual experience, this chapter describes the importance of concurrently optimizing both algorithm and architecture so that architecture information is back-annotated to the algorithm and is thus being considered as early as the algorithm design phase. We have also shown that advanced multicore platforms provide the best trade-off for higher performance per unit area (gate count), better flexibility, and lower power consumption. Reconfigurable architectures with lower granularity are also favored in applications for higher performance and lower power, but at the cost of less flexibility. AAC for multicore platforms and reconfigurable architectures will inevitably become a trend for implementing the coming generation of visual computing algorithms.

References

1. G. G. (Chris) Lee, M.-J. Wang, H.-T. Li, and H.-Y. Lin, A motion-adaptive deinterlacer via hybrid motion detection and edge-pattern recognition, *EURASIP Journal on Image and Video Processing*, 2008, 10, 2008.
2. G. G. Lee, H.-T. Li, M.-J. Wang, and H.-Y. Lin, Motion adaptive deinterlacing via edge pattern recognition, in *Proceedings of IEEE International Symposium on Circuits and Systems (ISCAS 2007)*, pp. 2662–2665, New Orleans, USA, May 2008.
3. G. G. Lee, H.-Y. Lin, M.-J. Wang, R.-L. Lai, C. W. Jhuo, and B.-H. Chen, Spatial–temporal content-adaptive deinterlacing algorithm, *IET Image Processing*, 2(6), 323–336, 2008.
4. G. G. Lee, H.-Y. Lin, M.-J. Wang, R.-L. Lai, and C.-W. Jhuo, A high-quality spatial–temporal content-adaptive deinterlacing algorithm, in *Proceedings of IEEE International Symposium on Circuits and Systems (ISCAS 2008)*, pp. 2594–2597, Seattle, USA, May 2008.
5. R. W. Schaefer and A. V. Oppenheim, *Discrete-Time Signal Processing*. Englewood Cliffs, NJ: Prentice-Hall, 1989.
6. S. Edeards, L. Lavagno, E. A. Lee, and A. Sangiovanni-Vincentelli. Design of embedded systems: Formal models, validation and synthesis. *Proceedings of the IEEE*, 85(3), 366–390, 1997.
7. G. G. Lee, E. S. Jang, M. Mattavelli, M. Raulet, C. Lucarz, H. Kim, S. Lee, H.-Y. Lin, J. Janneck, D. Ding, and C-J. Tsai, Text of ISO/IEC FDIS 23001-4: Codec Configuration Representation, Information technology—MPEG systems technologies—Part 4: Codec configuration representation, 2009.

8. H. S. Shin, Y.-S. Tung, C. Lucarz, K. Sugimoto, M. Raulet, Y. Yamada, H.-Y. Lin, Y.-L. Cheng, and M. Mattavelli, Text of ISO/IEC FDIS 23002-4: Video Tool Library, Information technology—MPEG video technologies—Part 4: Video tool library, 2010.

9. G. G. Lee, C. Lucarz, and H.-Y. Lin, Study of ISO/IEC 23002-4/FPDAM1 Video Tool Library Conformance and Reference Software, ISO/IEC JTC1/SC29/WG11 MPEG w10921, Xian, China, Oct. 2009. (will be an ISO/IEC international standard in October 2010).

10. V. Mathur and V. Prasanna, A hierarchical simulation framework for application development on system-on-chip architectures, in *Proceedings of 14th Annual IEEE International Applied Specific Integrated Circuit/System Chip Conference*, pp. 428–434, Brussels, Belgium, 2001.

11. L. Benini, D. Bertozzi, D. Bruni, N. Drago, F. Fummi, and M. Poncino, SystemC cosimulation and emulation of multiprocessor SoC designs, *Computer*, 36(4), 53–59, 2003.

12. L. Thiele, S. Chakraborty, M. Gries, and S. Künzli, A framework for evaluating design tradeoffs in packet processing architectures, in *Proceedings of Annual ACM IEEE Design Automation Conference*, pp. 880–885, New Orleans, LA, 2002.

13. G. Kahn, The semantics of simple language for parallel programming, in *IFIP Congress '74*, pp. 471–475, Stockholm, Sweden, 1974.

14. E. A. de Kock, W. J. M. Smits, P. van der Wolf, J.-Y. Brunel, W. M. Kruijtzer, P. Lieverse, K. A. Vissers, and G. Essink, YAPI: Application modeling for signal processing systems, in *Proceedings of Design Automation Conference*, pp. 402–405, Los Angeles, California, 2000.

15. T. Grotker, S. Liao, G. Martin, and S. Swan, *System Design with SystemC*, Philip Drive Norwell, Massachusetts: Kluwer Academic Publishers, 2002.

16. L. Cai and D. Gajski, Transaction level modeling: An overview, in *Proceedings of the International Conference on Hardware/Software Codesign and System Synthesis*, Newport Beach, CA, October 2003.

17. A. D. Booth, A signed binary multiplication technique, *Quarterly Journal of Mechanics and Applied Mathematics*, 4(2), 236–240, 1951.

18. M. Horowitz, A. Joch, F. Kossentini, and A. Hallapuro, H.264/AVC baseline profile decoder complexity analysis, *IEEE Transactions on Circuits and Systems for Video Technology*, 13(7), 704–716, 2003.

19. M. Ravasi and M. Mattavelli, High-abstraction level complexity analysis and memory architecture simulations of multimedia algorithms, *IEEE Transactions on Circuits and Systems for Video Technology*, 15(5), 673–684, 2005.

20. M. Ravasi and M. Mattavelli, High-level algorithmic complexity evaluation for system design, *Journal of Systems Architecture*, 48/13–15, 403–427, 2003.

21. G. M. Amdahl, Validity of single-processor approach to achieving large-scale computing capability, in *Proceedings of AFIPS Conference*, pp. 483–485, Atlantic City, New Jersey, 1967.

22. V. Escuder, R. Duran, and R. Rico, Quantifying ILP by means of graph theory, in *Proceedings of the 2nd International Conference on Performance Evaluation Methodologies and Tools*, pp. 317–322, San Francisco, California, 2007.

23. A. Prihozhy, M. Mattavelli, and D. Mlynek, Evaluation of the parallelization potential for efficient multimedia implementations: dynamic evaluation of algorithm critical path, *IEEE Transactions on Circuits and Systems for Video Technology*, 15(5), 593–608, 2005.

24. S. Y. Kung, *VLSI Array Processor*. Upper Saddle River, New Jersey: Prentice-Hall, 1988.

25. J. W. Janneck, D. Miller, and D. B. Parlour, Profiling dataflow programs, in *Proceedings of IEEE ICME 2008*, pp. 1665–1068, June 2008.

26. K. K. Parhi, *VLSI Digital Signal Processing Systems: Design and Implementation*. New York: Wiley, 1999.

27. H.-Y Lin and G. G. Lee, Quantifying Intrinsic parallelism via eigen-decomposition of dataflow graphs for algorithm/architecture co-exploration, in *Proceedings of IEEE SiPS 2010*, October 2010.

28. F. R. K. Chung, Spectral graph theory, *Regional Conferences Series in Mathematics*, No. 92, 1997.

29. M. Fiedler, Algebraic connectivity of graphs, *Czechoslovakia Mathematical Journal*, 23(2), 298–305, 1973.

30. M. C. McFarland, A. C. Parker, and R. Camposano, The high level synthesis of digital system, *Proceedings of IEEE*, 78(2), 301–318, 1990.

31. F. Balasa, F. Catthoor, and H. De. Man, Background memory area estimation for multidimensional signal processing systems, *IEEE Transactions on VLSI Systems*, 3(2), 157–172, 1995.

32. P. G. Kjeldsberg, F. Catthoor, and E. J. Aas, Storage requirement estimation for data intensive applications with partially fixed execution ordering, in *Proceedings of 8th International Workshop on Hardware/Software Codesign*, pp. 56–60, San Diego, CA, May 3–5, 2000.

33. J.-C. Tuan, T.-S. Chang, and C.-W. Jen, On the data reuse and memory bandwidth analysis for full-search block-matching VLSI architecture, *IEEE Transactions on Circuits and Systems for Video Technology*, 12(1), 61–72, 2002.

34. C.-Y. Chen, C.-T. Huang, Y.-H. Chen, and L.-G. Chen, Level C+ data reuse scheme for motion estimation with corresponding coding orders, *IEEE Transactions on Circuits and Systems for Video Technology*, 16(4), 553–555, 2006.

35. C. Loeffler, A. Lightenberg, and G. S. Moschytz, Practical fast 1-D DCT algorithms with 11-multiplications, in *Proceedings of IEEE International Conference on Acoustics, Speech, and Signal Processing*, Vol. 2, pp. 988–991, Glasgow, UK, May 1989.

36. J. E. Volder, The CORDIC trigonometric computing technique. *IRE Transactions on Computers*, EC-8, 330–334, 1959.

37. A. Segall and J. Zhao, Entropy slices for parallel entropy decoding, *ITU-T Q.6/SG16 VCEG, COM16-C405*, Geneva, CH, April 2008.

38. E. B. van der Tol, E. G. Jaspers, and R. H. Gelderblom, Mapping of H.264 decoding on a multiprocessor architecture, in *Proceedings of SPIE*, Vol. SPIE-5022, pp. 707–718, Bellingham, Washington, May 2003.

39. S. Dutta, D. Singh, and V. Mehra, Architecture and implementation of single-chip programmable digital television and media processor, in *Proceedings of IEEE SiPS99*, pp. 321–320, Taipei, Taiwan, 1999.

40. Y. Nie and K. Ma, Adaptive irregular pattern search with matching prejudgment for fast block-matching motion Estimation, *IEEE Transactions on Circuits and Systems for Video Technology*, 15(6), 789–794, 2005.

41. C. Zhu, X. Lin, L. Chau, and L. Po, Enhanced hexagonal search for fast block motion estimation, *IEEE Transactions on Circuits and Systems for Video Technology*, 14(10), 1210–1214, 2004.

42. Y. Murachi, K. Hamano, T. Matsuno, J. Miyakoshi, M. Miyama, and M. Yoshimoto, A 95 mW MPEG2 MP@HL motion estimation processor core for portable high-resolution video application, *IEICE Transactions of Fundamentals of Electronics, Communications and Computer Sciences*, E88-A, 3492–3499, 2005.

43. C. Zhu, X. Lin, and L.P. Chau, Hexagon-based search pattern for fast block motion estimation, *IEEE Transactions on Circuits and Systems for Video Technology*, 12, 349–355, 2002.

44. K. R. Namuduri, Motion estimation using spatio-temporal contextual information, *IEEE Transactions on Circuits and Systems for Video Technology*, 14, 1111–1115, 2004.

45. V. Bhaskaran and K. Konstantinides, *Image and Video Compression Standards Algorithms and Architectures*, 2nd edition, Philip Drive Norwell, Massachusetts: Kluwer Academic Publishers, 1997.

46. G. G. Lee, M.-J. Wang, H.-Y. Lin, Drew W. C. Su, and B.-Y. Lin, Algorithm/architecture co-design of 3D spatio-temporal motion estimation for video coding. *IEEE Transactions on Multimedia*, 9(3), 455–465, 2007.

47. C. Chen, S. Huang, Y. Chen, T. Wang, and L. Chen, Analysis and architecture design of variable block-size motion estimation for H.264/AVC, *IEEE Transactions on Circuits and Systems I*, 53(3), 578–593, 2006.

24

Dataflow-Based Design and Implementation of Image Processing Applications

Chung-Ching Shen, William Plishker, and Shuvra S. Bhattacharyya

CONTENTS

24.1 Dataflow Introduction

Model-based design has been explored extensively over the years in many domains of embedded systems. In model-based design, application subsystems are represented in terms of functional components that interact through formal models of computation (e.g., see [1]). By exposing and exploiting the high-level application structure that is often difficult to extract from platform-based design tools, model-based approaches facilitate systematic integration, analysis, synthesis, and optimization that can be used to exploit platform-based tools and devices more effectively.

Dataflow is a well-known computational model and is widely used for expressing the functionality of digital signal processing (DSP) applications, such as audio and video data stream processing, digital communications, and image processing (e.g., see [2–4]). These applications usually require real-time processing capabilities and have critical performance constraints. Dataflow provides a formal mechanism for describing specifications of DSP applications, imposes minimal data-dependency constraints in specifications, and is effective in exposing and exploiting task or data-level parallelism for achieving high-performance implementations.

Dataflow models of computation have been used in a wide variety of development environments to aid in the design and implementation of DSP applications (e.g., see [1,5–7]). In these tools, an application designer is able to develop complete functional specifications of model-based components, and functional validation or implementation on targeted platforms can be achieved through automated system simulation or synthesis processes.

Dataflow graphs are directed graphs, where vertices (*actors*) represent computational functions, and edges represent first-in-first-out (FIFO) channels for storing data values (*tokens*) and imposing data dependencies between actors. In DSP-oriented dataflow models of computation, actors can typically represent computations of arbitrary complexity as long as the interfaces of the computations conform to dataflow semantics. That is, dataflow actors produce and consume data from their input and output edges, respectively, and each actor executes as a sequence of discrete units of computation, called *firings*, where each firing depends on some well-defined amount of data from the input edges of the associated actor.

Retargetability is an important property to integrate into dataflow-based design environments for DSP. The need for efficient retargetability is of increasing concern due to the wide variety of platforms that are available for targeting implementations under different kinds of constraints and requirements. Such platforms include, for example, programmable digital signal processors, field programmable gate arrays (FPGAs), and graphics processing units (GPUs). For efficient exploration of implementation trade-offs, designers should be able to rapidly target these kinds of platforms for functional prototyping and validation. For example, Sen et al. [4] introduced a structured, dataflow-based design methodology for mapping rigid image registration applications onto FPGA platforms under real-time performance constraints. Shen et al. [8] presented a method to derive Pareto points in the design space that provide trade-offs between memory usage and performance based on different scheduling strategies. This approach to derive such Pareto points demonstrates a systematic, retargetable methodology based on high-level dataflow representations.

By following structured design methods enabled by dataflow modeling, designers can efficiently port implementations of dataflow graph components (actors and FIFOs) across different platform-oriented languages, such as C, C++, CUDA, MATLAB, SystemC, Verilog, and VHDL, while the application description can still be tied as closely as possible to the application domain, not the target platform. This makes a dataflow-based design highly portable while still structured enough to be optimized for. For example, in ref. [8], a dataflow-based design tool is introduced for design and analysis of embedded software for multimedia systems. In this tool, exploitation of data parallelism can be explored efficiently and associated performance metrics are evaluated for different dataflow graph components that are implemented using different platform-oriented languages.

By promoting formally specified component interfaces (in terms of dataflow semantics), and modular design principles, dataflow techniques also provide natural connections to powerful unit testing methodologies, as well as automated, unit-testing-driven correspondence checking between different implementations of the same component (e.g., see [9,10]).

To demonstrate dataflow-based design methods in a manner that is concrete and easily adapted to different platforms and back-end design tools, we present case studies based on the *lightweight dataflow* (LWDF) programming methodology [11]. LWDF is designed as a "minimalistic" approach for integrating coarse grain dataflow programming structures into arbitrary simulation- or platform-oriented languages, such as those listed above (i.e., C, C++, CUDA, etc.). In particular, LWDF requires minimal dependence on specialized tools or libraries. This feature—together with the rigorous adherence to dataflow principles throughout the LWDF design framework—allows designers to integrate and experiment with dataflow modeling approaches relatively quickly and flexibly into existing design methodologies and processes.

Thus, LWDF is well suited for presenting case studies in the context of this chapter, where our objective is to emphasize fundamental dataflow concepts and features, and their connection to developing efficient parallel implementations of image processing applications. In this chapter, we provide a background on relevant dataflow concepts and LWDF programming, and present LWDF-based case studies that demonstrate effective use of dataflow techniques for image processing systems. We focus specifically on case studies that demonstrate the integration of dataflow programming structures with C, Verilog, and CUDA for simulation, FPGA mapping, and GPU implementation, respectively.

24.2 Overview of Dataflow Models

In this section, we review a number of important forms of dataflow that are employed in the design and implementation of DSP systems.

Synchronous dataflow (SDF) [3] is a specialized form of dataflow that is useful for an important class of DSP applications, and is used in a variety of commercial design tools. In SDF, actors produce and consume constant amounts of data with respect to their input and output ports. Useful features of SDF include compile time, formal validation of deadlock-free operation and bounded buffer memory requirements; support for efficient static scheduling; and buffer size optimization (e.g., see [12]). *Cyclo-static dataflow* (CSDF) [13] is a generalization of SDF where the consumption or production rate of an actor port forms a periodic sequence of constant values. Both SDF and CSDF are static dataflow models in which production and consumption rates are statically known and data independent.

The *Parameterized synchronous dataflow* (PSDF) model of computation results from the integration of SDF with the metamodeling framework of *parameterized dataflow* [14]. PSDF expresses a wide range of dynamic dataflow behaviors while preserving much of the useful analysis and synthesis capability of SDF [14]. Based on the scheduling features provided by PSDF, low-overhead, quasi-static schedules can be generated for hand-coded implementation or software synthesis. Here, by a quasi-static schedule, we mean an ordering of execution for the dataflow actors whose structure is largely fixed at compile time, with a relatively small amount of decision points or symbolic adjustments evaluated at run-time based on the values of relevant input data.

Boolean dataflow (BDF) [15] allows data-dependent control actors to be integrated with SDF actors. In BDF, the consumption and production rates of an actor port can vary dynamically based on the values of tokens that are observed at a designated *control port* of the actor. Two basic control actors in the BDF model are the `switch` and `select` actors. The `switch` actor has two input ports—one for control and another for data—and two output ports. On each firing, the actor consumes one token from its data input, and one Boolean-valued token from its control input, and copies the value consumed from its data input onto one of its two output ports. The output port on which data are produced during a given firing is selected based on the Boolean value of the corresponding control input token.

On the contrary, the `select` actor has two data input ports, one control input port and one output port. On each firing, the actor consumes a Boolean-valued token from its control input, and selects a data input based on the value of this control token. The actor then consumes a single token from the selected data input, and copies this token onto the single actor output port.

Enable–invoke dataflow (EIDF) is a recently proposed dataflow model [16] that facilitates the design of applications with structured dynamic behavior. EIDF divides actors into sets of *modes*. Each mode, when executed, consumes and produces a fixed number of tokens. The fixed behavior of a mode provides a structure that can be exploited by analysis and optimization tools, while dynamic behavior can be achieved by switching between modes at run-time.

Each mode is defined by an *enable method* and an *invoke method*, which correspond, respectively, to testing for sufficient input data and executing a single quantum ("invocation") of execution for a given actor. After a successful invocation, the invoke method returns a set of admissible *next modes*, any of which may be then checked for readiness using the enable method and then invoked, and so on. By returning a *set* of possible next modes (as opposed to being restricted to a single next mode), a designer may model nondeterministic applications in which execution can proceed down many possible paths.

In the implementation of dataflow tools, functionalities corresponding to the enable and invoke methods are often interleaved—for example, an actor firing may have computations that are interleaved with blocking reads of data that provide successive inputs to those computations. In the contrary, there is a clean separation of enable and invoke capabilities in EIDF. This separation helps to improve the predictability of an actor invocation (since availability of the required data can be guaranteed in advance by the enable method), and in prototyping efficient scheduling and synthesis techniques (since enable and invoke functionalities can be called separately by the scheduler).

Dynamic dataflow behaviors require special attention in scheduling to retain efficiency and minimize loss of predictability. The enable function is designed so that if desired, one can use it as a "hook" for dynamic or quasi-static scheduling techniques to rapidly query actors at run-time to see if they are executable. For this purpose, it is especially useful to separate the enable functionality from the remaining parts of actor execution.

These remaining parts are left for the invoke method, which is carefully defined to avoid computation that is redundant with the enable method. The restriction that the enable method operates only on token counts within buffers and not on token values further promotes the separation of enable and invoke functionalities while minimizing redundant computation between them. At the same time, this restriction does not limit the overall expressive power of EIDF, which is Turing complete. This can be seen from our ability to formulate enabling and invoking methods for BDF actors [16]. Since BDF is known to be Turing complete, and EIDF is at least as expressive as BDF, EIDF can express any computable function, including conditional dataflow behaviors, and other important forms of dynamic dataflow.

The restrictions in EIDF can therefore be viewed as design principles imposed in the architecture of dataflow actors rather than restrictions in functionality. Such principles lay the foundation for optimization and synthesis tools to effectively target a diverse set of platforms including FPGAs, GPUs, and other kinds of multicore processors.

The LWDF programming approach, which is used as a demonstration vehicle throughout this chapter, is based on the *core functional dataflow* (CFDF) model of computation [16,17]. CFDF is a special case of the EIDF model. Recall that in EIDF, the invoking function in general returns a set of valid next modes in which the actor can subsequently be invoked. This allows for nondeterminism as an actor can be invoked in any of the valid modes within the next-mode set. In the deterministic CFDF model, actors must proceed deterministically to one particular mode of execution whenever they are enabled. Hence, the invoking function should return only a single valid mode of execution instead of a set of arbitrary cardinality. In other words, CFDF is the model of computation that results when EIDF is restricted so that the set of next modes always has exactly one element. With this restricted form of invoking function, only one mode can be meaningfully interrogated by the enabling function, ensuring that the application is deterministic.

24.3 Lightweight Dataflow

LWDF is a programming approach that allows designers to integrate various dataflow modeling approaches relatively quickly and flexibly into existing design methodologies and processes [11]. LWDF is designed to be minimally intrusive on existing design processes and requires minimal dependence on specialized tools or libraries. LWDF can be combined with a wide variety of dataflow models to yield a lightweight programming method for those models. In this chapter, the combination of LWDF and the CFDF model is used to demonstrate model-based design and implementation of image processing applications, and we refer this method to as *lightweight core functional dataflow* (LWCFDF) programming.

In LWCFDF, an actor has an *operational context* (OC), which encapsulates the following entities related to an actor implementation:

- Actor parameters
- Local actor variables—variables whose values store temporary data that do not persist across actor firings
- Actor state variables—variables whose values do persist across firings
- References to the FIFOs corresponding to the input and output ports (edge connections) of the actor as a component of the enclosing dataflow graph
- Terminal modes: a (possibly empty) subset of actor modes in which the actor cannot be fired

In LWCFDF, the OC for an actor also contains a *mode variable* whose value stores the next CFDF mode of the actor and persists across firings. The LWCFDF OC also includes references to the invoke function and enable function of the actor. The concept of terminal modes, defined above, can be used to model finite subsequences of execution that are "restarted" only through external control (e.g., by an enclosing scheduler). This is implemented in LWCFDF by extending the standard CFDF enable functionality such that it unconditionally returns `false` whenever the actor is in a terminal mode. An example of a terminal mode is given in Section 24.4 as part of the specification of a Gaussian filtering application.

24.3.1 Design of LWCFDF Actors

Actor design in LWCFDF includes four interface functions—*construct, enable, invoke,* and *terminate*. The construct function can be viewed as a form of object-oriented constructor, which connects an actor to its input and output edges (FIFO buffers), and performs any other pre-execution initialization associated with the actor. Similarly, the terminate function performs any operations that are required for "closing out" the actor after the enclosing graph has finished executing (e.g., deallocation of actor-specific memory or closing of associated files).

To describe LWCFDF operation (including the general operation of the underlying CFDF model of computation) in more detail, we define the following notation for dataflow graph actors and edges.

- *inputs(a)*: the set of input edges for actor a. If a is a source actor, then $inputs(a) = \emptyset$.
- *outputs(a)*: the set of output edges for actor a. If a is a sink actor, then $outputs(a) = \emptyset$.
- *population(e)*: the number of tokens that reside in the FIFO associated with e at a given time (when this time is understood from context).
- *capacity(e)*: the buffer size associated with e—that is, the maximum number of tokens that can coexist in the FIFO associated with e.
- *cons(a, m, e)*: the number of tokens consumed on input edge e in mode m of a given actor a.
- *prod(a, m, e)*: the number of tokens produced on output edge e in mode m of a given actor a.
- $\tau(a)$: the set of terminal modes of actor a.

In LWCFDF, a finite buffer size *capacity(e)* must be defined for every edge at any point during execution. Typically, this size remains fixed during execution, although in more complex or less predictable applications, buffer sizes may be varied dynamically. For simplicity, this optional time-dependence of *capacity(e)* is suppressed from our notation. Further discussion on dynamically varying buffer sizes is beyond the scope of this chapter.

In the enable function for an LWCFDF actor a, a `true` value is returned if

$$population(e) \geq cons(a, m, e) \text{ for all } e \in inputs(a); \tag{24.1}$$

$$population(e) \leq (capacity(e) - prod(a, m, e)) \text{ for all } e \in outputs(a); \text{ and} \tag{24.2}$$

$$m \notin \tau(a), \tag{24.3}$$

where m is the current mode of a.

In other words, the enable function returns `true` if the given actor is not in a terminal mode, and has sufficient input data to execute the current mode, and the output edges of the actor have sufficient data to store the new tokens that are produced when the mode executes. An actor can be invoked at a given point of time if the enable function is `true`-valued at that time.

In the invoke function of an LWCFDF actor a, the operational sequence associated with a single invocation of a is implemented. Based on CFDF semantics, an actor proceeds deterministically to one particular mode of execution whenever it is enabled, and in any given mode, the invoke method of an actor should consume data from at least one input or produce data on at least one output (or both) [16]. Note that in case an actor includes state, then the state can be modeled as a self-loop edge (a dataflow edge whose source and sink actors are identical) with appropriate delay, and one or more modes can be defined that produce or consume data only from the self-loop edge. Thus, modes that affect only the actor state (and not the "true" inputs or outputs of the actor) do not fundamentally violate CFDF semantics, and are therefore permissible in LWDF.

The enable and invoke functions of LWCFDF actors are executed by schedulers in the LWCFDF run-time environment. When an enclosing scheduler executes an LWCFDF application, each actor starts in an *initial mode* that is specified as part of the application specification. When the invoke function for an actor a completes, it returns the next mode for a to the runtime environment. This next mode information can then be used for subsequent checking of enabling conditions for a.

24.3.2   Design of LWCFDF FIFOs

FIFO design for dataflow edge implementation is orthogonal to the design of dataflow actors in LWDF. That is, by using LWDF, application designers can focus on design of actors and mapping of edges to lower-level communication protocols through separate design processes (if desired) and integrate them later through well-defined interfaces. Such design flow separation is useful due to the orthogonal objectives, which center around computation and communication, respectively, associated with actor and FIFO implementation.

FIFO design in LWDF typically involves different structures in software compared to hardware. For software design in C, tokens can have arbitrary types associated with them—for example, tokens can be integers, floating point values, characters, or pointers (to any kind of data). Such an organization allows for flexibility in storing different kinds of data values, and efficiency in storing the data values directly (i.e., without being encapsulated in any sort of higher-level "token" object).

In C-based LWCFDF implementation, FIFO operations are encapsulated by interface functions in C. These functions are referenced through function pointers so that they can be targeted to different implementations for different FIFO types while adhering to standard interfaces (polymorphism).

Standard FIFO operations in LWDF include operations that perform the following tasks:

- Create a new FIFO with a specified capacity.
- Read and write tokens from and to a FIFO.
- Check the capacity of a FIFO.
- Check the number of tokens that are currently in a FIFO.
- Deallocate the storage associated with a FIFO (e.g., for dynamically adapting graph topologies or, more commonly, as a termination phase of overall application execution).

For hardware design using hardware description languages (HDLs), a dataflow graph edge is typically mapped to a FIFO module, and LWCFDF provides designers with mechanisms for developing efficient interfaces between actor and FIFO modules. For maximum flexibility in design optimization, LWCFDF provides for retargetability of actor–FIFO interfaces across synchronous, asynchronous, and mixed-clock implementation styles.

24.4   Simulation Case Study

As an example of using the LWDF programming model based on the CFDF model of computation for DSP software design using C, a C-based LWCFDF actor is implemented as an abstract data type (ADT) to enable efficient and convenient reuse of the actor across arbitrary applications. Such ADTs provide object-oriented implementations in C. In particular, an actor context is encapsulated with a separation of interface and implementation. Furthermore, by building on memory layout conventions of C structures, an actor context inherits from an actor design template that contains common (or "base class") types that are shared across actors. Similarly, through the use of function pointers, different execution contexts for a related group of actors can share the same name (e.g., as a "virtual method").

In typical C implementations, ADT components include header files to represent definitions that are exported to application developers and implementation files that contain implementation-specific definitions. We refer to the ADT-based integration of LWCFDF and C as LWCFDF-C.

24.4.1 Application Specification

To demonstrate LWDF in a C-based design flow, we create a basic application centered around Gaussian filtering. Two-dimensional 2D Gaussian filtering is a common kernel in image processing that is used for preprocessing. Gaussian filtering can be used to denoise an image or to prepare for multiresolution processing. A Gaussian filter is a filter whose impulse response is a Gaussian curve, which in two dimensions resembles a bell.

For filtering in digital systems, the continuous Gaussian filter is sampled in a window and stored as coefficients in a matrix. The filter is convolved with the input image by centering the matrix on each pixel, multiplying the value of each entry in the matrix with the appropriate pixel, and then summing the results to produce the value of the new pixel. This operation is repeated until the entire output image has been created.

The size of the matrix and the width of the filter may be customized according to the application. A wide filter will remove noise more aggressively but will smoothen sharp features. A narrow filter will have less of an impact on the quality of the image, but will be correspondingly less effective against noise.

Figure 24.1 shows a simple application based on Gaussian filtering. It reads bitmap files in tile chunks, inverts the values of the pixels of each tile, runs Gaussian filtering on each inverted tile, and then writes the results to an output bitmap file. The main processing pipeline is single rate in terms of tiles, and can be statically scheduled, but after initialization and end-of-file behavior modeled, there is conditional dataflow behavior in the application graph, which is represented by square brackets in the figure.

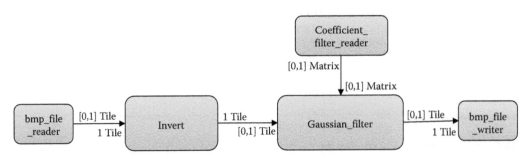

FIGURE 24.1
Dataflow graph of an image processing application for Gaussian filtering.

Such conditional behavior arises, first, because the Gaussian filter coefficients are programmable to allow for different standard deviations. The coefficients are set once per image—coefficient_filter_reader produces a coefficient matrix for only the first firing. To correspond to this behavior, the gaussian_filter actor consumes the coefficient matrix only once, and each subsequent firing processes tiles. Such conditional firing also applies to bmp_file_reader, which produces tiles until the end of the associated file is reached.

It should also be noted that the tiles indicated in Figure 24.1 do vary somewhat between edges. Gaussian filtering applied to tiles must consider a limited neighborhood around each tile (called a *halo*) for correct results. Therefore, tiles produced by bmp_file_reader overlap, while the halo is discarded after Gaussian filtering. As a result, nonoverlapping tiles form the input to bmp_file_writer.

24.4.2 Actor Design Using LWCFDF-C

As shown in Figure 24.1, our dataflow model of the image processing application for Gaussian filtering includes five actors: bmp_file_reader, coefficient_filter_reader, invert, gaussian_filter, and bmp_file_writer. The design and implementation of each actor using LWCFDF-C are described below.

The bmp_file_reader actor contains two modes: the process mode and the inactive mode. In the process mode, for each firing when the bmp_file_reader actor is enabled, it reads image pixels of the processing tile and the corresponding header information from a given bitmap file, and produces them to its output FIFOs. Then the actor returns the process mode as the mode for its next firing. This continues for each firing until all data have been read from the given bitmap file. After that, the actor returns the inactive mode, which is a terminal mode (see Section 24.3), and therefore indicates that the actor cannot be fired anymore until its current mode is first reset externally (e.g., by the enclosing scheduler). In other words, the inactive mode is in the terminal mode set of the bmp_file_reader.

The coefficient_filter_reader actor contains two modes: the process mode and the inactive mode. Again, the inactive mode is a terminal mode. For each firing when it is enabled (not in the inactive mode), the coefficient_filter_reader actor reads filter coefficients from a given file, stores them into a filter coefficient vector (FCV) array, and produces the coefficients onto its output FIFO. The FCV V has the form

$$V = (\text{sizeX}, \text{sizeY}, c_0, c_1, \dots, c_n), \tag{24.4}$$

where sizeX and sizeY denote the size of the FCV represented in 2D format; each c_i represents a coefficient value; and $n = \text{sizeX} \times \text{sizeY}$. After firing, the actor returns the process mode if there are data remaining in the input file; otherwise, the actor returns the inactive mode.

The bmp_file_writer actor contains only a single mode, called the process mode. Thus, the actor behaves as an SDF actor. For each firing when it is enabled, the bmp_file_writer actor reads the processed image pixels of the processing tile and the corresponding header information from its input FIFOs, and writes them to a bitmap file, which can later be used to display the processed results. The actor returns the process mode as the next mode for firing.

The invert actor contains only the process mode which makes the actor implemented as an SDF actor. For each firing when it is enabled, the invert actor reads the image pixels of the processing tile from its input FIFOs, inverts the color of the image pixels, and writes the processed result to its output FIFO. The actor returns the `process` mode as the next mode for firing.

The `gaussian_filter` actor contains two modes: the store coefficients STC (STore Coefficients) mode and the `process` mode. In the STC mode, for each firing when it is enabled, the `gaussian_filter` actor consumes filter coefficients from its coefficient input FIFO, caches them inside the actor for further reference, and then returns the `process` mode as the next mode for firing. In the `process` mode, for each firing when the `gaussian_filter` actor is enabled, image pixels of a single tile will be consumed from the tile input FIFO of the actor and the cached filter coefficients will be applied to these pixels. The results will be produced onto the tile output FIFO. The actor then returns the `process` mode as the next mode for firing. To activate a new set of coefficients, the actor must first be reset, through external control, back to the STC mode.

To demonstrate how actors are designed using LWCFDF-C, we use the `gaussian_filter` actor as a design example, and highlight the core implementation parts of the construct function and enable function in the following code segments.

```
/* actor operational context */
gfilter_cfdf_context_type *context = NULL;
context = util_malloc(sizeof(gfilter_cfdf_context_type));

/* actor enable function */
context->enable = (actor_enable_function_type)gfilter_cfdf_enable;

/* actor invoke function */
context->invoke = (actor_invoke_function_type)gfilter_cfdf_invoke;

context->mode = STC_MODE;      /* initial mode configuration */
context->filter = NULL;        /* pointer to filter coefficients */
context->filterX = 0;          /* size of filter coefficients X dimension */
context->filterY = 0;          /* size of filter coefficients Y dimension */
context->tileX = tileX;        /* length of the tile */
context->tileY = tileY;        /* height of the tile */
context->halo = halo;          /* halo padding around tile dimensions */
context->coef_in = coef_in;    /* coef input */
context->tile_in = tile_in;    /* tile input */
context->tile_out = tile_out;  /* tile output */
boolean result = FALSE;
switch (context->mode) {
    case STC_MODE:      /* Store Coefficients Mode */
        result = disps_fifo_population(context->coef_in) >= 1;
        break;
    case PROCESS_MODE:  /* Process Mode */
        result = disps_fifo_population(context->tile_in) >= 1;
        break;
    default:
        result = FALSE;
        break;
}
return result;
```

The invoke function implements the core computation of the `gaussian_filter` actor. The functionality of the corresponding modes is shown in the following code segments.

```
switch (context->mode) {
    case STC_MODE:
        fifo_read(context->coef_in, &fcv);

        /* first element of fcv stores the size of the filter. */
        context->filterY = fcv[0];
        context->filterX = fcv[1];
        context->filter = util_malloc(sizeof(float) * context->filterY *
                context->filterX);
        for (x = 0; x < context->filterY * context->filterX; x++) {
            context->filter[x] = fcv[x + 2];
        }
        sum_coefs = 0;
        for (x = 0; x < context->filterY* context->filterX; x++) {
            sum_coefs += context->filter[x];
        }
        for (x = 0; x < context->filterY * context->filterX; x++) {
            context->filter[x] /= sum_coefs;
        }
        context->mode = PROCESS_MODE;
        break;

case PROCESS_MODE:
    fifo_read(context->tile_in, &tile);

    /* form a new tile */
    newtile = malloc(sizeof(float) * (tileX) * (tileY));

    /* loop through the pixels in the tile */
    for (x = 0; x < tileX; x++) {
        for (y = 0; y < tileY; y++) {
            int yf, xf;
            newtile[(tileX) * y + x] = 0;

            /*loop through the coefs of the filter*/
            for (yf = 0; yf < context->filterY; yf++) {
                for (xf = 0; xf < context->filterX; xf++) {
                    newtile[(tileX) * y + x] +=
                            context->filter[yf * context->filterX + xf] *
                            tile[(tileX + 2 * halo) * (y + yf) + (x + xf)];
                }
            }
        }
    }
    fifo_write(context->tile_out, &newtile);
    context->mode = PROCESS_MODE;
    break;
```

Functional correctness of the LWCFDF design for the Gaussian filtering application can be verified by simulating its LWCFDF-C implementation using a simple scheduling strategy, which is an adaptation of the *canonical scheduling strategy* of the *functional Dataflow Interchange Format* (DIF) environment [17]. Since the semantics of LWCFDF dataflow guarantee deterministic operation (i.e., the same input/output behavior regardless of the schedule that is applied), validation under such a simulation guarantees correct operation regardless of the specific scheduling strategy that is ultimately used in the final implementation.

Such a simulation approach is therefore useful to orthogonalize functional validation of an LWDF design before exploring platform-specific scheduling optimizations—for example, optimizations that exploit the parallelism exposed by the given dataflow representation. This approach is also useful because it allows use of a standard, "off-the-shelf" scheduling strategy (the canonical scheduling adaptation described above) during functional validation so that designer effort on scheduling can be focused entirely on later phases of the design process, after functional correctness has been validated.

24.5 Background on FPGAs for Image Processing

24.5.1 FPGA Overview

An FPGA is a type of semiconductor device containing a regular matrix of user-programmable logic elements (LEs) whose interconnections are provided through a programmable routing network [18]. FPGAs can be easily configured to implement custom hardware functionalities. FPGAs provide for relatively low cost, and fast design time compared to implementation on application-specific integrated circuits (ASICs) with generally some loss in performance and energy efficiency. However, even when ASIC implementation is the ultimate objective, use of FPGAs is effective for rapid prototyping and early-stage design validation.

A typical FPGA architecture, which is shown in Figure 24.2, consists of various mixes of configurable logic blocks (CLBs), embedded memory blocks, routing switches, interconnects, and subsystems for high-speed I/O, and clock management.

CLBs or *LEs* are the main programmable LEs in FPGAs for implementing simple or complex combinatorial and sequential circuitry. A typical CLB element contains look-up tables for implementing Boolean functions; dedicated user-controlled multiplexers for combinational logic; dedicated arithmetic and carry logic; and programmable memory elements such as flip flops, registers, or RAMs. An example of a simplified Xilinx Virtex-6 FPGA CLB is shown in Figure 24.3.

Various FPGA's device families include significant support for computationally intensive DSP tasks. In addition to having higher-performance designs for major logic and memory elements and interconnects, dedicated hardware resources are provided in such FPGAs for commonly used DSP operations, such as multiply-accumulate (MAC) operations. Furthermore, performance-optimized intellectual property cores are also incorporated into FPGAs to perform specific DSP tasks, such as finite-impulse response (FIR) filtering and fast fourier transform (FFT) computation.

24.5.2 Image Processing on FPGAs

Image processing on FPGAs is attractive as many interesting applications, for example, in the domains of computer vision, and medical image processing, can now be implemented

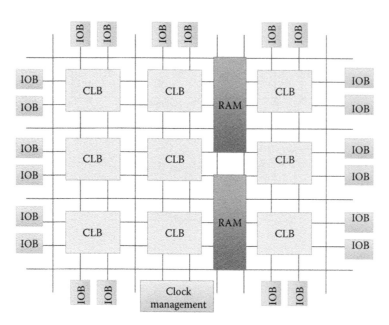

FIGURE 24.2
A typical FPGA architecture.

with high flexibility, relatively low cost, and high performance. This is because many common image processing operations can be mapped naturally onto FPGAs to exploit the inherent parallelism within them. Typical operations for image processing applications include image differencing, registration, and recognition.

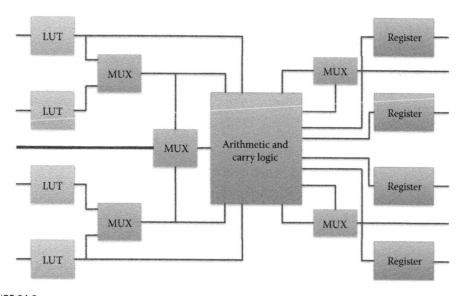

FIGURE 24.3
Simplified Xilinx Virtex-6 FPGA CLB. (From *Virtex-6 FPGA CLB User Guide, UG364 (v1.1)*, September 2009. With permission.)

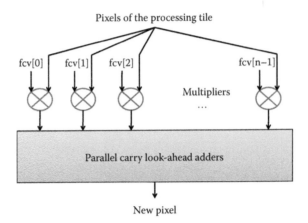

FIGURE 24.4
Parallel processing for tile pixels geared toward FPGA implementation.

Real-time image processing often requires DSP algorithms operating on multiple rows or columns of image pixels concurrently. FPGAs can provide extensive capabilities for high-speed, parallel processing of image pixels. For example, Xilinx's new FPGA, the Virtex-6, can deliver over 30 billion MAC operations per second for processing pixels in parallel. These elements can also be reconfigured to perform different tasks based on application requirements.

24.6   FPGA Design Case Study

To illustrate a dataflow-based design for FPGA implementation using the Verilog HDL, we design an LWCFDF actor as a Verilog module, where input and output ports of the actor are mapped to unidirectional module ports. Such an actor-level module can contain arbitrary submodules for complex actors, and the functionality of an actor can in general be described in either a behavioral or structural style. We refer to such an integration of LWCFDF with Verilog as *LWCFDF-V*.

Design of an LWCFDF-V actor module can be structured as a finite-state machine (FSM) for implementing the associated mode operations, an I/O controller for performing read and write operations on the FIFOs that are connected to the actor, and any number of submodules for concurrent execution of processing sub-tasks.

In the Gaussian filtering application based on the dataflow graph shown in Figure 24.1, only the `invert` and `gaussian_filtering` actors are synthesizable modules. The `bmp_file_read`, `coefficient_filter_reader`, and `bmp_file_writer` actors can be designed for verifying functional correctness of the application in Verilog simulations.

The `gaussian_filter` actor in LWCFDF-V has two modes: the `STC` and `process` modes. The actor implements the core computation of the Gaussian filtering application. In the `STC` mode, the actor reads coefficients and stores them into an internal memory. As described in Section 24.4, this operation under the `STC` mode is applied only once per image. In the `process` mode, the `gaussian_filter` actor incorporates a submodule to exploit parallelism associated with the convolution computations of the filter coefficients and image pixels. A design schematic of this parallel-processing component of the LWCFDF-V gaussian_filter actor module is shown in Figure 24.4.

As shown in Figure 24.4, the processing elements for the convolution computations basically consist of levels of multipliers and adders. In order to have efficient hardware implementations for both operations, we apply the *dadda tree multiplier* (DTM) [20] as the design method for designing the multipliers and the *carry look-ahead adder* (CLA) [21] for the adders. Both design methods provide speed improvements for the respective operations.

A Verilog specification of the gfilter_filter actor module is illustrated in the following code. This type of structural design provides a useful standard format for designing actor modules in LWCFDF-V.

```
/*********************************************************************
      Structural modeling for parallel processing
 *********************************************************************/
gfilter_process_module gpm(new_pixel, new_pixel_ready, pixel[0], pixel[1],
    ..., pixel[m-1], fcv[0], fcv[1], ..., fcv[n-1], mode, pixel_ready, clock
    , reset);
```

```
/*********************************************************************
      Behavioral modeling for FSM and I/O control
 *********************************************************************/
    always @(posedge clock or negedge reset) begin
        if (~reset) begin
            /* System reset */
        end
        else begin
            case (mode)
                `GFILTER_STC_MODE: begin
                    /* sequential operations and state transition */
                end
                `GFILTER_PROCESS_MODE: begin
                    /* sequential operations and state transition */
                end
            endcase
        end
    end

    always @(*) begin
        case (mode)
            `GFILTER_STC_MODE: begin
                /* combinational operations and state update */
            end
            `GFILTER_PROCESS_MODE: begin
                /* combinational operations and state update */
            end
        endcase
    end
```

Functional correctness of the LWCFDF-V design for the Gaussian filtering application can be verified by connecting all of the actor modules with the corresponding FIFO modules in a `testbench` module, and simulating the LWCFDF-V implementations based on a self-timed scheduling strategy. In such a scheduling strategy, an actor module fires whenever it has sufficient tokens (as required by the corresponding dataflow firing) available on its input FIFOs, and there is no central controller to drive the overall execution flow [22].

To experiment with the LWCFDF-V design, we used the Xilinx Virtex-5 FPGA device. We applied 256×256 images decomposed into 128×128 tiles and filtered with a 5×5 matrix of Gaussian filter coefficients. Based on these settings, 25 DTMs and 23 CLAs were instantiated for parallel processing. The synthesis result derived from the LWCFDF implementation of the `gaussian_filter` actor module achieves a clock speed of 500 MHz, and uses about 1% of the slice registers and 1% of the slice LUTs (LookUp Tables) on the targeted FPGA platform.

24.7 Background on GPUs for Image Processing

GPUs provide another class of high-performance computing platforms that can be leveraged for image processing. While they do not have the flexibility of FPGAs, GPUs are programmable with high-peak throughput capabilities. Typically a GPU architecture is structured as an array of hierarchically connected cores as shown in Figure 24.5. Cores tend to be lightweight as the GPU will instantiate many of them to support massively parallel graphics computation. Some memories are small and scoped for access to a few cores, but can be read or written in one or just a few cycles. Other memories are larger and accessible by more cores, but at the cost of longer read-and-write latencies.

Graphics processors began as dedicated coprocessors to CPUs for providing rendering capabilities. As graphics developers wanted more control over how the coprocessors behaved, GPU architectures introduced limited programmability to the processing pipeline. As GPUs became increasingly flexible to accommodate different rendering capabilities, developers began to realize that if a nongraphics problem could be cast as a graphics rendering problem, the cores on the GPU could be used for massive parallel processing. Initially, this required application developers to use a graphics language such as OpenGL [23] to describe their nongraphics problem, but modern programming interfaces to GPUs directly support general-purpose programming [24,25]. These general-purpose programming models have enabled GPU acceleration across a wide gamut of application areas [26].

Image processing is particularly well suited to acceleration from GPUs. If an image processing problem can be constructed as a per-pixel operation, then there is a good match to a GPU processing pipeline intended to render each pixel on the screen. Furthermore, GPUs typically have texture caches, which match well with 2D image tiles to be processed. Memory structures are tailored for contiguous access such as loading blocks of images. Furthermore, losing some floating point precision during the operation is also usually not critical.

This match between the application domain and the architecture has made GPUs popular in the image processing community. For example, GPUs are used extensively in computer vision, where computationally intensive kernels require acceleration to work in real time or on video streams [27]. Medical images may be enhanced through a variety of image

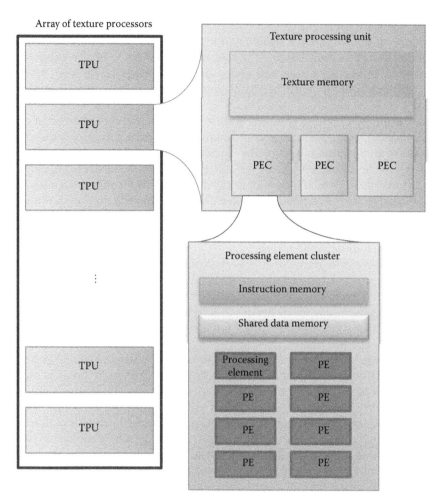

FIGURE 24.5
A typical GPU architecture.

processing techniques, and the GPU is leading to computationally intensive processing techniques that are clinically viable [28].

24.8   GPU Implementation Case Study

To examine concretely the problem of supporting dataflow applications on GPUs, we continue our development of the Gaussian filtering application, which was introduced in Section 24.4. Pipeline parallelism expressed by dataflow is generally not as applicable to GPUs as FPGAs. As GPU threads are executed with individual instruction streams and cannot coordinate at a coarse granularity, pipelines must be separated by distinct calls to the GPU. There are still benefits for GPU implementation to be derived from pipeline parallelism; however, in the context of GPU-based image processing, vectorization often leads to the most useful form of dataflow graph parallelism. In such vectorization, multiple firings of the same actor are executed concurrently to achieve parallel processing across different invocations of the actor. Each instance of a "vectorized actor" may be mapped to an individual thread or process, allowing the replicated instances to be executed in parallel.

Fundamentals of vectorized execution for dataflow actors have been developed by Ritz et al. [29], and explored further by Zivojnovic et al. [30], Lalgudi et al. [31], and Ko et al. [32].

Vectorized actor execution is simplest and most efficient when the actor in question is stateless. A stateful actor necessitates coordination among the duplicated instances of the actor. This does not prohibit a functionally correct vectorized realization, but the coordination step that must occur on every firing (and that potentially creates dependencies between actors) incurs overhead, which increases as the degree of duplication increases. Read-only data do not involve such overhead because read–write hazards do not exist and the data may be duplicated before invocation to alleviate memory access issues.

In our Gaussian filtering example, the core actor is stateless, and applies the same matrix to each tile and each pixel in the image.

An application developer may consider vectorization within and across actors while writing kernels for Compute Unified Device Architecture (CUDA) acceleration. In the context of LWCFDF-based design, the actor interface need not change as the vectorization degree changes, which makes it easy for designers to start with the programming framework provided by CUDA and wrap the resulting kernel designs in LWDF for integration at the dataflow graph level.

In the GPU-targeted version of our Gaussian filtering application, a CUDA kernel is developed to accelerate the core Gaussian filtering computation, and each thread is assigned to a single pixel, which leads to a set of parallel independent tasks. The threads are assembled into blocks to maximize data reuse. Each thread uses the same matrix for application to the local neighborhood, and there is significant overlap in the neighborhoods of the nearby pixels. To this end, the threads are grouped by tiles in the image. Once the kernel is launched, threads in a block cooperate to load the matrix, the tile to be processed, and a surrounding neighborhood of points. The image load itself is vectorized to ensure efficient bursting from memory. As CUDA recognizes the contiguous accesses across threads, the following code induces a vectorized access to global memory.

```
_ _global_ _ void gfilter_kernel(float *input, float *output,
    float *filter_in) {
    _ _shared_ _ float s_a[BLOCK_DIMX*BLOCK_DIMY];

    /* the global memory index for the pixel assigned to this thread */
    int input_ix = ...

    /* the local shared memory index for the pixel assigned to this thread */
    int local_ix = ...
    s_a[local_ix] = input[input_ix];

    ...
```

This loads only the core pixels of the tile to be processed, and so an additional code is needed to load the halo and the filter coefficients from the main memory. Computational vectorization occurs the same way with each thread in a block able to run simultaneously

TABLE 24.1

Runtime for 2 Different Gaussian Filter Implementations

Implementation	Time (s)
LWCFDF-C	3.793
LWCFDF-CUDA	0.995

with other threads to utilize the parallel datapaths in the GPU. For the core computation of Gaussian filtering, the iterations of the outer loop that index over a tile are controlled by the threads, where each thread executes the two inner loops that index over the filter, as shown in the following code.

```
float value = 0;
int yf, xf;
for (yf = 0;yf < FILTER_SIZE; yf++) {
    for(xf = 0;xf < FILTER_SIZE; xf++) {
        value += filter[yf][xf] * s_a[(TILE_PLUS_HALO) *
                (threadIdx.y+yf) + (threadIdx.x+xf)];
    }
}
output[output_ix] = value;
};
```

The original outer loops reveal exactly the style of parallelism we need to accelerate the actor using vectorization, but the inner loops could have been parallelized as well. However, these would return less performance benefits as the summation represents dependencies across threads. This would lead to further overhead to sum to the final result.

The outer-loop acceleration is accommodated by LWDF by simply augmenting the execution of a single mode of the original C actor. Since the actor interface need not change, all other aspects of the LWDF interface remain intact, opening the possibility of seamlessly using CUDA-accelerated actors in the same application with actors that employ other forms of acceleration. For example, if the designer had chosen the invert actor to also be CUDA accelerated, the actor's interfaces could have been altered to pass GPU pointers to image data instead of host pointers to image data. Then fewer loads to the device would be required, which would reduce execution time. If the GPU is not fully utilized, such that more blocks of threads are needed, vectorization of the actor (which is permissible because it is a stateless model) allows us to create more threads that may be spread across the GPU.

We examined software implementations of our Gaussian filtering application with and without GPU acceleration. For our experimental setup, we used 256×256 images decomposed into 128×128 tiles and filtered with a 21×21 matrix of Gaussian filter coefficients. We applied an NVIDIA GTX 285 running CUDA 3.1 and compared that to a C-only implementation running on 3 GHz Intel Xeons.

The total runtimes are showed in Table 24.1. While the GPU did not massively outperform the C-only implementation, it did show a marked improvement. The kernel itself was significantly faster, but because of transfer times to and from the GPU, only a modest overall speedup was observed.

24.9 Summary

In this chapter, we introduced dataflow-based methods for implementing efficient parallel implementations of image processing applications. Specially, we used an LWDF programming model to develop an application example for image processing and demonstrated implementations of this application using C, Verilog, and CUDA. These implementations are oriented, respectively, for fast simulation and, embedded software realization; FPGA mapping; and high-performance acceleration on multicore platforms that employ GPU.

Through these case studies on diverse platforms, we have demonstrated the utility of dataflow modeling in capturing high-level application structure, and providing design methods that are relevant across different implementation styles, and different forms of parallel processing for image processing. The systematic, dataflow-driven design methods illustrated in this chapter are more broadly applicable for improving the productivity of the design process; the agility with which designs can be retargeted across different platforms; and the application of high-level transformations for optimizing the implementation structure.

Acknowledgment

This work was sponsored in part by the Laboratory for Telecommunication Sciences, and the US National Science Foundation.

References

1. J. Eker, J. Janneck, E. A. Lee, J. Liu, X. Liu, J. Ludvig, S. Sachs, and Y. Xiong, Taming heterogeneity—The ptolemy approach, *Proceedings of the IEEE*, 91(1), 127–144, 2003.
2. S. S. Bhattacharyya, P. K. Murthy, and E. A. Lee, Synthesis of embedded software from synchronous dataflow specifications, *Journal of VLSI Signal Processing Systems for Signal, Image, and Video Technology*, 21(2), 151–166, 1999.
3. E. A. Lee and D. G. Messerschmitt, Synchronous dataflow, *Proceedings of the IEEE*, 75(9), 1235–1245, 1987.
4. M. Sen, Y. Hemaraj, W. Plishker, R. Shekhar, and S. S. Bhattacharyya, Model-based mapping of reconfigurable image registration on FPGA platforms, *Journal of Real-Time Image Processing*, 2008, 149–162.
5. G. Johnson, *LabVIEW Graphical Programming: Practical Applications in Instrumentation and Control.* McGraw-Hill, New York, NY, 1997.
6. J. L. Pino, K. Kalbasi, H. Packard, and E. Division, Cosimulating synchronous dsp applications with analog RF circuits, in *Proceedings of the IEEE Asilomar Conference on Signals, Systems, and Computers*, Pacific Grove, CA, pp. 1710–1714, November 1998.
7. C. Hsu, M. Ko, and S. S. Bhattacharyya, Software synthesis from the dataflow interchange format, in *Proceedings of the International Workshop on Software and Compilers for Embedded Systems*, Dallas, TX, September 2005, pp. 37–49.
8. C. Shen, H. Wu, N. Sane, W. Plishker, and S. S. Bhattacharyya, A design tool for efficient mapping of multimedia applications onto heterogeneous platforms, in *Proceedings of the IEEE International Conference on Multimedia and Expo*, Barcelona, Spain, July 2011, 6 pages in online proceedings.

9. W. Plishker, C. Shen, S. S. Bhattacharyya, G. Zaki, S. Kedilaya, N. Sane, K. Sudusinghe, T. Gregerson, J. Liu, and M. Schulte, Model-based DSP implementation on FPGAs, in *Proceedings of the International Symposium on Rapid System Prototyping*, Fairfax, Virginia, June 2010, invited paper, DOI 10.1109/RSP_2010.SS4, 7 pp.

10. P. Hamill, *Unit Test Frameworks*. O'Reilly & Associates, Inc., Sebastopol, CA, 2004.

11. C. Shen, W. Plishker, H. Wu, and S. S. Bhattacharyya, A lightweight dataflow approach for design and implementation of SDR systems, in *Proceedings of the Wireless Innovation Conference and Product Exposition*, Washington, DC, November 2010, pp. 640–645.

12. S. S. Bhattacharyya, R. Leupers, and P. Marwedel, Software synthesis and code generation for DSP, *IEEE Transactions on Circuits and Systems—II: Analog and Digital Signal Processing*, 47(9), 849–875, 2000.

13. G. Bilsen, M. Engels, R. Lauwereins, and J. A. Peperstraete, Cyclo-static dataflow, *IEEE Transactions on Signal Processing*, 44(2), 397–408, 1996.

14. B. Bhattacharya and S. S. Bhattacharyya, Parameterized dataflow modeling for DSP systems, *IEEE Transactions on Signal Processing*, 49(10), 2408–2421, 2001.

15. J. T. Buck and E. A. Lee, The token flow model, in *Advanced Topics in Dataflow Computing and Multithreading*, L. Bic, G. Gao, and J. Gaudiot, Eds. IEEE Computer Society Press, Los Alamitos, CA, 1993.

16. W. Plishker, N. Sane, M. Kiemb, K. Anand, and S. S. Bhattacharyya, Functional DIF for rapid prototyping, in *Proceedings of the International Symposium on Rapid System Prototyping*, Monterey, CA, June 2008, pp. 17–23.

17. W. Plishker, N. Sane, M. Kiemb, and S. S. Bhattacharyya, Heterogeneous design in functional DIF, in *Proceedings of the International Workshop on Systems, Architectures, Modeling, and Simulation*, Samos, Greece, July 2008, pp. 157–166.

18. W. Wolf, *FPGA-Based System Design*. Prentice-Hall, Englewood, NJ, 2004.

19. *Virtex-6 FPGA CLB User Guide, UG364 (v1.1)*, September 2009.

20. L. Dadda, Some schemes for parallel multipliers, *Alta Frequenza*, 34, 349–356, 1965.

21. R. Katz, *Contemporary Logic Design*. The Benjamin/Cummings Publishing Company, Redwood City, CA, 1994.

22. E. A. Lee and S. Ha, Scheduling strategies for multiprocessor real time DSP, in *Proceedings of the Global Telecommunications Conference*, Dallas, TX, pp. 1279–1283, November 1989.

23. D. Shreiner, M. Woo, J. Neider, and T. Davis, *OpenGL(R) Programming Guide: The Official Guide to Learning OpenGL(R), Version 2 (5th edition)*. Addison-Wesley Professional, Boston, MA, 2005.

24. *NVIDIA CUDA Compute Unified Device Architecture—Programming Guide*, 2007.

25. Khronos OpenCL Working Group, *The OpenCL Specification, version 1.0.29*, December 2008.

26. J. D. Owens, D. Luebke, N. Govindaraju, M. Harris, J. Krüger, A. Lefohn, and T. J. Purcell, A survey of general-purpose computation on graphics hardware, *Computer Graphics Forum*, 26(1), 80–113, 2007.

27. *Using graphics devices in reverse: GPU-based Image Processing and Computer Vision*, 2008.

28. W. Plishker, O. Dandekar, S. S. Bhattacharyya, and R. Shekhar, Utilizing hierarchical multiprocessing for medical image registration, *IEEE Signal Processing Magazine*, 27(2), 61–68, 2010.

29. S. Ritz, M. Pankert, and H. Meyr, Optimum vectorization of scalable synchronous dataflow graphs, in *Proceedings of the International Conference on Application Specific Array Processors*, Venice, Italy, pp. 285–296, October 1993.

30. V. Zivojnovic, S. Ritz, and H. Meyr, Retiming of DSP programs for optimum vectorization, in *Proceedings of the International Conference on Acoustics, Speech, and Signal Processing*, Adelaide, SA, Australia, April 1994, pp. 492–496.

31. K. N. Lalgudi, M. C. Papaefthymiou, and M. Potkonjak, Optimizing computations for effective block-processing, *ACM Transactions on Design Automation of Electronic Systems*, 5(3), 604–630, 2000.

32. M. Ko, C. Shen, and S. S. Bhattacharyya, Memory-constrained block processing for DSP software optimization, *Journal of Signal Processing Systems*, 50(2), 163–177, 2008.

25

Application-Specific Instruction Set Processors for Video Processing

Sung Dae Kim and Myung Hoon Sunwoo

CONTENTS

25.1 Introduction

With rapid progress in semiconductor technology, it is becoming feasible to have high-quality multimedia services using smart hand-held devices. Today, multimedia electronics like smart phones, smart pads, digital cameras, camcorders, smart TVs, and so on have become highly popular. Anyone can easily use these various multimedia equipments and can generate new creative multimedia contents by using them. Multimedia technologies have been developed with new applications. Various multimedia codecs have been standardized as MPEG-2, MPEG-4, H.261, H.263, H.264/AVC [1], high-performance video coding (HVC) [2], and so on. The purpose of multimedia codecs is to fulfill the requirements of various applications by improving the picture quality, increasing the coding efficiency, and obtaining more error robustness.

Application-specific integrated circuit (ASIC) designs can reduce the cost, size, and power consumption of systems. However, custom ASIC designs face several limitations such as lack of flexibility and high development costs. If a multimedia codec is implemented by ASIC, the design should be refabricated whenever algorithms, standards, and applications are changed. Because of soaring non-recurring engineering (NRE) cost, it is not suitable to deal with rapidly changing standards and technologies. On the other hand, programmable digital signal processor (DSP)-based designs can greatly reduce time to market and allow faster changes and upgrades. However, they require large die size and high power consumption. To compromise the advantages of custom ASIC designs and general DSP designs, application-specific instruction set processor (ASIP) can be a promising solution for the nanotechnology era and they can achieve high performance and low power of ASIC and flexibility of DSP. Moreover, ASIP can reduce NRE costs since they can support various standards and applications. Hence, ASIP-based implementation is becoming a key trend in SoC design [3].

Recently, the short design cycle requirements for embedded ASIPs led to the development of the new ASIP design methodology, that is, architecture description language (ADL)-based ASIP design methodology [4–6]. ASIP design is a difficult task that requires expert knowledge in different domains such as target applications, application profiling, architecture exploration, assembly instruction set design, programming tool-suites development, microarchitecture design, and so on. ADL is employed to model ASIP at a higher level of abstraction than the register transfer level, thereby allowing the designer to deal with much less complexity. ADL such as language for instruction set architecture (LISA) [4], nML [5], EXPRESSION [6], and so on offers ASIP designers quick and optimal design solutions by enabling rapid development and exploration of ASIP architectures and automatically generating software tool-suites. Therefore, application developers can concentrate on algorithm aspects during the ASIP development, and ASIP architecture designers using ADL can significantly reduce the efforts of designing new architectures.

In this chapter, we introduce the design issues and methodologies of ASIP for video processing. This chapter is organized as follows. Section 25.2 presents a brief introduction of ASIP and the roles of ASIP on SoC for video processing. Section 25.3 introduces the ADL-based ASIP design methodology. Section 25.4 is a comprehensive discussion about specific instructions and their architectures of recent ASIPs for video processing. Finally, Section 25.5 summarizes and presents conclusions.

25.2 Introduction of ASIP

In this section, we introduce ASIP, including a short definition and its differences from ASIC. Moreover, we discuss the roles of ASIP in video processing systems.

25.2.1 What Is ASIP?

H.264/AVC has been used in mobile communication standards, such as smart phones, smart pads, DMB, DVB-S2, and so on. Low power consumption and low cost are crucial for the implementation of H.264/AVC. To improve coding efficiency, H.264/AVC has adopted new features, such as multireference frames, variable block size, quarter pixel accuracy motion estimation (ME)/motion compensation (MC), intraprediction, in-loop deblocking filter, and so on. The hardware complexity of H.264/AVC main profile encoder is about 10 times more than that of MPEG-4 visual simple profile encoder [7]. ASIC design methods have been widely used for the implementation of H.264/AVC codec [8–10].

The recent SoC for hand-held devices consists of various parts such as CPUs, DSPs, memory, video/audio codecs, power management, peripheral, and so on as shown in Figure 25.1.

It requires the convergence of advanced technology, cores and IPs, methodology, system software, manufacturing, and so on. Among them, video/audio codecs have been implemented by using typical ASIC design methods since these codecs require low power and high performance. However, the short design cycle and the rapidly changing standards increase the requirement of flexibility, and thus, the ASIP design becomes a key trend in SoC. This section introduces the difference between ASIC design and ASIP design and the roles of ASIP on SoC for video systems.

As mentioned, ASIP compromises the advantages of ASIC and DSP designs. Hence, the understanding of ASIC and DSP designs should be preceded by the understanding of the ASIP design. Figure 25.2 shows the typical structure of ASIC, which can be divided into two parts, data processing part and control part. The data processing part covers computation units and storage elements to route the data. The control part is based on finite-state machine (FSM) and controls the data processing part. FSM represents the application-dependent schedule of operations. The operation schedules are fixed when the application is decided. Whenever applications are changed, the control part should be redesigned. Moreover, traditional ASIC combines the data processing part and the control part without clear interface. Even if only the control part is changed, ASIC has to be completely verified again.

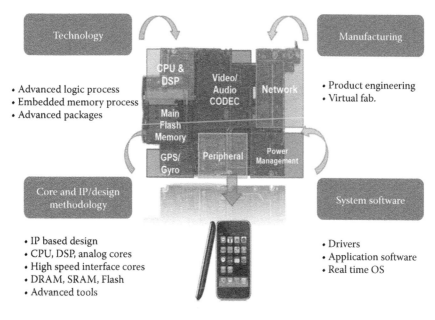

FIGURE 25.1
SoC components used in recent electronic device.

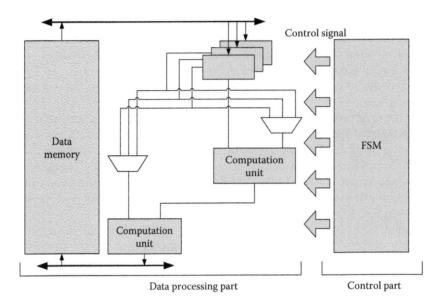

FIGURE 25.2
Typical structure of ASIC.

In contrast, a typical processor has a different control part compared to ASIC. The control part of the processor consists of instruction fetch logic, program memory, and decoder as shown in Figure 25.3. The designer defines the operations of the processor based on each instruction. The instruction fetch logic loads the instruction from program memory in the order of the program. The decoder generates the control signals using the fetched

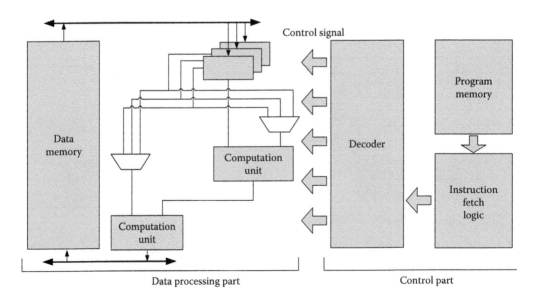

FIGURE 25.3
Typical structure of a processor.

instruction. The operation schedules are easily changed by changing the program. Every processor has its own instruction set, which is the main difference from ASIC.

There are various types of processors according to applications. General-purpose processors (GPPs) are designed to support any application such as data processing, program control, communication, multimedia processing, and so on. However, GPPs are not suitable for massive digital signal processing because of their architectures and performances. Hence, DSPs have been developed for digital signal processing. Early DSPs employ the architectures of GPPs with additional hardware and instruction set dedicated to digital signal processing. Therefore, early DSPs are more applicable to various digital signal processing applications than a particular application. As the variety of algorithms and the complexity of computations increase, these general-purpose DSPs have been diverging into various DSPs as shown in Figure 25.4. With the progress of DSPs, ASIP that has a specific instruction set for programming and shows comparable performance with ASIC is coming to the fore. Thus, ASIP can be defined as the optimized processor for a specific application. As shown in Figure 25.4, hundreds of DSPs and in-house ASIPs have been developed and widely used in commercial SoCs and products.

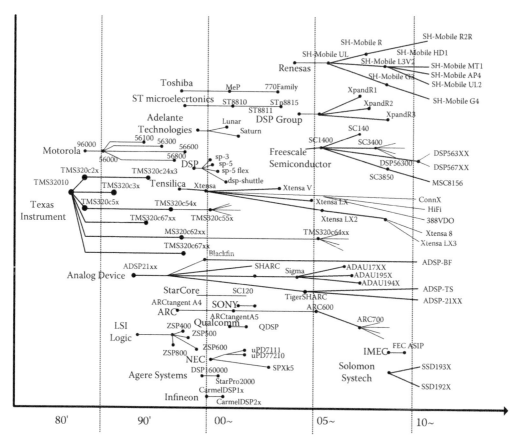

FIGURE 25.4
Progress of DSPs.

25.2.2 ASIP Roles on SoC for Video Processing

Video processing systems consist of several processors, hardware accelerators, memory, peripherals, bus, and so on. For real-time video processing, hardware accelerators are employed in video systems. As mentioned, such ASIC-based accelerators do not provide any flexibility while video processing standards and algorithms are rapidly changing. The systems should be redesigned to support new features for changing standards and algorithms. On the other hand, DSPs have several drawbacks such as low performance and high power consumption compared to hardware accelerators. The hardware accelerators and programmable cores in video processing systems can be replaced by ASIPs.

Several differences exist between ASIPs and general processors. ASIPs have the application-specific datapath, including specific H/W accelerators, special-purpose registers, tightly coupled local memories, and so on, as shown in Figure 25.5.

In addition, differences between DSPs and ASIPs also exist. DSPs can be defined as a specialized processor with an optimized architecture for various operations of digital signal processing. Hence, the optimized processors for video processing can be located between DSP and ASIP. In this chapter, DSP indicates only the general-purpose DSP that is not optimized for a specific application in contrast to ASIP.

ASIP can support various algorithms and standards by changing the program. Even after fabrication, other applications can be implemented by changing the software. For the computational-intensive parts of video processing, ASIP also adopts hardware accelerators by considering a trade-off between flexibility and performance. High performance is another reason why ASIP is suitable for video processing. ASIPs having specific instructions and their dedicated accelerators can show better performance than DSPs and can even be comparable with ASICs. In addition, ASIP supports instruction-level and data-level parallelism that are useful for target applications. Moreover, ASIP can achieve low power consumption. Program memory accesses account for a large percentage of power consumption. By designing application-specific instructions, the program code size can be

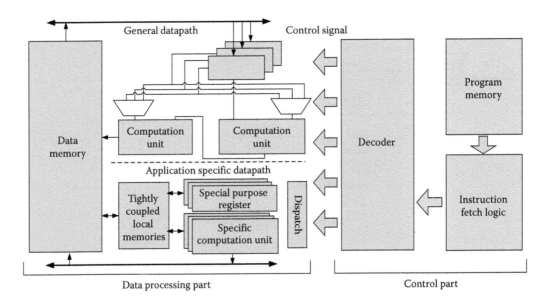

FIGURE 25.5
Typical structure of ASIP.

reduced, and thus, the number of program memory access can be dramatically reduced. Through these advantages, ASIP can be a promising solution for video processing.

25.3 ADL-Based ASIP Design Methodology

This section introduces ASIP design methodology, design flow, and design issues.

25.3.1 ASIP Design Methodology

There are two approaches to design ASIPs [11]. The first approach is to customize the instruction set of a predefined and preverified configurable processor [12–14]. These reconfigurable architectures cover a limited architectural design exploration but enable high design efficiency. In addition, they can provide the architectural flexibility even after fabrication.

The Xtensa configurable processor [12] can be customized for functions, registers, and datapath. The designer can select the pipeline stages and add functions for a specific application. The Xtensa processor provides direct connectivity with arbitrary widths and predictable latency without using the system bus. Figure 25.6 shows the data processing unit (DPU) architecture of the Xtensa LX3-configurable processor. It employs multi-issue Very Long Instruction Word (VLIW) using a Flexible Length Instruction eXtensions (FLIX) architecture. This solution includes the controllable function or adds a function for a specific application using the configuration block, optional block, and designer-defined extension. The designer-defined building blocks are specified in the Tensilica Instruction Extension (TIE) language. This Verilog-like specification language enables the designer to cover the hardware part, such as computational units and additional registers, and the software part, for example, instruction mnemonics and operand utilization.

Xtensa LX3 also provides a compiler named the Xtensa Processor Extension Synthesis (XPRES). This compiler analyzes C/C++ source codes and profiles the run-time application. Then, it automatically suggests configuration settings and new instructions to accelerate the application execution.

ARC cores [13] and CorExtend from MIPS [14] are other examples of configurable processors. However, this approach provides a limited set of atomic predefined options available to reduce processor complexity. These limited architectural flexibilities cause an upper bound to the performance increase.

The second approach is to design from scratch [15–17]. In other words, an entirely new instruction set is specifically designed for a target application. ADL such as LISA, nML, EXPRESSION, and so on is very helpful for the ASIP design approach. nML [5] originates from the Technical University of Berlin and targets the instruction set description. Based on nML, a retargetable C compiler (Chess) and an instruction set simulator and generator (Checkers) were developed by IMEC and commercialized by Target Compiler Technologies. EXPRESSION [6], developed at the University of California, Irvine, captures the structure and the instruction set of a processor as well as their mapping. A greater flexibility can be gained through the use of ADL. It also offers ASIP designers quick and optimal design solutions by enabling rapid development and exploration of ASIP architectures and automatically generating the software tool-suites.

LISA [4] is a representative and widely used ADL, which offers two main descriptions. One is the description of the ASIP structure, including registers, pipeline structure,

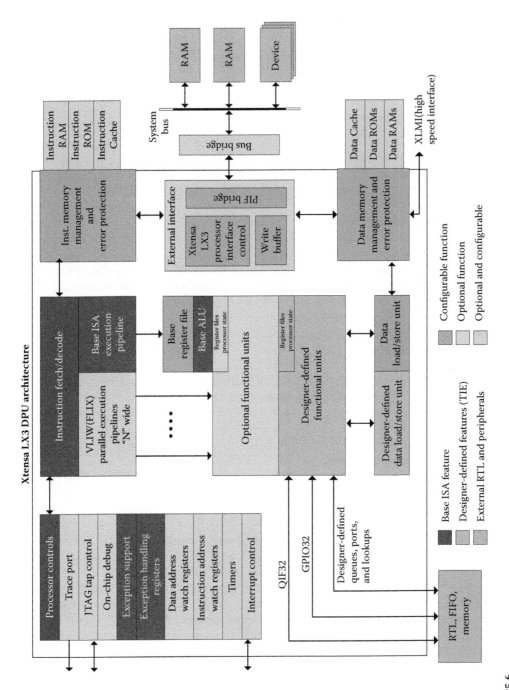

FIGURE 25.6
Xtensa LX3 DPU architecture.

instruction encoding, instruction syntax, and so on. The other is the description of the behavior of each instruction.

Figure 25.7 shows the design flow using LISA [4]. Once the processor is described in LISA, the processor designer framework can automatically generate the ASIP software development tools (C compiler, assembler, linker, disassembler, simulators, and debugger). The architecture exploration is iterative and repeated until a best fit between the selected architecture and the target application [11]. Every change to architecture specification requires a completely new set of software development tools. The development efforts can be saved by using ADL that can provide rapid exploration of ASIP architectures and can automatically generate the software tool-suites. ADL also supports verification and performance comparison phase. The generated simulators and debugger enable stimulation and debugging of the ASIP at a high level of abstraction. To perform comparison, simulations are running on different test conditions. The initial memory contents of each test condition are inserted in the assembly files in user-defined sections, which are defined in the linker command file.

In the abstraction level, the data flow is captured by the dependency of each operation. Data flow graph (DFG), which is a graphical representation of the data flow, is usually used for the design of ASIP. In LISA, the behavior section guides the data flow. Figure 25.8 shows the example of the DFG representation. The vertices in DFG are the basic operators for data manipulation and the edges represent the interconnections of inputs and outputs.

The ADL-based design methodology offers the possibility of fully customizing a processor to meet application-specific constraints [18], which includes specifying the size of various buses and data-path component widths to the bit-level granularity. The first step of

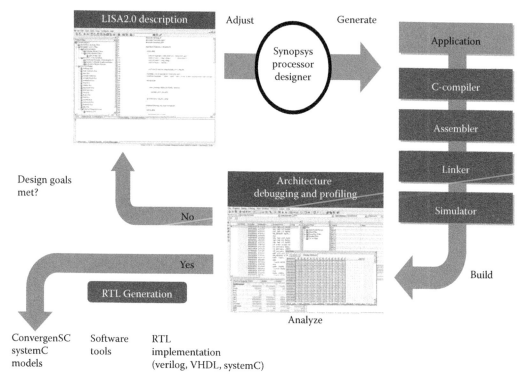

FIGURE 25.7
Design flow using LISA.

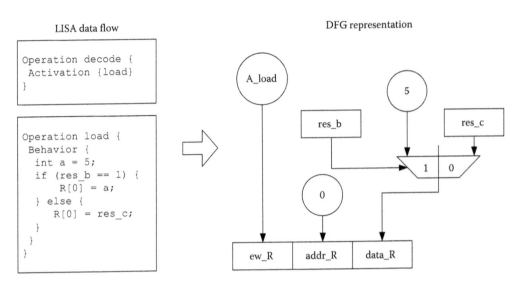

FIGURE 25.8
Example of DFG representation.

developing new instructions is separating a high-level source code into smaller functions. Special instructions and their hardware are then defined for each small function to reduce the number of cycles for executing the program. The custom instruction set is identified for critical parts of the application to increase computational performance.

25.3.2 ASIP Design Flow and Design Issues

The overall design flow of ASIP [19] is illustrated in Figure 25.9. First, the target application is chosen. ASIP can be designed to meet the target application having specific function blocks and instructions. Second, the application profiling is performed. At the abstraction level of arithmetic operations, the number of the required operations is measured. Profiling analyzes the target application to identify critical parts of the application. Criticality refers to computational intensive parts of the application. According to the profiling results, an ASIP designer divides the application into hardware implementation for high performance and software implementation for flexibility. Computation-intensive tasks are implemented by using H/W accelerators, while control-intensive tasks are usually performed by programmable processors. Hardware/software (HW/SW) codesign like ASIP involves various design problems, including HW/SW coverification and system synthesis. The well-defined interface between hardware accelerators and cores is essential for ASIP. Configurable parts of ASIP such as the instruction set, register file and its data transfer, cache size, application-specific hardware components, and so on can be implemented by using the ASIP design methodology.

There are three kinds of typical custom instruction set, including VLIW, single-instruction multiple data (SIMD) instruction set, and the operation fusion instruction set [20]. VLIW contains multiple execution slots for independent operations to execute in parallel. It greatly relies on a powerful compiler. SIMD targets at independent but identical operations for different data, and only duplicates function units for vector operands to accelerate program execution. The operation fusion, which refers to choosing sequential operations and replacing them by a single or multiple custom instructions to reduce the execution time, is

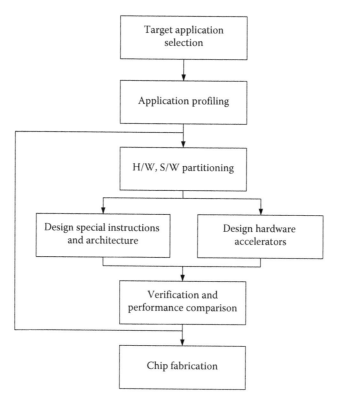

FIGURE 25.9
ADL-based ASIP design flow.

complementary to the above two techniques and has a broader application. Depending on the application, any of these different types of instruction set is selected. With the appropriate custom instructions selected, the instruction set architecture is extended, and the microarchitecture is augmented accordingly. ADL covers all phases of the design process from algorithmic specification of the application to implementation of the microarchitecture [21]. It improves the flexibility of modeling target architectures and significantly reduces description efforts. It provides a high level of flexibility to facilitate the description of various processors, such as VLIW, SIMD, and reduced instruction set computer (RISC) architectures.

To develop an ASIP, design space exploration is performed and the decision of an instruction set and parallelism are fixed at design time. ASIP cannot be altered after fabrication, which often compromises their performance for continually changing application areas. To alleviate this problem, reconfigurable ASIPs (rASIPs), which combine the programmability of ASIPs and the postfabrication hardware flexibility of reconfigurable structures, have been recently presented [22].

25.4 Recent ASIPs for Video Processing

Case studies of ASIP implementation for video processing are presented in this section. We explain specific instruction sets and architectures of ASIPs.

25.4.1 Video-Specific Instruction Set Processor

Video-specific instruction set processor (VSIP) [23] has been proposed for the implementation of mobile multimedia codec, especially H.264/AVC.

25.4.1.1 Overview of VSIP

Figure 25.10 shows the overall architecture of VSIP that consists of two parts, a programmable DSP part and a hardware accelerator part. The DSP part has a program control unit (PCU), a DPU, and an address unit (AU). The hardware accelerator part has an interprediction accelerator (IPA) and an entropy coding accelerator (ECA). IPA consists of an ME accelerator and an MC accelerator. ECA has a context-adaptive variable-length coding (CAVLC) accelerator and an exponential Golomb code (EGC) accelerator. The hardware accelerators can operate in parallel with the DSP unit.

PCU consists of a prefetch logic, a program counter, an instruction register, an FSM, a stack, and an interrupt controller. DPU consists of two multiply and accumulate (MAC) units for two 16-bit-by-16-bit multiplications and accumulations, two arithmetic logic units (ALU), a barrel shifter, and a register file. AU has two address generation units (AGU) for load and store. Each internal word length is 32 bits. The instruction pipeline consists of six stages, that is, prefetch, fetch, decode, execute1, execute2, and execute3. VSIP has 35 arithmetic instructions, 11 logical and shift instructions, 6 program control instructions, 4 move instructions, and 16 special instructions, including instructions for H.264/AVC, which are described next.

25.4.1.2 Specific Features in VSIP

The in-loop deblocking filter in H.264/AVC generates new pixel values according to the boundary strength (bS), which represents the difference of two neighboring blocks.

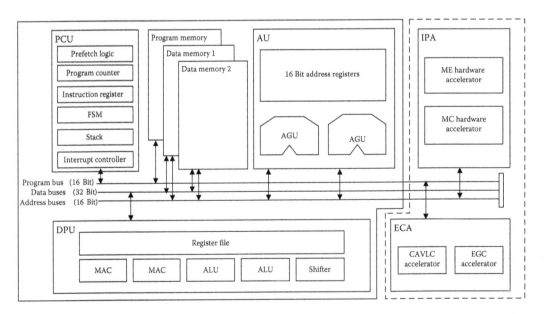

FIGURE 25.10
Overall VSIP architecture.

| q3 | q2 | q1 | q0 | p0 | p1 | p2 | p3 |

Boundary

FIGURE 25.11
Packed pixel data located in block boundary.

Figure 25.11 shows 8 pixels of neighboring 4 × 4 blocks. $p0 \sim p3$ are the packed data in a register and $q0 \sim q3$ are the packed data in another register.

Even though these computations are packed operations, the computations for the in-loop deblocking filter do not occur between two registers, but they occur between the packed data within the same register.

The intraprediction in H.264/AVC eliminates the redundancy of intraframe and interframe.

As the in-loop deblocking filter, the equations for the intraprediction also require the computation within the same register. Existing DSPs support only packed operations between two registers. A large number of instruction cycles is required to implement the in-loop deblocking filter and intraprediction. Hence, VSIP supports new instructions to execute packed operations within a register.

Figure 25.12 shows the three horizontal packed addition (HADD) instructions in VSIP. Three HADD instructions are as follows. The first *HADD* instruction in Figure 25.12a packs a 32-bit register into four 8-bit data, adds four packed data, and then saturates the result to 8-bit data. Figure 25.12b is similar to Figure 25.12a. However, the packed data, which is selected by a *mask*, is shifted by 1 bit to the left. In Figure 25.12c, *mask1* selects the data to be added, and *mask2* selects the data to be shifted. The in-loop deblocking filter and the intraprediction can be implemented using the proposed *HADD* instructions.

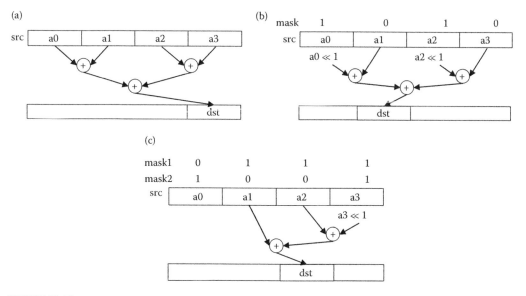

FIGURE 25.12
Horizontal packed addition instructions in VSIP. (a) dst = HADD(src). (b) dst = HADD(src:mask). (c) dst = HADD(src:mask1.mask2).

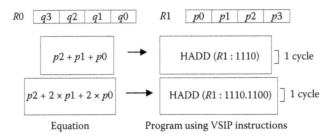

FIGURE 25.13
Assembly program of core block for in-loop deblocking filter.

Figure 25.13 shows the assembly program of the core block for the in-loop deblocking filter. $R0$ and $R1$ are general registers and the packed pixel data are stored in $R0$ and $R1$. Each result of the instruction can be obtained after one clock cycle. Hence, the proposed VSIP can execute equations for in-loop deblocking filter in one clock cycle.

Figure 25.14 shows the assembly program of the intraprediction. Acc represents an accumulator and the packed pixel data is stored in $R0$. $R1$ and $R2$ have offset values for rounding. The $ADDAR$ instruction in VSIP calculates an addition of two source data. After the addition, the result is shifted to the right by the immediate value in the instruction. Each result of the instruction can be obtained after one clock cycle. Hence, the proposed VSIP can execute these equations for intraprediction in two clock cycles.

Several core functions for generating the intrapredictor and the in-loop deblocking filter are coded using the proposed special instructions and the same blocks are also coded using the existing instructions of TMS320c64x. The proposed architecture can reduce the number of clock cycles for generating an intrapredictor of about 40% compared with TMS320c6x. Moreover, the total number of clock cycles to execute the in-loop deblocking filter can be reduced by about 20~25% than that of TMS320c6x. TMS320C64x supports the DOTPU4 instruction that executes packed multiplications of two registers and adds four results in four cycles. Other DSPs require more instructions since they do not support the special instructions.

Integer transform has a regular computation flow. The forward transform is executed with four rows of four packed data. Then, the forward transform is performed again with four

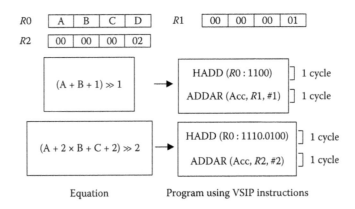

FIGURE 25.14
Assembly program of intraprediction.

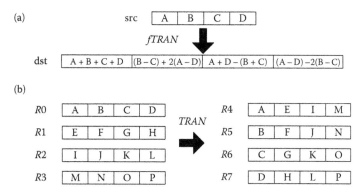

FIGURE 25.15
Operation flow of (a) *fTRAN* and (b) *TRAN* instruction in VSIP.

columns of four packed data to obtain the results of the 4×4 2-D integer transform. VSIP has novel instructions to efficiently execute the forward/inverse 4×4 integer transform. Figure 25.15a shows the operation flow of the *fTRAN* instruction. The *fTRAN* instruction reads a 32-bit general register in one register file, which consists of four 32-bit registers, and executes the forward transform. Then, the results are written in another register file consisting of four 32-bit registers. The *iTRAN* instruction performs a similar operation. These instructions can be implemented using the adders and eight additional 2×1 multiplexers. Figure 25.15b shows the operation flow of the *TRAN* instruction. The general register file has a 4×4 matrix whose elements are 8-bit pixel data. The *TRAN* instruction in VSIP executes the transpose of a 4×4 matrix.

4×4 integer transform can be easily programmed with the *fTRAN* and *TRAN* instructions. Table 25.1 shows the number of the required instructions for integer transform at 30 frames on existing DSPs [24,25] and VSIP. VSIP can be more efficient than the implementation using instructions of TMS320c55x (SW) and using the coprocessor of TMS320c55x (HW) for integer transform. TMS320c64x is a large VLIW architecture having eight function units while VSIP requires only two 32-bit adders.

VSIP has ME/MC and VLC hardware accelerators for real-time processing. ME/MC should frequently access memory. From a performance point of view and a low power point of view, it can be a serious problem. Thus, VSIP uses the sliding window method [26] to alleviate this problem. Figure 25.16 illustrates the operation flow of the ME hardware accelerator in VSIP. The ME architecture supports the $[+8, -7]$ search window. In the $[+8, -7]$ search window, 16 4×4 blocks exist in a row. In the first cycle, four sums of absolute difference (SADs) are simultaneously calculated as shown in Figure 25.16a. Next, the search window shifts right and each operation unit repeats the SAD calculation as shown in Figure 25.16b. The SADs of upper four pixels of every block in a row can be obtained after 4 cycles and 16 SADs are stored in buffers. The SADs of the second upper are calculated in

TABLE 25.1

Performance Comparisons of 4×4 Integer Transform

	TMS320c55x (SW) [24]	TMS320c55x (HW) [24]	TMS320c64x [25]	VSIP
MIPS	12.8	2.8	1.1	1.1

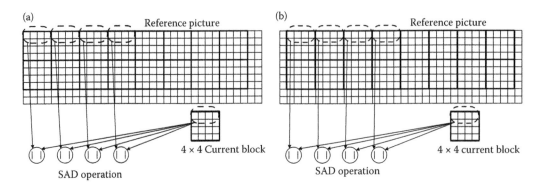

FIGURE 25.16
Operation flow of ME hardware accelerator in VSIP. (a) ME operation in the first cycle. (b) ME operation in the second cycle.

the same way, and the 16 SADs are accumulated with the 16 SADs in buffers, respectively. Then, after 16 cycles, the 16 SADs of 4×4 blocks can be obtained.

Table 25.2 shows the comparisons among Refs. [27] and [28] and VSIP ME architecture. The VSIP ME hardware accelerator can significantly reduce the gate counts compared to Refs. [27] and [28].

25.4.2 ASIPs for Motion Estimation

In most video coding standards, ME plays an important role in these hybrid video coding systems. To increase the accuracy of prediction, various features such as multireference, variable block size, fractional estimation, and so on are adopted in ME algorithms. In addition, fast search and early-termination schemes are used to reduce the computational complexity of ME. Hence, various ME algorithms according to the applications have been proposed [29–33]. ASIP is one of the promising solutions for the rapidly changing ME algorithms. Here are examples of three ASIPs for ME [34–36]. First, the ASIP [34] has been designed for supporting multiresolution ME. Second, the ASIP [35] has specific instructions for various ME algorithms including the adaptive ME algorithms [32,33]. Lastly, the ASIP [36] can control the number of function units according to the complexity of an ME algorithm.

25.4.2.1 ASIP for Multiresolution Motion Estimation

As known, ME consists of two steps: integer motion estimation (IME) and fractional motion estimation (FME) which searches around the best integer position obtained by IME. FME

TABLE 25.2

Performance Comparisons of ME Hardware Accelerator

	Clock Cycles/ Frame	Search Range	Supported Block Size	Gate Counts
[27]	405,603	[−16, +15]	Variable block support	154K
[28]	406,077	[−8, +7]	Variable block support	61K
VSIP ME accelerator	431,244	[−8, +7]	Variable block support	40K

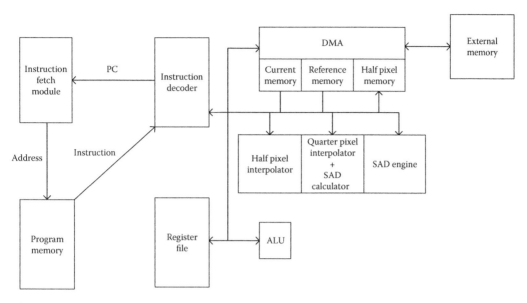

FIGURE 25.17
Architecture of the ASIP. (From S. H. Jeong, M. H. Sunwoo, and S. K. Oh, *EURASIP Journal on Applied Signal Processing*, 2005(16), 2655–2663, 2005. With permission.)

can significantly improve image quality. However, due to additional computations, such as interpolation and fractional search, FME sharply increases the computational complexity of the whole ME. As a result, ME occupies over 60% of the computation time of the whole encoder [37] and FME costs 45% of the ME time in H.264/AVC [38]. Therefore, reducing the computation complexity of fractional search becomes more important.

Fethi Tlili et al. proposed an ASIP [34] for H.264/AVC multiresolution ME. The different feature of the ASIP is to control the precision of ME. In other words, the ASIP can select interpolation depth (half, quarter, or none) by instructions so that users can control the computation complexity of applications. They proposed three efficient accelerators, SAD engine, half-pixel interpolator, and quarter-pixel interpolator, and they also proposed 12 instructions to provide coding flexibility. The instructions have 16-bit length and use operands stored in the register file.

Figure 25.17 shows the hardware architecture of the ASIP, which is composed of several modules. The accelerators are connected to the ALU and the internal memories to compute video processing algorithm. The instruction decoder controls functional units according to instructions and the instruction fetch module loads instructions from the program memory. The register file is used for storing the processed data.

This ASIP has several internal memories. The reference pixel memory is 2×18 Kb block RAM and the current pixel memory is 1×18 Kb block RAM. The internal memories have 8-bit width by implementation constraints. Data load to internal memory from external memory is managed by the direct memory access (DMA) controller. DMA enables data transfer when the CPU is running.

The half-pixel memory is used for storing results of half interpolation. The stored results should be used for further processing such as quarter interpolation. However, quarter pixels are not stored in this ASIP. Hence, the internal memory size can be reduced but it should compute the best matching pixels again for motion compensation (MC).

The ASIP has three special instructions for multiresolution ME. The *SAD4Pix* instruction computes the SAD value of 4 pixels for integer and half-pixel precision. However, the ASIP should fetch the instruction at every computation of 4 pixels, which may lead to an increase in instruction cycles and consequently power consumption. In addition, 1-D SAD computation per 4 pixels can be slow in high-resolution applications. Recently, 2-D SAD computation is widely used to satisfy high-resolution applications [39]. The *Interp4HafPix* instruction interpolates 4 half-pixels and stores the result in the internal memory. The interpolated pixels are stored in half-pixel memory. Then, *SAD4Pix* can be used to compute SAD of half-pixel precision. The *Interp4QpixSAD* instruction interpolates 4 quarter-pixels and computes SAD of quarter-pixel precision. This instruction returns SAD of each quarter-pixel and the ALU decides the minimum SAD and the best matching position.

This ASIP was implemented on Virtex II pro FPGA. The total area is about 61% of the FPGA Slices and 43% of the total LUTs is used. The implemented modules can be run on 172 MHz clock.

25.4.2.2 Adaptive Motion Estimation Processor

With the proliferation of autonomous and portable hand-held devices that support digital video coding, data-adaptive ME algorithms have been required to dynamically configure the search pattern not only to avoid unnecessary computations and memory accesses but also to save energy. ASIP, which has the specialized data path and the optimized instruction set, was introduced in Ref. [35]. The proposed ASIP can adapt the operation to the available energy level in runtime.

The authors classified the ME algorithms into two groups: (i) algorithms that treat each macro block independently and use the predefined patterns and (ii) algorithms that exploit interblock correlation to adapt the search patterns. The 3SS [29], 4SS [30], and DS [31] are well-known examples of the first group and MVFAST [32] and FAME [33] are examples of the second group that uses the information from adjacent blocks. The proposed ASIP covers both the ME algorithm groups.

Figure 25.18 shows the overall architecture of the ASIP. The architecture is based on a register to register architecture and provides only a reduced number of operations that focus on the most widely executed instructions in ME algorithms. The register file consists of 24 general-purpose registers and 8 special-purpose registers capable of storing one 16-bit word each. The control unit is very simple due to the adopted fixed instruction encoding format and a careful selection of the opcode for each instruction. The data path of complex and specific operations such as data load and sad calculation includes the *AGU* and the *SADU*, respectively.

The dedicated AGU fetches all pixel data for ME. To maximize the efficiency of data processing, AGU can work in parallel with the remaining functional units. *SADU* can execute the SAD operation in up to 16 clock cycles. The total instructions of the ASIP are only 8.

However, this design uses just one special instruction (*SAD16*), which lacks the flexibility to support various algorithms. Hence, users have to fetch the limited instruction repetitively to perform variable ME algorithms and this may lead to increase in the code length.

The performance of the ASIP was evaluated by implementing several ME algorithms such as FS, 4SS, DS, MVFAST, and FAME. These algorithms were programmed with the specific instruction set and the ASIP operation was simulated by using the developed software tools.

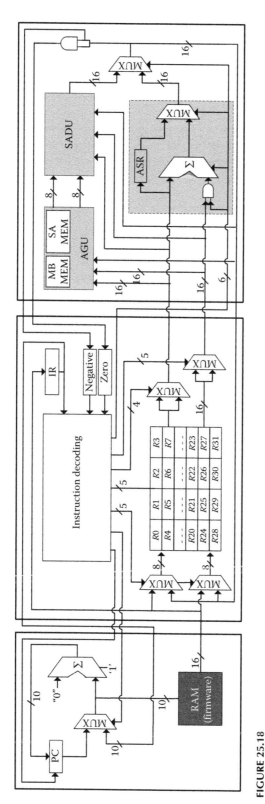

FIGURE 25.18
Architecture of the ASIP. (From S. H. Jeong, M. H. Sunwoo, and S. K. Oh, *EURASIP Journal on Applied Signal Processing*, 2005(16), 2655–2663, 2005. With permission.)

TABLE 25.3

Code Size of Each Algorithm

Algorithm	FS	4SS	DS	MVFAST	FAME
Code size	56	365	460	744	917

Table 25.3 shows the size of the program memory for various algorithms. The adaptive algorithms require more memory for storing the program. Moreover, the adaptive algorithms require additional memory to keep past information in memory.

25.4.2.3 Configurable and Programmable Motion Estimation Processor

The programmable and configurable ME processor for H.264/AVC was introduced in Ref. [36]. The ME processor can handle the requirements of high-definition (HD) video and is suitable for FPGA implementation. The specific instruction set has been proposed to satisfy both programmability and performance. Moreover, the ME processor can vary the number and type of execution units according to the application at compile time.

Figure 25.19 shows the microarchitecture of an example hardware configuration using four integer pel execution units (IPEU), one fractional pel execution unit (FPEU), and one interpolation execution unit (IEU). The main integer pel pipeline should be used to generate a valid processor configuration but the other units are optional. The number of IPEUs is configurable from one to the maximum limited only by the available resources. Each IPEU has its own copy of the point memory and processes 64 bits of data in parallel with the rest of IPEUs. However, IPEU with internal memory requires additional memory bandwidth. Moreover, this architecture cannot be used with data reuse algorithms since the reference data should be replicated to all IPEUs.

The point memories are 256×16 in size and contain the x and y offsets of the search patterns. IEU interpolates the 20×20 pixel area that contains the 16×16 macro block corresponding to the optimal integer motion vector. During interpolation, the integer pipeline is stalled, since the reference memory ports are in use. After the completion of interpolation, FME can start. There are nine specific instructions in this ASIP. All instructions execute in a single clock cycle except for the *check pattern* instruction which consumes a variable number of cycles depending on the complexity of the pattern and the number of available IPEUs. The *reconfiguration* instruction can reconfigure the processor state machine for the partition mode. The detailed operation of each unit and instructions are described in Ref. [36].

25.4.3 ASIP Design for Integer Transform

In hybrid video coding, discrete cosine transform (DCT) and quantization are effective algorithms that enhance the efficiency of the compression and decompression of data. Especially, the integer transform is adopted in H.264/AVC to avoid a mismatch problem and to reduce the computational complexity of floating-point arithmetic.

In H.264/AVC, the residual data can be coded using four different modes: 4×4 integer transform, 8×8 integer transform, and 4×4 and 2×2 Hadamard transforms. Each of the various transform operations has its own regular operation flow and such regularity can be a good feature to be implemented with ASIP. This is the reason why ASIP can be a quite suitable solution for integer transform. VSIP [23] has three special instructions such as *fTRAN*, *iTRAN*, and *TRAN* for integer transform already introduced in Section 25.4.1.

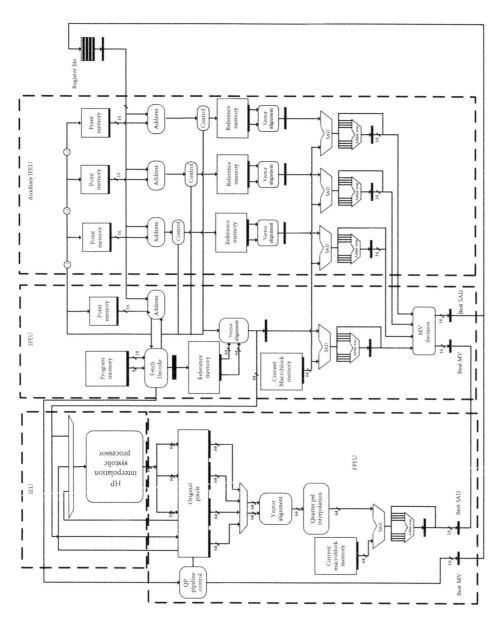

FIGURE 25.19

Architecture example of the ASIP [36] with 4 IPEU, 1 FPEU, and 1 IEU.

In this section, two integer transform ASIP designs are introduced. Through these examples, employing ASIP architecture on integer transform can be proved as a suitable alternative to ASIC solutions.

An ASIP for integer transform has been proposed in Ref. [40], which supports not only 4×4 and 8×8 inverse integer transforms but also 2×2 and 4×4 Hadamard transforms of DC coefficients. The ASIP and the inverse transform block are connected via the SoC-based Wishbone system bus. The top-level system architecture is shown in Figure 25.20.

The system is composed of four components: an ASIP core, an external memory, an inverse transform (IT) block with a direct memory access controller (DMAC), and an arbiter. The ASIP core controls the IT block connected via the system bus and it employs the conventional three-stage pipeline structure. With its 10 instructions, including a special instruction for IT, the ASIP can enable the entire system block to efficiently perform inverse integer transform. The special instruction IIT (inverse integer transform) performs inverse integer transform operations. In addition, IIT also has four operation modes such as 4×4 integer transform, 4×4 Hadamard transform, 2×2 Hadamard transform, and 8×8 integer transform.

An ASIP that fits the mobile environment and high scalability and flexibility has been proposed in Ref. [41]. The ASIP is an MIPS-based processor that has a five-stage pipeline: instruction fetch, decode, execution, memory, and write. In this ASIP, new SIMD instructions have been proposed and an SIMD-type register file is implemented to execute these instructions.

With its SIMD instructions, the ASIP can efficiently perform matrix multiplication. Through matrix multiplication, DCT and IDCT can be performed with 11% fewer instructions and the execution time can decrease up to about 15%. In addition, special SIMD instructions are employed to accelerate DCT and IDCT. Especially, its SI instruction operates as a special instruction, which is to accelerate transform matrix multiplication and this

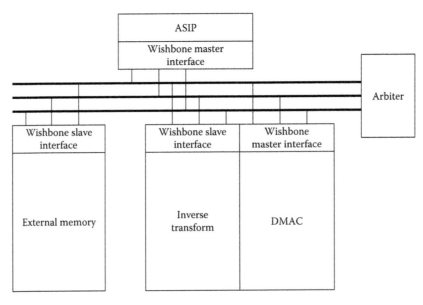

FIGURE 25.20
Top-level system architecture.

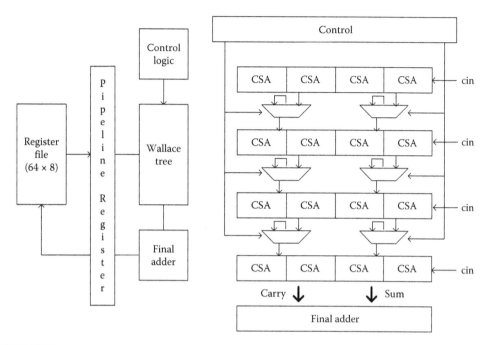

FIGURE 25.21
SIMD unit of the proposed ASIP.

instruction can reduce the nonefficient execution time incurred by the use of conventional instructions.

The ASIP also has its dedicated SIMD unit that can effectively process the special SIMD instruction SI, which can process four data of 16 bit or two data of 32 bit at the same time, and rapidly carry out the transform-specific instructions. Figure 25.21 shows the proposed SIMD unit and the Wallace Tree that are used for matrix computation. Comparing the register file size of Intel's Streaming SIMD Extensions (SSE), the chip area of the ASIP is very small so that it can be suitable for the embedded multimedia system.

25.5 Conclusions

Multimedia standards have been continuously developed to fulfill the requirements of various and new multimedia applications. Recently, due to soaring NRE costs in the nanotechnology era, ASIC design faces a major hurdle to implement the continuously changing multimedia standards, which has started a significant move toward the use of programmable solutions such as ASIP to implement various multimedia algorithms and standards.

Therefore, in this chapter, we have introduced what ASIP is and its advantages compared with ASIC and DSP. ASIP can give a better flexibility and competitive performance compared to ASIC but ASIP demands new design methodology and verification. This chapter introduced the design examples of ASIP and design methods, which include the method using a predefined reconfigurable architecture with a customized instruction set and the ADL-based method to design a system from scratch.

Through the ASIP design examples, this chapter presented an overview of how to design an efficient ASIP. Moreover, these design concepts can be applied to different applications. VSIP has been introduced as an example of ASIP designed for video codec, especially H.264/AVC. In addition, we have shown that ASIP can be a good solution for the implementation of various ME algorithms. We have discussed three ME ASIPs that have different specific instruction set and their dedicated architectures. Due to the regular data flow, ASIP is also quite suitable for the implementation of integer transform. Two ASIPs for integer transform have also been introduced.

Low-power reusable and reconfigurable ASIP solutions can significantly reduce the number of cycles, memory accesses, and computation complexity. Therefore, they can dramatically save power consumption. Moreover, ASIPs can reduce NRE costs since they can support various standards and applications. In summary, ASIP can be a promising low-power reusable solution for the nanotechnology era.

Acknowledgment

This work was supported by Mid-career Researcher Program through the National Research Foundation (NRF) grant of Korea funded by the MEST (20110016671), by the framework of international cooperation program managed by NRF of Korea (2011-0030930) and by IC Design Education Center (IDEC).

References

1. Draft ITU-T Recommendation and Final Draft International Standard of Joint Video Specification, ITU-T Rec. H.264 and ISO/IEC 14496-10 AVC, Joint Video Team (JVT) of ITU-T VCEG and ISO/IEC MPEG, Document JVT-G050, May 2003.
2. Working Draft 1 of High Efficiency Video Coding, Joint Collaborative Team on Video Coding (JCT-VC) of ITU-T SG16 WP3 and ISO/IEC JTC1/SC29/WG11, Document JCTVC-C403, October 2010.
3. V. R. Dodani, N. Kumar, U. Nanda, and K. Mahapatra, Optimization of an application specific instruction set processor using application description language, In *Proceedings of the IEEE International Conference on Industrial and Information Systems (ICIIS)*, Karnataka, India, July 2010, pp. 325–328.
4. Institute for Integrated Signal Processing Systems (ISS), RWTH Aachen University, Germany. Available: www.iss.rwth-aachen.de.
5. M. Freericks, The nML Machine Description Formalism, technical report, Technical University of Berlin, Department of Computer Science, 1993.
6. A. Halambi, P. Grun, V. Ganesh, A. Khare, N. Dutt, and A. Nicolau, EXPRESSION: A language for architecture exploration through compiler/simulator retargetability, In *Proceedings of the International Conference on Design Automation & Test in Europe (DATE)*, Munich, Germany, March 1999, pp. 485–490.
7. J. Ostermann, J. Bormans, P. List, D. Marpe, M. Narroschke, F. Pereira, T. Stockhammer, and T. Wedi, Video coding with H.264/AVC: Tools, performance, and complexity, *IEEE Circuits and Systems Magazine*, 4(1), 7–28, 2004.

8. T.-C. Chen, S. YiChien, Y. W. Huang, C. H. Tsai, C. Y. Chen, T. W. Chen, and L. G. Chen, Analysis and architecture design of an HDTV720p 30 frames/s H.264/AVC encoder, *IEEE Transactions on Circuits and Systems for Video Technology*, 16(6), 673–688, 2006.

9. Y. K. Lin, C. W. Ku, D. W. Li, and T. S. Chang, A 140-MHz 94K gates HD1080p 30-Frames/s intra only profile H.264 encoder, *IEEE Transactions on Circuits and Systems for Video Technology*, 19(3), 432–436, 2009.

10. K. Xu and C. S. Choy, A power-efficient and self-adaptive prediction engine for H.264/AVC decoding, *IEEE Transactions on Very Large Scale Integration Systems*, 16(3), 302–313, 2008.

11. M. Rashid, L. Apvrille, and R. Pacalet, Application specific processors for multimedia applications, In *Proceedings of the IEEE International Conference on Computational Science and Engineering*, Sao Paulo, Brazil, July 2008, pp. 109–116.

12. Xtensa LX3 processor. Available: http://www.tensilica.com.

13. ARC Configurable Cores. Available: http://www.synopsys.com/IP/ConfigurableCores/ARCProcessors/.

14. MIPS CorExtend. Available: www.mips.com.

15. J. H. Baek, S. D. Kim, and M. H. Sunwoo, SPOCS: Application specific signal processor for OFDM communication systems, *Journal of Signal Processing Systems*, 53(3), 383–397, 2008.

16. S. H. Yoon, J. H. Moon, and M. H. Sunwoo, Design of high-quality audio-specific DSP Core, In *Proceedings of the IEEE Workshop on Signal Processing Systems*, Athens, Greece, 2005, pp. 509–513.

17. S. H. Jeong, M. H. Sunwoo, and S. K. Oh, Bit manipulation accelerator for communication systems digital signal processor, *EURASIP Journal on Applied Signal Processing*, 2005(16), 2655–2663, 2005.

18. G. A. B. Ngoyi, J. M. Pierre Langlois, and Y. Savaria, Iterative design method for video processors based on an architecture design language and its application to ELA deinterlacing, In *Proceedings of the IEEE International Conference on Circuits and Systems and TAISA*, Montreal, Canada, June 2008, pp. 37–40.

19. M. K. Jain, M. Balakrishnam, and A. Kumar, ASIP design methodologies: Survey and issues, In *Proceedings of the Fourteenth International Conference on VLSI Design*, Bangalore, India, January 2001, pp. 76–81.

20. D. Goodwin and D. Petkov, Automatic generation of application specific processors, In *Proceedings of the IEEE International Conference on Compilers, Architecture, and Synthesis for Embedded Systems*, San Jose, USA, October 2003, pp. 137–147.

21. A. Chattopadhyay, D. Kammler, E. M. Witte, O. Schliebusch, H. Ishebabi, B. Geukes, R. Leupers, G. Ascheid, and H. Meyr, Automatic low power optimizations during ADL-driven ASIP design, In *Proceedings of the IEEE International Symposium on VLSI Design, Automation and Test*, Hsinchu, Taiwan, April 2006, pp. 1–4.

22. K. Karuri, A. Chattopadhyay, X. Chen, D. Kammler, L. Hao, R. Leupers, H. Meyr, and G. Ascheid, A design flow for architecture exploration and implementation of partially reconfigurable processors, *IEEE Transactions on Very Large Scale Integration (VLSI) Systems*, 16(10), 1281–1294, 2008.

23. S. D. Kim and M. H. Sunwoo, ASIP approach for implementation of H.264/AVC, *Journal of Signal Processing Systems*, 50(1), 53–67, 2008.

24. TMS320C55x *Hardware Extensions for Image/Video Applications Programmer's Reference*, Texas Instruments Inc., Dallas, TX, 2002.

25. TMS320C64x *Image/Video Processing Library*, Texas Instruments Inc., Dallas, TX, 2003.

26. T. Wiegand, X. Zhang, and B. Girod, Long-term memory motion-compensated prediction, *IEEE Transactions on Circuits and Systems for Video Technology*, 9(1), 70–84, 1999.

27. M. H. Kim, I. G. Hwang, and S. I. Chae, A fast VLSI architecture for full-search variable block size motion estimation in MPEG-4 AVC/H.264, In *Proceedings of Asia and South Pacific Design Automation Conference (ASP-DAC 2005)*, Shanghai, China, January 2005, pp. 631–634.

28. S. Y. Yap and J. V. McCanny, A VLSI architecture for variable block size video motion estimation, *IEEE Transactions on Circuits and Systems for Video Technology*, 51(7), 384–389, 2004.

29. R. Li, B. Zeng, and M. L. Liou, A new three-step search algorithm for block motion estimation, *IEEE Transactions on Circuits and Systems for Video Technology*, 4(4), 438–442, 1994.

30. L. M. Po and W. C. Ma, A novel four-step search algorithm for fast block motion estimation, *IEEE Transactions on Circuits and Systems for Video Technology*, 6(3), 313–317, 1996.
31. S. Zhu and K. K. Ma, A new diamond search algorithm for fast block matching motion estimation, *IEEE Transactions on Image Processing*, 9(2), 287–290, 2000.
32. A. M. Tourapis, O. C. Au, and M. L. Liou, Predictive motion vector field adaptive search technique (PMVFAST): Enhancing block-based motion estimation, In *Proceedings of Visual Communications and Image Processing*, San Jose, USA, January 2001, pp. 883–892.
33. I. Ahmad, W. Zheng, J. Luo, and M. Liou, A fast adaptive motion estimation algorithm, *IEEE Transactions on Circuits and Systems for Video Technology*, 16(3), 420–438, 2006.
34. F. Tliliand and A. Ghorbel, ASIP solution for implementation of H.264 multi resolution motion estimation, *International Journal of Communications, Network and System Sciences*, 3(5), 453–461, 2010.
35. T. Dias, S. Momcilovic, N. Roma, and L. Sousa, Adaptive motion estimation processor for autonomous video devices, *EURASIP Journal of Embedded Systems*, 2007(1), 10, 2007.
36. J. L. Nunez-Yanez, E. Hung, and V. Chouliaras, A configurable and programmable motion estimation processor for the H.264 video codec, In *Proceedings of the IEEE International Conference on Field Programmable Logic and Applications*, Heidelberg, Germany, September 2008, pp. 149–154.
37. Y. J. Wang, C. C. Cheng, and T. S. Chang, A fast fractional pel motion estimation algorithm for H.264/MPEG-4 AVC, In *Proceedings of the IEEE International Symposium on Circuits and Systems*, Island of Kos, Greece, May 2006, pp. 3974–3977.
38. T. C. Chen, Y. W. Huang, and L. G. Chen, Fully utilized and reusable architecture for fractional motion estimation of H.264/AVC, In *Proceedings of the IEEE International Conference on Acoustics, Speech, and Signal Processing*, Montreal, Canada, May 2004, pp. 9–12.
39. C. Y. Chen, S. Y. Chien, Y. W. Huang, T. C. Chen, T. C. Wang, and L. G. Chen, Analysis and architecture design of variable block-size motion estimation for H.264/AVC, *IEEE Transactions on Circuits and Systems I: Regular Papers*, 53(3), 578–593, 2006.
40. N. T. Ngo, T. T. T. Do, T. M. Le, Y. S. Kadam, and A. Bermak, ASIP-controlled inverse integer transform for H.264/AVC compression, In *Proceedings of the IEEE/IFIP International Symposium on Rapid System Prototyping*, Monterey, USA, 2008, pp. 158–164.
41. H. Kim, J. Kim, and Y. Lee, Low cost multimedia ASIP for transform and quantization of H.264, In *Proceedings of the International Conference on Electronics, Information and Communication*, Tashkent, Uzbekistan, June 2008, pp. 300–303.

Part VII

Methodology, Techniques, and Applications: Multimedia Systems and Applications

26

Interactive Multimedia Technology in Learning: Integrating Multimodality, Embodiment, and Composition for Mixed-Reality Learning Environments

David Birchfield, Harvey Thornburg, M. Colleen Megowan-Romanowicz, Sarah Hatton, Brandon Mechtley, Igor Dolgov, Winslow Burleson, and Gang Qian

CONTENTS

26.1 Introduction

Learning is often considered to be the process of transferring knowledge from an expert source (e.g., a teacher in a classroom) to the learner (e.g., a student). In addition to simply adding knowledge to the learner's memory, effective learning needs to be carefully designed so that knowledge acquired by the learner can be seamlessly integrated in the existing knowledge framework of the individual, and that the learner knows how to relate the knowledge to his/her everyday life and is able to apply such knowledge to solve practical problems.

Although teaching and learning have been practiced by mankind for thousands of years to transfer knowledge, education and cognitive scientists are still consistently in search of better theories, models, pedagogies, and tools to improve the effectiveness of teaching and learning. In the process, the advances in multimedia technology make it possible to implement, evaluate, and refine such theories for effective learning. Emerging research from interactive multimedia and human–computer interaction (HCI) has offered exciting new possibilities for the creation of transformative approaches to learning. Advanced interactive multimedia and HCI technology allows teachers and students to explore, create, and interact with course-content-related media in a more intuitive manner through movement and voice than traditional human–computer interfaces such as mouse click and key stroke. Current sensing, modeling, and feedback paradigms in advanced interactive multimedia and HCI can enrich collaborative learning, bridge the physical/digital realms, and prepare all students for the dynamic world they face. When grounded in contemporary research from the learning sciences, interactive multimedia and HCI approaches show great promise to redefine the future of learning and instruction through paradigms that cultivate students' sense of ownership and play in the learning process.

A convergence of recent trends across the education, interactive multimedia, and HCI research communities points to the promise of new learning environments that can realize this vision. In particular, many emerging technology-based learning systems are highly inquiry based, with the most effective being learner centered, knowledge centered, and assessment centered (Bransford et al. 2000). These systems are broadly termed as *student-centered learning environments* (SCLEs). Looking to the future of learning, we envision a new breed of SCLE that is rooted in contemporary education, interactive multimedia, and HCI research and is tightly coupled with appropriate curriculum and instruction design. Our research is focused on three concepts in particular: embodiment, multimodality, and composition, which we define in Section 26.2.

We begin with a discussion on these key concepts and situate them in the context of education, interactive multimedia, and HCI research. We present background theoretical work and examples of their application in a variety of learning contexts. We then present our work in the design and implementation of new platform for learning, the *Situated Multimedia Arts Learning Lab* (*SMALLab*). SMALLab (Figure 26.1) is a mixed-reality, interactive multimedia environment where students collaborate and interact with sonic and visual media through vocalization and full-body, 3D movements in an open, physical space. *SMALLab*

FIGURE 26.1
SMALLab mixed-reality learning environment.

emphasizes human-to-human interaction within a computational multimodal feedback framework that is situated within an open physical space. In collaboration with a network of school and community partners, we have deployed *SMALLab* in a variety of informal and formal educational settings and community-based contexts, impacting on thousands of students, teachers, and community members, many from underserved populations. We have developed innovative curricula in collaboration with our partner institutions. We summarize past deployments along with their supporting pilot studies, and present two recent examples as case studies of *SMALLab* learning. Finally, we present conclusions and describe our ongoing work and future plans.

26.2 Background

Recent research spanning education, interactive multimedia, and HCI has yielded three themes that inform our work across learning and play: *multimodality*, *embodiment*, and *composition*. Here, we define the scope of these terms in our research, and discuss their theoretical basis before presenting examples of prior related applications.

26.2.1 Multimodality

26.2.1.1 Learning Sciences

By *multimodality*, we mean interactions and knowledge representations that encompass students' full sensory and expressive capabilities, including visual, sonic, haptic, and kinesthetic/proprioceptive. Multimodality includes both student activities in *SMALLab* and the knowledge representations it enables.

The research of Jackendoff in cognitive linguistics suggests that information that an individual assimilates is encoded either as spatial representations (images) or as conceptual structures (symbols, words, or equations) (Jackendoff 1996). Traditional didactic approaches to teaching strongly favor the transmission of conceptual structures, and there is evidence that many students struggle with the process of translating these into spatial representations

(Megowan 2007). In contrast, information gleaned from the *SMALLab* environment is both propositional and imagistic as described above.

Working in *SMALLab*, students create multimodal artifacts such as sound recordings, videos, and digital images. They interact with computation using innovative multimodal interfaces such as 3D physical movements, visual programming interfaces, and audio capture technologies. These interfaces encourage the use of multiple modes of representation, which facilitates learning in general (Cunningham et al. 1993; Roth and McGinn 1998) and are robust to individual differences in students' optimal learning styles (Gardner 1993; Burt 2006), and can serve to motivate learning (Bransford et al. 2000).

26.2.1.2 Interactive Multimedia and HCI

Many recent developments in interactive multimedia and HCI research have emphasized the role of immersive, multisensory interaction through multimodal (auditory, visual, and tactile) interface design. This work can be applied in the design of new mixed-reality spaces. For example, in "Combining Audio and Video in Perceptive Spaces," Wren et al. (1999) describe their work in the development of environments utilizing unencumbered sensing technologies in situated environments. The authors present a variety of applications of this technology that span data visualization, interactive performance, and gaming. These technologies suggest powerful opportunities for the design of learning scenarios, but they have not yet been applied for this purpose.

Related work in arts and technology has influenced our approach to the design of mediated learning scenarios. Our work draws from extensive research in the creation of interactive sound environments (Bahn et al. 2001; Wessel and Wright 2001; Hunt et al. 2002). While much of this work is focused on applications in interactive computer music performance, the core innovations for interactive sound can be directly applied in our work with students. In addition, we are drawing from the 3D visualization community (Rheingans and Landreth 1995) in considering how to best apply visual design elements (e.g., color, lighting, and spatial composition) to render content in *SMALLab*.

There are many examples where interactive multimedia and HCI researchers are extending the multimodal tool set and are applying it to novel technologically mediated experiences for learning and play. Ishii's *Music Bottles* offer a multimodal experience through sound, physical interaction, and light color changes as different bottles are uncorked by the user to release sounds. The underlying sensor mechanism is a resonant RF coil that is modulated by an element in the cork. Edmonds has chronicled the significant contribution physiological sensors have made to the interactive computational media arts (Muller and Edmonds 2006). *RoBallet* uses laser beam-break sensors, such as those found in some elevators and garage doors, along with video and sonic feedback to engage students in interactive choreography and composition. Cavallo argues that this system would enable new forms of teaching not only music but math and programming as well (Cavallo et al. 2004). The work described in this paper builds from the research advanced in these examples and is, likewise, extending the tools and domains for multimodal HCI interfaces as they apply to learning and play.

26.2.1.3 Example: The MEDIATE Environment

One example of an immersive, multisensory learning environment that emphasizes multimodality is *MEDIATE*, an environment designed to foster a sense of agency and a capacity for creative expression in people on the autistic spectrum (PAS). Autism is a variable neuron-developmental disorder in which PAS are overwhelmed by excessive stimuli, noises, and colors that characterize interaction in the physical world (EU Community Report 2004; Pares et al. 2004, 2005). Perhaps as a result (although exact mechanisms and causes are unknown), PAS withdraw into their own world. They often find self-expression and even everyday social interaction difficult. *MEDIATE*, designed in collaboration with PAS, sets up an immersive 3D environment in which stimuli are quite focused and simplified, yet at the same time dynamic and engaging—capable of affording a wide range of creative expression. The *MEDIATE* infrastructure consists of a pair of planar screens alternating with a pair of tactile interface walls and completely surrounds the participant. On the screens are projected particle grids, a dynamic visual field that responds to the participant's visual silhouette, his/her vocalizations and other sounds, and his/her tactile interactions (Pares et al. 2004). A specially designed loudspeaker system provides immersive audio feedback that includes the subsonic range, and interface walls provide vibrotactile feedback.

Multimodality in *MEDIATE* is achieved through the integration of sonic, visual, and tactile interfaces in both sensing and feedback. The environment is particularly impressive in that it can potentially supplant the traditional classroom space with one that is much more conducive to learning in the context of PAS. However, *MEDIATE* remains specialized as a platform for PAS rehabilitation, and has not been generalized for use in everyday classroom instruction. In contrast, *SMALLab* emphasizes multimodality in the context of real-world classroom settings, where the immersive media coexists in the realm of everyday social interactions. *SMALLab* enables students and teachers to work together, physically interacting, face-to-face with one another and digital media elements. Thus, it facilitates the emergence of a natural zone of proximal development (Vygotsky 1978) where, on an informal basis, facilitators and student peer experts can interact with novices and increase what they are able to accomplish in the interaction.

Although *MEDIATE* was designed in collaboration with PAS (EU Community Report 2004), participants are not able to build in new modes of interaction or further customize the interface. This idea of *composition*, which comes from building, extending, and reconfiguring the interaction framework, is essential to engaging participants in more complex and targeted learning situations and has been integral to the design of *SMALLab*.

26.2.2 Embodiment

26.2.2.1 Learning Sciences

By *embodiment* we mean that *SMALLab* interactions engage students both in mind and in body, encouraging them to physically explore concepts and systems by moving within and acting upon an environment.

A growing body of evidence supports the theory that cognition is "embodied"—grounded in the sensorimotor system (Fauconnier and Turner 2002; Hutchins 2002; Hestenes 2006; Barsalou 2008). This research reveals that the way we think is a function of our body, its physical and temporal location, and our interactions with the world around us. In particular, the metaphors that shape our thinking arise from the body's experiences in our world and are, hence, embodied (Megowan 2007).

A recent study of the development of reading comprehension in young children suggests that when children explicitly "index" or map words to the objects or activities that represent them, either physically or imaginatively, their comprehension improves dramatically (Glenberg et al. 2004). This aligns well with the notion, advanced by Fauconnier and Turner (2002), that words can be thought of as form-meaning pairs. For example, when a reader encounters the lexical form "train" in a sentence, he can readily supply the sound form (trân). If he then maps it to the image of a train (a locomotive pulling cars situated on a track), we have a form-meaning pair that activates the student's mental model of trains, which he can then use to help him understand and interpret the sentence in which the word "train" appears (Megowan 2007).

SMALLab is a learning environment that supports and encourages students in this meaning-making activity by enabling them to make explicit connections between sounds, images, and movement. Abstract concepts can be represented, shared, and collaboratively experienced via physical interaction within a mixed-reality space.

26.2.2.2 Interactive Multimedia and HCI

Many emerging developments in interactive multimedia and HCI also emphasize the connections between physical activity and cognition (Suchman 1987; Varela et al. 1991; Hutchins 1995; Dourish 2001, 2004; Wilson 2003; Rambusch and Ziemke 2005), and the intimately embedded relationship between people and other entities and objects in the physical world (Gibson 1979; Norman 1988; Shepard 2001). The *embodied cognition* perspective (Wilson 2003; Rambusch and Ziemke 2005) argues, based on strong empirical evidence from psychology and neurobiology (Glenberg et al. 2004, 2005), that perception, cognition, and action, rather than being separate and sequential stages in human interaction with the physical world, in fact occur simultaneously and are closely intertwined. Dourish (2001, 2004) in particular emphasizes the importance of *context* in embodied interaction, which emerges from the interaction rather than being fixed by the system. As such, traditional interactive multimedia and HCI frameworks such as desktop computing (i.e., mouse/keyboard/screen) environments, which facilitate embodied interaction in a limited sense or not at all, risk binding the user to the system context, restricting many of his/her capacities for creative expression and free thought, which have proven so essential in effective learning contexts. From cognitive, ecological, and design psychology, Shepard (2001), Gibson (1979), Norman (1988), and Galperin (1992) further emphasize the importance of the embedded relationship between people and things, and the role that manipulating physical objects has in cognition. Papert, Resnick, and others (Papert 1980, 1993; Harel 1991; Resnick 1998) extend these approaches by explicitly stating their importance in educational settings. Design-based learning methodologies such as *Star Logo*, *Lego Mindstorms*, and *Scratch* (Papert 1980, 1991; Peppler and Kafai 2007) emphasize physical-digital simulation and thinking. These have proven quite popular and effective in fostering and orienting students' innate creativity toward specific learning goals.

In order for these tools to extend further into the physical world and to make use of the important connections provided by *embodiment*, they must include physical elements that afford embodied interactions. Ishii has championed the field of tangible media (Ishii and Ullmer 1997) and coined the term tangible user interfaces (TUIs vs. GUIs [graphical user interfaces]). His tangible media group has developed an extensive array of applications that pertain to enhancing not only productivity (e.g., Urban Simulation, SandScape) but also artistic expression and playful engagement in the context of learning (e.g., I/O Brush, Topobo, and Curlybot) (Ishii 2007).

Some prior examples of interactive multimedia and HCI systems that facilitate elements of embodiment and interaction with immersive environments include the Cave Automated Visualization Environment (CAVE) (Cruz-Neira et al. 1993). CAVEs typically present an immersive environment through the use of 3D glasses or some other head-mounted display (HMD) that enables a user to engage through a remote control joystick. A related environment, described as a step toward the holodeck, was developed by Johnson at

University of Southern California (USC) to teach topics ranging from submarine opera-tion to Arabic language training (Swartout et al. 2001). In terms of extending physical activity through nontraditional interfaces and applying them to collaboration and social engagement, the Nintendo Wii's recent impact on entertainment is the most pronounced. The Wii amply demonstrates the power of the body as a computing interface. Some learning environments that have made strides in this area include *Musical Play Pen*, *KidsRoom*, and *RoBallet* (Weinberg 1999; Bobick et al. 1999; Cavallo et al. 2004). These interfaces demonstrate that movement-based interactive multimedia and HCI can greatly impact on instructional design, play, and creativity.

26.2.2.3 Example

A particularly successful example of a learning environment that leverages embodiment in the context of instructional design is *River City* (Dede, Ketelhut, and Ruess 2002; Dede and Ketelhut 2003; Ketelhut 2007; Nelson et al. 2007). *River City* is a multiuser, online desktop vir-tual environment that enables middle-school children to learn about disease transmission. The virtual world in *River City* embeds a river in various types of terrain that influence water runoff and other environmental factors that in turn influence the transmission of disease through water, air, and/or insect populations. The factors affecting disease trans-mission are complex and have many causes, paralleling conditions in the physical world. Student participants are virtually embodied in the world, exploring through avatars that interact with each other, with facilitators' avatars, and with the auditory and visual stim-uli comprising the *River City* world. Participants can make complex decisions within this world by, for example, using virtual microscopes to examine water samples, and sharing and discussing their proposed solutions. In several pilot studies (Dede et al. 2002; Dede and Ketelhut 2003), the level of motivation, the diversity and originality of participants' solutions, and their overall content knowledge were found to increase with *River City* as opposed to a similar paper-based environment. Hence, the *River City* experience provides at least one successful example of how social embodiment through avatars in a multisensory world can result in learning gains.

However, a critical aspect of embodiment not addressed by *River City* is the bodily kines-thetic sense of the participant. Physically, participants interact with *River City* using a mouse and keyboard, and view 2D projections of the 3D world on a screen. The screen physically separates their body from the environment, which implies that perception and bodily action are not as intimately connected as they are in the physical world, resulting in embodiment in a lesser sense (Rambusch and Ziemke 2005). In *SMALLab*, multiple participants interact with the system and with each other via expressive, full-body movement. In *SMALLab*, there is no physical barrier between the participant and the audiovisual environment they manipulate. It has long been hypothesized (Gardner 1993) that bodily kinesthetic modes of representation and expression are an important dimension of learning and severely under-utilized in traditional education. Thus, it is plausible that an environment that affords full-body interactions in the physical world can result in even greater learning gains.

Another aspect of *SMALLab* not addressed by *River City* is *composition* (see Section 26.2.3). *River City* does not afford users the opportunity to conceive new disease scenarios, to design disease transmission models, or to test these new scenarios in the game environment.

26.2.3 Composition

Composition refers to reconfigurability, extensibility, and programmability of interaction tools and experiences. Specifically, we mean *composition* in two senses. First, students compose new interaction scenarios in the service of learning. Second, educators and mentors can extend the toolset to support new types of learning that is tailored to their students' needs.

26.2.3.1 Learning Sciences

In our design of the *SMALLab* learning experience, we proceed from the fundamentally constructivist view that knowledge must be actively constructed by the learner rather than passively received from the environment, and that the prior knowledge of the learner shapes the process of new construction (Dougiamas 1998). Drawing on the views of social constructivists (i.e., Vygotsky, Bruner, Bandura, Cobb, and Roth) we view the learning process as socially mediated, knowledge as socially and culturally constructed, and reality as not discovered but rather socially "invented" (Vygotsky 1962; Cobb 1994; Roth and McGinn 1998). We venture beyond constructivism in subscribing to the notion that teaching and learning should be centered on the construction, validation, and application of models—flexible, reusable knowledge structures that scaffold thinking and reasoning (Lesh and Doerr 2003; Hestenes 2006). This constructive activity of modeling lies at the heart of student activity in *SMALLab*.

In their seminal work describing the situated nature of cognition, Brown et al. (1989) observed that students in a classroom setting tend to acquire and use information in ways that are quite different from "just plain folks" (JPFs). They further revealed that the reasoning of experts and JPFs was far closer to one another than that of experts and students. They concluded that the culture of schooling, with its passive role for students and rule-based structure for social interactions, promotes decontextualization of information that leads to narrow procedural thinking and the inability to transfer lessons learned in one context to another. This finding highlights the importance of learning that is situated, both culturally and socially. *SMALLab* grounds students in a physical space that affords visual, haptic, and sonic feedback. The abstraction of conceptual information from this perceptual set is enabled through guided reflective practice of students' as they engage in the modeling process.

Student engagement in *SMALLab* experience is motivated both by the novelty of a learning environment that affords them some measure of control (Middleton and Toluk 1999) and by the opportunity to work collaboratively to achieve a specific goal, where the pathway they take to this goal *is not predetermined by the teacher or the curriculum*. Hence, *SMALLab* environment rewards originality and creativity with a unique digital-physical learning experience that affords new ways of exploring a problem space.

26.2.3.2 Interactive Multimedia and HCI

Compositional interfaces have a rich history in interactive multimedia and HCI research, as evidenced by Papert and Minsky's Turtle Logo, which fosters creative exploration and play in the context of a functional, Lisp-based programming environment (Papert 1991). More recent examples of interactive multimedia and HCI systems that incorporate compositional interfaces include novice-level programming tools such as *Star Logo*, *Scratch*, and *Lego Mindstorms*. Resnick extends these approaches through the *Playful Invention and Exploration* (PIE) museum network and the *Intel Computer Club Houses* (Resnick 2006), thus providing communities with tools for creative composition in rich, informal sociocultural contexts. Essentially, these interfaces create a "programming culture" at community technology centers, classrooms, and museums. Extensive research on programming languages for creative practitioners, including graphical programming environments for musicians and multimedia artists such as Max/MSP/Jitter, Reaktor, and PD have made significant contributions to extending the impact and viability of programming tools as compositional interfaces.

Embedding physical interactions into objects for composition is a strategy for advancing *embodied multimodal composition*. Ryokai's I/O Brush (Ryokai et al. 2004) is an example of a technology that encourages composition, learning, and play. This system enables capture from the physical world through a camera in the end of a paint brush that allows individuals to capture colors and textures from the physical world and compose with them in the digital world. It can even take video sequences such as a blinking eye that can then become part of the user's digital painting. Composition is a profoundly empowering experience and one that many learning environments are also beginning to emphasize to a greater extent.

26.2.3.3 Example: Scratch

The Scratch programming environment (Resnick 2007) emphasizes the power of compositional paradigms for learning. Scratch enables students to create games, interactive stories, animations, music, and art within a graphical programming environment. The interface extends the metaphor of LEGO bricks where programming functions snap together in a manner that prohibits programming errors and thus avoids the steep learning curve that can be a barrier to many students in traditional programming environments. The authors frame the goal of Scratch as providing "tinkerability" for learners that will allow them to experiment and redesign their creations in a manner that is analogous to physical elements, albeit with greater combinatorial sophistication.

Scratch has been deployed in a number of educational settings (Kafai et al. 2007; Peppler and Kafai 2007). In addition to focused research efforts to evaluate its impact, a growing Scratch community website, where authors can publish their work, provides mounting evidence that it is a powerful tool for fostering meaningful participation for a broad and diverse population.

Scratch incorporates multimodality through the integration of sound player modules within the primarily visual environment. However, it provides only a limited set of available tools for sound transformation (e.g., soundfile playback, speech synthesis) and as a consequence, authors are not able to achieve the multimodal sophistication that is possible within *SMALLab*. Similarly, Scratch addresses the theme of embodiment in the sense that authors and users can represent themselves as avatars within the digital realm. However, Scratch exists within the standard desktop computing paradigm and students cannot interact through other more physically embodied mechanisms.

26.2.4 Defining Play

With a focus on play in the context of games, Salen and Zimmerman (2004) summarize a multitude of definitions. First, they consider the diverse meanings and contexts of the very term "play." They further articulate multiple scopes for the term, proposing a hierarchy comprised of three broad types. The most open sense is "being playful," such as teasing or wordplay. Next is "ludic activity," such as playing with a ball, but without the formal structure of a game. The most focused type is "game play," where players adhere to rigid rules that define a particular game space.

Play and game play in particular have been shown to be an important motivational tool (Gee 2003), and as Salen and Zimmerman (2004) note, play can be transformative as, "it can overflow and overwhelm the more rigid structure in which it is taking place, generating emergent, unpredictable results." Our work is informed by these broad conceptions of play that are applied to the implementation of game-like learning scenarios for K-12 content learning.

Jenkins offers an expansive definition of play as "the capacity to experiment with one's surroundings as a form of problem-solving" (Jenkins et al. 2006). Students engaged in this notion of play exhibit the same transformative effects as described above, and we apply this definition of collaborative problem-solving as play in our work with students in formal learning contexts.

26.2.5   Toward a Theoretical and Technological Integration

As described above, there has been extensive theoretical and practice-based research across education, interactive multimedia, and HCI that is aimed at improving learning through the use of embodiment, multimodality, and compositional frameworks. We have described examples of prior projects, each of which strongly emphasizes one or two of these concepts. This prior work has yielded significant results that demonstrate the powerful impact of educational research that is aligned with emerging interactive media and HCI practices. However, while there are some prior examples of interactive platforms that integrate these principles (SmartUs 2008), there are few prior efforts to date that do so while leveraging the powerful affordances of mixed reality for content learning. As such, there is an important opportunity to improve upon prior work.

In addition, many technologically driven efforts are limited by the use of leading-edge technologies that are prohibitively expensive and/or too fragile for most real-world learning situations. As a consequence, many promising initiatives do not make a broad impact on students and cannot be properly evaluated owing to a failure to address the practical constraints of today's classrooms and informal learning contexts. Specifically, in order to see large-scale deployment on a 2–5-year horizon, learning environments must be inexpensive, mindful of typical site constraints (e.g., space, connectivity, and infrastructure support), robust, and easily maintainable. It is essential to reach a balance between reliance upon leading-edge technologies and consideration of the real-world context in order to collect longitudinal data over a broad population of learners that will demonstrate the efficacy of these approaches.

Our own efforts are focused on advancing research at the intersection of interactive multimedia, HCI, and education. We next describe a new mixed-reality environment for learning, a series of formative pilot studies, and two recent in-school programs that illustrate the implementation and demonstrate the impact of our work.

26.3   *SMALLab*: Integrated Interactive Multimedia for Learning

SMALLab represents a new breed of SCLE that incorporates multimodal sensing, modeling, and feedback while addressing the constraints of real-world classrooms. Figure 26.2 diagrams the full system architecture, and here we detail select hardware and software components.

Physically, *SMALLab* consists of a $15' \times 15' \times 12'$ portable, freestanding media environment (Birchfield et al. 2006). A cube-shaped trussing structure frames an open physical architecture and supports the following sensing and feedback equipment: a six-element

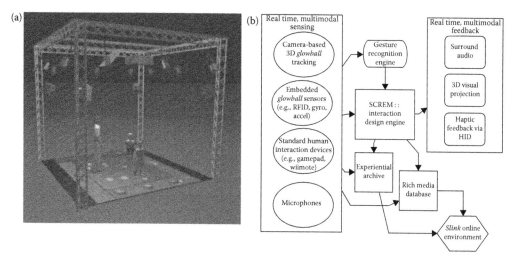

FIGURE 26.2
(See color insert.) (a) The *SMALLab* system with cameras, speakers, and project, and (b) *SMALLab* software architecture.

array of *Point Grey Dragonfly* firewire cameras (three color, three infrared) for vision-based tracking, a low-cost eight-camera OptiTrack marker-based motion capture system for accurate position and pose tracking, a top-mounted video projector providing real-time visual feedback, four audio speakers for surround sound feedback, and an array of tracked physical objects (custom-made *glowballs* and marker-attached *rigid objects*). A networked computing cluster with custom software drives the interactive system.

The open physical architecture of the space is designed to encourage human-to-human interaction, collaboration, and active learning within a computational framework. It can be housed in a large general-purpose classroom without the need for additional specialized equipment or installation procedures. The use of simple, unencumbered sensing technologies ensures that there is a minimal learning curve for interaction, yet it has been utilized in diverse educational contexts, including schools and museums.

With the exception of the *glowballs* and the marker-attached *rigid objects*, all *SMALLab* hardwares (e.g., audio speakers, cameras, multimedia computers, video projector, and support structure) are readily available off the shelf. This ensures that *SMALLab* can be easily maintained throughout the life of a given installation as all components can be easily replaced. Furthermore, the use of readily available hardware contributes to the overall low cost of the system. We have custom developed all *SMALLab* software, which is made freely available to our partner educational institutions.

SMALLab can be readily transported and installed in classrooms or community centers. We have previously disassembled, transported to a new site, reinstalled, and calibrated a functioning *SMALLab* system within one day's time.

26.3.1 Multimodal Sensing

Groups of students and educators interact in *SMALLab* together through the manipulation of up to five illuminated *glowball* objects, marker-attached rigid object, a set of standard HID devices, including wireless gamepads, Wii Remotes (Brain 2007; Shirai et al. 2007), and commercial wireless pointer/clicker devices.

The vision-based tracking system senses the real-time 3D position of these *glowballs* at a rate of 50–60 frames per second using robust multiview techniques (Jin et al. 2006; Jin and Qian 2008). Robust real-time object tracking in mediated environments with interfering visual projection in the background is challenging. To address interference from visual projection, each object is partially coated with a tape that reflects infrared light. Reflections from this tape can be picked up by the infrared cameras, while the visual projection cannot. To be specific, two major contributions have been made in our research to achieve robust object tracking. A reliable outlier rejection algorithm is developed using the epipolar and homography constraints to remove false candidates caused by interfering background projections and mismatches between cameras. To reliably integrate multiple estimates of the 3D object positions, an efficient fusion algorithm based on mean shift is used. This fusion algorithm can also reduce tracking errors caused by partial occlusion of the object in some of the camera views. Experimental results obtained in real-life scenarios demonstrate that the proposed system is able to achieve decent 3D object tracking performance in the presence of interfering background visual projection. Figure 26.3 shows the block diagram of

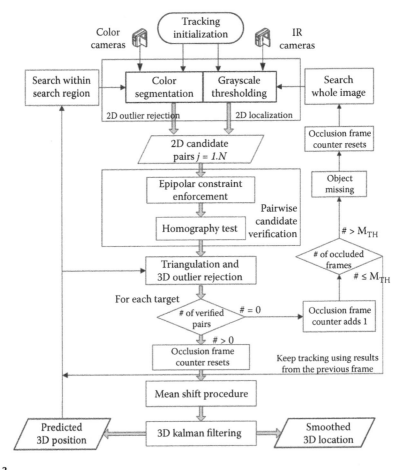

FIGURE 26.3
The block diagram of the object tracking system used in the multimodal sensing module of *SMALLab*. (From Jin, H. and G. Qian. 2008. *EURASIP Journal on Image and Video Processing*. 2008, Article ID 638073, 14 pp., With permission.)

the object tracking system. The tracking initialization is to manually obtain camera projection matrices, homogeneous plane coefficients, and the histogram of objects, and the gray-scale thresholds for IR cameras. It is a one-time process, and all the parameters can be saved for future use. Object histogram and gray-scale thresholds are used to locate the target in the color and IR camera view, respectively. The search region on the image plane predicted by Kalman filtering helps to remove 2D false candidates. A pair of 2D candidates from two different views are examined by the epipolar constraint and planar homography test. Label information is considered in this step. Each valid pair corresponds to a 3D triangulation result. The unlabeled pair from two different IR cameras might be obtained from 2D false candidates such as reflective or bright spots. The predicted 3D location and velocity from Kalman filtering help detect those false pairs. The distribution of all the objects being tracked is modeled as a mixture of Gaussian. Each mixture component corresponds to a target object. The kernel-based mean-shift algorithm is employed for each target to find the optimal 3D location. Each target is assigned a Kalman filter that plays the role of a smoother and predictor. The partial occlusions are alleviated by the mean-shift algorithm while the complete occlusions that occur in some camera views are automatically compensated by other nonoccluded camera views.

Object position data are routed to custom software modules (described below) that perform various real-time pattern analyses on these data, and in response, generate real-time interactive sound and visual transformations in the space. With this simple framework we have developed an extensible suite of interactive learning scenarios and curricula that integrate the arts, sciences, and engineering education.

26.3.2   Rich Media Database

SMALLab features an integrated and extensible rich media database that maintains multimodal content provided by students, teachers, and researchers. This is an important tool in support of multimodal knowledge representation in *SMALLab*. It manages audio, video, images, text, and 3D objects and enables users to annotate all media content with user-specific metadata and typed links between elements. The SCREM interface (described below) tightly integrates search and navigation tools so that scenario authors and students can readily access this media content.

26.3.3   SCREM

We apply the notion of composition at two levels. First, we have conceived of *SMALLab* as a modular framework to ensure that educators and administrators can continuously extend and improve it through the design and implementation of new scenarios. In this regard, *SMALLab* is not a one-size-fits-all solution, but rather, it enables an educator- and community-driven learning environment. Second, many *SMALLab* curricula emphasize learning through collaborative problem solving and open-ended design challenges. These approaches demand that students are able to readily design and deploy new interactive scenarios through the manipulation of powerful, yet easy-to-use interfaces—interfaces that provide both depth and breadth.

To this end we have developed an integrated authoring environment, the *SMALLab Core for Realizing Experiential Media* (SCREM). SCREM is a high-level object-oriented framework that is at the center of interaction design and multimodal feedback in *SMALLab*. It provides a suite of graphical user interfaces to either create new learning scenarios or modify existing frameworks. It provides integrated tools for adding, annotating, and linking content in the

SMALLab Media Content database. It facilitates rapid prototyping of learning scenarios, enables multiple entry points for the creation of scenarios, and provides age- and ability-appropriate authoring tools.

SCREM supports student and teacher composition at three levels. First, users can easily load and unload existing learning scenarios. These learning scenarios are stored in an XML format that specifies interactive mappings, visual and sonic rendering attributes, typed media objects, and metadata including the scenario name and date. Second, users can configure new scenarios through the reuse of software elements that are instantiated, destroyed, and modified via a graphical user interface. Third, developers can write new software code modules through a plug-in-type architecture that are then made available through the high-level mechanisms described above. Depending on developer needs, low-level *SMALLab* code can be written in a number of languages and media frameworks, including Max/MSP/Jitter, Javascript, Java, C++, Objective C, Open Scene Graph, and VR-Juggler.

26.3.4 *SLink* Web Portal

The *SMALLab Link*, or *SLink*, web portal (Birchfield et al. 2006) provides an online interface that enables teaching and learning to seamlessly span multiple *SMALLab* installations and to extend from the physical learning environment and into students' digital realms. It serves three functions as (1) a supportive technology, (2) a research tool, and (3) an interface to augment *SMALLab* learning.

As a supportive technology, *SLink* acts as a central server for all *SMALLab* media content and user data. It provides functionality to sync media content that is created at a given *SMALLab* site to all other sites while preserving unique metadata. Similarly, *SLink* maintains dynamic student and educator profiles that can be accessed by teachers and researchers online or in *SMALLab*.

SLink is a research tool and an important component of the learning evaluation infrastructure. Through a browser-based interface, educational researchers can submit, search, view, and annotate video documentation of *SMALLab* learning. Multiple annotations and annotator metadata is maintained for each documentation element.

SLink serves as a tool for students where they can access or contribute media content from any location through the web interface. This media content and metadata will sync to all *SMALLab* installations. In ongoing work, we are expanding the *SLink* web interface to provide greater functionality for students. Specifically, we are developing tools to search and render 3D *SMALLab* movement data through a browser-based application. Student audio interactions can be published as podcasts, and present visual interactions are presented as streaming movies. In these ways, *SLink* extends into the web our paradigms of multimodal interaction and learning through composition.

26.3.5 Experiential Activity Archive

All *glowball* position data, path shape quality information, SCREM interface actions, and projected media data are streamed in real time to a central archive application. Incoming data are timestamped and inserted into a MySQL database where it is made available in three ways. First, archived data can be replayed in real time such that they can be rerendered in *SMALLab* for the purpose of supporting reflection and discussion among students regarding their interactions. Second, archived data are made available to learners and researchers through the *SLink* web interface. Third, archived data can be later mined for the purposes of evaluation and assessment of *SMALLab* learning. We are currently developing a greatly expanded version of the activity archive that will include the archival of real-time video and audio streams, interfaces to create semantic links among entries, and tools to access the data from multiple perspectives.

26.4 Case Study: Earth Science Learning in *SMALLab*

Having presented a theoretical basis, and described the development and integration of various HCI technologies into a new mixed-reality environment, we now focus on the application and evaluation of *SMALLab* for learning. This research is undertaken at multiple levels, including focused user studies to validate subcomponents of the system (Tu et al. 2007a,b), and perception/action experiments to better understand the nature of embodied interaction in mixed-reality systems such as *SMALLab* (Dolgov et al. 2009). Over the past several years, we have reached over 25,000 students and educators through research and outreach in both formal and informal contexts that span the arts, humanities, and sciences (Cuthbertson et al. 2007; Birchfield et al. 2008; Hatton et al. 2008). This prior work serves as an empirical base that informs our theoretical framework. Here, we present a recent case study to illustrate our methodology and results.

26.4.1 Research Context

In the summer of 2007, we began a long-term partnership with a large urban high school in the greater Phoenix, Arizona metropolitan area. We have permanently installed *SMALLab* in a classroom and are working closely with teachers and students across the campus to design and deploy new learning scenarios. This site is typical of public schools in our region. The student demographic is 50% white, 38% Hispanic, 6% native American, 4% African American, and 2% others. 50% of students are on free or reduced lunch programs, indicating that many students are of low socioeconomic status. 11% are English language learners and 89% of these students speak Spanish at home. In this study, we are working with 9th-grade students and teachers from the school's Coronado Offers Reading, Writing, and Math for Everyone (C.O.R.E.) program for at-risk students. The C.O.R.E. program is a specialized "school within a school" with a dedicated faculty and administration. Students are identified for the program because they are studying at least two levels below their grade and have been recommended by their middle-school teachers and counselors. After almost a year of classroom observation by our research team, it is evident that students are tracked into this type of program not because they have low abilities but because they are often underserved by traditional instructional approaches and exhibit low motivation for learning. Our work seeks to address the needs of this population of students and teachers.

Throughout the year, we collaborated with a cohort of high-school teachers to design new *SMALLab* learning scenarios and curricula for language arts and science content learning. Embodiment, multimodality, and composition served as pillars to frame the formulation of new *SMALLab* learning scenarios, associated curricula, and the instructional design. In this context, we present one such teaching experiment. This case study illustrates the use of *SMALLab* for teaching and learning in a conventional K-12 classroom. It demonstrates the implementation of our theoretical framework around the integration of embodiment, multimodality, and composition in a single learning experience. Finally, we present empirical evidence of student learning gains as a result of the intervention.

26.4.2 Design and Teaching Experiment

The evolution of the earth's surface is a complex geological process that is affected by numerous interdependent processes. Geological evolution is an important area of study for high school students because it provides a context for the exploration of systems-thinking (Chen and Stroup 1993) that touches upon a wide array of earth science topics. Despite the nature of this complex, dynamic process, geological evolution is typically studied in a very static manner in the classroom. In a typical learning activity, students are provided with an image of the cross section of the earth's crust. Due to the layered structure of the rock formations, this is termed as a geological layer cake. Students are asked to annotate the image by labeling the rock layer names, ordering the layers according to which were deposited first, and identifying the evidence of uplift and erosion (Lutgens et al. 2004). Our partner teacher has numerous years of experiences with conventional teaching approaches in his classroom. Through preliminary design discussions with him, we identified a deficiency of this traditional instructional approach: *when students do not actively engage geological evolution as a time-based, generative process, they often fail to conceptualize the artifacts (i.e., cross sections of the earth's surface) as the products of a complex, dynamic system.* As a consequence, they struggle to develop robust conceptual models during the learning process.

For 6 weeks we collaborated with the classroom teacher, using the *SMALLab* authoring tools, to realize a new mixed-reality learning scenario to aid learning about geological evolution in a new way. Our three-part theoretical framework guided this work: embodiment, multimodality, and composition. At the end of this process, the teacher led a 3-day teaching experiment with 72 of his 9th-grade earth science students from the C.O.R.E. program. The goals for the teaching experiment were twofold. First, we wanted to advance participating students' understanding of earth science concepts relating to geological evolution. Second, we wanted to evaluate our theoretical framework and validate *SMALLab* as a platform for mixed-reality learning in a formal classroom-learning environment.

We identified four content learning goals for students: (1) understanding of the principle of *superposition*—that older structures exist below younger structures on the surface of the Earth; (2) understanding geological evolution as a complex system with interdependent relationships between surface conditions, fault events, and erosion forces; (3) understanding that geological evolution is a time-based process that unfolds over multiple scales; and (4) understanding how the fossil record can provide clues regarding the age of geological structures. These topics are central to high-school earth science learning and are components of the State of Arizona Earth and Space Science Standards (Arizona Department of Education 2005). We further stipulate that from a modeling instruction perspective (Hestenes 1992, 1996) students should demonstrate their mastery of these topics through evidence that they have consolidated robust conceptual models that integrate both descriptive and explanatory knowledge of these principles.

Our collaborative design process yielded three parts: (1) a new mixed-reality learning scenario, (2) a student participation framework, and (3) an associated curriculum. We now describe each of these parts, discussing how each tenet of our theoretical framework is expressed. We follow this with a discussion on the outcomes with respect to our goals.

26.4.2.1 Interactive Scenario: Layer Cake Builders

Figure 26.4 shows the visual scene that is projected onto the floor of *SMALLab*. Within the scene, the center portion is the layer cake construction area where students deposit sediment layers and fossils. Along the edges, students see three sets of images. At the bottom

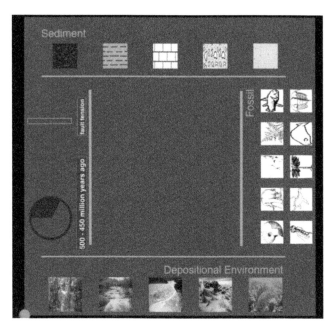

FIGURE 26.4
(**See color insert.**) Screen capture of projected *Layer Cake Builder* scene.

they see depictions of depositional environments. At the top are images that represent sedimentary layers. To the right they see an array of plant and animal images that represent the fossil record. Each image is an interactive element that can be selected by students and inserted into the layer cake structure. The images are iconic forms that students encounter in their studies outside of *SMALLab*. A standard wireless game pad controller is used to select the current depositional environments from among the five options. When a student makes a selection, they will *see* the image of the environment and *hear* a corresponding ambient soundfile. One *SMALLab* glowball is used to grab a sediment layer—by hovering above it—from among five options and drop it onto the layer cake structure in the center of the space. This action will insert the layer into the layer cake structure at the level that corresponds with the current time period. A second glowball is used to grab a fossil from among 10 options and drop it onto the structure. This action embeds the fossil in the current sediment layer. On the east side of the display, students see an interactive clock with geological time advancing to increment each new period. Three buttons on a wireless pointer device are used to pause, play, and reset geological time. A bar graph displays the current fault tension value in real time. Students use a Wii Remote game controller (Brain 2007; Shirai et al. 2007), with embedded accelerometers, to generate fault events. The more vigorously that a user shakes the device, the more the fault tension will increase. Holding the device still will decrease the fault tension. When a tension threshold is exceeded, a fault event (i.e., earthquake) will occur, resulting in uplift in the layer cake structure. Fault events can be generated at any time during the building process. Subsequently erosion occurs on the uplifted portion of the structure.

Figure 26.5 illustrates that in addition to the visual feedback present in the scene, students hear sound feedback with each action they take. A variety of organic sound events, including short clicks and ticks, accompany the selection and deposit of sediment layers and fossils. These events were created from field recordings of natural materials such as stones. This

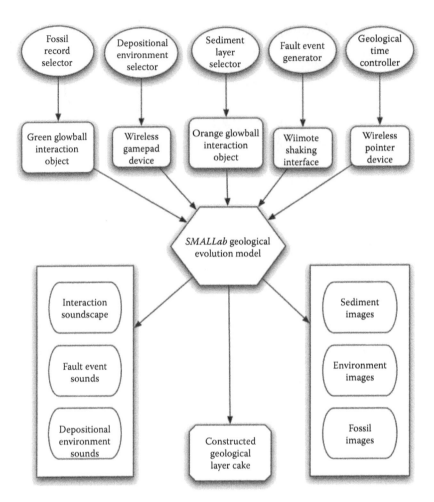

FIGURE 26.5
Layer Cake Builder interaction architecture schematic.

feedback contributes to an overarching soundscape that is designed to enrich students' sense of immersion in the earth science model. In addition, key earth science concepts and compositional actions are communicated to the larger group through sound. For example, the selection of a depositional environment is represented visually through an image, and sonically through looping playback of a corresponding soundfile. If a student selects the depositional environment of a fast-moving stream, all students will see an image of the stream, and hear the sound of fast-moving water. The multimodal display first alerts all students to be aware of important events in the compositional process. In addition, the dynamic nature of the fast-moving water sound communicates important features of the environment itself that are not necessarily conveyed through image alone. Specifically, a fast-moving stream is associated with the deposition of a conglomerate sediment layer that contains a mixture of large and small particles. The power of water to move large rocks and even boulders is conveyed to students through sound.

While students are engaged in the compositional process, sound is an important component to how they parse the activity and cue their own actions. Here we present a transcript from a typical layer cake build episode, demonstrating how sound helps students to orient themselves in the process. In the transcription, T is the teacher and FS indicates a member of the fossil selection team:

> (The student holding the controller from the depositional environment group selects an environment and the sound of ocean waves can be heard. Responding to the sound cue without even looking up at the image of the depositional environment highlighted, the student controlling the glowball for sediment layer team moves to select limestone.)
>
> *FS1: Shallow ocean.*
>
> *FS2: Wait, wait, wait.*
>
> (As the student holding the fossil glowball moves to make his selection, a fossil team member tells the boy with the glowball to wait because he could not see what sediment layer had been selected. After the sediment group and the fossil group made their selections, someone from the depositional environment team changes their selection. When the sound of a new environment is heard, the fossil team selector student (FS1) looks at the new environment and sees that the fossil he deposited is no longer appropriate for this environment. He picks up an image of a swimming reptile but then pauses uncertainly before depositing it.)
>
> *FS2: Just change it.*
>
> *T: Just change it to the swimming reptile*
>
> (The clock chimes the completion of one cycle at this point. The depositional environment team shifts their choice to desert and a whistling wind sound can be heard. Again, without even looking at the depositional environment image, the fossil group selector, FS2, quickly grabs a fossil and deposits it while the sediment layer girl runs back and forth above her 5 choices trying to decide which one to choose. She finally settles on one, picks and deposits it and then hands off the glowball and sits down. The next two selector students stand at the edge of the mat waiting for the clock to complete another cycle. The assessment team is diligently taking notes on what has been deposited. Another cycle proceeds as the sound of ocean waves can be heard. Students controlling the glowballs move quickly to make their selections without referring to the highlighted depositional environment.)

As shown in Figure 26.6, during the learning activities, all students are copresent in the space, and the scenario takes advantage of the embodied nature of *SMALLab*. For example, the concept of fault tension is embodied in the physical act of vigorously shaking the Wii Remote game controller. In addition, this gesture clearly communicates the user's intent to the entire group. Similarly, the deliberate gesture of physically stooping to select a fossil and carrying it across the space before depositing it in the layer cake structure allows all students to observe, consider, and act upon this decision as it is unfolding. Students might intervene verbally to challenge or encourage such a decision. Or they might coach a student who is struggling to take action. Having described the components of the system, we now narrate and discuss the framework that enables a class of over 20 students to participate in the scenario.

26.4.2.2 Participation Framework

The process of constructing a layer cake involves four lead roles for students: (1) the depositional environment selector, (2) the sediment layer selector, (3) the fossil record selector,

FIGURE 26.6
Students collaborating to compose a layer cake structure in *SMALLab*.

and (4) the fault event generator. In Figure 26.5, we diagram the relationship between each of these participant roles (top layer) and the physical interaction device (next layer down). The teacher typically assumes the role of geological time controller.

In the classroom, ~20–25 students are divided into four teams of five or six students each. Three teams are in active rotation during the build process, such that they take turns serving as the action lead with each cycle of the geological clock. These teams are the (1) depositional environment team and fault event team, (2) the sediment layer team, and (3) the fossil team. The remaining students constitute the evaluation team. These "evaluator" students are tasked to monitor the build process, record the activities of action leads, and to steer the discussion during the reflection process. Students are encouraged to verbally coach their teammates during the process.

There are at least two ways in which the build process can be structured. On the one hand, the process can be purely open ended, with the depositional environment student leading the process, experimenting with the outcomes, but without a specific constraint. This is an exploratory compositional process. Alternatively, the students can reference an existing layer cake structure as a script such as the one pictured in Figure 26.7. This second scenario is a goal-directed framing where only two students have access to the original

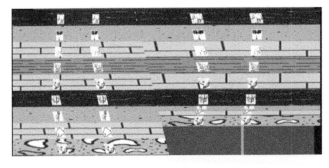

FIGURE 26.7
Layer cake structure created in *SMALLab*.

script, but all participants must work together to reconstruct the original. At the end of the build cycle, students compare their structure against the original. In this discussion we narrate the goal-directed build process.

At the beginning of each geological period, the lead "depositional environment" student examines the attributes of the source structure (e.g., Figure 26.7) and selects the appropriate depositional environment or surface condition on the earth. All students see an image and hear a sonic representation of the depositional environment. Based on that selected condition, another student grabs the appropriate sedimentary rock, and drops it onto the structure. While considering the current evolutionary time period and the current depositional environment, another student grabs a fossilized animal and lays it into the sedimentary layer. If students change their mind, sediment and fossil layers can be replaced by another element within a given geological time period. As the geological clock finishes a cycle, the next period begins. The action lead passes their interaction device to the next teammate, and these students collaborate to construct the next layer. The rotation continues in this fashion until the layer cake is complete. In this manner, the layer cake build process unfolds as a semistructured choreography of thought and action that is distributed across the four action leads and their teammates. The teams rotate their roles each time a new layer cake is to be constructed. The fossil students become evaluators, while the evaluators become the sediment layer team and so forth.

From a compositional perspective, this process is *open ended* and *improvisational*. By open ended we mean that any combination of depositional environments, sediment layers, fossils, and fault events can occur without constraint from the technology itself. By improvisational we mean that it unfolds in real time, and each participant acts with a clearly defined role, yet independently of the other students. The participation framework is analogous to a group of improvising jazz musicians. Students have individual agency to think and act freely. Yet they are bound by a constrained environment and driven by the shared goal of producing a highly structured outcome. Composition is distributed across multiple students where each has a clearly defined role to play and a distinct contribution to be made toward the collective goal. Collective success or failure depends on all participants. This process unfolds in real time with the expectation that there will be continuous face-to-face communication between participants.

This interaction model affords rich opportunities for whole group action and discussion about the relationship between in-the-moment events and the consequence of these decisions in the final outcome. For example, "fault event" students are free to generate earthquake after earthquake and explore the outcomes of this activity pattern, including its impact on students who are depositing sediment layers and fossils. Through this experimentation, students come to understand that in the real world, just as in the model, periods of numerous fault events are often interspersed with periods of little activity. This is a system-level understanding of geological evolution that must be negotiated by teams of students over the course of numerous cycles.

The learning activity is a form of structured play in two senses. Following Salen and Zimmerman's model, the layer cake build process unfolds in a structured manner as defined by the interaction framework. However, the play activity can take different forms according to the metarules set by the teacher. For example, during the open-ended compositional process, play is akin to "ludic activity" where a clear game space is articulated in *SMALLab*, but there are not clearly defined start and end conditions. When the activity is structured with a reference layer cake image and students are given the explicit goal to recreate that structure, the activity takes the form of goal-oriented "game play." Jenkins' notion of play also frames the learning activity as he defines play to be "the capacity to experiment with one's surroundings as a form of problem-solving." Again, in both the open-ended and structured forms, the layer cake build process is posed as a complex problem-solving activity that unfolds in real time. Importantly, individual participants must cooperatively integrate their thoughts and actions to achieve a shared success.

26.4.2.3 Curriculum

We collaborated with our partner teacher to design a curriculum that he implemented during a total of three, 45-min class periods across three consecutive days. The curriculum is informed by our overarching theoretical framework, and is designed to foster student-centered learning. Student activity is structured around a repeating cycle of *composition → reflection*. From a modeling instruction perspective (Hestenes 1992, 1996), this activity cycle supports students' underlying cognitive process that we term as *knowledge construction → consolidation*. During the first phase of the cycle (*composition/construction*), students devise conceptual models about the subject matter and the foundational systems. Teams of students work together in real time to create a layer cake structure. Through this hands-on, constructive activity, they are continuously forming, testing, and revising their conceptual models. This phase is immediately followed by a second stage (*reflection/consolidation*) when students discuss the construction activities, analyze any faults in decision-making, ask questions about various aspects of the layer cake structure, and challenge one another to justify their actions in the space. This reflective process leads to a consolidation of the conceptual models that spring from the construction phase. With every iteration of this cycle, new concepts are introduced and new knowledge is tested and consolidated, ultimately leading to a robust conceptual model of geological evolution.

As this was the first experience in *SMALLab* for most students, day 1 began with a brief introduction to the basic technology and an overview of expectation. The teacher then introduced the technological components of the learning scenario itself and students were divided into teams to begin creating layer cake structures in an open-ended, exploratory fashion. During this first day, the teacher structured the interactions, frequently pausing the scenario and prompting students to articulate their thinking before continuing the interaction. For example, he first started the geological clock and asked the depositional environment team to select an environment, leading to a discussion on the images and sounds, and what they represent. Once an environment was selected he would stop the geological clock and ask the sediment layer team to discuss the sediment icons and why a particular selection would be appropriate or not. Restarting geological time, the team selected their choice for the best sediment layer, placed it in the layer cake structure. Similar discussions and actions unfolded for the selection of an appropriate fossil. Over the course of the class period, the teacher intervened less and less as the students develop the ability to coordinate their activities and reason through the construction process on their own. Figure 26.7 shows an example of the outcome of a layer cake build cycle. During each reflection stage, we captured a screenshot of the layer cake structure and uploaded and annotated it in the *SLink* database for later reference.

During day 2, the teacher introduced the fault event interface and teams assumed this role in a similar manner as the exploration of day 1. Discussions regarding the selection of the fossil record grew more detailed as students were challenged to consider both the environmental conditions and the sequence of geological time in their selection process. For example, students reasoned through an understanding of why mammalian fossils should not appear early in the fossil record due to our understanding of the biological evolution of species. Midway through the class, the teacher moved students to the structured build process. He provided the "depositional environment" team with source images that show geological cross sections of the earth's crust such as the one pictured in Figure 26.7. These students must interpret the sequence of sediment layers and uplift/erosion evidence to properly initiate the environments and fault events that will cause the actions that follow to reproduce the source. Only the few students on the "depositional environment" team have access to this source image. Thus, all others' actions are dependent on their decision-making. For example, the "sediment" selection team could potentially add a rock layer that does not align with the source image for a particular geological period. While this could stem from a misunderstanding by their action lead, this deviation might follow the improper selection of a depositional environment from that team. Or, both the depositional environment and the sediment could be incorrect, causing a chain of deviations that would be unraveled at the end of the build. Students continued iterating through the *construction* → *reflection* process, rotating roles with each cycle, now structuring their interactions, and measuring their progress with the explicit goal of replicating the reference layer cake. The teacher at times guided the reflective process, but the student "evaluation" team members increasingly led these discussions.

On day 3, the teacher led a summative assessment activity. Prior to the session, he worked in *SMALLab* to create a set of four layer cake structures. He captured screenshots and printed images of these four structures. During class the students worked to recreate each of the structures in a similar manner as in day 2. At the end of each build process, the "evaluation" team reported any deviations from the reference structure, and the build teams were given the opportunity to justify and defend their actions. The teacher assigned a grade to each student at the end of the class period. This grade was a measure of their mastery of the build process as indicated by their ability to effectively contribute to the replication of the source structure and/or justify any deviations. Similar to days 1 and 2, team action leads rotated with each new geological period, and teams rotated through the different roles each time a new script was introduced. During this class session the teacher made very few interventions as students were allowed to reason through the building and evaluation process on their own.

26.4.3 Outcomes

During the final in-class assessment activity on day 3, all teams demonstrated an impressive ability to accurately reproduce the source structure. Collectively, the students composed 15 layer cakes during day 3. Eleven of the results were either a perfect match or within tolerable limits (e.g., only a slight deviation in the intensity of a fault event or no more than one incorrect sediment layer) of the source structure. Deviations typically stemmed from students' selection of alternate fossils in circumstances where there was room for interpretation or minor deviations in the magnitude of fault events within a given geological period. Students also exhibited improvement in their ability to justify their actions, developing arguments by the final day that suggest they quickly developed robust conceptual models of the underlying content.

For example, below is a transcript of the teacher and students in a typical cycle of *composition* → *reflection* from day 1 of the treatment. The teacher is controlling geological time during this episode. When the transcription begins, the students are in the middle of a layer cake build process and they have just completed discussion about creating one layer in the process. After his first comment, he starts the geological clock again, and the students commence constructing the next layer. In the transcriptions T is the teacher and students are identified by a first initial or S if the exact voice could not be identified.

T: *Alright, let's go one more time.*

(Sound of rushing water. The students with the glowballs pick a sediment layer (sandstone) and a fossil (fish) and lay them into the scenario. This takes less than 10 seconds. When they are done the teacher pauses the geologic clock to engage them in reflection).

T: *Alright, depositional environment—what are we looking at?*

Ss: *A river.*

T: *A river. Sandstone. Is that a reasonable choice for a type of rock that forms in a river? (Shrugs) Could be...is there any other types of rock over there that form in a river. Chuck. What's another rock over there that might form in a river?*

C: *In a river? I can't find one...*

T: *In a river. (there is a pause of several seconds)*

S: *Conglomerate.*

T: *Alright. Conglomerate is also an acceptable answer. Sandstone's not a bad answer. Conglomerate is pretty good...big chunks of rock that wash down in the river. So, what kind of fossil did you put in?*

S: *A fish.*

T: *A fish, okay. A fish in a stream makes good sense. Let's think about the fossils that we have in here. First we have a trilobite and then we had a jellyfish, then we had a fern and then we had a fish, alright? Is there anything wrong with the order of these animals so far?*

S: *They're aging.*

T: *What do you mean, 'they're aging'?*

S: *Evolution?*

T: *It's evolution so which ones should be the older fossils? (pause of several seconds)*

S: *...Trilobite?*

T: *Trilobite in this case...why the trilobite in this case? How do we know the trilobite's the oldest?*

S: *Because it's dead.*

T: *Just look at the picture. How do we know that the trilobite is oldest?*

S: *Because it's on the bottom?*

T: *We know that the oldest rocks are found...*

S: *On the bottom.*

T: *...on the bottom. So that's another thing that we want to make sure that we're keeping in check...we don't want to end up putting a whale on the bottom and a trilobite on top of a whale...because what kind of animal is a whale? (Pause) It's a mammal, alright? Mammals are relatively recently evolved. So let's pass off the spheres, guys. This next cycle I'm going to do a little different. I'm going to let two cycles go through without stopping you. Let's see how well we can do with the two cycles*

Now we present a brief transcription of a typical episode from day 3. Here, students have just finished building a complete layer cake. One student team controlled the depositional environment and faulting events, another team controlled sediment layers, a third team, controlled fossils, and a fourth team acted as evaluators, determining the plausibility of various elements used in the construction.

T: *Alright, JR, What's the first rock supposed to be?*

JR: *They got them all right.*

T: *All the rocks are correct?*

JR: *Yeah.*

T: *Ok. How about depositional environments, and Walt you're going to have to help her with this...do all the depositional environments match up with the rocks that were chosen?*

W: Yeah.

T: All the rocks match up…what about the fossils, A [student]?

A: They actually had some differences…

T: It doesn't have to exactly as it is on here. This is just a suggested order, right? What you need to do is figure out whether or not the ones they chose fit their environment.

W: Yeah. Well except for…

S1: Except for the fern in there…

W: Yeah, number 9 was supposed to be a fish, but it was a fern.

T: Ok, well, like I said, it doesn't necessarily have to be the fish that's there…is a fern possible as a formation of a fossil in a conglomerate, which is what type of depositional environment?

S1: Water…

S2: Stream…

S3: River…

T: A stream…is it possible for a fern to form a fossil in stream environment?

Many voices: yeah…no, no…no…yeah…

T: Alright. Bill says there is. Let's hear what you have to say Bill.

B: I just said that it can be.

T: Okay. How. How would that happen?

D: Cuz he thinks he knows everything.

T: David. Talk to Bill. I think you have a potential valid argument here but I want to hear it so we can make our…so we can judge.

D: That's cool.

B: Well…like, ferns grow everywhere, and if it lives near a river it could fall in…

T: Do ferns grow everywhere?

S4: No, not deserts.

T: Where do they typically grow? What do they need to grow?

Ss: Water.

T: Water. Would a fern growing next to a river make sense?

Ss: Yeah.

T: Do you think over the course of millions of years that one fern could end up preserved in a river environment?

B: Yeah. Fern plants could.

T: So since you guys over here are judging the fossils, Andy, do you accept his answer for why there's a fern there?

A: Yeah.

T: I would agree. I think that's an acceptable answer. It doesn't always have to be the way it pans out on the image here. Any other thing that you see? What about Allen, did they put the earthquake at the right point.

A: No. They're a little off.

T: How were they a little bit off?

A: They went, like, really long.

T: Could you be more specific. How did they go a little long?

A: She got excited. (Referring to the fact that she shook the Wii Remote hard for almost 10 seconds causing multiple faulting events.)

S1: I told you to stop.

S2: It's hard to do it right.

S1: Have an aneurism why don't you.

T: Okay. Allow me to just work it out. I'm mediating. I'm backing you guys up okay? So Allen, the important thing is, did it come at the right time?

A: Yeah. It was just too long.

T: Okay. That's more important. Maybe when you use the Wii controller sometimes it's hard to know when to stop.

S: Yeah.

T: Do you think that this is acceptable the way that they did it.

S: Yeah.

T: I would agree with you as well. So were there any points taken off for any decisions that were made in creating this geologic cross section?

S: No. Not really. It was all good.

T: It was all good...alright...awesome

These two transcripts demonstrate two important trends. First, there is a marked difference in the nature of the reflective discussion between the two days. The discussion on day 1 is exclusively led by the teacher as he prompts students to respond to direct questions. By day 3, while the teacher serves to moderate the discussion, he is able to steer the more free-flowing conversation in a way that encourages students to directly engage one another. Second, owing to the open-ended nature of the build process, students are by day 3 considering alternative solutions and deviations in the outcomes. They discuss the viability of different solutions and consider allowable tolerances. This shows that they are thinking of the process of evolution as a complex process that can have multiple "acceptable" outcomes so long as those outcomes align with their underlying conceptual models.

To assess individual students' content learning gains, we collaborated with the classroom teacher, to create a 10-item pencil and paper test to assess students' knowledge of earth science topics relating to geological evolution. Each test item included a multiple-choice concept question followed by an open-format question asking students to articulate an explanation for their answer. The content for this test was drawn from topics covered during a typical geological evolution curriculum and aligning with state and federal science standards. All test concepts were covered in the teacher's classroom using traditional instructional methods in the weeks leading up to the experiment. As such, at the time of the pretest, students had studied (and learned) all of the test material to the full extent that would be typically expected. To be clear, the 3-day teaching experiment did not introduce any new concepts, but rather only reinforced and reviewed previously studied topics. This concept test was administered 1 day before and then 1 day after the *SMALLab* treatment. Every student in our partner teacher's earth science classes participated in the teaching experiment and thus we were not able to administer the test to a control group.

Table 26.1 shows the pre- and posttest scores for the 72 participating students. The summary is divided into two categories for the multiple-choice items and corresponding open-answer explanation items. Open-answer questions were rated on a 0–2 scale where a score of 0 indicates a blank response or nonsense response. A score of 1 indicates a meaningful explanation that is incorrect or only partially accurate. A score of 2 indicates a well-formed and accurate explanation. We computed a percentage increase and the Hake gain for each category. A Hake gain is the actual percent gain divided by the maximum possible gain (Hake 1998). Participating students achieved a 22.6% overall percent increase in their multiple choice question scores, a 48% Hake gain ($P < 0.00002$, $r = 0.20$, $n = 72$, standard deviation = 1.9). They achieved a 40.4% overall percent increase in their explanation scores, a 23.5% Hake gain ($P < 0.000003$, $r = 0.60$, $n = 72$, standard deviation = 2.8). These results reveal that nearly all students made significant conceptual gains as measured by their ability to accurately respond to standardized-type test items and articulate their reasoning.

TABLE 26.1

Summary of Pre- and Posttreatment Geological Evolution Concept Test Results

		Scores
Multiple choice test items ($n = 72$)	Pretreatment multiple choice average score	6.82
	Posttreatment multiple choice average score	8.36
	Percent increase	22.6%
	Hake gain	48.5%
Free-response justifications ($n = 72$)	Pretreatment explanation average score	3.68
	Posttreatment explanation average score	5.17
	Percent increase	40.4%
	Hake gain	23.5%

We also observed that the student-centered, play-based nature of the learning experience had a positive impact on students. All participants were part of the school's C.O.R.E. program for at-risk students. While many of these students are placed in the program due to low academic performance, after 1 year of observation, we see that this is often not due to a lack of ability, but rather due to a lack of motivation to participate in the traditional culture of schooling. During our 3-day treatment we observed high motivation from students. Many students who might otherwise disengage from or even disrupt the learning process emerged as vocal leaders in this context. These students appeared intrinsically motivated to participate in the learning activity and displayed a sense of ownership for the learning process that grew with each day of the treatment. As evidence of the motivating impact of play, we informally observed a group of students from outside the teacher's regular classes. These students previously spoke with their peers about their in-class experience and subsequently visited *SMALLab* during their lunch hour to "play" in the environment. For nearly a full class period these students composed layer cake structures, working together, unsupervised by any teacher.

26.5   Conclusions and Future Work

We have presented theoretical research from interactive multimedia, HCI, and education that reveals a convergence of trends focused on *multimodality, embodiment,* and *composition.* While we have presented several examples of prior research that demonstrates the efficacy of learning in environments that align work in interactive multimedia, HCI, and education, there are few examples of large-scale projects that synthesize all three of these elements. We have presented our own efforts in this regard, using the integration of these three themes as a theoretical and technological framework that is informed by broad definitions of play. Our work includes the development of a new mixed-reality platform for learning that has been pilot tested and evaluated through diverse pedagogical programs, focused user studies, and perception/action experiments. We presented a recent high school earth science program that illustrates the application of our three-part theoretical framework in our mixed-reality environment. This study was undertaken with two primary goals: (1) to advance students' knowledge of earth science content relating to geological evolution, and (2) to evaluate our theoretical framework and validate *SMALLab* as a platform for mixed-reality learning in a formal classroom learning environment. Participating students demonstrated significant learning gains after only a 3-day treatment, and exhibited strong motivation for learning as a result of the integration of play in the scenario. This success demonstrates the feasibility of mixed-reality learning design and implementation in a mainstream formal school-based learning environment. Our preliminary conclusions suggest that there is great promise for the *SMALLab* learning technologies and our three-part theoretical base.

We are currently working to increase the scope and scale of the *SMALLab* platform and learning programs. Regarding the technological infrastructure, we are actively pursuing augmented sensing and feedback mechanisms to extend the system. This research includes an integrated framework for robotics, outfitting the tracked glowballs with sensors and wireless transmission capabilities, and integrating an active radio-frequency identification (RFID) system that will allow us to track participant locations in the space. We are extending the current multimodal archive to include real-time audio and video data that are interleaved with control data generated by the existing sensing and feedback structures. Furthermore, *SMALLab Learning* (http://smallablearning.com), a research spin-off company from ASU, has been founded in 2010 with the mission of advancing and disseminating embodied and mediated learning in schools and museums.

With regard to learning programs, we continue our collaboration with faculty and students at a regional high school. We are currently collecting data that will allow us to evaluate the long-term impact of *SMALLab* learning that is correlated across multiple content areas, grades, and instructional paradigms. Concurrently, we are developing a set of computationally based evaluation tools that will identify gains in terms of successful *SMALLab* learning strategies and the attainment of specific performance objectives. These tools will be applied to inform the design of *SMALLab* programs, support student-centered reflection, and communicate to the larger interactive multimedia, HCI, and education communities our successes and failures in this research.

Acknowledgments

We gratefully acknowledge that these materials document work supported by the National Science Foundation CISE Infrastructure grant under Grant No. 0403428 and IGERT Grant No. 0504647. We extend our gratitude to the students, teachers, and staff of the Herberger College for Kids, Herrera Elementary School, Whittier Elementary School, Metropolitan Arts High School, Coronado High School, and ASU Art Museum for their commitment to exploration and learning.

References

1. Arizona Department of Education. *Arizona Academic Standards: Science Strand 6—Earth and Space Science.* 2005. Available from http://www.ade.az.gov/standards/science/downloads/strand6.pdf.
2. Bahn, C., T. Hahn, and D. Trueman. 2001. Physicality and feedback: A focus on the body in the performance of electronic music. Paper read at *Multimedia Technology and Applications Conference,* Irvine, CA.

3. Barsalou, L. W. 2008. Grounded cognition, *Annual Review of Psychology*, 59(1), 617–645.
4. Birchfield, D., T. Ciufo, G. Minyard, G. Qian, W. Savenye, H. Sundaram, H. Thornburg, and C. Todd. 2006. SMALLab: A mediated platform for education. Paper read at *ACM SIGGRAPH*, Boston, MA.
5. Birchfield, D., B. Mechtley, S. Hatton, and H. Thornburg. 2008. Mixed-reality learning in the art museum context, in *Proceedings of ACM Multimedia*, Vancouver, Canada, October 27–31, 2008.
6. Bobick, A. F., S. S. Intille, J. W. Davis et al. 1999. The kidsRoom: A perceptually-based interactive and immersive story environment, *Presence: Teleoperators and Virtual Environments*, 8(4), 369–393.
7. Brain, Marshall. *How the Wii Works*. 2007 [Accessed July 2008]. Available from http://electronics.howstuffworks.com/wii.htm.
8. Bransford, J.D., A.L. Brown, and R.R. Cocking, eds. 2000. *How People Learn: Brain, Mind, Experience, and School*. Washington, DC: National Academy Press.
9. Brown, J.S., A. Collins, and P. Duguid. 1989. Situated cognition and the culture of learning. *Educational Researcher* 18(1):32–42.
10. Burt, G. 2006. Media effectiveness, essentiality, and amount of study: A mathematical model. *British Journal of Educational Technology* 37: 121–130.
11. Cavallo, D., A. Sipitakiat, A. Basu, S. Bryant, L. Welti-Santos, J. Maloney, S. Chen, E. Asmussen, C. Solomon, and E. Ackermann. 2004. RoBallet: Exploring learning through expression in the arts through constructing in a technologically immersive environment. *Proceedings of the 6th International Conference on Learning Sciences* pp. 105–112.
12. Chen, D., and W. Stroup. 1993. General systems theory: Toward a conceptual framework for science and technology education for all. *Journal for Science Education and Technology* 2(3): 447–459.
13. Cobb, P. 1994. Constructivism in math and science education. *Educational Researcher* 23(7):4.
14. Cruz-Neira, C., D.J. Sandin, and T.A. DeFanti. 1993. Surround-screen projection-based virtual reality: The design and implementation of the CAVE. *Proceedings of the 20th Annual Conference on Computer Graphics and Interactive Techniques* pp. 135–142.
15. Cunningham, D., T. Duffy, and R. Knuth. 1993. Textbook of the future. In *Hypertext: A Psychological Perspective*, C. McKnight, ed. London: Ellis Horwood.
16. Cuthbertson, A., S. Hatton, G. Minyard, H. Piver, C. Todd, and D. Birchfield. 2007. Mediated education in a creative arts context: Research and practice at Whittier Elementary School, in *Proceedings of the 6th International Conference on Interaction Design and Children (IDC '07)*, pp. 65–72, Aalborg, Danmark, June 2007.
17. Dede, C. and D.J. Ketelhut. 2003. Designing for motivation and usability in a museum-based multi-user virtual environment. Paper read at *American Educational Research Association Conference*, Chicago, IL.
18. Dede, C., D.J. Ketelhut, and K. Ruess. 2002. Motivation, usability and learning outcomes in a prototype museum-based multi-user virtual environment. Paper read at *5th International Conference of the Learning Sciences*, Seattle, WA.
19. Dolgov, I., D. Birchfield, M. McBeath, H. Thornburg, and C. Todd. 2009. Perception of approaching and retreating shapes in a large, immersive, multimedia learning environment. *Journal of Perceptual and Motor Skills*, 108(2):623–630.
20. Dougiamas, M. 1998. *A Journey into Constructivism*. Available at: http://dougiamas.com/writing/constructivism.html
21. Dourish, P. 2004. What we talk about when we talk about context. *Journal of Personal and Ubiquitous Computing* 8(1):19–30.
22. Dourish, P. 2001. *Where the Action Is: The Foundations of Embodied Interaction* Cambridge, MA: MIT Press.
23. EU Community Report, Information Society Technologies (IST) Programme. *A Multisensory Environment Design for an Interface between Autistic and Typical Expressiveness*. 2004.
24. Fauconnier, G. and M. Turner. 2002. *The Way We Think: Conceptual Blending and the Mind's Hidden Complexities*. New York: Basic Books.

25. Galperin, P.I. 1992. Stage-by-stage formation as a method of psychological investigation. *Journal of Russian and East European Psychology* 30(4):60–80.
26. Gardner, H. 1993. *Frames of Mind: The Theory of Multiple Intelligences*. New York: Basic Books.
27. Gee, J.P. 2003. *What Video Games Have to Teach Us about Learning and Literacy*. New York: Palgrave Macmillan.
28. Gibson, J.J. 1979. *The Ecological Approach to Visual Perception*. Boston, MA: Houghton Mifflin.
29. Glenberg, A.M., T. Gutierrez, and J. Levin. 2004. Activity and imagined activity can enhance young children's reading comprehension. *Journal of Educational Psychology* 96(3):424–436.
30. Glenberg, A.M., D. Havas, R. Becker, and M. Rinck. 2005. Grounding language in bodily states: The case for emotion. In *The Grounding of Cognition: The Role of Perception and Action in Memory, Language, and Thinking*, R. Zwann and D. Pecher, eds. Cambridge: Cambridge University Press.
31. Hake, R.R. 1998. Interactive-engagement vs traditional methods: A six-thousand-student survey of mechanics test data for introductory physics courses. *American Journal of Physics* 66: 64–74.
32. Harel, I. 1991. *Children Designers*, Ablex, Norwood, NJ, USA.
33. Hatton, S., D. Birchfield, and M. C. Megowan-Romanowicz. 2008. Learning metaphor through mixed-reality game design and game play, in *Proceedings of the ACM SIGGRAPH Symposium on Video Games (Sandbox '08)*, pp. 67–74, Los Angeles, California, USA, August 2008.
34. Hestenes, D. 1992. Modeling games in the Newtonian world. *American Journal of Physics* 60: 732–748.
35. Hestenes, D. 1996. Modeling methodology for physics teachers. Paper read at *International Conference on Undergraduate Physics*, College Park, MD.
36. Hestenes, D. 2006. Notes for a modeling theory of science cognition and instruction. In *GIREP Conference, Modeling and Physics and Physics Education*, Amsterdam, the Netherlands.
37. Hunt, A., M. Wanderley, and M. Paradis. 2002. The importance of parameter mapping in electronic instrument design. Paper read at *New Interfaces for Musical Expression*, Dublin, Ireland.
38. Hutchins, E. 1995. *Cognition in the Wild*, Vol. 1, MIT Press, Cambridge, Massachusetts, USA.
39. Hutchins, E. 2002. Material anchors for conceptual blends, in *Proceedings of the Symposium on Conceptual Blending*, Odense, Denmark.
40. Ishii, H. *Tangible User Interfaces*. 2007. Available from http://tangible.media.mit.edu.
41. Ishii, H. and B. Ullmer. 1997. Tangible bits: Towards seamless interfaces between people, bits and atoms. Paper read at *SIGCHI Conference on Human Factors in Computing Systems*, Atlanta, GA.
42. Jackendoff, R. 1996. The architecture of the linguistic spatial interface. In *Language and Space*, P. Bloom, M. A. Peterson, L. Nadel and M. F. Garrett, eds. Cambridge, MA: MIT Press.
43. Jenkins, H., R. Purushotma, M.Weigel, K. Clinton, and A. Robison. 2006. *Confronting the Challenges of Participatory Culture: Media Education for the 21st Century*. MacArthur Foundation, Cambridge, MA: MIT Press.
44. Jin, H., G. Qian, and S. Rajko. 2006. Real-time multi-view 3D object tracking in cluttered scenes. In *2nd International Symposium on Visual Computing*, Lake Tahoe, Nevada, USA.
45. Jin, H. and G. Qian. 2008. Robust real-time 3D object tracking with interfering background visual projections. *EURASIP Journal on Image and Video Processing* 2008, Article ID 638073, 14 pages, doi:10.1155/2008/638073.
46. Kafai, Y.B., K. Peppler, and G. Chin. 2007. High tech programmers in low income communities: Creating a computer culture in a community technology center. *Proceedings of the Third International Conference on Communities and Technology*, C. Steinfeld, B. Pentland, M. Achkermann, and N. Contractor, eds. New York: Springer.
47. Ketelhut, D. J. 2007. The impact of student self-efficacy on scientific inquiry skills: An exploratory investigation in River City, a multi-user virtual environment, *Journal of Science Education and Technology*, 16(1), 99–111.
48. Lesh, R. and H. Doerr. 2003. *Beyond Constructivism: Models and Modeling Perspectives on Mathematics Problem Solving, Learning, and Teaching*. Mahwah, NJ: Lawrence Erlbaum Associates.
49. Lutgens, F., E. Tarbuck, and D. Tasa. 2004. *Foundations of Earth Science (4th Edition)*. Upper Saddle River, NJ: Prentice Hall.

50. Megowan, M.C. 2007. *Framing Discourse for Optimal Learning in Science and Mathematics*, Tempe, AZ: College of Education, Division of Curriculum and Instruction, Arizona State University.
51. Middleton, J. and Z. Toluk. 1999. First steps in the development of an adaptive theory of motivation. *Educational Psychologist* 34(2):99–112.
52. Muller, L. and E. Edmonds. 2006. Living laboratories: Making and curating interactive art. In *ACM SIGGRAPH*. Boston, MA.
53. Nelson, B.C., D. J. Ketelhut, J. Clarke, E. Dieterle, C. Dede, and B. Erlandson. 2007. Robust design strategies for scaling educational innovations: The River City MUVE case study, in *The Design and Use of Simulation Computer Games in Education*, pp. 209–231, Sense Press, Rotterdam, the Netherlands.
54. Norman, D. 1988. *The Psychology of Everyday Things*. New York: Basic Books.
55. Papert, S. 1980. *Mindstorms: Children, Computers and Powerful Ideas*. New York: Basic Books.
56. Papert, S. and I. Harel, 1991. Situating constructionism. In *Constructionism*. Papert, S. and I. Harel. (Eds.) pp. 1–12. Ablex Publishing Corporation. Norwood, NJ.
57. Papert, S. 1993. *The Children's Machine*, Basic Books, New York, NY, USA.
58. Pares, N., A. Carreras, J. Durany, J. Ferrer, P. Freixa, D. Gomez, O. Kriglanski, R. Pares, J. Ribas, M. Soler, and A. Sanjurjo. 2004. MEDIATE: An interactive multisensory environment for children with severe autism and no verbal communication. *3rd International Workshop on Virtual Rehabilitation (IWVR'04)*, Lausanne, Switzerland.
59. Parés, N., P. Masri, G. vanWolferen, and C. Creed. 2005. Achieving dialogue with children with severe autism in an adaptive multisensory interaction: The "MEDIATE" project, *IEEE Transactions on Visualization and Computer Graphics*, 11(6), 734–742.
60. Peppler, K. and Y.B. Kafai. 2007. From SuperGoo to Scratch: Exploring media creative production in an informal learning environment. *Journal on Learning, Media and Technology* 32(2): 149–166.
61. Rambusch, J. and T. Ziemke. 2005. The role of embodiment in situated learning. *27th Annual Conference of the Cognitive Science Society* pp. 1803–1808.
62. Resnick, M. 1998. Technologies for lifelong kindergarten, *Educational Technology Research and Development*, 46(4), 43–55.
63. Resnick, M. 2006. Computer as paintbrush: Technology, play, and the creative society. In *Play=Learning: How Play Motivates and Enhances Children's Cognitive and Social-Emotional Growth*, D. Singer, R. Golikoff and K. Hirsh-Pasek, eds. Cambridge, UK: Oxford University Press.
64. Resnick, M. 2007. All I really need to know (about creative thinking) I learned (by studying how children learn) in kindergarten. Paper read at *SIGCHI Conference on Creativity and Cognition*, Washington, DC.
65. Rheingans, P. and C. Landreth. 1995. Perceptual principles for effective visualizations. In *Perceptual Issues in Visualization*. G. Grinstein and H. Levkowitz, Eds. pp. 59–74, Springer-Verlag Publishers, Berlin, Germany.
66. Roth, W.M. and M.K. McGinn. 1998. Inscriptions: Toward a theory of representing as social practice. *Review of Educational Research* 68(1):35–59.
67. Ryokai, K., S. Marti, and H. Ishii. 2004. I/O brush: Drawing with everyday objects as ink. *Proceedings of the 2004 Conference on Human Factors in Computing Systems* pp. 303–310.
68. Salen, K. and E. Zimmerman. 2004. *Rules of Play: Game Design Fundamentals* Cambridge, MA: MIT Press.
69. Shepard, R.N. 2001. Perceptual-cognitive universals as reflections of the world. *Behavioral & Brain Sciences* 24:581–601.
70. Shirai, A., E. Geslin, and S. Richir. 2007. WiiMedia: Motion analysis methods and applications using a consumer video game controller. Paper read at *Sandbox: ACM Siggraph Symposium on Video Games*, San Diego, CA.
71. SmartUs. *SmartUs: Games in Motion*. 2008. Available from http://www.smartus.com.
72. Suchman, L. 1987. *Plans and Situated Actions: The Problem of Human-Machine Communication*, Cambridge University Press, New York, NY, USA.

73. Swartout, W., R. Hill, J. Gratch, W.L. Johnson, C. Kyriakakis, C. LaBore, R. Lindheim, S. et al. 2001. Toward the Holodeck: Integrating graphics, sound, character and story. Paper read at *Autonomous Agents*, Montreal, Canada.
74. Tu, K., H. Thornburg, E. Campana, M. Fulmer, D. Birchfield, and A. Spanias. 2007a. Interaction and reflection via 3D path shape qualities in a mediated learning environment. *ACM Conference on Educational Multimedia and Multimedia Education*, Augsburg, Germany.
75. Tu, K., H. Thornburg, M. Fulmer, and A. Spanias. 2007b. Tracking the path shape qualities of human motion. In *International Conference on Acoustics, Speech, and Signal Processing*. Honolulu, HI.
76. Varela, F., E. Thompson, and E. Rosch. 1991. *The Embodied Mind: Cognitive Science and Human Experience*, MIT Press, Cambridge, Massachusetts, USA.
77. Vygotsky, L.S. 1962. *Thought and Language*. Cambridge, MA: MIT Press.
78. Vygotsky, L.S. 1978. *Mind in Society: The Development of Higher Psychological Processes*. Cambridge, MA: The Harvard University Press.
79. Weinberg, G. 1999. The Musical Playpen: an immersive digital musical instrument, *Personal and Ubiquitous Computing*, 3(3), 132–136.
80. Wessel, D. and M. Wright. 2001. Problems and prospects for musical control of computers. Paper read at *New Interfaces for Musical Expression*, Seattle, WA.
81. Wilson, M. 2003. Six views of embodied cognition. *Psychonomic Bulletin and Review* 9(4):625–636.
82. Wren, C.R., S. Basu, F. Sparacino, and A. Pentland. 1999. Combining audio and video in perceptive spaces. Paper read at *Managing Interactions in Smart Environments*, Dublin, Ireland.

27

Literature Survey on Recent Methods for 2D to 3D Video Conversion

Raymond Phan, Richard Rzeszutek, and Dimitrios Androutsos

CONTENTS

27.1 Introduction

Viewing 3D content, whether it be in the home, at the cinema, or in scientific and industrial environments, has become more popular over the last few years. Specifically, the sense of depth in a scene captured by conventional cameras, where virtually no depth was ever provided, can now be experienced. The end user can use this information to enhance their viewing experience, by means of this newly available depth perception, so that they can become truly immersed in the environment. In addition, a sense of depth can be used in many applications, ranging from training simulations, gaming, scientific model exploration, and in cinema. All these applications can use a sense of depth to get a sense of realism, and thus the users can truly appreciate the content that they are viewing.

27.1.1 Stereo Correspondence and Disparity Maps

In the past, much research focused on obtaining a sense of depth of a scene by using only two views, corresponding to the left and right eye viewpoints, known as *stereo correspondence*. Using a reference viewpoint, whether it is the left or right eye, a sense of depth is determined for each point in the reference, where points in one image are matched with their corresponding points in the other image. The amount of shift that the corresponding points undergo is known as *disparity*. Disparity and depth have an inverse relationship, where by the higher the horizontal shift, or disparity, the smaller the depth, and the closer the point is to the viewer. As an example, Figure 27.1 illustrates a sample stereo image pair—the Tsukuba image pair—from the standard Middlebury Stereo Database (http://vision.middlebury.edu/stereo). As can be seen in the figure, corresponding points between the two views that are closer to the camera have a larger shift, as evidenced by the bust, the desk, and the lamp. Corresponding points between the two views that are farther from the camera have a smaller shift, as evidenced by the bookshelf, the chalkboard, and the video camera.

By presenting these two views of a scene to a human observer, where one is slightly offset from the other, this generates the illusion of depth in an image, thus perceiving an image as 3D. The work is primarily done by the visual cortex in the brain, which processes the left and right views presented to the respective eyes. In the past, the disparities of each point between both viewpoints are collected into one coherent result, commonly known as a *disparity map*. These are the same size as either of the left and right images. The image is monochromatic (grayscale) in nature, and the brightness of each point is directly proportional to the disparity between the corresponding points. When a point is lighter, this corresponds to a higher disparity, and is closer to the viewer. Similarly, when a point is darker, this corresponds to a lower disparity, and is farther from the viewer. For scenes that have smaller disparities overall, the disparity maps are scaled accordingly, so that white corresponds to the highest disparity.

As an example, Figure 27.2 illustrates what a possible disparity map would look like for a given scene. Here, the disparity map is represented as an 8-bit monochromatic image, where 0 (black) denotes the farthest from the camera, and 255 (white) denotes the closest point to the camera. The house is closer to the camera, and the intensity is white, which means the disparity is greater. Similarly, the background is farther from the camera, and the intensity is black, which means the disparity is smaller.

(a)

(b)

FIGURE 27.1
(**See color insert.**) Tsukuba image pair: left view (a) and right view (b).

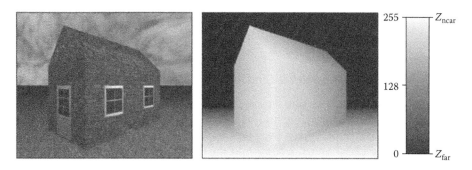

FIGURE 27.2
Disparity map example. (C. Fehn, R. de la Barre, and S. Pastoor, Interactive 3-DTV concepts and key technologies, *Proc. of the IEEE*, vol. 94, no. 3, pp. 524–538, March 2006. © (2006) IEEE.)

27.1.2 Viewing Stereoscopic and 3D Content with Specialized Hardware

Presently, to view 3D content, at least two views of the same scene are required. Traditionally, and as mentioned previously, only two views are required, but there are applications that require more than one view. To get a sense of depth in the scene, the existing 3D technology must be able to faithfully present different views of the scene to the user, where a view is presented to each eye without interference. Currently, there is a large amount of 3D display technology that is very successful in doing this. All of these consist of different lighting and electronics technologies that accomplish the common goal of depth perception. However, the amount of *content* for 3D display is still very scarce, and there are still efforts to acquire 3D video in the quickest and most efficient way possible. For the purpose of this chapter, we refer to 3D in terms of *stereoscopic content*, rather than the triangulation problem commonly seen in computer vision. There are two predominant methods for acquiring 3D video. The first is to take multiple cameras at different viewpoints, and obtain the raw video data directly.

One such example is the system designed by Mitsubishi Electrical Research Laboratories (MERL) [2], and is shown in Figure 27.3. Their 3DTV system has an array of 16 cameras

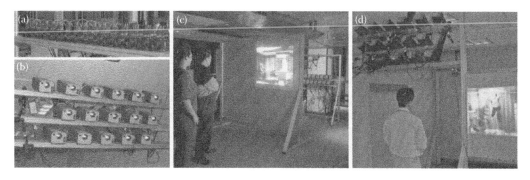

FIGURE 27.3
(**See color insert.**) 3DTV System by MERL. (a) Array of 16 cameras, (b) array of 16 projectors, (c) rear-projection 3D display with double-lenticular screen, and (d) front-projection 3D display with single-lenticular screen. (From W. Matusik and H. Pfister, 3D TV: A scalable system for real-time acquisition, transmission, and autostereoscopic display of dynamic scenes. *ACM Trans. on Graphics*, 24, 3, 2004. Available at: http://dl.acm.org/citation.cfm?id=1015805. With permission.)

and projectors, with a rear-projection 3D display with a double-lenticular screen, as well as a front-projection 3D display with a single-lenticular screen.

27.1.3 Viewing Stereoscopic and 3D Content Using Conventional 2D Video

With respect to the previous section, some postprocessing is required to get the views to be compatible with the target 3D technology, but requires only a slight amount of effort. However, filming directly in 3D can be rather expensive and difficult to set up, as can be clearly seen in Figure 27.3. In addition, should the user have an interest in viewing content in 3D that was only captured by one camera, this approach would be impossible to fill that need. As such, the second method, and the one we are ultimately interested in, is to take already-existing 2D content, and artificially produce the left and right views, or other views for multiview applications. Thus, the amount of material is endless, but the methods required to convert monocular video to 3D is a very difficult problem to solve, and still one of the most open-ended problems in computer vision that exist today. When a camera captures a scene, it performs a transformation from 3D to 2D coordinates, which is well understood. However, when considering the inverse situation, which is the one we are interested in, we are attempting to produce information from a model that lost the very information we are trying to obtain. Nevertheless, no matter the difficulty, the end result will be quite beneficial to anyone interested in the 3D experience, as conventional content can now be viewed in 3D, with a sense of depth that was never experienced before.

With respect to the second method of generating 3D content, the goal for the conversion of monocular video to its stereoscopic or multiview counterpart is to generate a set of disparity maps for each view that we wish to render. Each disparity map will determine the shift in pixels we need from the reference view, or a frame from the video sequence. The caveat here is that we must generate disparity maps for each frame of each view in the video sequence. In addition, most 2D to 3D conversion algorithms actually calculate a *depth* map, rather than the disparity map, but a simple conversion between the two can be achieved once a couple of the parameters of the camera are known.

Regardless, 2D to 3D conversion algorithms have been the subject of many useful applications which are targeted for the industry, as well as the end user. Specifically, 2D to 3D conversion techniques are of primary consideration in the latest 3DTVs. This is due to the fact that 3D technology is now widely available for the home, and they can now experience their existing 2D content in 3D. There have been many paradigms for the widespread of 3D technology to be experienced at home. As an example, Redert et al. [3] formed the Advanced Three-Dimensional Television System Technologies (ATTEST) group, which was the first ever attempt to introduce the 3DTV framework to the home. This included the paradigms for the transmission and reconstruction of the signals, the best way to capture stereoscopic content, and designing a compatible coding scheme for transmission. Similarly, Fehn et al. [1] discuss the overall process in greater detail. Figure 27.4 is a block diagram that illustrates their overall processing chain, from capturing content, to transmission, and ultimately to displaying this content to the user.

Unfortunately, this group currently no longer exists, but there have been other efforts that envision the same goal to deliver 3D to the end user. Specifically, the Mobile3DTV initiative [4], which started in 2008, was created to deliver 3D content to mobile phones in Europe. It encompasses the same objectives as the ATTEST project, but for more specialized mobile devices. This group is still conducting research, with many European universities in collaboration with each other. However, the method of delivering 3D content has a huge collaborative undertaking, and still does not solve how to deliver 2D content in 3D to the

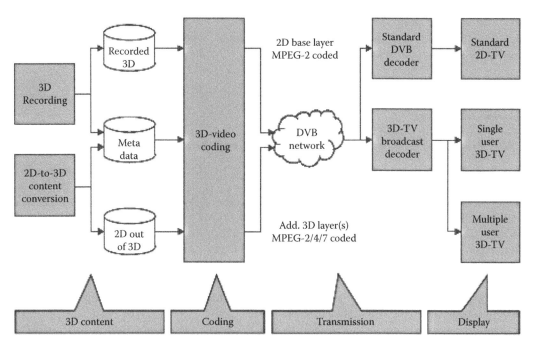

FIGURE 27.4
The ATTEST 3D-video processing chain. (A. Redert et al., ATTEST: Advanced three-dimensional television system technologies, *IEEE Intl. Symposium on 3D Data Processing Visualization and Transmission (3DPVT)*, pp. 313–319, Padova, Italy, 2002. © (2002) IEEE.)

home user. There has, however, been research to develop a full stereoscopic signal processing pipeline, in order to postprocess the 3D data received, such as the work done in Ref. [5], but we are thus brought to the same problem with the lack of 3D content available. There is a vast amount of 2D content available for user digest, which not only includes motion pictures and television shows, but standard computer vision and image processing applications can be augmented with a sense of depth, in order to give a more truer experience to the user.

27.1.4 Brief Overview of the Chapter

With the aforementioned, in this chapter, we present a literature survey that concentrates on converting conventional monocular video sequences, or those acquired by one camera, to produce two or more views of the same scene, to be presented for viewing using the end 3D technology. The conversion of this nature can be decomposed into different categories, and we will discuss the relevant and recent methods for each of these categories. We will discuss these methods in detail, citing works that are recent within the last few years. The categories that 2D to 3D conversion can be decomposed into are the following: using color information, using edge information, techniques from depth-based image rendering (DIBR), using motion, and analyzing scene features. However, the two most predominant techniques are using motion estimation and analyzing scene features.

To make this literature review more coherent, we will combine the discussion of color and DIBR into the analysis of scene features, as it technically falls into this category. The rest of this literature review will be formatted as follows. Section 27.2 discusses the relevant and recent 2D to 3D methods by using motion estimation as the underlying framework. Section 27.3 discusses the 2D to 3D methods using scene analysis, or other features extracted from the scene. Finally, Section 27.4 briefly summarizes the methods in each category, the disadvantages and drawbacks that can be encountered, leading to what can be formulated for future research in this area.

27.2 2D to 3D Conversion Methods Using Motion

The first of the most predominant techniques that is used for 2D to 3D conversion is using motion estimation to determine the depth or disparity of the scene. The underlying mechanism is that for objects that are closer to the camera, they should move faster, whereas for objects that are far, the motion should be slower. Thus, motion estimation can be used to determine correspondence between two consecutive frames, and can thus be used to determine what the appropriate shifts of pixels are from the reference view (current frame), to the target view (next frame). However, the use of the actual motion vectors themselves to generate a stereoscopic or multiview video sequence varies between the methods, but the underlying mechanism is the same. We will now begin our discussion on the use of motion estimation for 2D to 3D conversion.

We start with the method by Ideses et al. [6], where they concentrate on a real-time conversion from 2D to 3D. This work is arguably one of the methods that has established that using motion estimation is one of the key features to use when determining the depth map. Essentially, they use motion vectors from the MPEG4 decoder. They use the magnitude of the motion vector, by taking the Euclidean distance of the horizontal and vertical motion estimation components for each pixel. As a method of illustration, Figure 27.5 illustrates the basic methodology of how the system works. For the video sequence in question, each frame is decomposed into its RGB components. In parallel, the magnitude of the motion vector is determined. For display, anaglyph images are created, where the red channel of each frame is used, and the motion vector magnitudes are used to shift pixels to create the right image. The original image and modified red channel are thus merged to create the anaglyph image for each frame.

In a similar fashion, Huang et al. [7] determine depth maps by using both motion and scene geometry. Specifically, the motion vectors from the H.264 decoder are used to generate a motion-based depth map. In addition, a moving object detection algorithm is introduced, to diminish the block effect caused by the motion estimation in H.264. After, Gaussian mixture models are used to determine what the foreground objects are, and to modify the initial motion-based depth map with the H.264 decoder. Using the geometry of the scene, they extract vanishing lines and vanishing points by performing edge detection and the Hough transform. With the Hough transform, and the location of the vanishing lines, the location of points with respect to the vanishing lines—which can be transformed to vanishing planes—can be used to assign a depth value that is dependent on the location of the pixel point with respect to the vanishing lines or planes. With the combination of these two depth maps, a hybrid fusion is performed to merge the two depth maps together, and thus a depth map is created that handles both scene geometry, in addition to the pertinent objects shown in the scene. Figure 27.6 is the overall block diagram of this approach.

Pourazad et al. [8] use the H.264 motion vectors to produce 3D content. Specifically, camera motion correction is used when both the objects and the camera are moving, as this will inevitably produce depth ambiguity. When a camera is panning, the motion vector is most likely approximated by camera motion, and in the H.264 standard, these areas

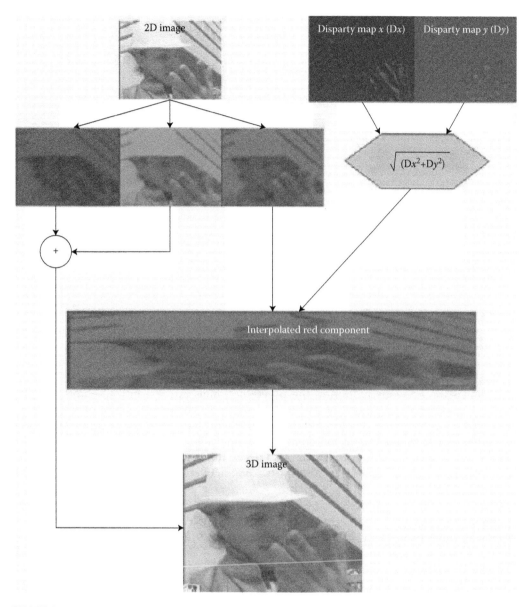

FIGURE 27.5
(**See color insert.**) Flow diagram of the algorithm by Ideses et al. (From I. Ideses, L. P. Yaroslavsky, and B. Fishbain, Real-time 2D to 3D video conversion, *J. Real-Time Image Processing*, 2, 3–9, 2007. With permission.)

are flagged by a "skip mode." This skip mode is used, in conjunction with calculating the histogram of the skipped blocks for the area of interest, to determine what the final motion of the camera is. This is subtracted from the motion vectors within the frame. Matching block correction is used, as matching blocks determined by a motion vector do not necessarily relate to the same part of an object in the scene. However, there is a case where the matching blocks do not represent the same part of an object, and thus do not form corresponding left and right areas when looking at this in a disparity point of view. Therefore, a block correction scheme is formulated so that the motion vector does point to the same area in

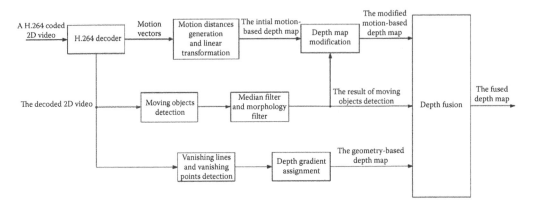

FIGURE 27.6
Block diagram of the algorithm by Huang et al. (X. Huang et al., A depth extraction method based on motion and geometry for 2D to 3D conversion, *Proc. IEEE Symp. on Intelligent Information Technology Applications*, Nanchang, China, 2009. © (2009) IEEE.)

the consecutive frame using simple statistical analysis and the difference image between the two consecutive frames. Finally, these are applied to the motion vectors, and some postprocessing is performed so that object border occlusions are minimized. The final depth map is obtained by using the preprocessed motion vectors from the previous three steps.

Chang et al. [9] use a simple difference equation to determine the motion estimation, which takes the previous motion in the previous frames into account to eliminate jitter. Next, a K-means color segmentation is performed, and a connected component algorithm is used to obtain connected color silhouettes of the objects. With the motion estimation, in combination with the connected components analysis, object segments are produced, and the object segments are converted into a depth map. They also continue this work in Ref. [10], where they use three different cues, in conjunction with the previous system they created, and fuse the three cues together to create a more accurate depth map. These three cues are using motion parallax, using texture, and depth from geometry. These cues are weighted, and an aggregation scheme is formulated to compute the best depth map that merges all three cues. The block diagram in Figure 27.7 illustrates their steps to achieve depth maps over 2D monocular video.

Cheng et al. [11] continue this work by using a bilateral filter to postprocess the depth maps, to flatten out possible occlusions. As will be mentioned later, there are problems when estimating the disparity, or depths at the edges of objects, as occlusions are inevitable. For boundary pixels, it is impossible to determine correspondence, as new pixels will emerge, which did not have any correspondences to begin with. Smoothing the depth map reduces the effect on occlusions, as well as minimizing what is known as the "cardboard" effect on rendered images. The cardboard effect is essentially viewing the objects stereoscopically, but they look like "cardboard" cutouts, as the depth values for each object are roughly the same for each corresponding object in the other frame. In order to have the maximum quality of depth perception, there should be some sort of gradient of depth values for each object. The bilateral filter is one example of minimizing occlusions and the cardboard effect influence.

Kim et al. [12] use MPEG-4 motion vectors to determine a sense of depth in a scene. They call this a motion vector field (MVF). With the MVF, a modified histogram is created that is dependent on the magnitude of the motion vector, as well as the angle it makes. The motion vectors are placed into bins accordingly using this principle. They call this motion vector

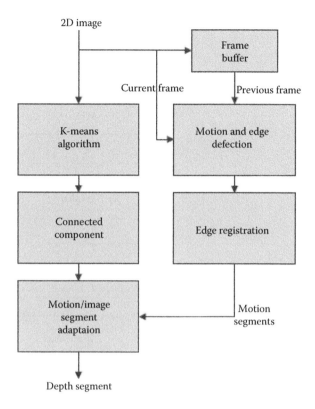

2D image

FIGURE 27.7
Block diagram of the algorithm by Chang et al. (Y.-L. Chang et al., Depth map generation for 2D-To-3D conversion by short-term motion assisted color segmentation, *Proc. IEEE Conf. on Multimedia and Expo (ICME)*, pp. 1958–1961, Beijing, China, 2007. © (2007) IEEE.)

binning (MVB). However, the bins have unequal ranges, or are not uniformly quantized. Within the MVB, each frame is classified as either being primary frame, where the motion information is highly reliable, or a secondary frame, where the information is unreliable, and its use is minimized. After this, confidence measures are used that are directly related to the modified histogram to the provided depth values. Figure 27.8 below illustrates a block diagram of this process.

Kim et al. [13], first use motion estimation by performing a color segmentation by the mean-shift algorithm [14], together with the Kanade–Lucas–Tomasi (KLT) feature tracker [15]. Color segmentation allows for estimating the depth and disparity of textureless patches. Here, it is assumed that each cluster of pixels that results after color segmentation shares the same motion. Feature points are selected by the KLT from the boundary pixels of each cluster. Bidirectional tracking is performed to increase the accuracy of feature tracking. Next, a motion-to-disparity calculation is performed, where a combination of three cues are used. The three cues for each pixel are multiplied together to create the final depth value of the pixel of interest. The three cues are: magnitude of the motion vector (or in this case, the optical flow vector), using the camera movement, and the scene complexity, or how much high-frequency content exists in the frame of interest. A more detailed version of their system can be found in Ref. [16].

Knorr et al. [17] perform a different approach, where not only do they determine stereoscopic video sequences from its monocular counterpart but they also generate multiple

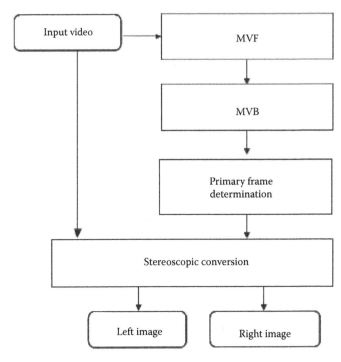

FIGURE 27.8
Block diagram of the algorithm by Kim et al. (From M. Kim et al., Automatic conversion of two-dimensional video into stereoscopic video, *Proc. SPIE Electronic Imaging—Three-Dimensional TV, Video and Display IV*, pp. 601610-1–601610-8, San Jose, California, 2006. With permission.)

views from the stereoscopic generated video sequence. In particular, they first employ structure from motion (SfM) to determine the external and internal parameters of the camera used to capture the scene. After the camera parameters are determined, multiple virtual cameras can be defined for each frame of the original video sequence, using their camera matrices, as well as the rough separation distance between the two eyes. Figure 27.9 illustrates this concept. First, single-view cameras are positioned throughout the scene of interest. If multiple cameras are not available, a single camera can be used. For a single camera, the camera films the scene, and then moves to another position and films the same scene again. However, the drawback with this single camera approach is that the scene has to be relatively static. Regardless, these camera positions are illustrated in the figure as blue shapes. For each original camera position, a corresponding multiview set is generated in order to create virtual camera positions, as illustrated by red shapes. This is done by estimating homographies to temporal neighboring views of the original camera path, where the original path is illustrated in gray shapes. The original camera path is not known *a priori*, but can be determined by the homographies calculated previously.

Knorr et al. extend their work to create superresolution stereoscopic videos, given that the original 2D video is of low resolution [18]. Essentially, a choosing mechanism that is based on bilinear interpolation is used. Similarly, Kunter et al. [19] use the SfM, and perform a segmentation of objects over a video sequence to help in the determination of novel views. They first determine what the background is by looking at multiple frames and determining homographies between the frames. Whatever is consistent over time most likely corresponds to background, and this is extracted. A simple foreground segmentation

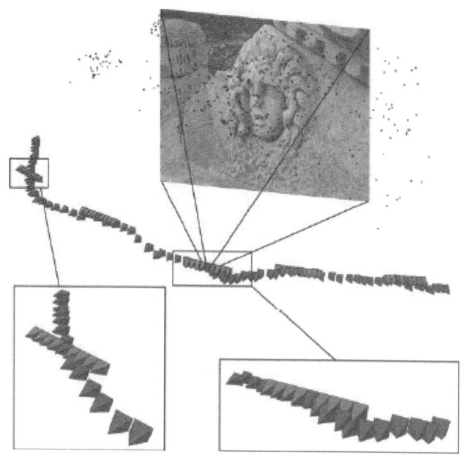

FIGURE 27.9

(See color insert.) Multiview synthesis using SfM and DIBR by Knorr et al. Gray: original camera path, red: virtual stereo cameras, blue: original camera of a multiview setup. (S. Knorr, A. Smolic, and T. Sikora, From 2D- to stereo- to multi-view video, *Proc. IEEE 3DTVCON*, pp. 1–4, Kos Island, Greece, 2007. © (2007) IEEE.)

algorithm is done, using gradients and edge information. In order to generate the different views, the background created with the SfM parameters is used to generate novel backgrounds, and the foreground object is converted to a mask, which is rendered differently for each of the rendered views.

Li et al. [20] incorporate both motion tracking and segmentation into their framework. Specifically, the KLT feature tracker is used, but is modified to promote efficiency. The difference frame between two consecutive frames is used, and those values not close to zero are examined further. These pixel locations are used for feature tracking via the KLT tracker, and the video volume is decomposed into chunks, where the KLT tracker is applied to each chunk separately. For the current chunk, the KLT feature vectors from the previous chunk of the last frame are used as the initial conditions. Next, feature pixels that have a similar motion are classified into one group as one object using flood fill. The flood fill is used not only for color similarity, but they also base it on motion similarity as well. For the pixels that were not feature tracked by not passing the difference frame test, Delaunay triangulation is used for interpolation. Next, the boundaries of each object are refined by

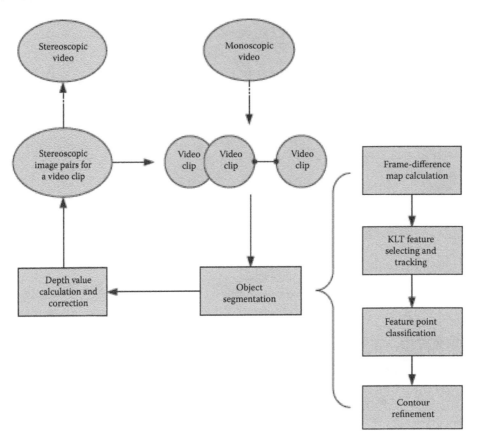

FIGURE 27.10
Block diagram of the algorithm by Li et al. (From T. Li, Q. Dai, and X. Xie, *Proc. IET Int. Conf. on Visual Information Engineering*, pp. 256–260, Xi'an, China, 2008. With permission.)

using active contours [21]. Once the motion vectors have been estimated, a simple motion to disparity operation is performed that depends on the motion vectors for each pixel. Figure 27.10 illustrates a block diagram of the aforementioned process.

Diplaris et al. [22] base their method on using the extended Kalman filter (EKF). First, the KLT tracker is used to find features in the first frame, and track them until the last one. Similar to Li et al.'s work, the features along the edges and object boundaries are tracked. These features are used in an SfM-like framework in conjunction with using the EKF, and the camera parameters are estimated, which are used to generate multiple views. This work is further extended in Ref. [23] where a bidirectional 2D motion estimation via the KLT feature tracker is performed, and is followed by efficiently tracking the object contours using the features from the KLT. Next, a Bayesian estimation framework is used to determine the depths at occluded points, and the rest of the algorithm remains the same, as seen in Ref. [22]. A more detailed version of this extended system can be found in Ref. [24].

Wang et al. [25] use the KLT feature tracker, as many others have done so far, but the system first determines the best candidate frame to perform matching on. Once the best candidate frame is determined, some postprocessing steps, such as frame shifting, transformation, and reshaping are performed to get the best experience. Wu et al. [26] design a 2D to 3D conversion algorithm by first segmenting pertinent objects from key frames using the lazy

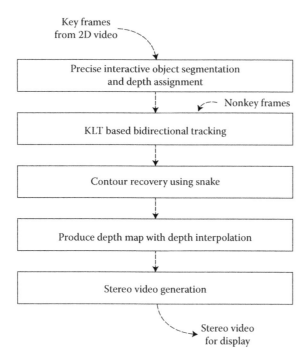

FIGURE 27.11
Block diagram of the algorithm by Wu et al. (C. Wu et al., A novel method for semi-automatic 2D to 3D video conversion, *Proc. IEEE 3DTVCON*, pp. 65–68, Istanbul, Turkey, 2008. © (2008) IEEE.)

snapping method [27]. Once the objects are segmented, they are assigned a rough depth value, which is user defined. The labels are normalized between [0, 255] where 0 denotes far and 255 denotes near. Next, for the nonkey frames, a bidirectional KLT tracker is used to track points along the contours of each object boundary, and the contours are refined via active contours. Once the contours have been refined, the depth values are estimated via bilinear interpolation for the other pixels that were not refined via active contours. Figure 27.11 illustrates a block diagram concerning this approach.

Xu et al. [28] first convert the frames to their monochromatic counterparts. In addition, they extract the optical flow (using the KLT) and perform a median filter on all of the vectors. Once they perform this, a least-squares like procedure is done to determine what the depth values are, given the monochromatic intensities, and the optical flow vectors. Figure 27.12 illustrates a block diagram concerning this approach.

Finally, Yan et al. [29] first use the scale invariant feature transform (SIFT) [30] between two consecutive frames, and then formulate a homography transformation between the frames. In addition, the current frame is oversegmented using the mean-shift [14] segmentation algorithm. After, the oversegmented result and the homography transformation are formulated into a graph cuts segmentation problem and the depths are solved in that fashion [31]. Essentially, homography estimation is used to determine the motion layers, or layers in the scene that roughly share the same motion. Motion vectors in the same motion layer most likely correspond to the same depth. Therefore, a region-based graph cuts segmentation is performed with the previous information. The result is a depth assignment that is coherent with respect to each motion layer. Figure 27.13 illustrates a block diagram describing this approach.

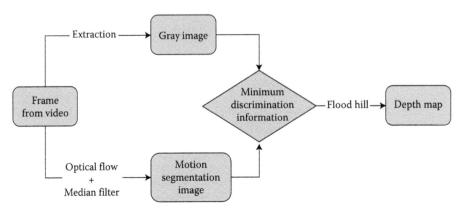

FIGURE 27.12
Block diagram of the algorithm by Xu et al. (F. Xu et al., 2D-to-3D Conversion based on motion and color mergence, *Proc. IEEE 3DTVCON*, 2008. © (2008) IEEE.)

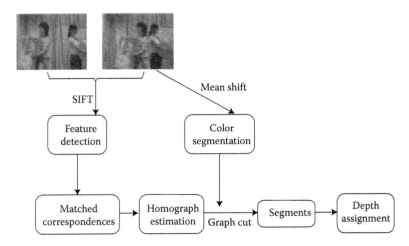

FIGURE 27.13
Block diagram of the algorithm by Yan et al. (Y. Yan et al., A novel method for automatic 2D-to-3D video conversion, *Proc. IEEE 3DTVCON*, pp. 1–4, Tampere, Finland, 2010. © (2010) IEEE.)

27.3 2D to 3D Conversion Methods Using Scene Features

In this section, we will discuss the second most popular method for converting 2D monocular video sequences into 3D, which is through analyzing the features of the scene of interest. Features such as shape, edges, color, or anything involving the direct use of features other than motion are of interest here. We will now begin our discussion for 2D to 3D conversion methods that are within this category.

27.3.1 Color

There have been attempts to use color information directly for determining depth maps from monocular video sequences. The most well-known one recently comes from Tam et al. [32]. Essentially, they decompose the video sequence into the YCbCr color space, and exclusively use the Cr color channel as a measure of depth. The reasoning is that different objects have different hues, and thus the Cr channel provides an approximate segmentation of the objects, which are characterized by different intensities in the Cr component image. Each intensity should belong to roughly the same kind of objects, and thus the same depth value. However, some tweaking is required, specifically when there is blue and green, as a blue sky has a lighter hue than a green tree, which would mean that the sky is closer to the viewer. There is some postprocessing that is involved here, but it is a very crude approach to determining a depth map.

27.3.2 Shape and Texture

Interestingly, some approaches to 2D to 3D conversion use shape and texture features to generate the stereoscopic content. One method to do so is the one by Feng et al. [33]. This is not a true 2D to 3D conversion, but it is of an interesting nature to note, and we shall cover this particular method in more detail. Specifically, there should be some stereoscopic pairs available throughout the video sequence, and the disparities for the rest of the monocular sequence are estimated accordingly. Shape features are used to estimate the disparity for the other 2D frames. Three features are extracted based on shape. These are the major axis (the straight line segment joining the two points of the shape on the boundary farthest away from each other), the minor axis (the line between the two points closest to each other on the boundary), and the center of mass of the object. This is essentially performing principal component analysis (PCA), as well as determining the first moment. After, dynamic programming is used to estimate the disparity map between those stereoscopic pairs and the disparities are propagated to the other 2D frames via the Hausdorff distance using the shape features. As an extension to this method, in Ref. [34], they created texture features based on the MPEG-4 decoder, and used these to generate the disparity maps.

27.3.3 Edges

Approaches have been formulated that use edge information to determine stereoscopic content from 2D data. Specifically, in the work by Cheng et al. [35], they first used a block-based algorithm to group regions of pixels together using edge information. Next, a depth from prior hypothesis step is used to generate the depth for each pixel. Specifically, in an aggregation window, the depth value is the gravitational center of the block. Then, bilateral filtering is performed to smooth the object boundaries, and thus minimize the cardboard effect. Figure 27.14 illustrates a block diagram of this approach.

Guo et al. [36] use 2D wavelet analysis to perform defocus estimation on edges. Essentially, the higher the wavelet components, the closer the pixel is to the viewer. Similar to using edges, blur information is used to determine the depth values. Essentially, depth from focus methods are used to determine the amount of defocusing, or blurring, in the frame, which is assumed that the blurring area is the result of convolving a properly focused area with an impulse response that is approximately a 2D Gaussian function. Because the values of the depth and the impulse response change linearly with the spatial position of the region, and getting close or far away from the camera plane, there exists a reasonable possibility to simulate the depth value by its corresponding focus value. Examples of this approach can be found in Feng et al. [37] and Valencia and Dagnino [38].

Finally, Li et al. [39] focus on 2D to 3D conversion by skeleton line tracking. Essentially, the foreground objects are segmented interactively, then a grass-fire algorithm is used to

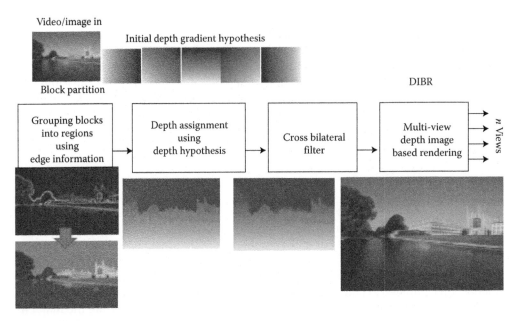

FIGURE 27.14
Block diagram of the algorithm by Cheng et al. (C.-C. Cheng, C.-T. Li, and L.-G. Chen, A 2D-to-3D conversion system using edge information, *Proc. IEEE Conf. on Consumer Electronics*, pp. 377–378, Las Vegas, Nevada, 2009. © (2009) IEEE.)

obtain the skeleton lines of the object. At the same time, optical flow is used to determine the motion vectors of the object. The skeleton lines and motion vectors are used to predict new skeleton lines of the objects for the next frame. While new skeleton lines are obtained, the lazy snapping method is used to recover the object without any interactivity. Next, depths are assigned so that the foreground is given a depth of intensity 255 (near), and the background a depth of intensity 0 (far). The rest of the depths are determined using a combination of the motion vectors and skeleton lines. Figure 27.15 illustrates a block diagram of this approach.

27.3.4 Filtering

There has been some research on applying filtering techniques for determining a sense of depth from monocular videos. Specifically, techniques that use edge-preserving smoothing or bilateral filtering have been investigated in the past for use in this framework. Specifically, we start with the work done by Angot et al. [40] who create an initial depth map, and refine based on the bilateral filter placed on a bilateral grid. The bilateral grid essentially exports the bilateral filtering problem to a 3D filtering problem, where the spatial and range variance are on the x and y axes, and the intensity is on the z axis. First, the frames are converted to their grayscale counterparts. Next, the frames are examined, and are categorized into one of six categories, where each category yields an initial depth map. Then, the bilateral filter is used to produce smooth depth gradients in the frames.

Chen et al. [41] propose a filter that works directly with edges, in order to solve the occlusion problem at object boundaries. Specifically, a 2D Gaussian-shaped filter, with its weights adjusted depending on the strength of the edges, is used to smooth the boundaries, and thus minimize occlusions. Jung et al. [42] propose a stereoscopic content delivery system

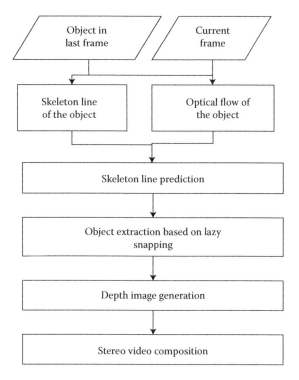

FIGURE 27.15
Block diagram of the algorithm by Li et al. (Z. Li, X. Xie, and X. Liu, An efficient 2D to 3D video conversion method based on skeleton line tracking, *Proc. IEEE 3DTVCON*, pp. 1–4, Potsdam, Germany, 2009. © (2009) IEEE.)

over T-DMB. However, this method does not specifically talk about how the content is converted from 2D to 3D, but the interesting thing to note here is the preprocessing step that is used. The depth images are smoothed by an adaptive smoothing filter based on the depth values. Essentially, the gradients (both magnitude and phase) are used as weights, which are then used to render the 3D content. A more detailed version of this system can be found in Ref. [43].

27.3.5 Segmentation

In addition to what was previously stated, there have been attempts to perform a segmentation of objects over a video sequence, and to assign depths for each of those objects. Specifically, we will talk about the work done by Ng et al. [44]. In their work, lazy snapping is used as a preprocessing step to extract objects from the background. An object tracking framework is formulated by partial differential equations, and a background inpainting algorithm is performed, to fill in the holes left by removing the objects from the scene. This algorithm is based on the work by Criminisi et al., where an intelligent hole-filling technique is formulated using nearby gradient and confidence measures that are based on texture and structure information [45]. As for the actual depth estimation itself, they refer to either using wavelet analysis, like in Ref. [38], or using short-term motion-assisted segmentation like in Ref. [9]. Figure 27.16 illustrates a block diagram that describes this process.

However, the most successful work in performing depth map estimation in a segmentation like framework is the work done by Guttmann et al. [46]. In this work, rough depth

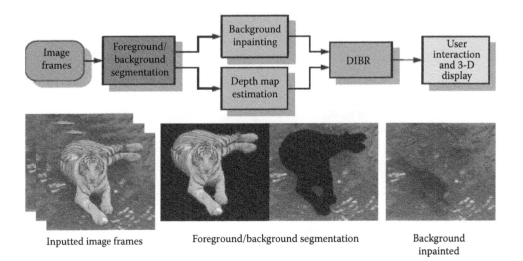

FIGURE 27.16
Block diagram of the algorithm by Ng et al. (K. T. Ng, Z. Y. Zhu, and S. C. Chan, An approach to 2D-to-3D conversion for multiview displays, *Proc. IEEE Int. Conf. on Information, Communications and Signal Processing*, pp. 1–5, Macau, 2009. © (2009) IEEE.)

values are marked by the user. Even though the true depth values are not known *a priori*, the authors argue that psychologically, the true depth values are not required, as long as they are relatively correct in assignment. For example, what the user thinks is very close to the camera should be marked as such, and what the user thinks is very far should be marked the same way. These depth values are initialized for the first and last frame. Next, a linear system is solved via least squares, where the system is decomposed into several features, such as the initial constraints, and spatial and time smoothness constraints. The solving of this system is essentially the solution to the random walks problem [47]. Then, the depth values are refined further using a support vector machine. The aforementioned method for converting from 2D to 3D is quite complex, with many processing steps required to obtain the final result. Noting this difficulty for this system, but knowing that random walks has the ability to respect weak boundaries, the work done by Phan et al. [48,49] merges the merits of random walks, together with the hard boundaries generated by the graph cuts segmentation paradigm, and the smoothing actions performed with random walks. However, the work in Ref. [46] performs the depth estimation over an entire video sequence, where in this work, only a single image is used. Essentially, the *a priori* labeling of the depth values is performed in the same fashion as Ref. [46], but the underlying mechanism to produce the final depth map is different, and more simplified.

27.3.6 Analysis of the Environment

In this category, depending on what the actual environment is, the depth map is calculated. Environments such as outdoors, ocean, indoors, portraits, mountains, land, and so on will inherently have different depth maps, as there are different vanishing points, horizon lines, and so on. The best example of this work was performed by Battiato et al. [50]. First, the scene is classified into one of the following categories: sky, farthest mountain, far mountain, Near mountain, land, and other. After an oversegmentation, either mean-shift or something related, is used to determine homogeneous regions of clusters. Next, an analysis of these clusters, in relation to randomly sampling N selected columns of the oversegmented image is performed, and a final classification step is performed using some predefined heuristics. Next, a vanishing lines detection algorithm was performed to determine the horizon. After, depths vary from the bottom of the frame, up to the vanishing point, and the depths vary as a decreasing gradient. The depths are assigned using predefined heuristics which are defined in their paper. A similar, but less complicated algorithm can be found in Ref. [51]. The next method that we will cover is the one by Cheng and Liang [52]. They assume that the classification of all possible images can be decomposed into portraits, outdoor images, and closeups. The scene classification is structured as follows. The image is transformed to the HSI color space, and a face detection algorithm is run. The face detection algorithm is based on the adaptive boosting (AdaBoost) algorithm used in the Viola and Jones framework [53]. If there is a face found, this is classified as belonging to the portrait class. If there is no face, then an object detection algorithm is run. Object detection is performed by analyzing the 2D wavelet coefficients, with some thresholding operations. After the object detection algorithm, a heuristic using the hue and saturation components determine whether or not there is a sky. If there is no sky, this is classified as a closeup, and if there is, we classify this as an outdoor image. The depth maps are then generated based on the category that the image was classified as. If it is a portrait, the face, as well as the body, is used to generate a depth gradient, using some other heuristics that are defined in the paper. If the scene was classified as an outdoor image, a logarithmic transformation is performed that uses the horizon line, and the boundaries of the image. Finally, if the scene is classified as a closeup, different heuristics are defined, but the same logarithmic transformation is used. Figure 27.17 illustrates a flow chart of the aforementioned method shown below.

Deng et al. [54] proposed a method for determining depth maps of indoor environments only. First, they extract the planes for each of the walls that are seen in the image. Next, a plane classification step is determined, in order to figure out which pixels belong to what plane. Once this is performed, for each pixel in the plane, a simple linear equation that is related to the area of each plane is used to determine the depth gradient from near to far. Similarly, Yamada and Suzuki [55] develop a real-time 2D to 3D conversion algorithm that converts to full 1080p HD resolution. Specifically, different depth models are created, which are ultimately what they postulate the different environments a scene could possibly be in. The first model is a spherical surface model, used for a generic scene with perspective geometry. The second model is a cylindrical and spherical model, and is used for scenes with an open sky, and the last model is used for a scene consisting of long range views, and flat grounds or water. With these models, they determine the amount of activity that is happening at the top half of the frame, as well as the bottom half of the frame. A blend between the two activities is used to determine what the final depth map is. Figure 27.18 illustrates a block diagram of the aforementioned approach.

27.3.7 Miscellaneous Methods

This last section details those algorithms that use features from the scene, but do not fall into any of the categories listed above. We will go through these methods in more detail. We start with the method performed by Harman et al. [56]. Essentially, the framework entails a machine learning algorithm, in order to determine what the depth values in the video sequence are. Essentially, some training samples, with user-defined depths are assigned to certain points in a frame. After, the machine learning algorithm determines what the rest of the depth values are. Machine learning algorithms, such as decision trees, support

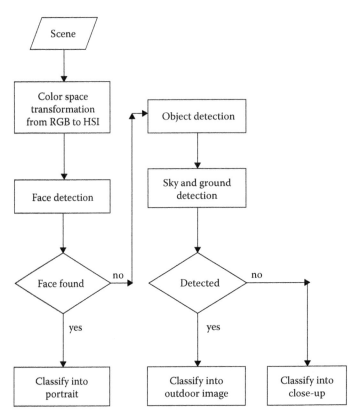

FIGURE 27.17
Flow chart of the algorithm by Cheng and Liang. (From F.-H. Cheng and Y.-H. Liang, Depth map generation based on scene categories, *SPIE J. of Electronic Imaging*, 18, 4, 043006-1–043006-6, 2009. With permission.)

vector machines, or neural networks, are used. The machine learning algorithm actually determines the depth values for only the frames that were marked. For the rest of the frames, consecutive keyframe depth maps are used to determine what the depth maps of the nonkey frames are. These are determined by a simple weighting scheme, with the depths at the consecutive keyframes used.

Jung and Ho [57] estimate the depth maps via a Bayesian learning algorithm. Training data of six different features are used, which are the horizontal line, the vanishing point, vertical lines, boundaries, the complexity of the wavelet coefficients, and the object size. This training data categorizes objects in the 2D image into four different types. According to the type, a relative depth value to each object is assumed, and a simple 3D model is generated. The four different types are: sky, ground, plane, and cubic. Sky and ground types are straightforward, whereas the plane type represents an object that is perpendicular to the optical axis of the camera and has a constant depth value. Things like a human and a tree are examples. Finally, the cubic type is regarded as an object that has different depth values, according to the distance from the vanishing point, and includes a building, a wall, and objects such as those. For the rest of the pixels that were not classified, a Bayesian classifier is formulated to choose the best depth values, given those previously mentioned six features.

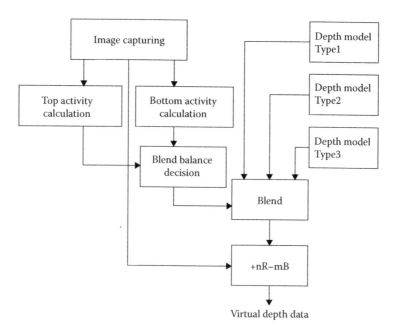

FIGURE 27.18

Flow chart of the algorithm by Yamada and Suzuki. (K. Yamada and Y. Suzuki, Real-time 2D-to-3D conversion at full HD 1080p resolution, *Proc. IEEE Int. Symp. on Consumer Electronics*, pp. 103–106, Kyoto, Japan, 2009. © (2009) IEEE.)

Kim et al. [58] formulate a method that is based on visual attention analysis. Essentially, a saliency map is created where objects that are of importance and attract the viewer visually are given a higher value, than those that do not attract any attention. A five-feature pyramidal scheme, consisting of color, luminance, orientation, texture, and motion, is considered. Next, the saliency map of each of these features is computed in another pyramidal scheme, and is thus fused to produce a final visual attention map. The visual attention map is thus used to determine what the relative depth of the pixels should be, as pixels of interest should be deemed closer, than pixels that are not as interesting.

In Ko et al. [59], the degree of focus of segmented regions are used to generate a stereoscopic image. A customized image segmentation algorithm is used to extract foreground objects. This is used to partition the image into homogeneous regions as well. Next, a higher-order statistics map is used to represent the spatial distribution of high-frequency components of the input image. The map is known to be well suited for detection and classification problems, as it can suppress Gaussian noise and preserve some of the non-Gaussian information. The relative depth map can be calculated with these two cues, and can be refined by postprocessing.

Finally, Rotem et al. [60] devise an automatic 2D to 3D conversion algorithm, which is based on calculating a planar transformation between images in the sequence and relies on the human capability to sense the residual parallax. Therefore, it does not depend on the calculation of the depth map. The system generates stereoscopic image pairs. Each pair consists of the original image and a transformed image. The transformed image is generated from another image of the original sequence. That image is selected by the algorithm such that a considerable parallax is developed between the two original frames, and the chosen image is then warped using the planar transformation.

TABLE 27.1

Summary of All Methods Discussed in This Chapter

Category	Major Characteristics	References
Motion	MPEG-4 Motion Vectors	[6,12]
Motion	H.264 Motion Vectors	[7,8]
Motion	Difference Frames and Scene Geometry	[9–11,25,26,28]
Motion	KLT	[12,13,20]
Motion	KLT and EKF	[22–24]
Motion	SfM	[17–19]
Motion	SIFT and Graph Cuts	[29]
Scene	Color	[32]
Scene	Shape and PCA	[33,34]
Scene	Edges	[35]
Scene	Wavelets	[36]
Scene	Skeleton Line Tracking	[39]
Scene	Filtering	[40–43]
Scene	Segmentation	[44,46,48,49]
Scene	Scene Classification	[50–52]
Scene	Indoor Environment Classification	[54]
Scene	Geometric Modelling	[55]
Miscellaneous	Machine Learning Algorithms	[56,57]
Miscellaneous	Visual Attention Analysis	[58]
Miscellaneous	Focus and Segmentation	[59]
Miscellaneous	Planar Transformations	[60]

27.4 Conclusion

In this literature review, we discussed the most recent undertakings in converting monocular video footage to their stereoscopic or multiview counterparts for display on 3D visualization technology. We have concentrated on the various recent methods to achieve this goal over the last few years, as research into this area has surfaced in higher volumes in this period. As previously mentioned, methods to convert from 2D to 3D can be further subdivided into two major categories: using motion and analyzing scene features. In order to fully illustrate and summarize all the methods we have discussed, Table 27.1 illustrates a summary of all of the methods we have discussed here, and the methods are grouped according to the major methods and features that they use. For brevity, we refer to each method by the reference number they are cited in, and the reader is referred to the reference list for the actual authors, and the titles of the papers.

For motion, the relative motion experienced throughout the frames can be used to determine a sense of depth in a scene, where higher motion denotes that objects are closer to the camera, and vice versa. However, there can be several disadvantages when using motion to determine a sense of depth. First, by directly using the motion vectors as an indicator of depth, it is possible that an object is moving quite fast, but the actual depth is quite far into the scene, while another object could be closer to the camera and could be moving at the same speed as the far object. Thus, this could be interpreted as both objects having the same depth, where they are clearly not in this case. In addition, motion vectors can be subject to estimation errors, especially when high motion is involved. This means that for objects that are moving quite fast, the motion vector, and hence the depth values, will inevitably be subject to errors as well. Finally, using motion is ultimately constrained to the window size that is used to estimate these vectors. Motion vectors are calculated by enclosing a window of interest around the center of the point to be motion estimated. Any motion occurring that is beyond this window size will not be captured, and it may be the case where objects are moving beyond this same window of calculation. Nevertheless, motion is a popular method, as they are already made available in the encoded video itself, and as can be evidenced by the amount of heavy research that was just discussed.

For analyzing scene features, prominent features that are observed in the frame, other than the use of motion, are exploited here. Color was one method that was discussed, where each pixel in the frame is decomposed into a more perceptual color space, and these components are directly used to determine a sense of depth. However, as mentioned previously, some tweaking is necessary for this to be used effectively, as the components themselves could misrepresent the depth of objects in certain situations. Another method was to use shape and texture, but as mentioned previously, this is not a true 2D to 3D conversion, as stereoscopic video sequences are required. Another feature that is commonly used are through edges, where they are exploited as a means of directly determining the amount of depth. Depth from defocus, using wavelet coefficients, or skeleton line tracking are such examples. However, the main drawback here is ultimately in selecting the right parameters in the way of exploiting those edges, and thus some tweaking is required. Similarly, with using filtering techniques, the same drawback can also be said here. By using segmentation techniques for estimating depth, the accuracy of the depth maps is ultimately constrained to the accuracy of the image or object segmentation methods that are used, and also require tweaking to get accurate results. For analyzing the environment, the depths are calculated based on the category of the scene, and will thus have different vanishing points, vanishing lines, and other areas of interest that make that scene unique, and and thus having a unique depth map. However, not all video footage acquired can compactly fit into this small set of categories, and it may be possible that a scene can be represented by more than one category. With this, it is inevitable that the parameters will be required to be adjusted to generate an accurate depth map.

Nevertheless, with the plethora of different research in this area that has surfaced, it has been demonstrated that it is quite an important topic for realizing a sense of depth in video sequences obtained by only a single camera, and is very useful in many aspects of the industry and to the end user. It is also important, as this is an alternative solution to producing 3D content to alleviate the lack of 3D content that exists, and to also provide a more cost-effective solution. Despite the difficulties experienced in each of the major categories that have been previously listed, there have been some successful results generated, and the future looks quite promising for the use of these techniques to realize 3D content.

References

1. C. Fehn, R. de la Barre, and S. Pastoor, Interactive 3-DTV concepts and key technologies, *Proc. of the IEEE*, vol. 94, no. 3, pp. 524–538, March 2006.
2. W. Matusik and H. Pfister, 3D TV: A scalable system for real-time acquisition, transmission, and autostereoscopic display of dynamic scenes, *ACM Trans. on Graphics*, 24, 3, 2004.

3. A. Redert, M. O. de Beeck, C. Fehn, W. Ijsselsteijn, M. Pollefeys, L. V. Gool, E. Ofek, I. Sexton, and P. Surman, ATTEST: Advanced three-dimensional television system technologies, *IEEE Intl. Symposium on 3D Data Processing Visualization and Transmission (3DPVT)*, pp. 313–319, Padova, Italy, 2002.

4. Mobile3DTV, Mobile 3DTV Content Delivery Optimization over DVB-H System, http://sp.cs.tut.fi/mobile3dtv/—Accessed on October 4th, 2010.

5. Z.-W. Gao, W.-K. Lin, and Y.-S. Shen, Design of signal processing pipeline for stereoscopic cameras, *IEEE Trans. Consumer Electronics*, 56, 2, pp. 324–331, 2010.

6. I. Ideses, L. P. Yaroslavsky, and B. Fishbain, Real-time 2D to 3D video conversion, *J. Real-Time Image Processing*, 2, 3–9, 2007.

7. X. Huang, L. Wang, J. Huang, D. Li, and M. Zhang, A depth extraction method based on motion and geometry for 2D to 3D conversion, *Proc. IEEE Symp. on Intelligent Information Technology Applications*, Nanchang, China, 2009.

8. M. Pourazad, P. Nasiopoulos, and R. K. Ward, Converting H.264-derived motion information into depth map, *Springer Science Advances in Multimedia Modeling*, 5371, 108–118, 2009.

9. Y.-L. Chang, C.-Y. Fang, L.-F. Ding, S.-Y. Chen, and L.-G. Chen, Depth map generation for 2D-To-3D conversion by short-term motion assisted color segmentation, *Proc. IEEE Conf. on Multimedia and Expo (ICME)*, pp. 1958–1961, Beijing, China, 2007.

10. Y.-L. Chang, C.-Y. Fang, L.-F. Ding, S.-Y. Chen, and L.-G. Chen, Priority Depth Fusion for the 2D-to-3D Conversion System, *Proc. SPIE Electronic Imaging: Three-Dimensional Image Capture and Applications*, Vol. 6805, pp. 680513-1–680513-8, San Jose, California, 2008.

11. C.-C. Cheng, C.-T. Li, P.-S. Huang, T.-K. Lin, Y.-M. Tsai, and L.-G. Chen, A block-based 2D-to-3D conversion system with bilateral filter, *Proc. IEEE Conf. on Consumer Electronics*, pp. 1–2, Kyoto, Japan, 2009.

12. M. Kim, S. Park, H. Kim, and I. Artem, Automatic conversion of two-dimensional video into stereoscopic video, *Proc. SPIE Electronic Imaging—Three-Dimensional TV, Video and Display IV*, 6016, pp. 601610-1–601610-8, San Jose, California, 2006.

13. D. Kim, D. Min, and K. Sohn, Stereoscopic video generation using motion analysis, *Proc. IEEE 3DTV Conf. (3DTVCON)*, pp. 1–4, Kos Island, Greece, 2007.

14. D. Comaniciu and P. Meer, Mean shift: A robust approach toward feature space analysis, *IEEE Trans. on Pattern Analysis and Machine Intelligence*, 24 (5), 603–619, 2002.

15. J. Shi and C. Tomasi, Good features to track, *Proc. IEEE Conf. on Computer Vision and Pattern Recognition*, pp. 593–600, Colorado Springs, Colorado, 1994.

16. D. Kim, D. Min, and K. Sohn, A stereoscopic video generation method using stereoscopic display characterization and motion analysis, *IEEE Trans. on Broadcasting*, 54 (2), 188–197, 2008.

17. S. Knorr, A. Smolic, and T. Sikora, From 2D- to stereo- to multi-view video, *Proc. IEEE 3DTVCON*, pp. 1–4, Kos Island, Greece, 2007.

18. S. Knorr, M. Kunter, and T. Sikora, Super-resolution stereo- and multi-view synthesis from monocular video sequences, *Proc. IEEE Intl. Conf. on 3-D Digital Imaging and Modeling (3DIM)*, pp. 55–64, Montreal, Quebec, Canada, 2007.

19. M. Kunter, S. Knorr, A. Krutz, and T. Sikora, Unsupervised object segmentation for 2D to 3D conversion, *Proc. SPIE Electronic Imaging: Stereoscopic Displays and Applications XX*, Vol. 7237, San Jose, California, 2009.

20. T. Li, Q. Dai, and X. Xie, An efficient method for automatic stereoscopic conversion, *Proc. IET Int. Conf. on Visual Information Engineering*, pp. 256–260, Xi'an, China, 2008.

21. M. Kass, A. Witkin, and D. Terzopoulos, Snakes: Active contour models, *Int. J. of Computer Vision*, 1 (4), 321–331, 1988.

22. S. Diplaris, N. Grammalidis, D. Tzovaras, and M. G. Strintzis, Generation of stereoscopic image sequences using structure and rigid motion estimation by extended Kalman filters, *Proc. IEEE ICME*, Vol. 2, pp. 233–236, Lausanne, Switzerland, 2002.

23. K. Moustakas, D. Tzovaras, and M. G. Strintzis, A noncausal Bayesian framework for object tracking and occlusion handling for the synthesis of stereoscopic video, *Proc. IEEE Int. Symp. on 3D Data Processing, Visualization and Transmission (3DPVT)*, Thessaloniki, Greece, 2004.

24. K. Moustakas, D. Tzovaras, and M. G. Strintzis, Stereoscopic video generation based on efficient layered structure and motion estimation from a monoscopic image sequence, *IEEE Trans. on Circuits and Systems for Video Technology*, 15 (8), 1065–1073, 2005.
25. H.-M. Wang, Y.-H. Chen, and J.-F. Yang, A novel matching frame selection method for stereoscopic video generation, *Proc. IEEE ICME*, pp. 1174–1177, New York, 2009.
26. C. Wu, G. Er, X. Xie, T. Li, X. Cao, and Q. Dai, A novel method for semi-automatic 2D to 3D video conversion, *Proc. IEEE 3DTVCON*, pp. 65–68, Istanbul, Turkey, 2008.
27. Y. Li, J. Sun, C.-K. Tang, and H.-Y. Shum, Lazy snapping, *ACM Trans. on Graphics*, 23, 3, 2004.
28. F. Xu, G. Er, X. Xie, and Q. Dai, 2D-to-3D Conversion based on motion and color mergence, *Proc. IEEE 3DTVCON*, pp. 205–208, Istanbul, Turkey, 2008.
29. Y. Yan, F. Xu, Q. Dai, and X. Liu, A novel method for automatic 2D-To-3D video conversion, *Proc. IEEE 3DTVCON*, pp. 1–4, Tampere, Finland, 2010.
30. D. G. Lowe, Distinctive image features from scale-invariant keypoints, *Int. J. of Computer Vision*, 60 (2), 91–110, 2004.
31. Y. Boykov and G. Funka-Lea, Graph cuts and efficient N-D image segmentation, *Int. J. of Computer Vision*, 2 (70), 109–131, 2006.
32. W. J. Tam, C. Vazquez, and F. Speranza, Three-dimensional TV: A novel method for generating surrogate depth maps using colour information, *Proc. SPIE Electronic Imaging—Stereoscopic Displays and Applications XX*, Vol. 7237, pp. 72371A-1–72371A-9, San Jose, California, 2009.
33. Y. Feng, J. Jiang, and S. S. Ipson, A shape-match based algorithm for pseudo-3D conversion of 2D videos, *Proc. IEEE Conf. on Image Processing (ICIP)*, Vol. 3, pp. 808–811, Genoa, Italy, 2005.
34. Y. Feng and J. Jiang, Pseudo-stereo conversion from 2D video, *Advanced Concepts in Intelligent Vision Systems*, 3708, 268–275, 2005.
35. C.-C. Cheng, C.-T. Li, and L.-G. Chen, A 2D-to-3D conversion system using edge information, *Proc. IEEE Conf. on Consumer Electronics*, pp. 377–378, Las Vegas, Nevada, 2009.
36. G. Guo, N. Zhang, L. Huo, and W. Gao, 2D to 3D conversion based on edge defocus and segmentation, *Proc. IEEE Int. Conf. on Acoustics, Speech and Signal Processing*, pp. 2181–2184, Las Vegas, Nevada, 2008.
37. Y. Feng, J. Jayaseelan, and J. Jiang, Cue based disparity estimation for possible 2D to 3D video conversion, *Proc. IET Conf. on Visual Information Engineering*, Bangalore, India, 2006.
38. S. A. Valencia and R. M. R. Dagnino, Synthesizing stereo 3D views from focus cues in monoscopic 2D images, *Proc. SPIE Electronic Imaging—Stereoscopic Displays and Virtual Reality Systems X*, Vol. 5006, pp. 377–388, San Jose, California, 2003.
39. Z. Li, X. Xie, and X. Liu, An efficient 2D to 3D video conversion method based on skeleton line tracking, *Proc. IEEE 3DTVCON*, pp. 1–4, Potsdam, Germany, 2009.
40. L. J. Angot, W.-J. Huang, and K.-C. Liu, A 2D to 3D video and image conversion technique based on a bilateral filter, *Proc. SPIE Electronic Imaging—Three-Dimensional Image Processing (3DIP) and Applications*, Vol. 7526, pp. 75260D-1–75260D-10, San Jose, California, 2010.
41. W.-Y. Chen, Y.-L. Chang, S.-F. Lin, L.-F. Ding, and L.-G. Chen, Efficient depth image based rendering with edge dependent depth filter and interpolation, *Proc. IEEE ICME*, pp. 1314–1317, Amsterdam, the Netherlands, 2005.
42. K. Jung, Y. K. Park, J. K. Kim, H. Lee, K. Yun, N. Hur, and J. Kim, Depth image based rendering for 3D data service over T-DMB, *Proc. IEEE 3DTVCON*, pp. 237–240, Istanbul, Turkey, 2008.
43. Y. K. Park, K. Jung, Y. Oh, S. Lee, J. K. Kim, G. Lee, H. Lee, K. Yun, N. Hur, and J. Kim, Depth image based rendering for 3DTV over T-DMB, *Elsevier Signal Processing: Image Communication*, 24(1–2), 122–136, 2009.
44. K. T. Ng, Z. Y. Zhu, and S. C. Chan, An approach to 2D-to-3D conversion for multiview displays, *Proc. IEEE Int. Conf. on Information, Communications and Signal Processing*, pp. 1–5, Macau, 2009.
45. A. Criminisi, P. Perez, and K. Toyama, Region filling and object removal by exemplar-based inpainting, *IEEE Trans. on Image Processing*, 13 (9), 1200–1212, 2004.
46. M. Guttmann, L. Wolf, and D. Cohen-Or, Semi-automatic stereo extraction from video footage, *Proc. IEEE Int. Conf. on Computer Vision (ICCV)*, Kyoto, Japan, 2009.
47. L. Grady, *Random Walks for Image Segmentation*, 28, 11, pp. 1768–1783, 2006.

48. R. Phan, R. Rzeszutek, and D. Androutsos, Semi-automatic 2D to 3D image conversion using a hybrid random walks and graph cuts based approach, *Proc. IEEE Int. Conf. on Acoustics, Speech and Signal Processing (ICASSP)*, pp. 897–900, Prague, Czech Republic, 2011.

49. R. Phan, R. Rzeszutek, and D. Androutsos, Semi-automatic 2D to 3D image conversion using scale-space random walks and a graph cuts based depth prior, Accepted to appear in *Proc. IEEE Int. Conf. on Image Processing (ICIP)*, Brussels, Belgium, 2011.

50. S. Battiato, A. Capra, S. Curti, and M. L. Cascia, 3D Stereoscopic image pairs by depth-map generation, *Proc. IEEE Int. Symp. 3DPVT*, pp. 124–131, Thessaloniki, Greece, 2004.

51. C. O. Yun, S. H. Han, T. S. Yun, and D. H. Lee, Development of stereoscopic image editing tool using image-based modeling, *IEEE Int. Conf. on Computer Graphics and Virtual Reality*, pp. 42–48, Las Vegas, Nevada, 2006.

52. F.-H. Cheng and Y.-H. Liang, Depth map generation based on scene categories, *SPIE J. of Electronic Imaging*, 18, 4, 043006-1–043006-6, 2009.

53. P. Viola and M. J. Jones, Robust real-time face detection, *Int. J. Computer Vision*, 57 (2), 137–154, 2004.

54. X.-L. Deng, X.-H. Jiang, Q.-G. Liu, and W.-X. Wang, Automatic depth map estimation of monocular indoor environments, *Proc. IEEE Int. Conf. on Multimedia and Information Technology*, pp. 646–649, Three Gorges, China, 2008.

55. K. Yamada and Y. Suzuki, Real-time 2D-to-3D conversion at full HD 1080p resolution, *Proc. IEEE Int. Symp. on Consumer Electronics*, pp. 103–106, Kyoto, Japan, 2009.

56. P. Harman, J. Flack, S. Fox, and M. Dowley, Rapid 2D to 3D conversion, *Proc. SPIE Electronic Imaging—Stereoscopic Displays and Virtual Reality Systems IX*, Vol. 4660, San Jose, California, 2002.

57. J.-I. Jung and Y.-S. Ho, Depth map estimation from single-view image using object classification based on bayesian learning, *Proc. IEEE 3DTVCON*, pp. 1–4, Tampere, Finland, 2010.

58. J. Kim, A. Baik, Y. J. Jung, and D. Park, 2D-to-3D conversion by using visual attention analysis, *Proc. SPIE Electronic Imaging—Stereoscopic Displays and Applications XXI*, Vol. 7524, pp. 752412-1–752412-12, San Jose, California, 2010.

59. J. Ko, M. Kim, and C. Kim, Depth-map estimation in a 2D single-view image, *Proc. SPIE Electronic Imaging—Applications of Digital Image Processing XXX*, Vol. 6696, pp. 66962A-1–66962A-9, San Jose, California, 2007.

60. E. Rotem, K. Wolowelsky, and D. Pelz, Automatic video to stereoscopic video conversion, *Proc. SPIE Electronic Imaging—Stereoscopic Displays and Virtual Reality Systems XII*, Vol. 5664, pp. 198–206, San Jose, California, 2005.

28

Haptic Interaction and Avatar Animation Rendering Centric Telepresence in Second Life

A. S. M. Mahfujur Rahman, S. K. Alamgir Hossain, and A. El Saddik

CONTENTS

> "Virtual Reality is electronic simulations of environments experienced via head-mounted eye goggles and wired clothing enabling the end user to interact in realistic three-dimensional situations."
>
> **George Coates, 1992 [8]**
> *Program from invisible site—a multimedia performance work*

28.1 Introduction

The key to defining virtual reality in terms of human experience rather than technological hardware is the concept of presence. Presence can be thought of as the experience of one's physical environment; it refers not only to the surroundings as they exist in the physical world, but also to the perception of those surroundings as mediated by both automatic and controlled mental processes (Gibson, 1979): Presence is defined as the sense of being in an environment.

Telepresence is defined as the experience of presence in an environment by means of a communication medium. In other words, presence refers to the natural perception of an environment, and telepresence refers to the mediated perception of an environment. This environment can be either temporally or spatially distant real environment, or an animated but nonexistent virtual world synthesized by a computer (for instance, the animated world created in a video game) [13]. A virtual reality is defined as a real or simulated environment in which a perceiver experiences telepresence. For example, reading a letter from a distant friend or colleague can evoke a sense of presence in the environment in which the letter was written, or can make the distant party seem locally present. This feeling can occur even when one is unfamiliar with the remote physical surroundings.

Interactivity, multimodality and haptic interaction play an important role in creating a vivid telepresence experience into the participating users. Vividness means the representational richness of a mediated environment as defined by its formal features; that is, the way in which an environment presents information to the senses. Vividness is stimulus driven, depending entirely upon the technical characteristics of a medium. Thereby, the haptic rendering, the ambient sound, 3D realistic animation and interaction with real-world user influences the user's senses and enriches the telepresence experience in the virtual environment. The sense of touch has much importance in technology-mediated human emotion communication and interaction. Emotional touches such as handshake, encouraging pat, kiss, tickle physical contacts are fundamental to mental and psychological development and they also enhance the users' interactive [4,11,30] and immersive experiences [9] with the virtual world. The haptic-based nonverbal modality can enhance social interactivity and emotional immersive experiences in a 3D multiuser virtual world that presents a 3D realistic environment, where people can enrol in an online virtual community [29]. One of the most popular and rapidly spreading examples of such systems is Linden Lab's Second Life (SL) [18]. In SL, similar to ActiveWorlds [1] and Sims [26], the users can view their avatars in a computer-simulated 3D environment and they can participate in real time in task-based games, play animation, communicate with other avatars through instant messaging and voice. The social communication aspect of SL is hugely popular and counts millions of users. Moreover, its open source viewer [27] provides a unique opportunity to extend it further and equip it with other interaction modality such as haptic.

In this pursuit, we explored the possibilities of integrating haptic interactions in SL [14,15]. We enhanced the open source SL viewer client and introduced a communication channel that provides physical and emotional intimacy to the remote users. In the prototype system, a user can take advantage of touch, tickle, and hug-type haptic commands in order to interact with the participating users by using visual, audio or text-based interface modalities. A haptic stimulation of touch and other touch-based interactions are rendered to the remote user on the contacted skin through our previously developed haptic jacket system [5] that is composed of an array of vibrotactile actuators. This chapter

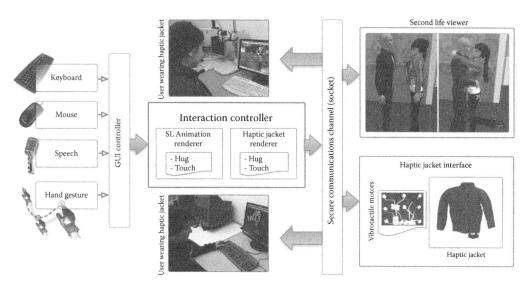

FIGURE 28.1
(**See color insert.**) A basic communication block diagram depicting various components of the SL interpersonal haptic communication system.

illustrates a preliminary prototype exploring the aforesaid haptic interactions between virtual and real environment actors. An overview of the system components is shown in Figure 28.1.

Our focus in this chapter is three-fold. First, in order to bridge the gap between virtual and real, we present an SL viewer add-on, where we provide haptic interaction opportunity between the real users and their respective virtual avatars through a 3D graphical-user interface (GUI) using speech, mouse, text, and gesture-based interaction modalities. Second, we discuss touch, hug, and tickle haptic features for the SL users through chat and GUI interactions. Third, we illustrate 3D annotation mechanism for the SL avatar so that user-dependent interpersonal haptic and animation interactions become possible.

The remainder of this chapter is organized as follows. At first, we present related work study in Section 28.2. In Section 28.3, we illustrate the various components of the system that facilitates the SL based interpersonal communication and provide a general overview of the system and its access mechanisms. Further in Section 28.4, we describe the implementation issues and development challenges of different modules. Also, in Section 28.5, we present response time comparisons for different haptic and animation data, accuracy of different interaction modalities and user study of the system. Section 28.6 concludes the chapter and states some possible future work directions.

28.2 Related Work

Thayer [28] states that touch as opposed to other forms of human-to-human communication will be more trusted by the person touched as a genuine reflection of emotion.

Especially in remote communication, touch is a unique channel in affect conveyance as the relation of touch to affect is immediate [30]. Haptic jacket [5] based rendering of touch has been incorporated previously into a conventional teleconferencing system to provide haptic interactions to the remote users. This approach uses marker-tracking technique to specify touchable parts of the user's body. The markers are further tracked using a dedicated camera. The system employs an expensive 3D camera in order to automatically create 3D touchable surface of the user.

In instant messaging, Rovers and van Essen [24] have provided a detailed study on the usage of hapticons that essentially are vibrotactile icons representing smileys. They incorporated six vibrotactile patterns that represent six associated smileys. These smileys could be triggered using mouse or keyboard-based interactions. In a 3D virtual environment, we attempted to employ a similar methodology. In our attempt, the smileys are replaced by different types of avatar animations such as hug, tickle, and touch that resemble the emotions that the user is trying to communicate to the other.

O'Brien and 'Floyd' Mueller [22] investigated an approach relating to intimate communication for couples. In this approach, a person could virtually hold hands by using their proposed probe to share tactile experiences with his or her partner's hand. They placed a small microchip inside a ball. When the ball is squeezed by a user the system sends vibrotactile data to the other ball that his or her partner is holding. For couples in long-distance relationships, these communication technologies may be a primary means of exchanging emotions [16]. In distance communication SL presents a multiuser communication framework that presents opportunities for interactions that connect people through a shared sense of place. Haptic-based input modes have been investigated in SL in order to assist blind people to be able to interact with the SL world [9]. The authors have implemented two new input modes that exploit the force feedback capabilities of haptic devices and allow the visually impaired users to navigate and explore the virtual environment. Recently, in SL, Tsetserukou et al. [29] have attempted to analyze the text conversations in SL chatting system. This system provides emotional haptic feedbacks to the users by using a specially designed wearable hardware. While the different hardware designs for HaptiTickler, HaptiHug, HaptiButterfly, and HaptiHeart are commendable, this approach does not seem to consider visual or pointer-based graphical interactions in the 3D environment other than the text-based conversation system. For example, it seems impossible to interact with specific parts of the virtual 3D avatar that can be used to generate haptic touch stimulation in that respective body part of the real user. Moreover, gesture [23] and audio-based interaction modalities can enhance the navigation and interaction experiences of the user in a 3D virtual gaming environment [12]. Hence, in this presented haptic communication framework we incorporated a flexible GUI-based multimodal interaction mechanism in order to provide more natural, easy, and accessible interactions in SL.

28.3 Haptic-Based Interpersonal Communication Framework

In this section we present the different components of our haptic-based interpersonal communication framework and their functional descriptions. The components of the system are depicted in Figure 28.1 as a block diagram.

28.3.1 SL Viewer

SL provides both commercial and open source versions of its client that are termed as viewer. The open source version of the viewer is called SnowGlobe [27] that provides the mechanism to handle different haptic responses and avatar animation sequences. We developed an add-on to communicate with the viewer and developed listeners to the SL communication channel. This coupling architecture provided the option to incorporate haptic interactions without affecting the functionality of the SL Communication system. In SL viewer, all messages are valid within a particular area, which is dependent on the avatars virtual 3D location. This area normally is defined as 10–30 m^2 centered on the user's virtual location. In our add-on, we develop a module that listens to the events that are generated from message transmissions in a SL component named Nearby Interaction Event Handler. The event handler performs actions by using text-based messaging protocol. A message contains event trigger data, animation data or simple communication data. The Message Transmission Module captures all the messages that are generated in the SL. By manipulating the 3D avatar the user triggers events in the 3D environment, for example, a collision event with other avatars or objects. The message transmission module captures those events and transfers the event messages to the nearby interaction event handler for further processing. The event handler module determines the particular event handling routine for a specific event and then packs the event-handling message with the handling routine. Afterwards, the handler sends the packet to the interaction event decoder. Message transmission module also receives animation data from the animation parcel manager and generates animation sequence for the avatars in the 3D virtual world.

28.3.2 Haptic Jacket Interface

Vibrotactile actuators communicate sound waves and create funneling illusion when it comes to physical contact with skin. The haptic jacket consists of an array of vibrotactile actuators that are placed in particular portions of the jacket and their patterned vibration can stimulate touch in the user's skin [3]. A series of small actuator motors are placed in a 2D plane in the jacket in a certain manner. An AVR Micro-controller controls the vibration of these actuators. We have configured the Micro-controller so that it processes input commands that are sent from the haptic interaction controller. In order to achieve the input command transmission the haptic interaction controller uses the Bluetooth communication channel. Figure 28.2 depicts the components of the jacket in more detail.

28.3.3 Interaction Controller

In representing the interaction controller, the engine of the system, we will introduce the avatar annotation procedure in Section 28.3.3.1. The annotation provides personalized animation and haptic feedback customization options. The way we provide security and authenticity in the avatar-based interactions are described in Section 28.3.3.2. We also present the avatar animations and define the associated haptic signal patterns in Sections 28.3.3.3 and 28.3.3.4, respectively.

28.3.3.1 Avatar Annotation

In our system we annotated visible body parts of the avatar in SL and specified the corresponding physical haptic actuators to render the haptic feedback. For each haptic signal

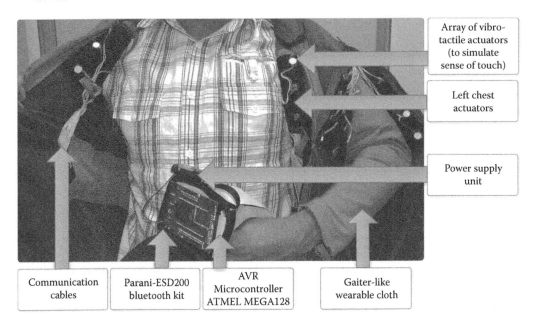

FIGURE 28.2
(**See color insert.**) The Haptic jacket controller and its hardware components. Array of vibro-tactile motors are placed in the gaiter-like wearable cloth in order to wirelessly stimulate haptic interaction.

we also annotated the avatar animation. Figure 28.3 depicts the geometric-based-avatar annotation scheme. We attached LSL scripts [18] in each of the annotated parts of the avatar that contain the haptic commands as well as the identification number of the animation sequences. For example, we annotated the 3D male avatar's left arm and specified particular

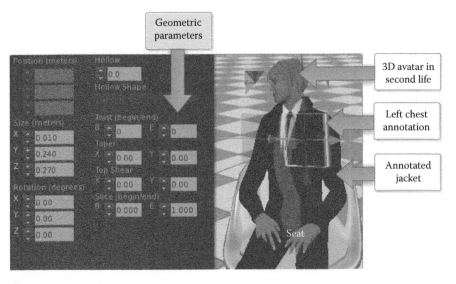

FIGURE 28.3
(**See color insert.**) The flexible avatar annotation scheme allows the user to annotate any part of the virtual avatar body with haptic and animation properties. When interacted by the other party, the user receives those haptic rendering on his/her haptic jacket and views the animation rendering on the screen.

vibrotactile actuator stimulation for it. Further, we specified the interacting animations for both the participating male and female virtual avatars. Afterwards, when the user representing the female avatar issues a GUI interaction command to the male avatar arm then the annotated haptic stimulation is rendered at the real male user's arm through the haptic jacket.

For intimate interactions such as a hug, we employed group-based annotation scheme. As evident, hugging with parents is different to that with a friend. Hence, we needed separate animation and haptic rendering for each type of hug, touch, and other interactions. We created groups and incorporated group-based annotation scheme of the 3D avatar. For each group we created different avatar animation and haptic rendering options. By using the script-based dialog interface any interacting contacts were then assigned to a group (default is formal). We provided four different groups namely family, friend, lovers, and formal. This group-based haptic interaction in SL further assisted the user to personalize his/her experience.

28.3.3.2 Access Control Scheme

In our prototype application we incorporated user profile specific access control mechanism in order to provide the participating users the means of authenticating and personalizing their interactions. For example, if user A issues a hug command to user B then the animation and haptic rendering takes place only if user B acknowledges the permission. A permission window is shown at user B's SL viewer for this purpose, where the interaction could be accepted or rejected. We used SL message notification and GUI to display the permission window in which the user is already adapted. In SL each user is associated with a string-based identification number. Message originated from a user's computer bears that identification number as a preamble to that message. Hence, in order to provide access control we compared the identification number with the list of contacts of the user and decided accordingly. In order to deliver user-specific haptic feedbacks to the user we used the group annotation. In any haptic interaction, the originator user information is mapped to obtain the group of the user. This phenomenon is depicted in Figure 28.4. In this approach the haptic renderer (HR) uses the group-specific avatar animation and haptic rendering data in order to deliver customized interactions to the users.

28.3.3.3 Avatar Animation

Animation helps the user to express the emotion in an intuitive manner (if compared to Instant Messaging). The animation rendering depicting a hug, for example, communicates the user's emotion directly when rendered with the hug haptic feedback. SL animation is a Biovision Hierarchy (BVH) file, which contains text data that describes each figure part's rotation and position along a time line. We controlled the avatar position or movement by triggering a message to animation parcel manager, which then executes the BVH animation file for that animation. We created these animation files for hug, touch, and tickle animations by using MilkShape 3D version 1.8.5 [21]. Both the participating male and female virtual avatars plays out the defined animation sequences in the SL viewer using their respective animation files. Empirically we created four different hug animations for the four groups of users in order to verify our concept that group-dependent animations were taking place in the SL viewer. In order to control the hug animations we used the SL scripting language of SL. In order to start the animation, user A issues a hug, touch, or tickle command, the participating second user B consents to it. Afterwards, in hug animation the two virtual

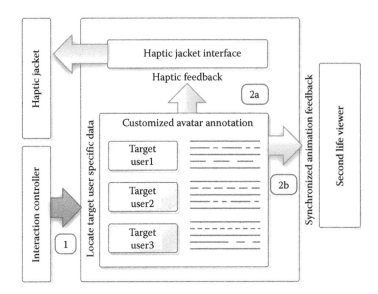

FIGURE 28.4
User-dependent haptic interaction access design. The haptic and animation data are annotated based on the target user groups such as family, friend, lovers, and formal.

avatars of user *A* and user *B* (who must be at the same virtual location in SL) comes closer by walking and holds each other closely. Similarly, for the touch animation, user *A* waves a hand emulating a touch activity and user *B* smiles or frowns (if done repeatedly) indicating that a touch has taken place. If user *A* touches stomach or neck of the avatar of user *B* then tickle animation for user *B* takes place. In tickle animation the avatar of user *B* moves awkwardly and laughs.

28.3.3.4 Haptic Renderer

The haptic jacket provides the funneling illusion-based touch haptic feedback. We leveraged the touch feature to create hug and tickle-based haptic feedbacks. We made careful observation about the real-life hug and noticed that when two people hug each other both feel a gradual touch feeling in some specific body parts. In a formal hug, a user receives touch feedbacks at the chest area and at the back shoulder area. Similarly, during our observations we noticed that in a tickle most users react to the random touch at the stomach area, at the underarm area, and sometimes at the neck area. Using these empirical parameters we constructed touch, hug, and tickle haptic feedbacks as the following:

- According to the virtual annotation the haptic touch sensation is delivered by incorporating the funneling illusion into the haptic jacket to stimulate real touch at the real user. When one person touches another person then both the participating users receive touch feelings.
- In order to create hug-type haptic feedbacks for the participating users we systematically increased the jacket's *leftChest*, *rightChest*, *neck*, *leftBackShoulder*, and *rightBackShoulder* motors intensity levels to produce the funneling illusion. The systematic control of the actuator intensity levels creates the touch effect in those areas and offers a hug-type haptic stimulation. The lover-type hug is different to that of the formal hug. In addition to the areas defined above we decided to add haptic

touch stimulation in the stomach area to emulate the joy emotion [10,29]. Hence, by following the laws of funneling illusion we activated the arrays of vibration motors attached to the abdomen area of a person.

- As described earlier, following our empirical study the tickle haptic feedback is evoked by incorporating random and unpredictable touch at the stomach area, at the underarm area and at the neck area provided that a GUI interaction at those virtual body places were performed.

28.3.4   GUI Controller

The Interaction Controller works as a core service and takes action according to the user inputs from the GUI Controller. The GUI controller enables the usage of keyboard, mouse, speech, and gesture-based inputs from the user. For example, a user representing a female avatar can point her mouse on a male avatar and produce a click event using the mouse. The GUI controller detects if the annotated body parts of the male avatar has received any GUI commands and sends the avatar body ID and type of action performed to the Interaction Controller. In our prototype the hug command is issued by using the speech, keyboard, and gesture-based interaction inputs. The GUI commands that were used in the various interaction inputs are discussed in the following:

- *Keyboard*: While processing the keyboard (text)-based inputs from the requester (sender) the controller analyzes the text messages sent to the jacket owner (receiver). The text message-based commands have certain preamble before the commands. Therefore, the interaction controller easily distinguishes the haptic commands that are issued based on the text inputs. The text command forms are HUG *username*, TOUCH *<username body parts>* and TICKLE *<username body parts>*, where *bodyparts = {leftChest, rightChest, stomach, leftShoulder, rightShoulder, leftBackShoulder, leftRightShoulder, leftArm, rightArm, neck}*.

- *Mouse*: It is extremely flexible to provide touch and tickle commands using a mouse. For each mouse click at the annotated body parts a touch command is issued. When the mouse click happens on the stomach, and neck area of the virtual avatar a tickle command is captured. In order to provide hug command the user clicks a GUI button on the screen and the nearest user is issued a hug command automatically.

- *Speech*: Similar to our previous speech-based interaction methodology [12] in virtual environment we processed the speech-based haptic commands from the user. The touch and tickle input commands are similar to that of keyboard interaction, where the user speaks out the type of interaction (touch, tickle) followed by body part names. In order to issue hug input command, the user simply speaks out hug and the nearest user is issued a hug command. User name recognition was not attempted in our approach.

- *Gesture*: In our previous work, we have proposed a novel motion path-based gesture interaction system [23]. This system allows the user to define a drawing symbol that can be associated with particular command. We tailored the motion path-based gesture interaction approach by introducing three main drawing symbols, for example, *h*, *T*, *k* representing hug, touch, and tickle commands, respectively. For each body parts we associated the following gesture commands, *bodyparts = {leftChest(L,C), rightChest (Γ,C), stomach (S), leftShoulder(L,S), rightShoulder(Γ,S), leftBackShoulder (L,b), leftRightShoulder (Γ,b), leftArm (L,m), rightArm (Γ,m), neck(n)}*.

The gesture drawing symbols were chosen based on their selection accuracy. For example, the gesture recognition rate for Γ is higher than R and since the selection.

28.4 Implementation of Interpersonal Communication Framework

In our prototype application, the interaction listener was developed as a service, which listens to a Communication Serial Port (COM). A Bluetooth device was connected with the PC's USB port, which was virtually configured with the COM port so that the Bluetooth device can send signals to the haptic jacket. For the haptic signal transmission, Bluetooth was configured at the PC COM port of the respective computers that interfaces with the hardware controller of the jacket. In the following, we present a detailed illustration of the various modules of our system.

28.4.1 Development of Different Modules

Here, we present the details of the implementation issues of different modules of our proposed system. We incorporated Microsoft Visual Studio 2005 IDE to develop our system and the primary language used was Visual C++. We adopted Microsoft Foundation Class library and asynchronous socket programming scheme to create a socket-based secure communication channel. In order to implement voice-based interaction we used Microsoft Speech SDK (SAPI version 5.1) [2]. We now briefly illustrate the development of different modules, which are SL Controller, Interaction Event Decoder, Permission Manager (PM), and HR. These modules and their inter message communications are depicted in Figure 28.5.

28.4.1.1 SL Controller Module

In order to develop the SL add-on, we locally build the SL open source viewer Snowglobe [27] version 1.3.2 by using the latest version of CMake (version 2.8.1) [7]. SL message transmission module is responsible for dispatching all the messages to handle virtual

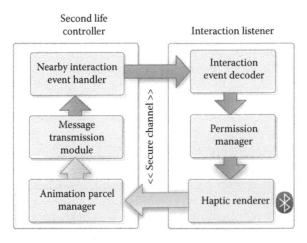

FIGURE 28.5
SL and haptic communication system block diagram.

environment. All the messages are in Extensible Markup Language (XML) format with detail avatar and virtual environment related data. All the event type messages are filtered by nearby interaction event handler.

28.4.1.2   Interaction Event Decoder Module

Interaction Event Decoder is a component in the interaction listener. It receives all the output messages from the Nearby Interaction Event Handler in an encrypted XML [31] format. The primary responsibilities of the module are to decrypt those received messages and transmit them further to the communication channel toward the PM.

28.4.1.3   PM Module

The PM looks up the user-dependent access control scheme and produces appropriate permission dialogues in SL viewer. The PM issues these dialogues by using SL script and receives appropriate permission parameters. Figure 28.6 shows the code snippet, that is, used to control user-dependent animation and the vibrotactile motors in the haptic jacket. As shown, before commencing avatar or haptic rendering functions, we call *llRequest-Permissions(key AvatarID, integer perm)* function. The function takes two parameters; the first parameter is the user's Avatar ID who requested an event. The second parameter *PERMISSION_TRIGGER_ANIMATION* is a permission type for that event.

28.4.1.4   HR Module

The HR operates the haptic jacket and notifies the Animation Parcel Manager for synchronized animation feedback. In order to control the jacket motors, it parses an XML file

```
state_entry()
{
    //Request animation permission to Second Life
    llRequestPermissions(llGetOwner(), PERMISSION_TRIGGER_ANIMATION);
    llSetTimerEvent(1.0);

    //Specify communication channel
    llListen(&iChannel,"",llGetOwner(),"");
}

touch_start(integer WHICH)
{
    //Obtain the calculated body part from the GUI interaction
    llGetAnnotationSelection(iChannel,&vBody_trg,
                        (string)llDetectedKey(iChannel));

    //User specific haptic feedback and animation
    aAnimations = SL_GetUserAnim(SL_GetUser(this),vBody_trg);
    hFeed = SL_GetUserFeed(SL_GetUser(this),iChannel,vBody_trg);

    //Render animation and haptic feedback
    llStartAnimation(llList2String(aAnimations,WHICH));
    SLStartFeed(llList2String(hFeed,WHICH));

    //Animation calibration and user interaction throttling
    if(WHICH++ >= TOTAL)
    {
        WHICH = INITIAL VALUE;
    }
```

FIGURE 28.6
A code snippet depicting portion of the Linden Script that allows customized control of the user interaction.

```
<?xml version="1.0"encoding="utf-8"?>
<InteractionRules>
    <hug userType="Friends"hapticFeedback="Yes"name="hug1">
        <animationModules>
            <UUID>6b61c8e8-4747-0d75-12d7-e49ff207a4ca</UUID>
            <animationBVH>hug.bvh</animationBVH>
            <animationPriority>MEDIUM</animationPriority>
            <animationLooped>YES</animationLooped>
            <animationSpeed>30</animationSpeed>
            <animationDuration>LOW</animationDuration>
            <animationScaleTo>0.75</animationScaleTo>
        </animationModules>
        <tactileModules>
            <module name="leftChest">
                <highestIntensity>15</highestIntensity>
                <lowestIntensity>0</lowestIntensity>
                <vibrationType>GRADUAL_INCREASE</vibrationType>
                <interactionTime>500</interactionTime>
                <numberOfRepetation>3</numberOfRepetation>
            </module>
            <module name="rightChest">...</module>
        </tactileModules>
    </hug>
    <hug userType="Family"hapticFeedback="Yes" name="hug2">...</hug>
    <touch>...</touch>
    <tickle>...</tickle>
```

FIGURE 28.7
An overview of the target user group specific interaction rules stored (and could be shared) in an XML file.

containing haptic patterns and sends a message to the microcontroller unit of the jacket accordingly. Portion of the XML file is shown in Figure 28.7. In our implementation the actuator motors have a total of 16 intensity levels from 0 to 15. Where, 0 means no vibration and 15 indicates the maximum vibration level. To repeat the vibration patterns we set the value for the *numberOfRepetation* attribute.

28.4.2   Processing Time of Different Modules

In order to ensure that the implemented interacting modules of our system performs on par with the interfacing modules of the SL controller we measured their performances with a set of haptic and animation data. The data size for the haptic and animation rendering in each step of the experiments were kept the same for all the components. In order to measure the performance metrics we embedded performance thread hooks in the components and recorded the responses of those. For each pair of haptic and animation rendering the experiment setups were repeated four times and later averaged. In Figure 28.8 the performance of the SL interfacing modules are depicted. Similarly for the same data the processing time of the implemented interacting modules are shown in Figure 28.9.

In order to compare the processing times we evaluated the component in the SL controller that required the highest processing time. As shown in Figure 28.8 in the experiment setup number 12, the Nearby Interaction Event Handler required >175 ms to compute the interaction and report that to the Interaction Event Decoder. However, the two core components, for example, the Interaction Event Decoder and the PM processed the message data in <65 ms and 10 ms, respectively, in all the experiments. This showed that the core components of our system were able to handle the message data as par to that of the SL controller. The two other components namely the Bluetooth Transmission module and the Haptic Rendering module render their operations locally and dependent on their hardwire processing time of the Bluetooth and the Haptic controller subsystem, respectively.

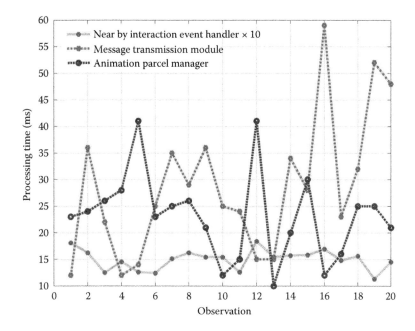

FIGURE 28.8
Processing time of different interfacing modules of the SL Controller. The figure depicts the modules that interface with our system.

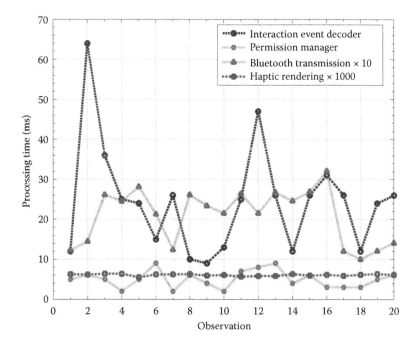

FIGURE 28.9
Processing time of the components of the implemented interaction controller with respect to different haptic and animation interactions.

28.5 Results

We present the different parameters that affect the transmission time of the haptic and animation data in our system in Section 28.5.1. In Section 28.5.3, we illustrate a detailed analysis of the impact of different interaction modalities. Results pertaining to the multiuser access performance and usage of the system are discussed further in Section 28.5.2. Further in Section 28.5.4 we describe the usability study setup and its analysis.

28.5.1 Response Time

We calculate the haptic transmission time from the sender machine to the receiver jacket by using Equation 28.1. Where user's average interaction time to interact with the SL viewer is I unit, average data transmission rate via the server is Π, n is the message size and the time for sending data from the receiver machine to the jacket actuators is β_1 unit.

$$R = I + \frac{n}{\Pi} + \beta_1 \tag{28.1}$$

After generating a haptic interaction the system approximately requires $R = (3775 + 270 + 344)$ ms to complete the transmission. Here, in our experiments the average of the interaction time is 3775 ms, network overhead is 270 ms, and β_1 is 344 ms. The haptic acknowledgement from the receiver machine to the sender jacket is represented by Equation 28.2. Here, n/Ω is the average time for transmitting n byte feedback message from the receiver machine to the sender machine. We assumed that the transmitted message and its acknowledgment were of the same length.

$$
\begin{aligned}
S &= I + \frac{n}{\Pi} + \beta_2 + \frac{n}{\Omega} \\
&= R + \frac{n}{\Omega} + (\beta_2 - \beta_1) \tag{28.2} \\
&= R + \frac{n}{\Omega}, \qquad (\beta_2 - \beta_1) \simeq 0
\end{aligned}
$$

On average, the time S is higher than R by n/Ω unit, which is the network transmission delay. In order to ensure that the difference between S and R does not affect the interaction experience of the participating users, the haptic rendering and animation rendering are synchronized locally in respective users' machines. Figure 28.10 depicts the processing times required to render different haptic and animation data. From the result we see that hug interaction needs more time than other interactions as for the hug-type rendering the system is required to process more data than the others.

However, the SL message transmission architecture also plays a role to introduce delay in the synchronized haptic and animation rendering thereby increasing the Interaction Processing Time. We present two main factors that we observed during our experiments. We found out that the Nearby Interaction Listener component introduces delay in its message processing when the server receives too many requests from the surrounding of the avatar. To measure the difference in the processing times we designed experiment sessions on five empirically selected time intervals during a day. We continued to sample the Interaction Response Time in three successive weeks by running the same set of experiments. We show the recorded data in the following Figure 28.11. As seen in the figure, the Interaction Response Time reached its peak during the weekends.

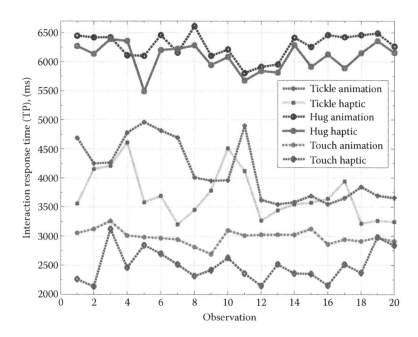

FIGURE 28.10
Haptic and animation rendering time over 18 samples. The interaction response time changes due to the network parameters of SL controller system.

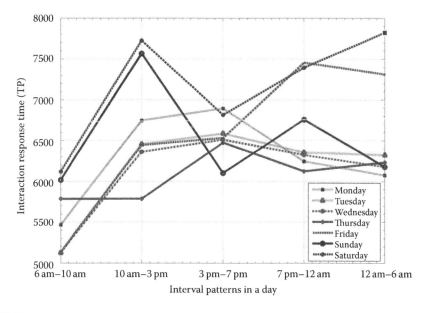

FIGURE 28.11
Average of the interaction response times that were sampled on particular time intervals. The data were gathered during three weeks experiment sessions and averaged. From our analysis, we observed that based on the server load the user might experience delay in their interactions.

TABLE 28.1

Interaction Response Time Based on Prim Size on Various SL Map Locations.

Location (Slurl) [19]	Area (m²)	Prims on Parcel (Object Density)	Interaction Response Time (TP), ms
Oak Grove/128/128/11	528	600	5120
Saint Lucia/128/128/22	1792	269	5469
Amberville/128/128/2	5472	1391	6089
Boreal/128/128/122	7168	1495	6201
Kissena Park/128/128/2	8192	1426	6213
Wichi/128/128/2	9408	2018	6213
Moose Beach/50/57/20	19,008	1382	6428
New York NYC/128/128/2	21,936	4341	6901
Zen Destani/128/128/18	28,672	2556	6310
Solace Beach/128/128/2	38,832	4911	6052
London UK/128/128/22	47,760	5460	7168

In SL the users can navigate to different map locations in the virtual world. A convenient method of specifying locations and teleporting to that location is achieved by using the slurls [19], which are hyperlinks that allows users login directly to that site or teleport to it if they are already inside SL. We noticed that different slurls have different 3D object density (Prims) and they are spanned in varying size. When then number of Prims increases the interaction complexity with the present virtual avatars in that area increases. These metrics therefore, influences the Interaction Response Time in our experiments as shown in Table 28.1.

The area of the slurl effectively creates different density of the avatars with particular prims in their surrounding virtual locations. In order to measure the effect of different density levels we teleported the avatars in locations with very low to very high density and determined their impact on the Interaction Response Time. Our findings are depicted in Figure 28.12.

We equipped the Interaction Controller to sense these network parameters. By sensing the density of the avatars, prims, nearby interaction message parsing frequency and day time metrics the Interaction Controller empirically calculates a threshold that can distinguish disruptive effects of network lag in the inter-personal communication. However, as our system extensively used the communication platform of the SL controller we tackled the lag at the network communication level by deciding to inform the participating users about the delay. The Interaction Controller incorporates the calculated lag threshold and provides color-coded decorators at the SL viewer's Heads Up Display (HUD). In this regard, we adopted the decorator scheme proposed in [25]. We developed a bar in the HUD that displays Green when the prims and avatars interaction messages do not create congestion. Similarly a Red bas is displayed reflecting the delay in the communication. Later, in our usability study we noticed that when the user was informed about possible interaction delays by using the decorators s/he accepted the lag with ease and reacted more intuitively.

28.5.2 Multiuser Haptic Interaction Response Time

The haptic and animation rendering-based communication is not specific to a pair of users. Rather, multiple users from different groups can interact with each other at the same time. This essentially extends the interpersonal interaction and provides option to leverage the

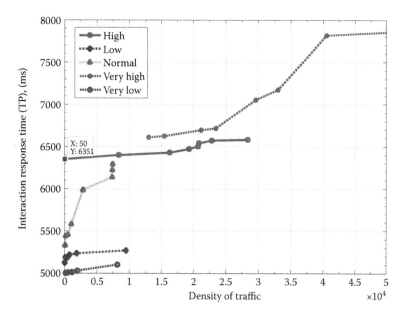

FIGURE 28.12
Interaction response time in varying density of traffic in the SL map location for the Nearby Interaction Handler.

framework in a group-specific interaction scenario. The developed Interaction Listener supports interaction requests from multiple users. For example, suppose user A receives interaction requests from users B and C then the system creates a queue of requests for user A and processes that on a first come first serve basis. In such cases the Interaction Response Time can be calculated by using the Little's Formula, which is a classical conservation equation in queuing theory.

$$E(T) = \frac{E(n)}{\lambda} \qquad (28.3)$$

Here, $E(T)$ is the average delay for a user request, that is, average throughput time, $E(n)$ is the average number of requests to the interaction listener. λ is the arrival rate of the requests. However,

$$E(n) = \frac{\rho}{1 - \rho} \quad \text{and} \quad \rho = \frac{\lambda}{\mu} \qquad (28.4)$$

where, ρ is the fraction of time the interaction listener requires to process the requests. Hence, combining Equations 28.3 and 28.4 we get,

$$E(T) = \frac{\rho}{\lambda(1 - \rho)} = \frac{\frac{\lambda}{\mu}}{\lambda(1 - \frac{\lambda}{\mu})} = \frac{1}{\mu - \lambda} \qquad (28.5)$$

Here, μ is request processing rate from the queue, and Equation 28.5 is the queuing delay. In this equation λ is the arrival rate of a interaction request by a user and μ is the average service rate. From Figures 28.9 and 28.8 we measured that the average service time is approximately $6.5\,s$. Hence, average service rate $\mu = \frac{1}{6.5} = 0.1538\,s$. For example, in case after every $60\,s$ an interaction request is triggered to the Interaction Listener then $\lambda = \dfrac{1}{60} = 0.0167$ request s. Therefore, from the Little's Formula (Equation 28.5), the total waiting time including the service time $= \dfrac{1}{0.1538 - 0.0167} = 7.29\,s$ (approx.).

28.5.3 Analysis of Different Interaction Modalities

A comparison of different interaction modalities used and their suitability for each haptic interactions are given in Table 28.2. The two other parameters are average time needed to produce the command and average accuracy, which are also listed. However, not all haptic input commands were convenient to use for each interaction modalities. For the keyboard (text) interaction modality we found that writing body parts names take time and often spelling mistakes impaired the accuracy of the command. Touch and tickle input commands were very easy to issue using the mouse-based modality. However, while issuing hug input command using the mouse, it became difficult to assign the command to a particular user, hence nearest user was selected automatically from the user group lists. Similar problem occurred while using speech and gesture-based interaction modalities as it became cumbersome to recognize the user names using either of those two approaches. From the table we see that the percentage of accuracy is highest for mouse-based interaction modality, which is 99.6%. This and its flexibility for usage in pointing and interacting with annotated body parts made it the ideal medium for haptic input command delivery in our system.

28.5.4 Usability Study

We have incorporated the usability evaluation guidelines [6] and designed our tests accordingly with the sensory analysis [17] of the system involving both the user and the targeted sensory communication modules. Before performing the usability test we designed a test plan where we defined our evaluation objectives, developed questions for the participants, identified the measurement criterion and decided upon the target users of the system. The test took place at a university laboratory with 16 participants comprising of different age groups. Five of the participants are in age group 13–18, eight of them are in age group 18–36 and the rest three are in age group 36+. Furthermore, in multiuser interpersonal communication setting the users was divided into two groups namely Groups A and Group B.

For the traditional experiments two users were chosen. In order to ensure that each communicating participant can converse with different age groups their selection was made randomly. Moreover to ensure the distributed communication behavior the physical

TABLE 28.2

A Comparison of Different Interaction Modality

Modality	Average Time	Accuracy	Suitability
Keyboard	5110 (ms)	85 (%)	Hug, touch, tickle
Mouse	2075 (ms)	99.6 (%)	Touch, tickle, hug
Speech	3790 (ms)	55 (%)	Hug, tickle, touch
Gesture	4125 (ms)	78.1 (%)	Hug, touch, tickle

TABLE 28.3

Usability Test Questions to the User

#	Question
Q1	Perceived system response was acceptable
Q2	The haptic feedbacks are realistic and/or acceptable
Q3	Consider using the system in SL
Q4	Perceived delay between haptic response and avatar rendering was tolerable
Q5	Easy to get familiar with

location of the users were separated. In a user's test machine the enhanced SL viewer was installed to provide animation and GUI-based interactions.

At a time, the selected volunteers were told to put on the haptic jackets and requested to use the prototype system by participating in certain haptic interaction-based tasks. Their activity was monitored throughout the experiment and recorded for analysis. Afterwards, based on their interaction experiences the users filled out a questionnaire where they were requested to provide ratings of their likeliness, familiarity, ease of usage, and so on, of the system (see Table 28.3).

The user responses are shown in Likert scale [20] in Figure 28.13. The ratings of the questionnaire were in the range of 1-5 (the higher the rating, the greater is the satisfaction). The average of the responses of the users were calculated in percentage form and measured after the usability tests. Figure 28.13 shows the user's responses for each given assertion. It is worth mentioning that >80% of the users would like to communicate using the enhanced system through haptic and animation interaction if they were available in SL. Overall the users were also satisfied with the synchronized animation and haptic rendering responses and 75% of users consented to that.

We conducted usability tests to evaluate the user's quality of experience with our proposed system and to measure the suitability of the approach. Table 28.4 summarizes the overall performance score of the users. The higher mean values of System response, Haptic feedbacks, and Ease of use represent a very satisfactory user response, while the moderate mean values of System in SL and Perceived delay show relatively good user satisfaction.

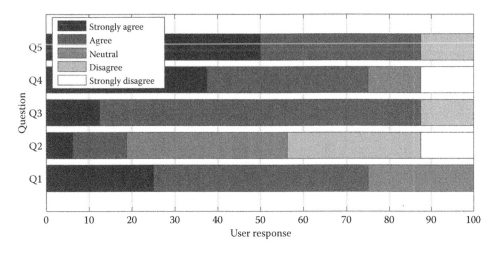

FIGURE 28.13
Usability study of the SL haptic interaction system.

TABLE 28.4

User Satisfaction on the Overall Evaluation in Likert Scale

	Mean	Std. Dev.	Mean Percentage
Acceptability	4	0.7303	80%
Haptic feedback	2.6875	1.0782	53.75%
Likeliness	3.8750	0.8062	77.5%
Delay	3.875	1.3102	77.5%
Ease of use	4.25	1	85%

In our study we also attempted to evaluate the acceptability of the system by the users from different genders, age groups and technical backgrounds. The result of these studies are depicted in Figure 28.14. We infer from Figure 28.14a that the female users gave more positive feedback in acceptability and likeness than the male users. However, the male users confirmed that it was easier for them to use the system after a couple of dry runs. Moreover, all the users favored that a refined haptic rendering is needed to make the interaction experience natural and realistic. In case of different age groups we divided the users into three age groups, namely *group*–1: ages 13–18, *group*–2: ages 18–36, and *group*–3: ages 36+ and recorder their responses in Figure 28.14b. In retrospect, compared to the older group of users, the users from the younger group seemed to be more attracted in using the system and wanted to participate in remote touch, hug, and tickle interactions. Also from Figure 28.14c we received favorable responses and recommendations from users with

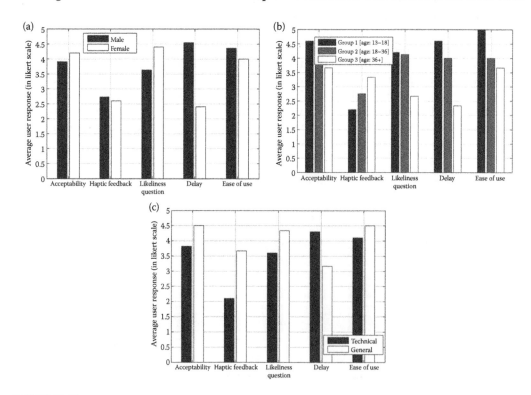

FIGURE 28.14

Comparison between the responses of users from different (a) gender, (b) age groups, and (c) technical background.

nontechnical background than that of the technical people. Although non-technical users were less happy with the Interaction Response Time of the system.

28.6 Conclusion

In this chapter, we presented a SL HugMe prototype system that bridges the gap between virtual and real-world events by incorporating interpersonal haptic communication system in SL. The developed system works as an add-on and is loosely coupled to the SL viewer. The haptic and animation data are annotated in the virtual 3D avatar body parts. We were able to successfully demonstrate the ability of the system to transfer a virtual touch and hug that had been generated in the virtual world and experienced that in the real world. Users defined the type of stimulus they wanted to translate to the real world and enjoyed them while they were happening in the virtual world. The 3D avatar and the annotated body parts representing a real user receive inputs when they are interacted through gesture, mouse, speech, or text-based input modalities and produces emotional feedbacks such as touch, tickle, and hug to the real user through the haptic jacket. We presented the implementation details of a preliminary prototype exploring the aforesaid haptic interactions in a real-virtual collaborative environment.

Glossary

Haptic Rendering: Haptic rendering allows users to 'feel' virtual objects in a simulated virtual environment.

Second Life (SL): Second Life is a 3D virtual world developed by Linden Lab that is accessible via the Internet. A free client program called the Second Life Viewer enables its users to interact with each other through virtual 3D avatars.

Gesture: A gesture is a motion of the body that conveys meaning information, which could be mapped to perform a pre-defined operation.

Gesture-Based Interaction: The process in which a gesture is sensed by an application to perform certain tasks could be defined as gesture based interaction.

Virtual Environment: Virtual environment is a computer generated, three-dimensional representation of a real world or an imaginary world that lacks the photorealism and a sense of total immersion unless augmented with other sensory modules.

Tele-Immersion: Tele-immersion or telepresence refers to a set of technologies that simulate the user with augmented sensory information to render the illusions that allow a user to feel as if they were present, or to have an effect, at a virtual or imaginary location other than their true physical location.

References

1. Activeworlds. Activeworlds inc. Technical report, http://www.activeworlds.com, 2010.
2. Microsoft Speech API. http://www.microsoft.com/speech/developers.aspx. Accessed 11 July 2010.

3. A. Barghout, Jongeun Cha, A. El Saddik, J. Kammerl, and E. Steinbach. Spatial resolution of vibro-tactile perception on the human forearm when exploiting funneling illusion. In *IEEE International Workshop on Haptic Audio Visual Environments and Games, 2009. HAVE 2009*, pp. 19–23, Politecnico di Milano, Lecco, Italy, 7–8 November, 2009.

4. S. Brave and A. Dahley. Intouch: A medium for haptic interpersonal communication. In *CHI '97: CHI '97 Extended Abstracts on Human Factors in Computing Systems*, pp. 363–364, New York, NY: ACM, 1997.

5. J. Cha, M. Eid, A. Barghout, A.S.M.M. Rahman, and A. El Saddik. Hugme: Synchronous haptic teleconferencing. In *MM '09: Proceedings of the Seventeen ACM international Conference on Multimedia*, pp. 1135–1136, New York, NY: ACM, 2009.

6. D. Chisnell. Usability testing: Taking the experience into account. *Instrumentation Measurement Magazine, IEEE*, 13(2): 13 –15, 2010.

7. CMake. Cross platform build system. Technical report, http://www.cmake.org/cmake/resources/software.html, Accessed Nov. 2008.

8. G. Coaets. Program from invisible site—A virtual show. Technical report, San Francisco, 1992.

9. M. de Pascale, S. Mulatto, and D. Prattichizzo. Bringing haptics to second life for visually impaired people. *Haptics: Perception, Devices and Scenarios*, 5024, 896–905, 2008.

10. C. DiSalvo, F. Gemperle, J. Forlizzi, and E. Montgomery. The hug: An exploration of robotic form for intimate communication. In *2003 IEEE International Workshop on Robot and Human Interactive Communication*, pp. 403–408, Millbrae, California, USA, Oct. 31–Nov. 2, 2003.

11. A. El Saddik. The potential of haptics technologies. *IEEE Instrumentation and Measurement*, 10(1): 10–17, 2007.

12. A. El Saddik, A.S.M.M. Rahman, and M. Anwar Hossain. Suitabiulity of searching and representing multimedia learning resources in a 3-d virtual gaming environment. *IEEE Transactions on Instrumentation and Measurement*, 57(9): 1830–1839, 2008.

13. J. J. Gibson. *The Ecological Approach to Visual Perception*. Technical report, Boston: Houghton Mifflin, 1979.

14. S.K. Alamgir Hossain, A.S.M.M. Rahman, and A. El Saddik. Interpersonal haptic communication in second life. In *2010 IEEE International Symposium on, Haptic Audio-Visual Environments and Games (HAVE)*, pp. 1–4, Phoenix, Arizona, USA, 16–17 October, 2010.

15. S.K. Alamgir Hossain, A.S.M.M. Rahman, and A. El Saddik. Haptic based emotional communication system in second life. In *2010 IEEE International Symposium on Haptic Audio-Visual Environments and Games (HAVE)*, pp. 1–1, Phoenix, Arizona, USA, 16–17 October, 2010.

16. S. King and J. Forlizzi. Slow messaging: Intimate communication for couples living at a distance. In *DPPI '07: Proceedings of the 2007 Conference on Designing Pleasurable Products and Interfaces*, pp. 451–454, New York, NY: ACM, 2007.

17. E.P. Kukula, M. J. Sutton, and S. J. Elliott. The human biometric-sensor interaction evaluation method: Biometric performance and usability measurements. *IEEE Transactions on, Instrumentation and Measurement*, 59(4): 784–791, 2010.

18. *Linden* Lab. Second life. Technical report, http://lindenlab.com/, 2010.

19. Linden Lab. Slurl: Location-based linking in second life. Technical report, http://slurl.com/, Accessed Nov. 2009.

20. Rensis Likert. A technique for the measurement of attitudes. *Archives of Psychology*, 140:1–55, 1932.

21. MilkShape-3D. Milkshape 3d, modelling and animation tool. Technical report, http://chumbalum.swissquake.ch/ms3d/index.html, Accessed Feb. 2009.

22. S. O'Brien and F. 'Floyd' Mueller. Holding hands over a distance: Technology probes in an intimate, mobile context. In *OZCHI '06: Proceedings of the 18th Australia Conference on Computer-Human Interaction*, pp. 293–296, New York, NY: ACM, 2006.

23. A.S.M.M. Rahman, M. A. Hossain, J. Parra, and A. El Saddik. Motion-path based gesture interaction with smart home services. In *Proceedings of the Seventeen ACM International Conference on Multimedia*, Beijing, China: ACM, 2009.

24. A. F. Rovers and H. A. van Essen. Him: A framework for haptic instant messaging. In *CHI '04: CHI '04 Extended Abstracts on Human Factors in Computing Systems*, pp. 1313–1316, New York, NY: ACM, 2004.

25. S. Shirmohammadi and N. Ho Woo. Shared object manipulation with decorators in virtual environments. In *Eighth IEEE International Symposium on, Distributed Simulation and Real-Time Applications, 2004. DS-RT 2004*, pp. 230–233, Budapest, Hungary, 21–23 October, 2004.

26. Sims. Electronic arts inc. Technical report, Sims Online, http://www.ea.com/official/thesims/thesimsonline, 2010.

27. SnowGlobe. Second life open source portal. Technical report, http://wiki.secondlife.com/wiki/Open_Source_Portal, Accessed 10 June 2010.

28. S. Thayer. *Social touching. In Schiff, W., Foulke, E. (eds.), Tactual Perception: A Sourcebook*, Cambridge: Cambridge University Press, 1982.

29. D. Tsetserukou, A. Neviarouskaya, H. Prendinger, M. Ishizuka, and S. Tachi. Ifeelim: Innovative real-time communication system with rich emotional and haptic channels. In *CHI EA '10: Proceedings of the 28th of the International Conference Extended Abstracts on Human Factors in Computing Systems*, pp. 3031–3036, New York, NY: ACM, 2010.

30. R. Wang and F. Quek. Touch and talk: Contextualizing remote touch for affective interaction. In *TEI '10: Proceedings of the Fourth International Conference on Tangible, Embedded, and Embodied Interaction*, pp. 13–20, New York, NY: ACM, 2010.

31. Wikipedia. Xml-encryption. Technical report, http://en.wikipedia.org/wiki/XML_Encryption, Accessed 10 July 2010.

Index